Fluid Mechanics

We work with leading authors to develop the
strongest educational materials in engineering,
bringing cutting-edge thinking and best learning
practice to a global market.

Under a range of well-known imprints, including
Prentice Hall, we craft high quality print and
electronic publications which help readers to
understand and apply their content,
whether studying or at work.

To find out more about the complete range of our
publishing please visit us on the World Wide Web at:
www.pearsoned.co.uk

Fluid Mechanics

Fourth Edition

JOHN F. DOUGLAS

M.Sc. (Eng.), Ph.D., A.C.G.I., D.I.C., C.Eng., M.I.C.E., M.I.Mech.E.
Formerly of Mechanical Engineering Department
University of the South Bank, London

JANUSZ M. GASIOREK

B.Sc. (Eng.), Ph.D., C.Eng., M.I.Mech.E., M.C.I.B.S.E.
Formerly of Mechanical Engineering Department
University of the South Bank, London

JOHN A. SWAFFIELD

B.Sc., M.Phil., Ph.D., C.Eng., M.R.Ae.S., F.C.I.W.E.M., M.C.I.B.S.E.
William Watson Professor of Building Engineering
Heriot-Watt University, Edinburgh

PEARSON
Prentice Hall

Harlow, England • London • New York • Boston • San Francisco • Toronto
Sydney • Tokyo • Singapore • Hong Kong • Seoul • Taipei • New Delhi
Cape Town • Madrid • Mexico City • Amsterdam • Munich • Paris • Milan

Pearson Education Limited
Edinburgh Gate
Harlow
Essex CM20 2JE
England

and Associated Companies throughout the world

Visit us on the World Wide Web at:
www.pearsoned.co.uk

First published by Pitman Publishing Limited 1979
Second Edition 1985
Third Edition published under the Longman imprint 1995
Fourth Edition 2001

ISBN 0 582 41476 8

British Library Cataloguing-in-Publication Data
A catalogue record for this book can be obtained from the British Library

Library of Congress Cataloging-in-Publication Data

Douglas, John F.
    Fluid mechanics / J. F. Douglas, J. M. Gasiorek, and J. A. Swaffield.—4th ed.
      p. cm.
    Includes bibliographical references and index.
    ISBN 0-582-41476-8
      1. Fluid mechanics.   I. Gasiorek, J. M. (Janusz Maria), 1927-   II. Swaffield, J. A., 1943-
III. Title.

TA357.D68 2000
620.1′06—dc21

00-033964

10 9 8 7 6 5 4 3
07 06 05 04

Set by 35 in 10/12pt Times Roman
Printed and bound by Mateu Cromo Artes Graficas, Madrid, Spain

Contents

PART I   ELEMENTS OF FLUID MECHANICS

PART III DIMENSIONAL ANALYSIS AND SIMILARITY 258

PART IV   BEHAVIOUR OF REAL FLUIDS   332

PART VI UNSTEADY FLOW IN BOUNDED SYSTEMS 592

18 Quasi-steady Flow 594

19 Unsteady Flow in Closed Pipeline Systems 606

20 Pressure Transient Theory and Surge Control 618

Appendix 1   Some Properties of Common Fluids   896

Appendix 2   Values of Drag Coefficient C_D for Various Body Shapes   900

Preface to the Fourth Edition

THE STUDY OF FLUID MECHANICS REMAINS WITHIN THE CORE OF ENGINEERING education. Advances in the media available for the delivery of that process provide exciting challenges to the academic. The availability of fast, powerful and inexpensive computing and the multi-various opportunities presented by the web and Internet have the potential to transform fluid mechanics education. Always an experimental subject that traditionally relied heavily on laboratory demonstrations and experience, the opportunities offered by validated simulations are particularly appropriate, extending the student's experience far beyond the constraints of laboratory space or equipment. While these changes are to be welcomed it remains essential that any fluid mechanics course or supporting text provides the fundamental underpinning that will allow the student, and later the practitioner, to recognize when a flow simulation, however 'sophisticated' the package, is less than accurate.

Advances in media and particularly computing have therefore provided both the challenges and the solutions necessary to ensure that fluid mechanics remains at the centre of engineering education. However, the objectives of that educational process have also changed, particularly in the UK where fundamental reassessments of the academic and practice levels necessary for professional recognition have introduced differentiated courses. Current Engineering Council regulations will progressively reduce the percentage of graduates reaching Chartered Engineer status, while the introduction of BEng and MEng course requirements, incorporating Matching Sections to allow those unable to progress directly to an MEng qualification the opportunity to reach chartered status, will inevitably determine the partition of fluid mechanics into fundamental principles required by all and a range of more specialist topics that may be covered in greater depth. The Matching Section approach will also inevitably lead to post-university courses taken in many instances by part-time or distance learning routes, again offering both tremendous challenges to the course provider and the opportunity to fully utilize the advantages of the computing and delivery media advances already mentioned.

The original aims of this text, dating back to the first edition and explicit by the third edition, clearly meet the needs of this changed educational landscape. The text has consistently emphasized the importance of a fundamental understanding of the principles of fluid mechanics, while at the same time providing specialist topics to be covered in greater depth, whether in the area of rotodynamic machinery or unsteady flow. The fundamental material may be seen as crossing the boundaries of the engineering disciplines committed to a coverage of fluid mechanics and it may be argued that the specialist areas chosen also meet this criterion, although in a more selective sense. The second and third editions experimented with the provision of computing simulations; however, the support infrastructure is only now fully available to allow the maximum benefit to be drawn from the provision, in this fourth edition, of a wider range of computing applications. In many ways the development of this provision, from a separately purchased BBC Basic floppy disk in 1984 to the opportunity in the current edition to utilise simulations of a complexity unavailable in the early 1980s, is an allegory for the development of computing over this period. Clearly the lifetime of this edition will see continued exponential change in delivery systems so that it is probably not practical to predict the format of the fifth edition. While the third edition provided a hardcopy solutions manual, this will now be available for downloading, making the 'Problem' sections provided more attractive as a basis for student tutorial activity or distance learning education; applications already available to students at Heriot-Watt University through the extensive distance learning Masters course provision from the Department of Building Engineering.

Thus the fourth edition retains the educational aims and objectives of earlier editions, while continuing to make full use of the available computing infrastructure. The content has been revised and extended – in the treatment of air and gas distribution networks to include time dependency, including the provision of simulations to extend any laboratory provision in this area, the inclusion of modelling and simulation considerations for both water and airborne pollution and groundwater seepage flow and the introduction of wind turbine coverage aimed at power generation. The impact of computing through computational fluid dynamics (CFD) is recognized with the emphasis placed firmly on the development of the fundamental principles, including the essential recognition that boundary equation definition, if not based on a full understanding of the flow condition, can lead to worthless predictions. Similarly the computational constraints defining stability have been reinforced. While all these are defined in terms of the exciting field of CFD, the text emphasizes that these considerations were always present in the simulations provided in this and earlier editions and may be found at the root of the cases described in the reworked coverage of unsteady flows across the whole spectrum of conditions from free surface wave attenuation to low-amplitude air pressure transient propagation and traditional waterhammer.

As in previous editions the text emphasizes the linkage between theory and practice; engineering is still fundamentally about changing the rules and making things work. Examples throughout the text illustrate the application of theory. All the computing is presented in terms of a description of the calculation or simulation, followed by an example and an invitation to consider further several linked problems. The programs provided with the text fall into a number of natural categories, firstly relatively simple calculations, for example friction factor, lift coefficient or free surface flow depths; then calculations designed to provide solutions for steady state system operation, for example fan or pump operating points, free surface gradually varied flow surface profiles or groundwater seepage flow nets beneath dams; and finally unsteady flow simulations, whether for air distribution and recirculation networks or for waterhammer in response to changes in system operating conditions. This content both extends that provided in previous editions and enhances its presentation.

The authors would again like to thank all their colleagues, both at Heriot-Watt University and at many other universities worldwide, who have contributed to the development of this series of editions directly or through informed comment, and all those students who have used the texts. In particular the authors wish to thank Dr Ruth Thomas and Dr Nils Tomes at Heriot-Watt University for the development of several of the third edition computer programs into a form suitable for both the Heriot-Watt distance learning MSc programme and for this text. From the Department of Building Engineering and Surveying Dr Ian McDougall contributed to the translation of the authors' Fortran-based simulations into a form suitable for the twenty-first century; Dr David Campbell is again to be thanked for the quality of the additional original artwork provided with the text and for collating the solutions manual into a form suitable for electronic transmission; and Dr Fan Wang provided the background to the CFD descriptions included in this fourth edition. Through their subtle but effective insistence Anna Faherty and Karen Sutherland, Commissioning Editors at Pearson Education, are ultimately responsible for the manuscript being produced almost to schedule and for this and the support of all our other colleagues, past and present, at Pitman, Longman and Pearson Education since 1974, we are grateful. Nevertheless any errors, factual or of understanding, remain the authors' responsibility. Fluid mechanics is the most fascinating and exciting of the engineering disciplines and one that impinges on all our lives in a multitude of ways both recognized and taken for granted. The authors hope that this text will communicate some of that excitement to the reader.

Finally it was with deep sadness that we learnt of the death in 1998 of John Douglas who initiated this series of editions in 1974. John was always committed to the educational concepts embodied in this text. His commitment to engineering education and his ability to introduce the fundamentals of fluids to students was exceptional, as evidenced by the parallel and highly successful 'Solving Problems in Fluid Mechanics' series. This text will we hope continue that commitment and is dedicated to his memory.

J. M. GASIOREK
J. A. SWAFFIELD
Edinburgh, January 2000

Preface to the Third Edition

THIS THIRD EDITION OF *Fluid Mechanics* HAS RETAINED THE AIMS OF THE ORIGINAL TEXT in providing a broad-based approach to the study of fluid flow together with a detailed and more advanced treatment of specialist topics that find wide application within the design and analysis of flow systems. The text repeats the previous mix of exposition and example shown to be successful by earlier editions.

The study of fluid flow is one of the few areas within engineering that truly crosses the boundaries between the various engineering disciplines. It is of equal importance to mechanical, civil, chemical and process, aeronautical and environmental and building services engineers and is to be found as a fundamental building block in the education and formation of these engineers. While this has remained true, the techniques available to enable students to achieve an understanding of fluid mechanics have been revolutionized by the readily available access to computing facilities; facilities that have already advanced immeasurably since the second edition of this text was published. The use of the computer to aid understanding through the provision of interactive simulations, including the use of multi-media packages, will undoubtedly advance even more rapidly during the lifetime of this edition. Whereas the second edition was accompanied by an optional floppy disk containing the programs presented in the text, this is no longer appropriate. Instead the program listings, including a number of new or enhanced programs, have been presented in a format that will make them readily scannable and so usable on a wide range of machines.

While the third edition text retains the philosophy and methodology introduced with the earlier editions, the content has been refined to both extend and, in the authors' view, improve the presentation of existing material. In particular the text has been reordered to present earlier the fundamentals of dimensional analysis and the laws of similarity. The treatment given to the steady flow energy equation has been extended, together with a general enhancement of the analysis of air and gas flow networks. In this context the coverage of fans within rotodynamic machinery has been strengthened and new material covering the use of fans in ventilation, and the ventilation of tunnels by jet fans in particular, has been presented. A new chapter dealing with the mechanisms of mechanical and natural ventilation has been added to provide both a treatment of this important topic and a background to one of the most common applications for fan technology.

As in previous editions current research has been utilized in the treatment of specialist topics, such as the jet fan tunnel ventilation and the unsteady flow analysis presentations. In the latter case the treatment presented in this edition seeks to emphasize the commonality of a range of unsteady flow analyses, from classical waterhammer to free surface waves and low-amplitude transient propagation in gas flows, by demonstrating the general development of the defining equations and the identical solution by finite difference techniques, allied to computer simulation, once the appropriate terms have been identified for each application.

Once again the authors would like to thank all their colleagues in the many universities in the UK and overseas who have contributed to this text by their support for, and comments on, the earlier editions. The authors are grateful to the staff at Longman, particularly Ian Francis and Chris Leeding, who have both supported us in completing this edition and shown considerable patience with the process. Nevertheless, any errors, factual or of understanding, remain ours. We have found fluid mechanics, in all its multi-disciplinary manifestations, to be the most stimulating of engineering areas; we hope that this text will communicate some of that experience and enthusiasm to students of this most demanding of engineering disciplines.

J. F. Douglas
J. M. Gasiorek
J. A. Swaffield
Edinburgh, December 1993

Preface to the Second Edition

IN THE PREPARATION OF THIS SECOND EDITION WE HAVE RETAINED THE AIMS OF THE ORIGINAL text, namely to provide a broad-based treatment of the essentials of fluid mechanics, while at the same time demonstrating the application of the subject, particularly to the study and solution of higher level problems in selected areas. In retaining this 'applications' approach we are both aware and pleased that this technique currently features in the UK Engineering Council statements on the training, education and 'formation' of engineers, strengthening our view that this is one of the most efficient and relevant methods of helping students in general to understand our subject. We believe that such an approach should also include the use of improved computer-based numerical solutions as these will become part of the engineer's everyday activities.

In the five years since the first edition was published there has been a significant change in the availability of and access to micro and other computers for both the student and the practising engineer. Computers and programs are of course not ends in themselves but rather they are powerful tools that we can utilize to dispense with many tedious and repetitive calculations, thereby allowing the study, within an educational framework, of problems of greater complexity and relevance, including time-dependent phenomena that were previously beyond our capability without recourse to simplifying assumptions. This second edition therefore includes a series of computer programs chosen to illustrate these aspects of computer application and to be of direct use to both student and practising engineer alike. While the programs have been written in BBC Basic they may be transferred with little difficulty to Apple, Commodore or Sinclair machines. A program cassette tape will also be available to support the text.

None of this of course removes the necessity to provide a thorough basis for the subject and this remains one of the text's main objectives. We have included new material in areas that have been found particularly interesting by our readers, as well as updating and refining the existing text. The treatment of incompressible flows around a body has been extended to include the study of wakes, while the coverage of fluid machinery has been strengthened by the inclusion of a major new chapter on positive displacement machines. The existing treatment of unsteady flow has been extended to allow the application of numerical modelling techniques to unsteady open channel or partially filled pipe flows. Taken together with the introduction of computing methods we view these additions as supporting and reaffirming the aims and objectives of the original text.

Once again we would like to thank all our colleagues in many universities and polytechnics in the UK and overseas who have encouraged us by their positive response to and constructive comments on the first edition. All have helped us to formulate this new edition which we hope will fulfil a useful role for both the student and the practising engineer.

J. F. DOUGLAS
J. M. GASIOREK
J. A. SWAFFIELD
London, May 1984

Preface to the First Edition

THIS IS A TEXTBOOK FOR ALL MANNER OF ENGINEERS. WHETHER THE READER IS CONCERNED with Civil, Mechanical or Chemical Engineering, Buiding Services or Environmental Engineering, the principles of fluid mechanics remain the same. Drawing on our joint experience of teaching students in all these disciplines, we have tried to set out these principles simply and clearly and to illustrate their application by examples drawn from the various branches of engineering.

In the planning of this book we are indebted to our colleagues in other colleges, polytechnics and universities for the opportunity to study their syllabuses and examination papers which has enabled us to cover the general requirements of the Honours Degree and Professional examinations. We have also deliberately dealt with the elementary aspects of the subject very fully and so the book will meet the requirements of those studying for the Higher National Diploma or for the Higher Diploma or Higher Certificate of the Business and Technician Education Council (B.T.E.C.).

For ease of reference the contents has been divided into Parts which are substantially self-contained and we hope that they will provide a convenient source of information for the practising engineer in his day to day activities.

J. F. DOUGLAS
J. M. GASIOREK
J. A. SWAFFIELD

List of Symbols

a	acceleration, area
A	area, constant
b	width, breadth
B	width, breadth, constant
c	chord length, velocity of sound
c_p	specific heat at constant pressure
c_v	specific heat at constant volume
C	constant
C_c	coefficient of contraction
C_d	coefficient of discharge
C_D	coefficient of drag
C_f	coefficient of friction
C_L	coefficient of lift
C_v	coefficient of velocity
d	diameter
D	drag, diameter, depth
e	base of natural logarithms
e	error, internal energy per unit mass
E	modulus of elasticity, energy
f	friction factor
$f(\ )$	reflected pressure wave
F	force, stress
$F(\ )$	pressure wave
$\boldsymbol{g}$	gravitational acceleration
h	vertical height, depth
h	head loss
H	head, enthalpy
i	hydraulic gradient
I	moment of inertia
k	constant, radius of gyration
K	bulk modulus
l	length
L	lift
m	mass, area ratio, doublet strength, hydraulic mean depth
M	molecular weight
n	number of, polytropic index
N	rotational speed
p	pressure
P	force, power, wetted perimeter

q	flow rate per unit width or unit depth
Q	volumetric flow rate
r	radius, radial distance
R	radius, reaction force
$\boldsymbol{R}$	gas constant
s	slope, distance, arbitrary coordinate within Cartesian system, slip
S	surface, entropy
t	time
T	temperature, torque surface width
u	velocity, peripheral blade velocity
U	internal energy, velocity
v, V	velocity
v_f	velocity of flow
v_r	relative velocity
v_x	velocity component in x direction
v_y	velocity component in y direction
v_z	velocity component in z direction
v_r	radial velocity
v_θ	tangential velocity
V	volume
w	specific weight
W	weight, work
x, y, z	orthogonal coordinates
y	gas content (per cent)
Z	potential head, depth
α	angle, angular acceleration
β	angle
γ	adiabatic index (c_p/c_v)
Γ	circulation
δ	difference, increment
Δ	change in
ε	absolute roughness, eddy viscosity
ζ	vorticity
η	efficiency
θ	angle
μ	coefficient of dynamic viscosity
v	coefficient of kinematic viscosity, Poisson's ratio
ρ	mass density
σ	relative density (specific gravity), surface tension
τ	shear stress
ϕ	shear strain, angle
Φ	velocity potential
Ψ	stream function
ω	angular (rotational) velocity, stage variable
Fr	Froude number
Ma	Mach number
Re	Reynolds number

Str	Strouhal number
We	Weber number
L	Dimensions of length
M	Dimensions of mass
T	Dimensions of time
Θ	Dimensions of temperature

List of Computer Programs

In each case the program background theory is presented, together with an application example and output and a series of suggested further investigations. In addition four further program listings are provided as solutions to end of chapter problems, namely:

Acknowledgements

We are grateful to the following for permission to reproduce copyright material:

Figures 5.20 and 5.21 reproduced with the permission of FLUENT Inc.; Figure 5.22 reproduced by permission of BRE Ltd.; Figure 5.23 reproduced with the permission of Computational Dynamics Ltd.; Figures 6.11 (a) and (b), 25.6, 25.12, 25.17, 25.28 and 25.35 reproduced with the permission of Woods Air Movement Ltd.; Figure 19.2 (a) reproduced with the permission of Thames Water; Figures 21.3, 21.4 and 21.5 (a) and (b) reproduced from *Pressure Surges in Pipe and Direct Systems* by J.A. Swaffield and A.P. Boldy, with kind permission from Ashgate Publishing Group and Adrian P. Boldy.

Parts I, V (page 473) and VI (page 593) photographs reproduced with the permission of Thames Water; Part II photograph reproduced with the permission of NEG Micon A/S © NEG Micon; Part V (p. 472) photograph reproduced with the permission of Scottish and Southern Energy plc; Part VI (page 592) photograph © Crown Copyright/MOD. Reproduced with the permission of Her Majesty's Stationery Office; Part VII photograph reproduced with the permission of Woods Air Movement Ltd.

Whilst every effort has been made to trace the owners of copyright material, in a few cases this has proved impossible and we take this opportunity to offer our apologies to any copyright holders whose rights we may have unwittingly infringed.

A Companion Web Site accompanies
Fluid Mechanics, 4e
by Douglas, Gasiorek and Swaffield

Visit the *Fluid Mechanics* Companion Web Site at www.booksites.net/douglas4 to find valuable teaching material including:

- Information about, and errata for, the book.
- Transparencies for selected figures.
- A password-protected area where a full Solutions Manual can be downloaded in pdf format.
- A syllabus manager that will build and host a course web page.

Elements of
Fluid Mechanics

Water effects, image
courtesy of Thames
Water

FLUID MECHANICS, AS THE NAME INDICATES, IS THAT branch of applied mechanics which is concerned with the statics and dynamics of liquids and gases. The analysis of the behaviour of fluids is based upon the fundamental laws of applied mechanics which relate to the conservation of mass–energy and the force–momentum equation, together with other concepts and equations with which the student who has already studied solid-body mechanics will be familiar. There are, however, two major aspects of fluid mechanics which differ from solid-body mechanics. The first is the nature and properties of the fluid itself, which are very different from those of a solid. The second is that, instead of dealing with individual bodies or elements of known mass, we are frequently concerned with the behaviour of a continuous stream of fluid, without beginning or end.

A further problem is that it can be extremely difficult to specify either the precise movement of a stream of fluid or that of individual particles within it. It is, therefore, often necessary – for the purpose of theoretical analysis – to assume ideal, simplified conditions and patterns of flow. The results so obtained may then be modified by introducing appropriate coefficients and factors, determined experimentally, to provide a basis for the design of fluid systems. This approach has proved to be reasonably satisfactory – in so far as the theoretical analysis usually establishes the form of the relationship between the variables; the experimental investigation corrects for the factors omitted from the theoretical model and establishes a quantitative relationship.

I

Fluids and their Properties

THIS CHAPTER WILL DEFINE THE NATURE OF FLUIDS, STRESSING BOTH THE COMMONALITY with concepts of applied mechanics applied to solid-body systems and the fundamental differences that arise from the nature of fluids. The appropriate physical properties that define these differences and allow the differentiation of fluids into gases and liquids, Newtonian and non-Newtonian, compressible and incompressible, will be identified. The application of the equation of state for perfect gases will be introduced. ● ● ●

1.1   FLUIDS

In everyday life, we recognize three states of matter: solid, liquid and gas. Although different in many respects, liquids and gases have a common characteristic in which they differ from solids: they are fluids, lacking the ability of solids to offer permanent resistance to a deforming force. Fluids *flow* under the action of such forces, deforming continuously for as long as the force is applied. A fluid is unable to retain any unsupported shape; it flows under its own weight and takes the shape of any solid body with which it comes into contact.

Deformation is caused by *shearing* forces, i.e. forces such as F (Fig. 1.1), which act tangentially to the surfaces to which they are applied and cause the material originally occupying the space ABCD to deform to AB′C′D. This leads to the definition:

FIGURE 1.1

Deformation caused by shearing forces

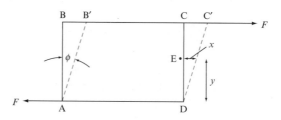

A fluid is a substance which deforms continuously under the action of shearing forces, however small they may be.

Conversely, it follows that:

If a fluid is at rest, there can be no shearing forces acting and, therefore, all forces in the fluid must be perpendicular to the planes upon which they act.

1.2   SHEAR STRESS IN A MOVING FLUID

Although there can be no shear stress in a fluid at rest, shear stresses *are* developed when the fluid is in motion, if the particles of the fluid move relative to each other so that they have different velocities, causing the original shape of the fluid to become distorted. If, on the other hand, the velocity of the fluid is the same at every point, no shear stresses will be produced, since the fluid particles are at rest relative to each other.

Usually, we are concerned with flow past a solid boundary. The fluid in contact with the boundary adheres to it and will, therefore, have the same velocity as the boundary. Considering successive layers parallel to the boundary (Fig. 1.2), the velocity of the fluid will vary from layer to layer as y increases.

If ABCD (Fig. 1.1) represents an element in a fluid with thickness s perpendicular to the diagram, then the force F will act over an area A equal to BC × s. The force per

FIGURE 1.2
Variation of velocity with
distance from a solid
boundary

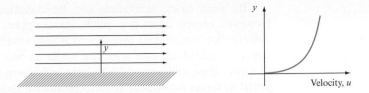

unit area F/A is the *shear stress* τ and the deformation, measured by the angle ϕ (the *shear strain*), will be proportional to the shear stress. In a solid, ϕ will be a fixed quantity for a given value of τ, since a solid can resist shear stress permanently. In a fluid, the shear strain ϕ will continue to increase with time and the fluid will flow. It is found experimentally that, in a true fluid, the rate of shear strain (or shear strain per unit time) is directly proportional to the shear stress.

Suppose that in time t a particle at E (Fig. 1.1) moves through a distance x. If E is a distance y from AD then, for small angles,

Shear strain, $\phi = x/y$,

Rate of shear strain $= x/yt = (x/t)/y = u/y,$

where $u = x/t$ is the velocity of the particle at E. Assuming the experimental result that shear stress is proportional to shear strain, then

$$\tau = \text{constant} \times u/y. \tag{1.1}$$

The term u/y is the change of velocity with y and may be written in the differential form $\mathrm{d}u/\mathrm{d}y$. The constant of proportionality is known as the *dynamic viscosity* μ of the fluid. Substituting into equation (1.1),

$$\tau = \mu \frac{\mathrm{d}u}{\mathrm{d}y}, \tag{1.2}$$

which is Newton's law of viscosity. The value of μ depends upon the fluid under consideration.

1.3 DIFFERENCES BETWEEN SOLIDS AND FLUIDS

To summarize, the differences between the behaviours of solids and fluids under an applied force are as follows:

1. For a solid, the strain is a function of the applied stress, providing that the elastic limit is not exceeded. For a fluid, the rate of strain is proportional to the applied stress.

2. The strain in a solid is independent of the time over which the force is applied and, if the elastic limit is not exceeded, the deformation disappears when the force is removed. A fluid continues to flow for as long as the force is applied and will not recover its original form when the force is removed.

In most cases, substances can be classified easily as either solids or fluids. However, certain cases (e.g. pitch, glass) appear to be solids because their rate of deformation under their own weight is very small. Pitch is actually a fluid which will flow and spread out over a surface under its own weight – but it will take days to do so rather than milliseconds! Similarly, solids will flow and become plastic when subjected to forces sufficiently large to produce a stress in the material which exceeds the elastic limit. They will also 'creep' under sustained loading, so that the deformation increases with time. A plastic substance does *not* meet the definition of a true fluid, since the shear stress must exceed a certain minimum value before flow commences.

1.4   NEWTONIAN AND NON-NEWTONIAN FLUIDS

Even among substances commonly accepted as fluids, there is a wide variation in behaviour under stress. Fluids obeying Newton's law of viscosity (equation (1.2)) and for which μ has a constant value are known as *Newtonian fluids*. Most common fluids fall into this category, for which shear stress is linearly related to velocity gradient (Fig. 1.3). Fluids which do not obey Newton's law of viscosity are known as non-Newtonian and fall into one of the following groups:

FIGURE 1.3

Variation of shear stress with velocity gradient

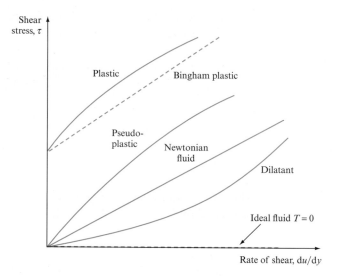

1.   *Plastic*, for which the shear stress must reach a certain minimum value before flow commences. Thereafter, shear stress increases with the rate of shear according to the relationship

$$\tau = A + B\left(\frac{du}{dy}\right)^{n},$$

where A, B and n are constants. If $n = 1$, the material is known as a Bingham plastic (e.g. sewage sludge).

2. *Pseudo-plastic*, for which dynamic viscosity decreases as the rate of shear increases (e.g. colloidal solutions, clay, milk, cement).

3. *Dilatant substances*, in which dynamic viscosity increases as the rate of shear increases (e.g. quicksand).

4. *Thixotropic substances*, for which the dynamic viscosity decreases with the time for which shearing forces are applied (e.g. thixotropic jelly paints).

5. *Rheopectic materials*, for which the dynamic viscosity increases with the time for which shearing forces are applied.

6. *Viscoelastic materials*, which behave in a manner similar to Newtonian fluids under time-invariant conditions but, if the shear stress changes suddenly, behave as if plastic.

The above is a classification of actual fluids. In analyzing some of the problems arising in fluid mechanics we shall have cause to consider the behaviour of an *ideal fluid*, which is assumed to have no viscosity. Theoretical solutions obtained for such a fluid often give valuable insight into the problems involved, and can, where necessary, be related to real conditions by experimental investigation.

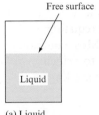

(a) Liquid

(b) Gas

FIGURE 1.4
Behaviour of a fluid in a container

1.5   LIQUIDS AND GASES

Although liquids and gases both share the common characteristics of fluids, they have many distinctive characteristics of their own. A liquid is difficult to compress and, for many purposes, may be regarded as incompressible. A given mass of liquid occupies a fixed volume, irrespective of the size or shape of its container, and a free surface is formed (Fig. 1.4(a)) if the volume of the container is greater than that of the liquid.

A gas is comparatively easy to compress. Changes of volume with pressure are large, cannot normally be neglected and are related to changes of temperature. A given mass of a gas has no fixed volume and will expand continuously unless restrained by a containing vessel. It will completely fill any vessel in which it is placed and, therefore, does not form a free surface (Fig. 1.4(b)).

1.6   MOLECULAR STRUCTURE OF MATERIALS

Solids, liquids and gases are all composed of molecules in continuous motion. However, the arrangement of these molecules, and the spaces between them, differ, giving rise to the characteristic properties of the three different states of matter. In solids, the molecules are densely and regularly packed and movement is slight, each molecule being restrained by its neighbours. In liquids, the structure is looser; individual molecules have greater freedom of movement and, although restrained to some degree by the surrounding molecules, can break away from this restraint, causing a change of structure. In gases, there is no formal structure, the spaces between molecules are large and the molecules can move freely.

The molecules of a substance exert forces on each other which vary with their intermolecular distance. Consider, for simplicity, a monatomic substance in which each molecule consists of a single atom. An idea of the nature of the forces acting may be formed from observing the behaviour of such a substance on a macroscopic scale.

FIGURE 1.5
(a) Variation of force with separation. (b) Variation of potential energy with separation

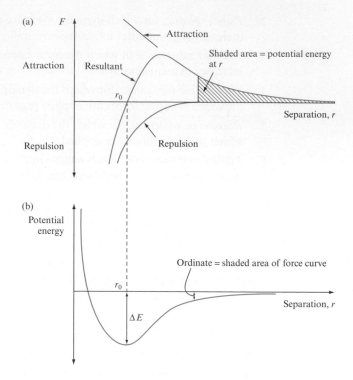

1.   If two pieces of the same material are far apart, there is no detectable force exerted between them. Thus, the forces between molecules are negligible when widely separated and tend to zero as the separation tends towards infinity.

2.   Two pieces of the same material can be made to weld together if they are forced into very close contact. Under these conditions, the forces between the molecules are attractive when the separation is very small.

3.   Very large forces are required to compress solids or liquids, indicating that a repulsive force between the molecules must be overcome to reduce the spacing between them.

It appears from these observations that interatomic forces vary with the distance of separation (Fig. 1.5(a)) and that there are two types of force, one attractive and the other repulsive. At small separations, the repulsive force is dominant; at larger separations, it becomes insignificant by comparison with the attractive force.

These conclusions can also be expressed in terms of the potential energy, defined as the energy required to bring one atom from infinity to a distance r from the second atom. The potential energy is zero if the atoms are infinitely far apart and is positive if external energy is required to move the first atom towards the second. Since Fig. 1.5(a) is the graph of the force F between the atom vs. the distance of separation, the potential energy curve (Fig. 1.5(b)) will be the integral of this curve from ∞ to r, which is the shaded area in Fig. 1.5(a).

At r_0, there is a condition of minimum energy, corresponding to $F = 0$ and representing a position of stable equilibrium, accounting for the inherent stability of solids and liquids in which the molecules are sufficiently densely packed for this condition to exist. Figure 1.5(b) also indicates that a pair of atoms can be separated completely, so that $r = \infty$, by the application of a finite amount of energy ΔE, which is called the dissociation or binding energy.

Considering a large number of particles of a substance, each particle will have kinetic energy $\frac{1}{2}mu^2$, where m is the mass of the particle and u its velocity. If a particle collides with a pair of particles, it will only cause them to separate if it can transfer to the pair energy in excess of ΔE. Thus, the possibility of stable pairs forming will depend on the average value of $\frac{1}{2}mu^2$ in relation to ΔE.

1. If the average value of $\frac{1}{2}mu^2 \gg \Delta E$, no stable pairs can form. The system will behave as a gas, consisting of individual particles moving rapidly with no apparent tendency to aggregate or occupy a fixed space.

2. If the average values of $\frac{1}{2}mu^2 \ll \Delta E$, no dissociation of pairs is possible and the colliding particle may be captured by the pair. The system has the properties of a solid, forming a stable conglomeration of particles which can only be dissociated by supplying energy from outside (e.g. by heating to produce melting and, subsequently, boiling).

3. If the average value of $\frac{1}{2}mu^2 \simeq \Delta E$, we have a system intermediate between (1) and (2), corresponding to the liquid state, since some particles will have values of $\frac{1}{2}mu^2 > \Delta E$, causing dissociation, while others will have values of $\frac{1}{2}mu^2 < \Delta E$ and will aggregate.

Summing up, in a solid, the individual molecules are close packed and their movement is restricted to vibrations of small amplitude. The kinetic energy is small compared with the dissociation energy, so that the molecules do not become separated but retain the same relative conditions.

In a liquid, the molecules are still close packed, but their movement is greater. Certain of the molecules will have sufficient kinetic energy to break through the surrounding molecules, so that the relative positions of the molecules can change from time to time. The material will cease to be rigid and can flow under the action of applied forces. However, the attraction between molecules is still sufficient to ensure that a given mass of liquid has a fixed volume and that a free surface will be formed.

In a gas, the spacing between molecules is some ten times as great as in a liquid. The kinetic energy is far greater than the dissociation energy. The attractive forces between molecules are very weak and intermolecular effects are negligible, so that molecules are free to travel until stopped by a solid or a liquid boundary. A gas will, therefore, expand to fill a container completely, irrespective of volume.

1.7   THE CONTINUUM CONCEPT OF A FLUID

Although the properties of a fluid arise from its molecular structure, engineering problems are usually concerned with the bulk behaviour of fluids. The number of molecules involved is immense, and the separation between them is normally negligible by comparison with the distances involved in the practical situation being studied. Under these conditions, it is usual to consider a fluid as a continuum – a hypothetical continuous substance – and the conditions at a point as the average of a very large number of molecules surrounding that point within a distance which is large compared with the mean intermolecular distance (although very small in absolute terms). Quantities such as velocity and pressure can then be considered to be constant at any point, and changes due to molecular motion may be ignored. Variations in such quantities can also be assumed to take place smoothly, from point to point. This assumption breaks down in the case of rarefied gases, for which the ratio of the mean free path of the molecules to the physical dimensions of the problem is very much larger.

In this book, fluids will be assumed to be continuous substances and, when the behaviour of a small element or particle of fluid is studied, it will be assumed that it contains so many molecules that it can be treated as part of this continuum.

Properties of fluids

The following properties of fluids are of general importance to the study of fluid mechanics. For convenience, a fuller list of the values of these properties for common fluids is given in Appendix 1, but typical values, SI units and dimensions in the MLT system (see Chapter 8) are given here.

1.8   DENSITY

The density of a substance is that quantity of matter contained in unit volume of the substance. It can be expressed in three different ways, which must be clearly distinguished.

1.8.1 Mass density

Mass density ρ is defined as the mass of the substance per unit volume. As mentioned above, we are concerned, in considering this and other properties, with the substance as a continuum and not with the properties of individual molecules. The mass density at a point is determined by considering the mass δm of a very small volume δV surrounding the point. In order to preserve the concept of the continuum, δV cannot be made smaller than x^3, where x is a linear dimension which is large compared with the mean distance between molecules. The density at a point is the limiting value as δV tends to x^3:

$$\rho = \lim_{\delta V \to x^3} \frac{\delta m}{\delta V}.$$

Units: kilograms per cubic metre (kg m^{-3}).
Dimensions: ML^{-3}.
Typical values at $p = 1.013 \times 10^5 \text{ N m}^{-2}$, $T = 288.15$ K: water, 1000 kg m^{-3}; air, 1.23 kg m^{-3}.

1.8.2 Specific weight

Specific weight w is defined as the weight per unit volume. Since weight is dependent on gravitational attraction, the specific weight will vary from point to point, according to the local value of gravitational acceleration g. The relationship between w and ρ can be deduced from Newton's second law, since

$$\text{Weight per unit volume} = \text{Mass per unit volume} \times g$$
$$w = \rho g.$$

Units: newtons per cubic metre (N m^{-3}).
Dimensions: ML^{-2}T^{-2}.
Typical values: water, 9.81×10^3 N m^{-3}; air, 12.07 N m^{-3}.

1.8.3 Relative density

Relative density (or specific gravity) σ is defined as the ratio of the mass density of a substance to some standard mass density. For solids and liquids, the standard mass density chosen is the maximum density of water (which occurs at 4 °C at atmospheric pressure):

$$\sigma = \rho_{\text{substance}}/\rho_{\text{H}_2\text{O at 4 °C}}.$$

For gases, the standard density may be that of air or of hydrogen at a specified temperature and pressure, but the term is not used frequently.
   Units: since relative density is a ratio of two quantities of the same kind, it is a pure number having no units.
   Dimensions: as a pure number, its dimensions are $M^0L^0T^0 = 1$.
   Typical values: water, 1.0; oil, 0.9.

1.8.4 Specific volume

In addition to these measures of density, the quantity specific volume is sometimes used, being defined as the reciprocal of mass density, i.e. it is used to mean volume per unit mass.

1.9 VISCOSITY

A fluid at rest cannot resist shearing forces, and, if such forces act on a fluid which is in contact with a solid boundary (Fig. 1.2), the fluid will flow over the boundary in such a way that the particles immediately in contact with the boundary have the same velocity as the boundary, while successive layers of fluid parallel to the boundary move with increasing velocities. Shear stresses opposing the relative motion of these layers are set up, their magnitude depending on the velocity gradient from layer to layer. For fluids obeying Newton's law of viscosity, taking the direction of motion as the x direction and v_x as the velocity of the fluid in the x direction at a distance y from the boundary, the shear stress in the x direction is given by

$$\tau_x = \mu\frac{\mathrm{d}v_x}{\mathrm{d}y}. \tag{1.3}$$

1.9.1 Coefficient of dynamic viscosity

The coefficient of dynamic viscosity μ can be defined as the shear force per unit area (or shear stress τ) required to drag one layer of fluid with unit velocity past another layer a unit distance away from it in the fluid. Rearranging equation (1.3),

$$\mu = \tau \bigg/ \frac{dv}{dy} = \frac{\text{Force}}{\text{Area}} \bigg/ \frac{\text{Velocity}}{\text{Distance}} = \frac{\text{Force} \times \text{Time}}{\text{Area}} \quad \text{or} \quad \frac{\text{Mass}}{\text{Length} \times \text{Time}}.$$

Units: newton seconds per square metre (N s m^{-2}) or kilograms per metre per second (kg m^{-1} s^{-1}). (But note that the coefficient of viscosity is often measured in poise (P); 10 P = 1 kg m^{-1} s^{-1}.)

Dimensions: ML^{-1}T^{-1}.

Typical values: water, 1.14×10^{-3} kg m^{-1} s^{-1}; air, 1.78×10^{-5} kg m^{-1} s^{-1}.

1.9.2 Kinematic viscosity

The kinematic viscosity v is defined as the ratio of dynamic viscosity to mass density:

$$v = \mu/\rho.$$

Units: square metres per second (m^2 s^{-1}). (But note that kinematic viscosity is often measured in stokes (St); 10^4 St = 1 m^2 s^{-1}.)

Dimensions: L^{2}T^{-1}.

Typical values: water, 1.14×10^{-6} m^2 s^{-1}; air, 1.46×10^{-5} m^2 s^{-1}.

1.10   CAUSES OF VISCOSITY IN GASES

When a gas flows over a solid boundary, the velocity of flow in the x direction, parallel to the boundary, will change with the distance y, measured perpendicular to the boundary. In Fig. 1.6, the velocity in the x direction is v_x at a distance y from the boundary and $v_x + \delta v_x$ at a distance $y + \delta y$. As the molecules of gas are not rigidly constrained, and cohesive forces are small, there will be a continuous interchange of molecules between adjacent layers which are travelling at different velocities. Molecules moving from the slower layer will exert a drag on the faster, while those moving from the faster layer will exert an accelerating force on the slower.

Assuming that the mass interchange per unit time is proportional to the area A under consideration, and inversely proportional to the distance δy between them,

$$\text{Mass interchange per unit time} = kA/\delta y,$$

where k is a constant of proportionality;

FIGURE 1.6

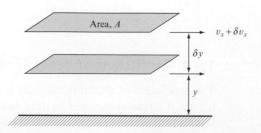

Change of velocity $= \delta v_x$;

Force exerted by one layer on the other

$\qquad$ = Rate of change of momentum

$\qquad$ = Mass interchange per unit time $\times$ Change of velocity

$$F = kA\frac{\delta v_x}{\delta y};$$

Viscous shear stress, $\tau = F/A = k\dfrac{\delta v_x}{\delta y}.$

Thus, from consideration of molecular mass interchange occurring in a gas, we arrive at Newton's law of viscosity.

If the temperature of a gas increases, the molecular interchange will increase. The viscosity of a gas will, therefore, increase as the temperature increases. According to the kinetic theory of gases, viscosity should be proportional to the square root of the absolute temperature; in practice, it increases more rapidly. Over the normal range of pressures, the viscosity of a gas is found to be independent of pressure, but it is affected by very high pressures.

1.11 CAUSES OF VISCOSITY IN A LIQUID

While there will be shear stresses due to molecular interchange similar to those developed in a gas, there are substantial attractive, cohesive forces between the molecules of a liquid (which are very much closer together than those of a gas). Both molecular interchange and cohesion contribute to viscous shear stress in liquids.

The effect of increasing the temperature of a fluid is to reduce the cohesive forces while simultaneously increasing the rate of molecular interchange. The former effect tends to cause a decrease of shear stress, while the latter causes it to increase. The net result is that liquids show a reduction in viscosity with increasing temperature which is of the form

$$\mu_T = \mu_0/(1 + A_1 T + B_1 T^2), \tag{1.4}$$

where μ_T is the viscosity at T °C, μ_0 is the viscosity at 0 °C and A_1 and B_1 are constants depending upon the liquid. For water, $\mu_0 = 0.0179$ P, $A_1 = 0.033\,68$ and $B_1 = 0.000\,221$. When plotted, equation (1.4) gives a hyperbola, viscosity tending to zero as temperature tends to infinity. An alternative relationship is

$$\mu/\mu_0 = A_2 \exp[B_2(1/T' - 1/T_0)], \tag{1.5}$$

where A_2 and B_2 are constants and T' is the *absolute* temperature.

High pressures also affect the viscosity of a liquid. The energy required for the relative movement of the molecules is increased and, therefore, the viscosity increases with increasing pressure. The relationship depends on the nature of the liquid and is exponential, having the form

$$\mu_p = \mu_0 \exp[C(p - p_0)], \tag{1.6}$$

where C is a constant for the liquid and μ_p is the viscosity at pressure p. For oils of the type used in oil hydraulic machinery, the increase in viscosity is of the order of 10 to 15 per cent for a pressure increase of 70 atm. Water, however, behaves rather differently from other fluids, since its viscosity only doubles for an increase in pressure from 1 to 1000 atm.

1.12 SURFACE TENSION

Although all molecules are in constant motion, a molecule within the body of the liquid is, on average, attracted equally in all directions by the other molecules surrounding it, but, at the surface between liquid and air, or the interface between one substance and another, the upward and downward attractions are unbalanced, the surface molecules being pulled inward towards the bulk of the liquid. This effect causes the liquid surface to behave as if it were an elastic membrane under tension. The surface tension σ is measured as the force acting across the unit length of a line drawn in the surface. It acts in the plane of the surface, normal to any line in the surface, and is the same at all points. Surface tension is constant at any given temperature for the surface of separation of two particular substances, but it decreases with increasing temperature.

The effect of surface tension is to reduce the surface of a free body of liquid to a minimum, since to expand the surface area molecules have to be brought to the surface from the bulk of the liquid against the unbalanced attraction pulling the surface molecules inwards. For this reason, drops of liquid tend to take a spherical shape in order to minimize surface area. For such a small droplet, surface tension will cause an increase of internal pressure p in order to balance the surface force.

Considering the forces acting on a diametral plane through a spherical drop of radius r, the force due to internal pressure $= p \times \pi r^2$, and the force due to surface tension around the perimeter $= 2\pi r \times \sigma$.

For equilibrium, $p\pi r^2 = 2\pi r\sigma$ or $p = 2\sigma/r$.

Surface tension will also increase the internal pressure in a cylindrical jet of fluid, for which $p = \sigma/r$. In either case, if r is very small, the value of p becomes very large. For small bubbles in a liquid, if this pressure is greater than the pressure of vapour or gas in a bubble, the bubble will collapse.

In many of the problems with which engineers are concerned, the magnitude of surface tension forces is very small compared with the other forces acting on the fluid and may, therefore, be neglected. However, these forces can cause serious errors in hydraulic scale models and through capillary effects. Surface tension forces can be reduced by the addition of detergents.

EXAMPLE 1.1

Air is introduced through a nozzle into a tank of water to form a stream of bubbles. If the bubbles are intended to have a diameter of 2 mm, calculate by how much the pressure of the air at the nozzle must exceed that of the surrounding water. Assume that $\sigma = 72.7 \times 10^{-3}$ N m^{-1}.

Solution

Excess pressure,

$$p = 2\sigma/r.$$

Putting $r = 1$ mm $= 10^{-3}$ m, $\sigma = 72.7 \times 10^{-3}$ N m^{-1}.
Excess pressure,

$$p = (2 \times 72.7 \times 10^{-3})/(1 \times 10^{-3}) = \mathbf{145.4 \ N \ m^{-2}}.$$

1.13 CAPILLARITY

If a fine tube, open at both ends, is lowered vertically into a liquid which wets the tube, the level of the liquid will rise in the tube (Fig. 1.7(a)). If the liquid does not wet the tube, the level of liquid in the tube will be depressed below the level of the free surface outside (Fig. 1.7(b)). If θ is the angle of contact between liquid and solid and d is the tube diameter (Fig. 1.7(a)),

FIGURE 1.7
Capillarity

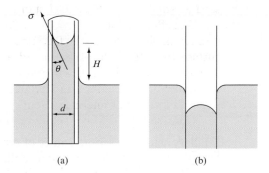

(a)   (b)

Upward pull due to surface tension	Component of surface tension acting upwards	Perimeter of tube		
	=	×	$= \sigma \cos\theta \times \pi d.$	(1.7)

The atmospheric pressure is the same inside and outside the tube, and, therefore, the only force opposing this upward pull is the weight of the vertical-sided column of liquid of height H, since, by definition, there are no shear stresses in a liquid at rest. Therefore, in Fig. 1.7, there will be no shear stress on the vertical sides of the column of liquid under consideration.

$$\text{Weight of column raised} = \rho g (\pi/4) d^2 H, \qquad (1.8)$$

where ρ is the mass density of the liquid.

Equating the upward pull to the weight of the column, from equations (1.7) and (1.8),

$$\sigma \cos\theta \times \pi d = \rho g (\pi/4) d^2 H,$$

Capillary rise, $H = 4\sigma \cos\theta / \rho g d.$

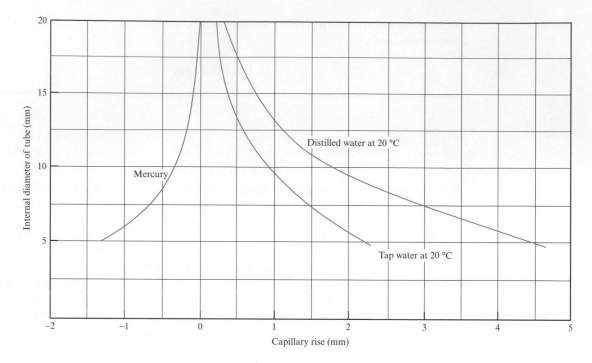

FIGURE 1.8
Capillary rise in glass tubes of circular cross-section

Capillary action is a serious source of error in reading liquid levels in fine-gauge tubes, particularly as the degree of wetting and, therefore, the contact angle θ are affected by the cleanness of the surfaces in contact. For water in a tube of 5 mm diameter, the capillary rise will be approximately 4.5 mm, while for mercury the corresponding figure would be −1.4 mm (Fig. 1.8). Gauge glasses for reading the level of liquids should have as large a diameter as is conveniently possible, to minimize errors due to capillarity.

1.14 VAPOUR PRESSURE

Since the molecules of a liquid are in constant agitation, some of the molecules in the surface layer will have sufficient energy to escape from the attraction of the surrounding molecules into the space above the free surface. Some of these molecules will return and condense, but others will take their place. If the space above the liquid is confined, an equilibrium will be reached so that the number of molecules of liquid in the space above the free surface is constant. These molecules produce a partial pressure known as the vapour pressure in the space.

The degree of molecular activity increases with increasing temperature, and, therefore, the vapour pressure will also increase. Boiling will occur when the vapour pressure is equal to the pressure above the liquid. By reducing the pressure, boiling can be made to occur at temperatures well below the boiling point at atmospheric pressure; for example, if the pressure is reduced to 0.2 bar (0.2 atm), water will boil at a temperature of 60 °C.

1.15 CAVITATION

Under certain conditions, areas of low pressure can occur locally in a flowing fluid. If the pressure in such areas falls below the vapour pressure, there will be local boiling and a cloud of vapour bubbles will form. This phenomenon is known as cavitation and can cause serious problems, since the flow of liquid can sweep this cloud of bubbles on into an area of higher pressure where the bubbles will collapse suddenly. If this should occur in contact with a solid surface, very serious damage can result due to the very large force with which the liquid hits the surface. Cavitation can affect the performance of hydraulic machinery such as pumps, turbines and propellers, and the impact of collapsing bubbles can cause local erosion of metal surfaces.

Cavitation can also occur if a liquid contains dissolved air or other gases, since the solubility of gases in a liquid decreases as the pressure is reduced. Gas or air bubbles will be released in the same way as vapour bubbles, with the same damaging effects. Usually, this release occurs at higher pressures and, therefore, before vapour cavitation commences.

1.16 COMPRESSIBILITY AND THE BULK MODULUS

All materials, whether solids, liquids or gases, are compressible, i.e. the volume V of a given mass will be reduced to $V - \delta V$ when a force is exerted uniformly all over its surface. If the force per unit area of surface increases from p to $p + \delta p$, the relationship between change of pressure and change of volume depends on the bulk modulus of the material:

Bulk modulus = Change in pressure/Volumetric strain.

Volumetric strain is the change in volume divided by the original volume; therefore,

$$\frac{\text{Change in volume}}{\text{Original volume}} = \frac{\text{Change in pressure}}{\text{Bulk modulus}},$$

$$-\delta V / V = \delta p / K,$$

the minus sign indicating that the volume decreases as pressure increases. In the limit, as $\delta p \to 0$,

$$K = -V \frac{\mathrm{d}p}{\mathrm{d}V}. \tag{1.9}$$

Considering unit mass of a substance,

$$V = 1/\rho. \tag{1.10}$$

Differentiating,

$$V\,\mathrm{d}\rho + \rho\,\mathrm{d}V = 0$$

$$\mathrm{d}V = -(V/\rho)\,\mathrm{d}\rho.$$

Substituting for V from equation (1.10),

$$\mathrm{d}V = -(1/\rho^2)\,\mathrm{d}\rho. \tag{1.11}$$

Putting the values of V and $\mathrm{d}V$ obtained from equations (1.10) and (1.11) in equation (1.9),

$$K = \rho\frac{\mathrm{d}p}{\mathrm{d}\rho}. \tag{1.12}$$

The value of K is shown by equation (1.12) to be dependent on the relationship between pressure and density and, since density is also affected by temperature, it will depend on how the temperature changes during compression. If the temperature is constant, conditions are said to be isothermal, while, if no heat is allowed to enter or leave during compression, conditions are adiabatic. The ratio of the adiabatic bulk modulus to the isothermal bulk modulus is equal to γ, the ratio of the specific heat of a fluid at constant pressure to that at constant volume. For liquids, γ is approximately unity and the two conditions need not be distinguished; for gases, the difference is substantial (for air, $\gamma = 1.4$).

The concept of the bulk modulus is mainly applied to liquids, since for gases the compressibility is so great that the value of K is not a constant, but proportional to pressure and changes very rapidly. The relationship between pressure and mass density is more conveniently found from the characteristic equation of a gas (1.13). For liquids, the value of K is high and changes of density with pressure are small, but increasing pressure does bring the molecules of the liquid closer together, increasing the value of K. For water, the value of K will double if the pressure is increased from 1 to 3500 atm. An increase of temperature will cause the value of K to fall.

For liquids, the changes in pressure occurring in many fluid mechanics problems are not sufficiently great to cause appreciable changes in density. It is, therefore, usual to ignore such changes and to treat liquids as incompressible. Where, however, sudden changes of velocity generate large inertial forces, high pressures can occur and compressibility effects cannot be disregarded in liquids (see Chapter 20). Gases may also be treated as incompressible if the pressure changes are very small, but, usually, compressibility cannot be ignored. In general, compressibility becomes important when the velocity of the fluid exceeds about one-fifth of the velocity of a pressure wave (e.g. the velocity of sound) in the fluid.

Units: since volumetric strain is the ratio of two volumes, the units of bulk modulus will be the same as those of pressure, newtons per square metre (N m^{-2}).

Dimensions: ML^{-1}T^{-2}.

Typical values: water, 2.05×10^9 N m^{-2}; oil, 1.62×10^9 N m^{-2}.

1.17   EQUATION OF STATE OF A PERFECT GAS

The mass density of a gas varies with its absolute pressure p and absolute temperature T. For a perfect gas,

$$p = \rho RT, \tag{1.13}$$

where R is the gas constant for the gas concerned. Most gases at pressures and temperatures well removed from liquefaction follow this characteristic equation closely, but it does not apply to vapours.

   Units: the gas contant is measured in joules per kilogram per kelvin (J kg^{-1} K^{-1}).
   Dimensions: $L^2 T^{-2} \Theta^{-1}$.
   Typical values: air, 287 J kg^{-1} K^{-1}; hydrogen, 4110 J kg^{-1} K^{-1}.

1.18   THE UNIVERSAL GAS CONSTANT

From equation (1.13) ρR is constant for a given value of pressure p and temperature T. By Avogadro's hypothesis, all pure gases have the same number of molecules per unit volume at the same temperature and pressure, so that ρ is proportional to the molar mass M (kg kmol^{-1}). Therefore, the quantity MR will be constant for all perfect gases, and is known as the universal gas constant R_0.

$$R_0 = MR = 8.314 \text{ kJ kmol}^{-1} \text{ K}^{-1}.$$

1.19   SPECIFIC HEATS OF A GAS

Since pressure, temperature and density of a gas are interrelated, the amount of heat energy H required to raise the temperature of a gas from T_1 to T_2 will depend upon whether the gas is allowed to expand during the process, so that some of the energy supplied is used in doing work instead of raising the temperature of the gas. Two different specific heats are, therefore, given for a gas, corresponding to the two extreme conditions of constant volume and constant pressure.

1.   *Specific heat at constant volume c_v.* For a temperature change from T_1 to T_2 at constant volume,

   Heat supplied per unit mass, $H = c_v(T_2 - T_1)$.

   Since there is no change in volume, no external work is done, so that the increase of internal energy per unit mass of gas is $c_v(T_2 - T_1)$ heat units.

2.   *Specific heat at constant pressure c_p.* If the pressure is kept constant, the gas will expand as the temperature changes from T_1 to T_2:

   Heat supplied per unit mass $= c_p(T_2 - T_1)$ heat units.

Only part of this energy is used to raise the temperature of the gas; the rest goes to external work.

Thus, $c_p > c_v$:

$$c_p(T_2 - T_1) = c_v(T_2 - T_1) + \text{External work (in heat units)}.$$

It can be shown that $R = (c_p - c_v)$, where R, c_p and c_v have the same units.
Units: specific heat is measured in joules per kilogram per kelvin, as is R.
Dimensions: $L^2 T^{-2} \Theta^{-1}$.
Typical values: air, $c_p = 1.005 \text{ kJ kg}^{-1} \text{ K}^{-1}$, $c_v = 0.718 \text{ kJ kg}^{-1} \text{ K}^{-1}$.

1.20 EXPANSION OF A GAS

When a gas expands, the amount of work done will depend upon the relationship between pressure and volume, which, in turn, depends upon whether the gas receives or loses heat during the process.

If a unit mass of a gas has a volume V_1 at pressure p_1 and volume V_2 at pressure p_2, as shown in Fig. 1.9, then,

$$\begin{matrix} \text{Work done} \\ \text{during expansion} \end{matrix} = \begin{matrix} \text{Area under } p - V \\ \text{curve between } V_1 \text{ and } V_2 \end{matrix} = \int_{V_1}^{V_2} p \, dV.$$

FIGURE 1.9
Expansion of a gas

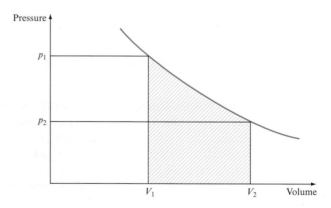

1. If the expansion is isothermal, the absolute temperature T (in kelvin) of the gas remains unchanged and the characteristic equation $p = \rho RT$ becomes $p/\rho = $ constant; or, putting $V = $ volume of unit mass $= 1/\rho$,

$$pV = \text{constant} = p_1 V_1 = RT, \tag{1.14}$$

$$p = p_1 V_1 (1/V).$$

From equation (1.14),

$$\text{Work done per unit mass} = p_1 V_1 \int_{V_1}^{V_2} \frac{dV}{V}$$

$$= p_1 V_1 \log_e(V_2/V_1)$$

$$= \boldsymbol{R} T \log_e(V_2/V_1).$$

2. For any known relationship between pressure and mass density of the form $p/\rho^n = \text{constant}$, putting $V = 1/\rho$,

$$pV^n = p_1 V_1^n = \text{constant.} \tag{1.15}$$

Therefore,

$$p = p_1 V_1^n V^{-n}.$$

$$\text{Work done by gas per unit mass} = \int_{V_1}^{V_2} p \; dV$$

$$= p_1 V_1^n \int_{V_1}^{V_2} V^{-n} \; dV$$

$$= [p_1 V_1^n/(1-n)][V_2^{(1-n)} - V_1^{(1-n)}]$$

$$= (1-n)^{-1}[p_1 V_1^n V_2^{(1-n)} - p_1 V_1],$$

or, since $p_1 V_1^n = p_2 V_2^n$,

$$\text{Work done by gas per unit mass} = (p_2 V_2 - p_1 V_1)/(1-n)$$

$$= (p_1 V_1 - p_2 V_2)/(n-1)$$

$$= \boldsymbol{R}(T_1 - T_2)/(n-1). \tag{1.16}$$

3. If the compression is carried out adiabatically, no heat enters or leaves the system. Now, for any mode of compression, considering unit mass,

Heat supplied = Change of internal energy + Work done (in heat units).

Change of internal energy $= c_v(T_2 - T_1)$.

Mechanical work done $= (p_2 V_2 - p_1 V_1)/(1-n)$.

Thus, in general, if H is the heat supplied,

$$H = c_v(T_2 - T_1) + (p_2 V_2 - p_1 V_1)/(1-n).$$

Now, $\boldsymbol{R} = (c_p - c_v)$ or $c_v = \boldsymbol{R}/(c_p/c_v - 1).$

Also $R(T_2 - T_1) = (p_2 V_2 - p_1 V_1)$.

Thus, $H = (p_2 V_2 - p_1 V_1)/(c_p/c_v - 1) + (p_2 V_2 - p_1 V_1)/(1 - n)$.

For an adiabatic change, $H = 0$, so that

$$(p_2 V_2 - p_1 V_1)/(c_p/c_v - 1) = -(p_2 V_2 - p_1 V_1)/(1 - n)$$
$$= (p_2 V_2 - p_1 V_1)/(n - 1),$$

and, therefore,

$$n = c_p/c_v = \gamma.$$

Thus, for an adiabatic change, the relationship between pressure and density is given by

$$pV^\gamma = p/\rho^\gamma = \text{constant}, \tag{1.17}$$

and, from (2),

$$\text{Work done by gas per unit mass} = (p_1 V_1 - p_2 V_2)/(\gamma - 1)$$
$$= R(T_1 - T_2)/(\gamma - 1). \tag{1.18}$$

Concluding remarks

The material presented in this chapter will be utilized in all sections of the text; in particular the influence of fluid viscosity will be of the utmost importance. The equation of state and the definition of compressible flows will also be used to differentiate flow conditions.

Summary of important equations and concepts

1. The relationship between shear stress, viscosity and velocity gradient, equation (1.3), will recur throughout the text. While this text will be concerned with Newtonian fluids, as defined in Section 1.4, the reader should be familiar with the differentiation between these and other non-Newtonian fluid types.

2. The fundamental fluid properties introduced must be understood and their dependence on temperature and pressure appreciated. In particular the defining differences between liquids and gases become essential in dealing with concepts of compressibility and time dependency, Section 1.5.

3. A range of properties are introduced in this chapter whose importance will be returned to later under particular flow conditions: for example surface tension and its effects on capillary action; vapour pressure and its role in pressure surge analysis; and cavitation as a limit to pump operation and propeller/turbine blade design.

4. The concept of compressibility, the differences between gas and liquid compressibility and the conditions under which flows may be considered incompressible must be understood, Section 1.16.

5. The gas laws are essential to the later development of the concepts of gaseous fluid flow, Sections 1.17 to 1.20.

2

Pressure and Head

THIS CHAPTER WILL CONSIDER AND INTRODUCE THE FORCES ACTING ON, OR GENERATED by, fluids at rest. In particular the concept of pressure will be introduced, including its variation with depth of submergence, via the hydrostatic equation, its unique value at any particular depth in a continuous fluid and direction of application at that depth. The concept that the atmosphere dictates that all activities on the Earth's surface are effectively carried out submerged in a fluid will be stressed and the pressure variations within, and stability of, the atmosphere will be treated. This understanding of pressure will be used to introduce methods of pressure measurement that will be essential to the treatment, in later chapters, of fluids in motion. ● ● ●

2.1   STATICS OF FLUID SYSTEMS

The general rules of statics apply to fluids at rest, but, from the definition of a fluid (Section 1.1), there will be no shearing forces acting and, therefore, all forces (such as *F* in Fig. 2.1(a)) exerted between the fluid and a solid boundary must act at right angles to the boundary. If the boundary is curved (Fig. 2.1(b)), it can be considered to be composed of a series of chords on each of which a force $F_1, F_2, \ldots, F_n$ acts perpendicular to the surface at the section concerned. Similarly, considering any plane drawn through a body of fluid at rest (Fig. 2.1(c)), the force exerted by one portion of the fluid on the other acts at right angles to this plane.

FIGURE 2.1

Forces in a fluid at rest

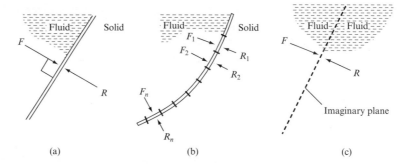

(a)   (b)   (c)

Shear stresses due to viscosity are only generated when there is relative motion between elements of the fluid. The principles of fluid statics can, therefore, be extended to cases in which the fluid is moving as a whole but all parts are stationary relative to each other.

In the analysis of a problem it is usual to consider an element of the fluid defined by solid boundaries or imaginary planes. A free body diagram can be drawn for this element, showing the forces acting on it due to the solid boundaries or surrounding fluid. Since the fluid is at rest, the element will be in equilibrium, and the sum of the component forces acting in any direction must be zero. Similarly, the sum of the moments of the forces about any point must be zero. It is usual to test equilibrium by resolving along three mutually perpendicular axes and, also, by taking moments in three mutually perpendicular planes.

Although a body or element may be in equilibrium, it can also be of interest to know what will happen if it is displaced from its equilibrium position. For example, in the case of a ship it is of the utmost importance to know whether it will overturn when it pitches or rolls or whether it will tend to right itself and return to its original position. There are three possible conditions of equilibrium:

1.   *Stable equilibrium.* A small displacement from the equilibrium position generates a force producing a righting moment tending to restore the body to its equilibrium position.

2. *Unstable equilibrium.* A small displacement produces an overturning moment tending to displace the body further from its equilibrium position.

3. *Neutral equilibrium.* The body remains at rest in any position to which it is displaced.

These conditions are typified by the three positions of a cone on a horizontal surface shown in Fig. 2.2.

FIGURE 2.2
Types of equilibrium

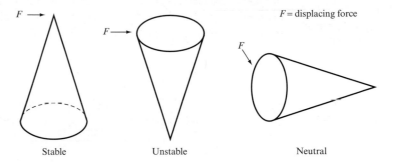

Stable   Unstable   Neutral

2.2   PRESSURE

A fluid will exert a force normal to a solid boundary or any plane drawn through the fluid. Since problems may involve bodies of fluids of indefinite extent and, in many cases, the magnitude of the force exerted on a small area of the boundary or plane may vary from place to place, it is convenient to work in terms of the *pressure p* of the fluid, defined as the force exerted per unit area. If the force exerted on each unit area of a boundary is the same, the pressure is said to be uniform:

$$\text{Pressure} = \frac{\text{Force exerted}}{\text{Area of boundary}} \quad \text{or} \quad p = \frac{F}{A}.$$

If, as is more commonly the case, the pressure changes from point to point, we consider the element of force δF normal to a small area δA surrounding the point under consideration:

$$\text{Mean pressure,} \ \ p = \frac{\delta F}{\delta A}.$$

In the limit, as $\delta A \to 0$ (but remains large enough to preserve the concept of the fluid as a continuum),

$$\text{Pressure at a point,} \ \ p = \lim_{\delta A \to 0} \frac{\delta F}{\delta A} = \frac{\mathrm{d}F}{\mathrm{d}A}.$$

Units: newtons per square metre (N m^{-2}). (Note that an alternative metric unit is the bar; 1 bar $= 10^5$ N m^{-2}.)
Dimensions: ML^{-1}T^{-2}.

2.3   PASCAL'S LAW FOR PRESSURE AT A POINT

By considering the equilibrium of a small fluid element in the form of a triangular prism surrounding a point in the fluid (Fig. 2.3), a relationship can be established between the pressures p_x in the x direction, p_y in the y direction and p_s normal to any plane inclined at any angle θ to the horizontal at this point.

FIGURE 2.3

Equality of pressure in all directions at a point

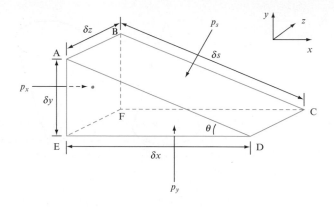

If the fluid is at rest, p_x will act at right angles to the plane ABFE, p_y at right angles to CDEF and p_s at right angles to ABCD. Since the fluid is at rest, there will be no shearing forces on the faces of the element and the element will not be accelerating. The sum of the forces in any direction must, therefore, be zero.

Considering the x direction:

$$\text{Force due to } p_x = p_x \times \text{Area ABFE} = p_x \delta y \delta z;$$

$$\text{Component of force due to } p_s = -(p_s \times \text{Area ABCD}) \sin \theta$$

$$= -p_s \delta s \delta z \frac{\delta y}{\delta s} = -p_s \delta y \delta z$$

(since $\sin \theta = \delta y / \delta s$). As p_y has no compound in the x direction, the element will be in equilibrium if

$$p_x \delta y \delta z + (-p_s \delta y \delta z) = 0,$$

$$p_x = p_s. \tag{2.1}$$

Similarly, in the y direction,

$$\text{Force due to } p_y = p_y \times \text{Area CDEF} = p_y \delta x \delta z;$$

$$\text{Component of force due to } p_s = -(p_s \times \text{Area ABCD}) \cos \theta$$

$$= -p_s \delta s \delta z \frac{\delta x}{\delta s} = -p_s \delta x \delta z$$

(since $\cos \theta = \delta x / \delta s$).

Weight of element = −Specific weight × Volume

$$= -\rho g \times \tfrac{1}{2}\,\delta x \delta y \delta z.$$

As p_x has no component in the y direction, the element will be in equilibrium if

$$p_y \delta x \delta z + (-p_s \delta x \delta z) + (-\rho g \times \tfrac{1}{2}\,\delta x \delta y \delta z) = 0.$$

Since δx, δy and δz are all very small quantities, $\delta x \delta y \delta z$ is negligible in comparison with the other two terms, and the equation reduces to

$$p_y = p_s. \tag{2.2}$$

Thus, from equations (2.1) and (2.2),

$$p_s = p_x = p_y. \tag{2.3}$$

Now p_s is the pressure on a plane inclined at *any* angle θ; the x, y and z axes have not been chosen with any particular orientation, and the element is so small that it can be considered to be a point. This proof may be extended to the z axis. Equation (2.3), therefore, indicates that the pressure at a point is the same in all directions. This is known as *Pascal's law* and applies to a fluid at rest.

If the fluid is flowing, shear stresses will be set up as a result of relative motion between the particles of the fluid. The pressure at a point is then considered to be the mean of the normal forces per unit area (stresses) on three mutually perpendicular planes. Since these normal stresses are usually large compared with shear stresses it is generally assumed that Pascal's law still applies.

2.4   VARIATION OF PRESSURE VERTICALLY IN A FLUID UNDER GRAVITY

Figure 2.4 shows an element of fluid consisting of a vertical column of constant cross-sectional area A and totally surrounded by the same fluid of mass density ρ. Suppose that the pressure is p_1 on the underside at level z_1 and p_2 on the top at level z_2. Since the fluid is at rest the element must be in equilibrium and the sum of all the vertical forces must be zero. The forces acting are:

Force due to p_1 on area A acting up $= p_1 A$,

Force due to p_2 on area A acting down $= p_2 A$,

Force due to the weight of the element $= mg$

$$= \text{Mass density} \times g \times \text{Volume} = \rho g A (z_2 - z_1).$$

Since the fluid is at rest, there can be no shear forces and, therefore, no vertical forces act on the side of the element due to the surrounding fluid. Taking upward forces as positive and equating the algebraic sum of the forces acting to zero,

$$p_1 A - p_2 A - \rho g A (z_2 - z_1) = 0,$$

FIGURE 2.4
Vertical variation
of pressure

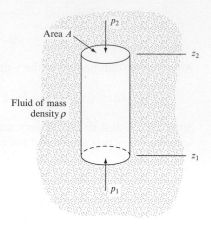

$$p_2 - p_1 = -\rho g (z_2 - z_1). \tag{2.4}$$

Thus, in any fluid under gravitational attraction, pressure decreases with increase of height z.

EXAMPLE 2.1

A diver descends from the surface of the sea to a depth of 30 m. What would be the pressure under which the diver would be working above that at the surface assuming that the density of sea water is 1025 kg m^{-3} and remains constant?

Solution

In equation (2.4), taking sea level as datum, $z_1 = 0$. Since z_2 is lower then z_1 the value of z_2 is −30 m. Substituting these values and putting $\rho = 1025$ kg m^{-3}:

$$\text{Increase of pressure} = p_2 - p_1$$
$$= -1025 \times 9.81 \, (-30 - 0) = \mathbf{301.7 \times 10^3 \, N \, m^{-2}}.$$

2.5   EQUALITY OF PRESSURE AT THE SAME LEVEL IN A STATIC FLUID

If P and Q are two points at the same level in a fluid at rest (Fig. 2.5), a horizontal prism of fluid of constant cross-sectional area A will be in equilibrium. The forces acting on this element horizontally are $p_1 A$ at P and $p_2 A$ at Q. Since the fluid is at rest, there will be no horizontal shear stresses on the sides of the element. For static equilibrium the sum of the horizontal forces must be zero:

$$p_1 A = p_2 A,$$

FIGURE 2.5

Equality of pressures at
the same level

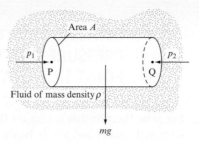

$$p_1 = p_2.$$

Thus, the pressure at any two points at the same level in a body of fluid at rest will be the same.

In mathematical terms, if (x, y) is the horizontal plane,

$$\frac{\partial p}{\partial x} = 0 \quad \text{and} \quad \frac{\partial p}{\partial y} = 0;$$

partial derivatives are used because pressure p could vary in three directions.

Pressures at the same level will be equal even though there is no direct horizontal path between P and Q, provided that P and Q are in the same continuous body of fluid. Thus, in Fig. 2.6, P and Q are connected by a horizontal pipe, R and S being two points at the same level at the entrance and exit to the pipe. If the pressure is p_P at P, p_Q at Q, p_R at R and p_S at S, then, since R and S are at the same level,

$$p_R = p_S; \tag{2.5}$$

also $p_R = p_P + \rho gz$ and $p_S = p_Q + \rho gz.$

Substituting in equation (2.5),

$$p_P + \rho gz = p_Q + \rho gz, \quad p_P = p_Q.$$

FIGURE 2.6

Equality of pressures in a
continuous body of fluid

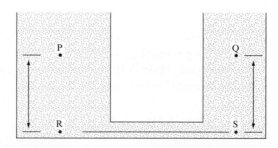

2.6 GENERAL EQUATION FOR THE VARIATION OF PRESSURE DUE TO GRAVITY FROM POINT TO POINT IN A STATIC FLUID

Let p be the pressure acting on the end P of an element of fluid of constant cross-sectional area A and $p + \delta p$ be the pressure at the other end Q (Fig. 2.7).

FIGURE 2.7

Variation of pressure in a stationary fluid

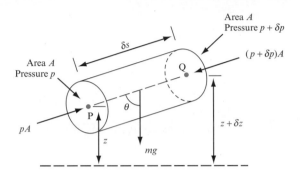

The axis of the element is inclined at an angle θ to the vertical, the height of P above a horizontal datum is z and that of Q is $z + \delta z$. The forces acting on the element are:

pA	acting at right angles to the end face at P along the axis of the element,
$(p + \delta p)A$	acting at Q along the axis in the opposite direction;
$m\mathbf{g}$	the weight of the element, due to gravity, acting vertically down

$$= \text{Mass density} \times \text{Volume} \times \text{Gravitational acceleration}$$
$$= \rho \times A\delta s \times \mathbf{g}.$$

There are also forces due to the surrounding fluid acting normal to the sides of the element, since the fluid is at rest, and, therefore, perpendicular to its axis PQ.

For equilibrium of the element PQ, the resultant of these forces in any direction must be zero. Resolving along the axis PQ,

$$pA - (p + \delta p)A - \rho g A \delta s \cos \theta = 0, \quad \delta p = -\rho g \delta s \cos \theta,$$

or, in differential form,

$$\frac{\mathrm{d}p}{\mathrm{d}s} = -\rho \mathbf{g} \cos \theta.$$

In the general three-dimensional case, s is a vector with components in the x, y and z directions. Taking the (x, y) plane as horizontal, if the axis of the element is also horizontal, $\theta = 90°$ and

$$\left(\frac{\mathrm{d}p}{\mathrm{d}s}\right)_{\theta=90°} = \frac{\partial p}{\partial x} = \frac{\partial p}{\partial y} = 0, \tag{2.6}$$

confirming the results of Section 2.5 that, in a static fluid, pressure is constant everywhere in a horizontal plane. It is for this reason that the free surface of a liquid is horizontal.

If the axis of the element is in the vertical z direction, $\theta = 0°$ and

$$\left(\frac{\mathrm{d}p}{\mathrm{d}s}\right)_{\theta=0°} = \frac{\partial p}{\partial z} = -\rho g,$$

and, since $\partial p/\partial x = \partial p/\partial y = 0$, the partial derivative $\partial p/\partial z$ can be replaced by the total differential $\mathrm{d}p/\mathrm{d}z$, giving

$$\frac{\mathrm{d}p}{\mathrm{d}z} = -\rho g, \tag{2.7}$$

which corresponds to the result obtained in Section 2.4.

Also, considering any two horizontal planes a vertical distance z apart,

Pressure at all points on lower plane $= p$,

Pressure at all points on upper plane $= p + z\dfrac{\partial p}{\partial z}$,

Difference of pressure $= z\dfrac{\partial p}{\partial z}$.

Since the planes are horizontal, the pressure must be constant over each plane; therefore, $\partial p/\partial z$ cannot vary horizontally. From equation (2.7), this implies that ρg shall be constant and, therefore, for equilibrium, the density ρ must be constant over any horizontal plane.

Thus, the conditions for equilibrium under gravity are:

1. The pressure at all points on a horizontal plane must be the same.
2. The density at all points on a horizontal plane must be the same.
3. The change of pressure with elevation is given by $\mathrm{d}p/\mathrm{d}z = -\rho g$.

The actual pressure variation with elevation is found by integrating equation (2.7):

$$\mathrm{d}p = -\int \rho g \, \mathrm{d}z \quad \text{or} \quad p_2 - p_1 = -\int_{z_1}^{z_2} \rho g \, \mathrm{d}z, \tag{2.8}$$

but this cannot be done unless the relationship between ρ and p is known.

2.7 VARIATION OF PRESSURE WITH ALTITUDE IN A FLUID OF CONSTANT DENSITY

For most problems involving liquids it is usual to assume that the density ρ is constant, and the same assumption can also be made for a gas if pressure differences are very small. Equation (2.8) can then be written

$$p = -\rho g \int dz = -\rho gz + \text{constant},$$

or, for any two points at altitude z_1 and z_2 above datum,

$$p_2 - p_1 = -\rho g(z_2 - z_1).$$

2.8 VARIATION OF PRESSURE WITH ALTITUDE IN A GAS AT CONSTANT TEMPERATURE

The relation between pressure, density and temperature for a perfect gas is given by the equation $p/\rho = \mathbf{R}T$. If conditions are assumed to be isothermal, so that temperature does not vary with altitude, ρ can be expressed in terms of p and the result substituted in equation (2.7):

$$\rho = \frac{p}{RT},$$

and, from equation (2.7),

$$\frac{dp}{dz} = -\rho g = -\frac{pg}{RT},$$

$$\frac{dp}{p} = -\frac{g}{RT} dz.$$

Integrating from $p = p_1$ when $z = z_1$, to $p = p_2$ when $z = z_2$,

$$\log_e(p_2/p_1) = -(g/RT)(z_2 - z_1),$$

$$p_2/p_1 = \exp[-(g/RT)(z_2 - z_1)].$$

EXAMPLE 2.2

At an altitude z, of 11 000 m, the atmospheric temperature T is −56.6 °C and the pressure p is 22.4 kN m⁻². Assuming that the temperature remains the same at higher altitudes, calculate the density of the air at an altitude of z_2 of 15 000 m. Assume $\mathbf{R} = 287$ J kg⁻¹ K⁻¹.

Solution

Let p_2 be the absolute pressure at z_2. Since the temperature is constant,

$$p_2/p_1 = \exp[-(g/RT)(z_2 - z_1)].$$

Putting $p_1 = 22.4$ kN m^{-2} = 22.4×10^3 N m^{-2}, $z_1 = 11\,000$ m, $z_2 = 15\,000$ m, $R = 287$ J kg^{-1} K^{-1}, $T = -56.6\,°C = 216.4$ K:

$$p_2 = 22.4 \times 10^3 \times \exp\left[-\frac{9.81(15\,000 - 11\,000)}{287 \times 216.4}\right]$$

$$= 22.4 \times 10^3 \times \exp(-0.632) = 11.91 \times 10^3 \text{ N m}^{-2}.$$

Also, from the equation of state for a perfect gas (see equation (1.13)), $p_2 = \rho_2 RT$ and so

$$\text{Density of air at } 15\,000 \text{ m}, \rho_2 = p_2/RT$$

$$= (11.91 \times 10^3)/(287 \times 216.4)$$

$$= \mathbf{0.192 \text{ kg m}^{-3}}.$$

2.9   VARIATION OF PRESSURE WITH ALTITUDE IN A GAS UNDER ADIABATIC CONDITIONS

If conditions are adiabatic, the relationship between pressure and density is given by $p/\rho^\gamma = \text{constant} = p_1/\rho_1^\gamma$, so that

$$\rho = \rho_1(p/p_1)^{1/\gamma}.$$

Substituting in equation (2.7),

$$\frac{\mathrm{d}p}{\mathrm{d}z} = -\frac{\rho_1 g}{p_1^{1/\gamma}}\, p^{1/\gamma},$$

$$\mathrm{d}z = -\left(\frac{p_1^{1/\gamma}}{\rho_1 g}\right) p^{-1/\gamma}\, \mathrm{d}p.$$

Integrating from $p = p_1$ when $z = z_1$, to $p = p_2$ when $z = z_2$,

$$z_2 - z_1 = -\frac{p_1^{1/\gamma}}{\rho_1 g}\left[\frac{p^{(\gamma-1)/\gamma}}{(\gamma-1)/\gamma}\right]_{p_1}^{p_2}$$

$$= -\left(\frac{\gamma}{\gamma-1}\right)\frac{p_1^{1/\gamma}}{\rho_1 g}(p_2^{(\gamma-1)/\gamma} - p_1^{(\gamma-1)/\gamma})$$

$$= -\left(\frac{\gamma}{\gamma-1}\right)\frac{p_1}{\rho_1 g}\left[\left(\frac{p_2}{p_1}\right)^{(\gamma-1)/\gamma} - 1\right],$$

or, since $p_1/\rho_1 = RT$, for any gas,

$$z_2 - z_1 = -\left(\frac{\gamma}{\gamma - 1}\right)\frac{RT_1}{g}\left[\left(\frac{p_2}{p_1}\right)^{(\gamma-1)/\gamma} - 1\right],$$

$$\left(\frac{p_2}{p_1}\right)^{(\gamma-1)/\gamma} = \frac{g(z_2 - z_1)}{RT_1}\left(\frac{\gamma - 1}{\gamma}\right) + 1,$$

$$\frac{p_2}{p_1} = \left[1 - \frac{g(z_2 - z_1)}{RT_1}\left(\frac{\gamma - 1}{\gamma}\right)\right]^{\gamma/(\gamma-1)}. \tag{2.9}$$

This can be extended to any isentropic process for which $p/\rho^n = $ constant, to give

$$\frac{p_2}{p_1} = \left[1 - \frac{g(z_2 - z_1)}{RT_1}\left(\frac{n - 1}{n}\right)\right]^{n/(n-1)}. \tag{2.10}$$

The rate of change of temperature with altitude – the temperature lapse rate – can also be found for adiabatic conditions. From the characteristic equation, $\rho = p/RT$ and, since, from equation (2.7),

$$dz = -dp/\rho g,$$

substituting for ρ,

$$dz = -(RT/gp)\,dp. \tag{2.11}$$

For adiabatic conditions,

$$p/\rho^\gamma = p_1/\rho_1^\gamma,$$

or, since $p/\rho = RT$,

$$p = p_1(T_1/T)^{\gamma/(1-\gamma)}$$

and, differentiating,

$$dp = -[\gamma/(1 - \gamma)]p_1 T_1^{\gamma/(1-\gamma)} T^{-1/(1-\gamma)}\,dT.$$

Substituting these values of p and dp in equation (2.11):

$$dz = -\frac{RT}{g}\frac{\{-[\gamma/(1 - \gamma)]p_1 T_1^{\gamma/(1-\gamma)} T^{-1/(1-\gamma)}\,dT\}}{p_1 T_1^{\gamma/(1-\gamma)} T^{-\gamma/(1-\gamma)}}$$

$$= [\gamma/(1 - \gamma)](R/g)\,dT.$$

Temperature gradient,

$$\frac{\mathrm{d}T}{\mathrm{d}z} = -[(\gamma-1)/\gamma](g/R). \tag{2.12}$$

EXAMPLE 2.3

Calculate the pressure, temperature and density of the atmosphere at an altitude of 1200 m if at zero altitude the temperature is 15 °C and the pressure 101 kN m^{-2}. Assume that conditions are adiabatic ($\gamma = 1.4$) and $R = 287$ J kg^{-1} K^{-1}.

Solution

From equation (2.9),

$$p_2 = p_1 \left[1 - \frac{g(z_2 - z_1)}{RT_1} \left(\frac{\gamma - 1}{\gamma} \right) \right]^{\gamma/(\gamma-1)}.$$

Putting $p_1 = 101 \times 10^3$ N m^{-2}, $z_1 = 0$, $z_2 = 1200$ m, $T_1 = 15\ °C = 288$ K, $\gamma = 1.4$, $R = 287$ J kg^{-1} K^{-1}:

$$p_2 = 101 \times 10^3 \left[1 - \frac{9.81 \times 1200}{287 \times 288} \left(\frac{0.4}{1.4} \right) \right]^{1.4/0.4} \ \mathrm{N\ m}^{-2}$$

$$= 87.33 \times 10^3\ \mathrm{N\ m}^{-2}.$$

From equation (2.12),

Temperature gradient,

$$\frac{\mathrm{d}T}{\mathrm{d}z} = -[(\gamma-1)/\gamma](g/R) = -(0.4/1.4) \times (9.81/287)$$

$$= -9.76 \times 10^{-3}\ \mathrm{K\ m}^{-1},$$

$$T_2 = T_1 + \frac{\mathrm{d}T}{\mathrm{d}z}(z_2 - z_1)$$

$$= 288 - (9.76 \times 10^{-3} \times 1200) = 276.3\ \mathrm{K} = 3.3\ °C.$$

From the equation of state,

Density at 1200 m,

$$\rho_2 = p_2/RT_2$$

$$= (87.33 \times 10^3)/(287 \times 276.3) = 1.101\ \mathrm{kg\ m}^{-3}.$$

2.10 VARIATION OF PRESSURE AND DENSITY WITH ALTITUDE FOR A CONSTANT TEMPERATURE GRADIENT

Assuming that there is a constant temperature lapse rate (i.e. $\mathrm{d}T/\mathrm{d}z = \text{constant}$) with elevation in a gas, so that its temperature falls by an amount δT for a unit change of elevation, then, if $T_1 = $ temperature at level z_1, $T = $ temperature at level z,

$$T = T_1 - \delta T(z - z_1). \tag{2.13}$$

From equation (2.7), $\mathrm{d}p/\mathrm{d}z = -\rho g$ and, since $p/\rho = RT$, putting $\rho = p/RT$,

$$\frac{\mathrm{d}p}{\mathrm{d}z} = -p\frac{g}{RT},$$

$$\frac{\mathrm{d}p}{p} = -\frac{g}{RT}\,\mathrm{d}z.$$

Substituting for T from equation (2.13),

$$\mathrm{d}p/p = -\{g/R[T_1 - \delta T(z - z_1)]\}\mathrm{d}z.$$

Integrating between the limits p_1 and p_2 and z_1 and z_2,

$$\log_e(p_2/p_1) = (g/R\delta T)\log_e\{[T_1 - \delta T(z_2 - z_1)]/T_1\},$$

$$p_2/p_1 = [1 - (\delta T/T_1)(z_2 - z_1)]^{g/R\delta T}. \tag{2.14}$$

Comparing this with the result obtained in Section 2.9, and putting

$$\delta T = -\frac{\mathrm{d}T}{\mathrm{d}z} = \left(\frac{n-1}{n}\right)\left(\frac{g}{R}\right),$$

we have

$$g/R\delta T = n/(n - 1).$$

Substituting in equation (2.14),

$$p_2/p_1 = [1 - (g/RT_1)(z_2 - z_1)(n - 1)/n]^{n/(n-1)}$$

which agrees with equation (2.10).

To find the corresponding change of density ρ, since $p/\rho = RT$,

$$\frac{\rho_2}{\rho_1} = \frac{p_2}{p_1}\times\frac{T_1}{T_2} = \frac{p_2}{p_1}\times\frac{T_1}{T_1 - \delta T(z_2 - z_1)}$$

and, substituting from equation (2.14) for p_2/p_1,

$$\rho_2/\rho_1 = [1 - (\delta T/T_1)(z_2 - z_1)]^{g/R\delta T}[1 - (\delta T/T_1)(z_2 - z_1)]^{-1}$$

$$= [1 - (\delta T/T_1)(z_2 - z_1)]^{(g/R\delta T)-1}. \tag{2.15}$$

EXAMPLE 2.4

Assuming that the temperature of the atmosphere diminishes with increasing altitude at the rate of 6.5 °C per 1000 m, find the pressure and density at a height of 7000 m if the corresponding values at sea level are 101 kN m^{-2} and 1.235 kg m^{-3} when the temperature is 15 °C. Take $R = 287$ J kg^{-1} K^{-1}.

Solution

From equation (2.14),

$$p_2 = p_1[1 - (\delta T/T_1)(z_2 - z_1)]^{g/R\delta T}.$$

Putting $p_1 = 101 \times 10^3$ N m^{-2}, $\delta T = 6.5$ °C per 1000 m = 0.0065 K m^{-1}, $T_1 = 15$ °C = 288 K, $(z_2 - z_1) = 7000$ m, $R = 287$ J kg^{-1} K^{-1}:

$$p_2 = 101 \times 10^3[1 - (0.0065/288) \times 7000]^{9.81/(287 \times 0.0065)}$$

$$= \mathbf{40.89 \times 10^3 \ N \ m^{-2}}.$$

From the equation of state,

$$\text{Density, } \rho_2 = p_2/RT_2 = p_2/R[T_1 - \delta T(z_2 - z_1)]$$

$$= 40.89 \times 10^3/287(288 - 0.0065 \times 7000) = \mathbf{0.588 \ kg \ m^{-3}}.$$

2.11   VARIATION OF TEMPERATURE AND PRESSURE IN THE ATMOSPHERE

A body of fluid which is of importance to the engineer is the atmosphere. In practice, it is never in perfect equilibrium and is subject to large incalculable disturbances. In order to provide a basis for the design of aircraft an International Standard Atmosphere has been adopted which represents the average conditions in Western Europe; the relations between altitude, temperature and density have been tabulated (Table 2.1).

Essentially, the standard atmosphere comprises the troposphere – extending from sea level to 11 000 m – in which the temperature lapse rate is constant at approximately 0.0065 K m^{-1} and the pressure–density relationship is $p/\rho^n = $ constant, where $n = 1.238$. Above 11 000 m lies the stratosphere, in which conditions are assumed to be isothermal, with the temperature constant at −56 °C. Figure 2.8 shows the variation of pressure with altitude in the International Standard Atmosphere. The atmospheric pressure at sea level is assumed to be equivalent to 760 mm of mercury, the temperature 15 °C and the density 1.225 kg m^{-3}.

In the real atmosphere, the troposphere extends to an average of 11 000 m, but can vary from 7600 m at the poles to 18 000 m at the equator. While the temperature, in general, falls steadily with altitude, meteorological conditions can arise in the lower layers which produce temperature inversion – the temperature increasing with altitude. In the stratosphere, the temperature remains substantially constant up to approximately 32 000 m; it then rises to about 70 °C before falling again. Figure 2.9 shows typical values for pressure and temperature.

TABLE 2.1

International Standard
Atmosphere

ALTITUDE ABOVE SEA LEVEL (m)	ABSOLUTE PRESSURE (bar)	ABSOLUTE TEMPERATURE (K)	MASS DENSITY (kg m^{-3})	KINEMATIC VISCOSITY (m^2 s^{-1})
0	1.013 25	288.15	1.225 0	1.461×10^{-5}
1 000	0.898 8	281.7	1.111 7	1.581
2 000	0.795 0	275.2	1.006 6	1.715
4 000	0.616 6	262.2	0.819 4	2.028
6 000	0.472 2	249.2	0.660 2	2.416
8 000	0.356 5	236.2	0.525 8	2.904
10 000	0.265 0	223.3	0.413 6	3.525
11 500	0.209 8	216.7	0.337 5	4.213
12 000	0.194 0	216.7	0.311 9	4.557
14 000	0.141 7	216.7	0.227 9	6.239
16 000	0.103 5	216.7	0.166 5	8.540
18 000	0.075 65	216.7	0.121 6	11.686
20 000	0.055 29	216.7	0.088 91	15.989
22 000	0.040 47	218.6	0.064 51	22.201
24 000	0.029 72	220.6	0.046 94	30.743
26 000	0.021 88	222.5	0.034 26	42.439
28 000	0.016 16	224.5	0.025 08	58.405
30 000	0.011 97	226.5	0.018 41	80.134
32 000	0.008 89	228.5	0.013 56	109.620

FIGURE 2.8

Variation of temperature
with altitude in the
International Standard
Atmosphere

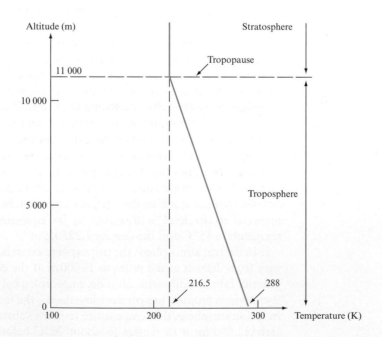

FIGURE 2.9

Variation of temperature and pressure in the real atmosphere

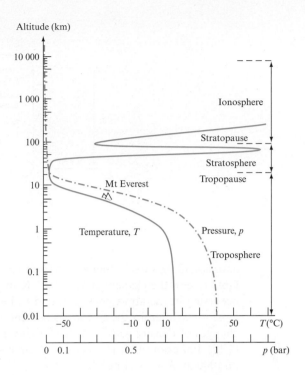

Because of vertical currents, the composition of the air remains practically constant in both the troposphere and the stratosphere, except that there is a negligible amount of water vapour in the stratosphere and a slight reduction in the ratio of oxygen to nitrogen above an altitude of 20 km. Nine-tenths of the mass of the atmosphere is contained below 20 km and 99 per cent below 60 km.

2.12 STABILITY OF THE ATMOSPHERE

We have seen that there are variations of density from point to point in the atmosphere when it is at rest. In practice, there are local disturbances due to air currents. There are also changes of density as a result of local thermal effects, which cause the movement of elements of air into regions where they are surrounded by air of slightly different density and temperature. If the density of the surrounding air is less than that of the newly arrived element, there is a tendency for the element to return to its original position – since the net upward force exerted by the surrounding fluid is less than the weight of the element. In Fig. 2.10, if ρ_1 is the density of the surrounding air and ρ_2 is the density of the air in the displaced element,

Weight of element, $mg = \rho_2 g A \delta z$.

Upward force due to surrounding fluid $= \delta p \times A = \rho_1 g \delta z A$, and there is, therefore, a net downward force of $(\rho_2 - \rho_1) g \delta z A$.

As an element of fluid rises in the atmosphere, its pressure and temperature fall. Air is a poor conductor, and conditions are, therefore, approximately adiabatic. From

FIGURE 2.10
Stability of the
atmosphere

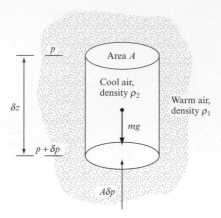

equation (2.9), substituting $\gamma = 1.414$ and $R = 287\ \text{J}\ \text{kg}^{-1}\text{K}^{-1}$, the adiabatic temperature lapse rate in the element is $\delta T = 0.01\ \text{K}\ \text{m}^{-1}$. The natural temperature lapse rate $\delta T'$ occurring in the atmosphere is found to be of the order of $0.0065\ \text{K}\ \text{m}^{-1}$. Since the lapse rates differ for the ascending element of air and the surrounding atmosphere the changes of density with altitude will also differ and can be calculated from equation (2.15). For example, if ρ_1 = density of air at sea level, ρ_2 = density of air at an elevation of 1000 m, $R = 287\ \text{J}\ \text{kg}^{-1}\text{K}^{-1}$, then:

1.   Assuming $\delta T = 0.01\ \text{K}\ \text{m}^{-1}$, for the air in the displaced element,

$$\frac{g}{R\delta T} - 1 = \frac{9.81}{287 \times 0.01} - 1 = 2.418,$$

and from equation (2.15),

$$\rho_2 = \rho_1\left(1 - \frac{0.01 \times 1000}{288}\right)^{2.48}, \quad \frac{\rho_2}{\rho_1} = 0.9181$$

for air in the element.

2.   Assuming $\delta T = 0.0065\ \text{K}\ \text{m}^{-1}$ for the surrounding atmosphere,

$$\frac{g}{R\delta T} - 1 = \frac{9.81}{287 \times 0.0065} - 1 = 4.258,$$

$$\frac{\rho_2}{\rho_1} = \left(1 - \frac{0.0065 \times 1000}{288}\right)^{4.258} = 0.9018$$

for the surrounding air.

Thus, the density of the ascending element expanding adiabatically decreases less rapidly than that of the surrounding air; the element eventually becomes denser than the surroundings and tends to fall back to its original level. The atmosphere, therefore, tends to be stable under normal conditions. If, however, the natural temperature lapse rate were to exceed the adiabatic lapse rate, equilibrium would be unstable and an element displaced upwards would continue to rise. Such conditions can arise in thundery weather.

2.13 PRESSURE AND HEAD

In a fluid of constant density, $dp/dz = -\rho g$ can be integrated immediately to give

$$p = -\rho g z + \text{constant}.$$

In a liquid, the pressure p at any depth z, measured downwards from the free surface so that $z = -h$ (Fig. 2.11), will be

$$p = \rho g h + \text{constant}$$

and, since the pressure at the free surface will normally be atmospheric pressure p_{atm},

$$p = \rho g h + p_{atm}. \tag{2.16}$$

It is often convenient to take atmospheric pressure as a datum. Pressures measured above atmospheric pressure are known as *gauge pressures*.

FIGURE 2.11
Pressure and head

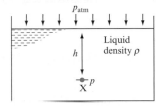

Since atmospheric pressure varies with atmospheric conditions, a perfect vacuum is taken as the absolute standard of pressure. Pressures measured above perfect vacuum are called *absolute pressures*:

Absolute pressure = Gauge pressure + Atmospheric pressure.

Taking p_{atm} as zero, equation (2.16) becomes

$$p = \rho g h, \tag{2.17}$$

which indicates that, if g is assumed constant, the gauge pressure at a point X (Fig. 2.11) can be defined by stating the vertical height h, called the *head*, of a column of a given fluid of mass density ρ which would be necessary to produce this pressure. Note that when pressures are expressed as head, it is essential that the mass density ρ is given or the fluid named. For example, since from equation (2.17) $h = p/\rho g$, a pressure of 100 kN m^{-2} can be expressed in terms of water ($\rho_{H_2O} = 10^3$ kg m^{-3}) as a head of $(100 \times 10^3)/(10^3 \times 9.81) = 10.19$ m of water. Alternatively, in terms of mercury (relative density 13.6) a pressure of 100 kN m^{-2} will correspond to a head of $(100 \times 10^3)/(13.6 \times 10^3 \times 9.81) = 0.75$ m of mercury.

EXAMPLE 2.5

A cylinder contains a fluid at a gauge pressure of 350 kN m^{-2}. Express this pressure in terms of a head of (a) water ($\rho_{H_2O} = 1000$ kg m^{-3}), (b) mercury (relative density 13.6).

What would be the absolute pressure in the cylinder if the atmospheric pressure is 101.3 kN m^{-2}?

Solution

From equation (2.17), head, $h = p/\rho g$.

(a) Putting $p = 350 \times 10^3$ N m^{-2}, $\rho = \rho_{H_2O} = 1000$ kg m^{-3},

$$\text{Equivalent head of water} = \frac{350 \times 10^3}{10^3 \times 9.81} = \textbf{35.68 m}.$$

(b) For mercury $\rho_{Hg} = \sigma\rho_{H_2O} = 13.6 \times 1000$ kg m^{-3},

$$\text{Equivalent head of water} = \frac{350 \times 10^3}{1.36 \times 10^3 \times 9.81} = \textbf{2.62 m},$$

$$\text{Absolute pressure} = \text{Gauge pressure} + \text{Atmospheric pressure}$$

$$= 350 + 101.3 = \textbf{451.3 kN m}^{-2}.$$

2.14 THE HYDROSTATIC PARADOX

From equation (2.17) it can be seen that the pressure exerted by a fluid is dependent only on the vertical head of fluid and its mass density ρ; it is not affected by the weight of the fluid present. Thus, in Fig. 2.12 the four vessels all have the same base area A and are filled to the same height h with the same liquid of density ρ.

Pressure on bottom in each case, $p = \rho gh$,

Force on bottom = Pressure × Area = $pA = \rho ghA$.

Thus, although the weight of fluid is obviously different in the four cases, the force on the bases of the vessels is the same, depending on the depth h and the base area A.

FIGURE 2.12
The hydrostatic paradox

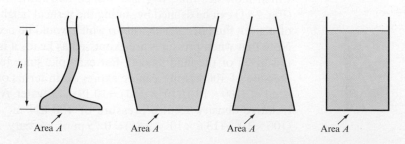

Area A Area A Area A Area A

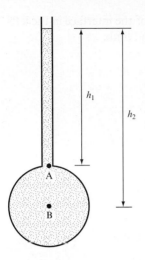

FIGURE 2.13
Pressure tube or piezometer

2.15   PRESSURE MEASUREMENT BY MANOMETER

The relationship between pressure and head is utilized for pressure measurement in the manometer or liquid gauge. The simplest form is the pressure tube or *piezometer* shown in Fig. 2.13, consisting of a single vertical tube, open at the top, inserted into a pipe or vessel containing liquid under pressure which rises in the tube to a height depending on the pressure. If the top of the tube is open to the atmosphere, the pressure measured is 'gauge' pressure:

Pressure at A = Pressure due to column of liquid of height h_1,

$$p_A = \rho g h_1.$$

Similarly,

Pressure at B = $p_B = \rho g h_2.$

If the liquid is moving in the pipe or vessel, the bottom of the tube must be flush with the inside of the vessel, otherwise the reading will be affected by the velocity of the fluid. This instrument can only be used with liquids, and the height of the tube which can conveniently be employed limits the maximum pressure that can be measured.

EXAMPLE 2.6

What is the maximum gauge pressure of water that can be measured by means of a piezometer tube 2 m high? (Mass density of water $\rho_{H_2O} = 10^3$ kg m^{-3}.)

Solution
Since $p = \rho g h$ for maximum pressure, put $\rho = \rho_{H_2O} = 10^3$ and $h = 2$ m, giving

Maximum pressure, $p = 10^3 \times 9.81 \times 2 = \mathbf{19.62 \times 10^3}$ N m^{-2}.

The U-tube gauge, shown in Fig. 2.14, can be used to measure the pressure of either liquids or gases. The bottom of the U-tube is filled with a manometric liquid Q which is of greater density ρ_{man} and is immiscible with the fluid P, liquid or gas, of

FIGURE 2.14
U-tube manometer

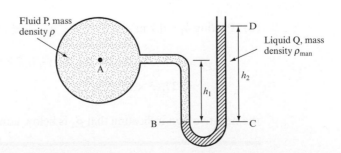

density ρ, whose pressure is to be measured. If B is the level of the interface in the left-hand limb and C is a point at the same level in the right-hand limb,

$$\text{Pressure } p_B \text{ at B} = \text{Pressure } p_C \text{ at C.}$$

For the left-hand limb,

$$p_B = \text{Pressure } p_A \text{ at A} + \text{Pressure due to depth } h_1 \text{ of fluid P}$$
$$= p_A + \rho g h_1.$$

For the right-hand limb,

$$p_C = \text{Pressure } p_D \text{ at D} + \text{Pressure due to depth } h_2 \text{ of liquid Q.}$$

But $p_D = \text{Atmospheric pressure} = \text{Zero gauge pressure,}$

and so $p_C = 0 + \rho_{man} g h_2.$

Since $p_B = p_C$,

$$p_A + \rho g h_1 = \rho_{man} g h_2,$$

$$p_A = \rho_{man} g h_2 - \rho g h_1. \tag{2.18}$$

EXAMPLE 2.7

A U-tube manometer similar to that shown in Fig. 2.14 is used to measure the gauge pressure of a fluid P of density $\rho = 800$ kg m^{-3}. If the density of the liquid Q is 13.6×10^3 kg m^{-3}, what will be the gauge pressure at A if (a) $h_1 = 0.5$ m and D is 0.9 m *above* BC, (b) $h_1 = 0.1$ m and D is 0.2 m *below* BC?

Solution

(a) In equation (2.18), $\rho_{man} = 13.6 \times 10^3$ kg m^{-3}, $\rho = 0.8 \times 10^3$ kg m^{-3}, $h_1 = 0.5$ m, $h_2 = 0.9$ m; therefore:

$$p_A = 13.6 \times 10^3 \times 9.81 \times 0.9 - 0.8 \times 10^3 \times 9.81 \times 0.5$$
$$= 116.15 \times 10^3 \text{ N m}^{-2}.$$

(b) Putting $h_1 = 0.1$ m and $h_2 = -0.2$ m, since D is below BC:

$$p_A = 13.6 \times 10^3 \times 9.81 \times (-0.2) - 0.8 \times 10^3 \times 9.81 \times 0.1$$
$$= -27.45 \times 10^3 \text{ N m}^{-2},$$

the negative sign indicating that p_A is below atmospheric pressure.

FIGURE 2.15
Measurement of
pressure difference

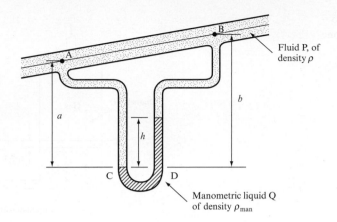

In Fig. 2.15, a U-tube gauge is arranged to measure the pressure difference between two points in a pipeline. As in the previous case, the principle involved in calculating the pressure difference is that the pressure at the same level CD in the two limbs must be the same, since the fluid in the bottom of the U-tube is at rest. For the left-hand limb,

$$p_C = p_A + \rho g a.$$

For the right-hand limb

$$p_D = p_B + \rho g(b - h) + \rho_{man} g h.$$

Since $p_C = p_D$,

$$p_A + \rho g a = p_B + \rho g(b - h) + \rho_{man} g h,$$

$$\text{Pressure difference} = p_A - p_B = \rho g(b - a) + h g(\rho_{man} - \rho). \tag{2.19}$$

EXAMPLE 2.8

A U-tube manometer is arranged, as shown in Fig. 2.15, to measure the pressure difference between two points A and B in a pipeline conveying water of density $\rho = \rho_{H_2O} = 10^3 \ \text{kg m}^{-3}$. The density of the manometric liquid Q is $13.6 \times 10^3 \ \text{kg m}^{-3}$, and point B is 0.3 m higher than point A. Calculate the pressure difference when $h = 0.7$ m.

Solution

In equation (2.19), $\rho = 10^3 \ \text{kg m}^{-3}$, $\rho_{man} = 13.6 \times 10^3 \ \text{kg m}^{-3}$, $(b - a) = 0.3$ m and $h = 0.7$ m.

$$\text{Pressure difference} = p_A - p_B$$
$$= 10^3 \times 9.81 \times 0.3 + 0.7 \times 9.81(13.6 - 1) \times 10^3 \ \text{N m}^{-2}$$
$$= \mathbf{89.467 \times 10^3 \ N \ m^{-2}}.$$

FIGURE 2.16
U-tube with one
leg enlarged

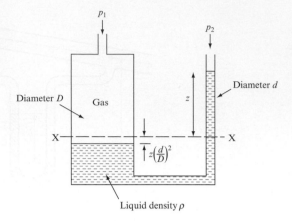

In both the above cases, if the fluid P is a gas its density ρ can usually be treated as negligible compared with ρ_{man} and the equations (2.18) and (2.19) can be simplified.

In forming the connection from a manometer to a pipe or vessel in which a fluid is flowing, care must be taken to ensure that the connection is perpendicular to the wall and flush internally. Any burr or protrusion on the inside of the wall will disturb the flow and cause a local change in pressure so that the manometer reading will not be correct.

Industrially, the simple U-tube manometer has the disadvantage that the movement of the liquid in both limbs must be read. By making the diameter of one leg very large as compared with the other (Fig. 2.16), it is possible to make the movement in the large leg very small, so that it is only necessary to read the movement of the liquid in the narrow leg. Assuming that the manometer in Fig. 2.16 is used to measure the pressure difference $p_1 - p_2$ in a gas of negligible density and that XX is the level of the liquid surface when the pressure difference is zero, then, when pressure is applied, the level in the right-hand limb will rise a distance z vertically.

Volume of liquid transferred from left-hand leg to right-hand leg $= z \times (\pi/4)d^2$;

Fall in level of the left-hand leg

$$= \frac{\text{Volume transferred}}{\text{Area of left-hand leg}} = \frac{z \times (\pi/4)d^2}{(\pi/4)D^2} = z\left(\frac{d}{D}\right)^2.$$

The pressure difference, $p_1 - p_2$, is represented by the height of the manometric liquid corresponding to the new difference of level:

$$p_1 - p_2 = \rho g[z + z(d/D)^2] = \rho g z[1 + (d/D)^2],$$

or, if D is large compared with d,

$$p_1 - p_2 = \rho g z.$$

If the pressure difference to be measured is small, the leg of the U-tube may be inclined as shown in Fig. 2.17. The movement of the meniscus along the inclined leg, read off on the scale, is considerably greater than the change in level z:

Pressure difference, $p_1 - p_2 = \rho g z = \rho g x \sin \theta.$

FIGURE 2.17
U-tube with inclined leg

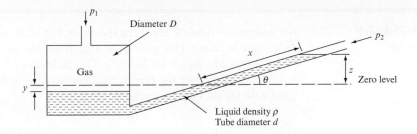

FIGURE 2.18
Inverted U-tube
manometer

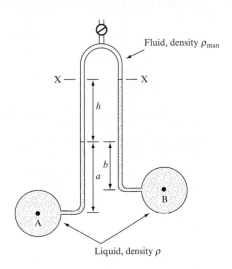

The manometer can be made as sensitive as may be required by adjusting the angle of inclination of the leg and choosing a liquid with a suitable value of density ρ to give a scale reading x of the desired size for a given pressure difference.

The *inverted U-tube* shown in Fig. 2.18 is used for measuring pressure differences in liquids. The top of the U-tube is filled with a fluid, frequently air, which is less dense than that connected to the instrument. Since the fluid in the top is at rest, pressures at level XX will be the same in both limbs.

For the left-hand limb,

$$p_{XX} = p_A - \rho ga - \rho_{man}gh.$$

For the right-hand limb,

$$p_{XX} = p_B - \rho g(b + h).$$

Thus $p_B - p_A = \rho g(b - a) + gh(\rho - \rho_{man}),$

or, if A and B are at the same level,

$$p_B - p_A = (\rho - \rho_{man})gh.$$

If the top of the tube is filled with air ρ_{man} is negligible compared with ρ and $p_B - p_A = \rho gh$. On the other hand, if the liquid in the top of the tube is chosen so that ρ_{man} is very nearly equal to ρ, and provided that the liquids do not mix, the result will be a very sensitive manometer giving a large value of h for a small pressure difference.

EXAMPLE 2.9

An inverted U-tube of the form shown in Fig. 2.18 is used to measure the pressure difference between two points A and B in an inclined pipeline through which water is flowing ($\rho_{H_2O} = 10^3$ kg m^{-3}). The difference of level $h = 0.3$ m, $a = 0.25$ m and $b = 0.15$ m. Calculate the pressure difference $p_B - p_A$ if the top of the manometer is filled with (a) air, (b) oil of relative density 0.8.

Solution

In either case, the pressure at XX will be the same in both limbs, so that

$$p_{XX} = p_A - \rho g a - \rho_{man} g h = p_B - \rho g (b + h),$$
$$p_B - p_A = \rho g (b - a) + g h (\rho - \rho_{man}).$$

(a) If the top is filled with air ρ_{man} is negligible compared with ρ. Therefore,

$$p_B - p_A = \rho g (b - a) + \rho g h = \rho g (b - a + h).$$

Putting $\rho = \rho_{H_2O} = 10^3$ kg m^{-3}, $b = 0.15$ m, $a = 0.25$ m, $h = 0.3$ m:

$$p_B - p_A = 10^3 \times 9.81(0.15 - 0.25 + 0.3)$$
$$= \mathbf{1.962 \times 10^3 \ N \ m^{-2}}.$$

(b) If the top is filled with oil of relative density 0.8, $\rho_{man} = 0.8\rho_{H_2O} = 0.8 \times 10^3$ kg m^{-3}.

$$p_B - p_A = \rho g (b - a) + g h (\rho - \rho_{man})$$
$$= 10^3 \times 9.81(0.15 - 0.25) + 9.81 \times 0.3 \times 10^3(1 - 0.8) \ N \ m^{-2}$$
$$= 10^3 \times 9.81(-0.1 + 0.06) = \mathbf{-392.4 \ N \ m^{-2}}.$$

The manometer in its various forms is an extremely useful type of pressure gauge, but suffers from a number of limitations. While it can be adapted to measure very small pressure differences, it cannot be used conveniently for large pressure differences – although it is possible to connect a number of manometers in series and to use mercury as the manometric fluid to improve the range. A manometer does not have to be calibrated against any standard; the pressure difference can be calculated from first principles. However, for accurate work, the temperature should be known, since this will affect the density of the fluids. Some liquids are unsuitable for use because they do not form well-defined menisci. Surface tension can also cause errors due to capillary rise; this can be avoided if the diameters of the tubes are sufficiently large – preferably not less than 15 mm diameter. It is difficult to correct for surface tension, since its effect will depend upon whether the tubes are clean. A major disadvantage of the manometer is its slow response, which makes it unsuitable for measuring fluctuating pressures. Even under comparatively static conditions, slight fluctuations of pressure can make the liquid in the manometer oscillate, so that it is difficult to get a precise reading of the levels of the liquid in the gauge. These oscillations can be reduced by putting restrictions in the manometer connections. It is also essential that the pipes connecting the manometer to the pipe or vessel containing the liquid under pressure should be filled with this liquid and that there should be no air bubbles in the liquid.

2.16   RELATIVE EQUILIBRIUM

If a fluid is contained in a vessel which is at rest, or moving with constant linear velocity, it is not affected by the motion of the vessel; but if the container is given a continuous acceleration, this will be transmitted to the fluid and affect the pressure distribution in it. Since the fluid remains at rest relative to the container, there is no relative motion of the particles of the fluid and, therefore, no shear stresses, fluid pressure being everywhere normal to the surface on which it acts. Under these conditions the fluid is said to be in relative equilibrium.

2.17   PRESSURE DISTRIBUTION IN A LIQUID SUBJECT TO HORIZONTAL ACCELERATION

Figure 2.19 shows a liquid contained in a tank which has an acceleration a. A particle of mass m on the free surface at O will have the same acceleration a as the tank and so will be subject to an accelerating force F. From Newton's law,

$$F = ma. \tag{2.20}$$

FIGURE 2.19
Effect of horizontal acceleration

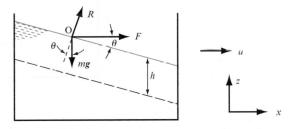

Also, F is the resultant of the fluid pressure force R, acting normally to the free surface at O, and the weight of the particle mg, acting vertically. Therefore,

$$F = mg \tan \theta. \tag{2.21}$$

Comparing equations (2.20) and (2.21),

$$\tan \theta = a/g \tag{2.22}$$

and is constant for all points on the free surface. Thus, the free surface is a plane inclined at a constant angle θ to the horizontal.

Since the acceleration is horizontal, vertical forces are not changed and the pressure at any depth h below the surface will be $\rho g h$. Planes of equal pressure lie parallel to the free surface.

2.18 EFFECT OF VERTICAL ACCELERATION

If the acceleration is vertical, the free surface will remain horizontal. Considering a vertical prism of cross-sectional area A (Fig. 2.20), subject to an upward acceleration a, then at depth h below the surface, where the pressure is p,

$$\text{Upward accelerating force, } F = \text{Force due to } p - \text{Weight of prism}$$

$$= p\text{A} - \rho gh A.$$

By Newton's second law,

$$F = \text{Mass of prism} \times \text{Acceleration} = \rho h A \times a.$$

Therefore,

$$pA - \rho gh A = \rho h A a,$$

$$p = \rho gh(1 + a/g). \tag{2.23}$$

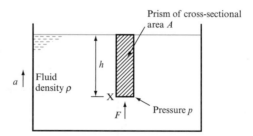

FIGURE 2.20
Effect of vertical acceleration

2.19 GENERAL EXPRESSION FOR THE PRESSURE IN A FLUID IN RELATIVE EQUILIBRIUM

If $\partial p/\partial x$, $\partial p/\partial y$ and $\partial p/\partial z$ are the rates of change of pressure p in the x, y and z directions (Fig. 2.21) and a_x, a_y and a_z the accelerations,

$$\text{Force in } x \text{ direction, } F_x = p\,\delta y\,\delta z - \left(p + \frac{\partial p}{\partial x}\delta x\right)\delta y\,\delta z$$

$$= -\frac{\partial p}{\partial x}\delta x\,\delta y\,\delta z.$$

By Newton's second law, $F_x = \rho\delta x\delta y\delta z \times a_x$; therefore,

$$-\frac{\partial p}{\partial x} = \rho a_x. \tag{2.24}$$

FIGURE 2.21
Relative equilibrium: the general case

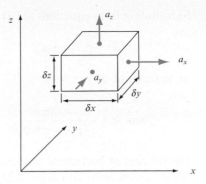

Similarly, in the y direction,

$$-\frac{\partial p}{\partial y} = \rho a_y.$$ (2.25)

In the vertical z direction, the weight of the element $\rho g \delta x \delta y \delta z$ must be considered:

$$F_z = p\,\delta x\,\delta y - \left(p + \frac{\partial p}{\partial z}\delta z\right)\delta x\,\delta y - \rho g\,\delta x\,\delta y\,\delta z$$

$$= -\frac{\partial p}{\partial z}\delta x\,\delta y\,\delta z - \rho g\,\delta x\,\delta y\,\delta z.$$

By Newton's second law,

$$F_z = \rho\,\delta x\,\delta y\,\delta z \times a_z;$$

therefore,

$$-\frac{\partial p}{\partial z} = \rho(g + a_z).$$ (2.26)

For an acceleration a_s in any direction in the $x - z$ plane making an angle ϕ with the horizontal, the components of the acceleration are

$$a_x = a_s \cos\phi \quad \text{and} \quad a_z = a_s \sin\phi.$$

Now $$\frac{dp}{ds} = \frac{\partial p}{\partial x}\frac{dx}{ds} + \frac{\partial p}{\partial z}\frac{dz}{ds}.$$ (2.27)

For the free surface and all other planes of constant pressure, $dp/ds = 0$. If θ is the inclination of the planes of constant pressure to the horizontal, $\tan\theta = dz/dx$. Putting $dp/ds = 0$ in equation (2.27),

$$\frac{\partial p}{\partial x}\frac{dx}{ds} + \frac{\partial p}{\partial z}\frac{dz}{ds} = 0$$

$$\frac{dz}{dx} = \tan\theta = -\frac{\partial p}{\partial x}\bigg/\frac{\partial p}{\partial z}.$$

Substituting from equations (2.24) and (2.26),

$$\tan \theta = -a_x/(g + a_z),$$ (2.28)

or, in terms of a_s,

$$\tan \theta = -\frac{a_s \cos \phi}{(g + a_s \sin \phi)}.$$ (2.29)

For the case of horizontal acceleration, $\phi = 0$ and equation (2.29) gives $\tan \theta = -a_s/g$, which agrees with equation (2.22). (Note effect of sign convention.) For vertical acceleration, $\phi = 90°$ giving $\tan \theta = 0$, indicating that the free surface remains horizontal.

Since, for the two-dimensional case,

$$dp = \frac{\partial p}{\partial x} dx + \frac{\partial p}{\partial z} dz.$$

the pressure at a particular point in the fluid can be found by integration:

$$p = \int dp = \int \frac{\partial p}{\partial x} dx + \int \frac{\partial p}{\partial z} dz.$$

Substituting from equations (2.24) and (2.26) and assuming that ρ is constant:

$$p = \int (-\rho a_x) \, dx + \int [-\rho(g + a_z)] \, dz + \text{constant}$$

$$= -\rho(x a_s \cos \phi - gz - za_s \sin \phi) + \text{constant},$$

or, since $x/z = \tan \theta$,

$$p = -z\rho(a_s \tan \theta \cos \phi - g - a_s \sin \phi) + \text{constant},$$ (2.30)

where z is positive measured upwards from a horizontal datum fixed relative to the fluid.

EXAMPLE 2.10

A rectangular tank 1.2 m deep and 2 m long is used to convey water up a ramp inclined at an angle ϕ of 30° to the horizontal (Fig. 2.22). Calculate the inclination of the water surface to the horizontal when (a) the acceleration parallel to the slope on starting from the bottom is 4 m s^{-2}, (b) the deceleration parallel to the slope on reaching the top is 4.5 m s^{-2}.

If no water is to be spilt during the journey what is the greatest depth of water permissible in the tank when it is at rest?

Solution

The slope of the water surface is given by equation (2.29).

(a) During acceleration, $a_s = +4$ m s^{-2}.

FIGURE 2.22
Acceleration up an
inclined plane

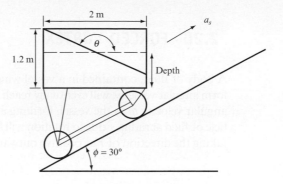

$$\tan \theta_A = \frac{-a_s \cos \phi}{g + a_s \sin \phi} = -\frac{4 \cos 30°}{9.81 + 4 \sin 30°}$$

$$= -0.2933$$

$$\theta_A = \mathbf{163° \ 39'}.$$

(b) During retardation, $a_s = -4.5$ m s^{-2}.

$$\tan \theta_R = -\frac{(-4.5) \cos 30°}{9.81 - 4.5 \sin 30°} = 0.5154$$

$$\theta_R = \mathbf{27° \ 16'}.$$

Since $180° - \theta_R < \theta_A$, the worst case for spilling will be during retardation. When the water surface is inclined, the maximum depth at the tank wall will be

$$\text{Depth} + \tfrac{1}{2}\text{Length} \times \tan \theta,$$

which must not exceed 1.2 m if the water is not to be spilt. Putting length = 2 m, $\tan \theta = \tan \theta_R = 0.5154$,

$$\text{Depth} + (2.0/2) \times 0.5154 = 1.2,$$

$$\text{Depth} = 1.2 - 0.5154$$

$$= \mathbf{0.6846 \ m}.$$

The equations derived in this section indicate:

1. if there is no horizontal acceleration, $a_x = 0$ and than $\theta = 0$ so that surfaces of constant pressure are horizontal;

2. in free space, g will be zero so that $\tan \theta = -a_x/a_z$ (surfaces of constant pressure will therefore be perpendicular to the resultant acceleration);

3. since free surfaces of liquids are surfaces of constant pressure, their inclination will be determined by equation (2.29); thus, if a_x and a_y are zero, the free surface will be horizontal.

2.20   FORCED VORTEX

A body of fluid, contained in a vessel which is rotating about a vertical axis with uniform angular velocity, will eventually reach relative equilibrium and rotate with the same angular velocity ω as the vessel, forming a forced vortex. The acceleration of any particle of fluid at radius r due to rotation will be $-\omega^2 r$ perpendicular to the axis of rotation, taking the direction of r as positive outward from the axis. Thus, from equation (2.24),

$$\frac{\mathrm{d}p}{\mathrm{d}r} = -\rho\omega^2 r.$$

Figure 2.23 shows a cylindrical vessel containing liquid rotating about its axis, which is vertical. At any point P on the free surface, the inclination θ of the free surface is given by equation (2.28):

FIGURE 2.23
Forced vortex

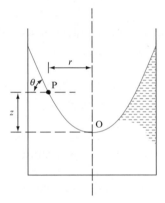

$$\tan\theta = -\frac{a_x}{g + a_z} = \frac{\omega^2 r}{g} = \frac{\mathrm{d}z}{\mathrm{d}r}. \tag{2.31}$$

The inclination of the free surface varies with r and, if z is the height of P above O, the surface profile is given by integrating equation (2.31):

$$z = \int_0^r \frac{\omega^2 r}{g}\,\mathrm{d}r = \frac{\omega^2 r^2}{2g} + \text{constant}. \tag{2.32}$$

Thus, the profile of the water surface is a paraboloid. Similarly other surfaces of equal pressure will also be paraboloids.

If the container is closed and the fluid has no free surface, the paraboloid drawn to represent the imaginary free surface represents the variation of pressure head with radius. Thus, the pressure p at radius r is given by equation (2.32) as

$$z = p/\rho g = \omega^2 r^2/2g + \text{constant},$$

$$p = \rho\omega^2 r^2/2 + \text{constant}. \tag{2.33}$$

Concluding remarks

The definitions of pressure and head presented in this chapter are fundamental to the study of fluid mechanics. In particular the development of the hydrostatic equation and the proof that pressure at a depth in a continuous fluid is constant was central to the development of both flow pressure and velocity measuring techniques due to the equation's application to the design and operation of pressure measuring manometers. The basic expressions derived for manometer use will be utilized later in the text in the treatment of flow measurement, Chapter 6. The treatment of the atmosphere stresses the obvious point that all our activities and structures must be seen as submerged within a fluid.

Summary of important equations and concepts

1. Pascal's law, equation (2.3), stating that pressures at a depth are equal in all coordinate directions, coupled with the hydrostatic equation (2.4) that links depth, gravitational acceleration, density and pressure are the fundamental underpinning of this chapter's treatment of hydrostatic forces.

2. Equation (2.8) underlines the importance of the density–pressure relationship identified in Chapter 1. This becomes essential to the understanding of the atmospheric variations of pressure and density with altitude, Sections 2.7 to 2.12.

3. Despite the predominance of electronic measurement of fluid flow conditions the application of hydrostatics to manometer measurement of pressure, and hence flow conditions, remains important. Sections 2.13 to 2.15 investigate these applications fully.

4. Relative equilibrium is introduced in Section 2.16 and is used to discuss the effects of acceleration and the generation of a forced vortex, concepts returned to in more detail in Chapter 6.

Problems

2.1 Calculate the pressure in the ocean at a depth of 2000 m assuming that salt water is (*a*) incompressible with a constant density of 1002 kg m^{-3}, (*b*) compressible with a bulk modulus of 2.05 GN m^{-2} and a density at the surface of 1002 kg m^{-3}.

[(*a*) 19.66 MN m^{-2}, (*b*) 19.75 MN m^{-2}]

2.2 What will be (*a*) the gauge pressure, (*b*) the absolute pressure of water at a depth of 12 m below the free surface? Assume the density of water to be 1000 kg m^{-3} and the atmospheric pressure 101 kN m^{-2}.

[(*a*) 117.72 kN m^{-2}, (*b*) 218.72 kN m^{-2}]

2.3 What depth of oil, specific gravity 0.8, will produce a pressure of 120 kN m^{-2}? What would be the corresponding depth of water?   [15.3 m, 12.2 m]

2.4 At what depth below the free surface of oil having a density of 600 kg m^{-3} will the pressure be equal to 1 bar?

[17 m]

2.5 What would be the pressure in kilonewtons per square metre if the equivalent head is measured as 400 mm of (*a*) mercury of specific gravity 13.6, (*b*) water, (*c*) oil of specific weight 7.9 kN m^{-3}, (*d*) a liquid of density 520 kg m^{-3}?

[(*a*) 53.4 kN m^{-2}, (*b*) 3.92 kN m^{-2},
(*c*) 3.16 kN m^{-2}, (*d*) 2.04 kN m^{-2}]

2.6 A mass of 50 kg acts on a piston of area 100 cm^{2}. What is the intensity of pressure on the water in contact with the underside of the piston, if the piston is in equilibrium?

[4.905 × 10^{4} N m^{-2}]

2.7   The pressure head in a gas main at a point 120 m above sea level is equivalent to 180 mm of water. Assuming that the densities of air and gas remain constant and equal to 1.202 kg m^{-3} and 0.561 kg m^{-3}, respectively, what will be the pressure head in millimetres of water at sea level?

[103 mm]

2.8   A manometer connected to a pipe in which a fluid is flowing indicates a negative gauge pressure head of 50 mm of mercury. What is the absolute pressure in the pipe in newtons per square metre if the atmospheric pressure is 1 bar?

[93.3 kN m^{-2}]

2.9   An open tank contains oil of specific gravity 0.75 on top of water. If the depth of oil is 2 m and the depth of water 3 m, calculate the gauge and absolute pressures at the bottom of the tank when the atmospheric pressure is 1 bar?

[44.15 kN m^{-2}, 144.15 kN m^{-2}]

2.10   A closed tank contains 0.5 m of mercury, 2 m of water, 3 m of oil of density 600 kg m^{-3} and there is an air space above the oil. If the gauge pressure at the bottom of the tank is 200 kN m^{-2}, what is the pressure of the air at the top of the tank?

[96 kN m^{-2}]

2.11   An inverted cone 1 m high and open at the top contains water to half its height, the remainder being filled with oil of specific gravity 0.9. If half the volume of water is drained away find the pressure at the bottom (apex) of the inverted cone.

[9033 N m^{-2}]

2.12   A hydraulic press has a diameter ratio between the two pistons of 8:1. The diameter of the larger piston is 600 mm and it is required to support a mass of 3500 kg. The press is filled with a hydraulic fluid of specific gravity 0.8. Calculate the force required on the smaller piston to provide the required force (*a*) when the two pistons are level, (*b*) when the smaller piston is 2.6 m below the larger piston.

[(*a*) 536 N, (*b*) 626.2 N]

2.13   Show that the ratio of the pressures (p_2/p_1) and densities (ρ_2/ρ_1) for altitudes h_2 and h_1 in an isothermal atmosphere is given by

$$\frac{p_2}{p_1} = \frac{\rho_2}{\rho_1} = e^{-g(h_2 - h_1)/RT}.$$

What increase in altitude is necessary in the stratosphere to halve the pressure? Assume a constant temperature of −56.5 °C and the gas constant $R = 287$ J kg^{-1} K^{-1}.

[4390 m]

2.14   From observation it is found that at a certain altitude in the atmosphere the temperature is −25 °C and the pressure is 45.5 kN m^{-2}, while at sea level the corresponding

values are 15 °C and 101.5 kN m^{-2}. Assuming that the temperature decreases uniformly with increasing altitude, estimate the temperature lapse rate and the pressure and density of the air at an altitude of 3000 m.

[6.37 °C per 1000 m, 70.22 kN m^{-2}, 0.91 kg m^{-3}]

2.15   Show that the ratio of the atmospheric pressure at an altitude h_1 to that at sea level may be expressed as $(p/p_0) = (T/T_0)^n$, a uniform temperature lapse rate being assumed. Find the ratio of the pressures and the densities at 10 700 m and at sea level taking the standard atmosphere as having a sea level temperature of 15 °C and a lapse rate of 6.5 °C per 1000 m to a minimum of −56.5 °C.   [0.2337, 0.3082]

2.16   The barometric pressure of the atmosphere at sea level is equivalent to 760 mm of mercury and its temperature is 288 K. The temperature decreases with increasing altitude at the rate of 6.5 K per 1000 m until the stratosphere is reached in which the temperature remains constant at 216.5 K. Calculate the pressure in millimetres of mercury and the density in kilograms per cubic metre at an altitude of 14 500 m. Assume $R = 287$ J kg^{-1} K^{-1}.

[97.52 mm, 0.209 kg m^{-3}]

2.17   In Fig. 2.24 fluid P is water and fluid Q is mercury. If the specific weight of mercury is 13.6 times that of water and the atmospheric pressure is 101.3 kN m^{-2}, what is the absolute pressure at A when $h_1 = 15$ cm and $h_2 = 30$ cm?

[59.8 kN m^{-2}]

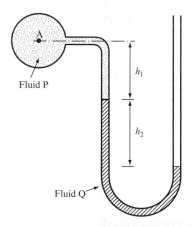

Fluid P

h_1

h_2

Fluid Q

FIGURE 2.24

2.18   A U-tube manometer (Fig. 2.25) measures the pressure difference between two points A and B in a liquid of density ρ_1. The U-tube contains mercury of density ρ_2. Calculate the difference of pressure if $a = 1.5$ m, $b = 0.75$ m and $h = 0.5$ m if the liquid at A and B is water and $\rho_2 = 13.6\rho_1$.

[54.4 kN m^{-2}]

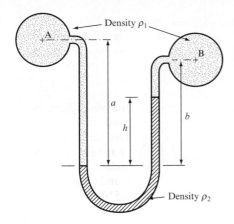

Density ρ_1

A

B

a

b

h

Density ρ_2

FIGURE 2.25

2.19 The top of an inverted U-tube manometer is filled with oil of specific gravity 0.98 and the remainder of the tube with water of specific gravity 1.01. Find the pressure difference in newtons per square metre between two points at the same level at the base of the legs when the difference of water level is 75 mm. [22 N m^{-2}]

2.20 An inclined manometer is required to measure an air pressure difference of about 3 mm of water with an accuracy of ±3 per cent. The inclined arm is 8 mm diameter and the enlarged end is 24 mm diameter. The density of the manometer fluid is 740 kg m^{-3}. Find the angle which the inclined arm must make with the horizontal to achieve the required accuracy assuming an acceptable readability of 0.5 mm.
[12° 39′]

2.21 An inclined tube manometer consists of a vertical cylinder of 35 mm diameter to the bottom of which is connected a tube of 5 mm diameter inclined upwards at 15° to the horizontal. The manometer contains oil of relative density 0.785. The open end of the inclined tube is connected to an air duct while the top of the cylinder is open to the atmosphere. Determine the pressure in the air duct if the manometer fluid moves 50 mm along the inclined tube.

What is the error if the movement of the fluid in the cylinder is ignored? [107.5 N m^{-2} 7.85 N m^{-2}]

2.22 A manometer consists of a U-tube, 7 mm internal diameter, with vertical limbs each with an enlarged upper end of 44 mm diameter. The left-hand limb and the bottom of the tube is filled with water and the top of the right-hand limb is filled with oil of specific gravity 0.83. The free surfaces of the liquids are in the enlarged ends and the interface between the oil and water is in the tube below the enlarged end. What would be the difference in pressures applied to the free surfaces which would cause the oil/water interface to move 1 cm? [21 N m^{-2}]

2.23 A vessel 1.4 m wide and 2.0 m long is filled to a depth of 0.8 m with a liquid of mass density 840 kg m^{-3}. What will be the force in N on the bottom of the vessel (*a*) when being accelerated vertically upwards at 4 m s^{-1}, (*b*) when the acceleration ceases and the vessel continues to move at a constant velocity of 7 m s^{-1} vertically upwards?
[(*a*) 25 985 N, (*b*) 18 458 N]

2.24 A pipe 25 mm in diameter is connected to the centre of the top of a drum 0.5 m in diameter, the cylindrical axis of the pipe and the drum being vertical. Water is poured into the drum through the pipe until the water level stands in the pipe 0.6 m above the top of the drum. If the drum and pipe are now rotated about their vertical axis at 600 rev min^{-1} what will be the upward force exerted on the top of the drum? [13.26 kN]

2.25 A tube ABCD has the end A open to the atmosphere and the end D closed. The portion ABC is vertical while the portion CD is a quadrant of radius 250 mm with its centre at B, the whole being arranged to rotate about its vertical axis ABC. If the tube is completely filled with water to a height in the vertical limb of 300 mm above C find (*a*) the speed of rotation which will make the pressure head at D equal to the pressure head at C, (*b*) the value and position of the maximum pressure head in the curved portion CD when running at this speed.

[(*a*) 84.6 rev min^{-1}, (*b*) 0.362 m of water and 0.12 m below point D]

2.26 A closed airtight tank 4 m high and 1 m in diameter contains water to a depth of 3.3 m. The air in the tank is at a pressure of 40 kN m^{-2} gauge. What are the absolute pressures at the centre and circumference of the base of the tank when it is rotating about its vertical axis at a speed of 180 rev min^{-1}? At this speed the water wets the top surface of the tank.

[17.01 m absolute, 21.53 m absolute]

3

Static Forces on Surfaces. Buoyancy

THE IMPLICATION OF THE HYDROSTATIC EQUATION, TOGETHER WITH THE REALIZATION that pressures at any equal depth in a continuous fluid are both equal and act equally in all directions, leads to the treatment of static forces on submerged surfaces. This also defines buoyancy and the stability of floating bodies. This chapter will introduce the techniques available to determine the forces acting on surfaces as a result of the applied fluid pressure and will stress the difference between pressure, which is a scalar quantity acting equally in all directions at a particular depth, and the associated force, which is a vector quantity possessing both magnitude and direction. ● ● ●

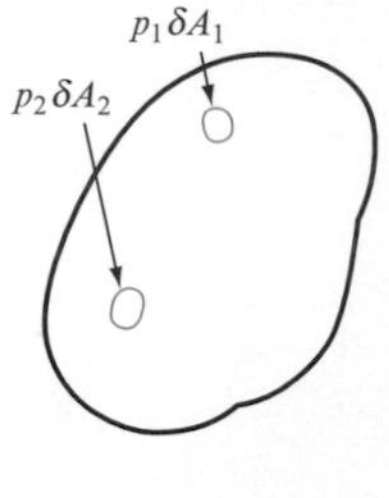

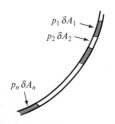

FIGURE 3.1

Forces on a plane surface

FIGURE 3.2

Forces on a curved
surface

FIGURE 3.3

Forces on a cylindrical
surface

3.1   ACTION OF FLUID PRESSURE ON A SURFACE

Since pressure is defined as force per unit area, when fluid pressure p acts on a solid boundary – or across any plane in the fluid – the force exerted on each small element of area δA will be $p\delta A$, and, since the fluid is at rest, this force will act at right angles to the boundary or plane at the point under consideration.

In a body of fluid, the pressure p may vary from point to point, and the forces on each element of area will also vary. If the fluid pressure acts on or across a plane surface, all the forces on the small elements will be parallel (Fig. 3.1) and can be represented by a single force, called the *resultant force*, acting at right angles to the plane through a point called the *centre of pressure*.

Resultant force, R = Sum of forces on all elements of area

$$R = p_1\,\delta A_1 + p_2\,\delta A_2 + \cdots + p_n\,\delta A_n = \Sigma p\,\delta A,$$

where Σ means 'the sum of'.

If the boundary is a curved surface, the elementary forces will act perpendicular to the surface at each point and will, therefore, not be parallel (Fig. 3.2). The resultant force can be found by resolution or by a polygon of forces, but will be less than $\Sigma p\delta A$. For example, in the extreme case of the curved surface of a bucket filled with water (Fig. 3.3), the elementary forces acting radially on the vertical wall will balance and the resultant force will be zero. If this were not so, there would be an unbalanced horizontal force in some direction and the bucket would move of its own accord.

3.2   RESULTANT FORCE AND CENTRE OF PRESSURE ON A PLANE SURFACE UNDER UNIFORM PRESSURE

The pressure p on a plane horizontal surface in a fluid at rest will be the same at all points, and will act vertically downwards at right angles to the surface. If the area of the plane surface is A,

Resultant force $= pA$.

It will act vertically downwards and the centre of pressure will be the centroid of the surface.

For gases, the variation of pressure with elevation is small and so it is usually possible to assume that gas pressure on a surface is uniform, even though the surface may not be horizontal. The resultant force is then pA acting through the centroid of the plane surface.

FIGURE 3.4
Resultant force on a
plane surface immersed
in a fluid

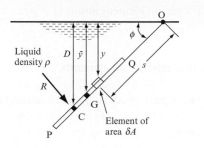
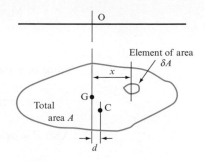

3.3   RESULTANT FORCE AND CENTRE OF PRESSURE ON A PLANE SURFACE IMMERSED IN A LIQUID

Figure 3.4 shows a plane surface PQ of any area A totally immersed in a liquid of density ρ and inclined at an angle ϕ to the free surface. Considering one side only, there will be a force due to fluid pressure p acting on each element of area δA. The magnitude of p will depend on the vertical depth y of the element below the free surface. Taking the pressure at the free surface as zero, from equation (2.4), and measuring y downwards, $p = \rho gy$; therefore,

$$\text{Force on element of area, } \delta A = p\delta A = \rho gy\delta A.$$

Summing the forces on all such elements over the whole surface, since these forces are all perpendicular to the plane PQ,

$$\text{Resultant force, } R = \Sigma\,\rho gy\,\delta A.$$

If we assume that ρ and g are constant,

$$R = \rho g\,\Sigma y\delta A. \tag{3.1}$$

The quantity $\Sigma y\delta A$ is the first moment of area under the surface PQ about the free surface of the liquid and is equal to $A\bar{y}$, where A = the area of the whole immersed surface PQ and $\bar{y}$ = the vertical depth to the centroid G of the immersed surface. Substituting in equation (3.1),

$$\text{Resultant force, } R = \rho gA\bar{y}. \tag{3.2}$$

This resultant force R will act perpendicular to the immersed surface at the centre of pressure C at some vertical depth D below the free surface, such that the moment of R about any point will be equal to the sum of the moments of the forces on all the elements δA about the same point. Thus, if the plane of the immersed surface cuts the free surface at O,

$$\text{Moment of } R \text{ about O} = \text{Sum of moments of forces on all elements}$$
$$\text{of area } \delta A \text{ about O,} \tag{3.3}$$

$$\text{Force on any small element} = \rho gy\delta A = \rho gs \sin\,\phi \times \delta A,$$

since $y = s \sin\,\phi$.

$$\text{Moment of force on element about } O = pgs \sin \phi \times \delta A \times s$$
$$= \rho g \sin \phi \times \delta A \times s^2.$$

Since ρ, g and ϕ are the same for all elements,

$$\text{Sum of the moments of all such forces about } O = \rho g \sin \phi \sum s^2 \delta A.$$

Also $R = \rho g A \bar{y}$; therefore,

$$\text{Moment of } R \text{ about } O = \rho g A \bar{y} \times OC = \rho g A \bar{y} (D/\sin \phi).$$

Substituting in equation (3.3),

$$\rho g A \bar{y} (D/\sin \phi) = \rho g \sin \phi \sum s^2 \delta A,$$
$$D = \sin^2 \phi (\sum s^2 \delta A)/A\bar{y},$$
$$\sum s^2 \delta A = \text{Second moment of area of the immersed surface}$$
$$\text{about an axis in the free surface through } O$$
$$= I_O = A k_O^2,$$

where k_O = the radius of gyration of the immersed surface about O. Therefore,

$$D = \sin^2 \phi (I_O/A\bar{y}) = \sin^2 \phi (k_O^2/\bar{y}). \tag{3.4}$$

The values of I_O and k_O^2 can be found if the second moment of area of the immersed surface I_G about an axis through its centroid G parallel to the free surface is known by using the parallel axis rule,

or, $$A k_O^2 = A k_G^2 + A(\bar{y}/\sin \phi)^2.$$

Thus $$D = \sin^2 \phi [k_G^2 + (\bar{y}/\sin \phi)^2/\bar{y}] = \sin^2 \phi (k_G^2/\bar{y}) + \bar{y}. \tag{3.5}$$

The geometrical properties of some common figures are given in Table 3.1.

From equation (3.5) it can be seen that the centre of pressure will always be below the centroid G except when the surface is horizontal ($\phi = 0°$). As the depth of immersion increases, the centre of pressure will move nearer to the centroid, since for the given surface the change of pressure between the upper and lower edge becomes proportionately smaller in comparison with the mean pressure, making the pressure distribution more uniform.

The lateral position of the centre of pressure can be found by taking moments about the line OG, which is the line of intersection of the immersed surface with a vertical plane through G:

$$R \times d = \text{Sum of moments of forces on small elements about } OG$$
$$= \sum \rho g \delta A y x.$$

Putting $$R = \rho g A \bar{y},$$
$$d = (\sum \delta A \pi x)/Ay.$$

TABLE 3.1
Geometrical properties
of some common figures

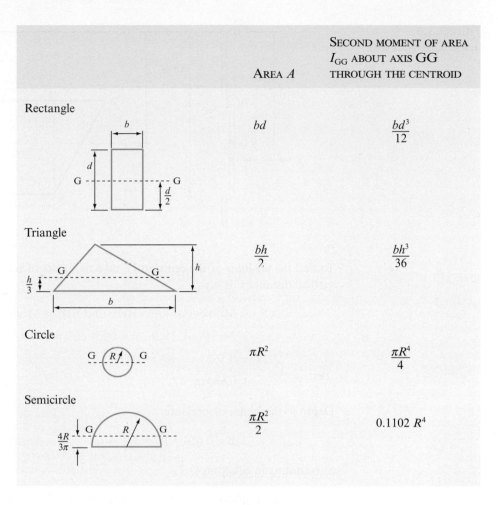

	AREA A	SECOND MOMENT OF AREA I_{GG} ABOUT AXIS GG THROUGH THE CENTROID
Rectangle	bd	$\dfrac{bd^3}{12}$
Triangle	$\dfrac{bh}{2}$	$\dfrac{bh^3}{36}$
Circle	πR^2	$\dfrac{\pi R^4}{4}$
Semicircle	$\dfrac{\pi R^2}{2}$	$0.1102\,R^4$

If the area is symmetrical about a vertical plane through the centroid G, the moment of each small element on one side is balanced by that due to a similar element on the other side so that $\sum \delta A y = 0$. Therefore, $d = 0$ and the centre of pressure will be on the axis of symmetry.

EXAMPLE 3.1

A trapezoidal opening in the vertical wall of a tank is closed by a flat plate which is hinged at its upper edge (Fig. 3.5). The plate is symmetrical about its centreline and is 1.5 m deep. Its upper edge is 2.7 m long and its lower edge is 1.2 m long. The free surface of the water in the tank stands 1.1 m above the upper edge of the plate. Calculate the moment about the hinge line required to keep the plate closed.

Solution

The moment required to keep the plate closed will be equal and opposite to the moment of the resultant force R due to the water acting at the centre of pressure C, i.e. $R \times$ CB. From equation (3.2), $R = \rho g A \bar{y}$.

Area of plate, $A = \frac{1}{2}(2.7 + 1.2) \times 1.5 = 2.925\ \text{m}^2$.

FIGURE 3.5
Trapezoidal sluice gate

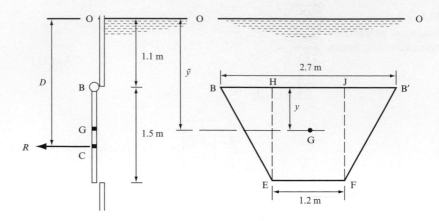

To find the position of the centroid G, take moments of area about BB′, putting the vertical distance GB = y:

$$A \times y = \text{Moment of areas BHE and FJB}' + \text{Moment of EFJH}$$

$$= 2 \times (\tfrac{1}{2} \times 1.5 \times 0.75) \times 0.5 + (1.2 \times 1.5) \times 0.75$$

$$2.925y = 0.5625 + 1.35 = 1.9125,$$

$$y = 0.654 \, \text{m}.$$

Depth to the centre of pressure,

$$\bar{y} = y + \text{OB} = 0.654 + 1.1 = 1.754 \, \text{m}.$$

Substituting in equation (3.2),

$$\text{Resultant force}, \ R = 10^3 \times 9.81 \times 2.925 \times 1.754 = \textbf{50.33 kN}.$$

From equation (3.4),

$$\text{Depth to centre of pressure C}, \ D = \sin^2 \phi (I_O/A\bar{y}).$$

Using the parallel axis rule for second moments of area,

$$I_O = \text{Second moment of EFJH about O} + \text{Second moment of BEH}$$
$$\text{and B}'\text{FJ about O}$$

$$= \left(\frac{1.2 \times 1.5^3}{12} + 1.2 \times 1.5 \times 1.85^2 \right) + \left(\frac{1.5 \times 1.5^3}{36} + 1.5 \times 0.75 \times 1.6^2 \right) \text{m}^4$$

$$= 9.5186 \, \text{m}^4.$$

As the wall is vertical, $\sin \phi = 1$; therefore,

$$\text{Depth to centre of pressure}, \ D = \frac{9.5186}{2.925 \times 1.754} = 1.8553 \, \text{m}.$$

$$\text{Moment about hinge} = R \times \text{BC} = 50.33(1.8553 - 1.1)$$

$$= \textbf{38.01 kN m}.$$

EXAMPLE 3.2

The angle between a pair of lock gates (Fig. 3.6) is 140° and each gate is 6 m high and 1.8 m wide, supported on hinges 0.6 m from the top and bottom of the gate. If the depths of water on the upstream and downstream sides are 5 m and 1.5 m, respectively, estimate the reactions at the top and bottom hinges.

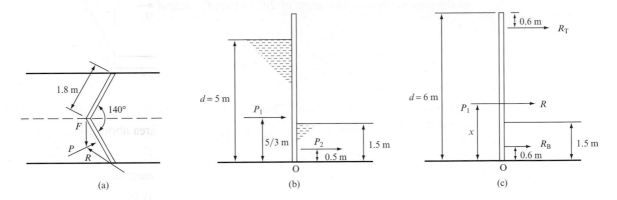

FIGURE 3.6
Lock gate

Solution

Figure 3.6(a) shows the plan view of the gates. F is the force exerted by one gate on the other and is assumed to act perpendicular to the axis of the lock if friction between the gates is neglected. P is the resultant of the water forces P_1 and P_2 (Fig. 3.6(b)) acting on the upstream and downstream faces of the gate, and R is the resultant of the forces R_T and R_B on the top and bottom hinges. Using equation (3.2)

Upstream water force,

$$P_1 = \rho g A_1 \bar{y}_1 = 10^3 \times 9.81 \times (5 \times 1.8) \times 2.5$$
$$= 220.725 \times 10^3 \, \text{N},$$

Downstream water force,

$$P_2 = \rho g A_2 \bar{z}_2 = 10^3 \times 9.81 \times (1.5 \times 1.8) \times 0.75$$
$$= 19.865 \times 10^3 \, \text{N},$$

Resultant water force on one gate,

$$P = P_1 - P_2 = (220.73 - 19.86) \times 10^3 \, \text{N}$$
$$= 200.87 \times 10^3 \, \text{N}.$$

The gates are rectangular, and so P_1 and P_2 will act at one-third of the depth of water (as shown in Fig. 3.6(b)), since, in equation (3.5), $\phi = 90°$, $\bar{y} = d/2$, $k_G^2 = d^2/12$, where d is the depth of the gate immersed (see also Section 3.4).

The height above the base at which the resultant force P acts can be found by taking moments. If P acts at a distance x from the bottom of the gate, then by taking moments about O,

$$Px = P_1 \times (5/3) - P_2 \times (1.5/3)$$
$$= (220.73 \times 5/3 - 19.86 \times 0.5) \times 10^3 = 357.95 \times 10^3 \, \text{N m}.$$
$$x = (357.95 \times 10^3)/(200.87 \times 10^3) = 1.782 \, \text{m}.$$

Assuming that F, R and P are coplanar, they will meet at a point, and, since F is assumed to be perpendicular to the axis of the lock on plan, both F and R are inclined to the gate as shown at an angle of $20°$ so that $F = R$ and

$$P = F \sin 20° + R \sin 20° = 2R \sin 20°,$$

$$R = \frac{P}{2 \sin 20°} = \frac{200.87 \times 10^3}{2 \times 0.342} = 293.65 \times 10^3 \, \text{N}.$$

If R is coplanar with P it acts at $1.78 \, \text{m}$ from the bottom of the gate. Taking moments about the bottom hinge,

$$4.8R_T = 1.18R$$

$$R_T = 1.18/4.8 \times 293.65 \times 10^3 = 72.2 \times 10^3 \, \text{N} = 72.2 \, \text{kN},$$

$$R_B = R - R_T = 293.65 - 72.2 = \textbf{221.45} \, \textbf{kN}.$$

3.4   PRESSURE DIAGRAMS

The resultant force and centre of pressure can be found graphically for walls and other surfaces of constant vertical height for which it is convenient to calculate the horizontal force exerted per unit width. In Fig. 3.7, ABC is the pressure diagram for the vertical wall of the tank containing a liquid, pressure being plotted horizontally against depth vertically. At the free surface A, the (gauge) pressure is zero. At depth y, $p = \rho g y$. The relationship between p and y is linear and can be represented by the triangle ABC. The area of this triangle will be the product of depth (in metres) and pressure (in newtons per square metre), and will represent, to scale, the resultant force R on unit width of the immersed surface perpendicular to the plane of the diagram (in newtons per metre).

$$\text{Area of pressure diagram} = \tfrac{1}{2} \text{AB} \times \text{BC} = \tfrac{1}{2} H \times \rho g H.$$

FIGURE 3.7

Pressure diagram for a vertical wall

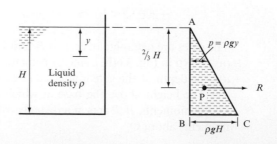

Therefore,

Resultant force, $R = \rho g H^2/2$ for unit width,

and R will act through the centroid P of the pressure diagram, which is at a depth of $\frac{2}{3} H$ from A.

This result could also have been obtained from equations (3.2) and (3.5), since, for unit width,

$$R = \rho g A \bar{y} = \rho g (H \times 1) \times \tfrac{1}{2} H = \rho g H^2/2,$$

and, in equation (3.5), $\phi = 90°$, $\sin \phi = 1$, $\bar{y} = H/2$, $k_G^2 = H^2/12$; therefore,

$$D = \frac{H^2/12}{H/2} + \frac{H}{2} = \frac{2}{3} H,$$

as before.

If the plane surface is inclined and submerged below the surface, the pressure diagram is drawn perpendicular to the immersed surface (Fig. 3.8) and will be a straight line extending from $p = 0$ at the free surface to $p = \rho g H$ at depth H. As the immersed surface does not extend to the free surface, the resultant force R is represented by the shaded area, instead of the whole triangle, and acts through the centroid P of this area.

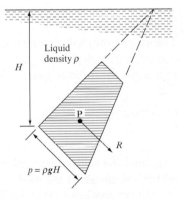

FIGURE 3.8

Pressure diagram for an inclined submerged surface

It is also possible to draw pressure diagrams in three dimensions for immersed areas of various shapes as, for example, the triangular sluice gate in Fig. 3.9. However, such diagrams do little more than provide assistance in visualizing the situation.

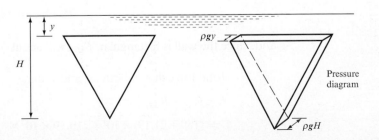

FIGURE 3.9

Pressure diagram for a triangular sluice gate

EXAMPLE 3.3

A closed tank (Fig. 3.10), rectangular in plan with vertical sides, is 1.8 m deep and contains water to a depth of 1.2 m. Air is pumped into the space above the water until the air pressure is 35 kN m^{-2}. If the length of one wall of the tank is 3 m, determine the resultant force on this wall and the height of the centre of pressure above the base.

FIGURE 3.10

Pressure diagram

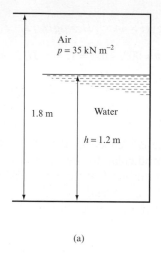

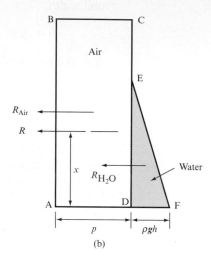

Solution

The air pressure will be transported uniformly over the whole of the vertical wall, and can be represented by the pressure diagram ABCD (Fig. 3.10(b)), the area of which represents the force exerted by the air per unit width of wall.

$$\text{Force due to air, } R_{Air} = (p \times AB) \times \text{Width}$$

$$= 35 \times 10^3 \times 1.8 \times 3 = 189 \times 10^3 \text{ N,}$$

and, since the wall is rectangular and the pressure uniform, R_{Air} will act at mid-height, which is 0.9 m above the base.

The pressure due to the water will start from zero at the free surface, corresponding to the point E, and reach a value DF equal to ρgh at the bottom. The area of the triangular pressure diagram EFD represents the force exerted by the water per unit width:

$$\text{Force due to water, } R_{H_2O} = \tfrac{1}{2} \times (\rho gh \times DE) \times \text{Width}$$

$$= \tfrac{1}{2} \times 10^3 \times 9.81 \times 1.2 \times 1.2 \times 3$$

$$= 21.19 \times 10^3 \text{ N,}$$

and, since the wall is rectangular, R_{H_2O} will act at $\tfrac{1}{3}h = 0.4$ m from the base.

Total force due to both air and water,

$$R = R_{Air} + R_{H_2O}$$

$$= (189 + 21.19) \times 10^3 = \mathbf{210.19 \times 10^3 \text{ N.}}$$

If x is the height above the base of the centre of pressure through which R acts,

$$R \times x = R_{Air} \times 0.9 + R_{H_2O} \times 0.4,$$

$$x = (189 \times 0.9 + 21 \times 0.4)/210.19 = \mathbf{0.85\,m}.$$

3.5 FORCE ON A CURVED SURFACE DUE TO HYDROSTATIC PRESSURE

If a surface is curved, the forces produced by fluid pressure on the small elements making up the area will not be parallel and, therefore, must be combined vectorially. It is convenient to calculate the horizontal and vertical components of the resultant force. This can be done in three dimensions, but the following analysis is for a surface curved in one plane only.

In Fig. 3.11(a) and (b), AB is the immersed surface and R_h and R_v are the horizontal and vertical components of the resultant force R of the liquid on one side of the surface. In Fig. 3.11(a) the liquid lies above the immersed surface, while in Fig. 3.11(b) it acts below the surface.

In Fig. 3.11(a), if ACE is a vertical plane through A, and BC is a horizontal plane, then, since element ACB is in equilibrium, the resultant force P on AC must equal the horizontal component R_h of the force exerted by the fluid on AB because there are no other horizontal forces acting. But AC is the projection of AB on a vertical plane; therefore,

FIGURE 3.11
Hydrostatic force on a curved surface

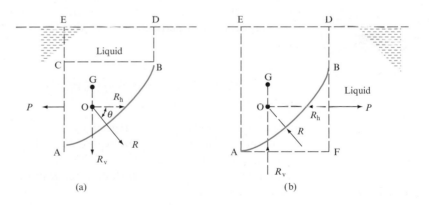

(a)   (b)

Horizontal component R_h = Resultant force on the projection of AB on a vertical plane.

Also, for equilibrium, P and R_h must act in the same straight line; therefore, the horizontal component R_h acts through the centre of pressure of the projection of AB on a vertical plane.

Similarly, in Fig. 3.11(b), element ABF is in equilibrium, and so the horizontal component R_h is equal to the resultant force on the projection BF of the curved surface AB on a vertical plane, and acts through the centre of pressure of this projection.

In Fig. 3.11(a), the vertical component R_v will be entirely due to the weight of the fluid in the area ABDE lying vertically above AB. There are no other vertical forces, since there can be no shear forces on AE and BD because the fluid is at rest. Thus,

$$\text{Vertical component, } R_v = \text{Weight of fluid vertically above AB,}$$

and will act vertically downwards through the centre of gravity G of ABDE.

In Fig. 3.11(b), if the surface AB were removed and the space ABDE filled with the liquid, this liquid would be in equilibrium under its own weight and the vertical force on the boundary AB. Therefore,

$$\text{Vertical component, } R_v = \text{Weight of the volume of the same fluid} \\ \text{which would lie vertically above AB,}$$

and will act vertically upwards through the centre of gravity G of this imaginary volume of fluid.

In the case of closed vessels under pressure, a free surface does not exist, but an imaginary free surface can be substituted at a level $p/\rho g$ above a point at which the pressure p is known, ρ being the mass density of the actual fluid.

The resultant force R is found by combining the components vectorially. In the general case, the components in three directions may not meet at a point and, therefore, cannot be represented by a single force. However, in Fig. 3.11, if the surface is of uniform width perpendicular to the diagram, R_h and R_v will intersect at O. Thus,

$$\text{Resultant force, } R = \sqrt{(R_h^2 + R_v^2)},$$

and acts through O at an angle θ given by $\tan\theta = R_v/R_h$.

In the special case of a cylindrical surface, all the forces on each small element of area acting normal to the surface will be radial and will pass through the centre of curvature O (Fig. 3.12). The resultant force R must, therefore, also pass through the centre of curvature O.

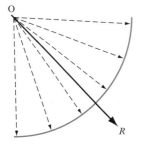

FIGURE 3.12

Resultant force on a cylindrical surface

EXAMPLE 3.4

A sluice gate is in the form of a circular arc of radius 6 m as shown in Fig. 3.13. Calculate the magnitude and direction of the resultant force on the gate, and the location with respect to O of a point on its line of action.

Solution

Since the water reaches the top of the gate,

FIGURE 3.13
Sector gate

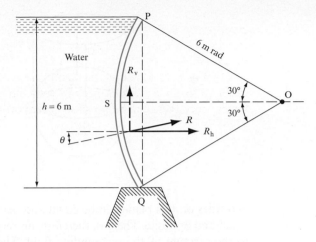

Depth of water, $h = 2 \times 6 \sin 30° = 6\,\text{m}$,

Horizontal component of force on gate = R_h per unit length

    = Resultant force on PQ per unit length

    = $\rho g \times h \times h/2 = \rho g h^2/2$

    = $(10^3 \times 9.81 \times 36)/2\,\text{N m}^{-1} = 176.58\,\text{kN m}^{-1}$,

Vertical component of force on gate = R_v per unit length

    = Weight of water displaced by segment PSQ

    = (Sector OPSQ − ΔOPQ)ρg

    = $[\,(60/360) \times \pi \times 6^2 - 6 \sin 30° \times 6 \cos 30°] \times 10^3 \times 9.81\,\text{N m}^{-1}$

    = $32.00\,\text{kN m}^{-1}$,

Resultant force on gate, $R = \sqrt{(R_h^2 + R_v^2)}$

    = $\sqrt{(176.58^2 + 32.00^2)} = \mathbf{179.46\ kN\,m^{-1}}$.

If R is inclined at an angle θ to the horizontal,

    $\tan \theta = R_v/R_h = 32.00/176.58 = 0.181\,22$

    $\theta = \mathbf{10.27°\ to\ the\ horizontal}$.

Since the surface of the gate is cylindrical, the resultant force R **must pass through O**.

3.6 BUOYANCY

The method of calculating the forces on a curved surface applies to all shapes of surface and, therefore, to the surface of a totally submerged object (Fig. 3.14). Considering any vertical plane VV through the body, the projected area of each of the

FIGURE 3.14
Buoyancy

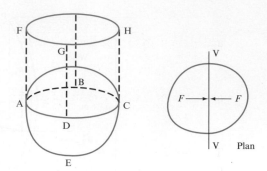

two sides on this plane will be equal and, as a result, the horizontal forces F will be equal and opposite. There is, therefore, no resultant horizontal force on the body due to the pressure of the surrounding fluid. The only force exerted by the fluid on an immersed body is vertical and is called the buoyancy or upthrust. It will be equal to the difference between the resultant forces on the upper and lower parts of the surface of the body. If ABCD is a horizontal plane,

Upthrust on body = Upward force on lower surface ADEC

– Downward force on upper surface ABCD

= Weight of volume of fluid AECDGFH

– Weight of volume of fluid ABCDGFH

= Weight of volume of fluid ABCDE,

Upthrust on body = Weight of fluid displaced by the body,

and will act through the centroid of the volume of fluid displaced, which is known as the *centre of buoyancy*. This result is known as Archimedes' principle. As an alternative to the proof given above, it can be seen that, if the body were completely replaced by the fluid in which it is immersed, the forces exerted on the boundaries corresponding to the original body would exactly maintain the substituted fluid in equilibrium. Thus, the upward force on the boundary must be equal to the downward force corresponding to the weight of the fluid displaced by the body.

If a body is immersed so that part of its volume V_1 is immersed in a fluid of density ρ_1 and the rest of its volume V_2 in another immiscible fluid of mass density ρ_2 (Fig. 3.15),

FIGURE 3.15
Body immersed in two fluids

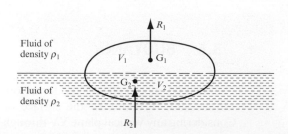

Upthrust on upper part, $R_1 = \rho_1 g V_1$

acting through G_1, the centroid of V_1,

Upthrust on lower part, $R_2 = \rho_2 g V_2$

acting through G_2, the centroid of V_2,

Total upthrust $= \rho_1 g V_1 + \rho_2 g V_2$.

The positions of G_1 and G_2 are not necessarily on the same vertical line, and the centre of buoyancy of the whole body is, therefore, not bound to pass through the centroid of the whole body.

EXAMPLE 3.5 A rectangular pontoon has a width B of 6 m, a length l of 12 m, and a draught D of 1.5 m in fresh water (density 1000 kg m^{-3}). Calculate (a) the weight of the pontoon, (b) its draught in sea water (density 1025 kg m^{-3}) and (c) the load (in kilonewtons) that can be supported by the pontoon in fresh water if the maximum draught permissible is 2 m.

Solution

When the pontoon is floating in an unloaded condition,

Upthrust on immersed volume = Weight of pontoon.

Since the upthrust is equal to the weight of the fluid displaced,

Weight of pontoon = Weight of fluid displaced,

$$W = \rho g B l D.$$

(a) In fresh water, $\rho = 1000$ kg m^{-3} and $D = 1.5$ m; therefore,

Weight of pontoon, $W = 1000 \times 9.81 \times 6 \times 12 \times 1.5$ N

$$= \mathbf{1059.5\,kN}.$$

(b) In sea water, $\rho = 1025$ kg m^{-3}; therefore,

Draught in sea water, $D = W/\rho g B l$

$$= \frac{1059.5 \times 10^3}{1025 \times 9.81 \times 6 \times 12} = \mathbf{1.46\ m}.$$

(c) For the maximum draught of 2 m in fresh water,

Total upthrust = Weight of water displaced $= \rho g B l D$

$$= 1000 \times 9.81 \times 6 \times 12 \times 2\ \text{N}$$

$$= 1412.6\,\text{kN},$$

Load which can be supported = Upthrust − Weight of pontoon

$$= 1412.6 - 1059.5 = 353.1\ \text{kN}.$$

3.7   EQUILIBRIUM OF FLOATING BODIES

When a body floats in vertical equilibrium in a liquid, the forces present are the upthrust R acting through the centre of buoyancy B (Fig. 3.16) and the weight of the body $W = mg$ acting through its centre of gravity. For equilibrium, R and W must be equal and act in the same straight line. Now, R will be equal to the weight of fluid displaced, $\rho g V$, where V is the volume of fluid displaced; therefore,

$$V = mg/\rho g = m/\rho.$$

As explained in Section 2.1, the equilibrium of a body may be stable, unstable or neutral, depending upon whether, when given a small displacement, it tends to return to the equilibrium position, move further from it or remain in the displaced position. For a floating body, such as a ship, stability is of major importance.

FIGURE 3.16
Body floating in equilibrium

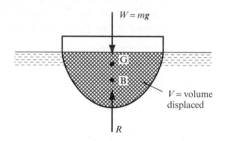

3.8   STABILITY OF A SUBMERGED BODY

For a body totally immersed in a fluid, the weight $W = mg$ acts through the centre of gravity of the body, while the upthrust R acts through the centroid of the body B, which is the centre of buoyancy. Whatever the orientation of the body, these two points will remain in the same positions relative to the body. It can be seen from Fig. 3.17 that a small angular displacement θ from the equilibrium position will generate a moment $W \times BG \times \theta$. If the centre of gravity G is below the centre of buoyancy B (Fig. 3.13(a)), this will be a righting moment and the body will tend to return to its equilibrium position. However, if (as in Fig. 3.17(b)) the centre of gravity is above the centre of buoyancy, an overturning moment is produced and the body is unstable. Note that, as

FIGURE 3.17
Stability of submerged bodies

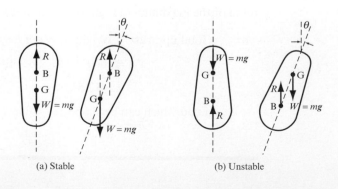

(a) Stable   (b) Unstable

the body is totally immersed, the shape of the displaced fluid is not altered when the body is tilted and so the centre of buoyancy remains unchanged relative to the body.

3.9 STABILITY OF FLOATING BODIES

Figure 3.18(a) shows a body floating in equilibrium. The weight $W = mg$ acts through the centre of gravity G and the upthrust R acts through the centre of buoyancy B of the displaced fluid in the same straight line as W. When the body is displaced through an angle θ (Fig. 3.18(b)), W continues to act through G; the volume of liquid remains unchanged since $R = W$, but the shape of this volume changes and its centre of gravity, which is the centre of buoyancy, moves relative to the body from B to B_1. Since R and W are no longer in the same straight line, a turning moment proportional to $W \times \theta$ is produced, which in Fig. 3.18(b) is a righting moment and in Fig. 3.18(d) is an overturning moment. If M is the point at which the line of action of the upthrust R cuts the original vertical through the centre of gravity of the body G,

$$x = GM \times \theta,$$

provided that the angle of tilt θ is small, so that $\sin \theta = \tan \theta = \theta$ in radians.

> The point M is called the *metacentre* and the distance GM is the *metacentric height*. Comparing Fig. 3.18(b) and (d) it can be seen that:
>
> 1. If M lies above G, a righting moment $W \times GM \times \theta$ is produced, equilibrium is stable and GM is regarded as positive.
> 2. If M lies below G, an overturning moment $W \times GM \times \theta$ is produced, equilibrium is unstable and GM is regarded as negative.
> 3. If M coincides with G, the body is in neutral equilibrium.

FIGURE 3.18
Stable and unstable
equilibrium

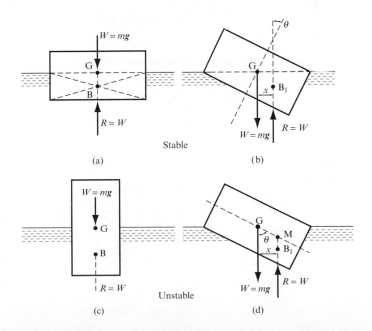

Since a floating body can tilt in any direction, it is usual, for a ship, to consider displacement about the longitudinal (rolling) and transverse (pitching) axes. The position of the metacentre and the value of the metacentric height will normally be different for rolling and pitching.

3.10 DETERMINATION OF THE METACENTRIC HEIGHT

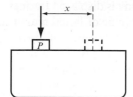

FIGURE 3.19
Determination of metacentric height

The metacentric height of a vessel can be determined if the angle of tilt θ caused by moving a load P (Fig. 3.19) a known distance x across the deck is measured.

$$\text{Overturning moment due to movement of load } P = Px. \tag{3.6}$$

If GM is the metacentric height and $W = mg$ is the total weight of the vessel including P,

$$\text{Righting moment} = W \times \text{GM} \times \theta. \tag{3.7}$$

For equilibrium in the tilted position, the righting moment must equal the overturning moment so that, from equations (3.6) and (3.7),

$$W \times \text{GM} \times \theta = Px,$$

$$\text{Metacentric height, GM} = Px/W\theta. \tag{3.8}$$

The true metacentric height is the value of GM as $\theta \to 0$.

3.11 DETERMINATION OF THE POSITION OF THE METACENTRE RELATIVE TO THE CENTRE OF BUOYANCY

For a vessel of known shape and displacement, the position of the centre of buoyancy B is comparatively easily found and the position of the metacentre M relative to B can be calculated as follows. In Fig. 3.20, AC is the original waterline plane and B the

FIGURE 3.20
Height of metacentre above centre of buoyancy

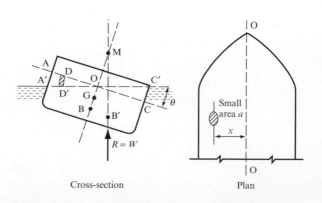

Cross-section   Plan

centre of buoyancy in the equilibrium position. When the vessel is tilted through a small angle θ, the centre of buoyancy will move to B′ as a result of the alteration in the shape of the displaced fluid. A′C′ is the waterline plane in the displaced position. For small angles of tilt,

$$BM = BB'/\theta.$$

The movement of the centre of buoyancy, which is the centre of gravity of the displaced fluid, from B to B′ is the result of the removal of a volume of fluid corresponding to the wedge AOA′ and the addition of a wedge COC′. The total weight of fluid displaced remains unchanged, since it is equal to the weight of the vessel; therefore,

$$\text{Weight of wedge AOA}' = \text{Weight of wedge COC}'.$$

If a is a small area in the waterline plane at a distance x from the axis of rotation OO, it will generate a small volume, shown shaded, when the vessel is tilted.

$$\text{Volume swept out by } a = DD' \times a = ax\theta.$$

Summing all such volumes and multiplying by the specific weight ρg of the liquid,

$$\text{Weight of wedge AOA}' = \sum_{x=0}^{x=AO} \rho g a x \theta. \tag{3.9}$$

Similarly,

$$\text{Weight of wedge COC}' = \sum_{x=0}^{x=CO} \rho g a x \theta. \tag{3.10}$$

Since there is no change in displacement, we have, from equations (3.9) and (3.10),

$$\rho g \theta \sum_{x=0}^{x=AO} ax = \rho g \theta \sum_{x=0}^{x=CO} ax,$$

$$\sum ax = 0.$$

But $\sum ax$ is the first moment of area of the waterline plane about OO; therefore the axis OO must pass through the centroid of the waterline plane.

The distance BB′ can now be calculated, since the couple produced by the movement of the wedge AOA′ to COC′ must be equal to the couple due to the movement of R from B to B′.

$$\text{Moment about OO of the weight of fluid swept out by area } a = \rho g a x \theta \times x.$$

$$\text{Total moment due to altered displacement} = \rho g \theta \sum ax^2.$$

Putting $\sum ax^2 = I = \text{Second moment of area of waterline plane about OO,}$

$$\text{Total moment due to altered displacement} = \rho g \theta I, \tag{3.11}$$

$$\text{Moment due to movement of } R = R \times BB' = \rho g V \times BB', \tag{3.12}$$

where V = volume of liquid displaced. Equating equations (3.11) and (3.12),

$$\rho g V \times BB' = \rho g \theta I,$$

$$BB' = \theta I / V, \tag{3.13}$$

giving $BM = BB'/\theta = I/V.$ (3.14)

The distance BM is known as the metacentric radius.

EXAMPLE 3.6

A cylindrical buoy (Fig. 3.21) 1.8 m in diameter, 1.2 m high and weighing 10 kN floats in salt water of density 1025 kg m^{-3}. Its centre of gravity is 0.45 m from the bottom. If a load of 2 kN is placed on the top, find the maximum height of the centre of gravity of this load above the bottom if the buoy is to remain in stable equilibrium.

FIGURE 3.21
Stability of a
cylindrical buoy

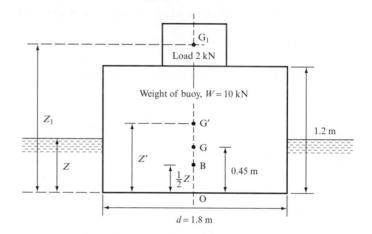

Solution

In Fig. 3.21, let G be the centre of gravity of the buoy, G_1 the centre of gravity of the load at a height Z_1 above the bottom, G' the combined centre of gravity of the load and the buoy at a height Z' above the bottom.

When the load is in position, let V be the volume of salt water displaced and Z the depth of immersion of the buoy.

$$\text{Buoyancy force} = \text{Weight of salt water displaced}$$

$$= \rho g V = \rho g (\pi/4) d^2 Z.$$

For equilibrium, the buoyancy force must equal the combined weight of the buoy and the load $(W + W_1)$; therefore,

$$W + W_1 = \rho g (\pi/4) d^2 Z,$$

Depth of immersion,

$$Z = 4(W + W_1)/\rho g \pi d^2$$

$$= 4(10 + 2) \times 10^3/(1025 \times 9.81 \times 1.8^2 \times \pi) = 0.47 \text{ m}.$$

The centre of buoyancy B will be at the centre of gravity of the displaced water, so that $OB = \frac{1}{2}Z = 0.235\,\text{m}$.

If the buoy and the load are just in stable equilibrium, the metacentre M must coincide with the centre of gravity G' of the buoy and load combined. The metacentric height G'M will then be zero and BG' = BM. From equation (3.14),

$$\text{BG}' = \text{BM} = \frac{I}{V} = \frac{\pi d^4/64}{\pi d^2 z/4} = \frac{1.8^2}{16 \times 0.47} = 0.431\ \text{m}.$$

Thus, the position of G' is given by

$$Z' = \tfrac{1}{2}Z + \text{BG}' = 0.235 + 0.431 = 0.666\,\text{m}.$$

The value of Z_1 corresponding to this value of Z' is found by taking moments about O:

$$W_1 Z_1 + 0.45 W = (W + W_1)Z'.$$

Maximum height of load above bottom,

$$Z_1 = \frac{(W + W_1)Z' - 0.45\,W}{W_1}$$

$$= \frac{12 \times 10^3 \times 0.666 - 0.45 \times 10 \times 10^3}{2 \times 10^3}\ \text{m} = \mathbf{1.746\ m}.$$

3.12 PERIODIC TIME OF OSCILLATION

The displacement of a stable vessel through an angle θ from its equilibrium position produces a righting moment T which, from equation (3.7), is given by $T = W \times \text{GM} \times \theta$, where $W = mg$ is the weight of the vessel and GM is the metacentric height. This will produce an angular acceleration $\text{d}^2\theta/\text{d}t^2$, and, if I is the mass moment of inertia of the vessel about its axis of rotation,

$$\frac{\text{d}^2\theta}{\text{d}t^2} = \frac{T}{I} = -\frac{W \times \text{GM} \times \theta}{(W/g)k^2} = -\frac{\text{GM} \times \theta g}{k^2},$$

where k is the radius of gyration from its axis of rotation. The negative sign indicates that the acceleration is in the opposite direction to the displacement. Since this corresponds to simple harmonic motion,

$$\text{Periodic time, } t = 2\pi\sqrt{\left(\frac{\text{Displacement}}{\text{Acceleration}}\right)} = 2\pi\sqrt{\left[\frac{\theta}{\text{GM} \times \theta \times (g/k^2)}\right]}$$

$$= 2\pi\sqrt{[k^2/(\text{GM} \times g)]}, \tag{3.15}$$

from which it can be seen that, although a large metacentric height will improve stability, it produces a short periodic time of oscillation, which results in discomfort and excessive stress on the structure of the vessel.

3.13   STABILITY OF A VESSEL CARRYING LIQUID IN TANKS WITH A FREE SURFACE

The stability of a vessel carrying liquid in tanks with a free surface (Fig. 3.22) is affected adversely by the movement of the centre of gravity of the liquid in the tanks as the vessel heels. Thus, G_1 will move to G_1' and G_2 to G_2'. The distance moved is calculated in the same way as the movement BB' of the centre of buoyancy, given by equation (3.13):

$$G_1G_1' = \theta I_1/V_1 \quad \text{and} \quad G_2G_2' = \theta I_2/V_2,$$

FIGURE 3.22
Vessel carrying liquid
in tanks

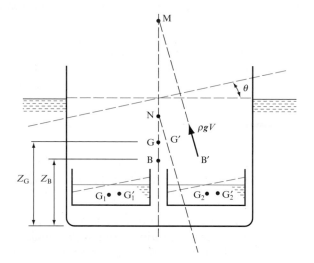

where I_1 and I_2 are the second moments of area of the free surfaces, and V_1 and V_2 the volumes, of the liquid in the tanks. As a result of the movement of G_1 and G_2, the centre of gravity G of the whole vessel and contents will move to G'. If V is the volume of water displaced by the vessel and ρ is the mass density of water,

$$\text{Weight of vessel and contents} = \text{Weight of water displaced}$$
$$= \rho g V.$$

If the volume of liquid of density ρ_1 in the tanks is V_1 and V_2,

$$\text{Weight of contents of the first tank} = \rho_1 g V_1,$$
$$\text{Weight of contents of the second tank} = \rho_1 g V_2.$$

Taking moments to find the change in the centre of gravity of the vessel and contents,

$$\rho g V \times GG' = \rho_1 g V_1 \times G_1 G_1' + \rho_1 g V_2 \times G_2 G_2'$$
$$= \rho_1 g V_1 \times \theta I_1 / V_1 + \rho_1 g V_2 \times \theta I_2 / V_2,$$
$$GG' = \frac{1}{V}(\rho_1/\rho)\theta(I_1 + I_2).$$

In the tilted position, the new vertical through B′ intersects the original vertical through G at the metacentre M, but the weight W acts through G′ instead of G and its line of action cuts the original vertical at N, reducing the metacentric height from GM to NM.

Effective metacentric height, $NM = Z_B + BM - (Z_G + GN)$,

and, since $BM = I/V$ and $GN = GG'/\theta = \frac{1}{V}(\rho_1/\rho)(I_1 + I_2)$,

$$NM = Z_B - Z_G + \frac{1}{V}[1 - (\rho_1/\rho)(I_1 + I_2)]. \tag{3.16}$$

Thus, the effect of the liquid in the tank is to reduce the effective metacentric height and impair stability, provided that the liquid in the tanks has a free surface so that its centre of gravity moves as the vessel tilts. Lateral subdivision of the tanks improves stability by reducing the sum of the second moments of area I_1, I_2, etc.

EXAMPLE 3.7

A barge (Fig. 3.23) has vertical sides and ends and a flat bottom. In plan view it is rectangular, 20 m long by 6 m wide, but with an additional semicircular portion of 3 m radius at one end. The empty barge weighs 200 kN and floats upright in fresh water. The part of the vessel which is rectangular in plan is divided by a wall into two

FIGURE 3.23
Barge containing liquids

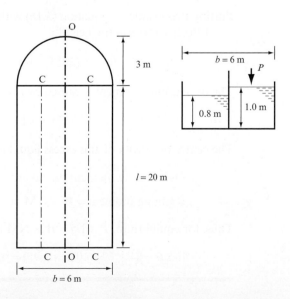

compartments 3 m wide by 20 m long. These compartments form open-top tanks which are partly filled with liquid of relative density 0.8 to a depth of 0.8 m in one tank and 1.0 m in the other. The vessel rolls about a horizontal axis, but the flat end remains in a vertical plane. Ignoring the thickness of the material of the barge structure and assuming that the centre of gravity of the barge and contents is 0.45 m above the bottom, find the angle of roll.

Solution

In order to be able to determine the angle of roll, we must first find the effective metacentric height from equation (3.16).

For the whole vessel

$$I_{OO} = \frac{lb^3}{12} + \frac{\pi b^4}{128} = 20 \times 6^3/12 + \pi \times 6^4/128 \text{ m}^4 = 391.9 \text{ m}^4.$$

For each tank

$$I_{CC} = l \times (\tfrac{1}{2} b)^3/12 = 20 \times 3^3/12 = 45 \text{ m}^4.$$

Weight of barge = 200 kN.

Weight of liquid load = $0.8 \times 10^3 \times 9.81(20 \times 3 \times 1 + 20 \times 3 \times 0.8)$ N

$$= 846 \text{ kN}.$$

Total weight of barge and contents = 1046 kN.

Area of waterline plane of vessel = $20 \times 6 + \tfrac{1}{2} \pi \times 3^2 = 134.1 \text{ m}^2$.

Volume of vessel submerged = Weight/(Density × g)

$$= 1046 \times 10^3/(10^3 \times 9.81) = 106.8 \text{ m}^3.$$

Depth submerged = 106.8/134.1 = 0.80 m.

Height of centre of buoyancy B above bottom = $\tfrac{1}{2}$ Depth submerged

$$= 0.4 \text{ m}.$$

Putting these values in equation (3.16) with $\rho_1/\rho = 0.8$,

Effective metacentric height,

$$\text{NM} = 0.4 - 0.45 + (391.9 - 0.8 \times 2 \times 45)/106.8 = 2.95 \text{ m}.$$

The overturning moment is caused by the weight of the excess liquid in one tank,

$$P = 0.8 \times 10^3 \times 9.81 \times 20 \times 3(1.0 - 0.8) = 94 \text{ kN}.$$

The centre of gravity of this excess liquid is 1.5 m from the centreline.

Overturning moment due to excess liquid = $P \times 1.5$,

Righting moment = $W \times \text{NM} \tan \theta$.

Thus, for equilibrium, $P \times 1.5 = W \times \text{NM} \tan \theta$,

$$\tan \theta = 94 \times 1.5/(1046 \times 2.95) = 0.0457, \quad \text{Angle of roll, } \theta = 2°37'.$$

Concluding remarks

Arising directly from the hydrostatic equation developed in Chapter 2, this chapter has demonstrated techniques available for determining the forces acting on submerged or partially submerged surfaces. The chapter has also stressed the relationship between pressure and force, and in particular has highlighted the fact that force is a vector quantity, calculated by reference to the applied pressure and the surface area normal to the force direction. This concept, although apparently obvious, will form the basis of later calculations of lift and drag on aerofoils, and the definition of appropriate lift and drag coefficients. The treatment of buoyancy and floating-body stability is a further demonstration of the application of the techniques appropriate to the analysis of solid-body mechanics.

Summary of important equations and concepts

1. The integration necessary to determine the resultant force acting on a surface is emphasized, Section 3.2, and most importantly the concept of a centre of pressure through which this force acts is introduced, Section 3.3. It will be shown later that centre of pressure movement as flow becomes supersonic can affect wing stability and introduce the need for remedial action, either by control surface activation or corresponding movement of the aircraft centre of gravity.

2. The treatment of a range of hydrostatic force and moment examples illustrates the interface between hydrostatics and mechanics, Sections 3.4 and 3.5.

3. The treatment of buoyancy and stability of floating bodies continues this linkage.

Problems

3.1 A circular lamina 125 cm in diameter is immersed in water so that the distance of its edge measured vertically below the free surface varies from 60 cm to 150 cm. Find the total force due to the water acting on one side of the lamina, and the vertical distance of the centre of pressure below the surface.   [12 639 N, 1.1 m]

3.2 One end of a rectangular tank is 1.5 m wide by 2 m deep. The tank is completely filled with oil of specific weight 9 kN m^{-3}. Find the resultant pressure on this vertical end and the depth of the centre of pressure from the top.
[27 kN, 1.33 m]

3.3 What is the position of the centre of pressure of a vertical semicircular plane submerged in a homogeneous liquid with its diameter d at the free surface?
[Depth $3\pi d/32$]

3.4 A culvert draws off water from the base of a reservoir. The entrance to the culvert is closed by a circular gate 1.25 m in diameter which can be rotated about its horizontal diameter. Show that the turning moment on the gate is independent of the depth of water if the gate is completely immersed and find the value of this moment.   [1177 N m]

3.5 A barge in the form of a closed rectangular tank 20 m long by 4 m wide floats in water. If the bottom is 1.5 m below the surface, what is the water force on one long side and at what level below the surface does it act?

If a uniform pressure of 50 kN m^{-2} gauge is applied inside the barge what will be the new magnitude and point of action of the resultant force on the side? The deck is 0.2 m above water level.
[220.73 kN, 1.0 m; 1479.27 kN, 0.6 m below surface]

3.6 A rectangular sluice door (Fig. 3.24) is hinged at the top at A and kept closed by a weight fixed to the door.

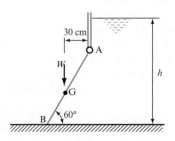

FIGURE 3.24

The door is 120 cm wide and 90 cm long and the centre of gravity of the complete door and weight is at G, the combined weight being 9810 N. Find the height of the water h on the inside of the door which will just cause the door to open.   [0.88 m]

3.7   A rectangular gate (Fig. 3.25) of negligible thickness, hinged at its top edge and of width b, separates two tanks in which there is the same liquid of density ρ. It is required that the gate shall open when the level in the left-hand tank falls below a distance H from the hinge. The level in the right-hand tank remains constant at a height y above the hinge. Derive an expression for the weight of the gate in terms of H, Y, y, b and $\boldsymbol{g}$. Assume that the weight of the gate acts at its centre of area.

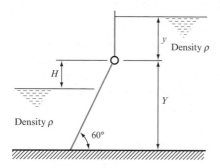

FIGURE 3.25

$$\left[W = 0.77\rho gb\left[\frac{3\,Y^2(y+H)-H^3}{Y}\right]\right]$$

3.8   A masonry dam 6 m high has the water level with the top. Assuming that the dam is rectangular in section and 3 m wide, determine whether the dam is stable against overturning and whether tension will develop in the masonry joints. Density of masonry 1760 kg m^{-3}.
                              [Stable, tension on the water face]

3.9   A pair of lock gates, each 3 m wide, have their lower hinges at the bottom of the gates and their upper hinges 5 m from the bottom. The width of the lock is 5.5 m. Find the reaction between the gates when the water level is 4.5 m above the bottom of one side and 1.5 m on the other. Assuming that this force acts at the same height as the resultant force due to the water pressure find the reaction forces on the hinges.   [331 kN; 107.6 kN, 223.4 kN]

3.10   A spherical container is made up of two hemispheres, the joint between the two halves being horizontal. The sphere is completely filled with water through a small hole in the top. It is found that 50 kg of water are required for this

purpose. If the two halves of the container are not secured together, what must be the mass of the upper hemisphere if it just fails to lift off the lower hemisphere?   [12.5 kg]

3.11   A sluice gate (Fig. 3.26) consists of a quadrant of a circle of radius 1.5 m pivoted at its centre O. Its centre of gravity is at G as shown. When the water is level with the pivot O, calculate the magnitude and direction of the resultant force on the gate due to the water and the turning moment required to open the gate. The width of the gate is 3 m and it has a mass of 6000 kg.

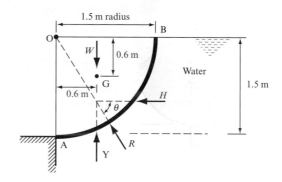

FIGURE 3.26

                              [61.6 kN, 57°31′, 35.3 kN m]

3.12   A sector-shaped sluice gate having a radius of curvature of 5.4 m is as shown in Fig. 3.27. The centre of curvature C is 0.9 m vertically below the lower edge A of the gate and 0.6 m vertically above the horizontal axis passing through O about which the gate is constructed to turn. The mass of the gate is 3000 kg per metre run and its centre of gravity is 3.6 m horizontally from the centre O. If the water level is 2.4 m above the lower edge of the gate, find per metre run (a) the resultant force acting on the axis at O, (b) the resultant moment about O.

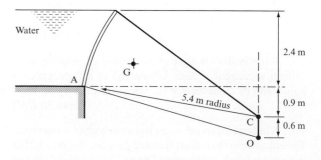

FIGURE 3.27

                              [(a) 39.2 kN, (b) 106 kN m]

3.13 The face of a dam (Fig. 3.28) is curved according to the relation $y = x^2/2.4$, where y and x are in metres. The height of the free surface above the horizontal plane through A is 15.25 m. Calculate the resultant force F due to the fresh water acting on unit breadth of the dam, and determine the position of the point B at which the line of action of this force cuts the horizontal plane through A.

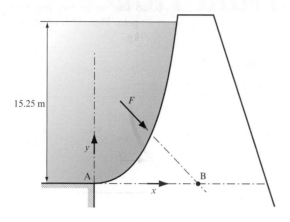

FIGURE 3.28

[1290 kN m^{-1}, 14.15 m]

3.14 A steel pipeline conveying gas has an internal diamter of 120 cm and an external diameter of 125 cm. It is laid across the bed of a river, completely immersed in water and is anchored at intervals of 3 m along its length. Calculate the buoyancy force in newtons per metre and the upward force in newtons on each anchorage. Density of steel = 7900 kg m^{-3}, density of water = 1000 kg m^{-3}.
[12 037 N m^{-1}, 13 742 N]

3.15 The ball-operated valve shown in Fig. 3.29 controls the flow from a tank through a pipe to a lower tank, in which it is situated. The water level in the upper tank is 7 m above the 10 mm diameter valve opening. Calculate the volume of the ball which must be submerged to keep the valve closed.

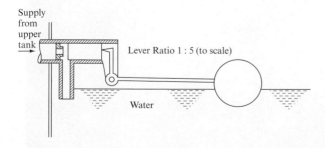

FIGURE 3.29

[110 cm^3]

3.16 The shifting of a portion of cargo of mass 25 000 kg through a distance of 6 m at right angles to the vertical plane containing the longitudinal axis of a vessel causes it to heel through an angle of 5°. The displacement of the vessel is 5000 metric tons and the value of I is 5840 m^4. The density of sea water is 1025 kg m^{-3}. Find (*a*) the metacentric height and (*b*) the height of the centre of gravity of the vessel above the centre of buoyancy. [(*a*) 0.342 m, (*b*) 0.849 m]

3.17 A buoy floating in sea water of density 1025 kg m^{-3} is conical in shape with a diameter across the top of 1.2 m and a vertex angle of 60°. Its mass is 300 kg and its centre of gravity is 750 mm from the vertex. A flashing beacon is to be fitted to the top of the buoy. If this unit has a mass of 55 kg what is the maximum height of its centre of gravity above the top of the buoy if the whole assembly is not to be unstable? (The centre of volume of a cone of height h is at a distance $\frac{3}{4}h$ from the vertex.) [1.25 m]

3.18 A rectangular pontoon 10 m by 4 m in plan weighs 280 kN. A steel tube weighing 34 kN is placed longitudinally on the deck. When the tube is in a central position, the centre of gravity for the combined weight lies on the vertical axis of symmetry 250 mm above the water surface. Find (*a*) the metacentric height, (*b*) the maximum distance the tube may be rolled laterally across the deck if the angle of heel is not to exceed 5°. [(*a*) 1.02 m, (*b*) 0.82 m]

3.19 A rectangular tank 90 cm long and 60 cm wide is mounted on bearings so that it is free to turn on a longitudinal axis. The tank has a mass of 68 kg and its centre of gravity is 15 cm above the bottom. When the tank is slowly filled with water it hangs in stable equilibrium until the depth of water is 45 cm after which it becomes unstable. How far is the axis of the bearings above the bottom of the tank? [0.21 m]

3.20 A cylindrical buoy 1.35 m in diameter and 1.8 m high has a mass of 770 kg. Show that it will not float with its axis vertical in sea water of density 1025 kg m^{-3}.

If one end of a vertical chain is fastened to the base, find the pull required to keep the buoy vertical. The centre of gravity of the buoy is 0.9 m from its base.
[GM = −0.42 m, 4632 N]

3.21 A solid cylinder 1 m in diameter and 0.8 m high is of uniform relative density 0.85. Calculate the periodic time of small oscillations when the cylinder floats with its axis vertical in still water. [2.90 s]

3.22 A ship has displacement of 5000 metric tons. The second moment of area of the waterline section about a fore and aft axis is 12 000 m^4 and the centre of buoyancy is 2 m below the centre of gravity. The radius of gyration is 3.7 m. Calculate the period of oscillation. Sea water has a density of 1025 kg m^{-3}. [10.94 s]

Concepts of Fluid Flow

Offshore wind turbines, photo courtesy of the British Wind Energy Association, © NEG Micon

THE STUDY OF FLUID MOTION IS COMPLICATED BY THE introduction of viscosity-dependent shear forces that were absent in the preceding treatment of stationary fluids. In the majority of flow cases analysis relies upon a body of empirical work, supported by the concepts of dimensional analysis and similarity. In this part of the text we will establish the analytical techniques that will later be combined with the empirical representation of frictional forces to allow the study of 'real' fluid behaviour.

In order to deal effectively with flowing fluids it is first necessary to identify flow categories, defined in predominantly mathematical terms, that will allow the appropriate analysis to be undertaken by identifying suitable and acceptable simplifications. Examples of the categories to be introduced include variation of the flow parameters with time (steady or unsteady) or variations along the flow path (uniform or non-uniform). Similarly, compressibility effects may be important in high-speed gas flows but may be ignored in many liquid flow situations.

In parallel to setting up these flow categories it is also necessary to develop a series of mathematically expressed principles that will allow the variations in flow parameters as a result of the motion of the fluid to be predicted.

The principles of continuity, energy and momentum are developed in this part of the text. The steady flow energy equation is introduced and will be utilized later to describe the behaviour of real fluids by the inclusion of an empirically based friction term. The momentum equation will be introduced and its application illustrated for both fluid-to-solid boundary transfers, such as the calculation of forces acting on moving vanes or pipe nozzles, and for other flow situations, such as the formation of hydraulic jumps in open-channel flows.

While the treatment of the behaviour of real fluid motion requires the introduction of viscous and, possibly, compressibility terms, the study of an ideal fluid freed from these constraints is useful and important, particularly in the consideration of flow patterns away from the influence of solid boundaries. Primarily a mathematical modelling tool, the study of ideal fluid flow has its roots in the work of eighteenth-century hydrodynamicists and has applications now in aerodynamics as it allows the introduction of a further flow classification, namely rotational or irrotational flow. The study of ideal flow allows flow patterns around aerofoil sections to be considered and therefore naturally leads to considerations of lift and vorticity.

Taken together with Part I, this portion of the text provides the foundation upon which the study and application of the behaviour of real fluids may be based.

4

Motion of Fluid Particles and Streams

THE PREDICTION OF THE CONDITIONS ENCOUNTERED BY, AND AS A RESULT OF, FLUIDS IN motion presents a range of problems that must be resolved by reference to the fundamental laws of physics, coupled with the particular fluid properties identified in Chapter 1. The treatment presented in this chapter will lay the foundations for later analysis in that the various fluid flow regimes, whether time dependent or determined by the shear forces assumed to act on the boundaries of the fluid flow or the compressibility of the fluid, will be identified, including the Reynolds number-dependent laminar and turbulent flow regimes. The presence of velocity profiles within any fluid flow will be emphasized, together with the importance of fluid viscosity in determining the detail conditions within the fluid flow. The application of the conservation of mass relationship across a control volume defined within a flowing fluid will be introduced and the relationship linking mass flow to local or mean flow velocity values will be detailed. ● ● ●

4.1   FLUID FLOW

The motion of a fluid is usually extremely complex. The study of a fluid at rest, or in relative equilibrium, was simplified by the absence of shear forces, but when a fluid flows over a solid surface or other boundary, whether stationary or moving, the velocity of the fluid in contact with the boundary must be the same as that of the boundary, and a velocity gradient is created at right angles to the boundary (see Section 1.2). The resulting change of velocity from layer to layer of fluid flowing parallel to the boundary gives rise to shear stresses in the fluid. Individual particles of fluid move as a result of the action of forces set up by differences of pressure or elevation. Their motion is controlled by their inertia and the effect of the shear stresses exerted by the surrounding fluid. The resulting motion is not easily analyzed mathematically, and it is often necessary to supplement theory by experiment.

If an individual particle of fluid is coloured, or otherwise rendered visible, it will describe a *pathline*, which is the trace showing the position at successive intervals of time of a particle which started from a given point. If, instead of colouring an individual particle, the flow pattern is made visible by injecting a stream of dye into a liquid, or smoke into a gas, the result will be a *streakline* or *filament line*, which gives an instantaneous picture of the positions of all the particles which have passed through the particular point at which the dye is being injected. Since the flow pattern may vary from moment to moment, a streakline will not necessarily be the same as a pathline. When using tracers or dyes it is essential to choose a material having a density and other physical properties as similar as possible to those of the fluid being studied.

In analyzing fluid flow, we also make use of the idea of a *streamline*, which is an imaginary curve in the fluid across which, at a given instant, there is no flow. Thus, the velocity of every particle of fluid along the streamline is tangential to it at that moment. Since there can be no flow through solid boundaries, these can also be regarded as streamlines. For a continuous stream of fluid, streamlines will be continuous lines extending to infinity upstream and downstream, or will form closed curves as, for example, round the surface of a solid object immersed in the flow. If conditions are steady and the flow pattern does not change from moment to moment, pathlines and streamlines will be identical; if the flow is fluctuating this will not be the case. If a series of streamlines are drawn through every point on the perimeter of a small area of the stream cross-section, they will form a *streamtube* (Fig. 4.1). Since there is no flow across a streamline, the fluid inside the streamtube cannot escape through its walls, and behaves as if it were contained in an imaginary pipe. This is a useful concept in dealing with the flow of a large body of fluid, since it allows elements of the fluid to be isolated for analysis.

FIGURE 4.1
A streamtube

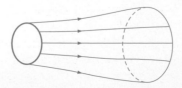

4.2 UNIFORM FLOW AND STEADY FLOW

Conditions in a body of fluid can vary from point to point and, at any given point, can vary from one moment of time to the next. Flow is described as *uniform* if the velocity at a given instant is the same in magnitude and direction at every point in the fluid. If, at the given instant, the velocity changes from point to point, the flow is described as *non-uniform*. In practice, when a fluid flows past a solid boundary there will be variations of velocity in the region close to the boundary. However, if the size and shape of the cross-section of the stream of fluid is constant, the flow is considered to be uniform.

A *steady* flow is one in which the velocity, pressure and cross-section of the stream may vary from point to point but do not change with time. If, at a given point, conditions do change with time, the flow is described as *unsteady*. In practice, there will always be slight variations of velocity and pressure, but, if the average values are constant, the flow is considered to be steady.

There are, therefore, four possible types of flow:

1. *Steady uniform flow.* Conditions do not change with position or time. The velocity and cross-sectional area of the stream of fluid are the same at each cross-section; for example, flow of a liquid through a pipe of uniform bore running completely full at constant velocity.

2. *Steady non-uniform flow.* Conditions change from point to point but not with time. The velocity and cross-sectional area of the stream may vary from cross-section to cross-section, but, for each cross-section, they will not vary with time; for example, flow of a liquid at a constant rate through a tapering pipe running completely full.

3. *Unsteady uniform flow.* At a given instant of time the velocity at every point is the same, but this velocity will change with time; for example, accelerating flow of a liquid through a pipe of uniform bore running full, such as would occur when a pump is started up.

4. *Unsteady non-uniform flow.* The cross-sectional area and velocity vary from point to point and also change with time; for example, a wave travelling along a channel.

4.3 FRAMES OF REFERENCE

Whether a given flow is described as steady or unsteady will depend upon the situation of the observer, since motion is relative and can only be described in terms of some frame of reference which is determined by the observer. If a wave travels along a channel, then to an observer on the bank the flow in the channel will appear to vary with time, and, therefore, be unsteady. If, however, the observer were travelling on the crest of the wave, conditions would not appear to the observer to change with time, and the flow would be steady according to the observer's frame of reference.

The frame of reference adopted for describing the motion of a fluid is usually a set of fixed coordinate axes, but the analysis of steady flow is usually simpler than that of unsteady flow and it is sometimes useful to use moving coordinate axes to convert an unsteady flow problem to a steady flow problem. The normal laws of mechanics will still apply, provided that the movement of the coordinate axes takes place with uniform velocity in a straight line.

4.4   REAL AND IDEAL FLUIDS

When a real fluid flows past a boundary, the fluid immediately in contact with the boundary will have the same velocity as the boundary. As explained in Section 4.1, the velocity of successive layers of fluid will increase as we move away from the boundary. If the stream of fluid is imagined to be of infinite width perpendicular to the boundary, a point will be reached beyond which the velocity will approximate to the free stream velocity, and the drag exerted by the boundary will have no effect. The part of the flow adjoining the boundary in which this change of velocity occurs is known as the *boundary layer*. In this region, shear stresses are developed between layers of fluid moving with different velocities as a result of viscosity and the interchange of momentum due to turbulence causing particles of fluid to move from one layer to another. The thickness of the boundary layer is defined as the distance from the boundary at which the velocity becomes equal to 99 per cent of the free stream velocity. Outside this boundary layer, in a real fluid, the effect of the shear stresses due to the boundary can be ignored and the fluid can be treated as if it were an *ideal fluid*, which is assumed to have no viscosity and in which there are no shear stresses. If the fluid velocity is high and its velocity low, the boundary layer is comparatively thin, and the assumption that a real fluid can be treated as an ideal fluid greatly simplifies the analysis of the flow and still leads to useful results.

Even in problems in which the effects of viscosity and turbulence cannot be neglected, it is often convenient to carry out the mathematical analysis assuming an ideal fluid. An experimental investigation can then be made to correct the theoretical analysis for the factors omitted and to bring the results obtained into agreement with the behaviour of a real fluid.

4.5   COMPRESSIBLE AND INCOMPRESSIBLE FLOW

All fluids are compressible, so that their density will change with pressure, but, under steady flow conditions and provided that the changes of density are small, it is often possible to simplify the analysis of a problem by assuming that the fluid is incompressible and of constant density. Since liquids are relatively difficult to compress, it is usual to treat them as if they were incompressible for all cases of steady flow. However, in unsteady flow conditions, high pressure differences can develop (see Chapter 20) and the compressibility of liquids must be taken into account.

Gases are easily compressed and, except when changes of pressure and, therefore, density are very small, the effects of compressibility and changes of internal energy *must* be taken into account.

4.6   ONE-, TWO- AND THREE-DIMENSIONAL FLOW

Although, in general, all fluid flow occurs in three dimensions, so that velocity, pressure and other factors vary with reference to three orthogonal axes, in some problems the major changes occur in two directions or even in only one direction. Changes along the other axis or axes can, in such cases, be ignored without introducing major errors, thus simplifying the analysis.

Flow is described as *one-dimensional* if the factors, or parameters, such as velocity, pressure and elevation, describing the flow at a given instant, vary only along the direction of flow and not across the cross-section at any point. If the flow is unsteady, these parameters may vary with time. The one dimension is taken as the distance along the central streamline of the flow, even though this may be a curve in space, and the values of velocity, pressure and elevation at each point along this streamline will be the average values across a section normal to the streamline. A one-dimensional treatment can be applied, for example, to the flow through a pipe, but, since in a real fluid the velocity at any cross-section will vary from zero at the pipe wall (Fig. 4.2) to a maximum at the centre, some correction will be necessary to compensate for this (see Chapter 10) if a high degree of accuracy is required.

In *two-dimensional* flow it is assumed that the flow parameters may vary in the direction of flow and in one direction at right angles, so that the streamlines are curves lying in a plane and are identical in all planes parallel to this plane. Thus, the flow over a weir of constant cross-section (Fig. 4.3) and infinite width perpendicular to the plane of the diagram can be treated as two-dimensional. A real weir has a limited width, but it can be treated as two-dimensional over its whole width and then an end correction can be introduced to modify the result to allow for the effect of the disturbance produced by the end walls (see Chapter 16).

FIGURE 4.2
Velocity profiles for one-dimensional flow

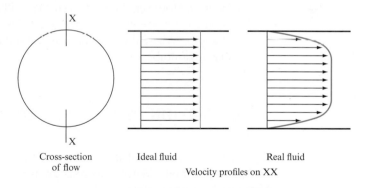

Cross-section
of flow

Ideal fluid

Real fluid

Velocity profiles on XX

FIGURE 4.3
Two-dimensional flow

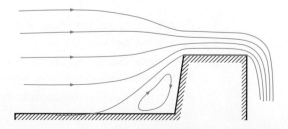

A special case of two-dimensional flow occurs when the cross-section of the flow is circular and the flow parameters vary symmetrically about the axis. For example, ideally the velocity distribution in a circular pipe will be the same across any diameter, the velocity varying from zero at the wall to a maximum at the centre. Referred to orthogonal coordinate axes (x in the direction of motion, y and z in the plane of the cross-section) the flow is three-dimensional, but, since it is axisymmetric, it can be reduced to two-dimensional flow by using a system of *cylindrical* coordinates (x in the direction of flow and r the radius defining the position in the cross-section).

4.7   ANALYZING FLUID FLOW

One difficulty encountered in deciding how to investigate the flow of a fluid is that, in the majority of problems, we are dealing with an endless stream of fluid. We have to decide what part of this stream shall constitute the element or system to be studied and what shall be regarded as the surroundings which act upon this system. There are two main alternatives:

1.   We can study the behaviour of a specific element of the fluid of fixed mass. Such an element constitutes a closed system. Its boundaries are a closed surface which may vary with time, but always contain the same mass of fluid. At any instant, a free body diagram can be drawn showing the forces exerted by the surrounding fluid and any solid boundaries on this element.

2.   We can define the system to be studied as a fixed region in space, or in relation to some frame of reference, known as a *control volume*, through which the fluid flows, forming, in effect, an open system. The boundary of this system is its control surface and its shape does not change with time. The control volume for a particular problem is chosen arbitrarily for reasons of convenience of analysis. However, the control surface will usually follow solid boundaries where these are present, and where it cuts the flow direction it will do so at right angles. Where there are no solid boundaries the control volume may form a streamtube.

4.8   MOTION OF A FLUID PARTICLE

Any particle or element of fluid will obey the normal laws of mechanics in the same way as a solid body. When a force is applied, its behaviour can be predicted from Newton's laws, which state:

1.   A body will remain at rest or in a state of uniform motion in a straight line until acted upon by an external force.

2.   The rate of change of momentum of a body is proportional to the force applied and takes place in the direction of action of that force.

3.   Action and reaction are equal and opposite.

Since momentum is the product of mass and velocity, for an element of fixed mass Newton's second law relates the change of velocity occurring in a given time (i.e. the acceleration) to the applied force. Working in a coherent system of units, such as SI, the proportionality becomes an equality and Newton's second law can be written

$$\text{Force} = \text{Mass} \times \frac{\text{Change of velocity}}{\text{Time}}$$

$$= \text{Mass} \times \text{Acceleration}.$$

The relationships between the acceleration a, initial velocity v_1, final velocity v_2 and the distance moved s in time t are given by the equations of motion:

$$v_2 = v_1 + at,$$

$$s = v_1 t + \tfrac{1}{2} at^2,$$

$$v_2^2 = v_1^2 + 2as.$$

In any body of flowing fluid, the velocity at a given instant will generally vary from point to point over any specified region, and if the flow is unsteady the velocity at each point may vary with time. In this field of flow, at any given time, a particle at point A will have a different velocity from that of a particle at point B. The velocities at A and B may also change with time. Thus the change of velocity δv, which occurs when a particle moves from A to B through a distance δs in time δt, is given by

$$\begin{array}{ccc} \text{Total change} \\ \text{of velocity} \end{array} = \begin{array}{c} \text{Difference of} \\ \text{velocity between} \\ \text{A and B at the} \\ \text{given instant} \end{array} + \begin{array}{c} \text{Change of} \\ \text{velocity at} \\ \text{B occurring} \\ \text{in time } \delta t. \end{array} \qquad (4.1)$$

The velocity v depends on both distance s and time t. The rate of change of velocity with position at a given time is, therefore, expressed by the partial differential $\partial v/\partial s$ and the rate of change of velocity with time at a given point is expressed by the partial differential $\partial v/\partial t$. Since A and B are a distance δs apart,

$$\text{Difference of velocity between A and B at the given instant} = \frac{\partial v}{\partial s}(\delta s).$$

Also,

$$\text{Change of velocity at B in time } t = \frac{\partial v}{\partial t}(\delta t).$$

Thus, in symbols, equation (4.1) is

$$\mathrm{d}v = \frac{\partial v}{\partial s}(\delta s) + \frac{\partial v}{\partial t}(\delta t). \qquad (4.2)$$

4.9   ACCELERATION OF A FLUID PARTICLE

The forces acting on a particle are related to the resultant acceleration $\delta v / \delta t$ of the particle by Newton's second law. From equation (4.2) in the limit at $\delta t \rightarrow 0$,

$$\text{Acceleration in the direction of flow, } a = \frac{dv}{dt} = \frac{\partial v}{\partial s}\frac{ds}{dt} + \frac{\partial v}{\partial t}.$$

To denote that the derivative dv/dt is obtained by following the motion of a single particle, it is written Dv/Dt, and since $ds/dt = v$,

$$a = \frac{Dv}{Dt} = v\frac{\partial v}{\partial s} + \frac{\partial v}{\partial t}. \tag{4.3}$$

The derivative D/Dt is known as the *substantive derivative*. The total acceleration, known as the *substantive acceleration*, is composed of two parts, as shown in equation (4.3):

1. the *convective acceleration* $v(\partial v/\partial s)$ due to the movement of the particle from one point to another point at which the velocity at the given instant is different;

2. the *local* or *temporal acceleration* $\partial v/\partial t$ due to the change of velocity at every point with time.

For steady flow, $\partial v/\partial t = 0$, while for uniform flow, $\partial v/\partial s = 0$.

We have so far assumed that the particle is accelerating in a straight line, but, if it is moving in a curved path, its velocity will be changing in direction and consequently there will be an acceleration perpendicular to its path, whether the velocity v is changing in magnitude or not. Figure 4.4 shows a particle moving from A to B along a curved path of length δs subtending a small angle $\delta\theta$ at the centre of curvature. The change of velocity δv_n will be perpendicular to the direction of motion and, from the velocity diagram,

$$\delta v_n = v\delta\theta = v\delta s/R.$$

FIGURE 4.4

Change of velocity for a circular path

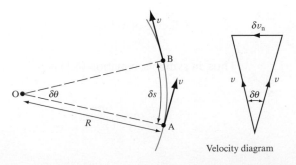

Velocity diagram

Dividing by δt, the time in which the change occurs, in the limit the acceleration perpendicular to the direction of motion is

$$a_n = \frac{\mathrm{d}v_n}{\mathrm{d}t} = \frac{v}{R}\frac{\mathrm{d}s}{\mathrm{d}t}$$

or, since

$$\frac{\mathrm{d}s}{\mathrm{d}t} = v,$$

$$a_n = v^2/R.$$

This is the convective term, and, if v has a component v_n towards the instantaneous centre of curvature, there will be a temporal term $\partial v_n/\partial t$ so that the substantial derivative is

$$a_n = \frac{v^2}{R} + \frac{\partial v_n}{\partial t}.$$

In general, the motion of a fluid particle is three-dimensional and its velocity and acceleration can be expressed in terms of three mutually perpendicular components. Thus, if v_x, v_y and v_z are the components of the velocity in the x, y and z directions, respectively, and a_x, a_y and a_z the corresponding components of acceleration, the velocity field is described by

$$v_x = v_x(x, y, z, t), \quad v_y = v_y(x, y, z, t), \quad v_z = v_z(x, y, z, t),$$

and the velocity $\mathbf{v}$ at any point is given by

$$\mathbf{v} = v_x\mathbf{i} + v_y\mathbf{j} + v_z\mathbf{k},$$

where $\mathbf{i}$, $\mathbf{j}$ and $\mathbf{k}$ are the unit vectors in the x, y and z directions.

The change of the component velocities in each direction as a particle moves in a fluid can now be calculated. Thus, in the x direction,

$$\delta v_x = \frac{\partial v_x}{\partial x}(\delta x) + \frac{\partial v_x}{\partial y}(\delta y) + \frac{\partial v_x}{\partial z}(\delta z) + \frac{\partial v_x}{\partial t}(\delta t),$$

and the acceleration in the x direction, in the limit as $\delta t \to 0$, will be

$$a_x = \frac{\mathrm{D}v_x}{\mathrm{D}t} = \frac{\partial v_x}{\partial x}\frac{\mathrm{d}x}{\mathrm{d}t} + \frac{\partial v_x}{\partial y}\frac{\mathrm{d}y}{\mathrm{d}t} + \frac{\partial v_x}{\partial z}\frac{\mathrm{d}z}{\mathrm{d}t} + \frac{\partial v_x}{\partial t}$$

or, since $\mathrm{d}x/\mathrm{d}t = v_x$, $\mathrm{d}y/\mathrm{d}t = v_y$, $\mathrm{d}z/\mathrm{d}t = v_z$,

$$a_x = \frac{\mathrm{D}v_x}{\mathrm{D}t} = v_x\frac{\partial v_x}{\partial x} + v_y\frac{\partial v_x}{\partial y} + v_z\frac{\partial v_x}{\partial z} + \frac{\partial v_x}{\partial t}. \tag{4.4}$$

Similarly,

$$a_y = \frac{\mathrm{D}v_y}{\mathrm{D}t} = v_x\frac{\partial v_y}{\partial x} + v_y\frac{\partial v_y}{\partial y} + v_z\frac{\partial v_y}{\partial z} + \frac{\partial v_y}{\partial t}, \tag{4.5}$$

$$a_z = \frac{\mathrm{D}v_z}{\mathrm{D}t} = v_x\frac{\partial v_z}{\partial x} + v_y\frac{\partial v_z}{\partial y} + v_z\frac{\partial v_z}{\partial z} + \frac{\partial v_z}{\partial t}. \tag{4.6}$$

The first three terms in each of equations (4.4) to (4.6) represent the convective acceleration and the final term the local or temporal acceleration.

4.10   LAMINAR AND TURBULENT FLOW

Observation shows that two entirely different types of fluid flow exist. This was demonstrated by Osborne Reynolds in 1883 through an experiment in which water was discharged from a tank through a glass tube (Fig. 4.5). The rate of flow could be controlled by a valve at the outlet, and a fine filament of dye injected at the entrance to the tube. At low velocities, it was found that the dye filament remained intact throughout the length of the tube, showing that the particles of water moved in parallel lines. This type of flow is known as *laminar*, *viscous* or *streamline*, the particles of fluid moving in an orderly manner and retaining the same relative positions in successive cross-sections.

FIGURE 4.5
Reynolds' apparatus

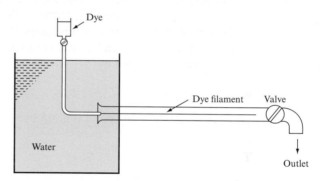

As the velocity in the tube was increased by opening the outlet valve, a point was eventually reached at which the dye filament at first began to oscillate and then broke up so that the colour was diffused over the whole cross-section, showing that the particles of fluid no longer moved in an orderly manner but occupied different relative positions in successive cross-sections. This type of flow is known as *turbulent* and is characterized by continuous small fluctuations in the magnitude and direction of the velocity of the fluid particles, which are accompanied by corresponding small fluctuations of pressure.

When the motion of a fluid particle in a stream is disturbed, its inertia will tend to carry it on in the new direction, but the viscous forces due to the surrounding fluid will tend to make it conform to the motion of the rest of the stream. In viscous flow, the viscous shear stresses are sufficient to eliminate the effects of any deviation, but in turbulent flow they are inadequate. The criterion which determines whether flow will be viscous or turbulent is therefore the ratio of the inertial force to the viscous force acting on the particle.

Suppose l is a characteristic length in the system under consideration, e.g. the diameter of a pipe or the chord of an aerofoil, and t is a typical time; then lengths, areas, velocities and accelerations can all be expressed in terms of l and t. For a small element of fluid of mass density ρ,

$$\text{Volume of element} = k_1 l^3,$$

$$\text{Mass of element} = k_1 \rho l^3,$$

$$\text{Velocity of element, } v = k_2 l/t,$$

$$\text{Acceleration of element} = k_3 l/t^2,$$

where k_1, k_2 and k_3 are constants. By Newton's second law,

$$\text{Inertial force} = \text{Mass} \times \text{Acceleration}$$

$$= k_1 \rho l^3 \times k_3 l/t^2$$

$$= k_1 k_3 \rho l^2 (l/t)^2$$

$$= (k_1 k_3 / k_2^2) \rho l^2 v^2.$$

Similarly,

$$\text{Viscous force} = \text{Viscous shear stress} \times \text{Area on which stress acts.}$$

From Newton's law of viscosity (equation (1.2)),

$$\text{Viscous shear stress} = \mu \times \text{Velocity gradient} = \mu(v/k_4 l),$$

where μ = coefficient of dynamic viscosity.

$$\text{Area on which shear stress acts} = k_5 l^2.$$

Therefore,

$$\text{Viscous force} = \mu(v/k_4 l) \times k_5 l^2 = (k_5/k_4)\mu v l.$$

The ratio

$$\frac{\text{Inertial force}}{\text{Viscous force}} = \frac{k_1 k_3 k_4}{k_2^2 k_5} \frac{\rho l^2 v^2}{\mu v l} = \text{constant} \times \frac{\rho v l}{\mu}.$$

Thus, the criterion which determines whether flow is viscous or turbulent is the quantity $\rho v l/\mu$, known as the Reynolds number. It is a ratio of forces and, therefore, a pure number and may also be written as vl/ν, where ν is the kinematic viscosity ($\nu = \mu/\rho$).

Experiments carried out with a number of different fluids in straight pipes of different diameters have established that if the Reynolds number is calculated by making l equal to the pipe diameter and using the mean velocity $\bar{v}$ (Section 4.11), then, below a critical value of $\rho \bar{v} d/\mu = 2000$, flow will normally be laminar (viscous), any tendency to turbulence being damped out by viscous friction. This value of the Reynolds number applies only to flow in pipes, but critical values of the Reynolds number can be established for other types of flow, choosing a suitable characteristic length such as the chord of an aerofoil in place of the pipe diameter. For a given fluid flowing in a pipe of a given diameter, there will be a *critical velocity* of flow $\bar{v}_c$ corresponding to the critical value of the Reynolds number, below which flow will be viscous.

In pipes, at values of the Reynolds number above 2000, flow will not necessarily be turbulent. Laminar flow has been maintained up to Re = 50 000, but conditions are unstable and any disturbance will cause reversion to normal turbulent flow. In straight pipes of constant diameter, flow can be assumed to be turbulent if the Reynolds number exceeds 4000.

4.11   DISCHARGE AND MEAN VELOCITY

The total quantity of fluid flowing in unit time past any particular cross-section of a stream is called the *discharge* or flow at that section. It can be measured either in terms of mass, in which case it is referred to as the mass rate of flow $\dot{m}$ and measured in units such as kilograms per second, or in terms of volume, when it is known as the volume rate of flow Q, measured in such units as cubic metres per second.

In an ideal fluid, in which there is no friction, the velocity u of the fluid would be the same at every point of the cross-section (Fig. 4.2). In unit time, a prism of fluid would pass the given cross-section and, if the cross-sectional area normal to the direction of flow is A, the volume passing would be Au. Thus

$$Q = Au.$$

In a real fluid, the velocity adjacent to a solid boundary will be zero or, more accurately, equal to the wall velocity in the flow direction, a condition known as 'no slip', which will be true as long as the flow does not separate from the wall. For a pipe, the velocity profile would be as shown in Fig. 4.6(a) for laminar flow and Fig. 4.6(b) for turbulent flow.

FIGURE 4.6
Calculation of discharge for a circular section, note 'no slip' at wall

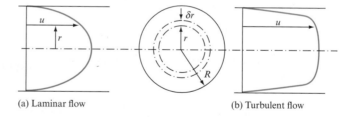

(a) Laminar flow   (b) Turbulent flow

If u is the velocity at any radius r, the flow δQ through an annular element of radius r and thickness δr will be

$$\delta Q = \text{Area of element} \times \text{Velocity}$$

$$= 2\pi r \delta r \times u,$$

and, hence,

$$Q = 2\pi \int_0^R ur\,dr. \tag{4.7}$$

If the relation between u and r can be established, this integral can be evaluated or the integration may be undertaken numerically, see Section 6.8, program VOLFLO.

In many problems, the variation of velocity over the cross-section can be ignored, the velocity being assumed to be constant and equal to the *mean velocity* $\bar{u}$, defined as volume rate of discharge Q divided by the area of cross-section A *normal* to the stream:

Mean velocity, $\bar{u} = Q/A$.

EXAMPLE 4.1

Air flows between two parallel plates 80 mm apart. The following velocities were determined by direct measurement.

Distance from one plate (mm)	0	10	20	30	40	50	60	70	80
Velocity (m s^{-1})	0	23	28	31	32	29	22	14	0

Plot the velocity distribution curve and calculate the mean velocity.

Solution

Figure 4.7 shows the velocity distribution curve. The area enclosed by the curve represents the product of velocity and distance, and since the two plates are parallel

$$\text{Mean velocity, } \bar{u} = \frac{\text{Discharge per unit width}}{\text{Distance between plates}}.$$

FIGURE 4.7
Velocity distribution curve

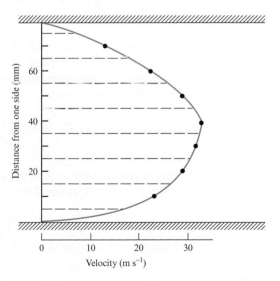

The area under the graph may be determined by the mid-ordinate method, taking values from Fig. 4.7,

$$\bar{u} = (\Sigma \text{Mid-ordinates}/8)$$

$$= (17.5 + 26.0 + 29.6 + 31.9 + 30.7 + 25.4 + 18.1 + 7.7)/8 = \textbf{23.36 m s}^{-1}.$$

4.12 CONTINUITY OF FLOW

Except in nuclear processes, matter is neither created nor destroyed. This principle of *conservation of mass* can be applied to a flowing fluid. Considering any fixed region in the flow (Fig. 4.8) constituting a control volume,

$$\text{Mass of fluid entering per unit time} = \text{Mass of fluid leaving per unit time} + \text{Increase of mass of fluid in the control volume per unit time.}$$

FIGURE 4.8
Continuity of flow

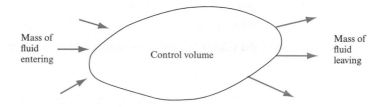

For steady flow, the mass of fluid in the control volume remains constant and the relation reduces to

$$\text{Mass of fluid entering per unit time} = \text{Mass of fluid leaving per unit time.}$$

Applying this principle to steady flow in a streamtube (Fig. 4.9) having a cross-sectional area small enough for the velocity to be considered as constant over any given cross-section, for the region between sections 1 and 2, since there can be no flow through the walls of a streamtube,

$$\text{Mass entering per unit time at section 1} = \text{Mass leaving per unit time at section 2.}$$

Suppose that at section 1 the area of the streamtube is δA_1, the velocity of the fluid u_1 and its density ρ_1, while at section 2 the corresponding values are δA_2, u_2 and ρ_2; then

$$\text{Mass entering per unit time at 1} = \rho_1 \delta A_1 u_1,$$

$$\text{Mass leaving per unit time at 2} = \rho_2 \delta A_2 u_2.$$

FIGURE 4.9
Continuous flow
through a streamtube

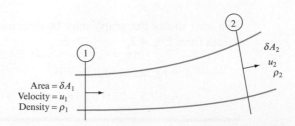

Then, for steady flow,

$$\rho_1 \delta A_1 u_1 = \rho_2 \delta A_2 u_2 = \text{Constant.} \tag{4.8}$$

This is the *equation of continuity* for the flow of a compressible fluid through a streamtube, u_1 and u_2 being the velocities measured at right angles to the cross-sectional areas δA_1 and δA_2.

For the flow of a real fluid through a pipe or other conduit, the velocity will vary from wall to wall. However, using the mean velocity $\bar{u}$, the equation of continuity for steady flow can be written as

$$\rho_1 A_1 \bar{u}_1 = \rho_2 A_2 \bar{u}_2 = \dot{m}, \tag{4.9}$$

where A_1 and A_2 are the total cross-sectional areas and $\dot{m}$ is the mass rate of flow.

If the fluid can be considered as incompressible, so that $\rho_1 = \rho_2$, equation (4.9) reduces to

$$A_1 \bar{u}_1 = A_2 \bar{u}_2 = Q. \tag{4.10}$$

The continuity of flow equation is one of the major tools of fluid mechanics, providing a means of calculating velocities at different points in a system.

The continuity equation can also be applied to determine the relation between the flows into and out of a junction. In Fig. 4.10, for steady conditions,

Total inflow to junction = Total outflow from junction,

$$\rho_1 Q_1 = \rho_2 Q_2 + \rho_3 Q_3.$$

FIGURE 4.10

Applications of the continuity equation

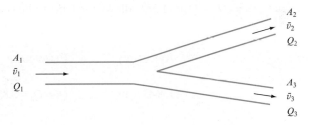

For an incompressible fluid, $\rho_1 = \rho_2 = \rho_3$ so that

$$Q_1 = Q_2 + Q_3$$

or $\quad A_1 \bar{v}_1 = A_2 \bar{v}_2 + A_3 \bar{v}_3.$

In general, if we consider flow towards the junction as positive and flow away from the junction as negative, then for steady flow at any junction the algebraic sum of all the mass flows must be zero:

$$\Sigma \rho Q = 0.$$

EXAMPLE 4.2

Water flows from A to D and E through the series pipeline shown in Fig. 4.11. Given the pipe diameters, velocities and flow rates below, complete the tabular data for this system.

FIGURE 4.11

Relations between discharge, diameter and velocity

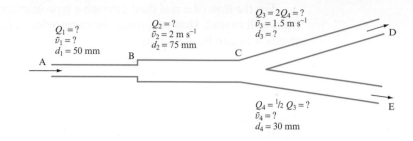

PIPE	DIAMETER (mm)	FLOW RATE ($m^3 s^{-1}$)	VELOCITY ($m s^{-1}$)
AB	$d_1 = 50$	$Q_1 = ?$	$\bar{v}_1 = ?$
BC	$d_2 = 75$	$Q_2 = ?$	$\bar{v}_2 = 2.0$
CD	$d_3 = ?$	$Q_3 = 2Q_4$	$\bar{v}_3 = 1.5$
DE	$d_4 = 30$	$Q_4 = 0.5Q_3$	$\bar{v}_4 = ?$

Solution

Adding area $A = (22/7)d^2/4$ to the data table and noting that $Q = A\bar{v}$ and that $Q_1 = Q_2 = (Q_3 + Q_4) = 1.5Q_3$ allows the table to be completed as (additions in **bold**),

DIAMETER (mm)	AREA (m^2)	FLOW RATE ($m^3 s^{-1}$)	VELOCITY ($m s^{-1}$)
$d_1 = 50$	$\mathbf{1.9643 \times 10^{-3}}$	$\mathbf{Q_1 = Q_2 = 8.839 \times 10^{-3}}$	$\mathbf{\bar{v}_1 = \bar{v}_2\, A_2/A_1}$ $\mathbf{= 2.0 \times 4.4196/1.9643}$ $\mathbf{= 4.27}$
$d_2 = 75$	$\mathbf{4.4196 \times 10^{-3}}$	$\mathbf{Q_2 = 2.0 \times 4.4196 \times 10^{-3}}$ $\mathbf{= 8.839 \times 10^{-3}}$	$\bar{v}_2 = 2.0$
$\mathbf{d_3 = [Q_3/(\bar{v}_3\, \pi/4)]^{0.5}}$ $\mathbf{= (5.893 \times 10^{-3}/1.5 \times 0.786)^{0.5}}$ $\mathbf{= 0.707}$		$\mathbf{Q_3 = Q_2/1.5}$ $\mathbf{= 5.893 \times 10^{-3}}$	$\bar{v}_3 = 1.5$
$d_4 = 30$	$\mathbf{0.707 \times 10^{-3}}$	$\mathbf{Q_4 = 0.5Q_3}$ $\mathbf{= 0.5 \times 5.893 \times 10^{-3}}$ $\mathbf{= 2.947 \times 10^{-3}}$	$\mathbf{\bar{v}_4 = Q_4/A_4}$ $\mathbf{= 2.947/0.7071}$ $\mathbf{= 4.17}$

(The calculation route is as follows: calculate areas where possible and then Q_2 and hence Q_1 and $\bar{v}_1$. From Q_2 calculate Q_3 and Q_4 and hence $\bar{v}_4$. Calculate d_3 from Q_3 and $\bar{v}_3$.)

4.13   CONTINUITY EQUATIONS FOR THREE-DIMENSIONAL FLOW USING CARTESIAN COORDINATES

The control volume ABCDEFGH in Fig. 4.12 is taken in the form of a small rectangular prism with sides δx, δy and δz in the x, y and z directions, respectively. The mean values of the component velocities in these directions are v_x, v_y and v_z. Considering flow in the x direction,

$$\text{Mass inflow through ABCD in unit time} = \rho v_x \delta_y \delta_z.$$

FIGURE 4.12
Continuity in three dimensions

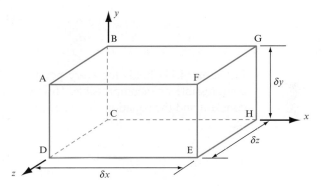

In the general case, both mass density ρ and velocity v_x will change in the x direction and so

$$\text{Mass outflow through EFGH in unit time} = \left[\rho v_x + \frac{\partial}{\partial x}(\rho v_x)\delta x \right]\delta y \,\delta z.$$

Thus,

$$\text{Net outflow in unit time in } x \text{ direction} = \frac{\partial}{\partial x}(\rho v_x)\delta x \,\delta y \,\delta z.$$

Similarly,

$$\text{Net outflow in unit time in } y \text{ direction} = \frac{\partial}{\partial y}(\rho v_y)\delta x \,\delta y \,\delta z,$$

$$\text{Net outflow in unit time in } z \text{ direction} = \frac{\partial}{\partial z}(\rho v_z)\delta x \,\delta y \,\delta z.$$

Therefore,

$$\text{Total net outflow in unit time} = \left[\frac{\partial}{\partial x}(\rho v_x) + \frac{\partial}{\partial y}(\rho v_y) + \frac{\partial}{\partial z}(\rho v_z) \right]\delta x \,\delta y \,\delta z.$$

Also, since $\partial \rho / \partial t$ is the change in mass density per unit time,

$$\text{Change of mass in control volume in unit time} = -\frac{\partial \rho}{\partial t} \delta x \, \delta y \, \delta z$$

(the negative sign indicating that a net outflow has been assumed). Then,

Total net outflow in unit time

= change of mass in control volume in unit time

$$\left[\frac{\partial}{\partial x}(\rho v_x) + \frac{\partial}{\partial y}(\rho v_y) + \frac{\partial}{\partial z}(\rho v_z) \right] \delta x \, \delta y \, \delta z = -\frac{\partial \rho}{\partial t} \delta x \, \delta y \, \delta z$$

or

$$\frac{\partial}{\partial x}(\rho v_x) + \frac{\partial}{\partial y}(\rho v_y) + \frac{\partial}{\partial z}(\rho v_z) = -\frac{\partial \rho}{\partial t}. \tag{4.11}$$

Equation (4.11) holds for every point in a fluid flow whether steady or unsteady, compressible or incompressible. However, for incompressible flow, the density ρ is constant and the equation simplifies to

$$\frac{\partial v_x}{\partial x} + \frac{\partial v_y}{\partial y} + \frac{\partial v_z}{\partial z} = 0. \tag{4.12}$$

For two-dimensional incompressible flow this will simplify still further to

$$\frac{\partial v_x}{\partial x} + \frac{\partial v_y}{\partial y} = 0. \tag{4.13}$$

EXAMPLE 4.3

The velocity distribution for the flow of an incompressible fluid is given by $v_x = 3 - x$, $v_y = 4 + 2y$, $v_z = 2 - z$. Show that this satisfies the requirements of the continuity equation.

Solution

For three-dimensional flow of an incompressible fluid, the continuity equation simplifies to equation (4.12):

$$\frac{\partial v_x}{\partial x} = -1, \quad \frac{\partial v_y}{\partial y} = +2, \quad \frac{\partial v_z}{\partial z} = -1,$$

and, hence,

$$\frac{\partial v_x}{\partial x} + \frac{\partial v_y}{\partial y} + \frac{\partial v_z}{\partial z} = -1 + 2 - 1 = 0,$$

which satisfies the requirement for continuity.

4.14 CONTINUITY EQUATION FOR CYLINDRICAL COORDINATES

The form of the continuity equation for a system of cylindrical coordinates r, θ and z, in which r and θ are measured in a plane corresponding to the x–y plane for Cartesian coordinates, can be found by using the relations between polar and Cartesian coordinates:

$$x^2 + y^2 = r^2, \quad (y/x) = \tan\theta,$$

$$v_x = v_r \cos\theta - v_\theta \sin\theta, \quad v_y = v_r \sin\theta + v_\theta \cos\theta,$$

$$\frac{\partial}{\partial x} = \frac{\partial}{\partial r}\frac{\partial r}{\partial x} + \frac{\partial}{\partial\theta}\frac{\partial\theta}{\partial x}, \quad \frac{\partial}{\partial y} = \frac{\partial}{\partial r}\frac{\partial r}{\partial y} + \frac{\partial}{\partial\theta}\frac{\partial\theta}{\partial y}.$$

This results in equation (4.12) becoming

$$\frac{1}{r}\left[\frac{\partial}{\partial r}(rv_r)\right] + \frac{1}{r}\frac{\partial v_\theta}{\partial\theta} + \frac{\partial v_z}{\partial z} = 0. \tag{4.14}$$

In the case of two-dimensional flow, this can be simplified further. Putting $\partial v_z/\partial z = 0$ and writing

$$\frac{\partial}{\partial r}(rv_r) = \left(r\frac{\partial v_r}{\partial r} + v_r\right),$$

equation (4.14) becomes

$$\frac{v_r}{r} + \frac{\partial v_r}{\partial r} + \frac{1}{r}\frac{\partial v_\theta}{\partial\theta} = 0.$$

Concluding remarks

This chapter has provided the frame of reference for much of the later material in this text. The classification of flows based on time or distance dependence, together with the influence of viscosity, via Reynolds number is fundamental. In defining the flow into laminar and turbulent regimes much of the observed fluid flow behaviour becomes understandable. The presence of velocity profiles across a fluid stream between boundaries is similarly fundamental as it heralds the later work on the development of the boundary layer and the recognition of the condition of 'no slip' at fluid/surface interfaces.

The introduction of the continuity equation, whether in its volumetric or its more widely applicable mass flow form, provides one of the recurring tools for fluid flow analysis, which, when the storage terms are introduced, will find application throughout this text for both compressible and incompressible flows under steady or unsteady conditions.

Summary of important equations and concepts

1. Chapter 4 introduces definitions of flow conditions, such as steady and unsteady, relating these to changes in flow condition, Section 4.2, together with the concept of the boundary layer, Section 4.4, that will become central to later chapters.

2. Movement in more than one dimension is introduced, with examples, emphasizing the simplifications possible if flow can be considered as one-dimensional, Sections 4.6 to 4.9.

3. The classification of flows into laminar and turbulent regimes, following Reynolds, is an essential concept and one that will be returned to continuously in later chapters. It is important to recognize that the ratio of forces represented by the Reynolds number applies to a whole range of flow geometries, not restricted to pipeflow, Section 4.10.

4. The concept of continuity of mass flow is established and shown in the special case of incompressible flow to reduce to volumetric flow continuity that allows mean velocities to be calculated in series and parallel pipe networks, equations (4.9) and (4.10).

Problems

4.1 The velocity of a fluid varies with time t. Over the period from $t = 0$ to $t = 8$ s the velocity components are $u = 0$ m s^{-1} and $v = 2$ m s^{-1}; while from $t = 8$ s to $t = 16$ s the components are $u = 2$ m s^{-1} and $v = -2$ m s^{-1}. A dye streak is injected into the flow at a certain point commencing at time $t = 0$ and the path of a particle of fluid is also traced from that point starting at $t = 0$. Draw to scale the streak-line, the pathline of the particle and the streamlines at time $t = 12$ s.

4.2 The velocity distribution for a two-dimensional field of flow is given by

$$u = \frac{2}{3+t} \text{ m s}^{-1} \quad \text{and} \quad v = 2 - \frac{t^2}{32} \text{ m s}^{-1}.$$

For the period of time from $t = 0$ to $t = 12$ s draw a streak-line for an injection of dye through a certain point A and a pathline for a particle of fluid which was at A when $t = 0$. Draw also the streamlines for $t = 6$ s and $t = 12$ s.

4.3 A nozzle is formed so that its cross-sectional area converges linearly along its length. The inside diameters are 75 mm and 25 mm at inlet and exit and the length of the nozzle is 300 mm. What is the convective acceleration at a section halfway along the length of the nozzle if the discharge is constant at 0.014 m^3 s^{-1}?   [337.94 m s^{-2}]

4.4 During a wind tunnel test on a sphere of radius $r = 150$ mm it is found that the velocity of flow u along the longitudinal axis of the tunnel passing through the centre of the sphere at a point upstream which is a distance x from the centre of the sphere is given by

$$u = U_0\left(1 - \frac{r^3}{x^3}\right)$$

where U_0 is the mean velocity of the undisturbed airstream. If $U_0 = 60$ m s^{-1} what is the convective acceleration when the distance x is (a) 300 mm, (b) 150 mm?
    [(a) 3937.5 m s^{-2}, (b) 0]

4.5 The velocity along the centreline of a nozzle of length L is given by

$$u = 2t\left(1 - 0.5\frac{x}{L}\right)^2$$

where u is the velocity in metres per second, t is the time in seconds from the commencement of flow, x is the distance from the inlet to the nozzle. Find the convective acceleration and the local acceleration when $t = 3$ s, $x = \frac{1}{2} L$ and $L = 0.8$ m.
    [18.99 m s^{-2}, 1.125 m s^{-2}]

4.6 Water flows through a pipe 25 mm in diameter at a velocity of 6 m s^{-1}. Determine whether the flow will be laminar or turbulent assuming that the dynamic viscosity of water is $1.30 \times 10^{-3}\,\mathrm{kg\,m^{-1}\,s^{-1}}$ and its density $1000\,\mathrm{kg\,m^{-3}}$. If oil of specific gravity 0.9 and dynamic viscosity $9.6 \times 10^{-2}\,\mathrm{kg\,m^{-1}\,s^{-1}}$ is pumped through the same pipe, what type of flow will occur?

[Turbulent, Re = 115 385; laminar, Re = 1406]

4.7   An air duct is of rectangular cross-section 300 mm wide by 450 mm deep. Determine the mean velocity in the duct when the rate of flow is $0.42\,\mathrm{m^3\,s^{-1}}$. If the duct tapers to a cross-section 150 mm wide by 400 mm deep, what will be the mean velocity in the reduced section assuming that the density remains unchanged?   [3.11 m s⁻¹, 7.0 m s⁻¹]

4.8   A sphere of diameter 300 mm falls axially down a 305 mm diameter vertical cylinder which is closed at its lower end and contains water. If the sphere falls at a speed of 150 mm s⁻¹, what is the mean velocity relative to the cylinder wall of the water in the gap surrounding the midsection of the sphere?   [4.46 m s⁻¹]

4.9   The air entering a compressor has a density of $1.2\,\mathrm{kg\,m^{-3}}$ and a velocity of 5 m s⁻¹, the area of the intake being 20 cm². Calculate the mass flow rate. If air leaves the compressor through a 25 mm diameter pipe with a velocity of 4 m s⁻¹, what will be its density?

[$12 \times 10^{-3}\,\mathrm{kg\,s^{-1}}$, 6.11 kg m⁻³]

4.10   Water flows through a pipe AB 1.2 m in diameter at 3 m s⁻¹ and then passes through a pipe BC which is 1.5 m in diameter. At C the pipe forks. Branch CD is 0.8 m in diameter and carries one-third of the flow in AB. The velocity in branch CE is 2.5 m s⁻¹. Find (a) the volume rate of flow in AB, (b) the velocity in BC, (c) the velocity in CD, (d) the diameter of CE.

[(a) 3.393 m³ s⁻¹, (b) 1.92 m s⁻¹, (c) 2.25 m s⁻¹, (d) 1.073 m]

4.11   A closed tank of fixed volume is used for the continuous mixing of two liquids which enter at A and B and are discharged completely mixed at C. The diameter of the inlet pipe at A is 150 mm and the liquid flows in at the rate of 56 dm³ s⁻¹ and has a specific gravity of 0.93. At B the inlet pipe is of 100 mm diameter, the flow rate is 30 dm³ s⁻¹ and the liquid has a specific gravity of 0.87. If the diameter of the outlet pipe at C is 175 mm, what will be the mass flow rate, velocity and specific gravity of the mixture discharged?

[78.18 kg s⁻¹, 3.58 m s⁻¹, 0.909]

4.12   In a 0.6 m diameter duct carrying air the velocity profile was found to obey the law $u = -5r^2 + 0.45\,\mathrm{m\,s^{-1}}$ where u is the velocity at radius r. Calculate the volume rate of flow of the air and the mean velocity.

[0.0636 m³ s⁻¹, 0.225 m s⁻¹]

4.13   During a test on a circular duct 2 m in diameter it was found that the fluid velocity was zero at the duct surface and 6 m s⁻¹ on the axis of the duct when the flow rate was 9 m³ s⁻¹. Assuming the velocity distribution to be given by

$$u = c_1 - c_2 r^n,$$

where u is the fluid velocity at any radius r, determine the values of the constants c_1, c_2 and n, specifying the units of c_1 and c_2.

   Evaluate the mean velocity and determine the radial position at which a Pitot tube must be placed to measure this mean velocity.

[6 m s⁻¹, 6 m⁻⁰·⁸²⁵ s⁻¹, 1.825; 2.86 m s⁻¹, 0.701 m]

4.14   Air flows through a rectangular duct which is 30 cm wide by 20 cm deep in cross-section. To determine the volume rate of flow experimentally the cross-section is divided into a number of imaginary rectangular elements of equal area and the velocity measured at the centre of each element with the following results:

DISTANCE FROM BOTTOM OF DUCT (cm)	DISTANCE FROM SIDE OF DUCT (cm)				
	3	9	15	21	27
18	1.6	2.0	2.2	2.0	1.7
14	1.9	3.4	6.9	3.7	2.0
10	2.1	6.8	10.0	7.0	2.3
6	2.0	3.5	7.0	3.8	2.1
2	1.8	2.0	2.3	2.1	1.9

Calculate the volume rate of flow and the mean velocity in the duct.   [0.202 m³ s⁻¹, 3.364 m s⁻¹]

4.15   If a two-dimensional flow field were to have velocity components

$$u = U(x^3 + xy^2) \quad \text{and} \quad v = U(y^3 + yx^2)$$

would the continuity equation be satisfied?   [Yes]

4.16   Determine whether the following expressions satisfy the continuity equation:

(a) $u = 10xt$, $v = -10yt$, $\rho = \text{constant}$ [Yes]

(b) $u = U(y/\delta)^{1/7}$, $v = 0$, $\rho = \text{constant}$.   [Yes]

The Momentum Equation
and its Applications

THE ANALYSIS OF FLUID FLOW PHENOMENA FUNDAMENTALLY DEPENDS UPON THE application of Newton's laws of motion, together with a recognition of the special properties of fluids in motion. The momentum equation relates the sum of the forces acting on a fluid element to its acceleration or rate of change of momentum in the direction of the resultant force. This relationship is, perhaps, when taken with the conservation of mass and the energy equation, the foundation upon which all fluid flow analysis is based. This chapter will introduce the application of the momentum equation to a range of fluid flow conditions, including forces exerted upon and by a fluid as a result of changes in direction and impact upon both stationary and moving surfaces, as well as introducing the application of the momentum equation to determine engine thrust as a result of changes to fluid momentum. The application of the momentum equation to the prediction of the rate of propagation of pressure or surface wave discontinuities will be presented. By utilizing the momentum equation, together with the conservation of mass, this chapter will also introduce Euler's equation for motion along a streamline under general conditions. Bernoulli's equation, the special form of Euler's equation applicable to incompressible inviscid flows, will be introduced and its application demonstrated. ● ● ●

5.1   MOMENTUM AND FLUID FLOW

In mechanics, the momentum of a particle or object is defined as the product of its mass m and its velocity v:

Momentum $= mv$.

The particles of a fluid stream will possess momentum, and, whenever the velocity of the stream is changed in magnitude or direction, there will be a corresponding change in the momentum of the fluid particles. In accordance with Newton's second law, a force is required to produce this change, which will be proportional to the rate at which the change of momentum occurs. The force may be provided by contact between the fluid and a solid boundary (e.g. the blade of a propeller or the wall of a bend in a pipe) or by one part of the fluid stream acting on another. By Newton's third law, the fluid will exert an equal and opposite force on the solid boundary or body of fluid producing the change of velocity. Such forces are known as dynamic forces, since they arise from the motion of the fluid and are additional to the static forces (see Chapter 3) due to pressure in a fluid; they occur even when the fluid is at rest.

To determine the rate of change of momentum in a fluid stream consider a control volume ABCD (Fig. 5.1). As the fluid flow is assumed to be steady and non-uniform in nature the continuity of mass flow across the control volume may be expressed as

FIGURE 5.1

Momentum in a flowing fluid

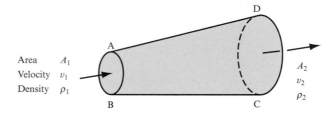

Area A_1
Velocity v_1
Density ρ_1

A_2
v_2
ρ_2

$$\rho_2 A_2 v_2 = \rho_1 A_1 v_1 = \dot{m}, \tag{5.1}$$

i.e. there is no storage within the control volume and $\dot{m}$ is the fluid mass flow.

The rate at which momentum exits the control volume across boundary CD may be defined as

$$\rho_2 A_2 v_2 v_2.$$

Similarly the rate at which momentum enters the control volume across AB may be expressed as

$$\rho_1 A_1 v_1 v_1.$$

Thus the rate of change of momentum across the control volume may be seen to be

$$\rho_2 A_2 v_2 v_2 - \rho_1 A_1 v_1 v_1 \tag{5.2}$$

or, from the continuity of mass flow equation,

$$\rho_1 A_1 v_1 (v_2 - v_1) = \dot{m}(v_2 - v_1)$$

$$= \text{Mass flow per unit time} \times \text{Change of velocity.} \qquad (5.3)$$

Note that this is the *increase* of momentum per unit time in the direction of motion, and according to Newton's second law will be caused by a force F, such that

$$F = \dot{m}(v_2 - v_1). \qquad (5.4)$$

This is the resultant force acting on the fluid element ABCD in the direction of motion. By Newton's third law, the fluid will exert an equal and opposite reaction on its surroundings.

5.2   MOMENTUM EQUATION FOR TWO- AND THREE-DIMENSIONAL FLOW ALONG A STREAMLINE

In Section 5.1, the momentum equation (5.4) was derived for one-dimensional flow in a straight line, assuming that the incoming and outgoing velocities v_1 and v_2 were in the same direction. Figure 5.2 shows a two-dimensional problem in which v_1 makes an angle θ with the x axis, while v_2 makes a corresponding angle ϕ. Since both momentum and force are vector quantities, they can be resolved into components in the x and y directions and equation (5.4) applied. Thus, if F_x and F_y are the components of the resultant force on the element of fluid ABCD,

FIGURE 5.2
Momentum equation for
two-dimensional flow

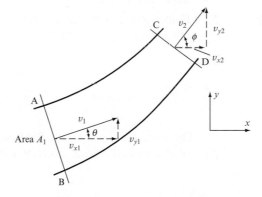

$$F_x = \text{Rate of change of momentum of fluid in } x \text{ direction}$$

$$= \text{Mass per unit time} \times \text{Change of velocity in } x \text{ direction}$$

$$= \dot{m}(v_2 \cos \phi - v_1 \cos \theta) = \dot{m}(v_{x2} - v_{x1}).$$

Similarly,

$$F_y = \dot{m}(v_2 \sin \phi - v_1 \sin \theta) = \dot{m}(v_{y2} - v_{y1}).$$

These components can be combined to give the resultant force,

$$F = \sqrt{(F_x^2 + F_y^2)}.$$

Again, the force exerted by the fluid on the surroundings will be equal and opposite. For three-dimensional flow, the same method can be used, but the fluid will also have component velocities v_{z1} and v_{z2} in the z direction and the corresponding rate of change of momentum in this direction will require a force

$$F_z = \dot{m}(v_{z2} - v_{z1}).$$

To summarize the position, we can say, in general, that

> Total force *exerted on* the fluid in a control volume in a given direction $=$ Rate of change of momentum in the given direction of the fluid passing through the control volume,
>
> $$F = \dot{m}(v_{\text{out}} - v_{\text{in}}).$$

The value of F is positive in the direction in which v is assumed to be positive.

For any control volume, the total force F which acts upon it in a given direction will be made up of three component forces:

$F_1 =$ Force exerted *in the given direction* on the fluid in the control volume by any *solid body* within the control volume or coinciding with the boundaries of the control volume.

$F_2 =$ Force exerted *in the given direction* on the fluid in the control volume by *body forces such as gravity.*

$F_3 =$ Force exerted *in the given direction* on the fluid in the control volume by the fluid outside the control volume.

Thus,

$$F = F_1 + F_2 + F_3 = \dot{m}(v_{\text{out}} - v_{\text{in}}). \tag{5.5}$$

The force R *exerted by* the fluid on the solid body inside or coinciding with the control volume in the given direction will be equal and opposite to F_1 so that $R = -F_1$.

5.3   MOMENTUM CORRECTION FACTOR

The momentum equation (5.5) is based on the assumption that the velocity is constant across any given cross-section. When a real fluid flows past a solid boundary, shear stresses are developed and the velocity is no longer uniform over the cross-section. In a pipe, for example, the velocity will vary from zero at the wall to a maximum at the centre. The momentum per unit time for the whole flow can be found by summing the momentum per unit time through each element of the cross-section, provided that these are sufficiently small for the velocity perpendicular to each element to be taken as uniform. Thus, if the velocity perpendicular to the element is u and the area of the element is δA,

Mass passing through element in unit time $= \rho \delta A \times u$,

$$\begin{matrix} \text{Momentum per unit time} \\ \text{passing through element} \end{matrix} = \text{Mass per unit time} \times \text{Velocity}$$

$$= \rho \delta A u \times u = \rho u^2 \delta A,$$

$$\begin{matrix} \text{Total momentum per unit} \\ \text{time passing whole} \\ \text{cross-section} \end{matrix} = \int \rho u^2 \, \mathrm{d}A. \tag{5.6}$$

To evaluate this integral, the velocity distribution must be known.

If we consider turbulent flow through a pipe of radius R (Fig. 5.3), the velocity u at any distance y from the pipe wall is given approximately by Prandtl's one-seventh power law:

FIGURE 5.3
Calculation of momentum correction factor

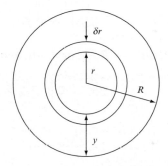

$$u = u_{\max}(y/R)^{1/7},$$

the maximum velocity, $u_{\max}$, occurring at the centre of the pipe. Since the velocity is constant at any radius $r = R - y$, it is convenient to take the element of area δA in equation (5.6) as an annulus of radius r and width δr,

$$\delta A = 2\pi r \delta r,$$

and, from equation (5.6), for the whole cross-section,

$$\text{Total momentum per unit time} = \int_0^R \rho u^2 \, \mathrm{d}A$$

$$= \int_0^R \rho u_{\max}^2 (y/R)^{2/7} 2\pi r \, \mathrm{d}r = \left(\frac{2\pi\rho}{R^{2/7}}\right) u_{\max}^2 \int_0^R y^{2/7} r \, \mathrm{d}r. \tag{5.7}$$

Since $r = R - y$, $\mathrm{d}r = -\mathrm{d}y$, and so, substituting for r and $\mathrm{d}r$ in equation (5.7) and changing the limits (because $y = 0$ when $r = R$),

$$\text{Total momentum per unit time } = \frac{2\pi\rho u_{max}^2}{R^{2/7}} \int_R^0 y^{2/7}(R-y)(-\mathrm{d}y)$$

$$= \frac{2\pi\rho u_{max}^2}{R^{2/7}} \int_R^0 (y^{9/7} - Ry^{2/7})\,\mathrm{d}y = \frac{2\pi\rho u_{max}^2}{R^{2/7}}\left(\frac{7}{16}y^{16/7} - \frac{7}{9}Ry^{9/7}\right)_R^0$$

$$= \frac{2\pi\rho u_{max}^2}{R^{2/7}} R^{16/7}\left(\frac{7}{9} - \frac{7}{16}\right) = \frac{49}{72}\pi\rho R^2 u_{max}^2. \tag{5.8}$$

In practice, it is usually more convenient to use the mean velocity $\bar{u}$ instead of the maximum velocity u_{max}:

$$\text{Mean velocity, } \bar{u} = \frac{\text{Total volume per unit time passing section}}{\text{Total area of cross-section}}$$

$$= \frac{1}{\pi R^2} \int_0^R u\,\delta A.$$

Putting $u = u_{max}(y/R)^{1/7}$ and $\delta A = 2\pi r\delta r$

$$\bar{u} = \frac{1}{\pi R^2} \int_0^R u_{max}\left(\frac{y}{R}\right)^{1/7} 2\pi r\,\mathrm{d}r = \frac{2u_{max}}{R^{15/7}} \int_0^R y^{1/7} r\,\mathrm{d}r.$$

Putting $r = R - y$, $\mathrm{d}r = -\mathrm{d}y$, and changing the limits,

$$\bar{u} = \frac{2u_{max}}{R^{15/7}} \int_R^0 y^{1/7}(R-y)(-\mathrm{d}y) = \frac{2u_{max}}{R^{15/7}} \int_R^0 (y^{8/7} - Ry^{1/7})\,\mathrm{d}y$$

$$= \frac{2u_{max}}{R^{15/7}}\left(\frac{7}{15}y^{15/7} - \frac{7}{8}Ry^{8/7}\right)_R^0 = \frac{49}{60}u_{max},$$

$$u_{max} = \frac{60}{49}\bar{u}. \tag{5.9}$$

Substituting from equation (5.9) in equation (5.8),

$$\text{Total momentum per unit time} = \frac{49}{72}\pi\rho R^2\left(\frac{60}{49}\right)^2\bar{u}^2 = 1.02\rho\pi R^2\bar{u}^2, \tag{5.10}$$

or, since $\rho\pi R^2\bar{u} = \text{mass per unit time}$,

$$\text{Momentum per unit time} = 1.02 \times \text{Mass per unit time} \times \text{Mean velocity}.$$

If the momentum per unit time of the stream had been calculated from the mean velocity without considering the velocity distribution, the value obtained would have been $\rho\pi R^2\bar{u}^2$. To take the velocity distribution into account, a momentum correction factor β must be introduced, so that, for the whole stream,

$$\text{True momentum per unit time} = \beta \times \text{Mass per unit time} \times \text{Mean velocity}.$$

The value of β depends upon the shape of the cross-section and the velocity distribution.

5.4   GRADUAL ACCELERATION OF A FLUID IN A PIPELINE NEGLECTING ELASTICITY

It is frequently the case that the velocity of the fluid flowing in a pipeline has to be changed, thus causing the momentum of the whole mass of fluid in the pipeline to change. This will require the action of a force, which can be calculated from the rate of change of momentum of the mass of fluid and is produced as a result of a change in the pressure difference between the ends of the pipeline. In the case of liquids flowing in rigid pipes, an approximate value of the change of pressure can be obtained by neglecting the effects of elasticity – providing that the acceleration or deceleration is small.

EXAMPLE 5.1

Water flows through a pipeline 60 m long at a velocity of 1.8 m s^{-1} when the pressure difference between the inlet and outlet ends is 25 kN m^{-2}. What increase of pressure difference is required to accelerate the water in the pipe at the rate of 0.02 m s^{-2}? Neglect elasticity effects.

Solution

Let A = cross-sectional area of the pipe, l = length of pipe, ρ = mass density of water, a = acceleration of water, δp = increase in pressure at inlet required to produce acceleration a.

As this is not a steady flow problem, consider a control mass comprising the whole of the water in the pipe. By Newton's second law,

$$\begin{array}{ll} \text{Force due to } \delta p \text{ in} & = \text{Rate of change of momentum of water in} \\ \text{direction of motion} & \quad \text{the whole pipe} \end{array}$$

$$= \text{Mass of water in pipe} \times \text{Acceleration}, \qquad \text{(I)}$$

$$\text{Force due to } \delta p = \text{Cross-sectional area} \times \delta p = A\delta p,$$

$$\text{Mass of water in pipe} = \text{Mass density} \times \text{Volume} = \rho A l.$$

Substituting in (I),

$$A\delta p = \rho A l a,$$

$$\delta p = \rho l a = 10^3 \times 60 \times 0.02 \text{ N m}^{-2}$$

$$= \textbf{1.2 kN m}^{-2}.$$

In this example, the change of pressure difference is small because the acceleration is small, but very large pressures can be developed by sudden accelerations or decelerations, such as may occur when valves are shut suddenly. The elasticity of the fluid and of the pipe must then be taken into account, as explained in Chapter 20.

5.5　FORCE EXERTED BY A JET STRIKING A FLAT PLATE

Consider a jet of fluid striking a flat plate that may be perpendicular or inclined to the direction of the jet, or indeed may be moving in the initial direction of the jet (Fig. 5.4).

A control volume encapsulating the approaching jet and the plate may be established, this control volume being fixed relative to the plate and therefore moving with it. It is helpful to consider components of the velocity and force vectors perpendicular and parallel to the surface of the plate.

In cases where the plate is itself in motion the most helpful technique is to reduce the plate, and therefore the associated control volume, to rest by the superimposition on the system of an equal but opposite plate velocity, as illustrated in Fig 5.4. This reduces all the cases illustrated to a simple consideration of a jet striking a stationary plain surface, the initial motion of the surface being reflected in the amendment of the jet velocity relative to the plate.

In each of the cases illustrated the impingement of the jet on the plate surface reduces the jet velocity component normal to the plate surface to zero. In general terms the jet velocity thus destroyed may be expressed as

$$v_{\text{normal}} = (v - u) \cos \theta.$$

FIGURE 5.4

Force exerted on a flat plate

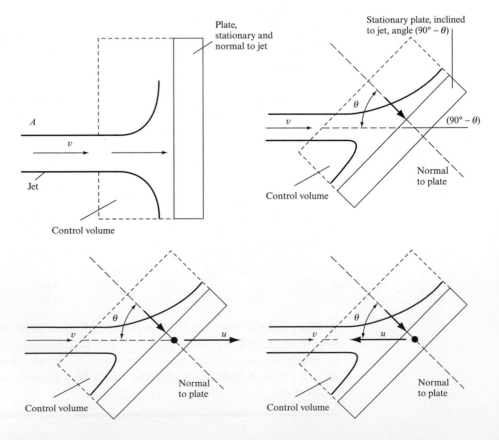

The mass flow entering the control volume is also affected by the superposition of a velocity equal and opposite to the plate velocity and may be expressed as

$$\dot{m} = \rho A(v - u)\cos\theta,$$

which reduces to

$$\dot{m} = \rho A v$$

if the plate is stationary. (Note that the sign convention adopted is positive in the direction of the initial jet velocity.)

Thus the rate of change of momentum normal to the plate surface is given by

$$\mathrm{dMomentum}/\mathrm{d}t = \rho A(v - u)(v - u)\cos\theta.$$

Clearly this expression reduces to

$$\mathrm{dMomentum}/\mathrm{d}t = \rho A v^2 \cos\theta$$

if the plate is stationary, and

$$\mathrm{dMomentum}/\mathrm{d}t = \rho A v^2$$

if the plate is both stationary and perpendicular to the initial jet direction. (Note that the plate velocity, represented by the superposition of an equal and opposite velocity on the system as a whole to bring the control volume to rest, appears in both the mass flow and relative jet velocity terms.)

There will therefore be a force exerted upon the plate equal to the rate of momentum destroyed normal to the plate, given in the general case by an expression of the form

$$\text{Force normal plate} = \rho A(v - u)(v - u)\cos\theta.$$

There will be an equal and opposite reaction force exerted on the jet by the plate.

In a direction parallel to the plate, the force exerted will depend upon the shear stress between the fluid and the surface of the plate. For an ideal fluid there would be no shear stress and hence no force parallel to the plate. The fluid would flow out over the plate so that the total momentum in unit time parallel to the plate remained unchanged.

EXAMPLE 5.2

A jet of water from a fixed nozzle has a diameter of 25 mm and strikes a flat plate inclined to the jet direction. The velocity of the jet is 5 m s^{-1}, and the surface of the plate may be assumed frictionless. (a) Indicate in tabular form the reduction in the force normal to the plate surface as the inclination of the plate to the jet varies from 90° to 0°. (b) Indicate in tabular form the force normal to the plate surface as the plate velocity changes from 2 m s^{-1} to −2 m s^{-1} in the direction of the jet, given that the plate is itself perpendicular to the approaching jet.

Solution

For each case the control volume taken is fixed relative to the plate. If the plate is in motion the control volume is brought to rest by the superposition of an equal and opposite velocity on the system as a whole. As a force applied normal to the plate is required in each case, the components of velocity and force normal and parallel to the plate surface are considered.

From equation (5.5) the gravity force is negligible and, if the fluid jet is assumed to be parallel sided and passing through a region at atmospheric pressure, there is no force exerted on the jet by fluid outside the control volume. Thus the force exerted normal to the plate in the general case is given by

$$\text{Force normal plate} = \rho A(v - u)(v - u) \cos \theta,$$

which may be utilized in the following tables drawn up to answer (a) and (b) above.

The cross-sectional area of the jet is $A = \pi\, 0.025^2/4 = 4.9 \times 10^{-4}$ and the density of the water jet $\rho = 1000.0$ kg m^{-3}. The jet velocity $v = 5.0$ m s^{-1}.

TABLE (A)
Variation of force exerted normal to the plate with plate angle

θ (deg)	$v \cos \theta$ (m s^{-1})	$\rho A v$ (kg s^{-1})	$\text{Force} = \rho A v^2 \cos \theta$ (N)
0	5.00	2.46	12.28
15	4.83	2.46	11.86
30	4.33	2.46	10.63
45	3.54	2.46	8.68
60	2.50	2.46	6.14
75	1.29	2.46	3.18
90	0.00	2.46	0.00

TABLE (B)
Variation of force exerted normal to the plate with plate velocity

θ (deg)	v (m s^{-1})	u (m s^{-1})	$v - u$ (m s^{-1})	$\rho A(v - u)$ (kg s^{-1})	$\text{Force} = \rho A(v - u)^2$ (N)
0	5.0	2.0	3.0	1.47	4.41
0	5.0	1.0	4.0	1.96	7.84
0	5.0	0.0	5.0	2.46	12.28
0	5.0	−1.0	6.0	2.94	17.64
0	5.0	−2.0	7.0	3.43	24.01

5.6   FORCE DUE TO THE DEFLECTION OF A JET BY A CURVED VANE

Both velocity and momentum are vector quantities and, therefore, even if the magnitude of the velocity remains unchanged, a change in direction of a stream of fluid will give rise to a change of momentum. If the stream is deflected by a curved vane (Fig. 5.5), entering and leaving tangentially without impact, a force will be exerted between the fluid and the surface of the vane to cause this change of momentum. It is usually convenient to calculate the components of this force parallel and perpendicular to the direction of the incoming stream by calculating the rate of change of momentum in these two directions. The components can then be combined to give the magnitude and direction of the resultant force which the vane exerts on the fluid, and the equal and opposite reaction of the fluid on the vane.

FIGURE 5.5

Force exerted on a curved vane

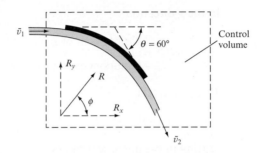

EXAMPLE 5.3

A jet of water from a nozzle is deflected through an angle $\theta = 60°$ from its original direction by a curved vane which it enters tangentially (see Fig. 5.5) without shock with a mean velocity $\bar{v}_1$ of 30 m s^{-1} and leaves with a mean velocity $\bar{v}_2$ of 25 m s^{-1}. If the discharge $\dot{m}$ from the nozzle is 0.8 kg s^{-1}, calculate the magnitude and direction of the resultant force on the vane if the vane is stationary.

Solution

The control volume will be as shown in Fig. 5.5. The resultant force *R exerted by the fluid* on the vane is found by determining the component forces R_x and R_y in the *x* and *y* directions, as shown. Using equation (5.5),

$$R_x = -F_1 = F_2 + F_3 - \dot{m}(v_{\text{out}} - v_{\text{in}})_x.$$

Neglecting force F_2 due to gravity and assuming that for a free jet the pressure is constant everywhere, so that $F_3 = 0$,

$$R_x = \dot{m}(v_{\text{in}} - v_{\text{out}})_x, \tag{I}$$

and, similarly,

$$R_y = \dot{m}(v_{\text{in}} - v_{\text{out}})_y. \tag{II}$$

Since the nozzle and vane are fixed relative to each other,

$$\begin{array}{l}\text{Mass per unit} \\ \text{time entering} \\ \text{control volume}\end{array} = \dot{m} = \begin{array}{l}\text{Mass per unit} \\ \text{time leaving} \\ \text{nozzle.}\end{array}$$

In the x direction,

$$v_{\text{in}} = \text{Component of } \bar{v}_1 \text{ in } x \text{ direction} = \bar{v}_1,$$

$$v_{\text{out}} = \text{Component of } \bar{v}_2 \text{ in } x \text{ direction} = \bar{v}_2 \cos\theta.$$

Substituting in (I),

$$R_x = \dot{m}(\bar{v}_1 - \bar{v}_2 \cos\theta). \qquad\qquad\text{(III)}$$

Putting $\dot{m} = 0.8$ kg s^{-1}, $\bar{v}_1 = 30$ m s^{-1}, $\bar{v}_2 = 25$ m s^{-1}, $\theta = 60°$,

$$R_x = 0.8(30 - 25 \cos 60°) = 14 \text{ N}.$$

In the y direction,

$$v_{\text{in}} = \text{Component of } \bar{v}_1 \text{ in } y \text{ direction} = 0,$$

$$v_{\text{out}} = \text{Component of } \bar{v}_2 \text{ in } y \text{ direction} = \bar{v}_2 \sin\theta.$$

Thus, from (II),

$$R_y = \dot{m}\bar{v}_2 \sin\theta. \qquad\qquad\text{(IV)}$$

Putting in the numerical values,

$$R_y = 0.8 \times 25 \sin 60° = 17.32 \text{ N}.$$

Combining the rectangular components R_x and R_y,

$$\begin{array}{l}\text{Resultant force exerted} \\ \text{by fluid on vane, } R\end{array} = \sqrt{(R_x^2 - R_y^2)}$$

$$= \sqrt{(14^2 + 17.32^2)} = \mathbf{22.27} \text{ N}.$$

This resultant force R will be inclined to the x direction at an angle $\phi = \tan^{-1}(R_y/R_x) = \tan^{-1}(17.32/14) = \mathbf{51°3'}$.

5.7   FORCE EXERTED WHEN A JET IS DEFLECTED BY A MOVING CURVED VANE

If a jet of fluid is to be deflected by a moving curved vane without impact at the inlet to the vane, the relation between the direction of the jet and the tangent to the curve of the vane at inlet must be such that the relative velocity of the fluid at inlet is tangential to the vane. The force in the direction of motion of the vane will be equal to the rate of change of momentum of the fluid in the direction of motion, i.e. the mass deflected per second multiplied by the change of velocity in that direction. The force at right angles to the direction of motion will be equal to the mass deflected per second times the change of velocity at right angles to the direction of motion.

EXAMPLE 5.4

A jet of water 100 mm in diameter leaves a nozzle with a mean velocity $\bar{v}_1$ of 36 m s^{-1} (Fig. 5.6) and is deflected by a series of vanes moving with a velocity u of 15 m s^{-1} in a direction at 30° to the direction of the jet, so that it leaves the vane with an absolute mean velocity $\bar{v}_2$ which is at right angles to the direction of motion of the vane. Owing to friction, the velocity of the fluid relative to the vane at outlet $\bar{v}_{r2}$ is equal to 0.85 of the relative velocity $\bar{v}_{r1}$ at inlet.

Calculate (a) the inlet angle α and outlet angle β of the vane which will permit the fluid to enter and leave the moving vane tangentially without shock, and (b) the force exerted on the series of vanes in the direction of motion u.

FIGURE 5.6

Force exerted on a series of moving vanes

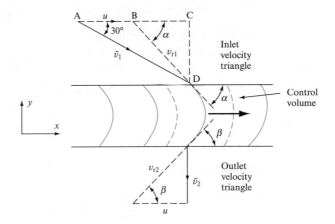

Solution

If the absolute velocity $\bar{v}_2$ is to be at right angles to the direction of motion, the vane must turn the fluid so that it leaves with a relative velocity $\bar{v}_{r2}$, which has a component velocity equal and opposite to u as shown in the outlet velocity triangle (Fig. 5.6).

(a) To determine the inlet angle α, consider the inlet velocity triangle. The velocity of the fluid relative to the vane at inlet, $\bar{v}_{r1}$, must be tangential to the vane and make an angle α with the direction of motion,

$$\tan \alpha = CD/BC = \bar{v}_1 \sin 30° / (\bar{v}_1 \cos 30° - u).$$

Putting $\bar{v}_1 = 36$ m s^{-1} and $u = 15$ m s^{-1},

$$\tan \alpha = 36 \times 0.5 / (36 \times 0.866 - 15) = 1.113,$$

$$\alpha = 48°3'.$$

To determine the outlet angle β, if $\bar{v}_2$ has no component in the direction of motion, the outlet velocity triangle is right angled, $\cos \beta = u/\bar{v}_{r2}$, but $\bar{v}_{r2} = 0.85\bar{v}_{r1}$ and, from the inlet triangle,

$$\bar{v}_{r1} = CD/\sin \alpha = \bar{v}_1 \sin 30° / \sin \alpha.$$

Therefore

$$\cos \beta = \frac{u \sin \alpha}{0.85 v_1 \sin 30°} = \frac{15 \times 0.744}{0.85 \times 36 \times 0.5} = 0.729,$$

$$\beta = 43°11'.$$

(b) Since the jet strikes a series of vanes, perhaps mounted on the periphery of a wheel, so that as each vane moves on its place is taken by the next in the series, the average length of the jet does not alter and the whole flow from the nozzle of diameter d is deflected by the vanes.

Neglecting the force due to gravity and assuming a free jet that does not fill the space between the vanes completely, so that the pressure is constant everywhere, the component forces in the x and y directions (Fig. 5.6) can be found from equation (5.5) putting $R = -F_1$ and $F_2 = F_3 = 0$. In the direction of motion, which is the x direction,

$$R_x = \dot{m}(v_{\text{in}} - v_{\text{out}})_x \tag{I}$$

$$\begin{array}{l} \text{Mass per unit time} \\ \text{entering control} \\ \text{volume} \end{array} = \dot{m} = \begin{array}{l} \text{Mass per unit time} \\ \text{leaving nozzle} \end{array}$$

$$= \rho(\pi/4)\mathrm{d}^2 \bar{v}_1,$$

$$v_{\text{in}} = \text{Component of } \bar{v}_1 \text{ in } x \text{ direction} = \bar{v}_1 \cos 30°,$$

$$v_{\text{out}} = \text{Component of } \bar{v}_2 \text{ in } x \text{ direction} = \bar{v}_2 \cos 90° = 0.$$

Substituting in (I),

$$\begin{array}{l} \text{Force on vanes in} \\ \text{direction of motion} \end{array} = R_x = \rho(\pi/4)\mathrm{d}^2 \bar{v}_1 \times \bar{v}_1 \cos 30°.$$

Putting in the numerical values,

$$\begin{array}{l} \text{Force on vanes in} \\ \text{direction of motion} \end{array} = 1000 \times (\pi/4)(0.1)^2 \times 36 \times 36 \times 0.866 \text{ N} = \mathbf{8816 \text{ N}}.$$

5.8   FORCE EXERTED ON PIPE BENDS AND CLOSED CONDUITS

Figure 5.7 shows a bend in a pipeline containing fluid. When the fluid is at rest, it will exert a static force on the bend because the lines of action of the forces due to pressures p_1 and p_2 do not coincide. If the bend tapers, the magnitude of the static forces will also be affected.

When the fluid is in motion, its momentum will change as it passes round the bend due to the change in its direction and, if the pipe tapers, any consequent change in magnitude of its velocity. There must, therefore, be an additional force acting between the fluid and the pipe.

EXAMPLE 5.5

A pipe bend tapers from a diameter of d_1 of 500 mm at inlet (see Fig. 5.7) to a diameter of d_2 of 250 mm at outlet and turns the flow through an angle θ of 45°. Measurements of pressure at inlet and outlet show that the pressure p_1 at inlet is 40 kN m^{-2} and the pressure p_2 at outlet is 23 kN m^{-2}. If the pipe is conveying oil which has a density ρ of 850 kg m^{-3}, calculate the magnitude and direction of the resultant force on the bend when the oil is flowing at the rate of 0.45 m^3 s^{-1}. The bend is in a horizontal plane.

FIGURE 5.7

Force on a tapering bend

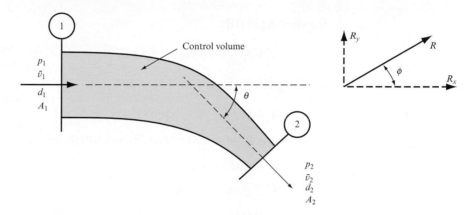

Solution

Referring to Fig. 5.7, take the x direction parallel to the incoming velocity $\bar{v}_1$ and the y direction as shown. The control volume is bounded by the inside wall of the bend and the inlet and outlet sections 1 and 2.

Mass per unit time entering control volume $= \rho Q$.

The forces *acting on the fluid* will be F_1 exerted by the walls of the pipe, F_2 due to gravity (which will be zero), and F_3 due to the pressures p_1 and p_2 of the fluid outside the control volume acting on areas A_1 and A_2 at sections 1 and 2. The force exerted by the fluid on the bend will be $R = -F_1$. Using equation (5.5), putting $F_2 = 0$ and resolving in the x direction:

$$(F_1 + F_3)_x = \dot{m}(v_{\text{out}} - v_{\text{in}})_x$$

and, since $R_x = -(F_1)_x$,

$$R_x = (F_3)_x - \dot{m}(v_{\text{out}} - v_{\text{in}})_x. \tag{I}$$

Now $(F_3)_x = p_1 A_1 - p_2 A_2 \cos\theta$,

$$v_{\text{out}} = \text{Component of } \bar{v}_2 \text{ in } x \text{ direction} = \bar{v}_2 \cos\theta,$$

$$v_{\text{in}} = \text{Component of } \bar{v}_1 \text{ in } x \text{ direction} = \bar{v}_1.$$

Substituting in (I),

$$R_x = p_1 A_1 - p_2 A_2 \cos\theta - \rho Q(\bar{v}_2 \cos\theta - \bar{v}_1). \tag{II}$$

Resolving in the y direction,

$$(F_1 + F_3)_y = \dot{m}(v_{\text{out}} - v_{\text{in}})_y$$

and, since $R_y = -(F_1)_y$,

$$R_y = (F_3)_y - \dot{m}(v_{out} - v_{in})_y. \tag{III}$$

Now, $(F_3)_y = 0 + p_2 A_2 \sin\theta$,

$$v_{out} = \text{Component of } \bar{v}_2 \text{ in } y \text{ direction} = -\bar{v}_2 \sin\theta,$$

$$v_{in} = \text{Component of } \bar{v}_1 \text{ in } y \text{ direction} = 0.$$

Substituting in (III),

$$R_y = p_2 A_2 \sin\theta + \rho Q \bar{v}_2 \sin\theta. \tag{IV}$$

For the given problem,

$$A_1 = (\pi/4)d_1^2 = (\pi/4)(0.5)^2 = 0.196\ 35\ \text{m}^2,$$

$$A_2 = (\pi/4)d_2^2 = (\pi/4)(0.25)^2 = 0.049\ 09\ \text{m}^2,$$

$$Q = 0.45\ \text{m}^3\ \text{s}^{-1},$$

$$\bar{v}_1 = Q/A_1 = 0.45/0.196\ 35 = 2.292\ \text{m s}^{-1},$$

$$\bar{v}_2 = Q/A_2 = 0.45/0.049\ 09 = 9.167\ \text{m s}^{-1}.$$

Putting $\rho = 850\ \text{kg m}^{-3}$, $\theta = 45°$, $p_1 = 40\ \text{kN m}^{-2}$, $p_2 = 23\ \text{kN m}^{-2}$, and substituting in equation (II)

$$R_x = 40 \times 10^3 \times 0.196\ 35 - 23 \times 10^3 \times 0.049\ 09 \cos 45°$$

$$- 850 \times 0.45(9.167 \cos 45° - 2.292)\ \text{N}$$

$$= 10^3(7.855 - 0.798 - 1.603)\ \text{N}$$

$$= \mathbf{5.454 \times 10^3\ N.}$$

Substituting in equation (IV)

$$R_y = 23 \times 10^3 \times 0.049\ 09 \sin 45° + 850 \times 0.45 \times 9.167 \sin 45°\ \text{N}$$

$$= 10^3(0.798 + 2.479)\ \text{N}$$

$$= \mathbf{3.277 \times 10^3\ N.}$$

Combining the x and y components,

$$\text{Resultant force on bend, } R = \sqrt{(R_x^2 + R_y^2)}$$

$$= \sqrt{(5.454^2 + 3.277^2)}\ \text{kN}$$

$$= \mathbf{6.362\ kN.}$$

The inclination of R to the x direction is given by

$$\phi = \tan^{-1}(R_y/R_x) = \tan^{-1}(3.277/5.454) = \mathbf{31°.}$$

5.9 REACTION OF A JET

Whenever the momentum of a stream of fluid is increased in a given direction in passing from one section to another, there must be a net force acting on the fluid in that direction, and, by Newton's third law, there will be an equal and opposite force exerted by the fluid on the system which is producing the change of momentum. A typical example is the reaction force exerted when a fluid is discharged in the form of a high-velocity jet, and which is applied to the propulsion of ships and aircraft through the use of propellers, pure jet engines and rocket motors. The propulsive force can be determined from the application of the linear momentum equation (5.5) to flow through a suitable control volume.

EXAMPLE 5.6

A jet of water of diameter $d = 50$ mm issues with velocity $\bar{v} = 4.9$ m s^{-1} from a hole in the vertical side of an open tank which is kept filled with water to a height of 1.5 m above the centre of the hole (Fig. 5.8). Calculate the reaction of the jet on the tank and its contents (a) when it is stationary, (b) when it is moving with a velocity $u = 1.2$ m s^{-1} in the opposite direction to the jet while the velocity of the jet relative to the tank remains unchanged. In the latter case, what would be the work done per second?

FIGURE 5.8
Reaction of a jet

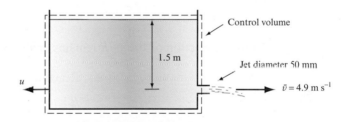

Control volume

1.5 m

Jet diameter 50 mm

u

$\bar{v} = 4.9$ m s^{-1}

Solution

Take the control volume shown in Fig. 5.8. In equation (5.5), the direction under consideration will be that of the issuing jet, which will be considered as positive in the direction of motion of the jet; therefore, $F_2 = 0$, and, if the jet is assumed to be at the same pressure as the outside of the tank, $F_3 = 0$.

Force exerted by
fluid system in $\quad = -F_1 = -\dot{m}(v_{\text{out}} - v_{\text{in}})$,
direction of motion, R

or in words,

Reaction force
in direction $\quad = \dfrac{\text{Mass discharged}}{\text{per unit time}} \times \dfrac{\text{Increase of velocity}}{\text{in direction of jet.}}$ 　　(I)
opposite to that
of the jet

In the present problem,

$$\begin{aligned}\text{Mass discharged} \atop \text{per unit time, } \dot{m} = \rho(\pi/4)d^2\bar{v} = 1000 \times (\pi/4)(0.05)^2 \times 4.9 \text{ kg s}^{-1}\end{aligned}$$

$$= 9.62 \text{ kg s}^{-1}.$$

(a) If the tank is stationary,

$$v_{\text{out}} = \bar{v} = 4.9 \text{ m s}^{-1},$$

$$v_{\text{in}} = \text{Component of velocity of the free surface in the direction of the jet} = 0.$$

Substituting in equation (I),

$$\text{Reaction of jet on tank} = 9.62 \times (4.9 - 0) \text{ N}$$

$$= \mathbf{47.14 \text{ N}}$$

in the direction opposite to that of the jet.

(b) If the tank is moving with a velocity u in the opposite direction to that of the jet, the effect is to superimpose a velocity of $-u$ on the whole system:

$$v_{\text{out}} = \bar{v} - u,$$

$$v_{\text{in}} = -u,$$

$$v_{\text{out}} - v_{\text{in}} = \bar{v}.$$

Thus, the reaction of the jet R remains unaltered at **47.14 N**.

$$\text{Work done per second} = \text{Reaction} \times \text{Velocity of tank}$$

$$= R \times u = 47.14 \times 1.2 = \mathbf{56.57 \text{ W}}.$$

A rocket motor is, in principle, a simple form of engine in which the thrust is developed as the result of the discharge of a high-velocity jet of gas produced by the combustion of the fuel and oxidizing agent. Both the fuel and the oxidant are carried in the rocket and so it can operate even in outer space. It does not require atmospheric air, either for combustion or for the jet to push against; the thrust is entirely due to the reaction developed from the momentum per second discharged in the jet.

EXAMPLE 5.7

The mass of a rocket m_r is 150 000 kg and, when ready to launch, it carries a mass of fuel m_{f0} of 300 000 kg. The initial thrust of the rocket motor is 5 MN and fuel is consumed at a constant rate $\dot{m}$. The velocity $\bar{v}_r$ of the jet relative to the rocket is 3000 m s^{-1}. Assuming that the flight is vertical, and neglecting air resistance, find (a) the burning time, (b) the speed of the rocket and the height above ground at the moment when all the fuel is burned, and (c) the maximum height that the rocket will reach. Assume that g is constant and equal to 9.81 m s^{-2}.

Solution

(a) From equation (I), Example 5.6,

$$\text{Initial thrust, } T = \dot{m}\bar{v}_r,$$

$$\text{Rate of fuel consumption, } \dot{m} = T/\bar{v}_r$$

$$= 5 \times 10^6/3000 = 1667 \text{ kg s}^{-1},$$

$$\text{Initial mass of fuel, } m_{f0} = 300\,000 \text{ kg,}$$

$$\text{Burning time} = m_{f0}/\dot{m} = 300\,000/1667 = \textbf{180 s.}$$

(b) If there is no air resistance, the forces acting on the rocket and the fuel which it contains during vertical flight are the thrust T acting upwards and the weight $(m_r + m_{ft})g$ acting downwards, where m_{ft} is the mass of the fuel in the rocket at time t.
From Newton's second law,

$$T - (m_r + m_{ft})g = (m_r + m_{ft})\frac{\mathrm{d}v_t}{\mathrm{d}t},$$

where v_t is the velocity of the rocket at time t.

$$\frac{\mathrm{d}v_t}{\mathrm{d}t} = \frac{T - (m_r + m_{ft})g}{m_r + m_{ft}}.$$

Since the fuel is being consumed at a rate $\dot{m}$,

$$\text{Mass of fuel at time } t, m_{ft} = m_{f0} - \dot{m}t.$$

Also, $T = \dot{m}\bar{v}_r$ and so

$$\frac{\mathrm{d}v_t}{\mathrm{d}t} = \frac{\dot{m}\bar{v}_r - (m_r + m_{f0} - \dot{m}t)g}{m_r + m_{f0} - \dot{m}t}.$$

Substituting numerical values,

$$\frac{\mathrm{d}v_t}{\mathrm{d}t} = \frac{1667 \times 3000 - (150\,000 + 300\,000 - 1667t) \times 9.81}{150\,000 + 300\,000 - 1667t}$$

$$= -9.81 + \frac{3000}{269.95 - t} \text{ m s}^{-2}.$$

Integrating,

$$v_t = -9.81t - 3000 \log_e(269.95 - t) + \text{constant.}$$

Putting $v_t = 0$ when $t = 0$, the value of the constant is $3000 \log_e 269.95$, giving

$$v_t = -9.81t - 3000 \log_e(1 - t/269.95). \tag{I}$$

From (a), all the fuel will be burnt out when $t = 180$ s. Substituting in equation (I),

$$v_t = -9.81 \times 180 - 3000 \log_e(1 - 180/269.95) \text{ m s}^{-1}$$

$$= -1765.8 + 3296.9 = \textbf{1531.2 m s}^{-1}.$$

The height at time $t = 180$ s is given by

$$Z_1 = \int_0^{180} v_t\, dt = -9.81 \int_0^{180} t\, dt - 3000 \int_0^{180} \log_e(1 - t/269.95)\, dt$$

$$= -(4.9t^2)_0^{180} + 3000\{269.95(1 - t/269.95)[\log_e(1 - t/269.95) - 1]\}_0^{180}$$

$$= -158\,760 + 243\,451.9 = 84\,691.9 \text{ m}$$

$$= \mathbf{84.692\ km}.$$

(c) When the fuel is exhausted, the rocket will have reached an altitude of 84 692 m and will have kinetic energy $m_r v_t^2/2g$. It will, therefore, continue to rise a further distance Z_2 until this kinetic energy has been converted into an increase of potential energy.

$$Z_2 = v_t^2/2g = 1531.2^2/(2 \times 9.81) = 119\,499 \text{ m},$$

$$\text{Maximum height reached} = Z_1 + Z_2 = 84\,692 + 119\,499 \text{ m}$$

$$= \mathbf{204.2\ km}.$$

For aircraft or missiles propelled in the atmosphere it is not necessary to employ a self-contained system, the propulsive force being obtained from the reaction of a jet of atmospheric air which is taken in and accelerated by means of a propeller, turboprop or jet engine and expelled at the rear of the craft.

In the case of the jet engine, air is taken in at the front of the engine and mixed with a small amount of fuel which, on burning, produces a stream of hot gas to be discharged at a much higher velocity at the rear. Figure 5.9(a) shows a jet engine moving through still air. It is convenient to take a control volume which is fixed relative to the engine and to reduce the system to a steady state by imposing a rearward velocity v upon it (Fig. 5.9(b)). Relative to the control volume,

FIGURE 5.9

Jet engine

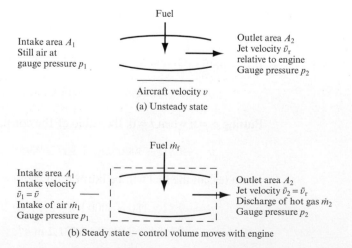

Fuel

Intake area A_1
Still air at
gauge pressure p_1

Outlet area A_2
Jet velocity $\bar{v}_r$
relative to engine
Gauge pressure p_2

Aircraft velocity v

(a) Unsteady state

Fuel $\dot{m}_f$

Intake area A_1
Intake velocity
$\bar{v}_1 = \bar{v}$
Intake of air $\dot{m}_1$
Gauge pressure p_1

Outlet area A_2
Jet velocity $\bar{v}_2 = \bar{v}_r$
Discharge of hot gas $\dot{m}_2$
Gauge pressure p_2

(b) Steady state – control volume moves with engine

Intake velocity, $\bar{v}_1 = v$, Jet velocity $= \bar{v}_r$,

Total force exerted on
fluid in the control $\quad$ Increase of momentum in
volume in the direction $=$ direction of the jet,
of the jet

$$F = \dot{m}_2 \bar{v}_2 - \dot{m}_1 \bar{v}_1.$$

Since the mass per unit time of the hot gases discharged will be greater than that of the air entering the control volume, owing to the addition of fuel,

$$\dot{m}_2 = \dot{m}_1 + \dot{m}_f.$$

Putting $\bar{v}_2 = \bar{v}_r$ and $\bar{v}_1 = v$,

$$F = (\dot{m}_1 + \dot{m}_f)\bar{v}_r - \dot{m}_1 v$$

$$= \dot{m}_1(\bar{v}_r - v) + \dot{m}_f \bar{v}_r.$$

If r is the ratio of the mass of fuel burned to the mass of air taken in,

$$F = \dot{m}_1[(1 + r)\bar{v}_r - v].$$

If T is the thrust exerted on the engine by the fluid, taken as positive in the direction of the jet, the force F_1 exerted by the engine on the fluid is equal to $-T$. There will be no gravity forces acting on the fluid in horizontal flight, but there will be a force $(p_1 A_1 - p_2 A_2)$ exerted on the fluid due to the fluid outside the control volume, so that

$$F = -T + (p_1 A_1 - p_2 A_2).$$

Substituting for F in the previous equation

$$T = (p_1 A_1 - p_2 A_2) - \dot{m}_1[(1 + r)\bar{v}_r - v],$$

Force on engine
in forward $\quad = -T = \dot{m}_1[(1 + r)\bar{v}_r - v] - (p_1 A_1 - p_2 A_2).$ $\qquad$ (5.11)
direction

EXAMPLE 5.8

A jet engine consumes 1 kg of fuel for each 40 kg of air passing through the engine. The fuel consumption is 1.1 kg s^{-1} when the aircraft is travelling in still air at a speed of 200 m s^{-1}. The velocity of the gases which are discharged at atmospheric pressure from the tailpipe is 700 m s^{-1} relative to the engine. Calculate (a) the thrust of the engine, (b) the work done per second, and (c) the efficiency.

Solution

(a) From equation (5.11), putting $r = 1/40$,

$$\dot{m}_1 = \dot{m}_f/r = 40 \times 1.1 = 44 \text{ kg s}^{-1};$$

$$v = 200 \text{ m s}^{-1}, \bar{v}_r = 700 \text{ m s}^{-1}, p_1 = p_2 = 0;$$

therefore,

$$\text{Thrust} = 44\left[\left(1 + \frac{1}{40}\right)700 - 200\right] = 22\,770 \text{ N}$$

$$= \textbf{22.77 kN.}$$

(b)   Work done per second = Thrust × Forward velocity

$$= T \times v = 22.77 \times 200 = \textbf{4554 kW.}$$

(c)   In addition to the useful work done on the aircraft, work is also done in giving the exhaust gases discharge from the tailpipe kinetic energy. Relative to the ground, the velocity of the air at outlet is $(\bar{v}_r - v)$, while at intake it is zero for still air. Since the mass discharge is $\dot{m}_1(1 + r)$,

$$\begin{aligned}\text{Loss of kinetic energy} \\ \text{per second}\end{aligned} = \tfrac{1}{2}\dot{m}_1(1 + r)(\bar{v}_r - v)^2$$

$$= \tfrac{1}{2} \times 44\left(1 + \frac{1}{40}\right)(700 - 200)^2 \text{ W}$$

$$= \textbf{5638 kW.}$$

$$\text{Efficiency} = \frac{\text{Work done per second}}{\text{Work done per second} + \text{Loss}}$$

$$= \frac{4554}{4554 + 5638} = 0.447 = \textbf{44.7 per cent}.$$

Jet propulsion can also be applied to boats. Water is taken in through an opening either in the bows of the vessel (Fig. 5.10(a)) or on either side (Fig. 5.10(b)) and pumped out of a jet pipe at the stern at high velocity. In both cases, the control volume taken for analysis is fixed relative to the vessel. The two cases differ in that the water entering at the bows has a velocity relative to the vessel in the direction of the jet equal to the absolute velocity of the vessel u, while in Fig. 5.10(b), for side intake, the water entering has no component velocity in the direction of the jet.

FIGURE 5.10

Jet propulsion of vessels.
(a) Intake in direction of
motion. (b) Intake in side
of vessel

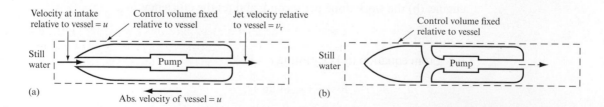

(a)

(b)

EXAMPLE 5.9

Derive a formula for the propulsion efficiency of a jet-propelled vessel in still water if u is the absolute velocity of the vessel, v_r the velocity of the jet relative to the vessel, when the intake is (a) at the bows facing the direction of motion, (b) amidships at right angles to the direction of motion.

Solution

(a) For intake in the direction of motion,

$$\text{Mass of fluid entering control volume in unit time} = \rho Q,$$

$$v_{\text{in}} = \text{Mean velocity of water at inlet in direction of motion relative to control volume} = u,$$

$$v_{\text{out}} = \text{Mean velocity of water at outlet in direction of motion relative to control volume} = v_r.$$

From equation (5.11), assuming that the pressure in the water is the same at outlet and inlet,

$$\text{Propelling force} = \rho Q(v_{\text{out}} - v_{\text{in}}) = \rho Q(v_r - u),$$

$$\text{Work done per unit time} = \text{Propelling force} \times \text{Speed of vessel}$$

$$= \rho Q(v_r - u)u.$$

In unit time, a mass of water ρQ enters the pump intake with a velocity u and leaves with a velocity v_r.

$$\text{Kinetic energy per unit time at inlet} = \tfrac{1}{2}\rho Q u^2,$$

$$\text{Kinetic energy per unit time at outlet} = \tfrac{1}{2}\rho Q v_r^2,$$

$$\text{Kinetic energy per unit time supplied by pump} = \tfrac{1}{2}\rho Q(v_r^2 - u^2),$$

$$\text{Hydraulic efficiency} = \frac{\text{Work done per unit time}}{\text{Energy supplied per unit time}}$$

$$= \rho Q(v_r - u)u / \tfrac{1}{2}\rho Q(v_r^2 - u^2),$$

$$= \mathbf{2u/(v_r + u)}.$$

(b) For intake at right angles to the direction of motion (Fig. 5.10(b)), the control volume used will be the same as in (a), as will the rate of change of momentum through the control volume, and therefore the propelling force. Hence,

$$\text{Work done per unit time} = \rho Q(v_r - u)u.$$

As, however, the intake to the pumps is at right angles to the direction of motion, the forward velocity of the vessel will not assist the intake of water to the pumps and, therefore, the whole of the energy of the outgoing jet must be provided by the pumps.

$$\text{Energy supplied per unit time} = \tfrac{1}{2}\rho Q v_r^2,$$

$$\text{Hydraulic efficiency} = \frac{\text{Work done per unit time}}{\text{Energy supplied per unit time}}$$

$$= \rho Q(v_r - u)u / \tfrac{1}{2}\rho Q v_r^2$$

$$= 2(v_r - u)u / v_r^2.$$

5.10 DRAG EXERTED WHEN A FLUID FLOWS OVER A FLAT PLATE

When a fluid flows over a stationary flat surface, such as the upper surface of the smooth flat plate shown in Fig. 5.11, there will be a shear stress τ_0 between the surface of the plate and the fluid, acting to retard the fluid. At a section AB of the flow well upstream of the tip of the plate O, the velocity will be undisturbed and equal to U. The fluid in contact with the surface of the plate will be at rest, and, at a cross-section such as CD, the velocity u of the adjacent fluid will increase gradually with the distance y away from the plate until it approximates to the free stream velocity at the outside of the *boundary layer* when $y = \delta$. The limit of this boundary layer, in which the drag of the stationary boundary affects the velocity of the fluid, is defined as the distance δ at which $u/U = 0.99$. The value of δ will increase from zero at the leading edge O, since the drag force D exerted on the fluid due to the shear stress τ_0 will increase as x increases. The value of D can be found by applying the momentum equation.

FIGURE 5.11
Drag on a flat plate

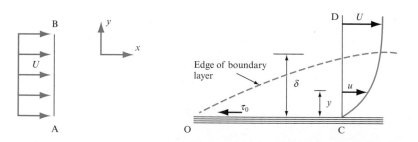

Consider a control volume PQSR (Fig. 5.12) consisting of a section of the boundary layer of length Δx at a distance x from the upstream edge of the plate. Fluid enters the control volume through section PQ and through the upper edge of the boundary layer QS, leaving through section RS. Applying the momentum equation,

Force acting on fluid   Rate of increase of momentum in
in control volume in = x direction of fluid passing
x direction   through control volume.

FIGURE 5.12
Momentum
equation applied to
a boundary layer

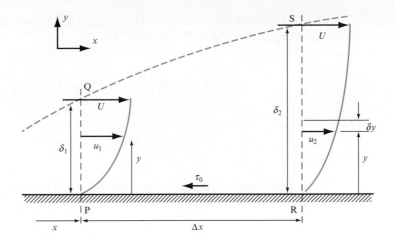

Since the velocity u in the boundary layer varies with the distance y from the surface of the plate, the momentum efflux through RS must be determined by integration. Consider an element of thickness δy, through which the velocity in the x direction is u_2. For a width B perpendicular to the diagram,

$$\frac{\text{Momentum per second}}{\text{passing through element}} = \text{Mass per second} \times \text{Velocity}$$

$$= \rho B \delta y u_2 \times u_2,$$

$$\frac{\text{Total momentum per second}}{\text{passing through RS}} = \rho B \int_0^{\delta_2} u_2^2 \, \mathrm{d}y. \tag{5.12}$$

Similarly, for the control surface PQ,

$$\frac{\text{Total momentum per second}}{\text{passing through PQ}} = \rho B \int_0^{\delta_1} u_1^2 \, \mathrm{d}y. \tag{5.13}$$

where u_1 is the velocity through PQ at a distance y from the surface. For the control surface QS, for continuity of flow,

$$\frac{\text{Rate of flow into}}{\text{the control volume, } Q} = \frac{\text{Rate of flow}}{\text{through RS}} - \frac{\text{Rate of flow}}{\text{through PQ}}$$

$$Q = B \int_0^{\delta_2} u_2 \, \mathrm{d}y - B \int_0^{\delta_1} u_1 \, \mathrm{d}y,$$

$$\frac{\text{Momentum in } x \text{ direction}}{\text{entering through QS}} = \rho Q U$$

$$= \rho B \left(\int_0^{\delta_2} u_2 \, \mathrm{d}y - \int_0^{\delta_1} u_1 \, \mathrm{d}y \right) U \tag{5.14}$$

$$\frac{\text{Force exerted on the fluid by}}{\text{the boundary in the } x \text{ direction}} = -\tau_0 B \Delta x. \tag{5.15}$$

Equating the force given by equation (5.15) with the sum of the x momenta from equations (5.12), (5.13) and (5.14),

$$-\tau_0 B \Delta x = \rho B \left[\int_0^{\delta_2} u_2^2 \, \mathrm{d}y - \int_0^{\delta_1} u_1^2 \, \mathrm{d}y - U \left(\int_0^{\delta_2} u_2 \, \mathrm{d}y - \int_0^{\delta_1} u_1 \, \mathrm{d}y \right) \right]$$

$$= \rho B \left[\int_0^{\delta_2} (u_2^2 - U u_2) \, \mathrm{d}y - \int_0^{\delta_1} (u_1^2 - U u_1) \, \mathrm{d}y \right].$$

The term in the square brackets is the difference between $\int_0^\delta (u^2 - Uu) \, \mathrm{d}y$ at sections RS and PQ, which can be written as $\Delta[\int_0^\delta u(u - U) \, \mathrm{d}y]$ so that

$$-\tau_0 B \Delta x = \rho B \Delta \left[\int_0^\delta u(u - U) \, \mathrm{d}y \right].$$

In the limit, as Δx tends to zero,

$$\tau_0 = \rho U^2 \frac{\mathrm{d}}{\mathrm{d}x} \int_0^\delta \frac{u}{U} \left(1 - \frac{u}{U} \right) \mathrm{d}y. \tag{5.16}$$

The drag D on one surface of the plate will be given by

$$D = B \int_0^x \tau_0 \, \mathrm{d}x.$$

If the fluid acts on both the upper and the lower surface of the plate, this force will of course be doubled.

5.11 ANGULAR MOTION

In Section 4.8, we set out the equations of motion for a particle or element of fluid moving in a straight line. If the particle or element is rotating about a fixed point, similar equations can be written to describe its angular motion. Angular displacement will be measured as the angle θ in radians through which the particle or element has moved about the centre measured from a reference direction. Angular velocity ω will be the rate of change of displacement θ with time, i.e.

$$\omega = \dot{\theta} = \frac{\mathrm{d}\theta}{\mathrm{d}t},$$

and the angular acceleration α will be the rate of change of ω with time, so that

$$\alpha = \ddot{\theta} = \frac{\mathrm{d}^2\theta}{\mathrm{d}t^2}.$$

The laws of angular motion will be similar to those for linear motion (see Section 4.8):

$$\omega_2 = \omega_1 + \alpha t,$$

$$\theta = \omega_1 t + \tfrac{1}{2}\alpha t^2,$$

$$\omega_2^2 = \omega_1^2 + 2\alpha\theta.$$

For a particle (Fig. 5.13) which at a given instant is rotating about a fixed point with angular velocity ω at a radius r,

Tangential linear velocity, $v_\theta = \omega r$,

Momentum of particle, $mv_\theta = m\omega r$.

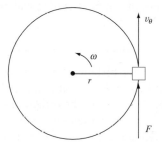

FIGURE 5.13
Angular motion

If the angular velocity changes from ω to zero in time t under the influence of a force F acting at radius r,

Rate of change of momentum of particle $= mr\omega/t$.

By Newton's second law,

$$F = mr\omega/t.$$

This force produces a turning moment or torque T about the centre of rotation,

$$T = Fr = mr^2\omega/t = \text{Angular momentum/Time}.$$

Now consider a particle moving in a curved path, so that in time t it moves from a position at which it has an angular velocity ω_1 at radius r_1 to a position in which the corresponding values are ω_2 and r_2. The effect will be equivalent to first applying a torque to reduce the particle's original angular momentum to zero, and then applying a torque in the opposite direction to produce the angular momentum required in the second position:

$$\text{Torque required to eliminate original angular momentum} = mr_1^2\omega_1/t,$$

$$\text{Torque required to produce new angular momentum} = mr_2^2\omega_2/t,$$

$$\text{Torque required to produce change of angular momentum} = (m/t)(\omega_2 r_2^2 - \omega_1 r_1^2)$$

$$= (m/t)(v_{\theta2} r_2 - v_{\theta1} r_1),$$

where v_θ = tangential velocity = ωr.

This analysis applies equally to a stream of fluid moving in a curved path, since m/t is the mass flowing per unit time, $\dot{m} = \rho Q$. The torque which must be acting on the fluid will be

$$T = \rho Q(v_{\theta2} r_2 - v_{\theta1} r_1), \tag{5.17}$$

and, of course, the fluid will exert an equal and opposite reaction.

EXAMPLE 5.10

A water turbine rotates at 240 rev min^{-1}. The water enters the rotating impeller at a radius of 1.2 m with an absolute mean velocity which has a tangential component of 2.3 m s^{-1} in the direction of motion and leaves with a tangential component of 0.2 m s^{-1} at a radius of 1.6 m. If the volume rate of flow through the turbine is 10 m^3 s^{-1}, calculate the torque exerted and the theoretical power output.

Solution

In equation (5.17), $\rho = 1000$ kg m^{-3}, $Q = 10$ m^3 s^{-1}, $\bar{v}_{\theta2} = 0.2$ m s^{-1}, $\bar{v}_{\theta1} = 2.3$ m s^{-1}, $r_2 = $ 1.6 m, $r_1 = 1.2$ m. Hence,

$$\text{Torque acting on fluid} = 1000 \times 10(0.2 \times 1.6 - 2.3 \times 1.2) \text{ N m}$$

$$= 10\,000(0.32 - 2.76) = -24\,400 \text{ N m}.$$

The torque exerted by the fluid on the rotor will be equal and opposite:

$$\text{Torque exerted by fluid} = 24\,400 \text{ N m}.$$

If n is the rotational speed in revolutions per second,

$$n = 240/60 = 4,$$

$$\text{Power output} = 2\pi nT = 2\pi \times 4 \times 24\,400 \text{ W}$$

$$= 613\,318 \text{ W} = \mathbf{613.32 \text{ kW}}.$$

5.12 EULER'S EQUATION OF MOTION ALONG A STREAMLINE

From consideration of the rate of change of momentum from point to point along a streamline and the forces acting due to the effects of the surrounding pressures and changes of elevation, it is possible to derive a relationship between velocity, pressure, elevation and density along a streamline.

Figure 5.14 shows a short section of a streamtube surrounding the streamline and having a cross-sectional area small enough for the velocity to be considered constant over the cross-section. AB and CD are two cross-sections separated by a short distance δs. At AB the area is A, velocity v, pressure p and elevation z, while at CD the corresponding values are $A + \delta A$, $v + \delta v$, $p + \delta p$ and $z + \delta z$. The surrounding fluid will exert a pressure p_{side} on the sides of the element and, if the fluid is assumed to be inviscid, there will be no shear stresses on the sides of the streamtube and p_{side} will act normally. The weight of the element mg will act vertically downward at an angle θ to the centreline.

FIGURE 5.14
Euler's equation

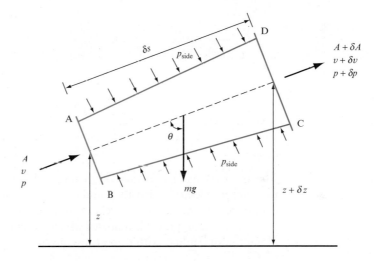

Mass per unit time flowing $= \rho A v$,

Rate of increase of momentum from AB to CD $= \rho A v[(v + \delta v) - v] = \rho A v \delta v.$ (5.18)

The forces acting to produce this increase of momentum in the direction of motion are

Force due to p in direction of motion $= pA$,

Force due to $p + \delta p$ opposing motion $= (p + \delta p)(A + \delta A)$,

Force due to p_{side} producing a component in the direction of motion $= p_{side} \delta A$,

Force due to mg producing a component opposing motion $= mg \cos \theta$,

Resultant force in the direction of motion $= \begin{aligned} &pA - (p + \delta p)(A + \delta A) \\ &+ p_{side} \delta A - mg \cos \theta. \end{aligned}$

The value of p_{side} will vary from p at AB to $p + \delta p$ at CD and can be taken as $p + k\delta p$, where k is a fraction,

$$\text{Weight of element, } mg = \rho g \times \text{Volume} = \rho g (A + \tfrac{1}{2}\delta A)\delta s,$$

$$\cos \theta = \delta z / \delta s,$$

$$\begin{array}{l}\text{Resultant force in the} \\ \text{direction of motion}\end{array} = \begin{array}{l} -p\delta A - A\delta p - \delta p \delta A + p\delta A + k\delta p \cdot \delta A \\ - \rho g (A + \tfrac{1}{2}\delta A)\delta s \cdot (\delta z / \delta s). \end{array}$$

Neglecting products of small quantities,

$$\text{Resultant force in the direction of motion} = -A\delta p - \rho g A \delta z. \tag{5.19}$$

Applying Newton's second law from equations (5.18) and (5.19),

$$\rho A v \delta v = -A\delta p - \rho g A \delta z.$$

Dividing by $\rho A \delta s$,

$$\frac{1}{\rho}\frac{\delta p}{\delta s} + v\frac{\delta v}{\delta s} + g\frac{\delta z}{\delta s} = 0, \tag{5.20}$$

or, in the limit as $\delta s \to 0$,

$$\frac{1}{\rho}\frac{\mathrm{d}p}{\mathrm{d}s} + v\frac{\mathrm{d}v}{\mathrm{d}s} + g\frac{\mathrm{d}z}{\mathrm{d}s} = 0. \tag{5.21}$$

This is known as *Euler's equation*, giving, in differential form, the relationship between pressure p, velocity v, density ρ and elevation z along a streamline for steady flow. It cannot be integrated until the relationship between density ρ and pressure p is known.

For an incompressible fluid, for which ρ is constant, integration of equation (5.21) along the streamline, with respect to s, gives

$$p/\rho + v^2/2 + gz = \text{constant}. \tag{5.22}$$

The terms represent energy per unit mass. Dividing by g,

$$p/\rho g + v^2/2g + z = \text{constant} = H, \tag{5.23}$$

in which the terms represent the energy per unit weight. Equation (5.23) is known as *Bernoulli's equation* and states the relationship between pressure, velocity and elevation for steady flow of a frictionless fluid of constant density. An alternative form is

$$p + \tfrac{1}{2}\rho v^2 + \rho gz = \text{constant}, \tag{5.24}$$

in which the terms represent the energy per unit volume.

These equations apply to a single streamline. The sum of the three terms is constant along any streamline, but the value of the constant may be different for different streamlines in a given stream.

If equation (5.21) is integrated along the streamline between any two points indicated by suffixes 1 and 2,

$$p_1/\rho g + v_1^2/2g + z_1 = p_2/\rho g + v_2^2/2g + z_2. \tag{5.25}$$

For a compressible fluid, the integration of equation (5.21) can only be partially completed, to give

$$\int \frac{\mathrm{d}p}{\rho g} + \frac{v^2}{2g} + z = H.$$

The relationship between ρ and p must then be inserted for the given case. For gases, this will be of the form $p\rho^n$ = constant, varying from adiabatic to isothermal conditions, while, for a liquid, $\rho(\mathrm{d}p/\mathrm{d}\rho) = K$, the bulk modulus.

5.13   PRESSURE WAVES AND THE VELOCITY OF SOUND IN A FLUID

In a real fluid, any change of pressure at a point or any cross-section will be associated with a change in density of the fluid, so that the particles of fluid will change their positions, moving closer together or further apart. Adjacent particles will, in turn, change their positions, and so the change of pressure and density will spread very rapidly through the fluid. Clearly, if the fluid were incompressible, every particle would have to change its position simultaneously and the speed of propagation of the disturbance or pressure wave would, theoretically, be infinite. However, the elasticity of a compressible fluid allows the particles to adjust their positions one after the other, so that the disturbance spreads with a finite velocity. The speed of propagation of a pressure change is very rapid and, in some problems, it is sufficient to assume that pressure changes are propagated instantaneously throughout the fluid. However, when studying abrupt changes of pressure, such as those occurring when a valve on a pipeline is closed suddenly, or when fluid velocities are high relative to a solid body (as in the case of aircraft in flight), the speed of propagation of pressure changes in the fluid can be a factor of major importance from the practical point of view.

In Fig. 5.15, a pressure wave is moving through a fluid from left to right with a velocity c relative to a stationary observer. The fluid to the right of the wavefront will not have been affected by the pressure wave and will have its original pressure p, velocity u relative to the observer and density ρ as indicated. To the left, the fluid behind

FIGURE 5.15

Pressure wave. Unsteady flow relative to a stationary observer

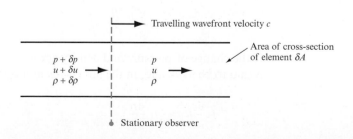

FIGURE 5.16
Pressure wave.
Steady state relative
to a moving observer

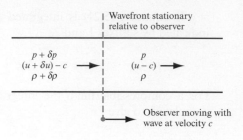

the wavefront will be at the new pressure $p + \delta p$, velocity $u + \delta u$ and density $\rho + \delta\rho$. From a terrestrial frame of reference, conditions in the fluid are not steady, since at a point fixed with reference to a stationary observer conditions will change with time. The usual equations for steady flow cannot, therefore, be applied. However, to an observer moving with the wave at velocity c, the wave will appear stationary; conditions will not change with time and the flow is steady and can be analyzed as such.

As shown in Fig. 5.16, the effect is equivalent to imposing a backward velocity c on the system from right to left. Considering an element of cross-sectional area δA perpendicular to the direction of flow,

$$\begin{array}{c}\text{Mass per unit time flowing} \\ \text{on the left of wavefront}\end{array} = \begin{array}{c}\text{Mass per unit time flowing} \\ \text{on the right of wavefront}\end{array}$$

$$(\rho + \delta\rho)(u + \delta u - c)\delta A = \rho(u - c)\delta A,$$

$$\rho\delta u + u\delta\rho + \delta u\delta\rho - c\delta\rho = 0,$$

$$(c - u)\delta\rho = (\rho + \delta\rho)\delta u. \tag{5.26}$$

Owing to the pressure difference δp across the wavefront, there will be a force acting to the right, in the direction of flow, which will cause an increase in momentum per unit time in this direction.

$$\text{Force due to } \delta p = \text{Increase of momentum per unit time to the right}$$

$$= \text{Mass per unit time} \times \text{Increase of velocity},$$

$$\delta p \times \delta A = \rho(u - c)\delta A \times [u - (u + \delta u)],$$

$$\delta p = \rho(u - c)(-\delta u),$$

$$\delta u = \delta p/\rho(c - u). \tag{5.27}$$

Substituting from equation (5.27) for δu in equation (5.26),

$$(c - u)\delta\rho = (\rho + \delta\rho)\delta p/\rho(c - u),$$

$$(c - u)^2 = \left(1 + \frac{\delta\rho}{\rho}\right)\frac{\delta p}{\delta\rho}. \tag{5.28}$$

If the change of pressure and density across the wavefront is small, the pressure wave is said to be *weak* and, in the limit as δp and $\delta\rho$ tend to zero, equation (5.28) gives

$$(c - u) = \sqrt{\left(\frac{\mathrm{d}p}{\mathrm{d}\rho}\right)}.$$

Now $(c - u)$ is the velocity of the wavefront relative to the fluid, so that

$$\begin{array}{l}\text{Velocity of propagation}\\\text{of a weak pressure wave}\end{array} = c - u = \sqrt{\left(\dfrac{\mathrm{d}p}{\mathrm{d}\rho}\right)}.$$

(5.29)

For a mass m of fluid of volume V and density ρ,

$$\rho V = m.$$

Differentiating,

$$\rho\,\mathrm{d}V + V\,\mathrm{d}\rho = 0,$$

$$\mathrm{d}\rho = -(\rho/V)\,\mathrm{d}V.$$

If K is the bulk modulus, then from equation (1.9),

$$K = -V\frac{\mathrm{d}p}{\mathrm{d}V} = \rho\frac{\mathrm{d}p}{\mathrm{d}\rho},$$

$$\frac{\mathrm{d}p}{\mathrm{d}\rho} = \frac{K}{\rho}.$$

Therefore,

$$\begin{array}{l}\text{Velocity of propagation of}\\\text{a weak pressure wave}\end{array} = c - u = \sqrt{(K/\rho)}.$$

(5.30)

This equation applies to solids, liquids and gases. Note, however, that when c represents the velocity of sound in still air, $u = 0$.

Since sound is propagated in the form of very weak pressure waves, equation (5.30) gives the velocity of sound or *sonic velocity*, with $u = 0$. In a gas, the pressure and temperature changes occurring due to the passage of a sound wave are so small and so rapid that the process can be considered as reversible and adiabatic, so that $p/\rho^{\gamma} = $ constant. Differentiating,

$$\frac{\mathrm{d}p}{\mathrm{d}\rho} = \frac{\gamma p}{\rho}$$

or, since $p/\rho = RT$,

$$\frac{\mathrm{d}p}{\mathrm{d}\rho} = \gamma RT.$$

Substituting in equation (5.29), with $u = 0$,

$$\text{Sonic velocity, } c = \sqrt{(\gamma p/\rho)} = \sqrt{(\gamma RT)}.$$

(5.31)

The above equations apply only to weak pressure waves in which the change of pressure is very small compared with the pressure of the fluid. The pressure change involved in the passage of a sound wave in atmospheric air, for example, varies from about 3×10^{-5} N m^{-2} for a barely audible sound to 100 N m^{-2} for a sound so loud that it verges on the painful. These are small in comparison with atmospheric pressure of 10^5 N m^{-2}. For relatively large pressure changes, the velocity of propagation of the pressure wave would be greater.

The sonic velocity is important in fluid mechanics, because when the velocity of the fluid exceeds the sonic velocity, i.e. becomes *supersonic*, small pressure waves cannot be propagated upstream. At *subsonic* velocities, lower than the sonic velocity, small pressure waves can be propagated both upstream and downstream. This results in the flow pattern around an obstacle, for example, differing for supersonic and subsonic flow with consequent differences in the forces exerted. The ratio of the fluid velocity u to the sonic velocity $c - u$ is known as the Mach number $\mathrm{Ma} = u/(c - u)$. If $\mathrm{Ma} > 1$, flow is supersonic; if $\mathrm{Ma} < 1$, flow is subsonic.

5.14　VELOCITY OF PROPAGATION OF A SMALL SURFACE WAVE

For flow with a free surface, as, for example, in open channels, the pressure cannot vary from point to point along the free surface. A disturbance in the fluid will be propagated as a surface wave rather than as a pressure wave. Using the approach adopted in Section 5.13, assume that a surface wave of height δZ (Fig. 5.17(a)) is being propagated from left to right in the view of a stationary observer. If this wave is brought to rest relative to the observer by imposing a velocity c equal to the wave velocity on the observer, conditions will now appear steady, as shown in Fig. 5.17(b). Considering a width B of the flow, perpendicular to the plane of the diagram,

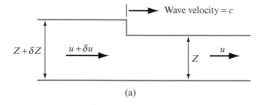 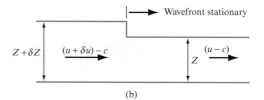

(a)　　　　　　　　　　　　　　　　　　　　(b)

FIGURE 5.17
Velocity of propagation of a small surface wave. (a) Unsteady flow as seen by a stationary observer (b) Steady flow as seen by a moving observer

Mass per unit time flowing on the left of the wavefront = Mass per unit time flowing on the right of the wavefront,

$$\rho \times B(Z + \delta Z) \times (u + \delta u - c) = \rho + B \times Z \times (u - c),$$

the mass density ρ being the same on both sides of the wavefront since the pressure is unchanged. Simplifying,

$$Z\delta u + u\delta Z + \delta Z \delta u - c\delta Z = 0,$$

$$(c - u)\delta Z = (Z + \delta Z)\delta u. \tag{5.32}$$

The change of momentum occurring as a result of the change of velocity across the wavefront is produced as a result of the hydrostatic force due to the difference of level δZ acting on the cross-sectional area BZ. By Newton's second law,

$$\frac{\text{Hydrostatic force}}{\text{due to } \delta Z} = \frac{\text{Mass per unit}}{\text{time}} \times \frac{\text{Change of}}{\text{velocity,}}$$

$$\rho g \delta Z B Z = B Z (u - c) \times (-\delta u),$$

$$\delta u = g \delta Z / (c - u).$$

Substituting from equation (5.32) for δu,

$$(c - u)\delta Z/(Z + \delta Z) = g\delta Z/(c - u),$$

$$(c - u)^2 = (Z + \delta Z)g$$

$$= gZ$$

if the wave height δZ is small.

$$\frac{\text{Velocity of propagation of the}}{\text{wave relative to the fluid}} = c - u = \sqrt{(gZ)}. \qquad (5.33)$$

Taking velocities in a downstream direction as positive, the wave velocity c relative to the bed of the channel is given by $\sqrt{(gZ)} + u$ if the wave is travelling downstream, and $\sqrt{(gZ)} - u$ if it is travelling upstream. Thus, if the stream velocity $u > \sqrt{(gZ)}$, the wave cannot travel upstream relative to the bed, while if $u < \sqrt{(gZ)}$ a surface wave will be propagated in both directions. The ratio of the stream velocity u to the velocity of propagation $c - u$ of the wave in the fluid is known as the *Froude number* Fr:

$$\mathrm{Fr} = u/(c - u) = u/\sqrt{(gZ)}. \qquad (5.34)$$

Thus the condition for a wave to be stationary is that the Froude number is unity, i.e. Fr = 1. The Froude number is also a criterion of the type of flow present in an open channel. If Fr $>$ 1 the flow is defined as supercritical, rapid or shooting, and is characterized by shallow and fast fluid motion. If Fr $<$ 1 the flow is defined as subcritical, tranquil or streaming, and is characterized, relative to supercritical flow, as slow and deep fluid motion. Analogies can be drawn between compressible flow and flow with a free surface, which can be utilized in experimental investigations of the former.

While it is convenient to develop free surface relationships with reference to rectangular channels, many free surface flow conditions occur in uniform but not rectangular channels, for example, the whole range of free surface flow conditions in partially filled pipes to be found in sewer and urban drainage and storm drainage applications. The fundamental concepts remain the same and it is possible to develop analogous expressions for wave speed. It may be shown that for a uniform channel of non-rectangular cross-section the wave speed as defined above is given by

$$c = \sqrt{(gA/T)} \qquad (5.35)$$

where A is the flow cross-sectional area of surface width T at any particular depth. (It may be noted that A/T is a form of 'average' depth relative to a rectangular channel, but this is a misleading interpretation.) The same criterion based on Froude number applies and the same flow-type nomenclature is utilized.

5.15 DIFFERENTIAL FORM OF THE CONTINUITY AND MOMENTUM EQUATIONS

The differential form of the continuity equation was developed in Section 4.13 as

$$\frac{\partial \rho}{\partial t} + \frac{\partial}{\partial x}(\rho v_x) + \frac{\partial}{\partial y}(\rho v_y) + \frac{\partial}{\partial z}(\rho v_z) = 0. \tag{5.36}$$

This expression is applicable to every point in a fluid flow, whether steady or unsteady, compressible or incompressible. The derivation as set out in Section 4.13 considered the continuity of flow and mass storage across an infinitesimal cuboid control volume within the flow. A similar approach may be taken to the determination of a differential form of the momentum equation that would be equally valid across the flow conditions listed above.

Figure 5.18 illustrates the flow in three dimensions through an element of the fluid, together with the forces acting on each surface of the element. Considering the x direction as an exemplar from which the other directional equations will be derived, the total acceleration in the x direction may be written as

$$\frac{\delta v_x}{\delta t} = v_x \frac{\partial v_x}{\partial x} + v_x \frac{\partial v_x}{\partial y} + v_x \frac{\partial v_z}{\partial z} + \frac{\partial v_x}{\partial t}, \tag{5.37}$$

from Section 4.9. The rate of change of momentum in the x direction may then be written as

$$\frac{\partial M_x}{\partial t} = \rho \delta x \delta y \delta z \left(v_x \frac{\partial v_x}{\partial x} + v_y \frac{\partial v_x}{\partial y} + v_z \frac{\partial v_x}{\partial z} + \frac{\partial v_x}{\partial t} \right). \tag{5.38}$$

FIGURE 5.18

Definition of coordinate axes and normal and shear stress notation. Stresses identified for the x axis derivation only

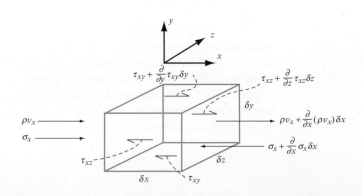

The force acting in the x direction may be determined from a summation of the normal stress, σ_x, on the element surfaces perpendicular to the x direction, having area $\delta y\delta z$, and the shear stresses, τ_{yx} and τ_{zx}, acting on the element surfaces parallel to the x direction, having areas $\delta z\delta x$ and $\delta x\delta y$, where the shear stress suffixes represent the flow direction considered and the separation of the two faces of the element.

In the x direction, therefore, the net force due to the normal stress on the perpendicular faces and the shear stress on the tangential faces of the element is

$$F_x = \left(\rho X - \frac{\partial\sigma_x}{\partial x} + \frac{\partial\tau_{yx}}{\partial y} + \frac{\partial\tau_{zx}}{\partial z}\right)\delta x\,\delta y\,\delta z, \tag{5.39}$$

where X is the body force in the x direction, comprising, for example, gravitational or Coriolis forces as appropriate, and the element is sufficiently small for the change of stress or mass flow with distance to be assumed linear. Therefore from equations (5.38) and (5.39) the general form of the momentum equation in each of the three dimensions may be written as

$$\rho X - \frac{\partial\sigma_x}{\partial x} + \frac{\partial\tau_{yx}}{\partial y} + \frac{\partial\tau_{zx}}{\partial z} = \rho\left(\frac{\partial v_x}{\partial t} + v_x\frac{\partial v_x}{\partial x} + v_y\frac{\partial v_x}{\partial y} + v_z\frac{\partial v_x}{\partial z}\right), \tag{5.40}$$

$$\rho Y + \frac{\partial\tau_{xy}}{\partial x} + \frac{\partial\sigma_y}{\partial y} + \frac{\partial\tau_{zy}}{\partial z} = \rho\left(\frac{\partial v_y}{\partial t} + v_x\frac{\partial v_y}{\partial x} + v_y\frac{\partial v_y}{\partial y} + v_z\frac{\partial v_y}{\partial z}\right), \tag{5.41}$$

$$\rho Z + \frac{\partial\tau_{xz}}{\partial x} + \frac{\partial\tau_{yz}}{\partial y} - \frac{\partial\sigma_z}{\partial z} = \rho\left(\frac{\partial v_z}{\partial t} + v_x\frac{\partial v_z}{\partial x} + v_y\frac{\partial v_z}{\partial y} + v_z\frac{\partial v_z}{\partial z}\right). \tag{5.42}$$

While these momentum equations are entirely general they cannot be integrated without reference to expressions defining the stresses assumed to act normal and tangential to the element surfaces.

In inviscid flow the shear stress terms disappear and the normal stress, σ, terms may be replaced by pressure, p terms, becoming the general form of the one-dimensional steady flow Euler equation presented in Section 5.12, equation (5.20), where the body force is gravitational.

Newtonian fluids, as previously defined in Section 1.4, display properties that allow stress to be related to velocity gradients, for both normal and shear components so that the viscous stresses are proportional to the rate of deformation, defined in terms of a linear deformation by the coefficient of dynamic viscosity, μ, and a second viscosity coefficient, λ, to cover volumetric deformation, defined as the sum of the velocity gradients along each of the three coordinate axes.

The stress velocity gradient expressions, known as the constitutive equations, may be defined as

$$\sigma_x = p - 2\mu\frac{\partial v_x}{\partial x} - \lambda\left(\frac{\partial v_x}{\partial x} + \frac{\partial v_y}{\partial y} + \frac{\partial v_z}{\partial z}\right), \quad \tau_{xy} = \mu\left(\frac{\partial v_x}{\partial y} + \frac{\partial v_y}{\partial x}\right), \tag{5.43}$$

$$\sigma_y = p - 2\mu\frac{\partial v_y}{\partial y} - \lambda\left(\frac{\partial v_x}{\partial x} + \frac{\partial v_y}{\partial y} + \frac{\partial v_z}{\partial z}\right), \quad \tau_{xz} = \mu\left(\frac{\partial v_x}{\partial z} + \frac{\partial v_z}{\partial x}\right), \tag{5.44}$$

$$\sigma_z = p - 2\mu\frac{\partial v_z}{\partial z} - \lambda\left(\frac{\partial v_x}{\partial x} + \frac{\partial v_y}{\partial y} + \frac{\partial v_z}{\partial z}\right), \quad \tau_{yz} = \mu\left(\frac{\partial v_y}{\partial z} + \frac{\partial v_z}{\partial y}\right). \tag{5.45}$$

The effect of the second viscosity coefficient, λ, is small in practice. A good approximation is to set $\lambda = -\frac{2}{3}\mu$, i.e. the Stokes hypothesis, and pressure may be seen to be the average of the three normal stresses from equations (5.43) to (5.45).

Using equation (5.40) as an exemplar it may be seen that for a homogeneous fluid, i.e. one where the properties are not affected by position, substitution for the normal and shear stress terms from equation (5.43) and using the Stokes hypothesis allows the left-hand side (LHS) of equation (5.40) to be recast as

$$\text{LHS} = \rho X - \frac{\partial p}{\partial x} + 2\mu\frac{\partial^2 v_x}{\partial x^2} - \frac{2}{3}\mu\frac{\partial}{\partial x}\left(\frac{\partial v_x}{\partial x} + \frac{\partial v_y}{\partial y} + \frac{\partial v_z}{\partial z}\right)$$

$$+ \mu\left[\frac{\partial}{\partial y}\left(\frac{\partial v_x}{\partial y} + \frac{\partial v_y}{\partial x}\right) + \frac{\partial}{\partial z}\left(\frac{\partial v_x}{\partial z} + \frac{\partial v_z}{\partial x}\right)\right],$$

$$\text{LHS} = \rho X - \frac{\partial p}{\partial x} + \mu\left(\frac{\partial^2 v_x}{\partial x^2} + \frac{\partial^2 v_x}{\partial y^2} + \frac{\partial^2 v_x}{\partial z^2}\right) - \frac{2}{3}\mu\frac{\partial}{\partial x}\left(\frac{\partial v_x}{\partial x} + \frac{\partial v_y}{\partial y} + \frac{\partial v_z}{\partial z}\right)$$

$$+ \mu\frac{\partial}{\partial x}\left(\frac{\partial v_x}{\partial x} + \frac{\partial v_y}{\partial y} + \frac{\partial v_z}{\partial z}\right),$$

so that if the right-hand side (RHS) of equation (5.40) is set to

$$\text{RHS} = \rho\frac{Dv_x}{Dt},$$

the expression in the x direction becomes

$$\rho X - \frac{\partial p}{\partial x} + \mu\left(\frac{\partial^2 v_x}{\partial x^2} + \frac{\partial^2 v_x}{\partial y^2} + \frac{\partial^2 v_x}{\partial z^2}\right) + \frac{1}{3}\mu\frac{\partial}{\partial x}\left(\frac{\partial v_x}{\partial x} + \frac{\partial v_y}{\partial y} + \frac{\partial v_z}{\partial z}\right) = \rho\frac{Dv_x}{Dt}, \quad (5.46)$$

with equivalent expressions for the y and z coordinate axes. If the flow is steady and incompressible then, by reference to the continuity equation, equation (5.46) may be reproduced in each of the three coordinate directions as

$$\rho X - \frac{\partial p}{\partial x} + \mu\left(\frac{\partial^2 v_x}{\partial x^2} + \frac{\partial^2 v_x}{\partial y^2} + \frac{\partial^2 v_x}{\partial z^2}\right) = \rho\frac{Dv_x}{Dt}, \quad (5.47)$$

$$\rho Y - \frac{\partial p}{\partial y} + \mu\left(\frac{\partial^2 v_y}{\partial x^2} + \frac{\partial^2 v_y}{\partial y^2} + \frac{\partial^2 v_y}{\partial z^2}\right) = \rho\frac{Dv_y}{Dt}, \quad (5.48)$$

$$\rho Z - \frac{\partial p}{\partial z} + \mu\left(\frac{\partial^2 v_z}{\partial x^2} + \frac{\partial^2 v_z}{\partial y^2} + \frac{\partial^2 v_z}{\partial z^2}\right) = \rho\frac{Dv_z}{Dt}. \quad (5.49)$$

Equations (5.47) to (5.49) are known as the Navier–Stokes equations following their independent derivation by these two nineteenth-century researchers. While in laminar flow the shear stress is proportional to the viscosity and the rate of shear strain, equation (1.3), turbulent stresses are complex and no wholly satisfactory model exists to be used in developing analogous forms of the Navier–Stokes equations. The introduction of an eddy viscosity, Section 11.4, which includes the turbulence effects and is based on the turbulence models discussed in Chapter 11, allows the Navier–Stokes equations to become central to the developing field of computational fluid dynamics.

5.16   COMPUTATIONAL TREATMENT OF THE DIFFERENTIAL FORMS OF THE CONTINUITY AND MOMENTUM EQUATIONS

The use of computer-based models and simulations to describe fluid flow conditions has numerous advantages for the designer and researcher. The development of computing capacity over the past decade has been exponential and has made possible the implementation of long recognized numerical solutions through the sledgehammer of fast computing. It is now possible to assess the likely effects of design changes without recourse to costly, both in time and resource, physical testing – a discipline with its own problems if not undertaken full scale. However, care must be exercised and computational fluid dynamics (CFD), as the application of computing power to the implementation of known numerical methods has become known, must be recognized as being still in the developmental stage itself. In particular the problem of turbulent flow description has not been wholly solved and care must be taken with the resulting simulation predictions.

While recognizing that caution, it is also necessary to recognize the benefits to be derived from the use of computational methods in dealing with flow conditions, both steady and transient, previously thought too complex, or at least too time consuming. Examples of quasi-steady and transient simulations will be developed later in Part VI of this text, while routine application of a computational approach will be found throughout the text.

The literature on CFD is now extensive and it would be inappropriate for this text to provide more than an introduction to the numerical methods employed. It must be stressed, however, that the availability of high-speed computing allows the time or distance grids used to become very small and this in turn leads to the application of relatively straightforward numerical methods to solve the governing equations for each case studied. However, the caution mentioned above must be exercised.

Within the rapidly growing application of computational methods three main approaches to CFD may be usefully identified, namely finite difference methods, the finite element method and the finite volume method, each with its own exponents and literature.

The *finite difference method* (FDM) utilizes a time–distance grid of nodes and a truncated Taylor series approach to determine the conditions at any particular node one time-step in the future based on the conditions at adjacent nodes at the current time. A brief coverage of the application of the Taylor series and the nodal grid will illustrate several points fundamental to flow simulation, points that will be considered again later in dealing with unsteady flow simulation.

Figure 5.19 illustrates a nodal grid superimposed on a duct or pipe, where the termination of the duct may be connection to a further duct, connection to some fitting, such as a damper, or energy source, such as a fan or pump, or connection to atmospheric or room conditions. The flow conditions along the duct are known at

FIGURE 5.19

Nodal grid basis for a finite difference method representation of flow conditions

Note: Nodes A, B, C and P are linked by explicit formulation, nodes A, O, B, P, C and Q by an implicit approach.

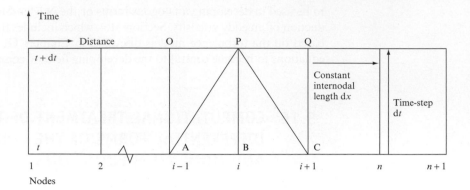

time zero. For full-bore flows these conditions might be zero flow and atmospheric pressure, while for partially filled pipe or channel flow the initial conditions could be a set uniform flow and depth.

At any time, t, the profile of variable y with x may be described by a truncated Taylor series as

$$y(x_0 \pm \Delta x) = y(x_0) \pm y'(x_0)\Delta x + y''(x_0)\frac{\Delta x^2}{2!} \pm y'''(x_0)\frac{\Delta x}{3!} + \cdots, \qquad (5.50)$$

where the value of x at node i is x_0 and the inclusion of the $\pm$ notation allows equation (5.50) to be used in either a forward or backward difference approach. From equation (5.50) it may be seen that the forward or backward first differential based on the first two terms only may be defined as

$$\text{Forward difference first order, } y'(x_0) = \frac{y(x_0 + \Delta x) - y(x_0)}{\Delta x}, \qquad (5.51)$$

$$\text{Backward difference first order, } y'(x_0) = -\frac{y(x_0 - \Delta x) - y(x_0)}{\Delta x}. \qquad (5.52)$$

Summation of the forward/backward application of equation (5.50) also yields a central difference expression for the first derivative of variable y at x_0:

$$\text{Central difference first order, } y'(x_0) = -\frac{y(x_0 + \Delta x) - y(x_0 - \Delta x)}{2\Delta x}. \qquad (5.53)$$

This approach is the equivalent of fitting a three-point parabolic curve through values of variable y at the nodes considered and has second order accuracy.

A more useful result is the summation of the second order forward/backward forms of the truncated Taylor series to determine the second differential of variable y at x_0:

$$y''(x_0) = -\frac{y(x_0 + \Delta x) - 2y(x_0) + y(x_0 - \Delta x)}{\Delta x^2}. \qquad (5.54)$$

Similarly the rate of change of variable y at x_0 with time may be written as

$$\frac{dy}{dt} = \frac{y(x_0, t+\Delta t) - y(x_0, t)}{\Delta t}. \tag{5.55}$$

Taken together these expressions allow conditions at nodes $i = 1$ to $i = n$ to be calculated one time-step in the future based on known conditions along the duct at time zero. Clearly on the next time-step this reduces to $i = 2$ to $i = n - 1$ unless some information is available as to the way in which conditions change at the boundary nodes, $i = 1$ and $i = n + 1$. Figure 5.19 illustrates this limitation and thereby introduces boundary equation considerations that will be returned to in more detail in the discussion of transient flow simulations. Steady boundary conditions such as atmospheric pressure may also be appropriate.

In order to demonstrate two solution techniques consider the solution of a general equation

$$\frac{\partial y}{\partial t} = K\frac{\partial^2 y}{\partial x^2}, \tag{5.56}$$

a common formulation describing, for example, unsteady conduction or contaminant decay.

Equations (5.54) to (5.55) may be reordered to demonstrate the *explicit* formulation

$$y(x_0, t+\Delta t) = y(x_0, t)(1 - 2F) + F[y(x_0 + \Delta x, t) + y(x_0 - \Delta x, t)],$$

$$F = \frac{K\Delta t}{\Delta x^2}, \tag{5.57}$$

where the required value of variable y at one time-step in the future is directly calculated from known current values at the three nodes considered. The value of the term F in equation (5.57) will be returned to later in a discussion of stability.

Alternatively, a time average value for the variable y at each of the three nodes, i.e. at x_0 and $x_0 \pm \Delta x$, would yield a form of equation (5.57)

$$y_{x_0}^{t+\Delta t}(1 + F) = y_{x_0}^t(1 - F) + \frac{F}{2}(y_{x_0+\Delta x}^{t+\Delta t} + y_{x_0+\Delta x}^t + y_{x_0-\Delta x}^{t+\Delta t} + y_{x_0-\Delta x}^t) \tag{5.58}$$

where direct solution for the value of y at x_0 at one time-step into the future is not possible as unknown values of y at nodes on either side are also present in the equation. Such a solution requires the simultaneous solution of all the equations of this form at each node along the duct at each time-step, including the termination nodes with boundary values provided via the appropriate boundary equation. This formulation is known as an *implicit* scheme.

Stability is a concern in finite difference solutions. Referring to the value of F in equation (5.57) common practice is to set this to one-quarter to prevent divergence of the simulation. This is based on an inspection of the equation, which would suggest values below one-half to prevent a change of sign for one of the y terms, and experience; for example, in the solution of the Fourier unsteady heat conduction equation. In addition while computing power has increased and accuracy improved there are still 'rounding errors' associated with each value used in the solution and truncation errors arising from the order of the Taylor series used to develop the simulation. Reducing the time-step or increasing the number of nodes will generally improve the situation but may also lead to other potential hazards, such as numerical dispersion or attenuation of a wave front. Reducing the time-step or increasing the number of nodes is not a panacea, as any advantage can be lost to increasingly important rounding errors. In the transient simulations discussed in Part VI, stability is determined by conformance to the Courant criterion, which links time-step and internodal distance to the local fluid velocity and wave propagation speed. Specialist CFD texts cover these areas in far more detail than would be appropriate here; however, for one group of partial differential equations, i.e. hyperbolic, it is possible to derive the Courant stability criterion.

Consider the reduced form of the continuity and momentum equations in one dimension, equations (5.36) and (5.47) for a frictionless horizontal pipeline where the convective terms $v_x \partial v_x / \partial x$ and $v_x \partial p / \partial x$ have been ignored as insignificant and the fluid bulk modulus K has been introduced from equation (1.12) and wave speed from equation (5.30):

$$\frac{\partial p}{\partial t} + \rho c^2 \frac{\partial v_x}{\partial x} = 0, \tag{5.59}$$

$$\frac{\partial p}{\partial x} + \rho \frac{\partial v_x}{\partial t} = 0, \tag{5.60}$$

Combining these equations involves introducing a multiplicative factor, F, so that

$$\left(\frac{\partial p}{\partial t} + F \frac{\partial p}{\partial x} \right) + F\rho \left(\frac{\partial v_x}{\partial t} + \frac{c^2}{F} \frac{\partial v_x}{\partial x} \right) = 0, \tag{5.61}$$

and, providing that $F = \mathrm{d}x/\mathrm{d}t = \pm c$, reduces to the total differential equation

$$\frac{\mathrm{d}p}{\mathrm{d}t} \pm \rho c \frac{\mathrm{d}v_x}{\mathrm{d}t} = 0, \tag{5.62}$$

capable of representation by a pair of finite difference equations, known as characteristics, linking conditions at the upstream and downstream nodes, $i + 1$ and $i - 1$, at time t to conditions at node i at time $t + \Delta t$, Fig. 5.19.

Thus in this case the stability question has a defined solution in that the relationship between time-step and internodal length is fixed.

Equation (5.62) may be seen to yield the Joukowsky pressure change when the velocity of flow in the x direction is changed:

$$p_i^{t+\Delta t} - p_i^t = \pm \rho c (v_i^{t+\Delta t} - v_i^t), \tag{5.63}$$

where the sign refers to the direction of wave propagation relative to the flow, i.e. a reduction in flow velocity generates a positive upstream surge but a negative downstream pressure change.

The form of the finite difference equations above also allows the wave reflection coefficients for constant pressure zones and dead ends to be defined, where it is assumed that the arriving wave results from a remote instantaneous stoppage of flow in a frictionless pipeline. A finite difference approach involving conditions at the boundary across one time-step and conditions upstream at the start of the time-step is sufficient. This approach will also be demonstrated for junctions or changes in pipeline properties.

EXAMPLE 5.11

Consider a simple horizontal frictionless pipeline where a remotely generated instantaneous pressure front approaches (a) a dead end where the fluid velocity will always be zero, (b) a constant pressure reservoir or some other constant pressure zone, such as a developed vapour cavity.

In each case determine the reflection coefficient applicable to the incoming pressure wave.

Solution

(a) Let the dead end be at node i, then from equation (5.61) with $F = +c$,

$$p_i^{t+\Delta t} - p_{i-1}^t + \rho c(v_i^{t+\Delta t} - v_{i-1}^t) = 0.$$

Adding and subtracting p_i^t yields

$$(p_i^{t+\Delta t} - p_i^t) - (p_{i-1}^t - p_i^t) + \rho c(v_i^{t+\Delta t} - v_{i-1}^t) = 0,$$

but as the velocity at the dead end is constantly zero

$$v_i^{t+\Delta t} = v_i^t = 0, \quad \rho c(v_i^{t+\Delta t} - v_{i-1}^t) = -(p_{i-1}^t - p_i^t),$$

hence

$$(p_i^{t+\Delta t} - p_i^t) = 2(p_{i-1}^t - p_i^t)$$

which for a frictionless pipeline yields a reflection coefficient of $+1$.

(b) Let the constant pressure zone be at node 1, then from equation (5.61) with $F = -c$,

$$p_1^{t+\Delta t} - p_2^t - \rho c(v_1^{t+\Delta t} - v_2^t) = 0.$$

Adding and subtracting p_1^t yields

$$(p_1^{t+\Delta t} - p_1^t) - (p_2^t - p_1^t) - \rho c(v_1^{t+\Delta t} - v_2^t) = 0,$$

but as the pressure at node 1 is constant

$$-\rho c(v_2^t - v_1^t) = \rho c(v_1^{t+\Delta t} - v_2^t),$$

and

$$(p_1^{t+\Delta t} - p_1^t) = -(p_2^t - p_1^t),$$

which for a frictionless pipeline yields a reflection coefficient of -1.

Boundary conditions: in simulations of networks, or in the heat transfer case referred to above involving multiple layers, the prime consideration is to ensure that the time-step utilized for the whole network remains constant. This allows boundary equations to be applied linking conditions in the constituent pipes or ducts. The introduction of the F term in equations (5.57) or (5.61) allows this to be achieved. Consider a network of ducts in which the wave speed, c, e.g. in equation (5.61) differs, either due to pipe cross-section or material properties. To retain a constant F while both c and Δt remain constant requires that Δx be varied between pipes, leading to possible computational difficulties as this results in an inordinate number of nodes in a particular pipe. However, this technique ensures a constant time-step and an orderly solution at the boundaries of each pipe length.

Boundary conditions range from the constant pressure or zero velocity cases already covered to continuity of mass flow or known variation in valve loss coefficients with time. Other boundary conditions may depend upon the pressure–time history at the boundary, for example, the onset of cavitation dependent on pressure level introduces a constant pressure vapour pocket boundary where at higher pressures the boundary might have been zero velocity at a dead end pipe termination.

For frictionless pipelines the reflection/transmission coefficient approach introduced above may be extended provided the appropriate boundary conditions of mass and pressure continuity are introduced, together with a constant F term, implying variable Δx and c between the joining pipes.

EXAMPLE 5.12

For a multiple pipe junction in a frictionless horizontal network subject to the arrival of a pressure surge due to some remote instantaneous flow stoppage, determine the junction reflection and transmission coefficients.

Solution

Assume that the surge arrives along pipe 1, that the positive flow direction is from the junction along pipe 1 and that the wave is transmitted into the remaining $n - 1$ pipes, where the positive flow direction was towards the junction. A reflection will be present in pipe 1.

At the junction the boundary conditions are that

$$A_1 v_{1,1}^{t+\Delta t} = \sum_{i=2}^{n} A_i v_{i,m}^{t+\Delta t},$$

where m is the final node of pipes $2 \to n$, and

$$p_{1,1}^{t+\Delta t} = (p_{i,m}^{t+\Delta t})_{(i=2 \to n)}.$$

For pipe 1, equation (5.62) may be written as

$$A_1 v_{1,1}^{t+\Delta t} = A_1 \left(v_{1,2}^t - \frac{1}{\rho c_1} p_{1,2}^t + \frac{1}{\rho c_1} p_{1,1}^{t+\Delta t} \right),$$

and for pipes 2 to n, equation (5.62) may be written as

$$A_i v_{i,m}^{t+\Delta t} = A_i \left(v_{i,m-1}^t + \frac{1}{\rho c_i} p_{i,m-1}^t - \frac{1}{\rho c_i} p_{i,m}^{t+\Delta t} \right).$$

From the continuity of flow boundary equation it follows that

$$A_1 v_{1,2}^t + \frac{A_1}{\rho c_1}(p_{1,1}^{t+\Delta t} - p_{1,2}^t) = -(p_{1,m}^{t+\Delta t} - p_{i,m-1}^t)\sum_{i=2}^{n} \frac{A_i}{\rho c_i} + \sum_{i=2}^{n} A_i v_{i,m-1}^t.$$

Adding and subtracting $A_1 v_{1,1}^t$ and noting that the continuity equation applies at time t and that for a steady frictionless flow state all nodes in one pipe display the same velocity, yields

$$-A_1(v_{1,1}^t - v_{1,2}^t) + \frac{A_1}{\rho c_1}(p_{1,1}^{t+\Delta t} - p_{1,2}^t) + (p_{i,m}^{t+\Delta t} - p_{1,1}^t)\sum_{i=2}^{n} \frac{A_i}{\rho c_i} = 0.$$

Noting that for a frictionless pipe system $p_{i,m-1}^t = p_{1,1}^t$, and adding and subtracting $p_{1,1}^t$ yields

$$(p_{1,1}^t - p_{1,2}^t)\left(\frac{2A_1}{\rho c_1}\right) + (p_{1,1}^{t+\Delta t} - p_{1,1}^t) + \left(\frac{A_1}{\rho c_1} + \sum_{i=2}^{n} \frac{A_i}{\rho c_i}\right) = 0,$$

and implies transmission and reflection coefficients of

$$C_T = \left(\frac{A_1}{c_1} - \sum_{i=2}^{n} \frac{A_i}{c_i}\right) \bigg/ \left(\sum_{i=1}^{n} \frac{A_i}{c_i}\right), \quad C_R = \left(\frac{2A_1}{c_1}\right) \bigg/ \left(\sum_{i=1}^{n} \frac{A_i}{c_i}\right).$$

5.17 COMPARISON OF CFD METHODOLOGIES

The *finite element method* (FEM) was initially developed for structural analysis but has been utilized for fluid flow predictions as it offers the advantage of a non-regular grid. This allows FEM simulations to address complex boundary geometries. FEM also has an advantage in that the base equations describing flow conditions within each 'cell' have a higher degree of accuracy than those used in FDM; however, the methodologies used are more complex than FDM, where as shown above, relatively easily understood techniques are applied.

The *finite volume method* draws together the best attributes of FDM and FEM in that it is capable of simulating complex boundary geometries and accurately modelling conservation for each cell while at the same time utilizing relatively straightforward finite difference relationships to represent the governing differential equations.

The physics of almost every fluid flow and heat transfer phenomenon is governed by three fundamental principles, namely mass conservation, momentum conservation (or Newton's second law) and energy conservation taken together with appropriate initial or boundary conditions. These three principles and conditions may be expressed mathematically, in most cases through integral or partial differential equations, whose close-form analytical solutions rarely exist.

The ability to seek the numerical solutions of these governing equations under a given set of boundary and initial conditions has led to the development of computational

FIGURE 5.20
The ability to model rotating components makes it possible to represent flow machines such as fans, pumps and compressors (CFD result courtesy of FLUENT Inc)

FIGURE 5.21
Modelling flow pathlines illustrates the vulnerability of a building ventilation intake to pollutants from a stack exhaust. Air flow around buildings is used to study pollutant transport from stacks or vents for a range of assumed wind conditions (CFD result courtesy of FLUENT Inc)

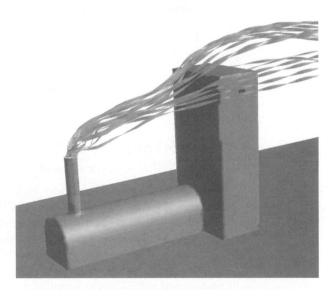

fluid dynamics (CFD), a new discipline in fluid dynamics that had to await the availability of computing power. Along with the traditional approaches of experimental and analytical fluid science, CFD is now widely used within a wide range of engineering applications, from fan designers concerned with detailed internal flow predictions to the description of air flow patterns around proposed building complexes. CFD is applied by all the fluid mechanics related disciplines from aeronautical/aerospace engineering where fluid dynamics is crucial to the industrial applications such as HVAC (heat, ventilation and air conditioning) design for buildings. Examples of the wide diversity of successful applications of CFD codes and packages are illustrated in Figs 5.20–5.24, ranging from the flow internal to a centrifugal fan, through the pollution spread from a process chimney to air flows within a building envelope and around a jet fighter.

FIGURE 5.22

Internal flow and
temperature modelling
used to investigate natural
ventilation in the Building
Research Establishment's
Environmental Building
(CFD application,
reproduced by permission
of BRE Ltd)

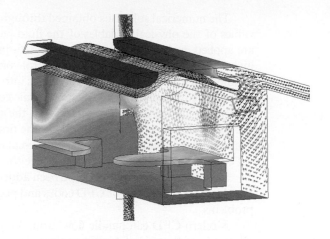

FIGURE 5.23

Air flow around a jet
fighter design illustrating
the application of CFD
to areas previously the
province of wind
tunnel testing (STAR-CD
from Computational
Dynamics Ltd)

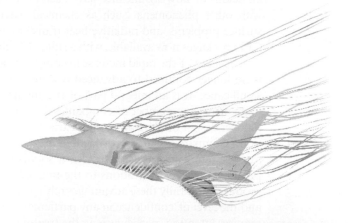

FIGURE 5.24

3 D Euler solution on an
unstructured mesh for
an Airbus research
configuration (courtesy
of Airbus UK Ltd,
Filton, Bristol)

The numerical solutions obtained through CFD for a flow problem represent the values of the physical variables of the fluid field. To achieve this, various techniques are applied, including manipulating the defining equations, dividing the fluid domain into a large number of small cells or control volumes (also called a mesh or grid), transforming the integrals or partial derivatives into discretized algebraic forms (so that the governing equations become linear algebraic equations in a discretized flow field) and, finally, solving the algebraic equations at the grid points. All these numerical algorithms were developed in the early 1970s, the first commercial CFD packages emerging some ten years later as computing power escalated. The exciting CFD history is briefly, yet colourfully, reviewed by Anderson, see the Further Reading section.

Since then new techniques have been added and new models have been developed to enhance the capability of CFD codes and packages to simulate more accurately 'real' problems.

Modern CFD can handle flow around a geometry of great complexity in which all details of flow significance have been faithfully represented. Fluid flow associated with other phenomena, such as chemical reactions, turbulence, multi-phase or free surface problems, and radiative heat transfer can all be simulated by the commercial CFD packages now available, with a suite of built-in models describing these processes.

Because of the rapid increase in computer power, memory and affordability, CFD is no longer confined to advanced research projects of great commercial or defence significance. It is an integral part of the engineering design and analysis environment in an increasing number of companies due to its ability to predict the performance of a novel design or simulate an industrial process before manufacture or implementation. Compared with experimental testing, computer modelling offers potential savings, removing the necessity for a sophisticated physical model and offering the possibility of fewer iterations to the final design with fewer expensive prototypes to produce. Naturally these advantages rely entirely on the validation of the CFD models and the level of confidence in any particular representation chosen – an area that still requires attention, particularly in the treatment of turbulence.

5.17.1 Structure of a CFD code

A CFD code has three basic components: pre-processor, solver and post-processor. The solver is the heart of the code, carrying out the major computations and providing the numerical solutions. The pre-processor and post-processor are the front and end of the code, providing the user/machine interface that allows a CFD operator to communicate with the solver: inputting data to define the problem to be simulated and commanding the solver to use certain models and schemes to carry out the simulation and, finally, presenting for study the computed results. Apart from these key elements, a commercial package aimed at multi-purpose modelling will have a suite of models for various flow problems, such as various turbulence models to cover a range of turbulence conditions and assumptions. The packages will also have a library of material properties for defining the fluid media and solid boundaries in the computational domain. Experience will guide the user in the choice of appropriate model and boundary condition.

The *pre-processor* has a number of functions that allow users to define a fluid domain, known as the *computational domain*, and build up the physical geometry of the zone considered, creating a mesh/grid system throughout the domain and tuning this mesh to improve computation quality, including accuracy and speed. The user may also specify the properties of the fluid and other materials in contact with the fluid at this stage.

The pre-processor allows the user to define the fluid flow phenomena to be modelled and choose from the appropriate physical or chemical models provided by the package and to select the numerical parameters and initial and boundary conditions for the computation. It is clearly apparent that experience will dictate success, as an inappropriate choice of boundary conditions or numerical parameters may lead to wholly erroneous results or computational instability.

All of these activities are crucial in ensuring a quality CFD modelling task. For example, the mesh/grid refinement in this stage affects directly the accuracy of the solution and the cost of computation, in terms of computation time and hardware requirement. An optimized mesh/grid system uses less computer memory space, requires less computing and, yet, gives satisfactory accuracy. The same applies to defining the computational domain – these two activities constitute more than half of all time spent in a CFD modelling exercise.

The *solver* is the heart of a CFD code, although it is very often treated as a 'black box' by many CFD operators. It is a collection of various algorithms and numerical techniques that perform the major computation tasks described above. It consists essentially of two components that provide a discretization for the defining equations and subsequent solution. The first component uses a discretizing scheme to express the governing equations in a discretized form for all the mesh/grid elements over the whole computational domain and discretizes the boundary equations appropriate at the boundary elements. In summary, this section converts the partial differential equations and boundary condition formulae into a group of algebraic equations. The second part uses an iterative procedure to find solutions that satisfy these boundary conditions for the algebraic equations defining the flow domain.

A solver in any CFD code is based on one of the available discretization methods. Currently there are three major methods, namely finite difference (FD), finite volume (FV) and finite element (FE). The finite element method was developed originally for structural stress analysis and is much more widely used in that area than in fluid dynamics. Over 90 per cent of CFD codes are based on either the finite difference or finite volume methods. The latter was developed as a special formulation of the former. As the finite volume method has been very well established and thoroughly validated, it is applied in most commercial CFD packages used worldwide, such as FLUENT, CFX, START-CD and FLOW-3D.

The distinguishing feature of the FV method is that it integrates the governing equations in the finite volumes (known as control volumes) over the whole computational domain. Hence there is one generic form of equation for one flow variable, ϕ, which could be a velocity component, enthalpy or species concentration:

$$\frac{\partial \rho \phi}{\partial t} = -\nabla(\rho V \phi) + \nabla(\Gamma_\phi \nabla \phi) + S_\phi, \tag{5.64}$$

where: ϕ = dependent variable
   Γ_ϕ = exchange coefficient (laminar + turbulent)
   V = flow velocity vector
   ρ = flow density
   S_ϕ = source of the variable.

This integrated expression has a clear physical meaning, namely that the rate of change of ϕ in the control volume with respect of time is equal to the sum of the net flux of ϕ due to convection into the control volume, the net flux of ϕ due to diffusion into the control volume and the net generation rate of ϕ inside the control volume.

Following integration the solver uses various approximations, which are based on an application of the finite difference method, to replace all the terms in the integrated equation, namely the time–change rate, the convective and diffusive fluxes and the source term. The process converts the equation into a set of algebraic equations, which are ready to be solved by an iterative method. The iteration is actually the third and last action the solver does in the whole computational task.

The output of the solver is a set of flow variable values at the mesh/grid nodes.

The *post-processor* allows the CFD operator to construct a picture of the simulated flow problem by displaying the geometry of the problem, the computational domain and the mesh/grid system. The post-processor may then display contour or iso-surface plots for the flow variables; including contours plotted over specified surfaces, such as on a solid/fluid interface or a iso-surface of a second flow variable. In many cases velocity vector plotting is important, as is streamline presentation or particle tracking. In some cases animation of the fluid flow or a flow process may be appropriate and, finally, it may be desirable to provide hardcopy printouts.

A complete CFD simulation often requires repeating the procedure a number of times. It is not rare to run a large number of trials before reaching a set of reasonable solutions. The procedure includes tuning the mesh system, adjusting boundary conditions, selecting numerical parameters, finding the right physical model, monitoring iteration and, finally, viewing the results. This again highlights the need for experience within the application of CFD code.

To support CFD predictions, model validation is the key issue in carrying out a CFD exercise. Considerable effort is required to ensure that the computer model developed is robust and that modelling quality is ensured. Validation normally has two parts: first, a mathematical/numerical verification, including convergence, grid-independence and stability test, and, second, physical validation through comparing predicted results with experimental data; this is the ultimate measure of model validation. However, very often it is hard to provide a physical validation for some obvious reasons: time, cost, or safety prohibitions to carrying out the experiment. This is often true in industrial applications, where the validation is limited to qualitative comparison with existing flow cases. The CFD operator's experience of various flow problems becomes important to ensuring a quick computer prediction with reasonable accuracy.

5.17.2 CFD model considerations

In addition to model validation, there are some other issues that need to be considered during a CFD exercise, some of these having already been mentioned. For example:

Explicit vs. implicit. These are two schemes of using finite difference type approximations to convert the governing partial differential equations into an algebraic format. Implicit methods allow arbitrarily large time-step sizes to be used in calculations so that the CPU time required can be reduced. Hence these methods are rather popular in many CFD codes. However, they require iterative solution methods that depend on the character of an under-relaxation in each iteration. This feature may introduce significant errors or very slow convergence in some circumstances, such as control volumes with large aspect ratios. In addition, the implicit methods are not accurate for convective processes. Explicit numerical methods, on the other hand, require less computational effort although there is a restriction on selection of time-step.

Their numerical stability requirements are equivalent to accuracy requirements. More detailed discussion on this account can be found in Anderson (1995).

Implicit methods gain their time-step independence by introducing diffusive effects into the approximating equations. The addition of numerical diffusion to physical diffusion, e.g. to heat conduction, may not cause a serious problem as it only modifies the diffusion rate. However, adding numerical diffusion to convective processes completely changes the character of the physical phenomena being modelled. In FLOW-3DX time-steps are automatically controlled by the program to insure time-accurate approximations.

Body-fitted coordinate vs. cartesian coordinate. In a cartesian coordinate system complex fluid/solid interfaces require very fine grid/mesh definition to reduce errors introduced by stepping the interface surface to approximate the actual surface profile. This also results in the necessity for non-uniform grid sizes. This results in extra computational time, storage and difficulties if it is necessary to transform the predictions for this grid into a regular or uniform one. In cases where the stepped surface approximation is not considered satisfactory or acceptable, a new mesh definition approach is required, such as body-fitted coordinates and unstructured coordinates as used in finite element analysis, allowing fluid phenomena such as heat transfer and shear stress to be satisfactorily modelled. These approaches do, however, require more computation than the regular grid technique and while this may not be a long-term problem in view of advances in computational power, elegant solutions require a minimization of such costs and effort. This issue is addressed in the literature, for example, Anderson (1995).

Relaxation and convergence criteria. Implicit schemes also need to select one or more numerical parameters to control convergence and relaxation. It is crucial to make a wise choice, as poor ones often lead to either divergence or slow convergence. To reduce the chances of making a poor selection some commercial CFD codes have reduced the range of choice or even developed devices to pre-select the parameters automatically.

Fluid/solid interface. Flow becomes more complex at the fluid/solid interface, as the variables are more likely to experience radical changes due to the presence of heat transfer, turbulence and wall shear stress at the surface and within the boundary layer. The size of the control volume can significantly affect the accuracy of the calculation, particularly in some CFD packages where a heat exchange coefficient is calculated locally. In addition, the mesh size also determines the estimation interface area so that total heat transfer prediction through the surface is influenced. Hence finer mesh/grid choices at the interface are reflected in the cost of increasing computation time.

Selecting a 'right physical model' and the 'right boundary conditions'. Commercial CFD codes always carry a collection of physical models and a suite of boundary conditions to make the codes user-friendly. It is important to select the right one that requires less computation and yet reveals enough details of any real, complex, flow phenomenon, e.g. there are many approximation equations developed to represent complex flow turbulence, and many new models are being formulated. A model that best suits one problem may be totally inadequate for another. Very often new equations are formulated to represent a complex flow phenomenon, the k–ε turbulence model is a typical example. Boundary conditions can be implemented differently from one code to another. Test runs and user manuals are essential to help find out the most suitable boundary condition for a particular problem.

The description above is clearly only the tip of the CFD iceberg. However, the issues raised may be seen to be echoed in all the computational examples included in this text, and particularly in the unsteady flow simulations included in Chapter 21 where the importance of transforming the governing equations into an algebraic format is emphasized, as is the stability conditions determining time-step size and the central importance of boundary condition selection. While commercial packages often major in the output data presentation phase, these constraints are common and have to be addressed if a realistic prediction of the flow condition is to be realized. The issue of validation is still not wholly resolved in many applications and hence user experience, grounded in a full understanding of the basics of fluid mechanics, becomes an imperative. In many cases, CFD appears more like art than science, as formulating mathematical equations, selecting numerical parameters and many other decisions made in an exercise are more likely to be based on experience or intuition, rather than scientific deduction. More practice makes a better CFD operator.

Concluding remarks

The application of the momentum equation to fluid flow situations provides the second of the fundamental tools to be used in understanding fluid flow. Its application to the derivation of Euler's equation has been demonstrated and the simplification of this equation for steady uniform inviscid flow into Bernoulli's equation linking flow pressure, velocity and elevation will be utilized continuously.

While the use of the momentum equation to determine the forces acting between a flow and its boundaries is important, including the application to turbines and the calculation of engine thrust, the momentum equation has a much wider range of application. Its use in determining the rate of propagation of pressure and surface wave discontinuities has been shown, and later in the text it will be used to calculate drag forces and to analyze unsteady flow situations that can lead to destructive situations.

The development of the Navier–Stokes equations has been included in this chapter and has led to the introduction of finite difference numerical solution techniques that will be of application later in the text. A brief description of the available CFD methodologies has also been included, although it is stressed that this subject area now has an extensive literature and application outwith the scope of this text, as represented by the Further Reading list presented here. However, the importance of a fundamental understanding of fluid flow mechanisms prior to embarking upon the use of CFD packages is reinforced by the discussion provided, as is the underlying commonality to the flow simulations presented later in the current text. Chapter 5 should be seen as a resource upon which later analysis and discussion will be based.

Summary of important equations and concepts

1. The statements of the momentum equation found in Section 5.2 and its application to a range of flow to structure interface applications in Sections 5.5 to 5.8 demonstrate the central importance of this chapter.

2. The use of the momentum equation to determine thrust, drag and torque, equations (5.11), (5.16) and (5.17) are all applications deriving directly from the momentum equation.

3. Euler's equation is developed in Section 5.12 as a form of the momentum equation and may be seen under certain flow conditions to be analogous to

Bernoulli's, which may also be derived by reference to the conservation of energy laws, Chapter 6. Equation (5.21) illustrates the importance of defining the flow as compressible or incompressible as in the latter case integration is possible, leading to a form of Bernoulli's equation.

4. The propagation of information through a fluid system at the appropriate wave or acoustic velocity depends upon the use of the momentum equation to define wave speed. Equations (5.31) and (5.35) demonstrate these relationships for compressible and free surface flows.

5. The Navier–Stokes equations derive from the momentum considerations introduced in this chapter. Modern computing, allied to finite difference schemes, allows numerical simulation and solution of a wide range of flow conditions. Sections 5.15 to 5.17 introduce these methodologies and reinforce the importance of user experience in defining the appropriate CFD model for any fluid flow phenomenon. The finite difference methods will be returned to in Chapters 18 to 21.

Further reading

Abbott, M. B. and Basco, D. R. (1989). *Computational Fluid Dynamics – An Introduction for Engineers.* Longman, Harlow.

Anderson Jr, J. D. (1995). *Computational Fluid Dynamics.* McGraw-Hill, New York.

Smith, G. D. (1985). *Numerical Solution of Partial Differential Equations: Finite Difference Methods*, 3rd edition. Clarendon Press, Oxford.

Versteeg, H. K. and Malalasekera, W. (1995). *An Introduction to Computational Fluid Dynamics, the Finite Volume Method.* Longman, Harlow.

Zeinkiewicz, O. C. and Taylor, R. L. (1991). *The Finite Element Method. Volume 2: Solid and Fluid Mechanics.* McGraw-Hill, New York.

Problems

5.1 Oil flows through a pipeline 0.4 m in diameter. The flow is laminar and the velocity at any radius r is given by $u = (0.6 - 15r^2)$ m s^{-1}. Calculate (a) the volume rate of flow, (b) the mean velocity, (c) the momentum correction factor.

[(a) 0.0377 m^3 s^{-1}, (b) 0.30 m s^{-1}, (c) 1.333]

5.2 A liquid flows through a circular pipe 0.6 m in diameter. Measurements of velocity taken at intervals along a diameter are:

Distance from wall m	0	0.05	0.1	0.2	0.3
Velocity m s^{-1}	0	2.0	3.8	4.6	5.0

Distance from wall m	0.4	0.5	0.55	0.6
Velocity m s^{-1}	4.5	3.7	1.6	0

(a) Draw the velocity profile, (b) calculate the mean velocity, (c) calculate the momentum correction factor.

[(b) 2.82 m s^{-1}, (c) 1.33]

5.3 Calculate the mean velocity and the momentum correction factor for a velocity distribution in a circular pipe given by $(v/v_0) = (y/R)^{1/n}$, where v is the velocity at a distance y from the wall of the pipe, v_0 is the centreline velocity, R the radius of the pipe and n an unspecified power.

$$\left[\frac{2v_0 n^2}{(n+1)(2n+1)}, \; \frac{(n+1)(2n+1)^2}{4n^2(n+2)} \right]$$

5.4 A pipeline is 120 m long and 250 mm in diameter. At the outlet there is a nozzle 25 mm in diameter controlled by a shut-off valve. When the valve is fully open water issues as a jet with a velocity of 30 m s^{-1}. Calculate the reaction of the jet.

If the valve can be closed in 0.2 s what will be the resulting rise in pressure at the valve required to bring the water in the pipe to rest in this time? Assume no change in density of the water and no expansion of the pipe.

[437.5 N, 180 kN m^{-2}]

5.5 A uniform pipe 75 m long containing water is fitted with a plunger. The water is initially at rest. If the plunger is forced into the pipe in such a way that the water is accelerated uniformly to a velocity of 1.7 m s^{-1} in 1.4 s what will be the increase of pressure on the face of the plunger assuming that the water and the pipe are not elastic?

If instead of being uniformly accelerated the plunger is driven by a crank 0.25 m long and making 120 rev min^{-1} so that the plunger moves with simple harmonic motion, what would be the maximum pressure on the face of the piston?

[91 kN m^{-2}, 2962.5 kN m^{-2}]

5.6 A flat plate is struck normally by a jet of water 50 mm in diameter with a velocity of 18 m s^{-1}. Calculate (a) the force on the plate when it is stationary, (b) the force on the plate when it moves in the same direction as the jet with a velocity of 6 m s^{-1}, (c) the work done per second and the efficiency in case (b).

[(a) 636.2 N, (b) 282.7 N, (c) 1696.2 W, 29.6 per cent]

5.7 A jet of water 50 mm in diameter with a velocity of 18 m s^{-1} strikes a flat plate inclined at an angle of 25° to the axis of the jet. Determine the normal force exerted on the plate (a) when the plate is stationary, (b) when the plate is moving at 4.5 m s^{-1} in the direction of the jet, and (c) determine the work done and the efficiency for case (b).

[(a) 269 N, (b) 151.2 N, (c) 287.55 W, 5 per cent]

5.8 A jet of water delivers 85 dm^3 s^{-1} at 36 m s^{-1} onto a series of vanes moving in the same direction as the jet at 18 m s^{-1}. If stationary, the water which enters tangentially would be diverted through an angle of 135°. Friction reduces the relative velocity at exit from the vanes to 0.80 of that at entrance. Determine the magnitude of the resultant force on the vanes and the efficiency of the arrangement. Assume no shock at entry.   [2546 N, 0.783]

5.9 A 5 cm diameter jet delivering 56 litres of water per second impinges without shock on a series of vanes moving at 12 m s^{-1} in the same direction as the jet. The vanes are curved so that they would, if stationary, deflect the jet through an angle of 135°. Fluid resistance has the effect of reducing the relative velocity by 10 per cent as the water traverses the vanes. Determine (a) the magnitude and direction of the resultant force on the vanes, (b) the work done per second by the vanes and (c) the efficiency of the arrangement.

[(a) 1632 N at 21°15′, (b) 18.25 kW, (c) 79.7 per cent]

5.10 Figure 5.25 shows a cross-section of the end of a circular duct through which air (density 1.2 kg m^{-3}) is discharged to atmosphere through a circumferential slot, the exit velocity being 30 m s^{-1}. Find the force exerted on the duct by the air if the gauge pressure at A is 2065 N m^{-2} below the pressure at outlet.   [720.7 N]

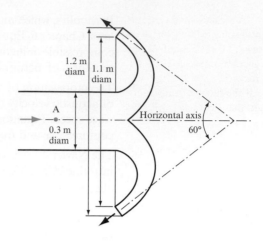

FIGURE 5.25

5.11 Water flows through the pipe bend and nozzle arrangement shown in Fig. 5.26 which lies with its axis in the horizontal plane. The water issues from the nozzle into the atmosphere as a parallel jet with a velocity of 16 m s^{-1} and the pressure at A is 128 kN m^{-2} gauge. Friction may be neglected. Find the moment of the resultant force due to the water on this arrangement about a vertical axis through the point X.   [65.4 N m counterclockwise]

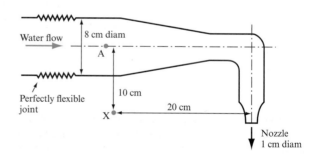

FIGURE 5.26

5.12 A ram-jet engine consumes 20 kg of air per second and 0.6 kg of fuel per second. The exit velocity of the gases is 520 m s^{-1} relative to the engine and the flight velocity is 200 m s^{-1} absolute. What is the power developed?

[1340 kW]

5.13 The resistance of a ship is given by $5.55u^6 + 978u^{1.9}$ N at a speed of u m s^{-1}. It is driven by a jet propulsion system with intakes facing forward, the efficiency of the jet drive being 0.8 and the efficiency of the pumps 0.72. The vessel is to be driven at 3.4 m s^{-1}. Find (a) the mass of water to be pumped astern per second, (b) the power required to drive the pump.   [(a) 10 928 kg s^{-1}, (b) 109.66 kW]

5.14 A rocket is fired vertically starting from rest. Neglecting air resistance, what velocity will it attain in 68 s if its initial mass is 13 000 kg and fuel is burnt at the rate of 124 kg s⁻¹, the gases being ejected at a velocity of 1950 m s⁻¹ relative to the rocket?

If the fuel is exhausted after 68 s what is the maximum height that the rocket will reach? Take $g = 9.8$ m s⁻².

[1372 m s⁻¹, 130.8 km]

5.15 A submarine cruising well below the surface of the sea leaves a wake in the form of a cylinder which is symmetrical about the longitudinal axis of the submarine. The wake velocity on the longitudinal axis is equal to the speed of the submarine through the water, which is 5 m s⁻¹, and decreases in direct proportion to the radius to zero at a radius of 6 m. Calculate the force acting on the submarine and the minimum power required to keep the submarine moving at this speed. Density of sea water = 1025 kg m⁻³.

[483 kN, 2415 kW]

5.16 If $u = ay + by^2$ represents the velocity of air in the boundary layer of a surface, a and b being constants and y the perpendicular distance from the surface, calculate the shear stress acting on the surface when the speed of the air relative to the surface is 75 m s⁻¹ at a distance of 1.5 mm from the surface and 105 m s⁻¹ when 3 mm from the surface. The viscosity of the air is 18×10^{-6} kg m⁻¹ s⁻¹.

[1.17 N m⁻²]

5.17 A lawn sprinkler consists of a horizontal tube with nozzles at each end normal to the tube but inclined upward at 40° to the horizontal. A central bearing incorporates the inlet for the water supply. The nozzles are of 3 mm diameter and are at a distance of 12 cm from the central bearing. If the speed of rotation of the tube is 120 rev min⁻¹ when the velocity of the jets relative to the nozzles is 17 m s⁻¹, calculate (*a*) the absolute velocity of the jets, (*b*) the torque required to overcome the frictional resistance of the tube and bearing.

[(*a*) 18.2 m s⁻¹, (*b*) 0.374 N m s⁻¹]

5.18 Derive an expression for the velocity of transmission of a pressure wave through a fluid of bulk modulus K and mass density ρ. What will be the velocity of sound through water if $K = 2.05 \times 10^9$ N m⁻² and $\rho = 1000$ kg m⁻³?

[1432 m s⁻¹]

5.19 Calculate the velocity of sound in air assuming an adiabatic process if the temperature is 20 °C, $\gamma = 1.41$ and $R = 287$ J kg⁻¹ K⁻¹.

[344.34 m s⁻¹]

5.20 Calculate the velocity of propagation relative to the fluid of a small surface wave along a very wide channel in which the water is 1.6 m deep. If the velocity of the stream is 2 m s⁻¹ what will be the Froude number?

[3.96 m s⁻¹, 0.505]

The Energy Equation and its Applications

WHILE CHAPTER 5 INTRODUCED THE MOMENTUM EQUATION, THE CONSIDERATION OF energy transfers within a flowing fluid is also fundamental to the study and prediction of fluid flow phenomena. This chapter will revisit the development of Bernoulli's equation and demonstrate that it is merely one special form of a more general energy equation that can accommodate apparent energy losses, due to frictional and separation effects, by application of the conservation of energy principle and the concept of changes in the internal energy of the flowing fluid. The transfer of energy into, or out of, a fluid flow system, by the introduction of mechanical devices such as fans, pumps or turbines, will be accommodated within the principle of conservation of energy across a predetermined control volume, leading to the introduction of the general steady flow energy equation. The representation of apparent energy losses due to friction and separation losses will be defined and the application of the energy equation to the measurement of fluid flow rate and fluid flow velocity demonstrated for a range of pipe flow and free surface flow conditions. A computer program designed to provide mass flow at a duct cross-section based on velocity traverse data is included. Finally, vortex flow will be introduced. ● ● ●

6.1   MECHANICAL ENERGY OF A FLOWING FLUID

An element of fluid, as shown in Fig. 6.1, will possess potential energy due to its height z above datum and kinetic energy due to its velocity v, in the same way as any other object. For an element of weight mg,

$$\text{Potential energy} = mgz,$$

$$\text{Potential energy per unit weight} = z, \tag{6.1}$$

$$\text{Kinetic energy} = \tfrac{1}{2}mv^2,$$

$$\text{Kinetic energy per unit weight} = v^2/2g. \tag{6.2}$$

FIGURE 6.1

Energy of a flowing fluid

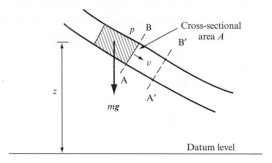

A steadily flowing stream of fluid can also do work because of its pressure. At any given cross-section, the pressure generates a force and, as the fluid flows, this cross-section will move forward and so work will be done. If the pressure at a section AB is p and the area of the cross-section is A,

$$\text{Force exerted on AB} = pA.$$

After a weight mg of fluid has flowed along the streamtube, section AB will have moved to A′B′:

$$\text{Volume passing AB} = mg/\rho g = m/\rho.$$

Therefore,

$$\text{Distance AA}' = m/\rho A,$$

$$\text{Work done} = \text{Force} \times \text{Distance AA}'$$

$$= pA \times m/\rho A,$$

$$\text{Work done per unit weight} = p/\rho g. \tag{6.3}$$

The term $p/\rho g$ is known as the flow work or pressure energy. Note that the term pressure energy refers to the energy of a fluid when *flowing* under pressure as part of a continuously maintained stream. It must not be confused with the energy stored in a fluid due to its elasticity when it is compressed. The concept of pressure energy is sometimes found difficult to understand. In solid-body mechanics, a body is free to change its velocity without restriction and potential energy can be freely converted to kinetic energy as its level falls. The velocity of a stream of fluid which has a steady volume rate of flow depends on the cross-sectional area of the stream. Thus, if the fluid flows, for example, in a uniform pipe and is incompressible, its velocity cannot change and so the conversion of potential energy to kinetic energy cannot take place as the fluid loses elevation. The surplus energy appears in the form of an increase in pressure. As a result, pressure energy can, in a sense, be regarded as potential energy in transit.

Comparing the results obtained in equations (6.1), (6.2) and (6.3) with equation (5.23) it can be seen that the three terms of Bernoulli's equation are the pressure energy per unit weight, the kinetic energy per unit weight and the potential energy per unit weight; the constant H is the total energy per unit weight. Thus, Bernoulli's equation states that, for steady flow of a frictionless fluid along a streamline, the total energy per unit weight remains constant from point to point although its division between the three forms of energy may vary:

$$\begin{matrix} \text{Pressure} & \text{Kinetic} & \text{Potential} \\ \text{energy per} & + \text{ energy per} & + \text{ energy per} & = \frac{\text{Total energy per}}{\text{unit weight}} = \text{constant,} \\ \text{unit weight} & \text{unit weight} & \text{unit weight} \end{matrix}$$

$$p/\rho g \times v^2/2g + z = H. \tag{6.4}$$

Each of these terms has the dimension of a length, or head, and they are often referred to as the pressure head $p/\rho g$, the velocity head $v^2/2g$, the potential head z and the total head H. Between any two points, suffixes 1 and 2, on a streamline, equation (6.4) gives

$$\frac{p_1}{\rho_1 g} + \frac{v_1^2}{2g} + z_1 = \frac{p_2}{\rho_2 g} + \frac{v_2^2}{2g} + z_2 \tag{6.5}$$

or

Total energy per unit weight at 1 = Total energy per unit weight at 2,

which corresponds with equation (5.25).

In formulating equation (6.5), it has been assumed that no energy has been supplied to or taken from the fluid between points 1 and 2. Energy could have been supplied by introducing a pump; equally, energy could have been lost by doing work against friction or in a machine such as a turbine. Bernoulli's equation can be expanded to include these conditions, giving

$$\begin{matrix} \text{Total energy} & \text{Total energy} & \text{Loss per} & \text{Work done} & \text{Energy} \\ \text{per unit} & = \text{per unit} & + \text{ unit} & + \text{ per unit} & - \text{ supplied} \\ \text{weight at 1} & \text{weight at 2} & \text{weight} & \text{weight} & \text{per unit} \\ & & & & \text{weight} \end{matrix}$$

$$\frac{p_1}{\rho_1 g} + \frac{v_1^2}{2g} + z_1 = \frac{p_2}{\rho_2 g} + \frac{v_2^2}{2g} + z_2 + h + w - q. \tag{6.6}$$

EXAMPLE 6.1

A fire engine pump develops a head of 50 m, i.e. it increases the energy per unit weight of the water passing through it by 50 N m N⁻¹. The pump draws water from a sump at A (Fig. 6.2) through a 150 mm diameter pipe in which there is a loss of energy per unit weight due to friction $h_1 = 5u_1^2/2g$ varying with the mean velocity u_1 in the pipe, and discharges it through a 75 mm nozzle at C, 30 m above the pump, at the end of a 100 mm diameter delivery pipe in which there is a loss of energy per unit weight $h_2 = 12u_2^2/2g$. Calculate (a) the velocity of the jet issuing from the nozzle at C and (b) the pressure in the suction pipe at the inlet to the pump at B.

FIGURE 6.2

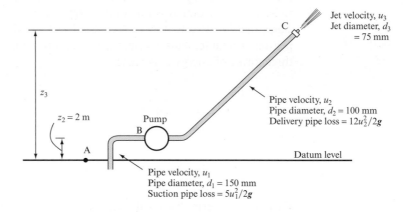

Solution

(a) We can apply Bernouilli's equation in the form of equation (6.6) between two points, one of which will be C, since we wish to determine the jet velocity u_3, and the other a point at which the conditions are known, such as a point A on the free surface of the sump where the pressure will be atmospheric, so that $p_A = 0$, the velocity v_A will be zero if the sump is large, and A can be taken as the datum level so that $z_A = 0$. Then,

$$
\begin{matrix}
\text{Total energy} \\
\text{per unit} \\
\text{weight at A}
\end{matrix}
=
\begin{matrix}
\text{Total energy} \\
\text{per unit} \\
\text{weight at C}
\end{matrix}
+
\begin{matrix}
\text{Loss in} \\
\text{inlet} \\
\text{pipe}
\end{matrix}
-
\begin{matrix}
\text{Energy per} \\
\text{unit weight} \\
\text{supplied by} \\
\text{pump}
\end{matrix}
+
\begin{matrix}
\text{Loss in} \\
\text{discharge} \\
\text{pipe,}
\end{matrix}
\qquad \textbf{(I)}
$$

$$
\begin{matrix}
\text{Total energy} \\
\text{per unit} \\
\text{weight at A}
\end{matrix}
= \frac{p_A}{\rho g} + \frac{v_A^2}{2g} + z_A = 0,
$$

$$
\begin{matrix}
\text{Total energy} \\
\text{per unit} \\
\text{weight at C}
\end{matrix}
= \frac{p_C}{\rho g} + \frac{u_3^2}{2g} + z_3,
$$

$$
p_C = \text{Atmospheric pressure} = 0,
$$

$$
z_3 = 30 + 2 = 32 \text{ m}.
$$

Therefore,

Total energy
per unit $= 0 + u_3^2/2g + 32 = u_3^2/2g + 32$ m.
weight at C

Loss in inlet pipe, $h_1 = 5u_1^2/2g$,

Energy per unit weight supplied by pump $= 50$ m,

Loss in delivery pipe, $h_2 = 12u_2^2/2g$.

Substituting in (I),

$$0 = (u_3^2/2g + 32) + 5u_1^2/2g - 50 + 12u_2^2/2g,$$

$$u_3^2 + 5u_1^2 + 12u_2^2 = 2g \times 18. \tag{II}$$

From the continuity of flow equation,

$$(\pi/4)d_1^2 u_1 = (\pi/4)d_2^2 u_2 = (\pi/4)d_3^2 u_3;$$

therefore,

$$u_1 = \left(\frac{d_3}{d_1}\right)^2 u_3 = \left(\frac{75}{150}\right)^2 u_3 = \frac{1}{4}u_3,$$

$$u_2 = \left(\frac{d_3}{d_2}\right)^2 u_3 = \left(\frac{75}{100}\right)^2 u_3 = \frac{9}{16}u_3.$$

Substituting in equation (II),

$$u_3^2 [1 + 5 \times (\tfrac{1}{4})^2 + 12 \times (\tfrac{9}{16})^2] = 2g \times 18,$$

$$5.109 u_3^2 = 2g \times 18$$

$$u_3 = \textbf{8.314 m s}^{-1}.$$

(b) If p_B is the pressure in the suction pipe at the pump inlet, applying Bernoulli's equation to A and B,

Total energy   Total energy   Loss in
per unit $=$ per unit $+$ inlet
weight at A   weight at B   pipe,

$$0 = (p_B/\rho g + u_1^2/2g + z_2) + 5u_1^2/2g,$$

$$p_B/\rho g = -z_2 - 6u_1^2/2g,$$

$$z_2 = 2 \text{ m}, u_1 = \tfrac{1}{4}u_3 = 8.314/4 = 2.079 \text{ m s}^{-1},$$

$$p_B/\rho g = -(2 + 6 \times 2.079^2/2g) = -(2 + 1.32) = -3.32 \text{ m},$$

$$p_B = -1000 \times 9.81 \times 3.32 = \textbf{32.569 kN m}^{-2} \textbf{ below atmospheric pressure.}$$

6.2   STEADY FLOW ENERGY EQUATION

Bernoulli's equation and its expanded form, as given in equation (6.6), were developed from Euler's equation (5.21) which, in turn, was derived from the momentum equation. It is possible to develop an energy equation for the steady flow of a fluid from the principle of conservation of energy, which states:

For any mass system, the net energy supplied to the system equals the increase of energy of the system plus the energy leaving the system.

Thus, if ΔE is the increase of energy of the system, ΔQ is the energy supplied to the system and ΔW the energy leaving the system, then, considering the energy balance for the system,

$$\Delta E = \Delta Q - \Delta W.$$

The energy of a mass of fluid will have the following forms:

1.   internal energy due to the activity of the molecules of the fluid forming the mass;
2.   kinetic energy due to the velocity of the mass of fluid itself;
3.   potential energy due to the mass of fluid being at a height above the datum level and acted upon by gravity.

Suppose that at section AA (Fig. 6.3) through a streamtube the cross-sectional area is A_1, the pressure p_1, velocity v_1, density ρ_1, internal energy per unit mass e_1 and height above datum z_1, while the corresponding values at BB are A_2, p_2, v_2, ρ_2, e_2 and z_2. The fluid flows steadily with a mass flow rate $\dot{m}$ and between sections AA and BB the fluid receives energy at the rate of q per unit mass and loses energy at the rate of w per unit mass. For example, q may be in the form of heat energy, while w might take the form of mechanical work.

$$\begin{array}{l}\text{Energy entering} \\ \text{at AA in unit} \\ \text{time, } E_1\end{array} = \underset{\text{energy}}{\text{Kinetic}} + \underset{\text{energy}}{\text{Potential}} + \underset{\text{energy}}{\text{Internal}} = \dot{m}(\tfrac{1}{2}v_1^2 + gz_1 + e_1),$$

$$\begin{array}{l}\text{Energy leaving} \\ \text{at BB in unit} \\ \text{time, } E_2\end{array} = \dot{m}(\tfrac{1}{2}v_2^2 + gz_2 + e_2).$$

FIGURE 6.3

Steady flow energy equation

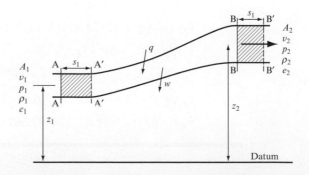

Therefore

$$\Delta E = E_2 - E_1 = \dot{m}[\tfrac{1}{2}(v_2^2 - v_1^2) + g(z_2 - z_1) + (e_2 - e_1)]. \tag{6.7}$$

This change of energy has occurred because energy has entered and left the fluid between AA and BB. Also, work is done on the fluid in the control volume between the two sections AA and BB by the fluid entering at AA and by the fluid in the control volume as it leaves at BB.

Energy entering per unit time between AA and BB $= \dot{m}q$,

Energy leaving per unit time between AA and BB $= \dot{m}w$.

As the fluid flows, work will be done by the fluid entering at AA since a force p_1A_1 is exerted on the cross-section by the pressure p_1 and, in unit time t, the fluid which was at AA will move a distance s_1 to A'A':

Work done in unit time on the fluid at AA $= p_1A_1s_1/t$.

But,

Mass passing per unit time, $\dot{m} = \rho_1 A_1 s_1/t$;

therefore,

$A_1 s_1 = \dot{m}/\rho_1$,

Work done per unit time on the fluid at AA $= p_1 \dot{m}/\rho_1$.

Similarly,

$$\begin{aligned}
\text{Change of energy of the system, } \Delta E ={}& \text{Work done on fluid at AA} \\
& - \text{Work done by fluid at BB} \\
& + \text{Energy entering between AA and BB} \\
& - \text{Energy leaving between AA and BB} \\
={}& \dot{m}p_1/\rho_1 - \dot{m}p_2/\rho_2 + \dot{m}q - \dot{m}w \\
={}& \dot{m}(q - w + p_1/\rho_1 - p_2/\rho_2). \tag{6.8}
\end{aligned}$$

Comparing equations (6.7) and (6.8),

$$\tfrac{1}{2}(v_2^2 - v_1^2) + g(z_2 - z_1) + (e_2 - e_1) = p_1/\rho_1 - p_2/\rho_2 + q - w.$$

Thus,

$$gz_1 + \tfrac{1}{2}v_1^2 + (p_1/\rho_1 + e_1) + q - w = gz_2 + \tfrac{1}{2}v_2^2 + (p_2/\rho_2 + e_2). \tag{6.9}$$

The terms $(p_1/\rho_1 + e_1)$ and $(p_2/\rho_2 + e_2)$ can be replaced by the enthalpies H_1 and H_2, giving

$$gz_1 + \tfrac{1}{2}v_1^2 + H_1 + q - w = gz_2 + \tfrac{1}{2}v_2^2 + H_2. \tag{6.10}$$

This steady flow energy equation can be applied to all fluids, real or ideal, whether liquids, vapours or gases, provided that flow is continuous and energy is transferred steadily to or from the fluid at constant rates q and w, conditions remaining constant with time and all quantities being constant across the inlet and outlet sections. In thermodynamics, it is usual to distinguish between heat and work and to treat q as the net inflow of heat and w as the net outflow of mechanical work per unit mass.

6.3   KINETIC ENERGY CORRECTION FACTOR

The derivation of Bernoulli's equation and the steady flow energy equation has been carried out for a streamtube assuming a uniform velocity across the inlet and outlet sections. In a real fluid flowing in a pipe or over a solid surface, the velocity will be zero at the solid boundary and will increase as the distance from the boundary increases. The kinetic energy per unit weight of the fluid will increase in a similar manner. If the cross-section of the flow is assumed to be composed of a series of small elements of area δA and the velocity normal to each element is u, the total kinetic energy passing through the whole cross-section can be found by determining the kinetic energy passing through an element in unit time and then summing by integrating over the whole area of the section.

$$\text{Mass passing through element in unit time} = \rho \delta A \times u,$$

$$\text{Kinetic energy per unit time passing through element} = \tfrac{1}{2} \times \text{Mass per unit time} \times (\text{Velocity})^2$$

$$= \tfrac{1}{2} \rho \delta A u^3,$$

$$\text{Total kinetic energy passing in unit time} = \int \tfrac{1}{2} \rho u^3 \, \delta A,$$

$$\text{Total weight passing in unit time} = \int \rho g u \, \delta A.$$

Thus, taking into account the variation of velocity across the stream,

$$\text{True kinetic energy per unit weight} = \frac{1}{2g} \frac{\int \rho u^3 \, \delta A}{\int \rho u \, \delta A}, \tag{6.11}$$

which is not the same as $\bar{u}^2/2g$, where $\bar{u}$ is the mean velocity:

$$\bar{u} = \int (u/A) \, \mathrm{d}A.$$

Thus,

$$\text{True kinetic energy per unit weight} = \alpha \bar{u}^2/2g, \tag{6.12}$$

where α is the kinetic energy correction factor, which has a value dependent on the shape of the cross-section and the velocity distribution. For a circular pipe, assuming Prandtl's one-seventh power law, $u = u_{\max}(y/R)^{1/7}$, for the velocity at a distance y from the wall of a pipe of radius R, the value of $\alpha = 1.058$.

6.4 APPLICATIONS OF THE STEADY FLOW ENERGY EQUATION

A comparison of equations (6.6) and (6.9) is helpful in applying the steady flow energy equation to a wide range of fluid flow conditions. Reference to the control volume AA'BB' in Fig. 6.3 allows the steady flow energy equation to be recast, from equation (6.9) for a constant density, i.e. incompressible flow, as

$$p_1 + \tfrac{1}{2}\rho v^2 + \rho g z_1 + \rho q - \rho w = p_2 + \tfrac{1}{2}\rho v^2 + \rho g z_2 + \rho \Delta e \qquad (6.13)$$

where, as in equation (5.24), the terms represent energy per unit volume. However, it is also clear that, in order to maintain dimensional homogeneity, each term in this representation of the steady flow energy equation has the dimensions of pressure. It will be shown how it is possible to utilize this particular form of the steady flow energy equation with remarkable ease in the definition of a wide range of fluid flow conditions.

The terms ρq and ρw may be identified, for example in Fig. 6.4, as the pressure rise across a pump or fan maintaining flow through a pipe or duct, or the pressure drop across a turbine.

Clearly each of these terms has values dependent upon the particular flow rate passing through the system, identified in this form of the steady flow energy equation by the mean flow velocity at the control volume boundaries AA' or BB'.

The term $\rho \Delta e$ represents an energy 'loss' due to frictional or separation losses between the boundaries of the control volume. As energy 'loss' cannot occur within the control volume it follows that this term represents a transfer of energy from one category to another, in this case into the fluid internal energy as identified earlier. Again the value of $\rho \Delta e$ will depend upon the flow rate in the system and on the fluid and conduit parameters; appropriate expressions defining these energy transfers in

FIGURE 6.4
Energy addition or extraction at rotodynamic machines

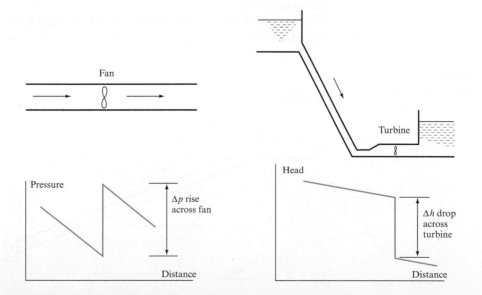

terms of the flow and pipe parameters will be developed later. It is sufficient at this stage to state that the pressure changes associated with these transfers, i.e. the $\rho\Delta e$ term, are dependent upon the square of flow velocity.

Thus the steady flow energy equation may be seen as an energy audit across a user-defined control volume. The appropriate choice of control volume makes the steady flow energy equation an immensely powerful tool in defining a wide range of flow conditions.

6.4.1 Choice of control volume boundary conditions for the steady flow energy equation

Referring to the general definition of the steady flow energy equation in Fig. 6.3, equation (6.13) may be written as

$$p_1 + \tfrac{1}{2}\rho v^2 + \rho g z_1 + \Delta p_{\text{input}} - \Delta p_{\text{out}} = p_2 + \tfrac{1}{2}\rho v^2 + \rho g z_2 + \Delta p_{\text{F+S}},$$

where Δp_{input} and Δp_{out} refer to the pressure rise experienced across a fan, or pump, and the pressure drop across a turbine, respectively. Suffixes 1 and 2 refer to the entry and exit boundary conditions of the control volume. The pressure loss experienced as a result of friction and separation of the flow from the walls of the conduit is encapsulated in the $\Delta p_{\text{F+S}}$ term and will be shown to be defined by a term of the form $\tfrac{1}{2}\rho K u^2$, where u is the local flow velocity and K is a constant dependent upon the conduit parameters, i.e. length, diameter, roughness or fitting type. The steady flow energy equation may thus be written as

$$p_1 + \tfrac{1}{2}\rho v^2 + \rho g z_1 + \Delta p_{\text{input}} - \Delta p_{\text{out}} = p_2 + \tfrac{1}{2}\rho v^2 + \rho g z_2 + \tfrac{1}{2}\rho K u^2,$$

where all terms are defined in the dimensions of pressure and hence are amenable to direct experimental measurement for any particular flow condition. The steady flow energy equation in this form may be easily applied to a range of flow conditions by 'dropping' terms that are irrelevant to the particular case to be studied. A range of common examples are presented below, many of which will be returned to later in the text in more detail.

The steady flow energy equation may be applied across a control volume whose boundaries may be taken as the water surfaces in each reservoir (Fig. 6.5).

FIGURE 6.5

Flow between two reservoirs open to atmosphere

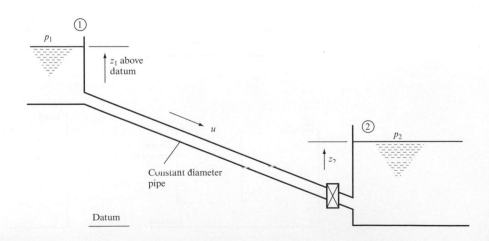

As there are no subtractions or additions of energy due to the presence of turbines or pumps in this system the steady flow energy equation reduces to

$$p_1 + \tfrac{1}{2}\rho v^2 + \rho g z_1 = p_2 + \tfrac{1}{2}\rho v^2 + \rho g z_2 + \tfrac{1}{2}\rho K u^2.$$

Further simplifications may be made by a careful study of the conditions at the system boundaries. In a gauge pressure frame of reference, the values of p_1 and p_2, the atmospheric pressure at the reservoir open surface, may be taken as zero. Further, as the surface areas of the two reservoirs may be assumed to be very large compared with the cross-sectional area of the connecting pipeline, it follows from an application of the continuity of flow equation between the two reservoirs that the values of the reservoir surface velocity, either v_1 vertically down at 1 or v_2 vertically up at 2, may be disregarded when compared with the flow velocity, u, in the actual pipeline. Therefore it is acceptable to neglect the surface kinetic energy terms in comparison with the combined friction and separation loss term, reducing the steady flow energy equation to

$$\rho g(z_1 - z_2) = \tfrac{1}{2}\rho K u^2,$$

i.e. the expected result that the difference in reservoir surface level, or potential energy, is solely responsible for overcoming the frictional and separation losses incurred in a flow between the two reservoirs. Therefore the choice of pipeline, in terms of its diameter, roughness or length, or the setting of any valves along the pipe, determines the throughflow – an expected result that conforms to our knowledge of the physical world.

If the upstream reservoir were to be replaced by a large pressurized tank at a pressure p_t above atmosphere, so that the continuity equation continued to support the dropping of the surface velocity terms, then the form of the steady flow energy equation would become

$$p_t + \rho g(z_1 - z_2) = \tfrac{1}{2}\rho K u^2,$$

and for any pipeline condition the flow delivered would rise compared with the open surface reservoir case – again a result that could be predicted.

An identical process may be seen to apply in the consideration of a simple ventilation system extracting air from a room at atmospheric pressure and discharging it to atmosphere at approximately the same elevation (Fig. 6.6).

If the boundaries of the control volume are positioned sufficiently far from the ductwork entry and exit grilles, then the local air velocity, and hence the associated kinetic energy, may be ignored in comparison with the ductwork air flow velocity and the associated friction and separation loss term. In this special case the pressure terms at the boundaries (points 1 and 2 in Fig. 6.6) may also be ignored as both are atmospheric, and as the fluid is air and the elevation difference across the control volume is stated to be small, the potential energy terms may also be dropped. Hence the steady flow energy equation reduces to the almost trivial

$$\Delta p_{input} = \tfrac{1}{2}\rho K u^2.$$

FIGURE 6.6
Room ventilation

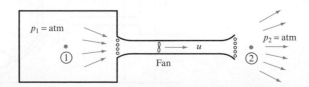

FIGURE 6.7
Pressurized room air supply and extract ventilation

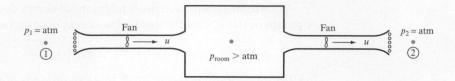

It will be appreciated that this form of the steady flow energy equation may also be utilized to represent the case where it is necessary either to supply or to extract air from a space held above atmospheric pressure. Examples of this would be clean rooms in electronics facilities or hospital operating theatres. In both cases it is required that any air leakage be out of the space (Fig. 6.7).

In the case where the fan is expected to supply air from atmosphere to a room held above atmospheric pressure, the steady flow energy equation becomes

$$\Delta p_{\text{input}} = p_{\text{room}} + \tfrac{1}{2}\rho K u^2,$$

and the extract fan receives support from the pressure gradient existing between the room and the external atmosphere, as illustrated by the appropriate form of the steady flow energy equation,

$$\Delta p_{\text{out}} = -p_{\text{room}} + \tfrac{1}{2}\rho K u^2.$$

6.5   REPRESENTATION OF ENERGY CHANGES IN A FLUID SYSTEM

The changes of energy, and its transformation from one form to another which occurs in a fluid system, can be represented graphically. In a real fluid system, the total energy per unit weight will not remain constant. Unless energy is supplied to the system at some point by means of a pump, it will gradually decrease in the direction of motion due to losses resulting from friction and from the disturbance of flow at changes of pipe section or as a result of changes of direction. In Fig. 6.8, for example, the flow of

FIGURE 6.8
Energy changes in a fluid system

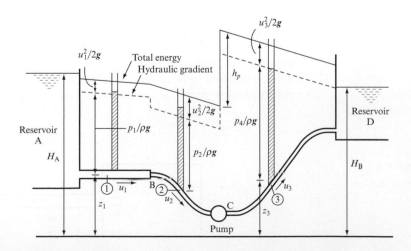

water from the reservoir at A to the reservoir at D is assisted by a pump which develops a head h_p, thus providing an addition to the energy per unit weight of h_p.

At the surface of reservoir A, the fluid has no velocity and is at atmospheric pressure (which is taken as zero gauge pressure), so that the total energy per unit weight is represented by the height H_A of the surface above datum.

As the fluid enters the pipe with velocity u_1, there will be a loss of energy due to disturbance of the flow at the pipe entrance and a continuous loss of energy due to friction as the fluid flows along the pipe, so that the total energy line will slope downwards. At B there is a change of section, with an accompanying loss of energy, resulting in a change of velocity to u_2. The total energy line will continue to slope downwards, but at a greater slope since u_2 is greater than u_1 and friction losses are related to velocity. At C, the pump will put energy into the system and the total energy line will rise by an amount h_p. The total energy line falls again due to friction losses and the loss due to disturbance at the entry to the reservoir, where the total energy per unit weight is represented by the height of the reservoir surface above datum (the velocity of the fluid being zero and the pressure atmospheric).

If a piezometer tube were to be inserted at point 1, the water would not rise to the level of the total energy line, but to a level $u_1^2/2g$ below it, since some of the total energy is in the form of kinetic energy. Thus, at point 1, the potential energy is represented by z_1, the pressure energy by $p_1/\rho g$ and the kinetic energy by $u_1^2/2g$, the three terms together adding up to the total energy per unit weight at that point.

Similarly, at points 2 and 3, the water would rise to levels $p_2/\rho g$ and $p_3/\rho g$ above the pipe, which are $u_2^2/2g$ and $u_3^2/2g$, respectively, below the total energy line. The line joining all the points to which the water would rise, if an open stand pipe was inserted, is known as the *hydraulic gradient*, and runs parallel to the total energy line at a distance below it equal to the velocity head.

If, as in Fig. 6.9, a pipeline rises above the hydraulic gradient, the pressure in the portion PQ will be below atmospheric pressure and will form a *siphon*. Under reduced pressure, air or other gases may be released from solution or a vapour pocket may form and interrupt the flow.

While earlier examples have concentrated on the application of the steady flow energy equation between the extremities of a system, in order to benefit from the resulting simplifications it is clear that the boundaries of the chosen control volume may be placed at any two points of interest along the system conduits. Figure 6.9 illustrates one example where this may be helpful and where a concentration on the extremities of the system may give quite misleading results. Application of the steady state flow energy equation between the open supply reservoir and the apex of the siphon allows the practicality of the siphon to be assessed.

FIGURE 6.9

Pipeline rising above hydraulic gradient

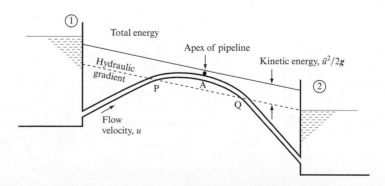

Between the system extremities, i.e. the open reservoir surfaces at 1 and 2, the steady flow energy equation implies that the flow is governed by the expression

$$\rho g(z_1 - z_2) = \tfrac{1}{2}\rho K u^2,$$

regardless of the intermediate elevation of the pipeline and its possible failure to operate under subatmospheric conditions. However, it is possible to apply the steady flow energy equation between the reservoir surface at 1 and the apex of the pipeline at A in order to assess the practicality of the siphon. Thus between 1 and A in Fig. 6.9, the steady flow energy equation becomes

$$p_1 + \tfrac{1}{2}\rho v^2 + \rho g z_1 = p_A + \tfrac{1}{2}\rho v_A + \rho g z_A + \tfrac{1}{2}\rho K u^2,$$

where the friction and separation loss term, $\tfrac{1}{2}\rho K u^2$, refers to the losses from 1 to A. The surface velocity v_1 may be neglected relative to the flow velocity in the pipeline, v_A. In this case the pipeline flow velocity u used in the loss calculation is identical to the local velocity at A, v_A, as the pipe up to A has been assumed to be of constant diameter. Thus the steady flow energy equation applied from the entry reservoir surface to the apex of the siphon becomes

$$p_A = \rho g(z_1 - z_A) - \tfrac{1}{2}\rho u^2 (1 + K).$$

Setting p_A to gas release pressure, fluid vapour pressure or indeed absolute zero, yields information as to the acceptability of any z_A or pipe length to the apex of the siphon value, the latter being contained in the loss coefficient K, which also depends on the other pipeline parameters mentioned previously, i.e. diameter and roughness.

6.6   THE PITOT TUBE

The Pitot tube is used to measure the velocity of a stream and consists of a simple L-shaped tube facing into the oncoming flow (Fig. 6.10(a)). If the velocity of the stream at A is u, a particle moving from A to the mouth of the tube B will be brought to rest so that u_0 at B is zero. By Bernoulli's equation,

FIGURE 6.10
Pitot tube

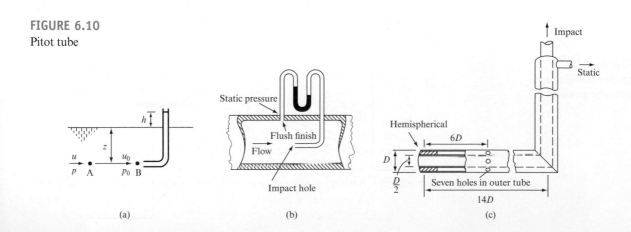

(a)   (b)   (c)

$$\begin{aligned}\text{Total energy per unit} \\ \text{weight at A}\end{aligned} = \begin{aligned}\text{Total energy per unit} \\ \text{weight at B,}\end{aligned}$$

$$u^2/2g + p/\rho g = u_0^2/2g + p_0/\rho g,$$

$$p_0/\rho g = u^2/2g + p/\rho g,$$

since $u_0 = 0$. Thus, p_0 will be greater than p. Now $p/\rho g = z$ and $p_0/\rho g = h + z$. Therefore,

$$u^2/2g = (p_0 - p)/\rho g = h,$$

$$\text{Velocity at A} = u = \sqrt{(2gh)}.$$

When the Pitot tube is used in a channel, the value of h can be determined directly (as in Fig. 6.10(a)), but, if it is to be used in a pipe, the difference between the static pressure and the pressure at the impact hole must be measured with a differential pressure gauge, using a static pressure tapping in the pipe wall (as in Fig. 6.10(b)) or a combined Pitot–static tube (as in Fig. 6.10(c)). In the Pitot–static tube, the inner tube is used to measure the impact pressure while the outer sheath has holes in its surface to measure the static pressure.

While, theoretically, the measured velocity $u = \sqrt{(2gh)}$, Pitot tubes may require calibration. The true velocity is given by $u = C\sqrt{(2gh)}$, where C is the coefficient of the instrument and h is the difference of head measured in terms of the fluid flowing. For the Pitot–static tube shown in Fig. 6.10(c), the value of C is unity for values of Reynolds number $\rho u D/\mu > 3000$, where D is the diameter of the tip of the tube.

6.7   DETERMINATION OF VOLUMETRIC FLOW RATE VIA PITOT TUBE

It will be shown in later chapters that relationships for the velocity distribution across fully developed pipe and duct flow may be utilized to determine the relationship between the velocity at the pipe centreline, or any other identified location, and the theoretical mean velocity in the conduit. Thus volumetric flow rate may be determined by a single Pitot tube or hot-wire/film anemometer. However, in practice this approach is flawed as it depends upon the flow conforming to a particular theoretical velocity distribution in circular cross-section flows and is particularly dubious for non-circular ducts.

A more common device utilized particularly in the study of fan characteristics is to mount a grid of Pitot tubes across the flow and to determine the volume flow by recording local velocities within preset areas; a subsequent summation yields the overall duct flow rate.

Figure 6.11 illustrates the guidance offered by a leading fan manufacturer for the flow integration in circular and rectangular ducts. A suitable method for circular ducts is to divide the duct cross-section into three or four concentric equal areas and to determine the velocity in each by averaging six velocity readings taken at 60° intervals round this annulus. Rectangular ducts should be divided into at least 25 equal rectangular areas by subdividing each side into five equal length increments. For 'long,

FIGURE 6.11

Location of velocity
measurements in
ducts (Woods Air
Movement Ltd)

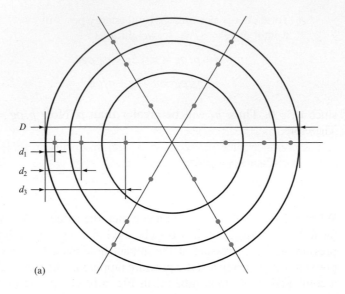

(a)

Circular ducts:
three zone, 18 point, traverse

$\dfrac{d_1}{D}$	$\dfrac{d_2}{D}$	$\dfrac{d_3}{D}$	
0.032	0.135	0.321	

four zone, 24 point, traverse

$\dfrac{d_1}{D}$	$\dfrac{d_2}{D}$	$\dfrac{d_3}{D}$	$\dfrac{d_4}{D}$
0.021	0.117	0.184	0.345

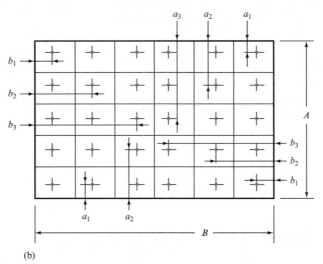

(b)

Rectangular ducts:
five zones per side, 25 points

$\dfrac{a_1}{A}$	$\dfrac{a_2}{A}$	$\dfrac{a_3}{A}$	
0.074	0.288	0.500	

6×5 zones, 30 points

$\dfrac{b_1}{B}$	$\dfrac{b_2}{B}$	$\dfrac{b_3}{B}$	
0.061	0.235	0.437	

7×5 zones, 35 points

$\dfrac{b_1}{B}$	$\dfrac{b_2}{B}$	$\dfrac{b_3}{B}$	$\dfrac{b_4}{B}$
0.053	0.203	0.366	0.500

thin' cross-sections better accuracy may be obtained by increasing the subdivision of the 'long' side to six or seven increments, yielding 30 or 35 areas across the flow. Flow velocity is then measured within each area.

The duct volumetric flow may then be calculated from the relationship

$$V_{\text{mean}} = \frac{\Sigma(V_{\text{local}} A_{\text{local}})}{A_{\text{duct}}}.$$

If the output of the Pitot tubes and the duct static pressure are recorded, either as pressures or as heights of manometer fluid, then this expression may be modified to yield volumetric flow directly from these readings by substituting for the local velocity. See Section 4.11 and Example 4.1.

6.8 COMPUTER PROGRAM VOLFLO

Program VOLFLO allows the determination of volumetric flow at a duct section from either Pitot–static pressure readings or velocity values following a traverse of the section. The program may use the duct area subdivision suggested in Fig. 6.11 or any other user-specified configuration of traverse measurement locations. The program handles both circular and rectangular section ducts and accepts velocity data in m s^{-1} or pressure data in mm of manometer fluid or N m^{-2}.

The data required for either circular or rectangular ducts are velocity or pressure, in the latter case either in mm or N m^{-2}. Constant static pressure at the traverse location is required if the traverse only records Pitot pressure at each location. The density of the flow and the manometer fluid may be required together with the dimensions of the duct and the number of sampling points and zones across the section, see Fig. 6.11.

6.8.1 Application example

For a rectangular section 0.3 m wide, 0.2 m deep, having five width and five depth increments and hence 25 traverse locations, velocity data are available as follows:

1.6	1.9	2.1	2.0	1.8
2.0	3.4	6.8	3.5	2.0
2.2	6.9	10.0	7.0	2.3
2.0	3.7	7.0	3.8	2.1
1.7	2.0	2.3	2.1	1.9

The VOLFLO determined flow rate is 0.202 m^3 s^{-1} with an average velocity of 3.36 m s^{-1}.

6.8.2 Additional investigations using VOLFLO

The computer program calculation may also be used to investigate the divergence in predicted volumetric flow rate when coarser grid settings are used for both rectangular or circular section ducts.

6.9 CHANGES OF PRESSURE IN A TAPERING PIPE

Changes of velocity in a tapering pipe were determined by using the continuity of flow equation (Section 4.12). Change of velocity will be accompanied by a change in the kinetic energy per unit weight and, consequently, by a change in pressure, modified by any change of elevation or energy loss, which can be determined by the use of Bernoulli's equation.

EXAMPLE 6.2

A pipe inclined at 45° to the horizontal (Fig. 6.12) converges over a length l of 2 m from a diameter d_1 of 200 mm to a diameter d_2 of 100 mm at the upper end. Oil of relative density 0.9 flows through the pipe at a mean velocity $\bar{v}_1$ at the lower end of 2 m s^{-1}. Find the pressure difference across the 2 m length ignoring any loss of energy, and the difference in level that would be shown on a mercury manometer connected

FIGURE 6.12

Pressure change in a
tapering pipe

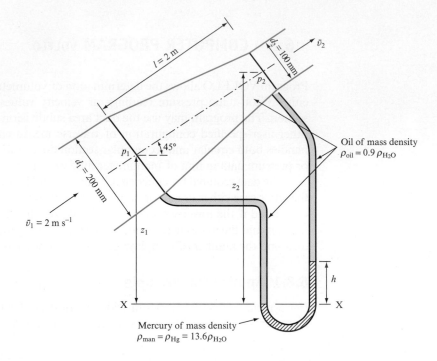

across this length. The relative density of mercury is 13.6 and the leads to the
manometer are filled with the oil.

Solution

Let A_1, $\bar{v}_1$, p_1, d_1, z_1 and A_2, $\bar{v}_2$, p_2, d_2, z_2 be the area, mean velocity, pressure, diameter
and elevation at the lower and upper sections, respectively. For continuity of flow,
assuming the density of the oil to be constant,

$$A_1\bar{v}_1 = A_2\bar{v}_2,$$

so that $\bar{v}_2 = (A_1/A_2)\bar{v}_1,$

where $A_1 = (\pi/4)d_1^2$ and $A_2 = (\pi/4)d_2^2.$

Thus, $\bar{v}_2 = (d_1/d_2)^2\bar{v}_1 = (0.2/0.1)^2 \times 2 = 8 \text{ m s}^{-1}.$

Applying Bernoulli's equation to the lower and upper sections, assuming no
energy losses,

$$\begin{array}{c}\text{Total energy per unit} \\ \text{weight at section 1}\end{array} = \begin{array}{c}\text{Total energy per unit} \\ \text{weight at section 2,}\end{array}$$

$$p_1/\rho_0 g + \bar{v}_1^2/2g + z_1 = p_2/\rho_0 g + \bar{v}_2^2/2g + z_2,$$

$$p_1 - p_2 = \tfrac{1}{2}\rho_0(\bar{v}_2^2 - \bar{v}_1^2) + \rho_0 g(z_2 - z_1). \tag{I}$$

Now, $z_2 - z_1 = l \sin 45° = 2 \times 0.707 = 1.414 \text{ m}$

and, since the relative density of the oil is 0.9, if ρ_{H_2O} = density of water, then $\rho_{oil} =$
$0.9\rho_{H_2O} = 0.9 \times 1000 = 900 \text{ kg m}^{-3}$. Substituting in equation (I),

$$p_1 - p_2 = \tfrac{1}{2} \times 900(8^2 - 2^2) + 900 \times 9.81 \times 1.414 \text{ N m}^{-2}$$

$$= 8829(3.058 + 1.414) = 39\,484 \text{ N m}^{-2}.$$

For the manometer, the pressure in each limb will be the same at level XX; therefore,

$$p_1 + \rho_{oil}gz_1 = p_2 + \rho_{oil}g(z_2 - h) + \rho_{man}gh,$$

$$(p_1 - p_2)/\rho_{oil}g + z_1 - z_2 = h(\rho_{man}/\rho_{oil} - 1),$$

$$h = \left(\frac{\rho_{oil}}{\rho_{man} - \rho_{oil}}\right)\left(\frac{p_1 - p_2}{\rho_{oil}g} + z_1 - z_2\right).$$

Putting $\rho_{oil} = 0.9\rho_{H_2O} = 900$ kg m^{-3} and $\rho_{man} = 13.6\rho_{H_2O}$,

$$h = [0.9/(13.6 - 0.9)][39\,484/(900 \times 9.81) - 1.414] = \mathbf{0.217\ m}.$$

6.10 PRINCIPLE OF THE VENTURI METER

As shown by equation (I) in Example 6.2, the pressure difference between any two points on a tapering pipe through which a fluid is flowing depends on the difference of level $z_2 - z_1$, the velocities $\bar{v}_2$ and $\bar{v}_1$, and, therefore, on the volume rate of flow Q through the pipe. Hence, the pressure difference can be used to determine the volume rate of flow for any particular configuration. The venturi meter uses this effect for the measurement of flow in pipelines. As shown in Fig. 6.13, it consists of a short converging conical tube leading to a cylindrical portion, called the throat, of smaller

FIGURE 6.13

Inclined venturi meter and U-tube

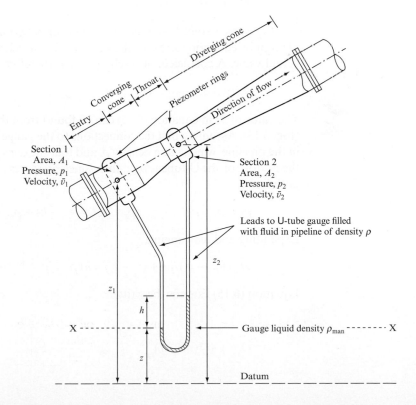

diameter than that of the pipeline, which is followed by a diverging section in which the diameter increases again to that of the main pipeline. The pressure difference from which the volume rate of flow can be determined is measured between the entry section 1 and the throat section 2, often by means of a U-tube manometer (as shown). The axis of the meter may be inclined at any angle. Assuming that there is no loss of energy, and applying Bernoulli's equation to sections 1 and 2,

$$z_1 + p_1/\rho g + v_1^2/2g = z_2 + p_2/\rho g + v_2^2/2g,$$

$$v_2^2 - v_1^2 = 2g[(p_1 - p_2)/\rho g + (z_1 - z_2)]. \tag{6.14}$$

For continuous flow,

$$A_1 v_1 = A_2 v_2 \quad \text{or} \quad v_2 = (A_1/A_2)v_1.$$

Substituting in equation (6.14),

$$v_1^2[(A_1/A_2)^2 - 1] = 2g[(p_1 - p_2)/\rho g + (z_1 - z_2)],$$

$$v_1 = \frac{A_2}{(A_1^2 - A_2^2)^{1/2}} \sqrt{\left[2g\left(\frac{p_1 - p_2}{\rho g} + z_1 - z_2\right)\right]}.$$

Volume rate of flow,

$$Q = A_1 v_1 = [A_1 A_2/(A_1^2 - A_2^2)^{1/2}]\sqrt{(2gH)},$$

where $H = (p_1 - p_2)/\rho g + (z_1 - z_2)$ or, if m = Area ratio = A_1/A_2,

$$Q = [A_1/(m^2 - 1)^{1/2}]\sqrt{(2gH)}. \tag{6.15}$$

In practice, some loss of energy will occur between sections 1 and 2. The value of Q given by equation (6.15) is a theoretical value which will be slightly greater than the actual value. A coefficient of discharge C_d is, therefore, introduced:

Actual discharge, $Q_{\text{actual}} = C_d \times Q_{\text{theoretical}}.$

The value of H in equation (6.15) can be found from the reading of the U-tube gauge (Fig. 6.13). Assuming that the connections to the gauge are filled with the fluid flowing in the pipeline, which has a density ρ, and that the density of the manometric liquid in the bottom of the U-tube is ρ_{man}, then, since pressures at level XX must be the same in both limbs,

$$p_X = p_1 + \rho g(z_1 - z) = p_2 + \rho g(z_2 - z - h) + \rho_{\text{man}} hg.$$

Expanding and rearranging,

$$H = (p_1 - p_2)/\rho g + (z_1 - z_2) = h(\rho_{\text{man}}/\rho - 1).$$

Equation (6.15) can now be written

$$Q = [A_1/(m^2 - 1)^{1/2}]\sqrt{\left[2gh\left(\frac{\rho_{\text{man}}}{\rho} - 1\right)\right]}. \tag{6.16}$$

Note that equation (6.16) is independent of z_1 and z_2, so that the manometer reading h for a given rate of flow Q is not affected by the inclination of the meter. If, however, the actual pressure difference ($p_1 - p_2$) is measured and equation (6.14) or (6.15) used, the values of z_1 and z_2, and, therefore, the slope of the meter, must be taken into account.

EXAMPLE 6.3

A venturi meter having a throat diameter d_2 of 100 mm is fitted into a pipeline which has a diameter d_1 of 250 mm through which oil of specific gravity 0.9 is flowing. The pressure difference between the entry and throat tappings is measured by a U-tube manometer, containing mercury of specific gravity 13.6, and the connections are filled with the oil flowing in the pipeline. If the difference of level indicated by the mercury in the U-tube is 0.63 m, calculate the theoretical volume rate of flow through the meter.

Solution

Using equation (6.16),

$$\text{Area at entry, } A_1 = (\pi/4)d_1^2 = (\pi/4)(0.25)^2 = 0.0491 \text{ m}^2,$$

$$\text{Area ratio, } m = A_1/A_2 = (d_1/d_2)^2 = (0.25/0.10)^2 = 6.25,$$

$$h = 0.63 \text{ m}, \quad \rho_{Hg} = \rho_{man} = 13.6 \times \rho_{H_2O}, \quad \rho_{oil} = 0.9\rho_{H_2O},$$

where ρ_{Hg} = density of mercury, ρ_{H_2O} = density of water and ρ_{oil} = density of oil. Substituting in equation (6.16),

$$Q = [0.0491/(6.25^2 - 1)^{1/2}]\sqrt{[2 \times 9.81 \times 0.63(13.6/0.9 - 1)]} = \mathbf{0.105 \text{ m}^3 \text{ s}^{-1}}.$$

6.11 PIPE ORIFICES

The venturi meter described in Section 6.10 operates by changing the cross-section of the flow, so that the cross-sectional area is less at the downstream pressure tapping than at the upstream tapping. A similar effect can be achieved by inserting an orifice plate which has an opening in it smaller than the internal diameter of the pipeline (as shown in Fig. 6.14). The orifice plate produces a constriction of the flow as shown, the cross-sectional area A_2 of the flow immediately downstream of the plate being

FIGURE 6.14
Pipe orifice meter

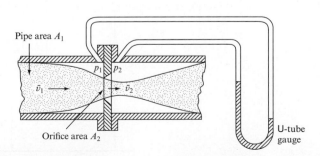

Pipe area A_1

p_1 p_2

$\bar{v}_1$ $\bar{v}_2$

Orifice area A_2

U-tube gauge

approximately the same as that of the orifice. The arrangement is cheap compared with the cost of a venturi meter, but there are substantial energy losses. The theoretical discharge can be calculated from equation (6.14) or (6.15), but the actual discharge may be as little as two-thirds of this value. A coefficient of discharge must, therefore, be introduced in the same way as for the venturi meter, a typical value for a sharp-edged orifice being 0.65.

6.12   LIMITATION ON THE VELOCITY OF FLOW IN A PIPELINE

Since Bernoulli's equation requires that the total energy per unit weight of a flowing fluid shall, if there are no losses, remain constant, any increase in velocity or elevation must be accompanied by a reduction in pressure. Furthermore, since the pressure can never fall below absolute zero, there will be a maximum velocity for a given configuration of a pipeline which cannot be exceeded. For a flowing liquid, the pressure will never fall to absolute zero since air or vapour will be released and form pockets in the flow well before this can occur.

6.13   THEORY OF SMALL ORIFICES DISCHARGING TO ATMOSPHERE

An orifice is an opening, usually circular, in the side or base of a tank or reservoir, through which fluid is discharged in the form of a jet, usually into the atmosphere. The volume rate of flow discharged through an orifice will depend upon the head of the fluid above the level of the orifice and it can, therefore, be used as a means of flow measurement. The term 'small orifice' is applied to an orifice which has a diameter, or vertical dimension, which is small compared with the head producing flow, so that it can be assumed that this head does not vary appreciably from point to point across the orifice.

Figure 6.15 shows a small orifice in the side of a large tank containing liquid with a free surface open to the atmosphere. At a point A on the free surface, the pressure p_A is atmospheric and, if the tank is large, the velocity v_A will be negligible. In the

FIGURE 6.15

Flow through a small orifice

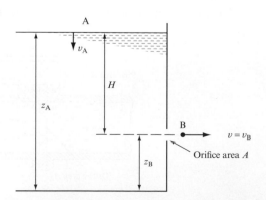

region of the orifice, conditions are rather uncertain, but at some point B in the jet, just outside the orifice, the pressure p_B will again be atmospheric and the velocity v_B will be that of the jet v. Taking the datum for potential energy at the centre of the orifice and applying Bernoulli's equation to A and B, assuming that there is no loss of energy,

Total energy per $=$ Total energy per
unit weight at A unit weight at B,

$$z_A + v_A^2/2g + p_A/\rho g = z_B + v_B^2/2g + p_B/\rho g.$$

Putting $z_A - z_B = H$, $v_A = 0$, $v_B = v$ and $p_A = p_B$,

Velocity of jet, $v = \sqrt{(2gH)}$. $\hfill$ (6.17)

This is a statement of *Torricelli's theorem*, that the velocity of the issuing jet is proportional to the square root of the head producing flow. Equation (6.17) applies to any fluid, H being expressed as a head of the fluid flowing through the orifice. For example, if an orifice is formed in the side of a vessel containing gas of density ρ at a uniform pressure p, the value of H would be $p/\rho g$. Theoretically, if A is the cross-sectional area of the orifice,

Discharge, Q = Area $\times$ Velocity = $A\sqrt{(2gH)}$. $\hfill$ (6.18)

In practice, the actual discharge is considerably less than the theoretical discharge given by equation (6.18), which must, therefore, be modified by introducing a *coefficient of discharge* C_d, so that

Actual discharge, $Q_{actual} = C_d Q_{theoretical} = C_d A\sqrt{(2gH)}$. $\hfill$ (6.19)

There are two reasons for the difference between the theoretical and actual discharges. First, the velocity of the jet is less than that given by equation (6.17) because there is a loss of energy between A and B:

Actual velocity at B = $C_v \times v = C_v\sqrt{(2gH)}$, $\hfill$ (6.20)

where C_v is a *coefficient of velocity*, which has to be determined experimentally and is of the order of 0.97.

Second, as shown in Fig. 6.16, the paths of the particles of the fluid converge on the orifice, and the area of the issuing jet at B is less than the area of the orifice A at C.

FIGURE 6.16

Contraction of issuing jet

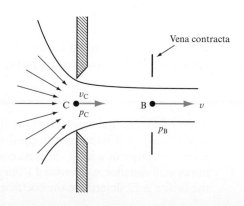

In the plane of the orifice, the particles have a component of velocity towards the centre and the pressure at C is greater than atmospheric pressure. It is only at B, a small distance outside the orifice, that the paths of the particles have become parallel. The section through B is called the *vena contracta*.

$$\text{Actual area of jet at B} = C_c A, \tag{6.21}$$

where C_c is the *coefficient of contraction*, which can be determined experimentally and will depend on the profile of the orifice. For a sharp-edged orifice of the form shown in Fig. 6.16, it is of the order of 0.64.

We can now determine the actual discharge from equations (6.20) and (6.21):

$$\text{Actual discharge} = \text{Actual area at B} \times \text{Actual velocity at B}$$

$$= C_c A \times C_v \sqrt{(2gH)}$$

$$= C_c \times C_v A \sqrt{(2gH)}. \tag{6.22}$$

Comparing equation (6.22) with equation (6.19), we see that the relation between the coefficients is

$$C_d = C_c \times C_v.$$

The values of the coefficient of discharge, the coefficient of velocity and the coefficient of contraction are determined experimentally and values are available for standard configurations in British Standard specifications.

To determine the coefficient of discharge, it is only necessary to collect, or otherwise measure, the actual volume discharged from the orifice in a given time and compare this with the theoretical discharge given by equation (6.18).

$$\text{Coefficient of discharge, } C_d = \frac{\text{Actual measured discharge}}{\text{Theoretical discharge}}.$$

Similarly, the actual area of the jet at the vena contracta can be measured,

$$\text{Coefficient of contraction, } C_c = \frac{\text{Area of jet at vena contracta}}{\text{Area of orifice}}.$$

In the same way, if the actual velocity of the jet at the vena contracta can be found,

$$\text{Coefficient of velocity, } C_v = \frac{\text{Velocity at vena contracta}}{\text{Theoretical velocity}}.$$

If the orifice is not in the bottom of the tank, one method of measuring the actual velocity of the jet is to measure its profile.

EXAMPLE 6.4

A jet of water discharges horizontally into the atmosphere from an orifice in the vertical side of a large open-topped tank (Fig. 6.17). Derive an expression for the actual velocity v of a jet at the vena contracta if the jet falls a distance y vertically in a horizontal distance x, measured from the vena contracta. If the head of water above the orifice is H, determine the coefficient of velocity.

FIGURE 6.17
Determination of the
coefficient of velocity

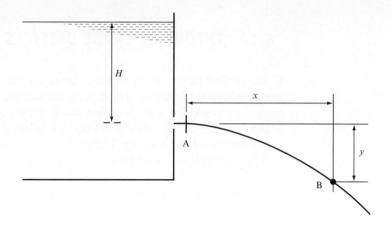

If the orifice has an area of 650 mm² and the jet falls a distance y of 0.5 m in a horizontal distance x of 1.5 m from the vena contracta, calculate the values of the coefficients of velocity, discharge and contraction, given that the volume rate of flow is 0.117 m³ and the head H above the orifice is 1.2 m.

Solution

Let t be the time taken for a particle of fluid to travel from the vena contracta A (Fig. 6.17) to the point B. Then

$$x = vt \quad \text{and} \quad y = \tfrac{1}{2}gt^2,$$

or $\quad v = x/t \quad \text{and} \quad t = \sqrt{(2y/g)}.$

Eliminating t,

Velocity at the vena contracta, $v = \sqrt{(gx^2/2y)}.$

This is the *actual* velocity of the jet at the vena contracta.
From equation (6.17),

Theoretical velocity $= \sqrt{(2gH)},$

$$\text{Coefficient of velocity} = \frac{\text{Actual velocity}}{\text{Theoretical velocity}} = v/\sqrt{(2gH)} = \sqrt{(x^2/4yH)}.$$

Putting $x = 1.5$ m, $y = 0.5$ m, $H = 1.2$ m and area, $A = 650 \times 10^{-6}$ m²,

Coefficient of velocity, $C_v = \sqrt{(x^2/4yH)} = \sqrt{[1.5^2/(4 \times 0.5 \times 1.2)]}$

$= \mathbf{0.968},$

Coefficient of discharge, $C_d = Q_{\text{actual}}/A\sqrt{(2gH)}$

$= (0.117/60)/[650 \times 10^{-6}\sqrt{(2 \times 9.81 \times 1.2)}]$

$= \mathbf{0.618},$

Coefficient of contraction, $C_c = C_d/C_v = 0.618/0.968 = \mathbf{0.639}.$

6.14 THEORY OF LARGE ORIFICES

If the vertical height of an orifice is large, so that the head producing flow is substantially less at the top of the opening than at the bottom, the discharge calculated from the formula for a small orifice, using the head h measured to the centre of the orifice, will not be the true value, since the velocity will vary very substantially from top to bottom of the opening. The method adopted is to calculate the flow through a thin horizontal strip across the orifice (Fig. 6.18), and integrate from top to bottom of the opening to obtain the theoretical discharge, from which the actual discharge can be determined if the coefficient of discharge is known.

FIGURE 6.18

Flow through a large orifice

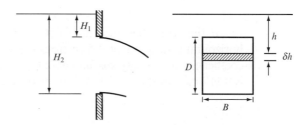

EXAMPLE 6.5

A reservoir discharges through a rectangular sluice gate of width B and height D (Fig. 6.18). The top and bottom of the opening are at depths H_1 and H_2 below the free surface. Derive a formula for the theoretical discharge through the opening.

If the top of the opening is 0.4 m below the water level and the opening is 0.7 m wide and 1.5 m in height, calculate the theoretical discharge (in cubic metres per second), assuming that the bottom of the opening is above the downstream water level.

What would be the percentage error if the opening were to be treated as a small orifice?

Solution

Since the velocity of flow will be much greater at the bottom than at the top of the opening, consider a horizontal strip across the opening of height δh at a depth h below the free surface:

$$\text{Area of strip} = B\delta h,$$

$$\text{Velocity of flow through strip} = \sqrt{(2gh)},$$

$$\text{Discharge through strip, } \delta Q = \text{Area} \times \text{Velocity} = B\sqrt{(2g)}h^{1/2}\delta h.$$

For the whole opening, integrating from $h = H_1$ to $h = H_2$,

$$\text{Discharge, } Q = B\sqrt{(2g)}\int_{H_1}^{H_2} h^{1/2}\,\mathrm{d}h$$

$$= \tfrac{2}{3}B\sqrt{(2g)}(H_2^{3/2} - H_1^{3/2}).$$

Putting $B = 0.7$ m, $H_1 = 0.4$ m, $H_2 = 1.9$ m,

$$\text{Theoretical discharge, } Q = \tfrac{2}{3} \times 0.7 \times \sqrt{(2 \times 9.81)}(1.9^{3/2} - 0.4^{3/2})$$
$$= 2.067(2.619 - 0.253) = \textbf{4.891 m}^3\,\textbf{s}^{-1}.$$

For a small orifice, $Q = A\sqrt{(2gh)}$, where A is the area of the orifice and h is the head above the centreline. Putting

$$A = BD = 0.7 \times 1.5 \text{ m}^2,$$
$$h = \tfrac{1}{2}(H_1 + H_2) = \tfrac{1}{2}(0.4 + 1.9) = 1.15 \text{ m},$$
$$Q = 0.7 \times 1.5\sqrt{(2 \times 9.81 \times 1.15)} = \textbf{4.988 m}^3\,\textbf{s}^{-1}.$$

This result is greater than that obtained by the large-orifice analysis.

$$\text{Error} = (4.988 - 4.891)/4.891 = 0.0198 = \textbf{1.98 per cent}.$$

6.15   ELEMENTARY THEORY OF NOTCHES AND WEIRS

A notch is an opening in the side of a measuring tank or reservoir extending above the free surface. It is, in effect, a large orifice which has no upper edge, so that it has a variable area depending upon the level of the free surface. A weir is a notch on a large scale, used, for example, to measure the flow of a river, and may be sharp edged or have a substantial breadth in the direction of flow.

   The method of determining the theoretical flow through a notch is the same as that adopted for the large orifice. For a notch of any shape (Fig. 6.19), consider a horizontal strip of width b at a depth h below the free surface and height δh.

FIGURE 6.19
Discharge from a notch

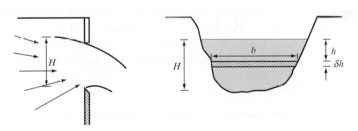

Area of strip $= b\delta h$,

Velocity through strip $= \sqrt{(2gh)}$,

Discharge through strip, $\delta Q = \text{Area} \times \text{Velocity} = b\delta h\sqrt{(2gh)}.$ (6.23)

Integrating from $h = 0$ at the free surface to $h = H$ at the bottom of the notch,

$$\text{Total theoretical discharge, } Q = \sqrt{(2g)}\int_0^H bh^{1/2}\,\mathrm{d}h. \tag{6.24}$$

Before the integration of equation (6.24) can be carried out, b must be expressed in terms of h.

FIGURE 6.20
Rectangular and
vee notches

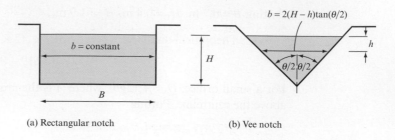

(a) Rectangular notch   (b) Vee notch

For a *rectangular notch* (Fig. 6.20(a)), put b = constant = B in equation (6.24), giving

$$Q = B\sqrt{(2g)} \int_0^H h^{1/2} \, dh = \tfrac{2}{3}B\sqrt{(2g)}H^{3/2}. \tag{6.25}$$

For a *vee notch* with an included angle θ (Fig. 6.20(b)), put $b = 2(H - h) \tan(\theta/2)$ in equation (6.24), giving

$$Q = \sqrt{(2g)} \tan(\theta/2) \int_0^H (H-h)h^{1/2} \, dh$$

$$= 2\sqrt{(2g)} \tan(\theta/2)(\tfrac{2}{3}Hh^{3/2} - \tfrac{2}{5}h^{5/2})_0^h$$

$$Q = \tfrac{8}{15}\sqrt{(2g)} \tan(\theta/2)H^{5/2}. \tag{6.26}$$

Inspection of equations (6.25) and (6.26) suggests that, by choosing a suitable shape for the sides of the notch, any desired relationship between Q and H could be achieved, but certain laws do lead to shapes which are not feasible in practice.

As in the case of orifices, the actual discharge through a notch or weir can be found by multiplying the theoretical discharge by a coefficient of discharge to allow for energy losses and the contraction of the cross-section of the stream at the bottom and sides.

EXAMPLE 6.6

It is proposed to use a notch to measure the flow of water from a reservoir and it is estimated that the error in measuring the head above the bottom of the notch could be 1.5 mm. For a discharge of 0.28 m³ s⁻¹, determine the percentage error which may occur, using a right-angled triangular notch with a coefficient of discharge of 0.6.

Solution

For a vee notch, from equation (6.26),

$$Q = C_d\tfrac{8}{15}\sqrt{(2g)} \tan(\theta/2)H^{5/2}.$$

Putting $C_d = 0.6$ and $\theta = 90°$,

$$Q = 0.6 \times \tfrac{8}{15} \times \sqrt{(19.62)} \times 1 \times H^{5/2} = 1.417H^{5/2}. \tag{I}$$

When $Q = 0.28$ m^3 s^{-1}, $H = (0.28/1.417)^{2/5} = 0.5228$ m. The error δQ in the discharge, corresponding to an error δH in the measurement of H, can be found by differentiating equation (I):

$$\delta Q = 2.5 \times 1.417 H^{3/2} \delta H = 2.5 Q \delta H / H,$$

$$\delta Q / Q = 2.5 \delta H / H.$$

Putting $\delta H = 1.5$ mm and $H = 0.5228$ m,

$$\text{Percentage error} = (\delta Q / Q) \times 100 = (2.5 \times 0.0015 / 0.5228) \times 100$$

$$= \textbf{0.72 per cent.}$$

In the foregoing theory, it has been assumed that the velocity of the liquid approaching the notch is very small so that its kinetic energy can be neglected; it can also be assumed that the velocity through any horizontal element across the notch will depend only on its depth below the free surface. This is a satisfactory assumption for flow over a notch or weir in the side of a large reservoir, but, if the notch or weir is placed at the end of a narrow channel, the *velocity of approach* to the weir will be substantial and the head h producing flow will be increased by the kinetic energy of the approaching liquid to a value

$$x = h + \alpha \bar{v}^2 / 2g, \tag{6.27}$$

where $\bar{v}$ is the mean velocity of the liquid in the approach channel and α is the kinetic energy correction factor to allow for the non-uniformity of velocity over the cross-section of the channel. Note that the value of $\bar{v}$ is obtained by dividing the discharge by the full cross-sectional area of the channel itself, not that of the notch. As a result, the discharge through the strip (shown in Fig. 6.19) will be

$$\delta Q = b \delta h \sqrt{(2gx)},$$

and, from equation (6.27), $\delta h = \delta x$, so that

$$\delta Q = b \sqrt{(2g)} x^{1/2} \, \mathrm{d}x. \tag{6.28}$$

At the free surface, $h = 0$ and $x = \alpha \bar{v}^2 / 2g$, while, at the sill, $h = H$ and $x = H + a \bar{v}^2 / 2g$. Integrating equation (6.28) between these limits,

$$Q = \sqrt{(2g)} \int_{\alpha \bar{v}^2 / 2g}^{(H + \alpha \bar{v}^2 / 2g)} b x^{1/2} \, \mathrm{d}x.$$

For a rectangular notch, putting $b = B = \text{constant}$,

$$Q = \tfrac{2}{3} B \sqrt{(2g)} H^{3/2} \left[\left(1 + \frac{\alpha \bar{v}^2}{2gH} \right)^{3/2} - \left(\frac{\alpha \bar{v}^2}{2gH} \right)^{3/2} \right]. \tag{6.29}$$

EXAMPLE 6.7

A long rectangular channel 1.2 m wide leads from a reservoir to a rectangular notch 0.9 m wide with its sill 0.2 m above the bottom of the channel. Assuming that, if the velocity of approach is neglected, the discharge over the notch, in SI units, is given by $Q = 1.84BH^{3/2}$, calculate the discharge (in cubic metres per second) when the head over the bottom of the notch H is 0.25 m (a) neglecting the velocity of approach, (b) correcting for the velocity of approach assuming that the kinetic energy correction factor α is 1.1.

Solution

(a) Neglecting the velocity of approach,

$$Q_1 = 1.84BH^{3/2}.$$

Putting $B = 0.9$ m and $H = 0.25$ m,

$$Q_1 = 1.84 \times 0.9 \times 0.25^{3/2} = \mathbf{0.207 \ m^3 \ s^{-1}}.$$

(b) Taking the velocity of approach into account, from equation (6.29) the correction factor k will be

$$k = [(1 + \alpha\bar{v}^2/2gH)^{3/2} - (\alpha\bar{v}^2/2gH)^{3/2}],$$

and the corrected value of Q will be $Q_2 = Q_1 \times k$, so that

$$Q_2 = 1.84BH^{3/2}\left[\left(1 + \frac{\alpha\bar{v}^2}{2gH}\right)^{3/2} - \left(\frac{\alpha\bar{v}^2}{2gH}\right)^{3/2}\right].$$

Putting $B = 0.9$ m, $H = 0.25$ m and $\alpha = 1.1$,

$$Q_2 = 1.84 \times 0.9 \times 0.25^{3/2}\left[\left(1 + \frac{1.1\bar{v}^2}{19.62 \times 0.25}\right)^{3/2} - \left(\frac{1.1\bar{v}^2}{19.62 \times 0.25}\right)^{3/2}\right]$$

$$= 0.207[(1 + 0.224\bar{v}^2)^{3/2} - (0.224\bar{v}^2)^{3/2}]. \tag{I}$$

Now,

$$V = \text{Velocity in approach channel} = \frac{\text{Discharge}}{\text{Area of channel}}$$

$$= Q_2/1.2(H + 0.2). \tag{II}$$

Using (II), the solution to (I) can be found by successive approximation, taking $\bar{v} = 0$ for the first approximation – which gives $Q = 0.207 \ m^3 \ s^{-1}$.

Inserting this value of Q in (II), with $H = 0.25$ m,

$$\bar{v} = 0.207/1.2 \times 0.45 = 0.3833 \ m \ s^{-1}.$$

Putting $\bar{v} = 0.3833 \ m \ s^{-1}$ in (I),

$$Q = 0.207[(1.0329)^{3/2} - (0.0329)^{3/2}] = 0.2161 \ m^3 \ s^{-1}.$$

For the next approximation,

$$\bar{v} = 0.2161/1.2 \times 0.45 = 0.4002 \text{ m s}^{-1},$$

giving

$$Q = 0.207[(1.0359)^{3/2} - (0.0359)^{3/2}] = 0.2168 \text{ m}^3 \text{ s}^{-1}.$$

A further approximation gives

$$\bar{v} = 0.2168/1.2 \times 0.45 = 0.4015 \text{ m s}^{-1}$$

and

$$Q = 0.207[(1.0360)^{3/2} - (0.0360)^{3/2}] = 0.2169 \text{ m}^3 \text{ s}^{-1}.$$

6.16 THE POWER OF A STREAM OF FLUID

In Section 6.1, it was shown that a stream of fluid could do work as a result of its pressure p, velocity v and elevation z and that the total energy per unit weight H of the fluid is given by

$$H = p/\rho g + v^2/2g + z.$$

If the weight per unit time of fluid flowing is known, the power of the stream can be calculated, since

$$\text{Power} = \text{Energy per unit time} = \frac{\text{Weight}}{\text{Unit time}} \times \frac{\text{Energy}}{\text{Unit weight}}.$$

If Q is the volume rate of flow, weight per unit time $= \rho g Q$,

$$\text{Power} = \rho g Q H = \rho g Q(p/\rho g + v^2/2g + z) = pQ + \tfrac{1}{2}\rho v^2 Q + \rho g Q z. \qquad (6.30)$$

EXAMPLE 6.8

Water is drawn from a reservoir, in which the water level is 240 m above datum, at the rate of 0.13 m³ s⁻¹. The outlet of the pipeline is at the datum level and is fitted with a nozzle to produce a high speed jet to drive a turbine of the Pelton wheel type. If the velocity of the jet is 66 m s⁻¹, calculate (a) the power of the jet, (b) the power supplied from the reservoir, (c) the head used to overcome losses and (d) the efficiency of the pipeline and nozzle in transmitting power.

Solution

(a) The jet issuing from the nozzle will be at atmospheric pressure and at the datum level so that, in equation (6.30), $p = 0$ and $z = 0$. Therefore,

$$\text{Power of jet} = \tfrac{1}{2}\rho v^2 Q.$$

Putting $\rho = 1000$ kg m^{-3}, $v = 66$ m s^{-1}, $Q = 0.13$ m^3 s^{-1},

Power of jet $= \frac{1}{2} \times 1000 \times 66^2 \times 0.13 = 283\ 140$ W $= \textbf{283.14 kW}$.

(b) At the reservoir, the pressure is atmospheric and the velocity of the free surface is zero so that, in equation (6.30), $p = 0$, $v = 0$. Therefore,

Power supplied from reservoir $= \rho g Q z$.

Putting $\rho = 1000$ kg m^{-3}, $Q = 0.13$ m^3 s^{-1}, $z = 240$ m,

Power supplied from reservoir $= 1000 \times 9.81 \times 0.13 \times 240$ W $= \textbf{306.07 kW}$.

(c) If $H_1 =$ total head at the reservoir, $H_2 =$ total head at the jet, and $h =$ head lost in transmission,

Power supplied from reservoir $= \rho g Q H_1 = 306.07$ kW,

Power of issuing jet $= \rho g Q H_2 = 283.14$ kW,

Power lost in transmission $= \rho g Q h = 22.93$ kW,

Head lost in pipeline $= h = \dfrac{\text{Power lost}}{\rho g Q}$

$$= \frac{22.93 \times 10^3}{1000 \times 9.81 \times 0.13} = \textbf{17.98 m}.$$

(d) Efficiency of transmission $= \dfrac{\text{Power of jet}}{\text{Power supplied by reservoir}}$

$= 283.14 / 306.07 = \textbf{92.5 per cent}$.

6.17 RADIAL FLOW

When a fluid flows radially inwards, or outwards from a centre, between two parallel planes as in Fig. 6.21, the streamlines will be radial straight lines and the streamtubes will be in the form of sectors. The area of flow will therefore increase as the radius increases, causing the velocity to decrease. Since the flow pattern is symmetrical, the total energy per unit weight H will be the same for all streamlines and for all points along each streamline if we assume that there is no loss of energy.

If v is the radial velocity and p the pressure at any radius r,

$$H = p/\rho g + v^2/2g = \text{constant.} \tag{6.31}$$

Applying the continuity of flow equation and assuming that the density of the fluid remains constant, as would be the case for a liquid,

Volume rate of flow, Q = Area × Velocity = $2\pi r b \times v$,

where b is the distance between the planes. Thus,

$v = Q/2\pi r b$

FIGURE 6.21
Radial flow

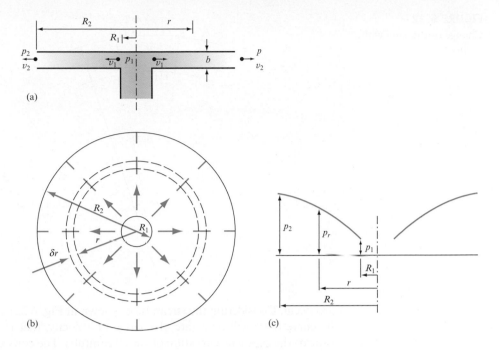

and, substituting in equation (6.31),

$$p/\rho g + Q^2/8\pi^2 r^2 b^2 g = H,$$

$$p = \rho g[H - (Q^2/8\pi^2 b^2 g) \times (1/r^2)]. \qquad (6.32)$$

If the pressure p at any radius r is plotted as in Fig. 6.21(c), the curve will be parabolic and is sometimes referred to as Barlow's curve.

If the flow discharges to atmosphere at the periphery, the pressure at any point between the plates will be below atmospheric; there will be a force tending to bring the two plates together and so shut off flow. This phenomenon can be observed in the case of a disc valve. Radial flow under the disc will cause the disc to be drawn down onto the valve seating. This will cause the flow to stop, the pressure between the plates will return to atmospheric and the static pressure of the fluid on the upstream side of the disc will push it off its seating again. The disc will tend to vibrate on the seating and the flow will be intermittent.

6.18 FLOW IN A CURVED PATH. PRESSURE GRADIENT AND CHANGE OF TOTAL ENERGY ACROSS THE STREAMLINES

Velocity is a vector quantity with both magnitude and direction. When a fluid flows in a curved path, the velocity of the fluid along any streamline will undergo a change due to its change of direction, irrespective of any alteration in magnitude which may

FIGURE 6.22

Change of pressure with radius

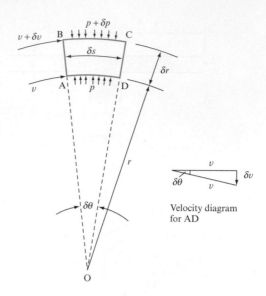

Velocity diagram for AD

also occur. Considering the streamtube (shown in Fig. 6.22), as the fluid flows round the curve there will be a rate of change of velocity, that is to say an acceleration, towards the centre of curvature of the streamtube. The consequent rate of change of momentum of the fluid must be due, in accordance with Newton's second law, to a force acting radially across the streamlines resulting from the difference of pressure between the sides BC and AD of the streamtube element.

In Fig. 6.22, suppose that the control volume ABCD subtends an angle $\delta\theta$ at the centre of curvature O, has length δs in the direction of flow and thickness b perpendicular to the diagram. For the streamline AD, let r be the radius of curvature, p the pressure and v the velocity of the fluid. For the streamline BC, the radius will be $r + \delta r$, the pressure $p + \delta p$ and the velocity $v + \delta v$, where δp is the change of pressure in a radial direction.

From the velocity diagram,

$$\text{Change of velocity in radial direction, } \delta v = v\delta\theta,$$

or, since $\delta\theta = \delta s/r$,

$$\text{Radial change of velocity between AB and CD} = v\frac{\delta s}{r},$$

$$\text{Mass per unit time flowing through streamtube} = \text{Mass density} \times \text{Area} \times \text{Velocity}$$

$$= \rho \times (b \times \delta r) \times v,$$

$$\text{Change of momentum per unit time in radial direction} = \text{Mass per unit time} \times \text{Radial change of velocity}$$

$$= \rho b \delta r v^2 \delta s/r. \tag{6.33}$$

This rate of change of momentum is produced by the force due to the pressure difference between faces BC and AD of the control volume:

$$\text{Force} = [(p + \delta p) - p]b\delta s. \tag{6.34}$$

Equating equations (6.33) and (6.34), according to Newton's second law,

$$\delta p b \delta s = \rho b \delta r v^2 \delta s / r,$$
$$\delta p / \delta r = \rho v^2 / r. \tag{6.35}$$

For an incompressible fluid, ρ will be constant and equation (6.35) can be expressed in terms of the pressure head h. Since $p = \rho g h$, we have $\delta p = \rho g \delta h$. Substituting in equation (6.35),

$$\rho g \delta h / \delta r = \rho v^2 / r,$$
$$\delta h / \delta r = v^2 / g r,$$

or, in the limit as δr tends to zero,

Rate of change of pressure head in radial direction

$$= \frac{\mathrm{d}h}{\mathrm{d}r} = \frac{v^2}{gr}. \tag{6.36}$$

To produce the curved flow shown in Fig. 6.22, we have seen that there must be a change of pressure head in a radial direction. However, since the velocity v along streamline AD is different from the velocity $v + \delta v$ along BC, there will also be a change in the velocity head from one streamline to another:

Rate of change of velocity head radially

$$= [(v + \delta v)^2 - v^2]/2g\delta r,$$

$$= \frac{v}{g} \frac{\delta v}{\delta r}, \text{ neglecting products of small quantities,}$$

$$= \frac{v}{g} \frac{\mathrm{d}v}{\mathrm{d}r}, \text{ as } \delta r \text{ tends to zero.} \tag{6.37}$$

If the streamlines are in a horizontal plane, so that changes in potential head do not occur, the change of total head H – i.e. the total energy per unit weight – in a radial direction, $\delta H/\delta r$, is given by

$$\delta H/\delta r = \text{Change of pressure head} + \text{Change of velocity head}.$$

Substituting from equations (6.36) and (6.37), in the limit,

$$\text{Change of total energy with radius, } \frac{\mathrm{d}H}{\mathrm{d}r} = \frac{v^2}{gr} + \frac{v}{g} \frac{\mathrm{d}v}{\mathrm{d}r}$$

$$\frac{\mathrm{d}H}{\mathrm{d}r} = \frac{v}{g} \left(\frac{v}{r} + \frac{\mathrm{d}v}{\mathrm{d}r} \right). \tag{6.38}$$

The term $(v/r + \mathrm{d}v/\mathrm{d}r)$ is also known as the *vorticity* of the fluid (see Section 7.2).

In obtaining equation (6.38), it has been assumed that the streamlines are horizontal, but this equation also applies to cases where the streamlines are inclined to the horizontal, since the fluid in the control volume is in effect weightless, being supported vertically by the surrounding fluid.

If the streamlines are straight lines, $r = \infty$ and $dv/dr = 0$. From equation (6.38) for a stream of fluid in which the velocity is uniform across the cross-section, and neglecting friction, we have $dH/dr = 0$ and the total energy per unit weight H is constant for all points on all streamlines. This applies whether the streamlines are parallel or inclined, as in the case of radial flow (Section 6.17).

6.19   VORTEX MOTION

In vortex motion, the streamlines form a set of concentric circles and the changes of total energy per unit weight will be governed by equation (6.38). The following types of vortex are recognized.

6.19.1 Forced vortex or flywheel vortex

The fluid rotates as a solid body with constant angular velocity ω, i.e. at any radius r,

$$v = \omega r \quad \text{so that} \quad \frac{dv}{dr} = \omega \quad \text{and} \quad \frac{v}{r} = \omega.$$

From equation (6.38),

$$\frac{dH}{dr} = \frac{\omega r}{g}(\omega + \omega) = \frac{2\omega^2 r}{g}.$$

Integrating,

$$H = \omega^2 r^2/g + C, \tag{6.39}$$

where C is a constant. But, for any point in the fluid,

$$H = p/\rho g + v^2/2g + z = p/\rho g + \omega^2 r^2/2g + z.$$

Substituting in equation (6.39),

$$p/\rho g + \omega^2 r^2/2g + z = \omega^2 r^2/g + C,$$

$$p/\rho g + z = \omega^2 r^2/2g + C. \tag{6.40}$$

If the rotating fluid has a free surface, the pressure at this surface will be atmospheric and therefore zero (gauge).

FIGURE 6.23
Forced vortex

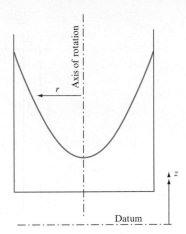

Putting $p/\rho g = 0$ in equation (6.40), the profile of the free surface will be given by

$$z = \omega^2 r^2/2g + C. \tag{6.41}$$

Therefore, the free surface will be in the form of a paraboloid (Fig. 6.23).

Similarly, for any horizontal plane, for which z will be constant, the pressure distribution will be given by

$$p/\rho g = \omega^2 r^2/2g + (C - z). \tag{6.42}$$

EXAMPLE 6.9

A closed vertical cylinder 400 mm in diameter and 500 mm high is filled with oil of relative density 0.9 to a depth of 340 mm, the remaining volume containing air at atmospheric pressure. The cylinder revolves about its vertical axis at such a speed that the oil just begins to uncover the base. Calculate (a) the speed of rotation for this condition and (b) the upward force on the cover.

Solution

(a) When stationary, the free surface will be at AB (Fig. 6.24), a height Z_2 above the base.

Volume of oil $= \pi r_1^2 Z_2$.

When rotating at the required speed ω, a forced vortex is formed and the free surface will be the paraboloid CDE.

Volume of oil = Volume of cylinder PQRS − Volume of paraboloid CDE

$$= \pi r_1^2 Z_1 - \tfrac{1}{2}\pi r_0^2 Z_1,$$

since the volume of a paraboloid is equal to half the volume of the circumscribing cylinder.

FIGURE 6.24

Forced vortex example

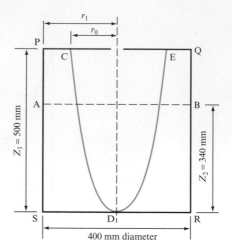

No oil is lost from the container; therefore,

$$\pi r_1^2 Z_2 = \pi r_1^2 Z_1 - \tfrac{1}{2}\pi r_0^2 Z_1,$$

$$r_0^2 = 2r_1^2(1 - Z_2/Z_1),$$

$$r_0 = r_1\sqrt{[2(1 - Z_2/Z_1)]} = r_1\sqrt{[2(1 - 340/500)]}$$

$$= 0.8r_1 = 0.8 \times 200 = 160 \text{ mm}.$$

Also, for the free surface of the vortex from equation (6.41),

$$z = \omega^2 r^2/2g + \text{constant},$$

or, between points C and D, taking D as the datum level,

$$Z_D = 0 \text{ when } r = 0 \quad \text{and} \quad Z_C = Z_1 \text{ when } r = r_0,$$

giving

$$Z_1 - 0 = \omega^2 r_0^2/2g,$$

$$\omega = \sqrt{(2gZ_1 r_0^2)}$$

$$= \sqrt{(2 \times 9.81 \times 0.5/0.16^2)} = \textbf{19.6 rad s}^{-1}.$$

(b)   The oil will be in contact with the top cover from radius $r = r_0$ to $r = r_1$. If p is the pressure at any radius r, the force on an annulus of radius r and width δr is given by

$$\delta F = p \times 2\pi r \delta r.$$

Integrating from $r = r_0$ to $r = r_1$,

$$\text{Force on top cover, } F = 2\pi \int_{r_0}^{r_1} pr\,\delta r. \tag{I}$$

From equation (6.42),

$$p/\rho g = \omega^2 r^2/2g + C.$$

Since the pressure at r_0 is atmospheric, $p = 0$ when $r = r_0$, so that

$$C = -\omega^2 r_0^2 / 2g,$$

and

$$p = \rho g \left(\frac{\omega^2 r^2}{2g} - \frac{\omega^2 r_0^2}{2g} \right) = \frac{\rho \omega^2}{2} (r^2 - r_0^2).$$

Substituting in (I),

$$F = 2\pi \frac{\rho \omega^2}{2} \int_{r_0}^{r_1} (r^2 - r_0^2) r \, \mathrm{d}r$$

$$= \rho \omega^2 \pi \int_{r_0}^{r_1} (r^3 - r_0^2 r) \, \mathrm{d}r$$

$$= \rho \omega^2 \pi (\tfrac{1}{4} r^4 - \tfrac{1}{2} r_0^2 r^2)_{r_0}^{r_1}$$

$$= \rho \omega^2 \pi (\tfrac{1}{4} r_1^4 - \tfrac{1}{4} r_0^4 - \tfrac{1}{2} r_0^2 r_1^2 + \tfrac{1}{2} r_0^4)$$

$$= \frac{\rho \omega^2 \pi}{4} (r_1^4 + r_0^4 - 2 r_0^2 r_1^2) = \frac{\pi}{4} \rho \omega^2 (r_1^2 - r_0^2)^2$$

$$= \frac{\pi}{4} \times (0.9 \times 1000) \times 19.6^2 \times (0.2^2 - 0.16^2)^2 \ \text{N} \ = \textbf{56.3 N}.$$

6.19.2 Free vortex or potential vortex

In this case, the streamlines are concentric circles, but the variation of velocity with radius is such that there is no change of total energy per unit weight with radius, so that $\mathrm{d}H/\mathrm{d}r = 0$. Substituting in equation (6.38),

$$0 = \frac{v}{g} \left(\frac{v}{r} + \frac{\mathrm{d}v}{\mathrm{d}r} \right),$$

$$\frac{\mathrm{d}v}{v} + \frac{\mathrm{d}r}{r} = 0.$$

Integrating,

$$\log_e v + \log_e r = \text{constant},$$

or $\quad vr = C,$

where C is a constant known as the *strength* of the vortex at any radius r;

$$v = C/r. \tag{6.43}$$

FIGURE 6.25
Free vortex

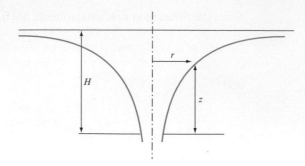

Since, at any point,

$$z + p/\rho g + v^2/2g = H = \text{constant},$$

substituting for v from equation (6.43)

$$z + p/\rho g + C^2/2gr^2 = H.$$

If the fluid has a free surface, $p/\rho g = 0$ and the profile of the free surface is given by

$$H - z = C^2/2gr^2, \tag{6.44}$$

which is a hyperbola asymptotic to the axis of rotation and to the horizontal plane through $z = H$, as shown in Fig. 6.25.

For any horizontal plane, z is constant and the pressure variation is given by

$$p/\rho g = (H - z) - C^2/2gr^2. \tag{6.45}$$

Thus, in the free vortex, pressure decreases and circumferential velocity increases as we move towards the centre.

EXAMPLE 6.10

A point A on the free surface of a free vortex is at a radius $r_A = 200$ mm and a height $z_A = 125$ mm above datum. If the free surface at a distance from the axis of the vortex, which is sufficient for its effect to be negligible, is 180 mm above datum, what will be the height above datum of a point B on the free surface at a radius of 100 mm?

Solution
For point A, from equation (6.44),

$$H - z_A = C^2/2gr_A^2;$$

therefore,

$$C^2/2g = r_A^2 (H - z_A).$$

Now H is the head above datum at an infinite distance from the axis of rotation, where the effect of the vortex is negligible, so that $H = 180$ mm $= 0.18$ m. Also $z_A = 0.125$ m and $r_A = 0.2$ m. Substituting,

$$\frac{C^2}{2g} = 0.2^2(0.18 - 0.125) = 2.2 \times 10^{-3} \text{ m}^3.$$

For point B,

$$H - z_B = C^2/2gr_B^2$$
$$z_B = H - C^2/2gr_B^2$$
$$= 0.18 - (2.2 \times 10^{-3})/0.1^2 = -0.04 = \textbf{40 mm below datum.}$$

6.19.3 Compound vortex

In the free vortex, $v = C/r$ and thus, theoretically, the velocity becomes infinite at the centre. The velocities near the axis would be very high and, since friction losses vary as the square of the velocity, they will cease to be negligible, and the assumption that the total head H remains constant will cease to be true. The central part of the vortex tends to rotate as a solid body, thus forming a forced vortex surrounded by a free vortex. Figure 6.26 shows the free surface profile of such a compound vortex, and also represents the variation of pressure with radius on any horizontal plane in the vortex. The velocity at the common radius R must be the same for the two vortices.

FIGURE 6.26
Compound vortex

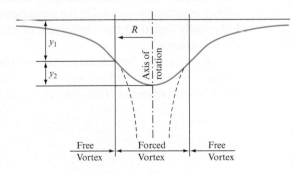

For the free vortex, if $y_1 =$ depression of the surface at radius R below the level of the surface at infinity,

$$y_1 = C^2/2gR^2 = v^2/2g = \omega^2 R^2/2g.$$

For the forced vortex, if y_2 = height of the surface at radius R above the centre of the depression,

$$y_2 = v^2/2g = C^2/2gR^2.$$

Thus,

$$\text{Total depression} = y_2 + y_1 = C^2/gR^2 = \omega^2 R^2/g. \tag{6.46}$$

For the forced vortex, the velocity at radius R is ωR, while for the free vortex, from equation (6.43), the velocity at radius R is C/R. Therefore, the common radius, at which these two velocities will be the same, is given by

$$\omega R = C/R \quad \text{or} \quad R = \sqrt{(C/\omega)}.$$

In Section 7.9 it will be shown that $C = \Gamma/2\pi$, where Γ is the circulation, so that

$$\text{Common radius, } R = \sqrt{(C/\omega)} = \sqrt{(\Gamma/2\pi\omega)}.$$

Concluding remarks

Chapters 4 and 5 introduced two of the fundamental relationships of fluid mechanics, namely the conservation of mass flow and the momentum equation. This chapter has introduced the third fundamental relationship, the steady flow energy equation, based on the conservation of energy principle. It is interesting to note the relationship between this expression and the simplified form of Euler's equation, Bernoulli's equation, generated from the application of the equation of momentum to steady, uniform, inviscid flow. The steady flow energy equation will be used throughout this text in the analysis of flow in pipes, ducts and open channels, as well as providing the basis for the study of the matching of machines with their fluid networks. Similarly the energy equation is essential in the measurement of flow velocity, this application being historically linked to the application of manometer techniques based on the hydrostatic equation.

An understanding of the manner in which the available energy in a fluid network changes in response to frictional or separation losses or to the presence of fans, pumps or turbines is essential. This chapter has laid the foundations for that understanding.

Summary of important equations and concepts

1. This chapter introduces the steady flow energy equation (SFEE), equation (6.13) and shows how it applies to a wide range of flow conditions, Section 6.4. This is probably the most important single equation to be met in the application of fluid theory to a wide range of actual applications, particularly when the frictional and separation loss terms are fully understood. The SFEE is an essential tool.

2. In addition to the application of the SFEE to conduit flow conditions, and its use in explaining the energy transfers within a fluid system, it also forms the basis for the measurement of flow velocity and flow rate, by Pitot–static tube or by venturi meter or orifice plate, Sections 6.5 through to 6.10.

3. Velocity profiles in real fluid flow situations make volumetric flow measurement difficult in some cases and the use of Pitot–static tube data in conjunction with an array and summing program are demonstrated, program VOLFLO.

4. The energy concepts demonstrated allow orifice flow to be summed and the various coefficients of velocity and discharge used to align 'real' flow with the theoretical to be introduced and defined.

5. This application of the energy equation to vortex flow allows the definition of both forced and free vortices, equations (6.42) and (6.45).

Problems

6.1 The suction pipe of a pump rises at a slope of 1 vertical in 5 along the pipe and water passes through it at 1.8 m s^{-1}. If dissolved air is released when the pressure falls to more than 70 kN m^{-2} below atmospheric pressure, find the greatest practicable length of pipe neglecting friction. Assume that the water in the sump is at rest.   [34.9 m]

6.2 A jet of water is initially 12 cm in diameter and when directed vertically upwards reaches a maximum height of 20 m. Assuming that the jet remains circular determine the rate of water flowing and the diameter of the jet at a height of 10 m.   [0.224 m^3 s^{-1}, 14.27 cm]

6.3 A pipe AB carries water and tapers uniformly from a diameter of 0.1 m at A to 0.2 m at B over a length of 2 m. Pressure gauges are installed at A, B and also at C, the midpoint of AB. If the pipe centreline slopes upwards from A to B at an angle of 30° and the pressures recorded at A and B are 2.0 and 2.3 bar, respectively, determine the flow through the pipe and pressure recorded at C neglecting all losses.   [0.0723 m^3 s^{-1}, 2.29 bar]

6.4 Air enters a compressor at the rate of 0.5 kg s^{-1} with a velocity of 6.4 m s^{-1}, specific volume 0.85 m^3 kg^{-1} and a pressure of 1 bar. It leaves the compressor at a pressure of 6.9 bar with a specific volume of 0.16 m^3 kg^{-1} and a velocity of 4.7 m s^{-1}. The internal energy of the air at exit is greater than that at entry by 85 kJ kg^{-1}. The compressor is fitted with a cooling system which removes heat at the rate of 60 kJ s^{-1}. Calculate the power required to drive the compressor and the cross-sectional areas of the inlet and outlet pipes.   [115.2 kW, 0.0664 m^2, 0.0170 m^2]

6.5 A Pitot–static tube is used to measure air velocity. If a manometer connected to the instrument indicates a difference in pressure head between the tappings of 4 mm of water, calculate the air velocity assuming the coefficient of the Pitot tube to be unity. Density of air = 1.2 kg m^{-3}.   [8.08 m s^{-1}]

6.6 A liquid flows through a circular pipe 0.6 m in diameter. Measurements of velocity taken at intervals along a diameter are:

Distance from the wall m	0	0.05	0.1	0.2	0.3
Velocity m s^{-1}	0	2.0	3.8	4.6	5.0

Distance from the wall m	0.4	0.5	0.55	0.6
Velocity m s^{-1}	4.5	3.7	1.6	0

Draw the velocity profile, calculate the mean velocity and determine the kinetic energy correction factor.   [2.82 m s^{-1}, 1.899]

6.7   A venturi meter with a throat diameter of 100 mm is fitted in a vertical pipeline of 200 mm diameter with oil of specific gravity 0.88 flowing upwards at a rate of 0.06 m³ s⁻¹. The venturi meter coefficient is 0.96. Two pressure gauges calibrated in kilonewtons per square metre are fitted at tapping points, one at the throat and the other in the inlet pipe 320 mm below the throat. The difference between the two gauge pressure readings is 28 kN m⁻².

Working from Bernoulli's equation determine the difference in level in the two limbs of a mercury manometer if it is connected to the tapping points and the connecting pipes are filled with the same oil. [202 mm]

6.8   An orifice plate is to be used to measure the rate of air flow through a 2 m diameter duct. The mean velocity in the duct will not exceed 15 m s⁻¹ and a water tube manometer, having a maximum difference between water levels of 150 mm, is to be used. Assuming the coefficient of discharge to be 0.64, determine a suitable orifice diameter to make full use of the manometer range. Take the density of air as 1.2 kg m⁻³. [1.31 m]

6.9   A sharp-edged orifice, 5 cm in diameter, in the vertical side of a large tank discharges under a head of 5 m. If $C_c = 0.62$ and $C_v = 0.98$, determine (a) the diameter of the jet at the vena contracta, (b) the velocity of the jet at the vena contracta and (c) the discharge in cubic metres per second. [(a) 3.94 cm, (b) 9.71 m s⁻¹, (c) 0.0118 m³ s⁻¹]

6.10   Find the diameter of a circular orifice to discharge 0.015 m³ s⁻¹ under a head of 2.4 m using a coefficient of discharge of 0.6. If the orifice is in a vertical plane and the jet falls 0.25 m in a horizontal distance of 1.3 m from the vena contracta, find the value of the coefficient of contraction. [6.82 cm, 0.715]

6.11   A tank has a circular orifice 20 mm diameter in the vertical side near the bottom. The tank contains water to a depth of 1 m above the orifice with oil of relative density 0.8 for a depth of 1 m above the water. Acting on the upper surface of the oil is an air pressure of 20 kN m⁻² gauge. The jet of water issuing from the orifice travels a horizontal distance of 1.5 m from the orifice while falling a vertical distance of 0.156 m. If the coefficient of contraction of the orifice is 0.65, estimate the value of the coefficient of velocity and the actual discharge through the orifice.
[0.97, 1.72 dm³ s⁻¹]

6.12   Water flows from a reservoir through a rectangular opening 2 m high and 1.2 m wide in the vertical face of a dam. Calculate the discharge in cubic metres per second when the free surface in the reservoir is 0.5 m above the top of the opening assuming a coefficient of discharge of 0.64. [8.16 m³ s⁻¹]

6.13   A vertical triangular orifice in the wall of a reservoir has a base 0.9 m long, 0.6 m below its vertex and 1.2 m below the water surface. Determine the theoretical discharge. [1.19 m³ s⁻¹]

6.14   A rectangular channel 1.2 m wide has at its end a rectangular sharp-edged notch with an effective width after allowing for side contractions of 0.85 m and with its sill 0.2 m from the bottom of the channel. Assuming that the velocity head averaged over the channel is $\alpha V^2/2g$ where V is the mean velocity and $\alpha = 1.1$, calculate the discharge in cubic metres per second when the head is 250 mm above the sill allowing for the velocity of approach.
[0.204 m³ s⁻¹]

6.15   In an experiment on a 90° vee notch the flow is collected in a 0.9 m diameter vertical cylindrical tank. It is found that the depth of water increases by 0.685 m in 16.8 s when the head over the notch is 0.2 m. Determine the coefficient of discharge of the notch. [0.613]

6.16   A pump discharges 2 m³ s⁻¹ of water through a pipeline. If the pressure difference between the inlet and the outlet of the pump is equivalent to 10 m of water, what power is being transmitted to the water from the pump?
[196.2 kW]

6.17   Inward radial flow occurs between two horizontal discs 0.6 m in diameter and 75 mm apart, the water leaving through a central pipe 150 mm in diameter in the lower disc at the rate of 0.17 m³ s⁻¹. If the absolute pressure at the outer edge of the disc is 101 kN m⁻² calculate the pressure at the outlet. Find also the resultant force on the upper disc.
[90 kN m⁻² abs, 373 N]

6.18   Two horizontal discs are 12.5 mm apart and 300 mm in diameter. Water flows radially outwards between the discs from a 50 mm diameter pipe at the centre of the lower disc. If the pressure at the outer edge of the disc is atmospheric, calculate the pressure in the supply pipe when the velocity of the water in the pipe is 6 m s⁻¹. Find also the resultant force on the upper disc, neglecting impact force.
[−17.5 kN m⁻², −92.4 N]

6.19 A hollow cylindrical drum with its axis vertical has an internal diameter of 600 mm and is full of water. A set of paddles 200 mm in diameter rotates concentrically with the axis of the drum at 120 rev min^{-1} and produces a compound vortex in the water. Assuming that all the water in the 200 cm core rotates as a forced vortex with the paddles and that the water outside this core moves as a free vortex, determine (*a*) the velocity of the water at 75 mm and 225 mm from the centre, and (*b*) the pressure head at these radii above the pressure head at the centre.

[(*a*) 0.943 m s^{-1}, 0.56 m s^{-1}, (*b*) 45.3 mm, 145 mm]

Two-dimensional Ideal Flow

AN IDEAL FLOW IS A PURELY THEORETICAL CONCEPT AS SUCH FLOWS POSSESS NO VISCOSITY, compressibility, surface tension or vaporization pressure limit. However, the mathematical treatment of such flows was fundamental in the development of modern fluid mechanics and finds application in the development of aerofoil lift, fan/pump blade design and groundwater flow predictions. This chapter will introduce the fundamental definitions of idealized flow, including circulation and vorticity, stream function, velocity potential and the techniques necessary for the generation of flow nets. The representation of flow conditions by a combination of rectilinear flows, sources and sinks will be demonstrated, leading to both the representation of flow over a cylinder, and, with the inclusion of circulation, the prediction of lift forces. A computer program to predict the lift coefficient and location of the stagnation point on a rotating cylinder is introduced. ● ● ●

In Chapter 4, a distinction was made between real and ideal fluids. The former exhibits the effects of viscosity and will be dealt with in the next part of the book, whereas the latter will be considered in this chapter.

An ideal fluid is a purely hypothetical fluid which is assumed to have no viscosity and no compressibility, and, in the case of liquids, no surface tension and no vaporization. The study of flow of such a fluid stems from the eighteenth-century hydrodynamics developed by mathematicians, who, by making the above assumptions regarding the fluid, aimed at establishing mathematical models for fluid flow. Although the assumptions of ideal flow appear to be very far fetched, the introduction of the boundary layer concept by Prandtl in 1904 enabled the distinction to be made between two regimes of flow: that adjacent to the solid boundary, in which viscosity effects are predominant and, therefore, the ideal flow treatment would be erroneous, and that outside the boundary layer, in which viscosity has negligible effect so that the idealized flow conditions may be applied. This argument is developed further in Chapter 12 when dealing with external flow.

The ideal flow theory may also be extended to situations in which fluid viscosity is very small and velocities are high, since they correspond to very high values of Reynolds number, at which flows are independent of viscosity. Thus, it is possible to see ideal flow as that corresponding to an infinitely large Reynolds number and to zero viscosity. The applications of ideal flow theory are found in aerodynamics, in accelerating flow, tides and waves.

The study of ideal flow provides mathematical expressions for streamlines in elementary or basic flow patterns. By combining these basic flow patterns in various ways, it is possible to obtain complex flow patterns which, in many cases, resemble remarkably closely the real situations outside the boundary layer and any associated wakes.

7.1   ROTATIONAL AND IRROTATIONAL FLOW

Considerations of ideal flow lead to yet another flow classification, namely the distinction between rotational and irrotational flow.

Basically, there are two types of motion: translation and rotation. The two may exist independently or simultaneously, in which case they may be considered as one super-imposed on the other. If a solid body is represented by a square, then pure translation or pure rotation may be represented as shown in Fig. 7.1(a) and (b), respectively.

FIGURE 7.1
Translation and rotation

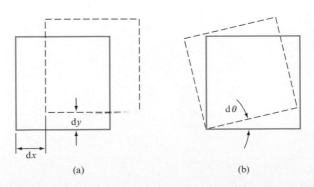

(a)   (b)

FIGURE 7.2
Linear and angular
deformation

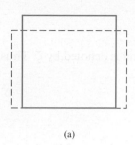
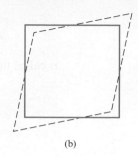

(a)   (b)

FIGURE 7.3
Rotation, translation and
deformation

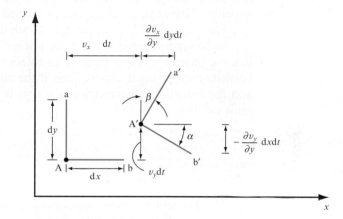

If we now consider the square to represent a fluid element, it may be subjected to deformation. This can be either linear or angular, as shown in Fig. 7.2(a) and (b), respectively.

Now, consider a motion of a fluid in which rotation of fluid elements is superimposed on their translation. In time dt, then, point A on the fluid element aAb moves to A′ and the element assumes position a′A′b′, as shown in Fig. 7.3. The two angles of rotation α and β will not be the same if deformation takes place and, therefore, the average rate of rotation in time dt will be

$$\omega = \frac{\alpha + \beta}{2} \times \frac{1}{dt} = \frac{1}{2} \frac{(\alpha + \beta)}{dt},$$

but, for small values and taking anticlockwise rotation as positive,

$$\alpha = \frac{\text{Arc}}{\text{Radius}} = \frac{\partial v_y}{\partial x} dx \, dt \frac{1}{dx} = \frac{\partial v_y}{\partial x} dt,$$

and $$\beta = -\frac{\partial v_x}{\partial y} dy \, dt \frac{1}{dy} = \frac{-\partial v_x}{\partial y} dt.$$

The rate of rotation about the z axis is, therefore,

$$\omega_z = \frac{1}{2}\left(\frac{\partial v_y}{\partial x} dt - \frac{\partial v_x}{\partial y} dt\right)\frac{1}{dt} = \frac{1}{2}\left(\frac{\partial v_y}{\partial x} - \frac{\partial v_x}{\partial y}\right).$$

The expression in brackets,

$$\frac{\partial v_y}{\partial x} - \frac{\partial v_x}{\partial y} = \zeta, \tag{7.1}$$

is called the *vorticity* and is denoted by ζ. Thus,

$$\zeta = 2\omega_z, \tag{7.2}$$

where ω_z is the angular velocity of the fluid elements about their mass centre in the x–y plane. In three-dimensional flow, ω_z would represent only one of three components of the angular velocity ω and vorticity would be equal to 2ω.

The expression (7.1) was obtained by stipulating rotation of fluid elements to exist and to be superimposed on their translation. Such a flow is known as *rotational*. It follows, therefore, that if there is no rotation, the expression (7.1) and, hence, the vorticity must be equal to zero. Thus, if the motion of particles is purely translational and the distortion is symmetrical, the flow is *irrotational* and the condition which it must satisfy is

$$\frac{\partial v_y}{\partial x} - \frac{\partial v_x}{\partial y} = 2\omega_z = 0. \tag{7.3}$$

The distinction between rotational and irrotational flow is important because, for example, it will be shown later that Bernoulli's equation derived for a streamline applies to all streamlines in the flow field only if the flow is irrotational. It will also be shown that the generation of lift by such surfaces as aerofoils is associated with irrotational flow. Also, a useful and practical procedure of determining 'flow nets' can only be applied to irrotational flow.

7.2 CIRCULATION AND VORTICITY

Consider a fluid element ABCD in rotational motion. Let the velocity components along the sides of the element be as shown in Fig. 7.4. Since the element is rotating,

FIGURE 7.4
Circulation

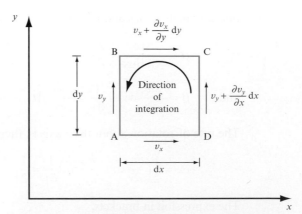

being part of rotational flow, there must be a 'resultant' peripheral velocity. However, since the centre of rotation is not known, it is more convenient to relate rotation to the sum of products of velocity and distance around the contour of the element. Such a sum is, of course, the line integral of velocity around the element and it is called *circulation*, denoted by Γ. Thus,

$$\text{Circulation, } \Gamma = \oint v_s \, \mathrm{d}s. \tag{7.4}$$

Circulation is, by convention, regarded as positive for anticlockwise direction of integration. Thus, for the element ABCD, starting from side AD,

$$\Gamma_{ABCD} = v_x \, \mathrm{d}x + \left(v_y + \frac{\partial v_y}{\partial x} \mathrm{d}x\right)\mathrm{d}y - \left(v_x + \frac{\partial v_x}{\partial y} \mathrm{d}y\right)\mathrm{d}x - v_y \, \mathrm{d}y$$

$$= \frac{\partial v_y}{\partial x} \mathrm{d}x \, \mathrm{d}y - \frac{\partial v_x}{\partial y} \mathrm{d}y \, \mathrm{d}x$$

$$= \left(\frac{\partial v_y}{\partial x} - \frac{\partial v_x}{\partial y}\right)\mathrm{d}x \, \mathrm{d}y,$$

but
$$\left(\frac{\partial v_y}{\partial x} - \frac{\partial v_x}{\partial y}\right) = \zeta$$

for two-dimensional flow in the x–y plane and, therefore, is the vorticity of the element about the z axis, ζ_z. The product $\mathrm{d}x \, \mathrm{d}y$ is the area of the element $\mathrm{d}A$. Thus,

$$\Gamma_{ABCD} = \left(\frac{\partial v_y}{\partial x} - \frac{\partial v_x}{\partial y}\right)\mathrm{d}x \, \mathrm{d}y = \zeta_z \, \mathrm{d}A.$$

It is seen, therefore, that the circulation around a contour is equal to the sum of the vorticities within the area of the contour. This is known as Stokes' theorem and may be stated mathematically, for a general case of any contour C (Fig. 7.5), as

$$\Gamma_C = \oint v \cos \theta \, \mathrm{d}s = \int_A \zeta \, \mathrm{d}A. \tag{7.5}$$

The concept of circulation is very important in the theory of lifting surfaces such as aerofoils, hydrofoils and blades of rotodynamic machines.

FIGURE 7.5
Circulation and vorticity

The above considerations indicate that, for irrotational flow, since vorticity is equal to zero, the circulation around a closed contour through which fluid is flowing must be equal to zero.

7.3   STREAMLINES AND THE STREAM FUNCTION

In Section 4.1, a distinction was made between streaklines, pathlines and streamlines. Of the three, the streamline is the one which is a purely theoretical line in space, defined as being tangential to instantaneous velocity vectors. From this definition of a streamline, it follows that there can be no flow across it, simply because a line cannot be tangential to a velocity vector which at the same time crosses it.

The concept of the streamline is very useful, especially in ideal flow, because it enables the fluid flow to be conceived as occurring in patterns of streamlines. These patterns may be described mathematically so that the whole system of analysis may be based on it. It requires, however, a mathematical definition of a streamline. Consider, in a two-dimensional case, the velocity and displacement vectors of a fluid at a point, together with their orthogonal components, as shown in Fig. 7.6.

FIGURE 7.6
Velocity and displacement components

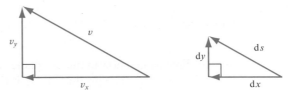

Since, by definition of a streamline, $\mathrm{d}s \parallel v$, it follows that

$$\mathrm{d}y \parallel v_y \quad \text{and} \quad \mathrm{d}x \parallel v_x.$$

Thus, the velocity triangle and the displacement triangle are similar and, therefore,

$$\frac{\mathrm{d}x}{v_x} = \frac{\mathrm{d}y}{v_y}. \tag{7.6}$$

This constitutes the equation of a streamline.

Since the streamlines in a flow pattern describe it, it is useful to label them by some numerical system. Furthermore, it is possible to relate the numerical labels to the flow rate of the pattern which is being described. Thus, let aa and bb be two streamlines in a flow bounded by solid boundaries AA and BB in Fig. 7.7 If the streamline aa is denoted by Ψ_a, which will be labelled by a numerical value representing the flow rate per unit depth between AA and the streamline aa, then,

$$\Psi_a = Q_{0c}$$

and, similarly, if

$$\Psi_b = Q_{0e},$$

FIGURE 7.7
The stream function

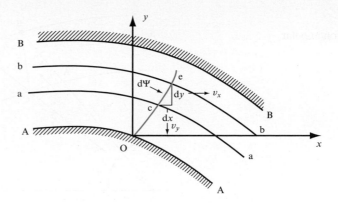

it follows that

$$\mathrm{d}\Psi = \Psi_\mathrm{b} - \Psi_\mathrm{a} = Q_\mathrm{ce},$$

so that

$$\mathrm{d}\Psi = v_x\,\mathrm{d}y - v_y\,\mathrm{d}x \qquad (7.7)$$

and Ψ, which is called the *stream function*, is given by

$$\Psi = \int v_x\,\mathrm{d}y - \int v_y\,\mathrm{d}x. \qquad (7.8)$$

Thus, the stream function depends upon position coordinates,

$$\Psi = f(x,\,y)$$

and, hence, the total derivative:

$$\mathrm{d}\Psi = \frac{\partial\Psi}{\partial x}\mathrm{d}x + \frac{\partial\Psi}{\partial y}\mathrm{d}y. \qquad (7.9)$$

Comparing equations (7.9) and (7.7), the relationships between the stream function and the velocity components are obtained:

$$v_x = \frac{\partial\Psi}{\partial y} \quad \text{and} \quad v_y = -\frac{\partial\Psi}{\partial x}. \qquad (7.10)$$

(*Note*: the sign convention adopted here is for the flow to be positive from left to right.) Since the value of a stream function represents the flow rate between a given streamline described by the stream function and a reference boundary, it follows that it must be constant for the given streamline in order to satisfy the continuity equation combined with the requirement of no flow across a streamline.

FIGURE 7.8
Stream function in polar
coordinates

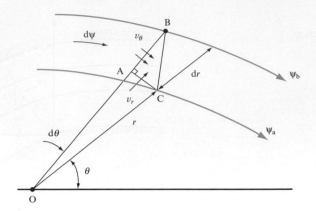

In some curved flows, it is more convenient to use the polar coordinates in mathematical analysis. In these, since

$$\Psi = f(r, \theta),$$

by differentiation,

$$d\Psi = \frac{\partial \Psi}{\partial r}\, dr + \frac{\partial \Psi}{\partial \theta}\, d\theta. \tag{7.11}$$

The sign convention in polar coordinates is that the tangential velocity is positive in the direction of positive θ, i.e. anticlockwise; the radial velocity is positive in the outward direction.

Consider, now, two curved streamlines Ψ_a and Ψ_b, as shown in Fig. 7.8. Assuming that $AB = dr$ when $d\theta = 0$ and applying the continuity equation, the following relationship is obtained:

$$v_r(r\, d\theta) - v_\theta\, dr = d\Psi. \tag{7.12}$$

Comparing equations (7.11) and (7.12), the following relationships are deduced:

$$v_\theta = -\frac{\partial \Psi}{\partial r} \quad \text{and} \quad v_r = \frac{1}{r}\frac{\partial \Psi}{\partial \theta}. \tag{7.13}$$

7.4   VELOCITY POTENTIAL AND POTENTIAL FLOW

In connection with flow nets mentioned earlier, an important concept is that of the *velocity potential*. It is defined as

$$\Phi = \int_A^B v_s\, ds, \tag{7.14}$$

FIGURE 7.9

Different paths in the
potential field

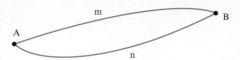

where A and B are two points in a potential field and v_s is the velocity tangential to
the elementary path s.

To understand the above concept, it is necessary to appreciate the meaning of the
term 'potential', so often used in mechanics as well as in other situations which satisfy
the same specific conditions. Consider, for example, points P and P' in a gravitational
field. If P' has a greater potential than P, the difference in their potential δW is defined
as the work required to move a particle from P to P' against the gravitational force. If
the distance between the points is δs and the force required is F then

$$\delta W = F\delta s,$$

or
$$W = \int_P^{P'} F\,\mathrm{d}s.$$
(7.15)

Clearly, the work done in such a case is independent of the path taken in doing the
work and this property is a characteristic of potential fields only. In any other case, say
against friction, the work done would depend upon the path taken. Therefore, not all
fields are potential, but only those in which the path taken is immaterial.

Returning to our velocity field, if points A and B in Fig. 7.9 belong to some
potential field, then the integral

$$\int_A^B v\,\mathrm{d}s$$

is independent of the path taken. Therefore,

$$\int_A^B v_\mathrm{m}\,\mathrm{d}s = \int_A^B v_\mathrm{n}\,\mathrm{d}s.$$
(7.16)

From the analogy with the gravitational field, it is apparent that the condition of
equation (7.16) will only be satisfied if the field is potential. Fluid flow in such a field
is known as *potential flow*.

Consider, now, the circulation around AnBm:

$$\Gamma_\mathrm{AnBm} = \int_A^B v_\mathrm{n}\,\mathrm{d}s + \int_B^A v_\mathrm{m}\,\mathrm{d}s$$

$$= \int_A^B v_\mathrm{n}\,\mathrm{d}s - \int_A^B v_\mathrm{m}\,\mathrm{d}s.$$
(7.17)

But, for the flow to be potential, the two integrals in equation (7.17) must be equal
and, therefore, $\Gamma_\mathrm{AnBm} = 0$. If the circulation is equal to zero, it follows that vorticity
must also be equal to zero and, therefore, the condition for potential flow is

$$\frac{\partial v_y}{\partial x} - \frac{\partial v_x}{\partial y} = 0, \tag{7.18}$$

which is identical with the condition for irrotational flow. Thus, potential flow is irrotational and vice versa.

In irrotational (potential) flow, therefore, the function (7.14)

$$\Phi = \int_A^B v_s \, ds$$

exists, from which it follows that, if v_x and v_y are the orthogonal components of v_s, then,

$$v_x = \frac{\partial \Phi}{\partial x} \quad \text{and} \quad v_y = \frac{\partial \Phi}{\partial y}, \tag{7.19}$$

so that

$$\Phi = \int v_x \, dx + \int v_y \, dy. \tag{7.20}$$

Similarly, in polar coordinates, if v_θ and v_r are tangential and radial components of v_s, then

$$\Phi = \int v_r \, dr + \int v_\theta r \, d\theta, \tag{7.21}$$

from which

$$v_r = \frac{\partial \Phi}{\partial r} \quad \text{and} \quad v_\theta = \frac{\partial \Phi}{r \partial \theta}. \tag{7.22}$$

It is now appropriate to consider the implications of potential flow to the applicability of Bernoulli's equation. It was originally derived in Section 5.12 to apply along a streamline, but not necessarily across streamlines, i.e. from one streamline to a neighbouring one. Furthermore, it was shown in Section 6.18 that in some curved flows the Bernoulli constant (or total head), defined as

$$H = p/\rho g + v^2/2g + z,$$

varies across streamlines, the variation in general being governed by equation (6.38), namely

$$\frac{dH}{dr} = \frac{v}{g}\left(\frac{v}{r} + \frac{dv}{dr}\right).$$

FIGURE 7.10
Circulation in polar
coordinates

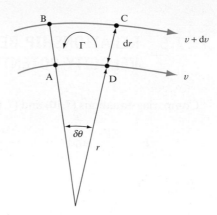

However, in some flows the Bernoulli constant is the same for all streamlines, so that Bernoulli's equation may be applied to any points in the flow field. Clearly, for this to happen dH/dr must be equal to zero. One such obvious case is when $dr \to \infty$, i.e. in the case of straight line flows. The other possibility is for the expression in brackets to be equal to zero, i.e.

$$\frac{v}{r} + \frac{dv}{dr} = 0. \tag{7.23}$$

Let us examine such a case by considering an element of fluid in a curved flow, as shown in Fig. 7.10. The circulation around the element ABCD is

$$\Gamma_{\text{ADCB}} = vr\delta\theta - (v + dv)(r + dr)\,\delta\theta = -v\,dr\delta\theta - dv\,dr\delta\theta - r\,dv\delta\theta.$$

Neglecting infinitesimals of the third order, this reduces to

$$\Gamma_{\text{ADCB}} = -v\,dr\delta\theta - r\,dv\delta\theta.$$

But the area of the element is $r\delta\theta\,dr$, so that the vorticity is given by

$$\zeta = \frac{\Gamma}{\text{Area}} = \frac{-v\,dr\delta\theta - r\,dv\delta\theta}{r\delta\theta\,dr}$$

$$= -\left(\frac{v}{r} + \frac{dv}{dr}\right), \tag{7.24}$$

which is the same as the left-hand side of equation (7.23). Thus, if the vorticity is zero, there is no variation of the Bernoulli constant. This condition applies to irrotational (potential) flow.

In potential flow, then, Bernoulli's equation applies to the whole flow field and is not limited to individual streamlines.

7.5   RELATIONSHIP BETWEEN STREAM FUNCTION AND VELOCITY POTENTIAL. FLOW NETS

Comparing equations (7.10) and (7.19), from (7.10)

$$v_x = \frac{\partial \Psi}{\partial y} \quad \text{and} \quad v_y = -\frac{\partial \Psi}{\partial x};$$

from (7.19),

$$v_x = \frac{\partial \Phi}{\partial x} \quad \text{and} \quad v_y = \frac{\partial \Phi}{\partial y}.$$

Thus, equating for v_x and v_y, we obtain

$$\frac{\partial \Psi}{\partial y} = \frac{\partial \Phi}{\partial x} \quad \text{and} \quad \frac{\partial \Phi}{\partial y} = -\frac{\partial \Psi}{\partial x}. \tag{7.25}$$

These equations are known as Cauchy–Riemann equations and they enable the stream function to be calculated if the velocity potential is known and vice versa in a potential flow.

It is now possible to return to the condition for potential flow and to restate it in terms of the stream function. The condition is

$$\frac{\partial v_y}{\partial x} - \frac{\partial v_x}{\partial y} = 0,$$

but

$$v_y = -\frac{\partial \Psi}{\partial x} \quad \text{and} \quad v_x = \frac{\partial \Psi}{\partial y},$$

so that, by substitution,

$$\frac{\partial}{\partial x}\left(-\frac{\partial \Psi}{\partial x}\right) - \frac{\partial}{\partial y}\left(\frac{\partial \Psi}{\partial y}\right) = 0$$

and

$$\frac{\partial^2 \Psi}{\partial x^2} + \frac{\partial^2 \Psi}{\partial y^2} = 0. \tag{7.26}$$

This is the Laplace equation for the stream function, which must be satisfied for the flow to be potential.

It is also interesting to note that the Laplace equation for the velocity potential must also be satisfied. This follows by substitution of equations (7.19) into the continuity equation for steady, incompressible, two-dimensional flow (equation (4.13)):

$$\frac{\partial v_x}{\partial x} + \frac{\partial v_y}{\partial y} = 0.$$

Substituting, now, for v_x and v_y from equations (7.19),

$$\frac{\partial}{\partial x}\left(\frac{\partial \Phi}{\partial x}\right) + \frac{\partial}{\partial y}\left(\frac{\partial \Phi}{\partial y}\right) = 0,$$

so that

$$\frac{\partial^2 \Phi}{\partial x^2} + \frac{\partial^2 \Phi}{\partial y^2} = 0. \tag{7.27}$$

Thus, the Laplace equation for the velocity potential must also be satisfied.

The fact that, for potential flow, both the stream function and the velocity potential satisfy the Laplace equation indicates that Ψ and Φ are interchangeable (Cauchy–Riemann equations) and that the lines of constant Ψ, i.e. streamlines, and the lines of constant Φ, called *equipotential* lines, are mutually perpendicular. This means that, if streamlines are plotted, points can be marked on them which have the same value of Φ and can be joined to form equipotential lines. Thus, a flow net of streamlines and equipotential lines is formed.

When the streamlines converge, the velocity increases and, therefore, for a given increment of $\delta\Psi$, the distance between the equipotential lines will also decrease.

The method of drawing a flow net consists of drawing by eye streamlines equispaced at $\delta\Psi$ at some section where the flow is rectilinear, such as AG or DE in Fig. 7.11, which shows an example of a network drawn for a rather unusual converging section, of which the upper half constitutes a sudden contraction but the lower half provides a smooth transition. The number of streamlines drawn, or rather, the size of intervals between them, depends upon the accuracy required. The more streamlines one uses, the more accurate will be the result, but, equally, the time spent in drawing the net will be greater. The set of equipotential lines is drawn next at intervals $\delta\Phi = \delta\Psi$ and in such a way that they cross each streamline at right angles. Thus a set of 'squares' is obtained. The process is done by eye and requires a series of successive adjustments to both streamlines and equipotential lines until a satisfactory network of 'squares' is achieved. As a final check, diagonals through the 'squares' may be drawn. They, too, should be smooth lines and should form a net of squares. A pair of such diagonals from A′ to F′ is shown in Fig. 7.11.

FIGURE 7.11

Example of a flow net

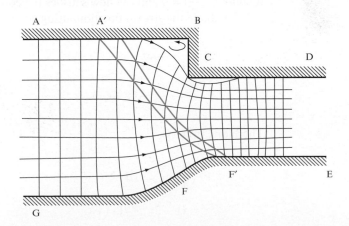

Where abrupt changes of the outer boundary occur, such as at points B and C, it can be seen that the streamline AA′D cannot follow the contour and separates from the boundary. At B, where the streamline turns towards the fluid, the velocity at the separation area will be zero and the fluid trapped there will be stagnant. At point C, the streamline turns away from the fluid, indicating high velocity in the separation bubble. This velocity is spent in rotation of considerable vigour. Certainly, therefore, the assumption of irrotational flow is not valid there. In general, then, wherever the streamlines diverge or converge abruptly, separation may occur. Because GFE is smooth and converging, no separation will occur there. Should, however, the flow direction be reversed, although the flow net would remain the same, separation might be expected downstream of F due to the divergence of flow. Separation phenomena are discussed fully in Chapter 11 in connection with boundary layer.

Constructing flow nets is a useful exercise which requires a lot of patience and experience. The alternative is to use precise mathematical expressions for stream function and velocity potential describing the flow from which a flow net can be plotted exactly. The following sections of this chapter deal with such mathematical expressions for some basic flows which may then be combined to represent more complex flow patterns.

EXAMPLE 7.1

In a two-dimensional incompressible flow the fluid velocity components are given by $v_x = x - 4y$ and $v_y = -y - 4x$. Show that the flow satisfies the continuity equation and obtain the expression for the stream function. If the flow is potential, obtain also the expression for the velocity potential.

Solution

For two-dimensional incompressible flow, the continuity equation is

$$\frac{\partial v_x}{\partial x} + \frac{\partial v_y}{\partial y} = 0,$$

but $v_x = x - 4y$ and $v_y = -y - 4x$,

and $\dfrac{\partial v_x}{\partial x} = 1, \quad \dfrac{\partial v_y}{\partial y} = -1$;

therefore, $1 - 1 = 0$ and the flow satisfies the continuity equation.

To obtain the stream function, using equations (7.10),

$$v_x = \frac{\partial \Psi}{\partial y} = x - 4y, \tag{I}$$

$$v_y = -\frac{\partial \Psi}{\partial x} = -(y + 4x). \tag{II}$$

Therefore, from (I),

$$\Psi = \int (x - 4y)\,dy + f(x) + C$$

$$= xy - 2y^2 + f(x) + C.$$

But, if $\Psi_0 = 0$ at $x = 0$ and $y = 0$, which means that the reference streamline passes through the origin, then $C = 0$ and

$$\Psi = xy - 2y^2 + f(x). \qquad \text{(III)}$$

To determine $f(x)$, differentiate partially the above expression with respect to x and equate to $-v_y$, equation (II):

$$\frac{\partial \Psi}{\partial x} = y + \frac{\partial}{\partial x} f(x) = y + 4x,$$

$$f(x) = \int 4x\,\mathrm{d}x = 2x^2.$$

Substitute into (III),

$$\Psi = 2x^2 + xy - 2y^2.$$

To check whether the flow is potential, there are two possible approaches:
(a) Since

$$\frac{\partial v_y}{\partial x} - \frac{\partial v_x}{\partial y} = 0,$$

but

$$v_y = -(4x + y) \quad \text{and} \quad v_x = (x - 4y),$$

therefore,

$$\frac{\partial v_y}{\partial x} = -4 \quad \text{and} \quad \frac{\partial v_x}{\partial y} = -4,$$

so that

$$\frac{\partial v_y}{\partial x} - \frac{\partial v_x}{\partial y} = -4 + 4 = 0$$

and the flow is potential.
(b) Laplace's equation must be satisfied:

$$\frac{\partial^2 \Psi}{\partial x^2} + \frac{\partial^2 \Psi}{\partial y^2} = 0,$$

$$\Psi = 2x^2 + xy - 2y^2.$$

Therefore,

$$\frac{\partial \Psi}{\partial x} = 4x + y \quad \text{and} \quad \frac{\partial \Psi}{\partial y} = x - 4y,$$

$$\frac{\partial^2 \Psi}{\partial x^2} = 4 \quad \text{and} \quad \frac{\partial^2 \Psi}{\partial y^2} = -4.$$

Therefore $4 - 4 = 0$ and so the flow is potential.
Now, to obtain the velocity potential,

$$\frac{\partial \Phi}{\partial x} = v_x = x - 4y;$$

therefore,

$$\Phi = \int (x - 4y)\, dx + f(y) + G.$$

But $\Phi_0 = 0$ at $x = 0$ and $y = 0$, so that $G = 0$. Therefore

$$\Phi = x^2/2 - 4yx + f(y).$$

Differentiating with respect to y and equating to v_y,

$$\frac{\partial \Phi}{\partial y} = -4x + \frac{d}{dy} f(y) = -(4x + y)$$

$$\frac{d}{dy} f(y) = -y \quad \text{and} \quad f(y) = -\frac{y^2}{2},$$

so that

$$\Phi = x^2/2 - 4yx - y^2/2$$

7.6   STRAIGHT LINE FLOWS AND THEIR COMBINATIONS

The simplest flow patterns are those in which the streamlines are all straight lines parallel to each other (Fig. 7.12).

FIGURE 7.12
Rectilinear flow

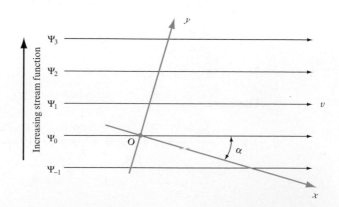

The convention for numbering the streamlines is that the stream function is considered to increase to the left of an observer looking downstream, i.e. in the direction of flow along the streamlines, as indicated in Fig. 7.12.

If the velocity of the rectilinear flow v is inclined to the x axis at an angle α, then its components are

$$v_x = v \cos \alpha \quad \text{and} \quad v_y = v \sin \alpha.$$

The stream function is obtained simply by substitution of the above expressions into

$$d\Psi = v_x \, dy - v_y \, dx,$$

whereupon

$$\Psi = \int v \cos \alpha \, dy - \int v \sin \alpha \, dx + \text{constant}.$$

Since in a uniform flow $v = \text{constant}$ and in a straight line flow α is also constant, the expression for the stream function becomes

$$\Psi = vy \cos \alpha - vx \sin \alpha + \text{constant}.$$

The constant of integration may be made zero by choosing the reference streamline $\Psi_0 = 0$ to pass through the origin, so that when $x = 0$ and $y = 0$ the stream function $\Psi = \Psi_0 = 0$. Thus

$$\Psi = v(y \cos \alpha - x \sin \alpha). \tag{7.28}$$

Since v_x and v_y are constant, then $\partial v_x / \partial y$ and $\partial v_y / \partial x$ are both zero and, therefore, the flow is potential.

The velocity potential is obtained from

$$d\Phi = \frac{\partial \Phi}{\partial x} dx + \frac{\partial \Phi}{\partial y} dy = v_x \, dx + v_y \, dy.$$

Therefore, by substitution and integration,

$$\Phi = \int v \cos \alpha \, dx + \int v \sin \alpha \, dy + \text{constant},$$

but, if $\Phi = \Phi_0 = 0$ at $x = 0$ and $y = 0$, then

$$\Phi = v(x \cos \alpha + y \sin \alpha). \tag{7.29}$$

Some simple straight line flows may be illustrated as follows.

1. Uniform, straight line flow in the direction Ox, velocity u, shown in Fig. 7.13. Let streamline $\Psi_0 = 0$ be along the x axis. Now,

$$v_x = \frac{\partial \Psi}{\partial y} = u = \text{constant}.$$

FIGURE 7.13
Straight line flow: $\Psi = uy$

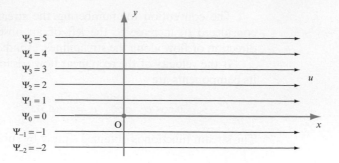

Therefore,

$$\partial\Psi = u\partial y.$$

Integrating,

$$\Psi = uy + \text{constant},$$

but $\Psi_0 = 0$ at $x = 0$ and $y = 0$ so that constant $= 0$, and the equation of the stream function becomes

$$\Psi = uy.$$

Alternatively, the volume flowing between the x axis and any streamline, per unit depth, is $q = uy$ and, therefore,

$$\Psi = uy.$$

   2. Uniform, straight line flow in the direction Oy, velocity v, shown in Fig. 7.14. Let streamline $\Psi_0 = 0$ be along the y axis. Now,

$$v_y = -\frac{\partial\Psi}{\partial y} = v = \text{constant}.$$

Therefore,

$$\partial\Psi = -v\,dx.$$

FIGURE 7.14
Straight line flow: $\Psi = -vx$

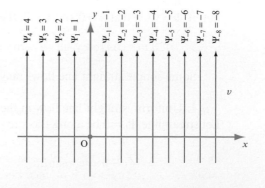

FIGURE 7.15

Combination of straight line flows

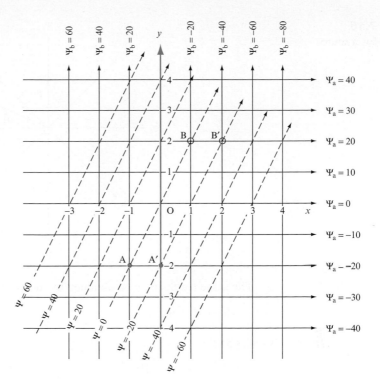

Integrating,

$$\Psi = -vx + \text{constant},$$

but $\Psi_0 = 0$ at $x = 0$ and $y = 0$ so that constant $= 0$, and the equation of the stream function is

$$\Psi = -vx.$$

3. Combined flow consisting of a uniform flow $u = 10 \text{ m s}^{-1}$ along Ox and uniform flow $v = 20 \text{ m s}^{-1}$ along Oy, shown in Fig. 7.15.

Choose a suitable scale for x and y, say 10 mm = 20 m. Draw horizontal streamlines $\Psi_0 = uy = 10y$ and label them. Draw vertical streamlines $\Psi_b = -vx = -20x$ and label them.

At point A the steam function due to v is $\Psi_b = 20$ and the stream function due to u is $\Psi_a = -20$. Therefore, the combined stream function (scalar quantity) is $\Psi = 20 - 20 = 0$. Similarly, it is zero at point B and the origin. Hence the stream function for the streamline passing through AOB is $\Psi = 0$.

By the same method, at point A′ the stream function due to v is $\Psi_b = 0$ and the stream function for the streamline due to u is $\Psi_a = -20$. Therefore, the combined stream function is $\Psi = 0 - 20 = -20$. Similarly, the combined stream function at B′ is also equal to -20. Thus, the straight line passing through A′B′ represents a streamline of the combined flow whose stream function is -20. By repeating the process of drawing lines through points at which the combined value of the stream function is the same, a new set of streamlines is obtained and it represents the combined flow pattern.

FIGURE 7.16
Radial flow: a source

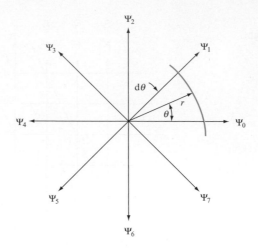

The same results may be obtained by the algebraic method. Since

$$\Psi = \Psi_a + \Psi_b,$$

it follows that

$$\Psi = uy - vx = 10y - 20x.$$

This equation represents a family of straight lines, each line being defined by the particular value of Ψ assigned to it.

The other basic flow patterns in which the streamlines are straight lines are those in which the fluid flows radially either outwards from a point, in which case it is know as a *source*, or inwards into a point, in which case it is known as a *sink*. A sink flow is simply treated as a negative source flow and, thus, the mathematics of both may be explained by considering only the source flow, which is shown in Fig. 7.16.

Radial flows and their applications were discussed in Section 6.17, but here we are concerned with the mathematical expressions for their stream function and velocity potential which lead to more complex and useful flow combinations.

In radial flows, it is seen that, since the velocity passes through the origin and is a function of θ only, the tangential component of velocity does not exist and

$$v = v_r.$$

Consider now a source of unit depth and let the steady rate of flow be q, known as the *strength* of the source. Then, at any radius r, the radial velocity is given by

$$v_r = q/2\pi r. \tag{7.30}$$

The stream function and the velocity potential are obtained in a similar manner as for the rectilinear flow, but polar coordinates are used. Since, from equation (7.12),

$$d\Psi = rv_r \, d\theta - v_\theta \, dr,$$

for radial flow $v_\theta = 0$, and for a source $v_r = q/2\pi r$, it follows that

$$d\Psi = r(q/2\pi r) \, d\theta = (q/2\pi) \, d\theta.$$

Integrating,

$$\Psi = q\theta/2\pi + \text{constant}.$$

If, however, $\Psi = \Psi_0 = 0$ when $\theta = 0$, the constant of integration becomes zero and

$$\Psi = q\theta/2\pi \quad \text{for a source,} \tag{7.31}$$

$$\Psi = -q\theta/2\pi \quad \text{for a sink.} \tag{7.32}$$

Similarly, it may be shown that

$$\Phi = (q/2\pi) \log_e r \quad \text{for a source,} \tag{7.33}$$

$$\Phi = -(q/2\pi) \log_e r \quad \text{for a sink.} \tag{7.34}$$

The simplest case of combining flow patterns is that in which a source is added to a uniform rectilinear flow. This is accomplished by the additions of the stream functions of the two types of flow. The stream function for a uniform rectilinear flow parallel to the x axis is

$$\Psi_R = v_0 y = v_0 r \sin\theta,$$

and that for a source is

$$\Psi_S = q\theta/2\pi.$$

Thus, the stream function for the combined flow is

$$\Psi = \Psi_R + \Psi_S = v_0 r \sin\theta + q\theta/2\pi. \tag{7.35}$$

Figure 7.17 shows graphically that this is the superposition of a system of radial streamlines onto a system of straight streamlines parallel to the x axis. By definition, a given streamline is associated with one particular value of the stream function and, therefore, if we join the points of intersection of the radial streamlines with the rectilinear streamlines where the sum of the stream functions is a given constant value, the resulting line will be one streamline of the combined flow pattern. If this procedure

FIGURE 7.17

Combination of rectilinear flow and a source

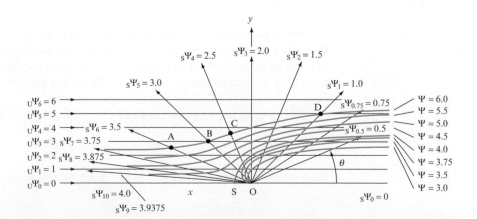

is repeated for a number of values of the combined stream function, the result will be a picture of the combined flow pattern. This is shown in Fig. 7.17, where numerical values were assigned to stream functions in order to illustrate the point. Observe, for example, the streamline of the combined flow $\Psi = 6$. It passes through points A, B, C and D such that:

$$\text{at A, } \Psi_U = 2.5 \quad \text{and} \quad \Psi_S = 3.5; \text{ therefore } \Psi = 2.5 + 3.5 = 6$$

$$\text{at B, } \Psi_U = 3.0 \quad \text{and} \quad \Psi_S = 3.0; \text{ therefore } \Psi = 3.0 + 3.0 = 6$$

$$\text{at C, } \Psi_U = 3.5 \quad \text{and} \quad \Psi_S = 2.5; \text{ therefore } \Psi = 3.5 + 2.5 = 6$$

$$\text{at D, } \Psi_U = 5.0 \quad \text{and} \quad \Psi_S = 1; \quad \text{therefore } \Psi = 5.0 + 1.0 = 6.$$

All streamlines of the combined flow are obtained in this manner.

It is interesting to note, in this particular flow pattern, that the resulting streamlines are grouped into two distinct sets. In one set all the streamlines emerge from the origin ($\Psi = 3, 3.5, 3.75$) and in the other they approach the rectilinear flow asymptotically at some distance upstream ($\Psi = 4.5, 5, 6$). The two sets are separated by the streamline $\Psi = 4$, which passes through point S. This point is a *stagnation point*, where the velocity from the source is equal to the uniform velocity of the parallel flow, so that the resultant velocity at S is zero. The distance OS $= a$ may, therefore, be determined by equating the uniform velocity to that from the source at radius a. Thus,

$$v_0 = q/2\pi a,$$

so that

$$a = q/2\pi v_0. \tag{7.36}$$

The value of the stream function for the streamline passing through point S is obtained by substituting $\theta = \pi$ and $r = a = q/2\pi v_0$ into (7.35). Thus,

$$\Psi_S = v_0(q/2\pi v_0) \sin \pi + q\pi/2\pi,$$

which simplifies to

$$\Psi_S = \tfrac{1}{2} q. \tag{7.37}$$

Since there can be no flow across a streamline, then the streamline Ψ_S passing through S may be replaced by a solid boundary of an object under investigation, such as a hill or the nose of an aerofoil. In the latter case, the flow pattern below the x axis must also be used as shown in Fig. 7.18. It is then known as a *half-body* or *Rankine body*.

The general equation for the streamline through point S is

$$\Psi = \tfrac{1}{2} q = v_0 r \sin \theta + q\theta/2\pi, \tag{7.38}$$

from which the radial distance r_S to any point on this streamline is

$$r_S = q(\pi - \theta)/2\pi v_0 \sin \theta, \tag{7.39}$$

FIGURE 7.18
Rankine body

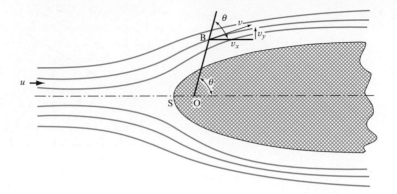

and it describes the contour of the Rankine body. It can be appreciated that as $x \to \infty$ this streamline becomes parallel to the x axis and, there, the perpendicular distance from the x axis to the streamline represents the maximum half-width of the Rankine body. The perpendicular distance is given by

$$y = r \sin \theta = q(\pi - \theta)/2\pi v_0.$$

But, as $x \to \infty$, the radius $r \to \infty$ and $\theta \to 0$, so that

$$y_{max} = q/2v_0. \tag{7.40}$$

EXAMPLE 7.2

In the ideal flow around a half-body, the free stream velocity is 0.5 m s^{-1} and the strength of the source is $2.0 \text{ m}^2 \text{ s}^{-1}$. Determine the fluid velocity and its direction at a point, $r = 1.0 \text{ m}$ and $\theta = 120°$.

Solution

The stream function for the flow around a half-body is given by

$$\Psi = v_0 r \sin \theta + q\theta/2\pi.$$

In this case, $v_0 = 0.5 \text{ m s}^{-1}$, $q = 2.0 \text{ m}^2 \text{ s}^{-1}$. To determine the fluid velocity and its velocity vector at a point, it is first necessary to determine its tangential and radial components. These are

$$v_r = \frac{1}{r}\frac{\partial \Psi}{\partial \theta} \quad \text{and} \quad v_\theta = -\frac{\partial \Psi}{\partial r}.$$

Therefore,

$$v_r = \frac{1}{r}\left(v_0 r \cos \theta + \frac{q}{2\pi}\right) = \frac{1}{1}\left(0.5 \times 1 \times \cos 120° + \frac{2}{2\pi}\right)$$

$$= -0.25 + 0.318 = 0.068 \text{ m s}^{-1},$$

$$v_\theta = -v_0 \sin \theta = -0.5 \sin 120° = -0.433 \text{ m s}^{-1},$$

FIGURE 7.19

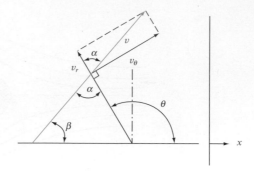

which is in the clockwise direction. Therefore,

$$v = \sqrt{(v_r^2 + v_\theta^2)} = \sqrt{(0.0047 + 0.188)} = \textbf{0.438 m s}^{-1}.$$

If β is the angle the velocity vector makes with the horizontal, as shown in Fig. 7.19, then

$$\beta = \theta - \alpha$$

and $\tan \alpha = v_\theta/v_r = 0.438/0.068 = 6.44.$

Therefore

$$\alpha = 81.2°, \text{ and } \beta = 120 - 81.2 = \textbf{38.8°}.$$

7.7   COMBINED SOURCE AND SINK FLOWS. DOUBLET

Let us consider the flow pattern resulting from the combination of a source and a sink of equal strength, which means that the flow rate from the source is equal to the flow rate into the sink. Also, let them be placed symmetrically about the origin and on the x axis, as shown in Fig. 7.20.

Let the stream function for the source be Ψ_1, for the sink be Ψ_2 and let the flow rate be q. Since the convention for stream functions is that they increase to the left while looking downstream, it follows that the stream functions for the source increase as the angle θ_1 increases and those for the sink decrease as angle θ_2 increases.

As discussed earlier, the value of a combined stream function is obtained by the addition of the values of stream functions at their intersection. For example, if the combined stream function is $\Psi = 5$, it will pass through points of intersection of Ψ_1 and Ψ_2 such that their values add up to 5 (e.g. $\Psi_1 = 1$ and $\Psi_2 = 4$ or $\Psi_1 = 2$ and $\Psi_2 = 3$ and so on, as shown in Fig. 7.20). Figure 7.20 also shows that the combined streamlines of a source and a sink of equal strength are circles passing through the point source and the point sink. Mathematically,

$$\Psi = \Psi_1 + \Psi_2 = q\theta_1/2\pi - q\theta_2/2\pi,$$
$$= (q/2\pi)(\theta_1 - \theta_2), \tag{7.41}$$

FIGURE 7.20
Source and sink

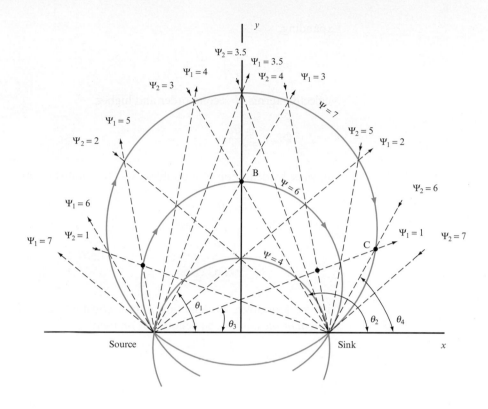

which, since Ψ and q are constant for any given streamline, is a condition satisfied by a circle.

Figure 7.20 shows only half of the flow pattern, the other half, below the x axis, being the mirror image of that above it.

The velocity potential for such a combined flow is also obtained by the addition of the velocity potentials for the source and the sink. Thus

$$\Phi = \Phi_{source} + \Phi_{sink} = (q/2\pi) \log_e r_1 - (q/2\pi) \log_e r_2,$$

$$\Phi = (q/2\pi)(\log_e r_1 - \log_e r_2).$$

(7.42)

7.7.1 Doublet

If a sink and a source of equal strength are brought together in such a way that the product of their strength and the distance between them remain constant, the resulting flow pattern is known as a *doublet*.

Consider point P (Fig. 7.21) on the velocity potential of a doublet. Let the velocity potential for the doublet be Φ_D; then

$$\Phi_D = (q/2\pi) \log_e(r + dr) - (q/2\pi) \log_e r$$

$$= \frac{q}{2\pi} \log_e \left(\frac{r + dr}{r}\right) = \frac{q}{2\pi} \log_e \left(1 + \frac{dr}{r}\right).$$

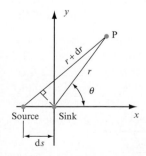

FIGURE 7.21
Source and sink

Expanding,

$$\log_e\left(1+\frac{\mathrm{d}r}{r}\right) = \frac{\mathrm{d}r}{r} - \frac{1}{2}\left(\frac{\mathrm{d}r}{r}\right)^2 + \cdots.$$

Neglecting terms of second order and higher,

$$\Phi_D = \frac{q}{2\pi}\frac{\mathrm{d}r}{r},$$

but, since by definition of a doublet $\mathrm{d}s \to 0$, it follows that

$$\mathrm{d}r \simeq \mathrm{d}s \cos\theta$$

and

$$\Phi_D = (q/2\pi r)\, \mathrm{d}s \cos\theta.$$

Also for a doublet, by definition $q\, \mathrm{d}s = $ constant. Let this constant, known as the *strength of the doublet*, be denoted by m; then

$$m = q\, \mathrm{d}s$$

and

$$\Phi_D = (m/2\pi r)\cos\theta \tag{7.43}$$

or, in rectangular coordinates,

$$\Phi_D = \frac{m}{2\pi}\left(\frac{x}{x^2 + y^2}\right). \tag{7.44}$$

From the above equations, the expressions for the stream function may be obtained, namely

$$\Psi_D = -(m/2\pi r)\sin\theta = -(m/2\pi)[y/(x^2 + y^2)]. \tag{7.45}$$

Note that the above equations were derived for a doublet in which the source and the sink were placed on the x axis. Such a doublet is shown in Fig. 7.22(a). If, however, the

FIGURE 7.22
Doublets

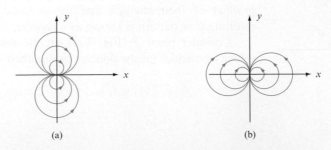

(a)   (b)

source and the sink are placed on the y axis, the resulting doublet is oriented as in Fig. 7.22(b) and the expressions for the stream function and the velocity potential become

$$\Phi_{D(yy)} = (m/2\pi r) \sin\theta = (m/2\pi)[y/(x^2 + y^2)], \qquad (7.46)$$

$$\Psi_{D(yy)} = -(m/2\pi r) \cos\theta = -(m/2\pi)[x/(x^2 + y^2)]. \qquad (7.47)$$

The flow is always from the source to the sink, so that, if they are placed as shown in Fig. 7.21, the flow in the doublet is as shown in Fig. 7.22(a). If, however, the positions of the source and the sink are reversed, the flow directions are also reversed, which means that the expressions for the stream function and the velocity potential change signs.

EXAMPLE 7.3

A source of strength 10 m² s⁻¹ at (1, 0) and a sink of the same strength at (−1, 0) are combined with a uniform flow of 25 m s⁻¹ in the −x direction. Determine the size of Rankine body formed by the flow and the difference in pressure between a point far upstream in the uniform flow and the point (1, 1).

Solution

For the source (Fig. 7.23),

$$\Psi_{\text{source}} = \frac{q}{2\pi}\theta = \frac{10}{2\pi}\tan^{-1}\left(\frac{y}{x-1}\right);$$

for the sink,

$$\Psi_{\text{sink}} = -\frac{q}{2\pi}\theta = -\frac{10}{2\pi}\tan^{-1}\left(\frac{y}{x+1}\right);$$

for the uniform flow,

$$\Psi_{\text{U}} = -25y.$$

Thus, the combined flow is represented by the stream function

$$\Psi = \frac{10}{2\pi}\left[\tan^{-1}\left(\frac{y}{x-1}\right) - \tan^{-1}\left(\frac{y}{x+1}\right)\right] - 25y.$$

FIGURE 7.23

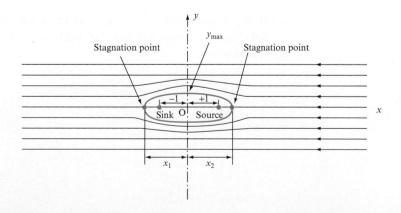

To obtain stagnation points, $v_x = 0$. Thus,

$$v_x = \frac{\partial \Psi}{\partial y} = \frac{10}{2\pi}\left[\frac{x-1}{(x-1)^2+y^2} - \frac{x+1}{(x+1)^2+y^2}\right] - 25.$$

Now, $v_x = 0$ at $y = 0$, i.e. on the x axis; therefore,

$$\frac{10}{2\pi}\left(\frac{1}{x-1} - \frac{1}{x+1}\right) = 25,$$

$$(x+1) - (x-1) = 5\pi(x-1)(x+1),$$

$$2/5\pi = x^2 - 1,$$

$$x^2 = 2/5\pi + 1 = 1.127,$$

$$x_{12} = \pm 1.062 \text{ m},$$

and the length of the Rankine body is

$$l = x_1 + x_2 = \textbf{2.124 m}.$$

To obtain the width of the Rankine body, it is necessary to determine the maximum value of y on the contour of the body, i.e. on Ψ_0. This will occur because of the symmetry at $v_y = 0$:

$$v_y = -\frac{\partial \Psi}{\partial x} = -\frac{10}{2\pi}\left[\frac{y}{(x-1)^2+y^2} - \frac{y}{(x+1)^2+y^2}\right] = 0.$$

Therefore,

$$\frac{y}{(x-1)^2+y^2} = \frac{y}{(x+1)^2+y^2},$$

but, since $y \neq 0$,

$$(x-1)^2 + y^2 = (x+1)^2 + y^2,$$

$$x^2 - 2x + 1 = x^2 + 2x + 1.$$

$$-4x = 0,$$

$$x = 0,$$

which is expected from the symmetry of the source and sink about the origin. To find the value of y_{max}, which will give the width of the body, substitute the above value of $x = 0$ into $\Psi = 0$:

$$0 = \frac{10}{2\pi}[\tan^{-1}(-y) - \tan^{-1} y] - 25 y_{max},$$

but, since $|y| = |-y|$,

$$\frac{25 \times 2\pi}{10} y_{max} = 2 \tan^{-1} y,$$

$$y_{max} = 0.127 \tan^{-1} y,$$

from which $y_{max} = 0.047$ m and the width of the Rankine body is $2y_{max} = \mathbf{0.094}$ m.
At point (1, 1),

$$v_x = \frac{10}{2\pi}(-\tfrac{2}{5}) - 25 = -25.63 \text{ m s}^{-1},$$

$$v_y = -\frac{10}{2\pi}(1 - \tfrac{1}{5}) = -1.27 \text{ m s}^{-1}.$$

Therefore,

$$v = (v_x^2 + v_y^2)^{1/2} = 25.66 \text{ m s}^{-1}.$$

Since the flow is potential, Bernoulli's equation may be applied to any two points, such as one in the free stream ($v_\infty = 25$ m s^{-1}) and point (1, 1):

$$p_\infty/\rho g + v_\infty^2/2g = p_{(1, 1)}/\rho g + v^2/2g,$$

and, hence,

$$p_{(1, 1)} - p_\infty = (\rho/2)(v^2 - v_\infty^2) = (\rho/2)33.43 \text{ N m}^{-2}.$$

7.8   FLOW PAST A CYLINDER

A flow pattern equivalent to that of an ideal fluid passing a stationary cylinder, with its axis perpendicular to the direction of flow, is obtained by combining a doublet with rectilinear flow. Figure 7.24 shows the resulting streamlines and the stagnation points S which are formed. The combined stream function and the velocity potential are

$$\Psi_c = \Psi_D + \Psi_R = -(m/2\pi r) \sin \theta + v_0 r \sin \theta$$
$$\Psi_c = (v_0 r - m/2\pi r) \sin \theta, \tag{7.48}$$

and $$\Phi_c = (v_0 r + m/2\pi r) \cos \theta. \tag{7.49}$$

Since the flow pattern corresponds to that around a cylinder, it is of interest to obtain an expression for the radius of this cylinder. Since the distance between the two

FIGURE 7.24
Flow around a cylinder

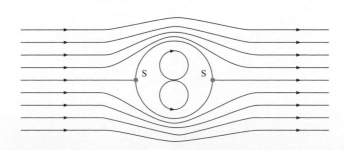

stagnation points is the diameter of this cylinder, say $2a$, and the flow at a stagnation point is zero, it follows that the streamline passing through S is $\Psi_0 = 0$. Thus

$$\Psi_0 = (v_0 a - m/2\pi a) \sin \theta = 0,$$

so that $v_0 a = m/2\pi a$

and $a = \sqrt{(m/2\pi v_0)}.$ (7.50)

For a given velocity of the uniform flow and a given strength of the doublet, the radius a is constant, which proves that the body so derived is circular. It is also possible to plot the flow pattern around a cylinder of radius a with uniform velocity v_0. From equation (7.50), the strength of the doublet is

$$m = 2\pi v_0 a^2,$$

and the combined stream function becomes

$$\Psi_c = (v_0 r - 2\pi v_0 a^2/2\pi r) \sin \theta$$
$$= v_0(r - a^2/r) \sin \theta.$$ (7.51)

Similarly, the velocity potential is

$$\Phi_c = v_0(r + a^2/r) \cos \theta.$$ (7.52)

EXAMPLE 7.4

If a 40 mm diameter cylinder is immersed in a stream having a velocity of 1.0 m s^{-1}, determine the radial and normal components of velocity at a point on a streamline where $r = 50$ mm and $\theta = 135°$, measured from the positive x axis. Assume flow to be ideal. Also determine the pressure distribution with radial distance along the y axis.

Solution

By equations (7.13), the velocity components are given by

$$v_r = \frac{1}{r}\frac{\partial \Psi}{\partial \theta} \quad \text{and} \quad v_\theta = -\frac{\partial \Psi}{\partial r},$$

but, for the ideal flow around a cylinder, the stream function is

$$\Psi_c = v_0\left(r - \frac{a^2}{r}\right)\sin \theta.$$

Therefore,

$$v_r = \frac{1}{r}\frac{\partial}{\partial \theta}\left[v_0\left(r - \frac{a^2}{r}\right)\sin \theta\right] = \frac{v_0}{r}\left(r - \frac{a^2}{r}\right)\cos \theta$$

$$= \left(1 - \frac{a^2}{r^2}\right)v_0 \cos \theta,$$

and $\quad v_\theta = -\dfrac{1}{\partial r}\left[v_0\left(r - \dfrac{a^2}{r}\right)\sin\theta\right] = -v_0\left(1 + \dfrac{a^2}{r^2}\right)\sin\theta.$

Substituting the numerical values $a = 2\,\text{cm}$, $r = 5\,\text{cm}$, $v_0 = 1.0\,\text{m s}^{-1}$, $\theta = 135°$, the following values for velocity components are obtained:

$$v_r = \left(1 - \frac{4}{25}\right)\cos 135° = -\frac{21}{25}\frac{1}{\sqrt{2}} = -0.594\,\text{m s}^{-1},$$

$$v_\theta = -\left(1 + \frac{4}{25}\right)\sin 135° = -\frac{29}{25}\frac{1}{\sqrt{2}} = -0.820\,\text{m s}^{-1}.$$

Remembering the sign convention for cylindrical coordinates, these components are as shown in Fig. 7.25.

FIGURE 7.25

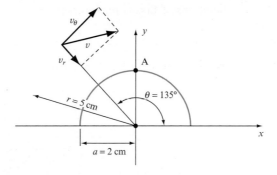

To obtain the pressure distribution along Ay it is first necessary to determine the velocity variation so that it may be used in applying Bernoulli's equation. Since for Ay, $\theta = 90°$, it follows that $v_r = 0$, and, hence,

$$v_\theta = -(1 + a^2/r^2)v_0,$$

which is the required velocity distribution for Ay. Now, applying Bernoulli's equation to a point far upstream where the velocity is v_0, the pressure is p_0, and to the section in the equation

$$p_0/\rho + v_0^2/2 = p/\rho + v_\theta^2/2,$$

gives $\quad p - p_0 = (\rho/2)(v_0^2 - v_\theta^2) = (\rho/2)[v_0^2 - (1 + a^2/r^2)^2 v_0^2]$

$$= (\rho/2)v_0^2[1 - (1 + 2a^2/r^2 + a^4/r^4)]$$

$$= -(\rho v_0^2/2)(2a^2/r^2 + a^4/r^4).$$

This equation shows that when $r \to \infty$, the pressure p approaches p_0, but at the surface of the cylinder (at point A), where $r = a$, the pressure is lower than that upstream by an amount equal to $\frac{3}{2}\rho v_0^2$.

7.9   CURVED FLOWS AND THEIR COMBINATIONS

The previous sections of this chapter dealt with flows whose basic components were straight line flows, either rectilinear or radial. The third basic type of flow is such that the streamlines are concentric circles, as shown in Fig. 7.26. Such flows are known as vortex flows. Their characteristic is that the radial component of velocity $v_r = 0$. This is so because, of course, there cannot be any flow across streamlines and, since in vortex flows they are circular, the flow must be confined to purely circular paths. Thus in any vortex flow

$$v_r = 0 \quad \text{and} \quad v = v_\theta. \tag{7.53}$$

There are two fundamental types of vortex flow distinguished by the nature of flow, namely rotational and irrotational. From these two basic types, various combinations of flows are possible.

FIGURE 7.26
Vortex flow

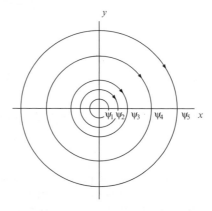

Vortex flows and their applications were discussed in Section 6.19. They were not, however, defined there with respect to rotational or irrotational flow, neither were the mathematics of their stream function and velocity potential appropriately discussed.

Let us first consider the irrotational vortex flow, which is known as the *free vortex*. Because it is irrotational, the vorticity and circulation across the stream must be equal to zero. Consider, in a free vortex flow, an element of fluid between streamlines Ψ and $(\Psi + d\Psi)$, as shown in Fig. 7.27. The circulation round the element, starting from A in the anticlockwise direction, is

$$\Gamma_{\text{ABCD}} = 0 - (v_\theta + dv_\theta)(r + dr)\, d\theta + 0 + v_\theta r\, d\theta,$$

and, neglecting infinitesimals of the third order,

$$\Gamma_{\text{ABCD}} = -(v_\theta\, dr + r\, dv_\theta)\, d\theta.$$

This, by the definition of irrotational flow, must be equal to zero. Therefore,

$$-(v_\theta\, dr + r\, dv_\theta)\, d\theta = 0,$$

FIGURE 7.27
An element of vortex flow

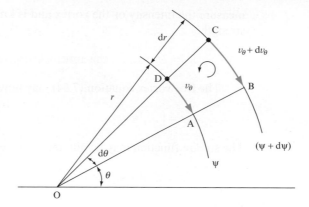

so that

$$v_\theta\, dr + r\, dv_\theta = 0,$$

but this is a differential of a product,

$$d(rv_\theta) = 0,$$

which, when integrated, gives

$$rv_\theta = \text{constant.} \tag{7.54}$$

This equation defines the relationship between the velocity and radius for a free vortex. It shows that the velocity increases towards the centre of the vortex and tends to infinity when the radius tends to zero. The velocity decreases as the radius increases and tends to zero as the radius tends to infinity. One practical example of this type of vortex flow is the emptying of a container through a central hole.

The constant in equation (7.54) may be established by making use of the singularity which exists in the free vortex flow, namely the infinite velocity at the centre of the vortex which we mentioned above. At this point, the vorticity which is given (equation (7.24)) by

$$-\left(\frac{\partial v_\theta}{\partial r} + \frac{v_\theta}{r}\right)$$

becomes indeterminate on substitution of $r \to 0$ and $v_\theta \to \infty$. It can, however, be determined by evaluating the circulation around the centre, i.e. along any of the concentric streamlines. This does not violate the condition for irrotational flow, by which the free vortex is defined, because the condition states that vorticity (and circulation) must be zero for any closed loop across the flow (see Section 7.1). The circulation around a circular streamline:

$$\Gamma_C = \oint v\, ds = \text{Circumference} \times \text{Tangential velocity} = 2\pi r v_\theta,$$

but, since $v_\theta r = \text{constant}$, it follows that this particular circulation is constant for any streamline and, therefore, for the whole vortex field. It may, therefore, be used to measure the intensity of the vortex and is known as the *vortex strength*. Thus, vortex strength

$$\Gamma_C = 2\pi r v_\theta \quad \text{for the anticlockwise vortex.}$$

The free vortex equation (7.54) may now be rewritten as

$$v_\theta r = \Gamma_C / 2\pi. \tag{7.55}$$

The stream function may be obtained from equation (7.12):

$$d\Psi = v_r r \, d\theta - v_\theta \, dr.$$

Since $v_r = 0$,

$$\Psi = -\int v_\theta \, dr,$$

and, substituting for v_θ from equation (7.55),

$$\Psi = -\int \frac{\Gamma_C}{2\pi r} \, dr = -\frac{\Gamma_C}{2\pi} \log_e r + \text{constant.}$$

The constant of integration is made zero by taking $\Psi = 0$ at $r = 1$, so that, finally,

$$\Psi = -(\Gamma_C / 2\pi) \log_e r \tag{7.56}$$

for anticlockwise rotation. The sign of the above expression becomes positive for clockwise rotation.

The velocity potential follows (equation (7.21)) from

$$d\Phi = v_r \, dr + r v_\theta \, d\theta,$$

whereas, upon substitution and making $\Phi_0 = 0$ at $\theta = 0$,

$$\Phi = \int \frac{\Gamma_C}{2\pi} \, d\theta = \frac{\Gamma_C}{2\pi} \theta. \tag{7.57}$$

Since the free vortex is irrotational, the Bernoulli constant remains the same for all streamlines.

EXAMPLE 7.5

A two-dimensional fluid motion takes the form of concentric, horizontal, circular streamlines. Show that the radial pressure gradient is given by

$$\frac{dp}{dr} = \rho \frac{v^2}{r},$$

where ρ = density, v = tangential velocity, r = radius. Hence, evaluate the pressure gradient for such a flow defined by $\Psi = 2 \log_e r$, where Ψ = stream function, at a radius of 2 m and fluid density of 10^3 kg m^{-3}.

Solution

For two concentric streamlines the variation of total head or the Bernoulli constant is, in general, given by

$$\frac{dH}{dr} = \frac{v_\theta}{g}\left(\frac{dv_\theta}{dr} + \frac{v_\theta}{r}\right),$$

but, for horizontal flow, $z = 0$ and, for vortex flow, $v = v_\theta$, so that

$$H = p/\rho g + v_\theta^2/2g;$$

therefore, differentiating,

$$\frac{dH}{dr} = \frac{1}{\rho g}\frac{dp}{dr} + \frac{v_\theta}{g}\frac{dv_\theta}{dr}.$$

Equating the two equations,

$$\frac{v_\theta}{g}\left(\frac{dv_\theta}{dr} + \frac{v_\theta}{r}\right) = \frac{1}{\rho g}\frac{dp}{dr} + \frac{v_\theta}{g}\frac{dv_\theta}{dr},$$

from which

$$\frac{1}{\rho}\frac{dp}{dr} = v_\theta\frac{dv_\theta}{dr} + \frac{v_\theta^2}{r} - v_\theta\frac{dv_\theta}{dr} = \frac{v_\theta^2}{r}.$$

Therefore,

$$\frac{dp}{dr} = \rho\frac{v_\theta^2}{r}.$$

The stream function $\Psi = 2\log_e r$ represents a free vortex, for which

$$\Psi = (\Gamma_C/2\pi)\log_e r,$$

and, hence,

$$\Gamma_C/2\pi = 2,$$

but, for a free vortex,

$$\Gamma_C/2\pi = v_\theta r,$$

so that

$$v_\theta r = 2 \quad \text{and} \quad v_\theta^2 = \frac{4}{r^2}$$

and, therefore,

$$\frac{dp}{dr} = \rho\frac{4}{r^3} = 10^3 \times \frac{4}{2^3} = 500\ \mathrm{N\,m^{-3}}.$$

The most common example of a rotational vortex, which is considered of fundamental importance, is a *forced vortex*. It will be shown in what follows that in a forced vortex the fluid rotates as a solid body with a constant rotational velocity.

Consider circulation around a segmental element (such as in Fig. 7.27) of a forced vortex, remembering that $v_r = 0$:

$$\Gamma_{ABCD} = -(v_\theta \, dr + r \, dv_\theta) \, d\theta.$$

The area of the element is

$$A = r \, d\theta \, dr,$$

so that vorticity (as already shown in Section 7.2, equation (7.5)) is given by

$$\zeta = \frac{\Gamma_{ABCD}}{dA} = -\left(\frac{v_\theta}{r} + \frac{dv_\theta}{dr}\right),$$

but if, for a solid body, rotation ω is the angular velocity which at any radius r is related to the tangential velocity v_θ by

$$\omega = v_\theta/r, \tag{7.58}$$

it follows, therefore, that for a forced vortex the vorticity

$$\zeta = -2\omega \tag{7.59}$$

and is constant for a given vortex.

The flow is rotational and there is, therefore, variation of the Bernoulli constant with radius. Using equation (6.38),

$$\frac{dH}{dr} = \frac{v_\theta}{g}\left(\frac{v_\theta}{r} + \frac{dv_\theta}{dr}\right) = v_\theta \times \frac{2\omega}{g}$$

and, since $v_\theta = \omega r$,

$$\frac{dH}{dr} = \frac{2\omega^2 r}{g}. \tag{7.60}$$

In order to determine the pressure distribution or the surface gradient in a forced vortex, the above expression must be used in conjunction with Bernoulli's equation, as was shown in Section 6.19.

The stream function for a forced vortex is obtained in the same manner as for the free vortex, but using the appropriate relationship (equation (7.58)), namely that

$$v_\theta = \omega r,$$

which yields

$$\Psi = -\int \omega r \, dr = -\tfrac{1}{2}\omega r^2 + \text{constant}.$$

But for $\Psi = 0$ at $r = 0$,

$$\Psi = -\tfrac{1}{2}\,\omega r^2 \tag{7.61}$$

for anticlockwise rotation.

FIGURE 7.28
Free spiral vortex

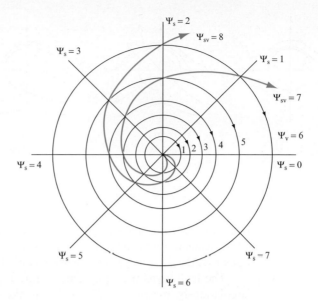

Since the forced vortex is rotational, there is no velocity potential corresponding to it and the Laplace equations are not satisfied.

A free spiral vortex, mentioned in Section 6.19, is, in contrast to a forced vortex, irrotational and represents a potential flow. The free spiral vortex is the combination of a free vortex and radial flow. It is, therefore, obtained by superposition of the stream functions of a free vortex with either a sink or a source flow depending upon the direction of the radial flow.

For outward flow using a source and for a clockwise vortex,

$$\Psi_{sv} = \Psi_{source} + \Psi_{free\ vortex}$$

$$= q\theta/2\pi + (\Gamma_C/2\pi)\log_e r$$

$$= (1/2\pi)(q\theta + \Gamma_C \log_e r) \tag{7.62}$$

and $$\Phi_{sv} = \Phi_{source} + \Phi_{free\ vortex}$$

$$= (q/2\pi)\log_e r + (\Gamma_C/2\pi)\theta$$

$$= (1/2\pi)(q\log_e r + \Gamma_C\theta). \tag{7.63}$$

The resulting flow is shown in Fig. 7.28.

7.10   FLOW PAST A CYLINDER WITH CIRCULATION. KUTTA–JOUKOWSKY'S LAW

Flow past a stationary cylinder may be obtained by superposition of a parallel flow and a doublet. This was discussed in Section 7.8. However, in the first half of the nineteenth century the German physicist H. G. Magnus observed experimentally that

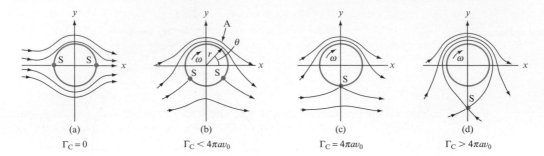

(a)   (b)   (c)   (d)

$\Gamma_C = 0$ $\Gamma_C < 4\pi a v_0$ $\Gamma_C = 4\pi a v_0$ $\Gamma_C > 4\pi a v_0$

FIGURE 7.29

if the cylinder in a parallel flow stream is rotated about its axis, a transverse force, which tends to move the cylinder across the parallel flow stream, is generated. This is known as the *Magnus effect* or aerodynamic lift.

The hydrodynamic equivalent of rotating a cylinder in a flow stream is to add circulation by means of a free vortex to the doublet in a parallel flow. Such a flow pattern can be obtained directly from the results of Sections 7.8 and 7.9 by adding the stream function of a free vortex (equation (7.56)) for clockwise rotation to the stream function of flow past a cylinder, given by equation (7.51). Thus, the combined stream function is

$$\Psi = v_0(r - a^2/r) \sin \theta + (\Gamma_C/2\pi) \log_e r. \tag{7.64}$$

The addition of circulation to the ideal flow past a cylinder gives rise to an asymmetric flow pattern, as shown in Fig. 7.29. There is an increase of velocity on one side of the cylinder and a decrease on the other. In consequence of this, the stagnation points move from the axis of the parallel flow. Their positions depend upon the magnitude of the circulation and can be determined using equation (7.64), remembering that at stagnation points, $v_\theta = 0$. Thus, the tangential velocity is given by

$$v_\theta = -\frac{\partial \Psi}{\partial r} = -v_0\left(1 + \frac{a^2}{r^2}\right) \sin \theta - \frac{\Gamma_C}{2\pi r} = 0.$$

Also, on the contour of the cylinder, $r = a$; therefore,

$$-2v_0 \sin \theta - \Gamma_C/2\pi a = 0 \quad \text{or} \quad \sin \theta = -\Gamma_C/4\pi a v_0. \tag{7.65}$$

The negative sign indicates that for the positive parallel flow (from left to right) and clockwise circulation, the stagnation points lie below the x axis. Furthermore, if the value of circulation $\Gamma_C < 4\pi a v_0$, then $0 > \sin \theta > -1$ and the stagnation points will lie in positions such as those shown in Fig. 7.29(b). For $\Gamma_C = 4\pi a v_0$, the stagnation points merge on the negative y axis, as shown in Fig. 7.29(c), and, finally, for $\Gamma_C > 4\pi a v_0$, the stagnation points will be as shown in Fig. 7.29(d). Since $\sin \theta$ cannot be greater than one, this is only possible if a is increased, which means that the stagnation point moves away from the surface of the cylinder.

In order to establish the magnitude of the transverse force acting on the cylinder and mentioned earlier, it is necessary to obtain first the pressure distribution around the cylinder and then the forces arising from it. Since the flow is irrotational, Bernoulli's equation may be applied to a point some distance upstream in the parallel flow and to a point on the surface of the cylinder. Thus,

$$p/\rho + v_\theta^2/2 = p_0/\rho + v_0^2/2,$$

where p is the pressure on the cylinder and it varies with θ, and p_0 is the pressure in the parallel flow some distance upstream, where the velocity is v_0. Rearranging and solving for pressure difference:

$$p - p_0 = (\rho/2)(v_0^2 - v_\theta^2) = (\rho v_0^2/2)(1 - v_\theta^2/v_0^2),$$

but $v_\theta = -2v_0 \sin \theta - \Gamma_C/2\pi a,$

so that

$$p - p_0 = (\rho v_0^2/2)[1 - (-2 \sin \theta - \Gamma_C/2\pi a v_0)^2]$$
$$= (\rho v_0^2/2)(1 - 4 \sin^2 \theta - 2\Gamma_C \sin \theta/\pi a v_0 - \Gamma_C^2/4\pi^2 a^2 v_0^2). \tag{7.66}$$

Consider, now, an element of the cylinder's surface. The force due to pressure acting on it is $(p - p_0)a \, d\theta$, and it may be resolved into vertical and horizontal components. The transverse force will, in our case, be the sum of the vertical components. Thus, the upward force

$$L = -\int_0^{2\pi} (p - p_0) a \sin \theta \, d\theta.$$

In equation (7.66), let

$$1 - \Gamma_C^2/4\pi^2 a^2 v_0^2 = A;$$

then $p - p_0 = (\rho v_0^2/2)(A - 4 \sin^2 \theta - 2\Gamma_C \sin \theta/\pi a v_0)$

and $L = -\int_0^{2\pi} (\rho v_0^2/2)(A - 4 \sin^2 \theta - 2\Gamma_C \sin \theta/\pi a v_0) a \sin \theta \, d\theta$

$$= -\int_0^{2\pi} (\rho v_0^2 a/2)(A \sin \theta - 4 \sin^3 \theta - 2\Gamma_C \sin^2 \theta/\pi a v_0) \, d\theta.$$

But $\int_0^{2\pi} \sin \theta \, d\theta = 0$ and $\int_0^{2\pi} \sin^3 \theta \, d\theta = 0,$

so that

$$L = \frac{\rho a v_0^2}{2} \int_0^{2\pi} \frac{2\Gamma_C \sin^2 \theta}{\pi a v_0} \, d\theta = \frac{\rho v_0 \Gamma_C}{\pi} \left(\frac{1}{2}\theta - \frac{\sin 2\theta}{4} \right)_0^{2\pi}$$

$$L = \rho v_0 \Gamma_C. \tag{7.67}$$

Thus, the force perpendicular to the direction of the parallel flow, or a main free stream, which in general is known as the *lift*, is, for a rotating cylinder of infinite length, independent of the diameter of the cylinder and equal to the product of fluid density, free stream velocity and circulation. This statement is known as *Kutta–Joukowsky's law* and gives theoretical justification for the experimentally observed Magnus effect.

It is interesting to note that the horizontal component of the force on the cylinder due to pressure, which in general is called the *drag*, and for our case is given by

$$D = \int_0^{2\pi} (p - p_0)a \cos\theta \, d\theta,$$

is equal to zero. This result, obtained on the assumption of the ideal flow, is not supported by experiments. This is so because in real fluids viscous friction provides resistance to flow and separation may occur.

7.11 COMPUTER PROGRAM ROTCYL

Program ROTCYL determines the angular positions of the stagnation points on the surface of a cylinder rotating in a uniform fluid stream and calculates the lift coefficient and the values of the pressure coefficient along the flow axis upstream of the cylinder, Fig. 7.29. If there is only one stagnation point then the program calculates its distance from the cylinder surface. Calculations are based upon potential flow theory developed in this chapter, in particular equations (7.65), (7.66) and (7.67), together with the definitions of lift coefficient, equation (12.5) and pressure coefficient (9.1).

The required data are cylinder diameter, D (mm), cylinder rotational speed, N (rev min^{-1}) and free stream velocity, U (m s^{-1}).

7.11.1 Application example

For a 60 mm diameter cylinder rotating at 1245 rev min^{-1} in a free stream of upstream velocity 20 m s^{-1}, the stagnation points on the cylinder surface are located at $-5.61°$ and $185.61°$ and the lift coefficient C_L is 1.23.

The values of pressure coefficient along the negative x axis are:

X (mm)	30	150	270	390	510	630
C_P	-0.962	-7.7E-2	-2.4E-2	-1.2E-2	-7E-3	-4E-3

7.11.2 Additional investigations using ROTCYL

ROTCYL may also be used to investigate the effect of cylinder rotational speed, diameter and free stream velocity on stagnation point location and both the lift and pressure coefficient values.

Concluding remarks

While the concept of an ideal fluid is purely theoretical it will be demonstrated that the techniques introduced in this chapter do contribute to the understanding of fluid flow, particularly under flow conditions where the viscosity effects are minimal, namely at high Reynolds numbers and well away from any boundary layer or wake effect. The approach developed may therefore be of use in the study of external flows, in the study of forces acting on aerofoil shapes (Chapter 12), or in the study of rotodynamic machinery (Chapter 22). The mathematical techniques introduced to allow the combination of sources, sinks and rectilinear flow are often helpful in determining the actual flow under complex conditions, provided that the restrictions mentioned above are recognized.

Summary of important equations and concepts

1. Chapter 7 defines an ideal fluid flow as having no viscosity, compressibility, surface tension or vaporization limits and shows that the theoretical study of such flows was fundamental to the development of modern fluid mechanics.

2. Definitions of vorticity, equation (7.2), and irrotational flow, equation (7.3) are presented, together with definitions of circulation, equation (7.4), and Stokes' theorem, equation (7.5).

3. The equation of the streamline and the definition of stream function are presented in Section 7.3, while Section 7.4 addresses velocity potential and potential flow and its implications for the application of Bernoulli's equation to straight line and curved flows, Fig. 7.10.

4. Flow nets are introduced in Section 7.5, together with the Cauchy–Riemann and Laplace equations (7.25) and (7.26), and the requirement that the Laplace equation be satisfied for velocity potential.

5. Combinations of straight line flows are introduced in Section 7.6 and this concept is then expanded to include sources and sinks, Fig. 7.23; doublets are introduced to illustrate the generation of stagnation points for a body in a uniform flow, Fig. 7.23. Flow past a cylinder is introduced in Section 7.8.

6. Combinations of curved flows are introduced in Section 7.9 with reference to forced vortices. Lift on a cylinder with rotation and the Kutta–Joukowsky law is introduced in Section 7.10, with the lift force defined by equation (7.67). A computer program, ROTCYL, is included to determine lift and stagnation point positions for a rotating cylinder in a uniform flow.

Problems

7.1   The x and y components of fluid velocity in a two-dimensional flow field are $u = x$ and $v = -y$ respectively.

(a)   Determine the stream function and plot the streamlines $\Psi = 1, 2, 3$.

(b)   If a uniform flow defined by $\Psi = y$ is superimposed on the above flow, plot the resulting streamlines and label them with Ψ values.

(c)   Determine the stream function and the velocity potential for the above combined flow.

$$[(a),\ \Psi = xy,\ (c)\ \Psi = y + xy,\ \Phi = x^2/2 + x - y^2/2]$$

7.2   The stream function for the two-dimensional flow of a liquid is given by $\Psi = 2xy$. In the range of values of x and y between 0 and 5 plot the streamlines and equipotential lines passing through coordinates (1, 1), (1, 2), (2, 2). Also determine the velocity in magnitude and direction at the point (1, 2).

$$[4.47, 63.4°]$$

7.3   A flow has a potential function Φ given by

$$\Phi = V(x^3 - 3xy^2).$$

Derive the corresponding stream function Ψ and show that some of the streamlines are straight lines passing through the origin of coordinates. Find the inclinations of these lines. Evaluate also the magnitude and direction of the velocity at an arbitrary point x, y.

$$[\pm 60°]$$

7.4   A source of strength 30 m^2 s^{-1} is located at the origin, and another source of strength 20 m^2 s^{-1} is located at (1, 0). Find the velocity components u and v at (−1, 0) and (1, 1). Also, if the dynamic pressure at infinity is zero for $\rho = 2.0$ kg m^{-3} calculate the dynamic pressure at the above points.

$$[40.58\ \text{N m}^{-2},\ 36.74\ \text{N m}^{-2}]$$

7.5   A source of strength m at the origin and a uniform flow of 15 m s^{-1} are combined in two-dimensional flow so that a stagnation point occurs at (1, 0). Obtain the velocity potential and stream function for this case.

$$[\Psi = +15\ \tan^{-1}(y/x) - 15y,$$
$$\Phi = +7.5\ \log_e(x^2 + y^2) - 15x]$$

7.6   A source discharging 20 m^3 s^{-1} is located at (−1, 0) and a sink of twice the strength is located at (2, 0). For the pressure at the origin of 100 N m^{-2} and density of 1.8 kg m^{-3}, find the velocity and pressure at points (0, 1) and (1, 1).

$$[4.15\ \text{m s}^{-1},\ 84.5\ \text{N m}^{-2},\ 5.14\ \text{m s}^{-1},\ 76.2\ \text{N m}^{-2}]$$

7.7   Show that the potential function for the flow generated by a source in a two-dimensional system is $a\ \ln(x^2 + y^2)$ where a is a constant. Hence derive an expression for the potential function for a doublet and show that the streamlines in a flow generated by a doublet are circular. Sketch these streamlines.

7.8   Show that the potential function $\Phi = ax/(x^2 + y^2)$ represents the flow generated by a doublet. In which direction is the doublet oriented? A cylinder of radius 4 cm is held with its centre at the point (0, 0) in a fluid stream. At large distances from the cylinder the fluid velocity is constant at 30 m s^{-1} parallel to the x axis and in the direction of x increasing. Calculate the components of the fluid velocity at the point $x = -4$ cm, $y = 1$ cm.

$$[-5.08\ \text{m s}^{-1},\ -17.85\ \text{m s}^{-1}]$$

7.9   Under what circumstances does potential flow analysis give an accurate prediction of the flow of real fluids? Show that the potential function

$$\Phi = U\left(x + \frac{a^2 x}{x^2 + y^2}\right)$$

gives the potential flow around a cylinder of radius a. A small particle whose velocity is at all times equal to that of the fluid immediately surrounding it passes through the point $(-3a, 0)$ at time $t = 0$. At what time will it pass through the point $(-2a, 0)$?

$$[1.2203a/U_0]$$

7.10   Show that a free vortex is an example of irrotational motion. A hollow cylinder 1 m diameter, open at the top, spins about its axis which is vertical, thus producing a forced vortex motion of the liquid contained in it. Calculate the height of the vessel so that the liquid just reaches the top when the minimum depth is 15 cm at 150 rev min^{-1}.

$$[3.29\ \text{m}]$$

7.11   Prove that, in the forced vortex motion of a liquid, the rate of increase of pressure p with respect to the radius r at a point in the liquid is given by

$$dp/dr = \rho \omega^2 r$$

in which ω is the angular velocity of liquid and ρ its density. What will be the thrust on the top of a closed vertical cylinder of 15 cm diameter, if it rotates about its axis at 400 rev min^{-1} and is completely filled with water?

$$[43.5\ \text{N}]$$

7.12 A compound vortex in a large tank of water comprises a forced vortex core surrounded by a free vortex. Determine the depth of water at the centre of the core below the free vortex level if the velocity is 2.5 m s^{-1} at the common radius of 18 cm.   [0.636 m]

7.13 Define vorticity and discuss the significance of irrotational motion. Give the vorticity at 1 m and 3 m radius in a vortex whose speed is 1 m s^{-1} throughout, and calculate the difference in pressure between these two places, if the axis of rotation is (*a*) vertical, and (*b*) horizontal.

[20.72 kN m^{-2}]

Dimensional Analysis and Similarity

Wind tunnel evaluation of aircraft model performance, photo courtesy of the Low Speed Wind Tunnel Facility, British Aerospace, Filton, Bristol

THE APPLICATION OF FLUID MECHANICS IN DESIGN, perhaps more than most engineering subjects, relies on the use of empirical results built up from an extensive body of experimental research. In many areas empirical data are supplied in the form of tables and charts that the designer may apply directly, an example being the values of friction factor for pipe flow and separation loss coefficients for duct and pipe fittings. However, even here, the tables and the underlying experimental work become too unwieldy and time consuming if no way can be found to replace the relationship between any two variables by generalized groupings. It is, therefore, in the organization of experimental work and the presentation of its results that dimensional analysis plays such an important role. This technique, which is dealt with first in this part of the text, commences with a survey of all the likely variables affecting any phenomenon, and, to the experienced researcher, then suggests the formation of groupings of more than one variable. Experimental work may then be based on these groups rather than on individual variables, considerably reducing the testing programme and leading to simplified design guides, such as the Moody charts mentioned in Chapter 10.

The application of results from one test series, involving say a particular pipe flow situation, to another case, depends on the full understanding of the principles of geometric and dynamic similarity which are covered in the second part of this section. Although similarity is inherent in the formation of relationships such as the Moody chart, it is more commonly associated with the use of models and model testing techniques. Examples of such applications as wind tunnel tests and river and harbour models are mentioned; however, the basic principles depend upon the equivalence of variable groupings formed initially by the use of dimensional analysis. Again, it will be appreciated that mathematics alone is not sufficient in the application of the similarity laws; in many cases it will be found that total equivalence of all the dimensionless groupings will be mutually impossible and here the experience of the researcher will be called upon, examples being found in the cases of ship model tests and pump or turbine modelling techniques utilizing gas in place of water.

Together, dimensional analysis, similarity and model testing techniques allow the design engineer to predict accurately and economically the performance of the prototype system, whether it is an aircraft wing, ship hull, dam spillway or harbour construction. The basis of these interactive techniques is presented in this part.

8

Dimensional Analysis

FLUID MECHANICS IS ESSENTIALLY AN EMPIRICALLY BASED DISCIPLINE THAT HAS RELIED upon the utilization of model flow representation to determine the likely performance of prototypes, whether aerofoils, ships or flow over and through buildings. In order to allow the results of model testing to be meaningful it is essential that the parameters determining the flow are identified and organized so that testing may be both directed and of value. This chapter will introduce the mathematical techniques of dimensional analysis whereby the parameters considered to be likely to affect the flow can be combined into a number of dimensionless groupings, thereby facilitating testing and reducing the overall test programme. The chapter will concentrate upon the intro-duction of parameter dimensions and the techniques available for parameter com-bination. Buckingham's π theorem will be introduced and shown to have general application. However, for the majority of fluid mechanics applications not involving heat transfer or temperature changes, a system based on mass, length and time will be shown to be appropriate. ● ● ●

8.1   DIMENSIONAL ANALYSIS

This is a useful technique for the investigation of problems in all branches of engineering and particularly in fluid mechanics. If it is possible to identify the factors involved in a physical situation, dimensional analysis can usually establish the form of the relationship between them. At first sight, the technique does not appear to be as precise as the usual algebraic analysis which seems to provide exact solutions, but these are usually obtained by making a series of simplifying assumptions which do not always correspond with the real facts. The qualitative solution obtained by dimensional analysis can usually be converted into a quantitative result, determining any unknown factors experimentally.

8.2   DIMENSIONS

Any physical situation, whether it involves a single object or a complete system, can be described in terms of a number of recognizable properties which the object or system possesses. For example, a moving object could be described in terms of its mass, length, area or volume, velocity and acceleration. Its temperature or electrical properties might also be of interest, while other properties – such as the density and viscosity of the medium through which it moves – would also be of importance, since they would affect its motion. These measurable properties used to describe the physical state of the body or system are known as its *dimensions*.

8.3   UNITS

To complete the description of the physical situation, it is also necessary to know the magnitude of each dimension. It is not usually sufficient to know, for example, that a body has the dimension of length; we also need to know the magnitude of this length. For this purpose we use agreed *units* of measurement. A length would be measured in terms of a standardized unit of length, such as the metre. Similarly, other agreed units are used to measure other dimensions. There is more than one system of units in common use, but of course the system of units chosen does not affect the real size of the body or system – only the numerical value of its measurements: 1 foot is precisely the same length as 0.3048 metres. The distinction between units and dimensions is that dimensions are properties that can be measured and units are the standard elements in terms of which these dimensions can be described quantitatively and assigned numerical values.

8.4   DIMENSIONAL REASONING

In analyzing any physical situation, it is necessary to decide what factors are involved and then to try to determine a quantitative relationship between them. The factors involved can often be assessed from observation, experiment or even, perhaps, intuition. However, it is sometimes difficult to establish precise quantitative relationships, because we cannot specify the conditions which exist exactly or the way in which the various factors interact. As a result, no suitable mathematical model can be constructed without making substantial assumptions to simplify the problem. A qualitative solution to the problem can sometimes be obtained by dimensional reasoning, and subsequent experimental investigation based on this analysis can frequently lead to a complete solution of the real problem.

In dimensional analysis, we are concerned only with the nature of the factors involved in the situation and not with their numerical values. The notation adopted to indicate this is to enclose the name or symbol of the quantity in square brackets: thus [length] means the dimension of length and not a particular length with a definite numerical value. For conciseness length is abbreviated to L and the dimension of length is written [L]. Similarly [M] is used for the dimension of mass, [T] for the dimension of time, [F] for the dimension of force, [Θ] for the dimension of temperature and so on. Whenever a property, in words or symbols, is enclosed in a square bracket it indicates that we are concerned with it only dimensionally and qualitatively, not quantitatively.

Dimensional reasoning is based on the proposition that, for an equation describing a physical situation to be true, the two sides must be equal both numerically and dimensionally. An equation must compare like with like. The simple equation $1 + 3 = 4$ is numerically correct but, in physical terms, may be entirely untrue, depending upon the nature of each term. Thus, it would be *untrue* to say that

1 elephant + 3 aeroplanes = 4 days,

since elephants, aeroplanes and days are not the same sort of things; it would be *true* to say that

1 metre + 3 metres = 4 metres,

provided that our concern is restricted to the study of length, since each term has the dimensions of length.

An equation describing a physical situation will only be true if all the terms are of the same kind and have the same dimensions. The equation is then said to be *dimensionally homogeneous*, and is valid only in relation to these dimensions. If an equation does not compare like with like, it will be physically meaningless, even though it may balance numerically. In general any equation of the form

$$a_1^{m_1} b_1^{n_1} c_1^{p_1} + a_2^{m_2} b_2^{n_2} c_2^{p_2} + \cdots = X$$

will be physically true if, in addition to being numerically correct, the terms are dimensionally the same so that

$$[a_1^{m_1} b_1^{n_1} c_1^{p_1}] = [a_2^{m_2} b_2^{n_2} c_2^{p_2}] = \cdots = [X],$$

where $[a_1^{m_1} b_1^{n_1} c_1^{p_1}]$ means the dimensions of $a_1^{m_1} b_1^{n_1} c_1^{p_1}$.

8.5   DIMENSIONLESS QUANTITIES

In describing an object or system, we sometimes use quantities which are non-dimensional or abstract. For example, the shape of an ellipse is defined by the ratio of the major and minor axes. Other non-dimensional quantities are relative density, strain and angle measured in radians. Such quantities are ratios comparing one quantity with another of the same kind and their numerical values are independent of the system of units employed. For example, tensile strain is defined as extension divided by original length. Since both these quantities have the dimension [L],

$$[\text{Strain}] = \frac{[\text{Extension}]}{[\text{Original length}]} = \frac{[\text{L}]}{[\text{L}]} = [\text{L}^0] = [1],$$

indicating that the dimension of strain is a pure number and strain is, therefore, dimensionless.

8.6   FUNDAMENTAL AND DERIVED UNITS AND DIMENSIONS

It would be possible to give independent units and dimensions to every physical property, but it would not help our understanding of their interrelationship if we did so. It is therefore desirable to select a number of fundamental dimensions and to express the dimensions of all physical properties in terms of these. Length is perhaps an obvious choice for one fundamental dimension. An area is defined as the product of two lengths. The area of a rectangle, having sides a and b, is given by the equation

Area of rectangle = Length $a \times$ Length b.

For this to be true, it must be dimensionally homogeneous, so that

$$[\text{Area}] = [\text{Length}] \times [\text{Length}] = [\text{L}^2],$$

i.e. area has the dimensions of $[\text{L}^2]$. The corresponding unit of area will be the unit of length squared, for example m^2 in SI units.

Similarly, if a tank has sides of length a, b and c,

Volume of tank = Length $a \times$ Length $b \times$ Length c

and so $[\text{Volume}] = [\text{L}] \times [\text{L}] \times [\text{L}] = [\text{L}^3],$

so that volume has the dimensions of $[\text{L}^3]$.

In kinematics, the dimension of time [T] is required. The dimensions of other quantities can be expressed in terms of length and time using the established definitions and relationships, together with the principle of dimensional homogeneity. For linear motion,

Velocity = Distance/Time

and so $[\text{Velocity}] = [\text{Distance}]/[\text{Time}] = [\text{L}]/[\text{T}] = [\text{LT}^{-1}].$

Similarly,

$$[\text{Acceleration}] = [\text{Velocity}]/[\text{Time}] = [LT^{-1}]/[T] = [LT^{-2}].$$

For angular motion, if we define the angle as being measured in radians,

$$\text{Angle} = \text{Length of arc}/\text{Length of radius}$$

$$[\text{Angle}] = [\text{Length}]/[\text{Length}] = [L]/[L] = [L^0],$$

which, dimensionally, is unity, indicating that angle is dimensionless:

$$[\text{Angular velocity}] = [\text{Angle}]/[\text{Time}] = 1/[T] = [T^{-1}]$$

and $$[\text{Angular acceleration}] = [\text{Angular velocity}]/[\text{Time}]$$

$$= [T^{-1}]/[T] = [T^{-2}].$$

In dynamics, an additional fundamental dimension is needed, since we are concerned with force and mass. Newton's second law provides the necessary relationship, which can be stated in the form

$$\text{Force} \propto \text{Mass} \times \text{Acceleration}.$$

In practice, for any given system of units, the constant of proportionality is made unity, so that

$$\text{Force} = \text{Mass} \times \text{Acceleration}.$$

For this equation to be valid, it must be dimensionally homogeneous and, therefore,

$$[\text{Force}] = [\text{Mass}] \times [\text{Acceleration}].$$

Writing $[F]$ for force, $[M]$ for mass, $[T]$ for time and $[L]$ for length, then, since $[\text{Acceleration}] = [LT^{-2}]$,

$$[F] = [M][LT^{-2}].$$

Thus, in dynamics, we can select $[L]$ and $[T]$ as fundamental dimensions together with either force $[F]$ or mass $[M]$, the remaining quantity being regarded as a derived dimension and expressed in terms of the three fundamental dimensions. In this book, we shall normally select the dimension of mass $[M]$ as the third fundamental dimension and treat force as a derived dimension, having the dimensions $[MLT^{-2}]$. If we were to select $[F]$, $[L]$ and $[T]$ as fundamental dimensions, mass would have the dimensions of $[FL^{-1}T^2]$.

The dimensions of other quantities can be expressed in terms of the fundamental dimensions using known relationships, definitions and equations and the principle of homogeneity of dimensions. For example,

$$\text{Pressure} = \text{Force}/\text{Area},$$

$$[\text{Pressure}] = [F]/[L^2] \quad \text{or} \quad [MLT^{-2}]/[L^2]$$

$$= [FL^{-2}] \quad \text{or} \quad [ML^{-1}T^{-2}].$$

Similarly,

$$\text{Mass density} = \text{Mass/Volume},$$

$$[\text{Mass density}] = [M]/[L^3] \quad \text{or} \quad [FL^{-1}T^2]/[L^3]$$

$$= [ML^{-3}] \quad \text{or} \quad [FL^{-4}T^2].$$

The identification of the relevant dimensions for power will also exemplify the regression through the appropriate equations necessary:

$$\text{Power} = \text{Rate of doing work} = \text{Work/Time}$$

$$= \text{Force} \times \text{Distance/Time}$$

$$= \text{Mass} \times \text{Acceleration} \times \text{Distance/Time}$$

$$= \text{Mass} \times \text{Distance/Time}^2 \times \text{Distance/Time}$$

$$= ML^2T^{-3} \text{ as shown in Table 8.1.}$$

The dimensions of quantities commonly occurring in mechanics are given in Table 8.1.

The system of fundamental dimensions chosen for problems involving the thermodynamics of fluids will depend upon whether it is desirable to separate thermal quantities, such as heat and temperature, from other related properties. Since heat is a form of energy it could be expressed dimensionally in the same way as energy which, in terms of fundamental dimensions of mass, length and time, is $[ML^2T^{-2}]$. We also have the relationship

$$H = cm\theta,$$

where H is the quantity of heat required to raise a mass m of a substance of specific heat c through a temperature difference of θ. If c is treated as a ratio and, therefore, is dimensionless,

$$[\theta] = [Hm^{-1}] = [ML^2T^{-2}][M^{-1}] = [L^2T^{-2}].$$

The dimensions of other thermal quantities can be derived in the same way and are shown in Table 8.2, column 1.

If the thermal aspects of a problem are of particular interest, it is useful to introduce at least one additional fundamental dimension which is thermal in character, such as temperature Θ. In the resulting MLTΘ system, heat energy may be treated thermally as $[M\Theta]$, in which case the dimensions of other quantities are as shown in column 2 of Table 8.2, or heat energy can be expressed in mechanical terms as $[ML^2T^{-2}]$ with the results shown in column 3.

The quantity of heat H might also be treated as an additional fundamental quantity. The dimensions of other quantities using the HLTΘ system are shown in Table 8.2, column 4, and for the HMLTΘ system in column 5. The units used for each system would, of course, differ. Heat might be measured in calories, if considered as $[M\Theta]$, or in joules, if regarded as $[ML^2T^{-2}]$.

TABLE 8.1
Dimensions of
quantities in mechanics
(based on Newton's
second law)

QUANTITY	DEFINING EQUATION	DIMENSIONS, MLT SYSTEM
Geometrical		
Angle	Arc/radius (a ratio)	$[M^0 L^0 T^0]$
Length	(Including all linear measurement)	$[L]$
Area	Length × Length	$[L^2]$
Volume	Area × Length	$[L^3]$
First moment of area	Area × Length	$[L^3]$
Second moment of area	Area × Length^2	$[L^4]$
Strain	Extension/Length	$[L^0]$
Kinematic		
Time	–	$[T]$
Velocity, linear	Distance/Time	$[LT^{-1}]$
Acceleration, linear	Linear velocity/Time	$[LT^{-2}]$
Velocity, angular	Angle/Time	$[T^{-1}]$
Acceleration, angular	Angular velocity/Time	$[T^{-2}]$
Volume rate of discharge	Volume/Time	$[L^3T^{-1}]$
Dynamic		
Mass	Force/Acceleration	$[M]$
Force	Mass × Acceleration	$[MLT^{-2}]$
Weight	Force	$[MLT^{-2}]$
Mass density	Mass/Volume	$[ML^{-3}]$
Specific weight	Weight/Volume	$[ML^{-2}T^{-2}]$
Specific gravity	Density/Density of water	$[M^0 L^0 T^0]$
Pressure intensity	Force/Area	$[ML^{-1}T^{-2}]$
Stress	Force/Area	$[ML^{-1}T^{-2}]$
Elastic modulus	Stress/Strain	$[ML^{-1}T^{-2}]$
Impulse	Force × Time	$[MLT^{-1}]$
Mass moment of inertia	Mass × Length^2	$[ML^2]$
Momentum, linear	Mass × Linear velocity	$[MLT^{-1}]$
Momentum, angular	Moment of inertia × Angular velocity	$[ML^2T^{-1}]$
Work, energy	Force × Distance	$[ML^2T^{-2}]$
Power	Work/Time	$[ML^2T^{-3}]$
Moment of a force	Force × Distance	$[ML^2T^{-2}]$
Viscosity, dynamic	Shear stress/Velocity gradient	$[ML^{-1}T^{-1}]$
Viscosity, kinematic	Dynamic viscosity/ Mass density	$[L^2T^{-1}]$
Surface tension	Energy/Area	$[MT^{-2}]$

TABLE 8.2   Dimensions of common quantities in thermodynamics

| | | | DIMENSIONS | | | |
| | | | MLTΘ SYSTEMS | | | |
QUANTITY	DEFINING EQUATION	MLT SYSTEM	THERMAL	DYNAMIC	HLTΘ SYSTEM	HMLTΘ SYSTEM
Temperature, θ		$[L^2T^{-2}]$	$[\Theta]$	$[\Theta]$	$[\Theta]$	$[\Theta]$
Heat quantity, H		$[ML^2T^{-2}]$	$[M\Theta]$	$[ML^2T^{-2}]$	$[H]$	$[H]$
Enthalpy		$[ML^2T^{-2}]$	$[M\Theta]$	$[ML^2T^{-2}]$	$[H]$	$[H]$
Entropy, S	$dS = dH/\theta$	$[M]$	$[M]$	$[ML^2T^{-2}\Theta^{-1}]$	$[H\Theta^{-1}]$	$[H\Theta^{-1}]$
Coeff. of thermal expansion	Change of length/unit length/degree	$[L^{-2}T^2]$	$[\Theta^{-1}]$	$[\Theta^{-1}]$	$[\Theta^{-1}]$	$[\Theta^{-1}]$
Thermal capacity	Heat required per degree temp. rise	$[M]$	$[M]$	$[ML^2T^{-2}\Theta^{-1}]$	$[H\Theta^{-1}]$	$[H\Theta^{-1}]$
Specific heat	Thermal capacity per unit mass	$[M^0 L^0 T^0]$	$[M^0 L^0 T^0 \Theta^0]$	$[L^2T^{-2}\Theta^{-1}]$	—	$[HM^{-1}\Theta^{-1}]$
Specific heat ratio		$[M^0 L^0 T^0]$	$[M^0 L^0 T^0 \Theta^0]$	$[M^0 L^0 T^0 \Theta^0]$	$[H^0 L^0 T^0 \Theta^0]$	$[H^0 M^0 L^0 T^0 \Theta^0]$
Thermal conductivity	Time rate of heat transmission per unit area and temp. gradient	$[ML^{-1}T^{-1}]$	$[ML^{-1}T^{-1}]$	$[MLT^{-3}\Theta^{-1}]$	$[HL^{-1}T^{-1}\Theta^{-1}]$	$[HL^{-1}T^{-1}\Theta^{-1}]$
Gas constant, R	Energy/Mass $\times$ Temp.	$[M^0 L^0 T^0]$	$[M^0 L^0 T^0 \Theta^0]$	$[L^2T^{-2}\Theta^{-1}]$	—	$[L^2T^{-2}\Theta^{-1}]$
Coeff. of heat transfer		$[ML^{-2}T^{-1}]$	$[ML^{-2}T^{-1}]$	$[MT^{-3}\Theta^{-1}]$	$[HL^{-2}T^{-1}\Theta^{-1}]$	$[HL^{-2}T^{-1}\Theta^{-1}]$
Mechanical equivalent of heat		$[M^0 L^0 T^0]$	$[L^2T^{-2}\Theta^{-1}]$	$[M^0 L^0 T^0]$	—	$[H^{-1}ML^2T^{-2}]$

8.7 DIMENSIONS OF DERIVATIVES AND INTEGRALS

The dimensions which should be assigned to partial or total derivatives can be determined easily, if it is remembered that, by definition, dy/dx is the limiting value of the ratio $\delta y/\delta x$ as δx tends to zero, where δy is the finite value of the increment of y which corresponds to a finite increment δx of x.

Clearly, the dimensions of δy are the same as those of y and the dimensions of δx are the same as those of x; thus, dimensionally,

$$\left[\frac{dy}{dx}\right] = \left[\frac{\delta y}{\delta x}\right] = \left[\frac{y}{x}\right].$$

Similarly,

$$\frac{d^2y}{dx^2} = \frac{d}{dx}\left(\frac{dy}{dx}\right) = \frac{\text{Increment of } dy/dx}{\text{Increment of } x},$$

so that, dimensionally,

$$\left[\frac{d^2y}{dx^2}\right] = \left[\frac{dy/dx}{x}\right] = \left[\frac{y/x}{x}\right] = \left[\frac{y}{x^2}\right]$$

or, in general,

$$\left[\frac{d^ny}{dx^n}\right] = \left[\frac{y}{x^n}\right].$$

The dimensions of integrals are found in the same way. The term

$$\int_b^a y\,dx$$

means the limit of the sum of all the products of $y\delta x$ between $x = a$ and $x = b$. Thus, the dimensions of

$$\int_b^a y\,dx$$

will be the same as those of $y\delta x$ and, since $[\delta x] = [x]$,

$$\left[\int_a^b y\,dx\right] = [yx].$$

Similarly, a double integral $\iint a\,dz_1\,dz_2$ means the limit sum of the products $a\delta z_1\delta z_2$. Since $[\delta z_1] = [z_1]$ and $[\delta z_2] = [z_2]$,

$$\left[\iint a\,dz_1\,dz_2\right] = [az_1z_2].$$

The dimensions of any multiple integral are found in the same way.

8.8   USE OF DIMENSIONAL REASONING TO CHECK CALCULATIONS

The principle of dimensional homogeneity requires that any equation which fully and correctly represents a possible physical situation must be dimensionally homogeneous, so that each term has the same dimensional formula. It is, therefore, possible to make a rapid check of any algebraic analysis of such a situation by substituting the dimensions of the quantities in each term and checking that, in the resultant dimensional equation, each fundamental dimension appears to the same power in each term. So long as symbols are not replaced by numerical values, each line of working can be checked in this way, and it is, therefore, often advantageous to delay the insertion of numerical values until the last possible moment.

Many engineering formulae in day-to-day use are committed to memory and, sometimes, there may be a doubt as to whether they have been recalled correctly. A quick dimensional check will establish the correct form immediately, although, of course, it will not indicate whether the values of any numerical constants are correct.

8.9   UNITS OF DERIVED QUANTITIES

The units in which a derived quantity can be measured may be determined directly from its dimensional formula by substituting the appropriate unit for each fundamental quantity. Thus, the SI unit of dynamic viscosity is obtained by substituting the kilogram for mass, the metre for length and the second for time in the dimensional formula $ML^{-1}T^{-1}$ and will be the kilogram per metre per second.

The SI system is a rationalized system of metric units, in which the units for all physical quantities can be derived from six basic, arbitrarily defined units, which are:

Length	metre;
Mass	kilogram;
Time	second;
Electric current	ampere;
Absolute temperature	kelvin;
Luminous intensity	candela.

The product or quotient of any two units is the unit of the resultant quantity. Certain of these derived units have been given special names. For example, force has the dimensions MLT^{-2} and is measured, therefore, in kg m s^{-2}. For convenience, this unit is referred to as the newton and abbreviated to N. Similarly, pressure is defined as force per unit area and measured in N m^{-2}, but this unit is sometimes called the pascal. Details of the basic and derived SI units are given in Table 8.3.

TABLE 8.3
SI units

QUANTITY	UNIT	SYMBOL
BASIC UNITS		
Length	metre	m
Mass	kilogram	kg
Time	second	s
Electric current	ampere	A
Absolute temperature	kelvin	K
Luminous intensity	candela	cd
Geometry		
Angle,		
plane	radian	rad
solid	steradian	sr
Area	square metre	m^2
Volume	metre cubed	m^3
First moment of area	metre cubed	m^3
Second moment of area	metre to fourth power	m^4
DERIVED UNITS		
Mechanics		
Frequency	hertz	Hz
Velocity,		
linear	metre per second	$m\,s^{-1}$
angular	radian per second	$rad\,s^{-1}$
Acceleration,		
linear	metre per second squared	$m\,s^{-2}$
angular	radian per second squared	$rad\,s^{-2}$
Force	newton (= kilogram metre per second squared)	$N\,(= kg\,m\,s^{-2})$
Density, mass	kilogram per metre cubed	$kg\,m^{-3}$
Specific weight	newton per metre cubed	$N\,m^{-3}$
Momentum,		
linear	kilogram metre per second	$kg\,m\,s^{-1}$
angular	kilogram metre squared per second	$kg\,m^2\,s^{-1}$
Moment of inertia	kilogram metre squared	$kg\,m^2$
Moment of force	newton metre	N m
Pressure or stress (intensity)	pascal (= newton per metre squared)	$Pa\,(= N\,m^{-2})$
Viscosity,		
dynamic	newton second per metre squared (= 10 poise)	$N\,s\,m^{-2}$
kinematic	metre squared per second	$m^2\,s^{-1}$
Surface tension	newton per metre	$N\,m^{-1}$
Energy, work	joule (= newton metre)	$J\,(= N\,m)$
Power	watt (= joule per second)	$W\,(= J\,s^{-1})$
Heat		
Temperature interval	degree kelvin	K
Linear expansion coefficient	expansion per unit length per degree kelvin	K^{-1}
Heat quantity	joule	J
Heat flow rate	watt	W
Entropy	joule per degree kelvin	$J\,K^{-1}$
Thermal capacity	joule per degree kelvin	$J\,K^{-1}$
Thermal conductivity	watt per metre per degree kelvin	$W\,m^{-1}\,K^{-1}$
Coefficient of heat transfer	watt per metre squared per degree kelvin	$W\,m^{-2}\,K^{-1}$

8.10 CONVERSION FROM ONE SYSTEM OF UNITS TO ANOTHER

It is intended that SI units should (at the earliest reasonable date) be used everywhere, but other systems are currently in use in many parts of the world. These can be divided into the so-called absolute systems – based on mass, length and time – and technical systems – based on force, length and time.

There are a number of units in common use which are not part of coherent systems. In some English-speaking countries, speeds on the roads are measured in miles per hour, weights are sometimes measured in tons, areas by the acre, volume by the Imperial gallon or the US gallon, and lengths by the chain or fathom, to name but a few. Similar local units are used in other countries and it seems probable that it will be a very long time before a single international system of coherent units will replace all others. In the meantime, the problem of converting from one system to another will remain important and dimensional reasoning can be of help in this process.

Clearly the true physical value of any quantity must be independent of the system of units by which it is measured. If a quantity Q is found to have a numerical value n_1 when measured in units of size u_1,

$$\text{Physical value of } Q = n_1 u_1.$$

Similarly, if the same quantity Q when measured in units of size u_2 has a numerical value of n_2,

$$\text{Physical value of } Q = n_2 u_2.$$

Since the value of Q must remain unchanged,

$$Q = n_1 u_1 = n_2 u_2.$$

Thus, the numerical value of a quantity is inversely proportional to the size of the units in which it is measured.

If the quantity Q has a dimensional formula $[M^a L^b T^c]$, the unit of measurement will be a derived unit. Suppose that the units of mass, length and time are m_1, l_1 and t_1 in the first system and m_2, l_2 and t_2 in the second system; then the derived units in these systems, based on the dimensional formulae, are

$$u_1 = m_1^a l_1^b t_1^c \quad \text{and} \quad u_2 = m_2^a l_2^b t_2^c.$$

The fundamental units in the two systems will also be related: the size of the unit of mass m_1 will be equal to $k_m m_2$, that of l_1 is $k_l l_2$ and that of t_1 is $k_t t_2$, where k_m, k_l and k_t are numerical constants.

$$Q = N_1 u_1 = N_1 m_1^a l_1^b t_1^c,$$

and so, replacing m_1 by $k_m m_2$, l_1 by $k_l l_2$ and t_1 by $k_t t_2$,

$$N_1 u_1 = N_1 (k_m^a m_2^a)(k_l^b l_2^b)(k_t^c t_2^c)$$
$$= N_1 k_m^a k_l^b k_t^c \times m_2^a l_2^b t_2^c,$$

But $N_1 u_1 = N_2 u_2 = Q$ and $m_2^a l_2^b t_2^c = u_2$; therefore,

$$N_2 u_2 = N_1 (k_m^a k_l^b k_t^c) u_2,$$

$$N_2 = N_1 k_m^a k_l^b k_t^c,$$

where k_m, k_l and k_t are the numbers of units in the second system required to make the corresponding fundamental units in the first system.

EXAMPLE 8.1

An engine produces 57 horsepower. What is the corresponding value in kilowatts and what is the conversion factor?

Solution

The horsepower and the kilowatt are multiples of the basic units for power in the British technical system and the SI system, respectively. Thus, if n_1 is the horsepower and N_1 is the corresponding number of British technical units (ft lbf/sec)

$$N_1 = 550 n_1.$$

Similarly, since the basic unit of power is the watt in the SI system,

$$N_2 = 1000 n_2.$$

The dimensional formula for power is $[ML^2T^{-3}]$. Using suffix 1 for British technical units and suffix 2 for SI units,

$$N_2 = N_1 k_m k_l^2 k_t^{-3},$$

where k_m, k_l and k_t are the ratios of the units of mass, length and time in system 1 to the corresponding units in system 2, as tabulated below:

Quantity	System 1 (FPS technical)	System 2 (SI)	Ratio (1)/(2)
Mass	slug	kilogram	14.6
Length	foot	metre	0.3048
Time	second	second	1

Therefore,

$$N_2 = N_1 \times 14.6 \times 0.3048^2 \times 1 = 1.356 N_1$$

or $1000 n_2 = 550 n_1 \times 1.356$

so that $n_2 = 0.746 n_1.$

Therefore,

Value in kilowatts $= 0.746 \times$ Value in horsepower.

Putting $n_1 = 57$ horsepower,

$$\text{Output} = 0.746 \times 57 = \textbf{42.5 kW},$$

$$\text{Conversion factor} = n_2/n_1 = \textbf{0.746}.$$

8.11　CONVERSION OF DIMENSIONAL CONSTANTS

Many equations commonly used by engineers in practice contain numerical constants. In some cases, these are pure numbers and are, therefore, unaffected by the system of units employed. In other cases, these numbers are not dimensionless and their numerical values will depend on the system of units being used. Conversion is carried out in the same way as has been explained in Section 8.10.

8.12　CONSTRUCTION OF RELATIONSHIPS BY DIMENSIONAL ANALYSIS USING THE INDICIAL METHOD

If the factors involved in any real physical situation can all be identified, the form of the equation relating them can be largely determined by dimensional reasoning. This is the case because the requirement that the resulting equations must be dimensionally homogeneous means that the variables can only be combined in a very limited number of ways, and in some cases this may lead to a unique solution. Since pure numbers are dimensionless, a complete numerical solution of the type provided by rigorous algebraic reasoning cannot be obtained by dimensional reasoning. However, the values of the missing numbers can usually be established with comparative ease by experiment. It should also be remembered that it is often necessary, when making an algebraic analysis, to simplify the situation by making a number of assumptions, some of which are not strictly correct. The choice is, therefore, between a complete answer to an idealized situation or the partial answer to the real problem provided by dimensional analysis.

EXAMPLE 8.2

The thrust F of a screw propeller is known to depend upon the diameter d, speed of advance v, fluid density ρ, revolutions per second N, and the coefficient of viscosity μ of the fluid. Find an expression for F in terms of these quantities.

Solution

The general relationship must be $F = \phi(d, v, \rho, N, \mu)$, which can be expanded as the sum of an infinite series of terms giving

$$F = A(d^m v^p \rho^q N^r \mu^s) + B(d^{m'} v^{p'} \rho^{q'} N^{r'} \mu^{s'}) + \cdots,$$

where A, B, etc., are numerical constants and m, p, q, r, s are unknown powers. Since, for dimensional homogeneity, all terms must be dimensionally the same, this can be reduced to

$$F = Kd^m v^p \rho^q N^r \mu^s,$$ (I)

where K is a numerical constant.

The dimensions of the dependent variable F and the independent variables d, v, ρ, N and μ are

$$[F] = [\text{Force}] = [\text{MLT}^{-2}],$$

$$[d] = [\text{Diameter}] = [\text{L}],$$

$$[v] = [\text{Velocity}] = [\text{LT}^{-1}],$$

$$[\rho] = [\text{Mass density}] = [\text{ML}^{-3}],$$

$$[N] = [\text{Rotational speed}] = [\text{T}^{-1}],$$

$$[\mu] = [\text{Dynamic viscosity}] = [\text{ML}^{-1}\text{T}^{-1}].$$

For convenience, these can be set out in the form of a table, or dimensional matrix, in which a column is provided for each variable and the power of each fundamental dimension in its dimensional formula is inserted in the corresponding row:

	F	d	v	ρ	N	μ
M	1	0	0	1	0	1
L	1	1	1	-3	0	-1
T	-2	0	-1	0	-1	-1

Substituting the dimensions for the variables in (I),

$$[\text{MLT}^{-2}] = [\text{L}]^m [\text{LT}^{-1}]^p [\text{ML}^{-3}]^q [\text{T}^{-1}]^r [\text{ML}^{-1}\text{T}^{-1}]^s.$$

Equating powers of [M], [L] and [T]:

$$[\text{M}], \quad 1 = q + s;$$ (II)

$$[\text{L}], \quad 1 = m + p - 3q - s;$$ (III)

$$[\text{T}], \quad -2 = -p - r - s.$$ (IV)

Since there are five unknown powers and only three equations, it is impossible to obtain a complete solution, but three unknowns can be determined in terms of the remaining two. If we solve for m, p and q, we get

$$q = 1 - s \quad \text{from (II)}$$

$$p = 2 - r - s \quad \text{from (IV)}$$

$$m = 1 - p + 3q + s = 2 + r - s \quad \text{from (III)}.$$

Substituting these values in (I),

$$F = Kd^{2+r-s}v^{2-r-s}\rho^{1-s}N^r\mu^s.$$

Regrouping the powers,

$$F = K\rho v^2 d^2 (\rho v d/\mu)^{-s}(dN/v)^r.$$

Since s and r are unknown this can be written

$$F = \rho v^2 d^2 \phi(\rho v d/\mu, dN/v), \tag{V}$$

where ϕ means 'a function of'. At first sight, this appears to be a rather unsatisfactory solution, but (V) indicates that

$$F = C\rho v^2 d^2, \tag{VI}$$

where C is a constant to be determined experimentally and the value of which is dependent on the values of $\rho v d/\mu$ and dN/v. Equation (VI) could also be used to calculate the thrust of a full-size propeller from experiments on a model, providing that the values of $\rho v d/\mu$ and dN/v were made the same for both and, therefore, C would have the same value in the two cases. Then we would be able to state that

$$F_{\text{model}}/F_{\text{full size}} = (\rho v^2 d^2)_{\text{model}}/(\rho v^2 d^2)_{\text{full size}}.$$

Equation (V) could have been written

$$F/\rho v^2 d^2 = \phi(\rho v d/\mu, dN/v)$$

or $$\phi(F/\rho v^2 d^2, \rho v d/\mu, dN/v) = 0. \tag{VII}$$

Each of the terms in the bracket forms a dimensionless group since

$$F/\rho v^2 d^2 = [MLT^{-2}]/[ML^{-3}][L^2T^{-2}][L^2] = [1]$$

$$\rho v d/\mu = [ML^{-3}][LT^{-1}][L]/[ML^{-1}T^{-1}] = [1]$$

$$dN/v = [L][T^{-1}]/[LT^{-1}] = [1].$$

8.13   DIMENSIONAL ANALYSIS BY THE GROUP METHOD

The indicial method is rather lengthy if there are a large number of variables. It was therefore necessary to develop a more generalized methodology that would lead directly to a set of dimensionless groupings whose number could be determined in advance by a scrutiny of the matrix formed from the variables considered to be relevant to the investigation and the relevant dimensions necessary to describe those variables. Such a technique was developed in the early years of the twentieth century and is known as the Buckingham π method.

The initial step in the application of the Buckingham π method is to list the variables considered to be significant and to form a matrix with their dimensions. Example 8.2 illustrated this technique with the six variables and three dimensions, M, L, T, forming the matrix illustrated. It is then necessary to determine the number of dimensionless groups into which the variables may be combined. This number may be found by application of the Buckingham π method, which states that

> The number of dimensionless groups arising from a particular matrix formed from n variables in m dimensions is n − r, where r is the largest non-zero determinant that can be formed from the matrix, and therefore the equation relating the variables will be of the form
>
> $$\phi(\pi_1, \pi_2, \pi_3, \ldots, \pi_{n-r}) = 0.$$

While this form of Buckingham's theorem is correct, it is often simpler, in the treatment of fluid conditions that only involve the dimensions of mass, length and time, to state that the number of dimensionless groups formed from n variables in three dimensions is $n - m$, or $n - 3$. While this is also correct and widely quoted, care must be taken in applying this rule outside the confines of a strictly three-dimensional problem. The introduction of heat and temperature as dimensions in the study of thermodynamic and heat transfer phenomena increases the number of possible dimensions to five; however, in some cases the correct value of r will be less than this. As fluid mechanics and thermodynamics/heat transfer often share common courses, care should be taken in the acceptance of the simplified rule.

The importance of determining the correct order for the largest non-zero determinant may be demonstrated by three examples taken from the area of fluid flow over a surface with and without heat transfer by either forced or natural convection.

EXAMPLE 8.3

Determine the number of dimensionless groups expected to be formed from the variables involved in the flow of fluid external to a solid body, extending this analysis to include both forced and natural convection from a surface.

Solution

Case 1: no heat transfer condition. The force acting, F, may be expected to be a function of flow velocity v, density ρ, dynamic viscosity μ, and body characteristic length L. The matrix formed in the applicable dimensions of mass, length and time from the five applicable variables is as follows:

	M	L	T
F	1	1	−2
v	0	1	−1
ρ	1	−3	0
μ	1	−1	−1
L	0	1	0

To determine the highest order of non-zero determinant it is necessary to refer to some simple laws governing determinants, for example:

1. The order of a determinant is defined by the equal number of rows and columns displayed.
2. The value of a determinant is unchanged if the order of its rows and columns is changed.
3. The value is unchanged if the rows are changed to columns.
4. If two rows or columns are identical then the value of the determinant is zero.
5. If any row or column is multiplied by a constant then the value of the determinant is its previous value multiplied by that constant.

The above 'rules' allow the largest non-zero determinant to be recognized in any dimensional matrix.

In the first case the highest order determinant possible would be third order, provided scrutiny of the rules above did not reduce this value. It will be seen by inspection that none applies and that the number of dimensionless groups expected in this case would be $(5 - 3)$ or two.

Case 2: forced convection. Considering the flow of fluid over a surface with a temperature difference between the fluid and the surface requires the introduction of two further dimensions, namely quantity of heat energy, H, and temperature, Θ, as set out in Table 8.2. The eight applicable variables may be formed into the dimensional matrix below:

VARIABLE	M	L	T	H	Θ
Force, F	1	1	−2	0	0
Length, L	0	1	0	0	0
Density, ρ	1	−3	0	0	0
Viscosity, μ	1	−1	−1	0	0
Heat capacity, c_p	−1	0	0	1	−1
Thermal conductivity, k	0	−1	−1	1	−1
Velocity, v	0	1	−1	0	0
Heat transfer coeff., h	0	−2	−1	0	−1

By inspection it is clear that the highest possible order would have been 5, however inspection of the H and Θ columns indicates that these columns are −1 multiples of each other, thus any fifth order determinant formed from the matrix would be zero.

Dropping either the H or Θ column and reapplying the rules set out above indicates that a fourth order non-zero determinant is possible and the number of dimensionless groups to be expected in this case is thus $(8 - 4)$ or four.

Case 3: natural convection. Here it is again necessary to introduce dimensions of heat and temperature. In this case, however, it is also necessary to introduce gravitational forces, represented by g, and a temperature difference, ΔT, to cater for the density–temperature buoyancy effects driving the process.

The ten applicable variables may be formed into the dimensional matrix below:

VARIABLE	M	L	T	H	Θ
Force, F	1	1	−2	0	0
Length, L	0	1	0	0	0
Density, ρ	1	−3	0	0	0
Viscosity, μ	1	−1	−1	0	0
Heat capacity, c_p	−1	0	0	1	−1
Thermal conductivity, k	0	−1	−1	1	−1
Coeff. of fluid thermal expansion, β	0	0	0	0	−1
Gravitational acceleration, g	0	1	−1	0	0
Temperature diff., ΔT	0	0	0	0	1
Heat transfer coeff., h	0	−2	−1	0	−1

By inspection it is clear that the highest possible order would have been 5. In this case none of the rules indicated would reduce the order and, therefore, the number of dimensionless groups expected would be $(10 − 5)$ or five.

Taken together these three examples illustrate the importance of careful inspection of the dimensional matrix in order to determine the applicable number of dimensionless groups to be expected in any investigation.

Referring back to Example 8.2, it may be seen that as there are six variables, F, d, v, ρ, N, μ, and three fundamental dimensions, M, L, T, the value of r would be 3, and hence the number of dimensionless groups would be 3.
The solution therefore contains the following groups:

$$\pi_1 = F/\rho v^2 d^2, \quad \pi_2 = \rho v d/\mu, \quad \pi_3 = dN/v.$$

Independent dimensionless groups are defined as those which can be formed from any particular number of quantities, but are independent of each other in the sense that none of them can be formed by any combination of the others.
In any particular problem, having determined the number of dimensionless groups as described above, the next step is to combine the variables to form the desired groupings. The following points are useful indicators of the best approach to follow:

1. From the independent variables thought to describe the fluid flow condition select certain variables to act as repeating variables. These variables may appear in all or some groups. The number of repeating variables is therefore $[n − (n − r)]$, i.e. the total number of variables minus the number of groups. The choice of these repeating variables is not arbitrary but should be guided by the following rules:

   (a) The repeating variables as a combination must include all the dimensions taken to describe the system; thus in Example 8.2 they must include M, L, T. This does not mean that each repeating variable includes all dimensions, but seen as a group this must be the case.

   (b) The repeating variables should be chosen with some regard for the practicality of any experimental investigation: they should be easily measurable or set by the investigator. Similarly, where the results of a dimensional analysis are to be the basis for a later design methodology, the repeating variable should be of prime interest to the designer. For example, it is more sensible to define pipe type in terms of pipe diameter than surface roughness as a repeating parameter and density is perhaps better than viscosity as a descriptor of fluid type.

2. Combine the repeating variables with the remaining independent variables to form the required number of groups. It follows that each of the remaining variables now only appears in one group and these groups are often referred to by that variable name.

3. A variable that is considered to be of minor significance will, as a result of (1) and (2) above, only appear in one group. The influence of this group will be negligible if this variable is truly inconsequential. This raises one of the interesting points in dimensional analysis, namely that there really are no 'wrong' answers, only answers that are more useful than others. The inclusion of a number of variables that have little or no effect on the flow phenomena will result in dimensionless groupings which will be shown by experimental investigation to be of no significance. Similarly, so long as the repeating variables chosen conform to the rule that they represent all the dimensions of the problem, the choice of repeating variable may also be arbitrary. This approach will result in the problem being defined in terms of a series of correctly dimensionless groups which, due to the poor choice of repeating variable, are of little use to the investigator. However, even in this case not all is lost, owing to the following rules that apply to the groups in an equation of the form

$$\phi(\pi_1, \pi_2, \pi_3, \ldots, \pi_{n-r}) = 0.$$

(a) Any number of dimensionless groups may be combined by multiplication or division to form a new valid group. Thus π_1 and π_2 may be combined to form $\pi_{1'} = \pi_1/\pi_2$ and the defining equation becomes

$$\phi(\pi_{1'}, \pi_2, \pi_3, \ldots, \pi_{n-r}) = 0.$$

(b) The reciprocal of any dimensional group remains valid. An example of this will be met in the later treatment of fans and pumps where a reciprocal form of the Reynolds number will be recognizable.

(c) Any dimensional group may be raised to any power and remain valid.

(d) Any dimensional group may be multiplied by a constant and remain valid. This is useful in relating a particular group to an easily measured quantity, e.g. pressure coefficients. Groups often include the combination ρv^2 as the non-dimensioning denominator, while in fact the use of $\frac{1}{2}\rho v^2$ would allow direct use of the flow kinetic energy term, in itself readily measured by use of a Pitot–static tube. Thus the general form of a dimensional relationship could appear as

$$\phi[\pi_{1'}, 1/\pi_2, (\pi_3)^i, \ldots, \tfrac{1}{2}\pi_{n-r}] = 0$$

and remain valid.

EXAMPLE 8.4

The variables controlling the motion of a floating vessel through the surrounding fluid are the drag force F, the vessel's speed v, its length l, the density ρ and dynamic viscosity μ of the fluid and the acceleration due to gravity g. Derive an expression for F by dimensional analysis.

Solution

The resistance to motion will be partly due to skin friction and partly due to wave resistance, thus involving both viscous and gravitational forces. The general form of the expression may be written as

$$F = \phi(v, l, \rho, \mu, g). \tag{I}$$

The dimensions of the variables involved may be tabulated as follows:

	F	v	l	ρ	μ	g
Mass M	1	0	0	1	1	0
Length L	1	1	1	−3	−1	1
Time T	−2	−1	0	0	−1	−2

As the number of variables n is six and the number of dimensions m is three, it follows that the number of dimensionless groups will be $(n - m)$ or 3. Thus there will be three repeating variables chosen to non-dimensionalized groups that feature three variables of prime interest. From an experimental or design perspective it would be useful to relate the force needed to the design speed of a vessel of a given length in a particular fluid so it would be reasonable to choose v, l and ρ as the repeating variables, leaving force F, viscosity μ, and gravitational acceleration as the independent variables. (While gravity is sensibly a constant, it appears as an independent variable as it is unlikely to be a variable that could be controlled easily in an experiment.)

The required solution will be

$$\pi_1 = \phi(\pi_2, \pi_3). \tag{II}$$

Let the first group in F be defined as

$$\pi_1 = \frac{F}{v^a l^b \rho^c}. \tag{III}$$

Equating powers of M, L and T results in a set of three simultaneous equations that may be solved for the indices a, b and c.

For M $\quad 1 = \qquad\quad c$
For L $\quad 1 = a + b - 3c$
For T $-2 = -a$

so that

$$a = 2,$$
$$b = 2,$$
$$c = 1,$$

and the first group becomes

$$\pi_1 = \frac{F}{v^2 l^2 \rho}. \qquad \text{(IV)}$$

(Note that this could be rearranged as

$$\pi_1 = \frac{F}{0.5 \rho v^2 l^2}, \qquad \text{(V)}$$

which has the form of a force coefficient incorporating a kinetic energy term and an area term.)

The second group will be formed around viscosity as the independent variable as follows:

$$\pi_2 = \frac{\mu}{v^a l^b \rho^c}, \qquad \text{(VI)}$$

where the indices have been retained as a, b and c, although they will now have different numerical values from those determined for the force group in order to demonstrate a useful 'short cut' in more complex cases.

Equating powers of M, L and T results in a set of three simultaneous equations that may be solved for the indices a, b and c.

For M $\quad 1 = \qquad\qquad c$
For L $-1 = \quad a + b - 3c$
For T $-1 = -a$

so that

$$a = 1,$$
$$b = 1,$$
$$c = 1,$$

and the second group becomes

$$\pi_2 = \frac{\mu}{v l \rho}, \qquad \text{(VII)}$$

which will be recognized as the invert form of Reynolds number so that the viscosity group may be rewritten as

$$\pi_2 = \frac{v l \rho}{\mu}. \qquad \text{(VIII)}$$

The third group will be formed around gravity as the independent variable as follows:

$$\pi_3 = \frac{g}{v^a l^b \rho^c}, \tag{IX}$$

where again the indices have been retained as a, b and c, but will have different numerical values from those determined for the force group to demonstrate a useful 'short cut'.

Equating powers of M, L and T results in a set of three simultaneous equations that may be solved for the indices a, b and c.

For M $0 = \qquad\qquad c$
For L $1 = a + b - 3c$
For T $-2 = -a$

so that

$$a = 2,$$
$$b = -1,$$
$$c = 0,$$

and the third group becomes

$$\pi_3 = \frac{g}{v^2 l^{-1}} = \frac{lg}{v^2}, \tag{X}$$

which may be recognized as the reciprocal of the square of the Froude number, so that the third group may be conventionally rewritten as

$$\pi_3 = \frac{v}{\sqrt{(lg)}}. \tag{XI}$$

Thus this example illustrates all the 'rules' set out in Section 8.13 and in addition illustrates one important time saver in these solutions – it will have been noted that the right-hand side of the M, L and T equations was *identical* for each group developed, leading to a tabular form of solution in cases such as this, for example

	GROUPS IN			
	F	μ	g	
For M	1	1	$0 =$	c
For L	1	-1	$1 =$	$a + b - 3c$
For T	-2	-1	$-2 = -a$	

Thus the force necessary to propel the vessel may be expressed as a non-dimensional force coefficient, which includes kinetic energy and area terms, found to be dependent on the viscous and gravitational forces present, defined in terms of the Reynolds and Froude numbers, respectively

$$\frac{F}{0.5 \rho v^2 l^2} = \phi(\text{Re, Fr}) \tag{XII}$$

EXAMPLE 8.5

The variables governing the resistance to flow, or surface shear stress τ_0 in a closed conduit are believed to include the flow mean velocity v, the conduit diameter D, its surface roughness k and the density ρ and dynamic viscosity μ of the fluid. In addition if the surface of the conduit is itself in motion then the surface velocity V_s may also be a factor.

For both cases utilize a tabular format to determine the likely dimensionless groups.

Solution

The general expression for the dependence of shear stress may be expressed as

$$\tau_0 = \phi(v,\, D,\, k, \rho,\, \mu,\, V_s). \tag{I}$$

The dimensions of the variables involved may be tabulated as follows:

	τ_0	v	D	k	ρ	μ	V_s
Mass, M	1	0	0	0	1	1	0
Length, L	1	1	1	1	-3	-1	1
Time, T	-2	-1	0	0	0	-1	-1

The number of variables is seven and the number of dimensions is three so the number of maximum groups, if V_s is included, is four, with three repeating variables. The choice of repeating variables is in this case straightforward as the investigator would wish to control the test variables of flow velocity, conduit diameter and fluid type, best described by density.

With repeating variables of v, D and ρ it follows that it will be necessary to seek dimensionless groups featuring τ_0, k, μ and V_s as follows:

$$\pi_1 = \frac{\tau_0}{v^a D^b \rho^c}, \tag{II}$$

$$\pi_2 = \frac{k}{v^a D^b \rho^c}, \tag{III}$$

$$\pi_3 = \frac{\mu}{v^a D^b \rho^c}, \tag{IV}$$

$$\pi_4 = \frac{V_s}{v^a D^b \rho^c}. \tag{V}$$

(Note that as before the index values a, b and c for each group will have different numerical values.)

The following tabular layout for the M, L and T equations may be developed as in Example 8.4:

	GROUPS IN			
	τ_0	k	μ	V_s
For M	1	0	1	$0 = \qquad\qquad c$
For L	-1	1	-1	$1 = \quad a + b - 3c$
For T	-2	0	-1	$-1 = -a$

resulting in the following dimensionless groups:

$$\pi_1 = \frac{\tau_0}{\frac{1}{2}\rho v^2}, \tag{VI}$$

$$\pi_2 = \frac{k}{D}, \tag{VII}$$

$$\pi_3 = \frac{\mu}{v D \rho} = \frac{\rho v D}{\mu} = \text{Re}, \tag{VIII}$$

$$\pi_4 = \frac{V_s}{v}, \tag{IX}$$

which include a stress coefficient and a form of Reynolds number. The wall roughness and the wall velocity groups could have been determined by inspection as both allow single variable non-dimensionality.

Examples 8.4 and 8.5 illustrate the 'mathematical' application of Buckingham's π theorem, however, its application to real investigations requires an understanding of the significance of the various groups, many of which recur throughout fluid analysis, as well as an appreciation of the impact of careful selection of repeating variables on the experimental process. These issues will be addressed in Sections 8.13 and 8.14 and in Chapter 9.

8.14   THE SIGNIFICANCE OF DIMENSIONLESS GROUPS

By definition, a dimensionless group is the ratio of two similar physical quantities. In Example 8.4, we obtain a dimensionless group $F/\rho v^2 l^2$. The numerator F is the force required to overcome the drag on the vessel and, therefore, the denominator should also represent a force existing in the system. Examining the term $\rho v^2 l^2$, it can be written $\rho l^3 \times v^2/l$. Taking l as some typical dimension, ρl^3 is the product of density and a typical volume and so represents the mass of a typical element of fluid. Similarly, the velocity v can be expressed as l/t, where t is the time required to traverse a typical distance l. Thus, $v^2/l \propto l^2/lt^2 \propto l/t^2$, which is a measure of acceleration. Thus, we have

$$\rho v^2 l^2 = \rho l^3 \times l/t^2$$

$$= \text{Mass} \times \text{Acceleration} = \text{Inertial force.}$$

The dimensionless group $F/\rho v^2 l^2$ is therefore the ratio

Drag force/Inertial force.

In the same way, the Reynolds number $\rho v l/\mu$ can also be expressed as the ratio of two forces, since it can be written $\rho v^2 l^2/\mu(v/l)l^2$ and we have already seen that $\rho v^2 l^2$ represents an inertial force. By definition, the coefficient of dynamic viscosity μ is the shear force per unit area produced by unit velocity gradient in the fluid. The denominator is therefore equivalent to a viscous force since v/l is the velocity gradient and l^2 represents a typical area. The dimensionless group is, therefore, the ratio

Inertial force/Viscous force.

The final group v^2/lg can also be expressed as a ratio of forces, since it can be written $\rho v^2 l^2/\rho l^3 g$. The denominator is the inertial force, as before, and, since ρl^3 is a typical mass and g the gravitational acceleration, $\rho l^3 g$ is a gravitational force. The dimensionless group v^2/lg is the ratio

Inertial force/Gravitational force.

Other dimensionless groups can be shown to be related to other physical properties of the system. For example, the Mach number $\bar{v}/\sqrt{(k/\rho)}$ is the ratio of the velocity of flow to the velocity of propagation of a pressure wave in the fluid, and is of importance in the study of compressible flow.

These and other dimensionless groups are useful means of defining the conditions which exist in a physical system and indicating which properties are of importance. As shown above, the Reynolds number is the ratio of inertial force to viscous force and is a measure of the importance of viscous resistance in controlling the flow of a fluid. A low Reynolds number indicates that viscosity is a dominant factor; a high Reynolds number indicates that its effect is small. For example, the flow in a circular pipe will be laminar for values of Reynolds number below 2000 and will be turbulent for higher values, so that, by stating the Reynolds number, we can specify the type of flow. As different laws apply to the two types of flow, it is a matter of practical importance to be able to distinguish between them.

8.15   THE USE OF DIMENSIONLESS GROUPS IN EXPERIMENTAL INVESTIGATION

Dimensional analysis can be of assistance in experimental investigation by reducing the number of variables in the problem. The result of the analysis is to replace an unknown relation between n variables by a relationship between a smaller number, $n - m$, of dimensionless groups. Any reduction in the number of variables greatly reduces the labour of experimental investigation. A moment's consideration will show how great this saving can be. A function of one variable can be plotted as a single curve constructed from a relatively small number of experimental observations, perhaps six if the relation is simple, or the results can be presented as a single table which might require just one page.

A function of two variables will require a chart consisting of a family of curves, one for each value of the second variable, or, alternatively, the information can be presented in the form of a book of tables. A function of three variables will require a set of charts or a shelf-full of books of tables.

As the number of variables increases, the number of observations to be taken grows so rapidly that the situation soon becomes impossible. Any reduction in the number of variables is, therefore, extremely important.

It is also, sometimes, simpler to alter the value of a dimensionless group than to alter the value of a single quantity. For example, the value of the Reynolds number $\rho \bar{v} d / \mu$ can be changed by altering any one or more of the quantities ρ, $\bar{v}$, d and μ. While the properties of individual fluids and the availability of pipes of a given diameter would considerably restrict the range of values available in each individual quantity, alteration of several variables making up the Reynolds number would provide a much wider range of values for the group. Thus, a low value of Reynolds number could be obtained by using a small-diameter pipe or by reducing the velocity in a given pipe or by changing the fluid passing through the given pipe.

FIGURE 8.1

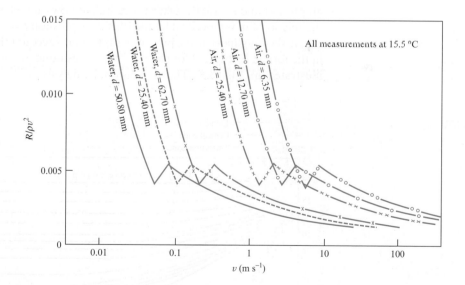

Following on from Example 8.5, for the stationary wall case it may be seen that for a smooth pipe, i.e. $k = 0.0$, the shear resistance to fluid flow, measured as $\tau_0 / (\rho v^2)$, would be expected to depend at any flow velocity, and any fluid type, on pipe diameter. Figure 8.1 illustrates some typical experimental results to illustrate this dependence.

Example 8.5 has, however, shown that the diameter and flow velocity variables may be combined through the introduction of a Reynolds number. A replotting of the data in Fig. 8.1, with the Reynolds number as the x axis, and the y axis provided by the Darcy friction factor f defined as

$$f = \tau_0/(0.5\rho v^2),$$

shown in Fig. 8.2, reduces the data to a single curve that also includes fluid type through the density and viscosity term in the Reynolds number. (Note the log axis in each case.)

FIGURE 8.2

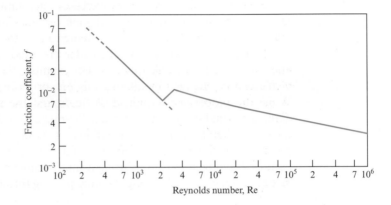

For low values of Reynolds number, when the flow is laminar, the slope of the graph is -1 and $f = 16/\text{Re}$, while for turbulent flows at higher Reynolds number values the relationship is described by the Blasius expression $f = 0.0791/\text{Re}^{0.25}$.

If the roughness of the pipe wall is taken into account then it is necessary to replot in the format of Fig. 8.2 for each k/d value, resulting in the well-known Moody chart illustrated in Fig. 8.3. The dimensional analysis presented and its experimental

FIGURE 8.3

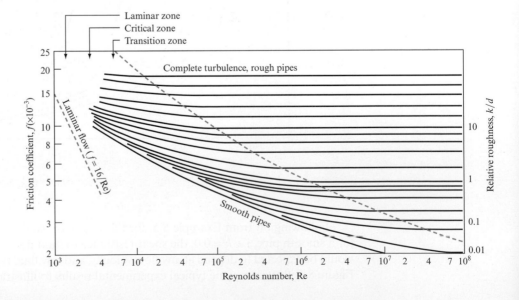

verification also leads to the Colebrook–White equation for friction factor that includes the groups discussed above, see Chapter 10.

In the laminar flow region, there is a single curve regardless of roughness, indicating that roughness is not significant and that the relationship remains $f = \phi(\rho \bar{v} d / \mu)$ in this region. At higher Reynolds numbers, when the flow is turbulent, we obtain a family of curves, one for each value of the relative roughness k/d, indicating that the relationship is of the form $f = \phi(\rho \bar{v} d / \mu, k/d)$. At high Reynolds numbers, when the flow is fully turbulent, the curves become a set of straight lines parallel to the Reynolds number axis, one for each value of k/d, showing that f is no longer a function of Reynolds number but is given by $f = \phi(k/d)$. From this example, it can be seen that while a dimensional analysis may suggest that certain groups affect the relationship, this effect may not necessarily be significant and must be verified experimentally.

One further advantage of the dimensionless presentation of experimental data is that it is independent of the units employed and should, therefore, be internationally intelligible and convenient to use. Care should, however, be taken when results are given in terms of named dimensionless groups, as the names and meanings of dimensionless groups have not been internationally agreed in all cases. It is desirable to ascertain the exact definition of the groups concerned and the precise quantities used in calculating them.

Concluding remarks

Dimensional analysis, as presented in this chapter, will be seen to be a mathematical technique that, in its simplified application to the three-dimensional system which is often sufficient in the study of fluid flows with no temperature-dependent or heat transfer effects, allows the enlightened design of experimental investigations. As mentioned, fluid mechanics depends heavily on empirical data, whether friction factors, life coefficients or machine performance data, and therefore a systematic empirical approach is essential. However, this chapter has also stressed that dimensional analysis alone cannot solve or define fluid flow problems; it can only suggest suitable groupings of variables that will allow the investigator to proceed. Chapter 9 will provide the basis for the use of dimensional analysis by introducing the laws of similarity. When combined with, or applied to, the groups suggested by dimensional analysis, similarity will allow the investigator to infer the performance of a prototype based upon the behaviour of a model, that behaviour being defined by the values and interrelationship of the variable groupings suggested by the dimensional analysis, and confirmed or modified by the experience of the investigator.

Summary of important equations and concepts

1. Units are based on choice; dimensions are fundamental, Sections 8.2 and 8.3.

2. Dimensional homogeneity is a requirement of any equation, Section 8.4.

3.   Mass, length, time, heat energy and temperature are relevant dimensions for fluid and thermodynamic analysis, Section 8.6.

4.   Identification of the relevant dimensions for any variable requires a regression to the fundamantal equations of motion, Section 8.6.

5.   The number of groups formed from n variables will be $(n-r)$, where r is the highest order non-zero determinant formed from the dimensional matrix. Note that for most fluid mechanics applications that do not feature energy or temperature dimensions r is 3.

6.   Dimensional analysis is only a tool to guide an investigation – the choice of repeating variables is determined by the investigator to be the most suitable.

7.   A range of groupings will recur and these should be sought in any analysis, for example Reynolds number relating viscous to inertia forces, Froude number to represent gravitational forces. Similarly, pressure coefficients based on the flow kinetic energy and force coefficients incorporating kinetic energy and area.

Further reading

Barr, D. I. H. (1983). A survey of procedures for dimensional analysis. *Int. J. Mech. Eng. Educ.*, **11**(3), 147–159.

Buckinkham, E. (1914, 1915). Model experiments and the form of empirical equations. *Phys. Rev.*, **2**, 345, (1915 *Trans. ASME*, **37**, 263–96).

Kline, S. J. (1965, reprinted 1986). *Similitude and Approximation Theory*. McGraw-Hill, New York.

Langhaar, H. L. (1980). *Dimensional Analysis and the Theory of Models*. Robert E. Kreiger, Florida.

Novak, P. and Cabelka, J. (1981). *Models in Hydraulic Engineering*. Pitman, London.

Sedov, L. I. (1959, reprinted 1996). *Similarity and Dimensional Methods in Mechanics*. Academic Press, London.

Problems

8.1   Show that the frictional torque T required to rotate a disc of diameter d at an angular velocity ω in a fluid of density ρ is given by

$$T = d^5 \omega^2 \rho \, \phi(\rho d^2 \omega / \mu),$$

and identify the Reynolds number group. $[\rho d(d\omega)/\mu]$

8.2   Develop an expression for the power P, developed by a hydraulic turbine, diameter d, at speed of rotation n, operating in a fluid of density ρ with available head h.

$$[P = \rho n^3 d^5 \phi(n^2 d^2 / gh)]$$

8.3   Determine the dependence of the force F acting on a sphere moving at a constant velocity through a fluid on the fluid density and viscosity. $[F = \rho D^2 v^2 \phi(\mathrm{Re})]$

8.4   If a circular cylinder of given length to diameter ratio, l/d, is rotated about its geometric axis at an angular velocity ω at right angles to and in a uniform fluid stream of velocity u, show that the power required to rotate the cylinder is given by $P = \rho v^3 / d\phi(ud/v, \, \omega d/v)$, where v is the fluid kinematic viscosity and ρ is the fluid density.

8.5 Show that the drag force on a body is a function of both Reynolds and Mach numbers in situations where viscous resistance and compressibility effects are major factors.

8.6 For a journal bearing of diameter d, length l, radial clearance c and eccentricity e, show that the load W that can be supported by the oil film of viscosity μ is given by

$$W/\mu n d^2 = \phi(c/d, e/d, l/d),$$

when the speed of rotation of the bearing is n.

8.7 Show that the rate of flow Q over a vee-notch of included angle θ may be expressed as $Q/(gh^5)^{0.5} = \phi[(gh^3)^{0.5}/v,\ gh^2\rho/\tau,\ \theta]$, where h is the head above the notch vertex, v is the fluid kinematic viscosity, τ is the fluid surface tension and g is acceleration due to gravity.

9

Similarity

THE TECHNIQUES OF DIMENSIONAL ANALYSIS INTRODUCED IN CHAPTER 8 ARE OF USE to indicate the parameter groups that may determine flow under any particular set of prevailing conditions. Model testing, based upon a systematic variation of these groups, is necessary to determine their relative importance. Once this has been established the rules of geometric and dynamic similarity must be invoked to allow design decisions to be taken based upon test results. This chapter defines both the most common dimensionless groupings found to determine flow conditions and their zones of influence, and introduces the laws of similarity necessary to translate a model into a prototype in each flow regime. Examples, drawn from a wide range of design cases, are presented, including internal pipe flows, external flows and fan/pump design, and free surface flows, including river and harbour models. ● ● ●

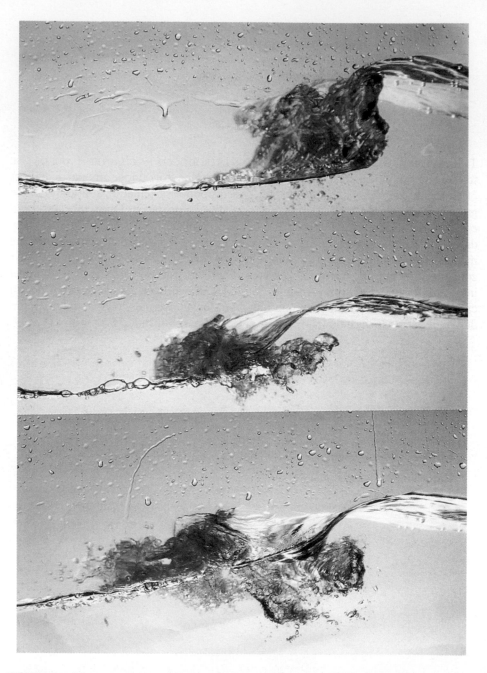

FIGURE 9.1(a)   Free surface waves in a laser wave tank experiment. The study of wave propagation and attenuation will be treated in Chapter 21. (Photograph courtesy of Professor Ian Grant, Fluid Loading and Instrumentation Centre, Heriot-Watt University, Edinburgh)

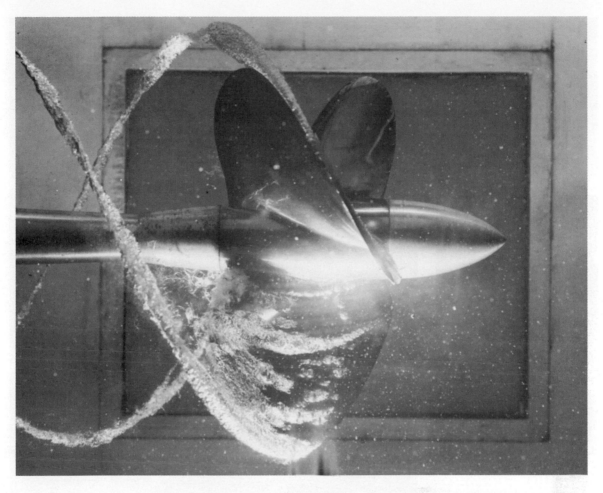

FIGURE 9.1(b)   Cavitation bubbles from a propeller under test in a cavitation tunnel. (Photograph courtesy of Professor Ian Grant, Fluid Loading and Instrumentation Centre, Heriot-Watt University, Edinburgh, and the Defence Research Agency, Haslar, UK)

Whenever the design engineer needs to take decisions at the design stage of a project, it will probably be necessary to initiate some form of model test programme. The basis of any such test series depends on the accurate use of instrumentation systems and the correct application of the theories of similarity. This, in turn, involves the application of dimensional analysis and the utilization of dimensionless groups such as the Reynolds, Froude or Mach numbers.

Model testing occurs in all areas of engineering based on fluid mechanics. Wind tunnel tests on aircraft, cars and trains, towing tests on ships and submarines, river/harbour tests carried out using models of high levels of intricacy to simulate tidal flow: all serve to illustrate such use of models. A recent application has arisen from the problems of air flow around buildings, which may be studied in wind tunnels and by examining smoke generation and propagation through building models. Model tests, then, depend on two basic types of similarity, which may be considered separately: geometric and dynamic similarity. Figure 9.1(a–f) illustrates a range of model testing.

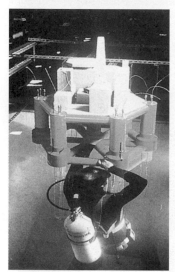

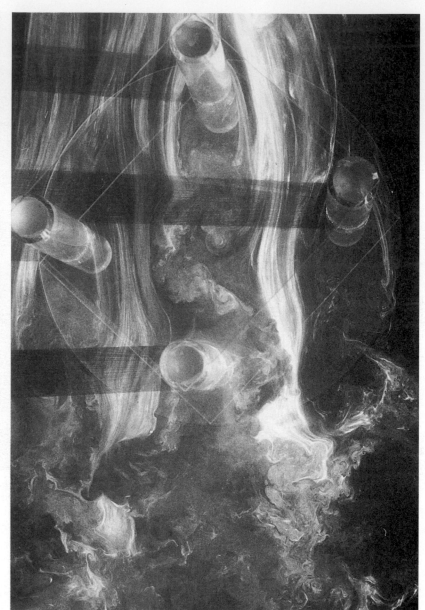

FIGURE 9.1(c)

Model testing applied to the design of a North Sea oil platform – including dynamic force measurement in an environmental wind tunnel, wave interaction tests in a long-wave channel and flow visualization studies using pulsed laser of the flow through the legs of a tension leg platform. (Photograph courtesy of Professor Ian Grant, Fluid Loading and Instrumentation Centre, Heriot-Watt University, Edinburgh)

FIGURE 9.1(d)
Flow visualization
utilizing smoke and
particle image
velocimetry applied to the
flow over a cylinder, an
obstruction in a flow field
and the shedding of
vortices from an 8 m
diameter wind turbine
blade. Vortex shedding is
addressed in Chapter 12.
(Photograph courtesy of
Professor Ian Grant,
Fluid Loading and
Instrumentation Centre,
Heriot-Watt University,
Edinburgh)

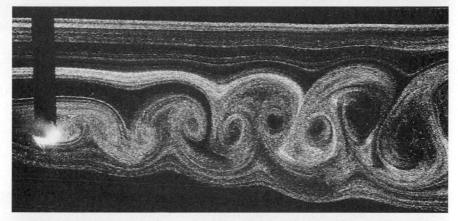

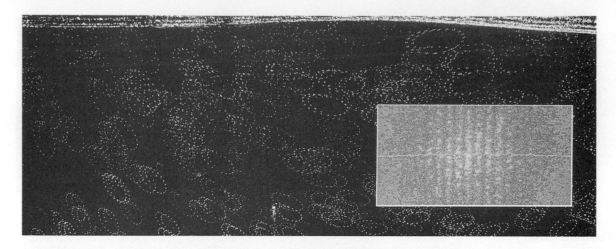

FIGURE 9.1(e)   Particle image velocimetry (PIV) applied to the flow field under surface waves in an open channel; the insert represents a PIV fringe analysis. The establishment of velocity profiles in free surface flows is addressed in Chapter 10. (Photograph courtesy of Professor Ian Grant, Fluid Loading and Instrumentation Centre, Heriot-Watt University, Edinburgh)

FIGURE 9.1(f)   Particle image velocimetry applied to yield instantaneous measurement of two components of velocity over an entire field by pulsed laser. (Photograph courtesy of Professor Ian Grant, Fluid Loading and Instrumentation Centre, Heriot-Watt University, Edinburgh)

9.1 GEOMETRIC SIMILARITY

The first requirement for model testing is a strict adherence to the principle of geometric similarity, i.e. the model must be an exact geometric replica of the prototype. Thus, for an aerofoil model of 1/10 scale, say, both the span and chord must be exactly 1/10 of the full-scale dimensions. This principle is, however, not fully applied to river models, where distortion of the vertical scale is necessary to obtain meaningful results because it is necessary to keep the relationship between wave properties and depth correct. Generally, it may be assumed that geometric similarity is achieved in model testing.

9.2 DYNAMIC SIMILARITY

The definition of dynamic similarity is that the forces which act on corresponding masses in the model and prototype shall be in the same ratio throughout the area of flow modelled. If this similarity is achieved, then it follows that the flow pattern will be identical for both the model and the prototype flow fields. Before moving on to the consideration of particular flow situations, it is worthwhile to restate the derivation of the most common dimensionless groups whose respective values govern model testing. Consider a general, hypothetical flow situation where the pressure change Δp between two points is dependent on mean velocity $\bar{v}$, length l, density ρ and viscosity μ, bulk modulus K, surface tension σ and gravitational acceleration g:

$$\Delta p = f(\bar{v}, l, \rho, \mu, K, \sigma, g).$$

With eight variables, defined by three dimensions, M, L and T, five dimensionless groups may be expected,

$$\frac{\Delta p}{\frac{1}{2}\rho v^2} = f_1[\rho v l/\mu,\ v/\sqrt{(K/\rho)},\ \rho v^2/\sigma,\ v/\sqrt{(lg)}]$$

recognizable as the pressure coefficient, $C_p = \Delta p/(\frac{1}{2}\rho v^2)$, and the dimensionless groups Re (Reynolds number), Ma (Mach number), We (Weber number), Fr (Froude number),

$$C_p = \frac{\Delta p}{\frac{1}{2}\rho v^2} = f_1(\mathrm{Re, Ma, We, Fr}). \qquad (9.1)$$

Equation (9.1) indicates that the pressure coefficient is dependent upon the other dimensionless groups and is defined if the other groups are defined. Thus if the values of the Re, Ma, We and Fr groups were identical for a prototype and its scale model, i.e. thereby conforming to geometric similarity, it would follow that the pressure coefficient would also be equal for the model and the prototype. This equivalence would therefore allow the model-generated results to be utilized to predict, for example, the lift and drag forces on aerofoils. This mathematical result follows directly from the development of dimensional analysis already presented.

However, it is also possible to confirm this conclusion by a parallel and independent analysis of the relative forces acting upon the model and prototype, remembering that the dimensionless groups included in equation (9.1) are in fact ratios of the applicable forces, the significance of each group depending upon the importance of those forces to the flow condition.

Two examples will be considered, namely the forces acting on the airstream over an aerofoil and the forces acting on an element of free surface flow. In both cases the forces acting will be shown to be generated due to gravity, F_g, pressure, F_P, and viscous, F_v, effects, although the significance of each will vary depending upon the example. The resultant force, F_R, will act on the fluid element in each case and accelerate it in accordance with Newton's second law. As the force polygons in the prototype, i.e. real flow situation, and the model will be similar, in a strictly geometric rather than a general sense, in both cases the magnitudes of the forces on the prototype fluid element will be in the same ratio to each other as the forces acting in the model flow. Thus, as the resultant force may be defined in terms of the fluid element mass and acceleration, it follows that

$$(F_R)_p/(F_R)_m = (ma)_p/(ma)_m = (F_P)_p/(F_P)_m = (F_g)_p/(F_g)_m$$

where the suffixes m and p refer to the model and prototype flow condition, respectively.

Now mass $m \propto \rho l^3$, acceleration $a \propto v/t$ and $F_g \propto \rho g l^3$. Thus by substitution

$$(v/gt)_m = (v/gt)_p$$

and, as $v \propto l/t$, it can be arranged that

$$t_m/t_p \propto (l_m/v_m)/(l_p/lv_p),$$

so that

$$(v^2/gl)_m = (v^2/gl)_p.$$

However, this relationship conforms to that already defined as the Froude number (note the squared form of the result compared with earlier definitions); thus the force conditions inherent in dynamic similarity have been shown to imply an equivalence of a variable group identified by the earlier dimensional analysis procedure.

Considering the viscous forces, then, if $F_v \propto \mu v l$, it follows that

$$(ma)_p/(ma)_m = (F_P)_p/(F_P)_m$$

or $$(\rho l^3 v/t)_p/(\rho l^3 v/t)_m = (\mu v l)_p/(\mu v l)_m,$$

where further simplification yields

$$(\rho l^2)/(\mu t)_p = (\rho l^2)/(\mu t)_m$$

and, as $t = l/v$, it follows that

$$(\rho v l)/(\mu)_p = (\rho v l)/(\mu)_m.$$

It will be appreciated that this equality is identical to the equivalence of Reynolds numbers already identified by dimensional analysis as a requirement for dynamic similarity. Thus a definition of dynamic similarity as requiring that the forces acting upon the model and prototype remain in the same ratio to each other may be developed to the point where the necessary equivalence of terms confirms the predictions of a dimensional analysis.

Considering the pressure force illustrated for both cases in Fig. 9.2(a) and (b) it follows that

$$(F_P)_p/(F_P)_m = (ma)_p/(ma)_m,$$

where $F_P \propto \Delta p l^2$, which leads to the conclusion that

$$(C_P)_m = (C_P)_p.$$

An inspection of the force polygons for both the air flow and free surface flow examples indicates that one of the forces could be determined if the other three were known. Thus the pressure force could be construed as dependent upon the viscous, gravitational and reaction forces shown. This would imply that the pressure coefficient depended upon the other parameters. Thus, if the Reynolds and Froude numbers are identical between the model and the prototype, it follows that the pressure coefficient will be equal for the model and the prototype. This is the same conclusion reached by the dimensional analysis approach, but arrived at independently by an analysis of the forces acting in each flow condition.

While these two cases have shown that equivalence of the Reynolds and Froude numbers is necessary, it is also clear that had the flow conditions around the model and the prototype involved other forces, such as surface tension or elasticity/compressibility effects, then the analysis presented could have been extended to include additional variables that would have confirmed the importance of other dimensionless groups, such as the Weber or Mach numbers.

Although the examples chosen are general it must be appreciated that in some flow conditions some forces predominate and the equivalence of the dimensionless groups representing these forces becomes imperative, while the equivalence of other groups becomes less significant. As was stressed in the development of the dimensional analysis approach, that methodology cannot itself inform the engineer as to the relative importance of each group; it was pointed out that in dimensional analysis, provided the mathematical rules are obeyed, there are no wrong answers, only more useful ones. Clearly for the case of air flow over an aerofoil the gravity forces become insignificant and the viscous forces predominate. Thus it is Reynolds number that must be held constant; the effect of Froude number may be neglected. Conversely, in the free surface flow case illustrated it is the gravity forces that determine the flow condition over the channel surface and thus Froude number equivalence is essential. In adopting this approach care must be taken to ensure that choice of model size takes account of other forces that might become important on the model. For example, if the spillway model is too small, i.e. the scale ratio is too small, then it is possible for viscous and surface tension forces to become important, thus rendering the dynamic similarity void as these forces do not appear in the prototype flow regime.

As a result of the independent confirmation provided by this approach it is now possible to state that dynamic similarity between a model and prototype requires that the significant dimensionless groups have the same values for both. It will be found that the dimensionless groups necessary in a wide range of flow conditions have been identified by engineers. A brief listing follows:

FIGURE 9.2
Forces acting on a fluid
element. (a) Passing over
an aerofoil. (b) In free
surface flow

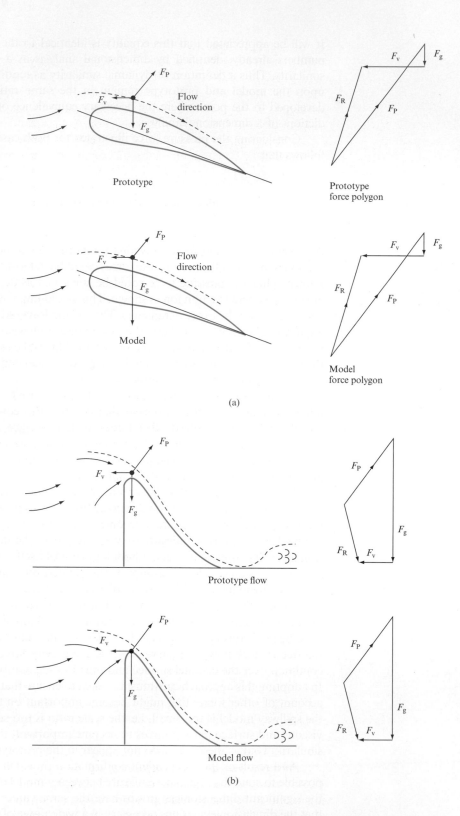

Prototype

Flow
direction

Prototype
force polygon

Model

Flow
direction

Model
force polygon

(a)

Prototype flow

Model flow

(b)

Reynolds number, Re: Reynolds number has already been defined as the ratio of inertia and viscous forces. It may be defined generally as $\mathrm{Re} = \rho V L / \mu$, where V is a characteristic velocity and L is a characteristic length. This general definition is important in understanding the wide influence of Reynolds number; in many cases Reynolds number is erroneously only associated with pipe flow where V is the mean flow velocity and L the pipe diameter. In open-channel flows the characteristic length L may be seen to be the hydraulic mean depth; in aerofoil theory it may be the wing chord; while in fan or pump analysis it may be the blade diameter.

Froude number, Fr: In flows dominated by gravitational effects, notably free surface flows, it has been shown above that dynamic similarity requires that the ratio of inertia to gravitational forces remains constant, and this scales to the square of Froude number, defined as $\mathrm{Fr} = V / (Lg)^{1/2}$, where V and L are characteristic values chosen as appropriate for the particular flow condition; for example, L could be depth in a rectangular cross-section channel but hydraulic mean depth in non-rectangular channels. Care should be taken with the definition of Froude number in texts as some authors utilize the squared term, which clearly differs for all values other than unity. A comparison of Froude and Reynolds numbers indicates immediately that if the same fluid is used for both prototype and model, it is impossible to have equivalence of both at the same time between a model and a prototype flow condition.

Mach number, Ma: Mach number is defined as the ratio of fluid velocity to the local sonic velocity. If the flow results in compressibility effects, i.e. the assumption of constant density is no longer supportable, it becomes important to include the effects of the elastic forces acting, and therefore the appropriate force ratio would be between inertia and elasticity. This ratio would therefore be proportional to $\rho L^2 V^2 / K L^2$ or $\rho V^2 / K$, where K is the bulk modulus of elasticity; this relationship is referred to as the Cauchy number. However, this is often simplified by noting that the wave propagation velocity in an isentropic gas or liquid may be expressed as $(K/\rho)^{1/2}$, and the ratio of inertia to elastic forces becomes $(V/c)^2$, where c is the wave speed. This reduces the force ratio to the simpler and widely used Mach number, defined simply as V/c. This makes Mach number equivalence a necessity in modelling air flows where compressibility effects are important.

It is also interesting to relate Mach number to an alternative definition of Froude number. The term $(Lg)^{1/2}$ may be shown to be the speed of propagation of a surface wave in a free surface or open-channel condition. Thus this form of Froude number becomes a form of Mach number. At first sight this might appear to be interesting but not relevant. However, this is not the case, as in open-channel flows where the Froude number exceeds unity, i.e. $V > (Lg)^{1/2}$, surface waves cannot propagate upstream; these cases are known as supercritical flow. This is analogous to the supersonic definition where sound waves cannot propagate ahead of the object. Again the characteristic length L may be taken as the channel hydraulic mean depth in non-rectangular channels.

Weber number, We: Weber number is defined as the ratio of inertia to surface tension forces. The definition of Weber number often varies between texts and care must be exercised in the use of tables. The expression following from the ratio quoted above and generally accepted is $\mathrm{We} = V(\rho L / \sigma)^{1/2}$, but the square and the reciprocal of this expression may be found in the literature.

Pressure, stress and force coefficients: e.g. pressure C_p, lift C_L, drag C_D, skin friction C_f. As already mentioned in the development of the dimensional analysis methodology, it is useful to be able to relate pressure differences or flow-induced forces to the flow parameters via a series of non-dimensional groupings. It therefore follows that it is necessary to 'non-dimensionalize' the pressure or force in terms of readily measured or defined variables. The pressure or stress coefficients may be 'non-dimensionalized' by use of the flow kinetic energy term. Thus

$$C_p = \Delta p / \tfrac{1}{2} \rho V^2, \qquad\qquad (9.2)$$

where V is a chosen characteristic velocity and the strictly not required constant of $\tfrac{1}{2}$ was historically included to allow direct use of experimental flow measurements. Force coefficients require that an area term be introduced, so that lift or drag coefficients are defined as

$$C_L = \Delta p / \tfrac{1}{2} \rho V^2 L^2. \qquad\qquad (9.3)$$

It will be appreciated that these coefficients are dependent upon other variable groupings as already illustrated.

Power coefficients: Power is simply a rate of doing work, which is in turn definable as a force moving through a given distance. Thus it is possible to define power in the three dimensions of mass, length and time. Power coefficients, useful in the definition of pump, fan and turbine characteristics in terms of other flow variables included in the other common dimensionless groupings, may thus be expected to have a form $P/\rho V^2 L^3$.

9.3   MODEL STUDIES FOR FLOWS WITHOUT A FREE SURFACE. INTRODUCTION TO APPROXIMATE SIMILITUDE AT HIGH REYNOLDS NUMBERS

Free surface effects are absent in bounded flows, this definition including both pipes or ducts flowing full and the flow around submerged bodies, such as aircraft, submarines or buildings. If the flows involved are low then compressibility effects may be ignored and the requirement to hold Mach number constant between the prototype and model may be relaxed. In these cases, therefore, the analysis presented would dictate an equivalence of Reynolds number. However, strict adherence to Reynolds number equivalence may prove inappropriate and unduly costly. The following examples will illustrate the difficulty.

EXAMPLE 9.1

A submarine-launched missile, 2 m in diameter and 10 m long, is to be tested in a water tunnel to determine the forces acting on it during its underwater launch. The maximum speed during this initial part of the missile's flight is 10 m s^{-1}. Determine the mean water tunnel flow velocity if a 1/20 scale model is employed and dynamic similarity is achieved.

Solution

To comply with dynamic similarity the Reynolds numbers must be identical for both the model and the prototype:

$$\text{Re}_m = \text{Re}_p,$$

$$(\rho VL/\mu)_m = (\rho VL/\mu)_p.$$

The model flow velocity is thus given by

$$V_m = V_p(L_p/L_m)(\rho_p/\rho_m)(\mu_m/\mu_p),$$

but as $\rho_p = \rho_m$ and $\mu_m = \mu_p$ it follows that

$$V_m = 10 \times 20 \times 1 \times 1 = \mathbf{200\ m\ s^{-1}}.$$

This is a high velocity and illustrates why few model tests are made will completely equal Reynolds numbers. At high Reynolds numbers, however, it will be shown that a relaxation of strict equivalence is acceptable.

EXAMPLE 9.2

An airship of 6 m diameter and 30 m length is to be studied in a wind tunnel. The airship speed to be investigated is at the docking end of its range, a maximum of $3\ m\ s^{-1}$. Determine the mean model wind tunnel air flow velocity if the model is made to a 1/30 scale, assuming the same sea level air pressure and temperature conditions for the model and the prototype.

Solution

Dynamic similarity requires the equivalence of Reynolds number; clearly there will be no compressibility effects on the prototype. Thus as before

$$V_m = V_p(L_p/L_m)(\rho_p/\rho_m)(\mu_m/\mu_p)$$

and substituting $\rho_p = \rho_m$ and $\mu_m = \mu_p$ it follows that

$$V_m = 3 \times 30 \times 1 \times 1 = \mathbf{90\ m\ s^{-1}}.$$

It is worth noting that at sea level the local sonic velocity may be taken as $340\ m\ s^{-1}$ and therefore the prototype Mach number is 3/340, approximately 0.01. The same calculation for the model indicates a Mach number of 90/340, approaching 0.3, a value at which Mach number effects could be present.

The objective of dimensional analysis and the development of the theories relating to dynamic similarity have been to enable the establishment of testing techniques that would allow engineers to predict from model tests the flow condition to be encountered in a prototype; by definition these techniques should be both reliable and affordable. While the definition of affordable is obviously a variable dependent upon the particular application, it is clear from the examples above that strict adherence to Reynolds number equivalence can be problematic. In both cases the power required to drive the water and wind tunnels is considerable, and in the air flow case the flow velocity necessary would introduce compressibility effects that would not be present on the prototype. Therefore an alternative approach has to be found that conforms to the engineer's understanding of both the forces acting on the model and the prototype and their relative importance. It has already been indicated that for pipe flows, at high Reynolds number values, frictional effects tend to become independent of Reynolds number. This effect will also be demonstrated later for drag coefficients for cylinders and spheres. Thus for model tests to be applicable it may be sufficient to ensure that both the model and prototype flows possess Reynolds number values well into this range. The precise Reynolds number values at which this approach is acceptable are dependent upon the flow condition being investigated. However, checks may be carried out based around measurements of C_p values as model Reynolds number rises: once the C_p values become independent of Re the 'safe' model flow condition has been reached. Guidance can also be obtained by reference to current practice in that branch of wind or water tunnel testing.

Alternative approaches involve retaining the Reynolds number equivalence by a change of fluid between the prototype and the model, e.g. the use of compressed air as a fluid in the model testing of hydroelectric turbines.

9.4   ZONE OF DEPENDENCE OF MACH NUMBER

Mach number becomes significant in flow situations where the ratio of flow velocity to sonic velocity exceeds about 0.25 to 0.3. It is normally difficult to satisfy both Reynolds number and Mach number equality simultaneously and so it is important that testing decisions are made based on previous experience of the type of flow to be investigated. For example, if the viscous motion of a fluid close to a boundary in supersonic flow is to be investigated, then Reynolds number equivalence would be the criterion, but if the object of the investigation is the flow conditions through the shock wave pattern around a body, then Mach number equivalence is paramount.

Mach and Reynolds number equivalence may be achieved if it is possible to vary the fluid density conditions between the model and prototype, as illustrated below.

EXAMPLE 9.3

In order to undertake predictions of the lift and drag force on a scale model of an aircraft during a section of its operational envelope involving sea level flight at 100 m s^{-1}, where the speed of sound may be taken as 340 m s^{-1}, it is proposed to utilize a cryogenic wind tunnel with nitrogen at 5 atmospheres of pressure and a temperature of −90 °C, conditions at which the nitrogen density and viscosity may be taken as 7.7 kg m^{-3} and 1.2×10^{-5} N s, respectively. The speed of sound in nitrogen at this temperature is 295 m s^{-1}. Determine the wind tunnel flow velocity, the scale of the model to ensure full dynamic similarity and the ratio of forces acting on the model and the prototype.

Solution

In order to provide an equivalence of Mach number it follows that the wind tunnel velocity must be given by

$$V_m = c_m(V_p/c_p),$$

where c is the appropriate local sonic velocity; hence

$$V_m = 295 \times 100/340 = \textbf{86.76 m s}^{-1}.$$

At this wind tunnel flow velocity it is possible, from the data given, to determine the necessary scale for which Reynolds number equivalence is achieved as

$$\mathrm{Re_m = Re_p,}$$
$$(\rho VL/\mu)_m = (\rho VL/\mu)_p,$$
$$(\rho_m/\rho_p)(V_m/V_p)(L_m/L_p)(\mu_p/\mu_m) = 1.$$

Thus

$$(7.7/1.2)(86.76/100)(L_m/L_p)(1.8 \times 10^{-5}/1.2 \times 10^{-5}) = 1$$
$$L_m/L_p = 1.0/(6.4 \times 0.868 \times 1.5)$$
$$= \mathbf{0.12}.$$

The ratio of forces acting on the model and prototype may be determined by noting that the force coefficients will be equal if both the Mach and Reynolds numbers are equal. Thus

$$(F_m/\rho V^2 L^2)_m = (F_p/\rho V^2 L^2)_p,$$

where F represents a typical force acting on the model and the prototype and L^2 represents an area, for this case the projected wing area.

Thus the ratio of forces becomes

$$F_m/F_p = (\rho_m/\rho_p)(V_m/V_p)^2(L_m/L_p)^2,$$
$$F_m/F_p = (7.7/1.2)(86.76/100)^2(0.12)^2 = \mathbf{0.0696 \text{ or } 6.96 \text{ per cent}}.$$

9.5 SIGNIFICANCE OF THE PRESSURE COEFFICIENT

Referring back to equation (9.1), it will be seen that the pressure coefficient was defined by the other dimensionless groups judged to define the flow condition. Therefore it follows that, if dynamic similarity is achieved, the values of pressure coefficient measured in model tests will also apply to the prototype. This is essential in relating both pressure changes in the model and prototype flow conditions and determining the forces acting on the prototype. This second calculation involves multiplying the pressure coefficient by an appropriate area, e.g. the projected wing area in an aircraft model investigation where the model lift and drag forces would have been measured via the lift and drag component balances of the wind tunnel instrumentation.

Thus in Example 9.1, if the pressure difference between two points on the surface of the missile had been 5.0 N m⁻², then the pressure difference on the model would have been given by

$$(C_p)_m = (C_p)_p,$$
$$(\Delta p/\tfrac{1}{2}\rho v^2)_m = (\Delta p/\tfrac{1}{2}\rho v^2)_p,$$
$$\Delta p_m = \Delta p_p(\rho_m/\rho_p)(V_m/V_p)^2,$$
$$\Delta p_m = 5 \times 1 \times (200/10)^2 = 2 \text{ kN m}^{-2}.$$

The importance of the pressure coefficient may also be appreciated if the case of pipe flow modelling is considered.

EXAMPLE 9.4

Flow through a heat exchanger tube is to be studied by means of a 1/10 scale model. If the heat exchanger normally carries water, determine the ratio of pressure losses between the model and the prototype if (a) water is used in the model, (b) air at normal temperature and pressure is used in the model.

Solution

For dynamic similarity the Reynolds numbers must be constant; hence

$$Re_m = Re_p,$$
$$V_m/V_p = (L_p/L_m)(\rho_p/\rho_m)(\mu_m/\mu_p).$$

If the Reynolds numbers are equal, then so must be the pressure coefficients; therefore,

$$(C_p)_m = (C_p)_p,$$
$$(\Delta p/\tfrac{1}{2}\rho v^2)_m = (\Delta p/\tfrac{1}{2}\rho v^2)_p,$$
$$\Delta p_m = \Delta p_p(\rho_m/\rho_p)(V_m/V_p)^2,$$
$$\frac{\Delta p_m}{\Delta p_p} = (\rho_m/\rho_p)(V_m/V_p)^2 = (\rho_p/\rho_m)(\mu_m/\mu_p)^2(L_p/L_m)^2.$$

(a) In the water model case, as the model and prototype fluid densities and viscosities are the same it follows that

$$(\Delta p)_m/(\Delta p)_p = 10^2 \times 1^2 \times 1 = \mathbf{100}.$$

(b) If air is used as the model fluid then the full form of the pressure coefficient equivalence must be used:

$$\rho_p/\rho_m = 1000/1.23,$$
$$\mu_m/\mu_p = 1.8 \times 10^{-5}/1 \times 10^{-3} = 1.8 \times 10^{-2},$$
$$(\Delta p)_m/(\Delta p)_p = 10^2 \times (1000/1.23) \times (1.8 \times 10^{-2})^2 = \mathbf{26.34}.$$

9.6　MODEL STUDIES IN CASES INVOLVING FREE SURFACE FLOW

In free surface model studies the effect of gravity becomes important and the governing parameter is Froude number. Generally the prototypes, i.e. large spillways, have Reynolds numbers large enough to be operating out of the range of dependence on Re; however, the model may be of such a size that, when Froude number equivalence is set up, the model Reynolds number is small enough to produce viscous effects not representative of the prototype. For this reason, the model must be large enough to place its Reynolds number above the viscous loss dependence level. One problem with free surface flow cases is that, generally, the same fluid is used for the model as for the prototype, so that the convenient expedient of substituting air for water in internal flows cannot be copied.

EXAMPLE 9.5

A 1/50 scale model of a proposed power station tailrace is to be used to predict prototype flow. If the design load rejection bypass flow is 1200 m^3 s^{-1}, what water flow rate should be used on the model?

Solution

Equating Froude numbers,

$$Fr_m = Fr_p,$$

where $Fr = \bar{v}/\sqrt{(lg)}$. Therefore,

$$\bar{v}_m/\bar{v}_p = \sqrt{(l_m/l_p)}.$$

Flow rate may be determined by introducing the area ratio $A_m/A_p = 1/2500 = (Scale)^2$. Hence,

$$\frac{Q_m}{Q_p} = \frac{A_m\bar{v}_m}{A_p\bar{v}_p} = \frac{l_m^2}{l_p^2}\sqrt{\left(\frac{l_m}{l_p}\right)},$$

$$Q_m = Q_p(l_m/l_p)^{2.5} = 1200 \times (1/50)^{2.5}$$

$$= 0.067 \text{ m}^3 \text{ s}^{-1}.$$

This relatively simple approach is complicated for the case of ship resistance testing, as the phenomenon is made up of two factors, namely the surface resistance of the hull, dependent on Reynolds number, and the wave resistance, which is Froude number dependent (see Section 12.4).

Consider the case of a model to be towed at a speed such that the Froude number is satisfied:

$$Fr_m = Fr_p,$$

$$v_m/\sqrt{(l_m g)} = v_p/\sqrt{(l_p g)},$$

$$v_m/v_p = \sqrt{(l_m/l_p)}.$$

Now consider the same model and equate Reynolds numbers:

$$Re_m = Re_p,$$

$$(\rho vl/\mu)_m = (\rho vl/\mu)_p,$$

$$v_m/v_p = (l_p/l_m)(\rho_p/\rho_m)(\mu_m/\mu_p) = l_p/l_m.$$

if $\rho_m = \rho_p$ and $\mu_m = \mu_p$. Obviously, then, the two criteria cannot be satisfied simultaneously and the approach followed is to equate Froude number to model wave resistance forces as these are the more difficult to analyze. Viscous hull resistance is then calculated by analytical techniques and added to the wave resistance measured.

While free surface flows are normally found in open channels or partially filled pipes, there exists an interesting subgroup of flow conditions where free surface flow analysis can be applied within closed conduits, namely annular gravity-driven downflow where the ratio of annular film thickness to pipe diameter is sufficiently large to allow the flow to be considered analogous to free surface flow in a shallow, rectangular cross-section open channel, see Chapters 16 and 21. Figure 9.3(a) illustrates the measurement of the unsteady annular downflow velocity by means of a laser anemometer in such a flow Q_w where, under terminal conditions, the gravity forces are balanced by the surface resistance between the falling fluid and the pipe walls, pipe diameter D. The falling annular water film entrains an air core flow which in turn is responsible for pressure fluctuations within the vertical pipe, as discussed in more detail in Section 21.8. Under these conditions the film terminal velocity V_t may be shown to be

$$V_t = K\left(\frac{Q_w}{D}\right)^{0.4}, \tag{9.4}$$

(a)   (b)

FIGURE 9.3 **(a)** Laser measurement of the annular water film thickness under unsteady flow conditions. The treatment of annular water downflows and the accompanying entrained air flow is discussed in Chapter 21. (Photograph courtesy of Professor J. A. Swaffield, Heriot-Watt University, Edinburgh.) **(b)** Definition of terms in annular downflow/entrained air relationship

where the coefficient K may be shown to be dependent on pipe roughness. Values of K may be determined from $K = 0.63/n^{0.6}$ (Chapter 16), where n is the appropriate Manning roughness coefficient for that pipe material. Coefficient K has values in the range 12 to 15.

The suction available to entrain an airflow in the stack may be considered to be dependent on the following parameters

$$\Delta p = \Phi(\rho_a, \rho_w, Q_a, V_w, D, t, \mu_a, \mu_w, g, H_w), \tag{9.5}$$

where ρ_a = air density,
  ρ_w = water density,
  Q_a = entrained air core volumetric flow rate,
  V_w = annular water film terminal velocity,
  D = stack diameter,
  t = annular water flow terminal thickness,
  μ_a = air viscosity
  μ_w = water viscosity
  g = acceleration due to gravity
  H_w = height of the vertical pipe carrying an annular water film.

A traditional dimensional analysis utilizing Buckingham's π theorem reduces this dependency to eight dimensionless groupings

$$\frac{\Delta p D^4}{\rho_w Q_w^2} = \Phi_1\left(\frac{\rho_a}{\rho_w}, \frac{t}{D}, \frac{Q_a}{Q_w}, \frac{\mu_w}{\mu_a}, \text{Re}_a, \frac{gD}{V_w^2}, \frac{H_w}{D}\right) \tag{9.6}$$

where Re_a is the entrained airflow Reynolds number, defined as $(\rho_a Q_a)/(D\mu_a)$ if the assumption that $t \ll D$ is complied with. The $(\Delta p D^4)/(\rho_w Q_w^2)$ term may be recognized as a form of pressure coefficient, non-dimensionalized in terms of the water downflow rate.

It will be seen that the annular thickness t may be expressed in terms of the water downflow rate Q_w, the pipe diameter D, and the terminal water velocity V_t. Thus the t/D dimensionless group may be recast by substitution:

$$\frac{t}{D} = \frac{Q_w}{\pi D V_t} \frac{1}{D} = \frac{Q_w}{\pi K D^2 (Q_w/D)^{0.4}}.$$

Substituting for K and ignoring the constant terms reduces the dimensionless group to

$$\frac{t}{D} = \frac{Q_w}{(D^2/n^{0.6})(Q_w/D)^{0.4}} = \frac{(Q_w n)^{0.6}}{D^{1.6}} = \frac{Q_w^3 n^3}{D^8},$$

where the Manning roughness coefficient n has dimensions of $L^{-1/3}T^1$.

The $(gD)^{0.5}/V_w$ term may be identified as a form of Froude number, expected due to the free surface nature of the flow. The H_w/D term recognizes the influence of water to air interface vertical length on the total shear force available to entrain the air flow Q_a and may perhaps be more usefully expressed in terms of a pseudo-pressure coefficient $gH_w D^4/Q_a^2$.

Thus the principles discussed earlier apply to this particular flow condition and allow suction pressure and entrained flow rates for different diameter pipes and water downflow rates to be determined.

$$\frac{\Delta p D^4}{\rho_w Q_w^2} = \Phi_2\left(\frac{\rho_a}{\rho_w}, \frac{Q_w^3 n^3}{D^8}, \frac{Q_a}{Q_w}, \frac{\mu_w}{\mu_a}, Re_a, \frac{gD}{V_w^2}, \frac{gH_w D^4}{Q_a^2}\right).$$ (9.7)

The dimensionless groups identified by the process demonstrated above form the basis for an empirical evaluation of the air entrainment within vertical stacks carrying a falling annular water film. These expressions will be utilized later, in Section 21.8, to develop a numerical simulation capable of modelling the air pressure transients generated within such systems by changes in the driving water flow rate.

9.7   SIMILARITY APPLIED TO ROTODYNAMIC MACHINES

Application of the techniques of dimensional analysis to fans and pumps yields relationships of the form

$$P/N^3 D^5 \rho = f(Q/ND^3, \mu/\rho ND^2, k/D, a/D, b/D, c/D),$$ (9.8)

$$P_s/\rho N^2 D^2 = f(Q/ND^3, \mu/\rho ND^2, k/D, a/D, b/D, c/D),$$ (9.9)

where P is shaft power, Q is volume flow rate, P_s is the pressure rise across the unit rotating at speed N, and of diameter D. The fluid type is defined by density ρ and viscosity μ, while the detail dimensions of the machine are a, b, c with surface roughness k. For geometrically similar machines operating at high Reynolds numbers, so that the term $\mu/ND^2\rho = Re$ becomes irrelevant, the expressions reduce to

$$P/N^3 D^5 \rho = f_2(Q/ND^3) \quad \text{and} \quad P_s/\rho N^2 D^2 = f_1(Q/ND^3).$$

Thus, for model testing to be valid, each of these groups should have identical values for the model and the prototype.

Model testing is of particular value in the design and manufacture of the larger-scale fans, pumps and turbines, to which these relationships also apply, except that the power terms relate to power generated rather than power supplied.

Generally, the model scale is arranged so that the impeller diameters are less than 0.5 m and the same fluid is usually employed in the model tests as for the prototype. However, due to the lack of effect of Reynolds number, provided the flow is well into the fully turbulent region, it is possible to employ air or pressurized gas in order to obtain more manageable flow rates or machine scales.

Denoting the model by m and the full-size machine by p, the following relations can be proved:

1.   Flow,

$$Q_m/Q_p = (ND^3)_m/(ND^3)_p.$$ (9.10)

2. Pressure rise (pumps) and pressure drop (turbines),

$$(P_s)_m/(P_s)_p = (\rho N^2 D^2)_m/(\rho N^2 D^2)_p, \tag{9.11}$$

where the density ratio may vary from unity.

3. Power supplied (pumps) and power generated (turbines),

$$P_m/P_p = (N^3 D^5 \rho)_m/(N^3 D^5 \rho)_p. \tag{9.12}$$

It will be appreciated that these relations are all independent of operating pressure, so that, in theory, any convenient operating test rig head may be employed. In practice, this is not entirely true as the onset of cavitation is dependent on absolute pressures and, for pumps and turbines, its occurrence is of major importance, so that pressure levels should be as close as possible to the full-scale installation values.

Theoretically, the efficiency of model and prototype should be the same. However, there will be some excess inefficiency in the model due to scale effects relating to leakage flow, roughness variations and manufacturing constraints (see Section 23.5).

EXAMPLE 9.6

A ventilation system fan is to be exported to a high-altitude region with an air density of 0.92 kg m^{-3} and is expected to deliver 2 m^3 s^{-1} at a pressure differential of 200 N m^{-2}.

If the fan is to be driven at 1400 rev min^{-1} on installation, calculate the flow rate and pressure rise required on test at sea level, air density 1.3 kg m^{-3}, and the appropriate fan test speed if conditions of dynamic similarity are to be achieved. (Assume no change in air viscosity with the change in altitude involved.)

Solution

The fan test speed may be determined by equating the fan Reynolds numbers:

$$(\mu/\rho N D^2)_S = (\mu/\rho N D^2)_A,$$

where the suffixes A and S refer to altitude and sea level conditions, respectively. Thus

$$N_S = (\rho N D^2)_A/(\rho D^2)_S$$
$$= 1400 \times (0.92/1.3) \times 1^2 = \mathbf{990 \ rev \ min^{-1}}.$$

Note that the geometric scale is unity as the fan is its own model and that the viscosity ratio has been set to unity also.

The sea level flow rate and pressure expected would thus follow as

$$Q_S = Q_A N_S/N_A$$
$$= 2 \times 990/1400$$
$$= \mathbf{1.41 \ m^3 \ s^{-1}},$$
$$\Delta p_S = \Delta p_A (\rho_S/\rho_A)(N_S/N_A)^2(D_S/D_A)^2$$
$$= 200 \times (1.3/0.92) \times (990/1400)^2 \times 1^2$$
$$= \mathbf{141 \ N \ m^{-2}}.$$

9.8   RIVER AND HARBOUR MODELS

River and harbour engineering projects are costly undertakings and, as analytical techniques only provide approximate predictions of the likely effects of any river widening or harbour improvement, the use of models at an early stage in the design has many advantages. The problems arise when a suitable scale is to be chosen for the model; the adoption of a scale that will give reasonable channel depths will usually result in a model too large to be practical, while choosing a scale on area/cost criteria yields channel depths that are very small.

The problems of shallow model channels are: (1) accuracy in level and level change measurement becomes impossible to achieve; (2) the surface roughness of the channel beds would be impractically small and there is even a probability that channel flow would be laminar rather than turbulent, as normally found in practice.

In order to provide a solution to these problems, distorted scaling is adopted, vertical scales of 1/100 and larger being typical, while horizontal scales vary from 1/200 to 1/500. Distortion of this sort is suitable if the overall discharge characteristics of a long length of river are being studied. However, it should be appreciated that the micro-situation is not well modelled, and situations such as the effects of breakwater positioning should be studied on as large and as undistorted a scale as possible.

In models of this type, strict geometric similarity is not achieved. However, if the mean flow velocity $\bar{v}$ and depth Z are arranged so that there is an equivalence of Froude number ($\bar{v}/\sqrt{gZ}$) between model and river, then it will be expected that the flow type, i.e. fast or slow, will be the same at corresponding points on the model and river. Thus,

$$\bar{v}_m^2/Z_m = \bar{v}_r^2/Z_r,$$

and, hence,

$$\bar{v}_m/\bar{v}_r = \sqrt{(Z_m/Z_r)}, \tag{9.13}$$

where m denotes model and r denotes full-scale river. The discharge Q through the model must depend on both vertical and horizontal scales; thus $Q = \bar{v}lZ$ and

$$Q_m/Q_r = (Z_m/Z_r)^{3/2}(l_m/l_r), \tag{9.14}$$

where l_m/l_r is the horizontal scale, 1:x, and Z_m/Z_r is the vertical scale, 1:y.

In order to manufacture models, it is necessary to have information on the effect of surface roughness, particularly the effects of model roughness size, which is often dictated by the manufacturing process chosen. From the Manning coefficient of surface roughness n and the Manning equation applied to both model and river,

$$\frac{\bar{v}_m}{\bar{v}_r} = \left(\frac{n_r}{n_m}\right)\left(\frac{Z_m}{Z_r}\right)^{2/3}\left(\frac{Z_m/l_m}{Z_r/l_r}\right)^{1/2},$$

$$n_m = n_r x^{1/2}/y^{2/3}. \tag{9.15}$$

Full-scale river beds have values of surface roughness defined by $n > 0.03$ and as the normal model surface finish, obtained by use of cement mortar for model surfaces, is of the order of $n \simeq 0.012$, artificial roughening is normally necessary. The maximum value of n obtainable by artificial means, i.e. adding wire mesh, gravel and even small rods to the bed, is of the order of 0.04. As mentioned, the micro-flow situation, i.e. eddies of local currents, will differ between model and river, so adjustment by surface roughness only is not a reasonable course of action. Figure 9.4(a–c) illustrates a typical model and the artificial channel bed roughening used.

Before it is used to predict the effects of any modifications to the river channels, the model should first be checked for discharge and depth accuracy. Measurements of full-scale depths and discharges should be checked against model discharge and depth by use of equation (9.14), relating total flow rates, and the vertical scale. Model discharge rates are produced using recirculating pump circuits and orifice plate or notch flow measuring instrumentation. If the model values are acceptable, then testing can continue. If the depth and flow rates are in poor agreement with the full-scale results, then the model surface roughness should be adjusted or the scales altered.

It should be noted that the above analysis is based on the premise that the Reynolds number of both model and full-scale river is such that the flow is totally turbulent and that variations in Reynolds number are not important. It is good practice, however, to check the values of model and channel Reynolds number to ensure that this simplification is valid, i.e. $\text{Re} > 2000$.

Estuary models may be constructed using the same general principles as outlined for river models; distorted scales are typically 1/50 to 1/150 vertical and 1/300 and 1/2500 horizontal. As the available data on tidal velocity distributions are likely to be sketchy, it is necessary to incorporate in the model the whole tidal channel system, as well as a substantial portion of adjacent coastline.

Although surface roughness is not so critical in estuary models, the speed of propagation of the tide becomes an important design criterion as the tidal period governs the time available for result recording. The tidal period of the model is, thus

$$\frac{\text{Time in model}}{\text{Time in estuary}} = \frac{(\text{Distance}/\text{Velocity})_{\text{model}}}{(\text{Distance}/\text{Velocity})_{\text{estuary}}}.$$

Now, as the Froude numbers are equal (as for river models) it follows from equation (9.13) that

$$(\text{Time})_m/(\text{Time})_e = (l_m/l_e)(\bar{v}_e/\bar{v}_m) = (l_m/l_e)\sqrt{(Z_e/Z_m)}. \tag{9.16}$$

Thus, for a tidal period of 12.4 h and a model with 1/50, 1/500 scales, the model tidal period is 10.5 min.

While equation (9.16) refers to horizontal fluid velocity, the scale factor for vertical silt particle velocity is also of interest and may be derived in the same manner:

$$\frac{\text{Rate of fall in model}}{\text{Rate of fall in estuary}} = \frac{\text{Depth scale}}{\text{Time scale}}$$

$$= (Z_m/Z_e)\sqrt{[(Z_m/Z_e)(l_e/l_m)]}$$

$$= (Z_m/Z_e)^{3/2}(l_e/l_m). \tag{9.17}$$

A further complication in estuary model studies, particularly of silting phenomena, is stratification effects due to density variations between salt and fresh water. If the use of saline solutions in the model testing is undesirable for corrosion reasons, then a stable clay solution may be employed.

Estuary models are now commonly used to investigate the effects of discharge of power station or industrial cooling water flows and, here, density variations may again

FIGURE 9.4(a) Full-scale model of the entrance channel to Dar-es-Salaam harbour

FIGURE 9.4(b) Empty scale model of the entrance channel to Dar-es-Salaam harbour. Note channel roughening in the view of the empty model

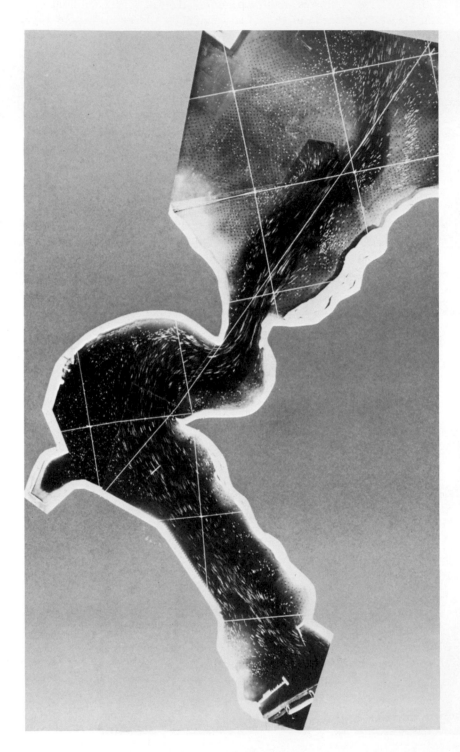

FIGURE 9.4(c)   A bird's eye view of the Dar-es-Salaam model showing flow patterns at ebb tide. The effect is obtained by scattering confetti on the water and then photographing it with a time exposure. Where the water is moving fastest, the confetti shows as a streak – the longer the streak, the faster the current. By studying results from a series of photographs representing various states of the tide, one is able to calibrate the model so that water flows in the model correspond closely to those recorded on site. (Courtesy of UN Development Program, East African Harbours Corp., International Bank for Reconstruction and Development, Bertlin and Partners, Redhill, and The Model Laboratory, Wimpey Laboratory, Middlesex)

have to be modelled, based on temperature. Other uses of estuary models include silting and erosion studies and the spread and deposit of effluent discharged into the sea. Generally, the role of the estuary model should be seen more as method of comparing the attributes of various design solutions, rather than as an accurate method of predicting the effects of one design.

Harbour and coastal models require the inclusion of wave effects, and these are reproducible by means of mechanical wave-making devices. However, the type of wave motion encountered in coastal engineering is dependent on both water depth and wavelength for its propagation velocity, so that the degree of scale distortion acceptable in river and estuary models can no longer be applied. The best model studies are carried out with equal scaling; however, distortion of the vertical scale up to two or three times the vertical has been used.

EXAMPLE 9.7

It is proposed to construct a model of 18 km length of river, for which the first 8 km are tidal. The normal discharge of the river is known to be in the region of 300 m³ s⁻¹, the average width and depth of the channel being 3 m and 65 m, respectively. Given a laboratory of 30 m length propose suitable scales and calculate the tidal period.

Solution

(i) The largest scale possible would be $30/18 \times 1000 = 1/600$ for the horizontal distances. (ii) As the river is tidal, scale distortions of around 6 to 10 are acceptable, so a vertical scale of $1/60$ could be employed. (iii) The model will be constructed to conform with these scales. However, in doing so, the effect of Reynolds number is assumed negligible. It is good practice to check the Reynolds number.

$$\text{Average river velocity} = 300 \text{ m}^3 \text{ s}^{-1}/(3 \times 65) \text{ m}^2 = 1.54 \text{ m s}^{-1}.$$

From equation (9.13),

$$\bar{v}_m = \bar{v}_r \sqrt{(Z_m/Z_r)} = 1.54 \times \sqrt{(1/60)}$$

$$= 0.199 \text{ m s}^{-1}.$$

$$\text{Re}_m = \rho v m/\mu,$$

where $m =$ Hydraulic mean depth $=$ Area/Perimeter flow cross-section

$$= (3 \times \tfrac{1}{60} \times 65 \times \tfrac{1}{600})/(\tfrac{65}{600} + \tfrac{6}{60})$$

$$= 5.4 \times 10^{-3}/0.208 = 0.026 \text{ m}.$$

Thus, $\text{Re}_m = 1000 \times 0.199 \times 0.026/1.14 \times 10^{-3} = 4532,$

which is sufficiently turbulent to allow Re effects to be ignored.

(iv) The tidal period can be calculated from the time scale (equation (9.16)):

$$\frac{(\text{Time})_m}{(\text{Time})_r} = \frac{l_m}{l_r}\sqrt{\left(\frac{Z_r}{Z_m}\right)} = \frac{1}{600} \times \sqrt{(60)} = 0.0129.$$

Therefore,

$$\text{Tidal period of model} = 12.4 \times 60 \times 0.0129 = \textbf{9.6 min}.$$

9.9   GROUNDWATER AND SEEPAGE MODELS

While the area of groundwater flow and its implications within soil mechanics, reservoir design and runoff predictions are outside the scope of this text, the development of groundwater and seepage modelling dependent upon both the fundamentals of dimensional analysis and similarity, and the solution of the Laplace equation introduced in Chapter 7, are of interest.

Water flow through the small passages or pores that exist within soils is known as seepage flow. The prediction of the forces consequent upon such flows is important as they can be of sufficient magnitude to be destructive. Similarly the prediction of seepage flow rate following rainstorms is essential in the design of reservoir catchments, the estimation of well yields and the provision of land drainage. In the study of flow through porous or granular media it is usual to exclude the effect of capillary action and to concentrate upon gravity-driven flow. Under this constraint the seepage may be investigated through a dimensional analysis based around the following parameters

$$V = \phi(\rho, d, \mu, g, n, S), \tag{9.18}$$

where V is the seepage velocity based on the flow divided by the seepage area, n is the void ratio, which in real situations will vary across the flow zone, d is the assumed particle size, which may also be variable in 'real' situations, ρ and μ are the fluid density and viscosity, and S is the hydraulic gradient driving the flow, $-\partial h/\partial x$. A standard analysis yields the following group relationships

$$Sgd/V^2 = \phi'(n, \text{Re}). \tag{9.19}$$

Manipulating equation (9.19) at low Reynolds numbers when inertial forces are negligible reduces to the Darcy law

$$V = (\rho g d^2/\mu)(1/\phi_1 n)\, S = -k\partial h/\partial x, \tag{9.20}$$

where k is the coefficient of permeability and is dependent upon soil porosity, particle size and orientation and the level of saturation. Equation (9.20) is only valid at Re values below 1, i.e. laminar filtration.

When inertial forces cannot be ignored, e.g. for flow through coarse sand or gravel, the seepage flow may be non-linear. Again if viscous forces are ignored then equation (9.19) may be expressed as

$$V^2 = Sgd\,\phi_2(n) = -C\partial h/\partial x, \tag{9.21}$$

where C is dependent on Reynolds number. Equation (9.21) is valid under turbulent conditions at Re values above 10^4.

Empirical results due to Yalin lead to an alternative formulation of equation (9.19):

$$Sgpd/(2\rho V^2) = (1/n^6)\,(0.01 + 1/\text{Re}). \tag{9.22}$$

Equation (9.18) identifies the dynamic similarity requirements for a model to determine seepage flows.

The coefficient of permeability can be determined through laboratory tests under a fixed head difference Δh, where the flow Q passing through a layer of thickness L and cross-section area A determines k as

$$k = LQ/(A\Delta h). \tag{9.23}$$

Alternatively in site operations the value of k is best determined through well pumping where it is assumed that the radial flow to the well identifies a k value as

$$k = [Qln(r_2/r_1)]/[\pi(h_2^2 - h_1^2)] \tag{9.24}$$

where $h_{1,2}$ refers to the water table elevation above an assumed impermeable strata observed at radial distances $r_{1,2}$ from the abstraction point under steady flow conditions in an unconfined homogeneous layer.

While both site measurement and model testing are options in the assessment of groundwater flows it is also attractive to consider numerical modelling of the flow by reference to the application of the stream function and velocity potential theory introduced in Chapter 7. The Laplace equations (7.26) and (7.27) apply to the flow conditions in a two-dimensional flow element in a soil of known permeability. As discussed in Chapter 7 the solution of the Laplace equations allows lines of constant stream function and velocity potential to be plotted in an x–y plane. The constant stream function lines represent streamlines, while equipotential lines define the distribution of pressure throughout the flow field.

From the Darcy law for seepage flow the velocity potential may be expressed as

$$\Phi = -kh, \tag{9.25}$$

while the volumetric flow may be determined from the flow area represented by the distance between adjacent streamlines. The resultant mesh of equipotential and streamlines is known as a flow net, as discussed in Section 7.5. Returning to the well abstraction rate measurement of permeability it will be appreciated that the flow streamlines will be radial to the well while the lines of equipotential will form concentric circles, centred on the well.

For flow through a porous medium assume for a unit thickness that adjacent streamlines are separated by a distance b, while adjacent lines of equipotential, differing by a head difference Δh, are separated by a distance L. Thus the local pressure gradient is $\Delta h/L$ and the flow is $b \times$ unit thickness. Hence from Darcy's law

$$\Delta q = -bk\Delta h/L \tag{9.26}$$

allowing a determination of the seepage flow.

Section 7.5 details the steps necessary to develop flow nets and defines the significance of changes to the net mesh size as flow velocity changes. These processes will not be repeated here; however, it is useful to develop a numerical approach to the derivation of appropriate flow nets using the finite difference techniques, already introduced in Chapter 5, to solve the Laplace equations, drawing also upon the discussion of the importance of defining network boundary conditions and calculation step size. A general description will be presented followed by a particular example.

The groundwater flow may be considered to occur in an x–y plane bounded by ground/air or ground/water boundaries, or by boundaries representing impenetrable layers or assumed boundaries so remote from the source of the groundwater flow that

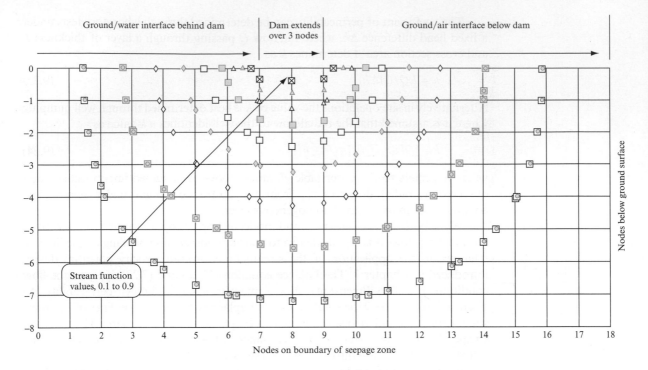

FIGURE 9.5(a)

Intersection points for stream function values on grid represeting seepage flow zone below dam

the boundary may be considered impenetrable. The x–y zone may be considered to be represented by a grid of side dimensions $\Delta x, \Delta y$, so that any zone consists of a network of nodal points, some at the boundaries to the zone and some internal to it – as shown in Fig. 9.5(a).

At each internal node, i.e. nodes not lying on a boundary, the Laplace equation for stream function, equations (7.26) must be satisfied:

$$\partial^2 \psi / \partial x^2 + \partial^2 \psi / \partial y^2 = 0. \tag{9.27}$$

As introduced in Chapter 5, this equation may be expressed in finite difference form through the application of Taylor's series, as the summation of the second order forward/backward forms of the truncated Taylor series may be used to determine the second differential of variable ψ at x_0 and y_0 within the grid. Hence in the x direction:

$$\psi''(x_0) = -\frac{\psi(x_0 + \Delta x) - 2\psi(x_0) + \psi(x_0 - \Delta x)}{\Delta x^2}, \tag{9.28}$$

and similarly in the y direction

$$\psi''(y_0) = -\frac{\psi(y_0 + \Delta y) - 2\psi(y_0) + \psi(y_0 - \Delta y)}{\Delta y^2}. \tag{9.29}$$

As suggested in Chapter 7 it is appropriate to set values of Δx and Δy equal to unity so that substituting for the second differentials of stream function in the Laplace equation yields the following expression for the stream function at the internal node of interest, located at x_0, y_0

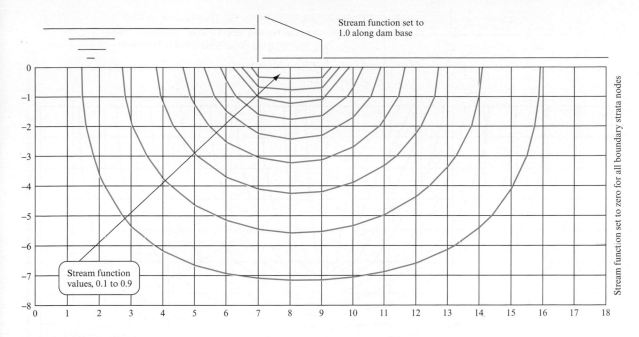

FIGURE 9.5(b)
Streamlines representing
seepage flow below dam

$$\psi_{x_0,y_0} = \frac{[\psi(x_0 - \Delta x) + \psi(x_0 + \Delta x) + \psi(y_0 - \Delta y) + \psi(y_0 + \Delta y)]}{4}. \tag{9.30}$$

Clearly an identical relationship may be obtained for velocity potential.

Boundaries formed by an impenetrable layer are clearly represented by zero flow streamlines while those between ground and air or ground and water are more difficult to define. However, it is clear that both of these may be represented by lines of equipotential, or velocity potential, as the pressure is constant. Also it is known from Chapter 7 that the streamlines, or lines of constant stream function, must intersect these boundaries at right angles, hence it is possible to represent these boundaries as the intersection between the zone of interest and a dummy symmetrical zone, see Fig. 9.5. Values of stream function at the nodes forming the boundary may then be determined from equation (9.30) with the values at the 'imaginary' node $y_0 + \Delta y$ and 'real' node $y_0 - \Delta y$ taken as equal.

It is clearly possible to develop and write a simple computer program based on the discussion above to simulate the groundwater flow beneath a dam of arbitrary thickness built upon soil subject to seepage but contained within an impenetrable layer that extends at a constant depth to points remote from the dam walls. Values of stream function are assigned to the boundaries to the zone and all internal nodes are assigned an arbitrary value. Equation (9.30) is used systematically to determine the appropriate nodal values on a square mesh until the difference between successive approximations is at an acceptable level. The solid boundary and impenetrable layer boundary values are retained throughout. The values at the ground/water and ground/air interfaces are determined from equation (9.30), with the 'dummy' node having the same value as the corresponding node Δy below the interface. A simple program may then be written to determine the intersection points on the grid for any series of stream function

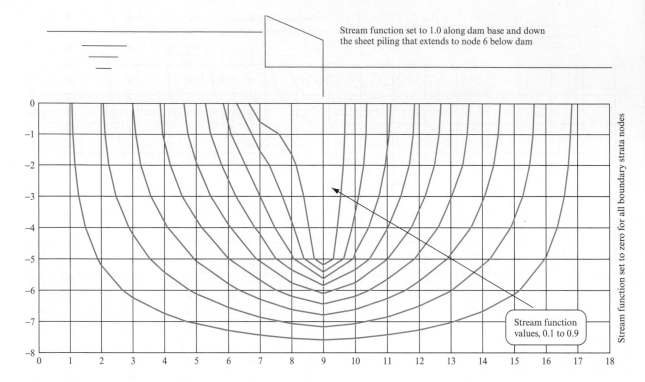

FIGURE 9.6

Streamlines representing seepage flow below dam with sheet piling added, cf. Fig. 9.5(b)

values; these values may then be sorted sequentially to allow graphical representation, Fig. 9.5(a) and (b).

Once the streamlines are established it is traditional to draw onto the figure the location of lines of equipotential using two simple rules, namely that the equipotential lines cross the streamlines at right angles and the zones enclosed by the intersection of adjacent streamlines and lines of equipotential are curvilinear squares. As it is assumed in the development of flow nets, see Chapter 7, that $\Delta\Psi = \Delta\Phi$, it is possible to draw the lines of equipotential and hence to determine the number of equipotential drops along the streamline network. Taken together with the number of 'flow passages', defined as the number of zones between adjacent streamlines, a calculation of total seepage flow q may be undertaken, as from the Darcy law

$$q = kHN_f/N_d, \tag{9.31}$$

where k is the coefficient of permeability, H is the pressure head difference across the system, N_f is the number of flow passages and N_d is the number of equipotential head drops estimated from drawing in the curvilinear squares between stream function and equipotential lines.

Many of the criteria established in the discussion of flow modelling in Chapter 5 come into play in this example. The choice of grid size initially determines the accuracy of the stream function array. Numerical methods improve on the accuracy of the stream function array from which the streamlines may be plotted, and clearly a user graphical interface to introduce lines of equipotential would be advantageous.

Figure 9.5(b) illustrates the flow under a simple dam. If a sheet pile curtain wall is introduced at the dam leading edge than the streamline contours change, as illustrated by Fig. 9.6. Here the sheet pile is included in the boundary representing the under-surface of the dam, namely a stream function value of unity. The resulting streamlines demonstrate the lack of symmetry to be expected.

The inclusion of ground water seepage modelling, even at this level, illustrates many of the advantages of numerical modelling as well as some of the pitfalls. The necessity to determine a suitable grid mesh is important as is the clear definition of the boundary equations or conditions to be utilized. A suitable Fortran code is presented in the solution to Problem 9.18, the program presented being capable of simple modification to extend the range of boundary conditions, e.g. in Fig. 9.6 the sheet pile-affected nodes were identified and values of stream function left unchanged within the internal node loop, activating equation (9.30).

9.10 POLLUTION DISPERSION MODELLING, OUTFALL EFFLUENT AND STACK PLUMES

Environmental concerns have led to increased interest in the modelling of pollution dispersion, either in applications such as sewage treatment outfalls or airborne pollution. These studies are beyond the scope of the current text; however, it is useful to review some of the modelling considerations that make this a highly interesting area of research.

In the case of effluent dispersion from sewer outfalls there are a number of distinct mechanisms and zones of application of the expected dimensionless groupings, such as the Reynolds and Froude numbers, with as might be expected some groups predominating within each zone. Figure 9.7 identifies these zones, based on the classification proposed by Ackers and Jaffrey in 1972, for a submerged effluent outfall in a tidal region. Sequentially from the outfall discharge the following zones may be considered:

1. Initially turbulent entrainment predominates, density differences are not important and the mechanism is governed by the discharge jet momentum. Successful modelling relies on geometric similarity, Reynolds number values above 2000 to avoid viscous effects and the Froude law.

2. The effluent discharge jet level within the recipient flow will depend upon buoyancy effects, with mixing at the boundaries being dependent upon turbulence

FIGURE 9.7

Dispersion of an effluent jet and identification of the relevant zones to be modelled, after Ackers and Jaffrey, 1972

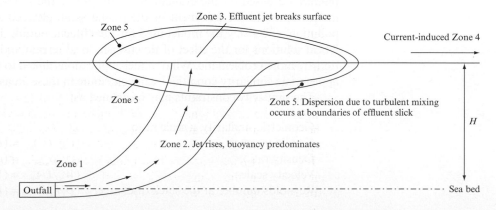

levels in the flow. The governing conditions again include geometric similarity and high Reynolds number but the applicable Froude number is now identified as the Froude densimetric number, defined as $V/(gh\Delta\rho/\rho)^{0.5}$.

3. Once the jet breaks the surface of the recipient flow the resulting convective spread depends upon the density differences between the point at which the jet breaks the surface and the recipient flow. The densimetric Froude number is again the predominant term with the characteristic length defined as the thickness of the buoyant layer. In addition Barr (1963) concludes that the model must be subject to a limiting value of the densimetric Reynolds number, defined as $(g\Delta\rho/\rho)^{0.5}H^{2.5}/(Lv) > 150$, where L is the distance travelled by the convective effluent spread at a velocity that satisfies the Froude law – a restriction that leads to a vertically distorted model, Section 9.8.

4. The modelling of the subsequent mass transport of the effluent via tidal or other currents requires the standard open channel conditions, namely Froude law, high Reynolds number and appropriate friction loss representation.

5. Geometric similarity is required for the general dispersion of the effluent due to turbulence interaction. An undistorted model is appropriate.

6. Finally, it may be necessary to model the rate of heat loss from the spreading effluent. Heat loss depends on area, time, temperature difference and a surface heat transfer coefficient, while temperature drop depends on mass and specific heat. Imposing equal model and prototype air temperatures leads to equal fluid temperatures between the model and the prototype to allow equal temperature drop so that the Froude number-dictated timescale leads to a distorted model scale.

To summarize these sequential mechanisms it may be seen that zones 1, 2 and 5 require an undistorted scale model, while zones 3 and 6 require a distorted model, as explained for river and harbour models in Section 9.8. Zone 4 may be undertaken with either. The actual bacterial and biological decay, in practice, is not normally considered. Overall the modelling of effluent outfalls has inherent difficulties, requiring at least two models with an interface. In addition, modelling the dispersion and diffusion of the effluent is problematic. While computational fluid dynamics offers apparent alternative approaches to this problem it is imperative to recall the importance of the definition of boundary conditions, no easier mathematically than described above through the zones identified by Ackers and Jaffrey.

Analytical methods for the prediction of dispersion of pollution from gaseous plumes are based on the evaluation of the rise of the plume due to buoyancy and momentum and the treatment of the plume as an elevated source of non-buoyant pollution, effectively the model used for the effluent outfall, Fig. 9.7. However, analytical solutions for the effect of downwash, local terrain and temperature stratification are more problematic. Wind tunnel applications linked to the identification of the appropriate similarity conditions can contribute in these areas.

The similarity constraints may be defined as:

Geometric similarity, a scale ratio	D_m/D_p,
Froude number,	$(D\mathbf{g}/U^2)_m - (D\mathbf{g}/U^2)_p$
Density ratio,	$(\rho_s/\rho_a)_m = (\rho_s/\rho_a)_p$
Velocity scale,	$(W_s/U)_m = (W_s/U)_p$
Reynolds number at the stack discharge,	$(DW_s/v_s)_m = (DW_s/v_s)_p$

where D is the stack diameter, U is the ambient mean wind velocity, ρ_s and ρ_a are the stack discharge and atmospheric air densities and W_s is the jet velocity on exit from the stack.

From these similarity considerations geometric scales are usually in the region 1/400 to 1/1000; velocity scales vary from 1/20 to 1/30 resulting in scale velocities for winds up to 20 m s^{-1} of as low as 0.15 to 1.0 m s^{-1}, which is very low and leads to modelling difficulties. Approximate solutions to deal with the low air speeds involve relaxing the density criterion and introducing the densimetric Froude number, already referred to in the discussion of effluent outfall studies, defined as $Fr_d = (\rho_s - \rho_a)Dg/(\rho_a U^2)$.

Use of a gas such as helium to represent the plume dispersion increases the velocity scale by approximately 50 per cent, thus making wind tunnel modelling practical.

Accurate Reynolds number similarity in the stack is not possible, particularly with helium, as the model stack internal flow remains laminar while in practice the stack discharge will be turbulent. This exaggerates the plume exit velocity and momentum. Generally the problem of similar Reynolds number has little effect on the modelling of the exhaust as buoyancy forces predominate, however, it is of importance in the study of any downwash effects as the initial trajectory of the plume is momentum dominated. Maintenance of a densimetric Froude number leads to a correct relative scaling of the plume buoyancy and inertia forces. Relaxing the density ratio criterion results in differing mixing rates between the model and prototype and precludes the consistent scaling of the plume properties.

The plume dispersion has been found to be highly sensitive to the presence of other buildings and local topography. The influence of building wakes in generating downwash forces on the plume has important consequences for the dispersion of stack plumes, leading to severe concerns as to the degree of ground level pollution. Figure 9.8 illustrates the likely effect of building wake on the dispersion of a stack plume and the effect of increasing stack height to limit the downwash experienced. In the first case, Fig. 9.8(a), the stack is relatively short compared with the height of the building. In this case the pollution is entrained within the wake and pollution levels are high downstream of the building, only decreasing due to natural buoyancy further downstream. The interaction of the dispersing plume with the downstream buildings has also been the subject of computational analysis, as illustrated by Fig. 5.21. In comparison increasing the stack height so that the plume is not entrained by the wake results in a naturally dispersing plume that limits ground level pollution.

In terms of the modelling of gaseous plume dispersion in a wind tunnel the same fundamental groupings apply as for the effluent outfall case. The exact modelling of the stack discharge velocity will only become important if downwash is a consideration. The dispersion of the plume is highly dependent upon entrainment and the presence of adjacent buildings and topographical features. Modelling the density ratio correctly is important in the prediction of eventual dispersion and approximations, or a failure to do so adequately will lead to an under-estimate of the ground level pollution levels. The similarities between the air- and water-based examples chosen are self-evident, despite the obvious differences in fluid densities and the application sites.

FIGURE 9.8(a)

Plume dispersion in presence of building wake-generated downwash, illustrating high pollution concentrations at ground level

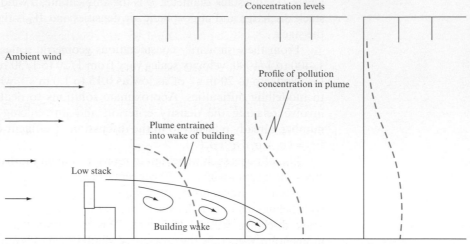

FIGURE 9.8(b)

Plume dispersion with an increased height of stack effectively removing plume from the effect of the building wake-generated downwash, illustrating lower pollution concentrations at ground level

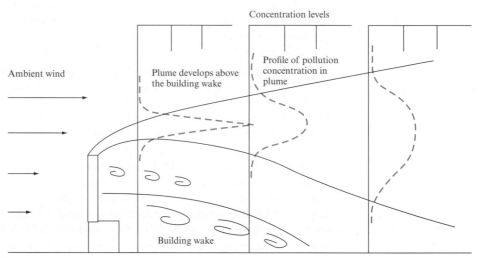

Concluding remarks

Chapter 9 has introduced a range of applications of dimensional analysis and similarity in the design of fluid machinery, aerofoils, channels and harbours and the entrainment of air flows by annular water flow. These should be seen as merely examples of the application of similarity and dimensional analysis across the whole spectrum of fluid mechanics applications. While the examples given are necessarily constrained to fluid situations, the technique applies equally to flow conditions involving temperature change and heat transfer. The approach demonstrated will be returned to throughout the text, for example in the treatment of frictional losses in Chapter 10, the analysis of fluid machinery in Chapters 22 and 23, the study of free surface flows in Chapters 15 and 16, and elsewhere. A thorough understanding of the application of dimensional analysis and similarity is essential to this text and to the general treatment of fluid mechanics.

Summary of important equations and concepts

1. This chapter defines the importance of both geometric and dynamic similarity. In particular it reinforces the role of experience in the choice of defining dimensionless groups where apparent incompatibilities arise.

2. The zones of influence of the major dimensionless groupings, Reynolds, Froude and Mach, are identified and the force ratios they represent defined, Section 9.2.

3. The use of non-dimensional coefficients to define lift and drag and other forces is discussed and the application of the similarity laws to a wide range of testing conditions introduced, from machinery to open channels and harbours. In the latter case the necessity to consider separate vertical and horizontal distance scales is explained.

Further reading

Ackers, P. and Jaffrey, L. J. (1972). The application of hydraulic models to pollution studies. *Symposium on Mathematical Modelling of Estuarine Pollution*, Paper 16, Water Pollution Research Laboratory, Stevenage.

Allen, J. (1952). *Scale Models in Hydraulic Engineering*. Longman, London.

Bain, D. C. *et al.* (1971). *Wind Tunnels, An Aid to Engineering Structure Design*. British Hydraulics Research Association, Cranfield.

Barr, D. I. H. (1963, 1967). *Densimetric Exchange Flow in Rectangular Channels*. La Houille Blanche, pp. 739–66 (1963); pp. 619–32 (1967).

Bradshaw, P. (1964). *Experimental Fluid Mechanics*. Pergamon, Oxford.

BS 848 Part I (1963). *Fan Testing for General Purposes*.

Cermak, J. E. (ed.) (1979). Wind engineering. In *Proc. 5th Int. Conf.*, Fort Collins, Colorado.

Kline, S. J. (1965). *Similitude and Approximation Theory*. McGraw-Hill, New York.

Langhaar, H. L. (1951). *Dimensional Analysis and Theory of Models*. Wiley, New York.

Novak, P. and Cabelka, J. (1981). *Models in Hydraulic Engineering*. Pitman, London.

Sedov, L. I. (1959). *Similitude and Dimensional Analysis in Mechanics*. Academic Press, New York.

Streeter, V. L. (ed.) (1961). *Handbook of Fluid Dynamics*. McGraw-Hill, New York.

Wood, I. R., Bell, R. G. and Wilkinson, D. L. (1995). Ocean disposal of wastewater. *Advanced Series on Ocean Engineering*, Volume 8, World Scientific, London.

Yallin, M. S. (1971). *Theory of Hydraulic Models*. Macmillan, London.

Problems

9.1   Water at 20 °C flows at 4 m s^{-1} in a 200 mm smooth pipe. Calculate the air velocity in a 100 mm pipe at 40 °C if the two flows are dynamically similar.   [135.0 m s^{-1}]

9.2   A spherical balloon to be used in air at 20 °C is tested by towing a 1/3 model submerged in a water tank. If the model is 1 m in diameter and the drag force is measured as 200 N at a model speed of 1.2 m s^{-1}, what would be the expected prototype drag if the water temperature was 15 °C and dynamic similarity is assumed?   [42.2 N]

9.3   Determine the relationship between model and prototype kinematic viscosity if both Reynolds number and Froude number are to be satisfied.
[Linear scale to power 3/2]

9.4   A large venturi meter is calibrated by means of a 1/10 scale model using the same fluid as the prototype. Calculate the discharge ratio between model and prototype for dynamic similarity.   [1/10]

9.5   The velocity at a point in a model spillway for a dam is 1 m s^{-1}. For a scale of 1/10 calculate the corresponding velocity in the prototype.   [3.16 m s^{-1}]

9.6   If the scale ratio between a model spillway and its prototype is 1/25 what velocity and discharge ratio should apply between model and prototype? If the prototype discharge is 3000 m^3 s^{-1} what is the model discharge?
[1/5, 1/3125, 0.96 m^3 s^{-1}]

9.7   A 1/5 scale model of a missile has a drag coefficient of 3 at Mach number 2. What would the model/prototype drag ratio be in air at the same temperature and one-third the density for the prototype at the same Mach number?   [1/25]

9.8   A ship model, scale 1/50, has a wave resistance of 30 N at its design speed. Calculate the prototype wave resistance.   [3750 kN]

9.9   A ship having a length of 200 m is to be propelled at 25 km h^{-1}. Calculate the prototype Froude number and the scale of a model to be towed at 1.25 m s^{-1}.   [0.157, 1/30.8]

9.10   A fan running at 8 rev s^{-1} delivers 2.66 m^3 s^{-1} at a fan total pressure of 418 N m^{-2}, the air having a temperature of 0 °C and 101.325 kN m^{-2} pressure. Given that the fan

efficiency is 69 per cent, calculate the air quantity delivered, the fan total pressure and the fan power when the air temperature is increased to 60 °C and the barometric pressure falls to 95 kN m^{-2}.   [2.66 m^3 s^{-1}, 322 N m^{-2}, 1.24 kW]

9.11   An axial flow water pump is to deliver 15 m^3 s^{-1} against a head of 20 m water. Calculate the air flow delivery rate and pressure rise for a 1/3 scale model using air at 1.3 kg m^{-3} density if the model and prototype are driven at the same speed.   [0.55 m^3 s^{-1}, 2.90 mm H$_2$O]

9.12   A model turbine employs 2 m^3 s^{-1} water flow when simulating a full-scale prototype designed to be served by a 15 m^3 s^{-1} flow. If the scale is 1/5 calculate the speed ratio and the shaft-delivered power ratio.   [16.66, 1.48]

9.13   In a test on a centrifugal fan it was found that the discharge was 2.75 m^3 s^{-1} and the total pressure 63.5 mm water column. The shaft power was 1.7 kW. If a geometrically similar fan having dimensions 25 per cent smaller but having twice the rotational speed was used, calculate the output, pressure generated and shaft power required. The air conditions are the same in both cases.
[2.32 m^3 s^{-1}, 142.9 mm H$_2$O, 3.2 kW]

9.14   A 1/5 scale model of the piping system of a water pumping station is to be tested to determine overall pressure losses. Air at 27 °C, 100 kN m^{-2} absolute, is available. For a prototype velocity of 0.45 m s^{-1} in a 4.25 m diameter duct section with water at 15 °C, determine the air velocity and quantity needed to model the situation.
[31.39 m s^{-1}, 17.83 m^3 s^{-1}]

9.15   The torque delivered by a water turbine depends upon discharge Q, head H, density ρ, angular velocity ω and efficiency η. Determine the form of the equation for torque.

$$\left[\rho g H^4 \cdot f\left(\omega \frac{H^3}{Q}, \eta \right) \right]$$

9.16   A 20 km length of river is to be modelled in a laboratory having only 12.5 m of available length. The river discharge is known to be in the range 400–500 m^3 s^{-1} and the average length and width are 3.5 and 55 m, respectively. Propose suitable scales.   [1/1600, 1/100]

9.17   If the model in Problem 9.16 is tidal calculate the tidal period on the model.   [4.65 min]

9.18   Develop and write a simple computer program to simulate the groundwater flow beneath a dam of arbitrary thickness built upon soil subject to seepage but contained within an impenetrable layer that extends at a constant depth to points remote from the dam walls. Utilize the program to draw the groundwater flow streamlines beneath the structure.

9.19   Use the program listing provided in the solution to Problem 9.18 to investigate the effect of grid size and further investigate the streamline formation for the following cases:

(*a*)   rectangular seepage zone beneath a dam, cf. Fig. 9.5;

(*b*)   rectangular seepage zone beneath a dam with a centrally placed sheet pile;

(*c*)   seepage flow beneath a dam where the rectangular zone is stepped in from the left-hand side.

Behaviour of Real Fluids

Weir and spillway flow from a reservoir, photo courtesy of West of Scotland Water

IN EARLIER CHAPTERS, THE BASIC EQUATIONS OF continuity, energy and momentum were introduced and applied to fluid flow cases where the assumption of frictionless flow was made. The analysis presented in the following chapters will introduce concepts necessary to extend the previous work to real fluids in which viscosity is accepted, and hence leads to situations where frictional effects cannot be ignored. The concept of Reynolds number as an indication of flow type will be used extensively and the fluid boundary layer, already introduced in Chapter 5, which lies between the free stream and the surface passed by the fluid and in which all the flow resistance is concentrated, will be expanded.

It will be necessary to distinguish between two different situations: namely, that in which the fluid moves inside a pipe or duct or in a channel so that it is guided by a boundary surrounding the fluid; and that in which the fluid flows around a solid body. In the first case, the flow is sometimes referred to as bounded flow and in the second case as external flow. Examples of the latter are fluid flow around a bridge pier or flow of wind around a house. Also to this category belong all the cases of solid objects moving through a stationary fluid, because it is the relative velocity between the fluid and the object that really matters. Thus an aeroplane in flight or a sailing ship are examples of such situations.

The bounded flow and the external flow around a body are both governed by the same basic principles. In all cases the fluid velocity at the boundary, i.e. where the fluid meets the solid surface, is equal to zero. This condition is sometimes referred to as the 'no-slip' condition. The velocity then increases with distance perpendicular to the boundary, the rate of increase being governed by the particular law applicable to the type of flow, which may be either laminar or turbulent.

In the external flow, the fluid velocity at some distance away from the boundary reaches a free stream velocity, which is the velocity of undisturbed (by the solid object) fluid, usually taken some distance upstream of the object. Thus for a bridge pier the fluid velocity at its surface will be zero and will increase away from it until it reaches the velocity of the undisturbed river. For a ship the velocity of the fluid at its surface will be equal to that of the ship and will diminish down to zero at some distance away from the ship as the water of the sea may be taken as stationary.

For bounded flow, such as in a pipe, the velocity of the fluid is zero at the wall and increases to a maximum at the centre of the pipe where the boundary layers, starting from the diametrically opposite points on the wall, meet.

In all the above cases, there is a velocity gradient and, thus, shear stresses in the fluid. In order to maintain flow this shear stress must also be maintained and this can only be achieved by additional forces doing work on the fluid. In other words, there must be a continuous supply of energy for the flow to exist. This energy, supplied solely to maintain flow in a bounded system, is usually expressed per unit weight of the fluid flowing and thus is in units of fluid head.

$$\frac{\text{Energy supplied per unit time}}{\text{Weight of fluid flowing}}$$

$$= \frac{\text{Force} \times \text{Distance}/\text{Time}}{\text{Specific weight} \times \text{Discharge}}$$

$$= \frac{pa \times s/t}{\rho g Q} = \frac{pav}{\rho g Q} = \frac{pQ}{\rho g Q} = \frac{p}{\rho g} = h.$$

This head (or energy) is considered as lost because it cannot be used for any other purpose than to maintain flow and hence it is called *head loss*. Such losses will be discussed in detail in Chapters 10 and 11.

In external flows, the forces required to maintain the velocity gradient in the boundary layer and energy dissipation in separation wakes are called the *drag* and will be discussed fully in Chapters 12 and 13. ● ● ●

Laminar and
Turbulent Flows
in Bounded Systems

THE FLOW OF REAL FLUIDS EXHIBITS VISCOUS EFFECTS. THIS CHAPTER WILL IDENTIFY THESE effects for both laminar and turbulent incompressible flow conditions. The relationships defining fluid friction will be developed and will be shown to be applicable to laminar, turbulent, free surface or full-bore flow situations, provided that due consideration is given to the appropriate value of flow friction factor. The Hagen–Poiseuille, Darcy and Chezy equations linking flow velocity to frictional loss will be developed. The chapter will draw on earlier material describing both the application of the momentum equation, in order to identify frictional forces, and dimensionless analysis and similarity, in order to identify the dependence of flow frictional losses on other flow parameters. The empirical nature of frictional loss prediction under turbulent flow conditions will be stressed and both the Moody chart and the Colebrook–White equation methodologies for loss determination introduced. Losses arising from flow disturbance caused by changes of direction, changes in flow cross-section or interaction with flow control devices will be introduced and characterized as separation losses, quantifiable via empirical coefficients based upon the flow kinetic energy. Expressions defining the velocity distribution across both laminar and turbulent fully developed pipe flow will be demonstrated. The calculation of flow friction factor based on the Colebrook–White equation is introduced as a computer program. ● ● ●

10.1 INCOMPRESSIBLE, STEADY AND UNIFORM LAMINAR FLOW BETWEEN PARALLEL PLATES

Consider first the case of steady laminar flow between inclined parallel plates, one of which is moving at a velocity U (Fig. 10.1) in the flow direction. It is required to calculate the velocity profile between the plates and hence the flow through the system.

FIGURE 10.1
Laminar flow between parallel plates

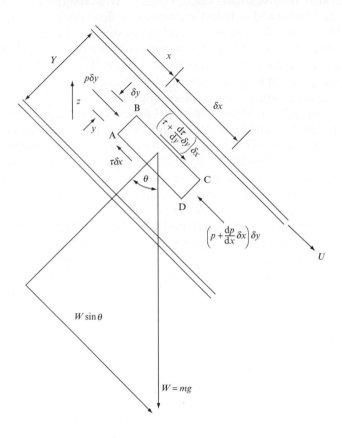

This flow condition may be analyzed by application of the momentum equation to an element of the flow – ABCD in Fig. 10.1 – and by consideration of the constraints imposed on the flow by limiting the analysis to steady, uniform, laminar flow. The momentum equation may be stated as

Resultant force in flow direction = Rate of change of momentum in flow direction.

However, as the flow is restricted to the steady uniform case, then the acceleration is zero. (If the acceleration of the flow is described by the equation

$$\frac{\mathrm{d}\bar{v}}{\mathrm{d}t} = \frac{\partial \bar{v}}{\partial t} + \frac{\partial \bar{v}}{\partial x} \cdot \frac{\partial x}{\partial t},$$

then for steady flow $\partial \bar{v}/\partial t$ is zero and for uniform flow $\partial \bar{v}/\partial x$ is zero; hence the zero value of $d\bar{v}/dt$.) Thus, the resultant force acting on the fluid element ABCD is zero and the flow is in a state of equilibrium under the action of the forces illustrated.

If it is assumed that the plates are sufficiently wide to make edge effects negligible, then the resultant force, in the flow direction, on the fluid element may be expressed, for unit width of plate, as

$$p\,\delta y - \left(p + \frac{\mathrm{d}p}{\mathrm{d}x}\,\delta x\right)\delta y + W\sin\theta - \tau\delta x + \left(\tau + \frac{\mathrm{d}\tau}{\mathrm{d}y}\,\delta y\right)\delta x = 0, \tag{10.1}$$

where p is the static pressure of the flow, τ is the shear stress, θ is the plate inclination and $W = \rho g \delta x \delta y$ per unit width. Therefore,

$$-\frac{\mathrm{d}p}{\mathrm{d}x}\,\delta x\delta y + W\sin\theta + \frac{\mathrm{d}\tau}{\mathrm{d}y}\,\delta y\delta x = 0.$$

If z is the elevation of the system above some horizontal datum, then

$$\sin\theta = -\frac{\mathrm{d}z}{\mathrm{d}x}$$

and, hence, by substitution for W and $\sin\theta$,

$$-\frac{\mathrm{d}p}{\mathrm{d}x}\,\delta x\delta y + \rho g\delta x\delta y\left(-\frac{\mathrm{d}z}{\mathrm{d}x}\right) + \frac{\mathrm{d}\tau}{\mathrm{d}y}\,\delta x\delta y = 0,$$

so that

$$\frac{\mathrm{d}\tau}{\mathrm{d}y} = \frac{\mathrm{d}}{\mathrm{d}x}(p + \rho gz), \tag{10.2}$$

where $(p + \rho gz)$ is the *piezometric pressure*, denoted by p^*.

As previously stated, Section 1.9.1, the shear stress in laminar flow may be expressed in the terms of fluid viscosity and the velocity gradient as

$$\tau = \mu\frac{\mathrm{d}u}{\mathrm{d}y}. \tag{10.3}$$

Hence, integrating equation (10.2) and substituting for τ yields an expression in terms of the velocity gradient:

$$\tau = \mu\frac{\mathrm{d}u}{\mathrm{d}y} = y\left[\frac{\mathrm{d}}{\mathrm{d}x}(p + \rho gz)\right] + C_1. \tag{10.4}$$

This integration was possible as $(p + \rho gz)$ is assumed to vary only in the x direction. Integration of equation (10.4) with respect to y will yield an equation for the velocity distribution between the plates in the form $u = f(y)$, in terms of fluid viscosity, piezometric head and two constants of integration that may be evaluated by consideration of the system boundary conditions at $y = 0$ and $y = Y$:

$$u = \frac{1}{\mu}\frac{\mathrm{d}}{\mathrm{d}x}(p + \rho g z)\frac{y^2}{2} + y\frac{C_1}{\mu} + C_2. \tag{10.5}$$

At the interface between the fluid and the plates at $y = 0$ and $y = Y$, the relative velocity of the fluid to the plate is zero, i.e. the condition of no slip. Thus, at $y = 0$ it follows that the fluid velocity $u = 0$ as this plate is itself stationary. At $y = Y$ the fluid velocity relative to the plate is zero; however, as the plate is moving at a velocity U in the flow direction, the value of u at $y = Y$ must similarly be $u = U$. Substituting these two boundary conditions in turn into equation (10.5) yields:

$$y = 0, u = 0; \quad \text{therefore } C_2 = 0;$$

$$y = Y, u = U; \quad \text{therefore } C_1 = \mu\frac{U}{Y} - \frac{Y}{2}\frac{\mathrm{d}}{\mathrm{d}x}(p + \rho g z);$$

or

$$u = \frac{y}{Y}U - \frac{1}{2\mu}\frac{\mathrm{d}}{\mathrm{d}x}(p + \rho g z)(Yy - y^2). \tag{10.6}$$

Equation (10.6) represents the velocity profile across the gap between the two plates, and is a general equation from which a number of restricted cases may be considered. For example,

1. Horizontal plates with no movement of the upper plates, i.e. $U = 0$, $\sin\theta = 0$; hence $\mathrm{d}z/\mathrm{d}x = 0$ and

$$u = -\frac{1}{2\mu}\frac{\mathrm{d}p}{\mathrm{d}x}(Yy - y^2). \tag{10.7}$$

Note equation (10.7) represents a parabolic velocity profile, and the negative sign recognizes that $\mathrm{d}p/\mathrm{d}x$ itself will be negative as the pressure drops in the flow direction.

2. Horizontal plates with upper plate motion:

$$u = \frac{y}{Y}U - \frac{1}{2\mu}\frac{\mathrm{d}p}{\mathrm{d}x}(Yy - y^2). \tag{10.8}$$

Equation (10.8) indicates that fluid flow may occur even if there is no pressure gradient in the x direction, provided the plates are in motion. In this case $u = (y/Y)U$, a straight line velocity distribution. This phenomenon is known as Couette flow.

The volume flow rate Q may be calculated for any of the above cases by integrating the expression for δQ, the flow through an element δy of the plate separation and of unit width, between the system boundary at $y = 0$ and $y = Y$.

Generally, $\delta Q = u \delta y$ per unit width; hence

$$Q = \int_{y=0}^{Y} u \, \mathrm{d}y.$$

For the general case, illustrated in Fig. 10.1, Q per unit width becomes

$$Q = \int_{y=0}^{Y} \left[\frac{y}{Y} U - \frac{1}{2\mu} \frac{\mathrm{d}}{\mathrm{d}x} (p + \rho g z)(Yy - y^2) \right] \mathrm{d}y$$

$$= \left(\frac{U}{Y} \frac{y^2}{2} \right)_0^Y - \frac{1}{2\mu} \frac{\mathrm{d}}{\mathrm{d}x} (p + \rho g z) \left(Y \frac{y^2}{2} - \frac{y^3}{3} \right)_0^Y.$$

Therefore,

$$Q = \frac{UY}{2} - \frac{1}{2\mu} \frac{\mathrm{d}}{\mathrm{d}x} (p + \rho g z) \frac{Y^3}{6}. \tag{10.9}$$

For flow between stationary horizontal plates this reduces to

$$Q = \frac{1}{12\mu} \frac{\mathrm{d}p}{\mathrm{d}x} Y^3 \text{ per unit width.} \tag{10.10}$$

EXAMPLE 10.1

Laminar flow of a fluid of viscosity $\mu = 0.9$ N s m^{-2} and density $\rho = 1260$ kg m^{-3} occurs between a pair of parallel plates of extensive width, inclined at 45° to the horizontal, the plates being 10 mm apart. The upper plate moves with a velocity 1.5 m s^{-1} relative to the lower plate and in a direction opposite to the fluid flow. Pressure gauges, mounted at two points 1 m vertically apart on the upper plate, record pressures of 250 kN m^{-2} and 80 kN m^{-2}, respectively. Determine the velocity and shear stress distribution between the plates, the maximum flow velocity and the shear stress on the upper plate (Fig. 10.2).

Solution

Flow direction from direction of pressure gradient.

At (1), $p_1 + \rho g z_1 = 250 + 9.81 \times 1.0 \times \dfrac{1260}{1000} = 262.36$ kN m^{-2}.

At (2), $p_2 + \rho g z_2 = 80$ kN m^{-2}

FIGURE 10.2

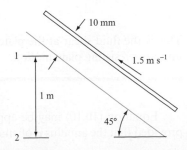

as $z = 0$ if datum taken at (2). Flow is down slope, upper plate moves 'up' slope. Pressure gradient

$$\frac{\mathrm{d}p^*}{\mathrm{d}x} = -\frac{(262.36 - 80)}{1 \cdot \sqrt{2}} = -182.36/\sqrt{2}$$

$$p^* = (p + \rho gz) = -128.95 \text{ kN m}^{-2} \text{ per metre, i.e. } z = 1.$$

From equation (10.6),

$$u = y\frac{U}{Y} - \frac{1}{2\mu}\frac{\mathrm{d}p^*}{\mathrm{d}x}(Yy - y^2),$$

where $U = -1.5 \text{ m s}^{-1}$, $Y = 0.01$ m and u is the local velocity at a point y above the lower plate. Thus the velocity profile is

$$u = \frac{-1.5}{0.01}y + \frac{128.95 \times 10^3}{2 \times 0.9}(0.01y - y^2)$$

$$= -150y + 716.4y - 71.64 \times 10^3 y^2,$$

$$u = \mathbf{566.4}y - \mathbf{71.64 \times 10^3}y^2.$$

Shear stress distribution is given by

$$\tau_y = \mu\left(\frac{\mathrm{d}u}{\mathrm{d}y}\right)_y,$$

$$\frac{\mathrm{d}u}{\mathrm{d}y} = 566.4 - 143.28 \times 10^3 y,$$

$$\tau_y = \mathbf{509.76 - 128.95 \times 10^3}y;$$

$u_{\max}$ occurs where $\mathrm{d}u/\mathrm{d}y = 0$, $y = 566.4 \times 10^{-3}/143.28 = 0.395 \times 10^{-2}$.
    Hence,

$$u_{\max} = 566.4 \times 0.003\,95 - 71.64 \times 10^3 \times 0.003\,95^2$$

$$= 2.24 + 1.117 = \mathbf{3.36 \text{ m s}^{-1}}.$$

Shear stress on upper plate is given by

$$\tau_Y = \mu\left(\frac{\mathrm{d}u}{\mathrm{d}y}\right)_{y=Y} = 509.76 - 128.95 \times 10^3 \times 0.01 = \mathbf{0.78 \text{ kN m}^{-2}}.$$

This is the fluid shear at the plate; hence, the shear force on the plate is 0.78 kN per unit area resisting plate motion.

Equation (10.10) may be applied to laminar flow between concentric cylinders, provided that the annulus is of small dimensions compared with the cylinder diameter.

FIGURE 10.3

Leakage flow past a
piston within a cylinder

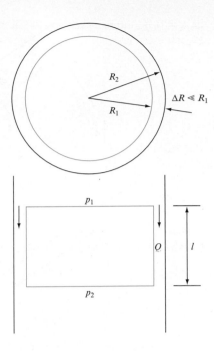

An example of this case would involve the leakage past a piston within a cylinder, as shown in Fig. 10.3. Hence, the leakage flow becomes

$$Q = \frac{1}{12\mu} \frac{p_1 - p_2}{l} (\Delta R)^3 2\pi R_1, \qquad (10.11)$$

where $\Delta R = R_1 - R_2$ and is the piston/cylinder separation, and the total width of the 'parallel plates' is given by the piston circumference. In this case, it will be seen that plate width edge effects can be ignored as the 'parallel plates' are effectively continuous.

10.2   INCOMPRESSIBLE, STEADY AND UNIFORM LAMINAR FLOW IN CIRCULAR CROSS-SECTION PIPES

Steady, uniform, laminar flow in a circular cross-section pipe or annulus may be treated in the same manner as described for laminar flow between parallel plates. The analysis rests on the same basic principles, namely the application of the momentum equation to an element of flow within the conduit; the application of the shear stress–velocity gradient relationship (10.3); and the knowledge of the flow condition at the pipe wall, which allows the constants of integration to be evaluated, namely the no-slip condition.

Consider an annular element in the flow of internal radius r and radial thickness δr, as shown in Fig. 10.4, in an inclined tube, of radius R, carrying a fluid under laminar flow conditions. Applying the momentum equation to the situation illustrated in Fig. 10.4 yields an expression

FIGURE 10.4
Forces acting on an
annular element in a
laminar pipe flow
situation

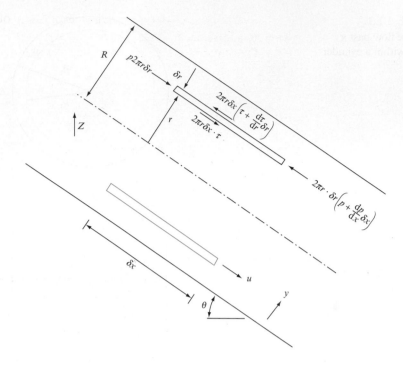

$$p2\pi r\delta r - \left(p + \frac{\mathrm{d}p}{\mathrm{d}x}\delta x\right)2\pi r\delta r + \tau 2\pi r\,\mathrm{d}x$$

$$- \left[2\pi r\tau\delta x + \frac{\mathrm{d}}{\mathrm{d}r}(2\pi r\tau\,\mathrm{d}x)\delta r\right] + W\sin\theta = 0, \tag{10.12}$$

where p is the flow static pressure, $W = mg$ is the element weight and τ is the shear stress at radius r. Owing to the assumption of steady uniform conditions, the flow acceleration is zero and, hence, the resultant force on the element is zero. Putting $W = 2\pi r\delta r\delta x\rho g$ and $\sin\theta = -\mathrm{d}z/\mathrm{d}x$, where z is the elevation of the pipe above some horizontal datum, reduces expression (10.12) to

$$-\frac{\mathrm{d}p}{\mathrm{d}x} - \frac{1}{r}\frac{\mathrm{d}}{\mathrm{d}r}(r\tau) - \rho g\frac{\mathrm{d}z}{\mathrm{d}x} = 0$$

by dividing by $2\pi r\delta r\delta x$. Rearranging,

$$\frac{\mathrm{d}}{\mathrm{d}x}(p + \rho gz) + \frac{1}{r}\frac{\mathrm{d}}{\mathrm{d}r}(r\tau) = 0. \tag{10.13}$$

The term $(p + \rho gz)$ is the flow piezometric pressure and is independent of r, enabling equation (10.13) to be integrated with respect to r. Hence,

$$\frac{r^2}{2}\frac{\mathrm{d}}{\mathrm{d}x}(p + \rho gh) + r\tau + C_1 = 0.$$

If conditions at the pipe centreline are substituted into the above expression, then $C_1 = 0$ as $r = 0$.

The shear stress–velocity gradient expression of equation (10.3) may be employed in a modified form to take note of the direction of measurement of distance r from the centre of the pipe, rather than the use of y measured from the pipe wall; hence

$$\tau = \mu\frac{\mathrm{d}u}{\mathrm{d}y} = -\mu\frac{\mathrm{d}u}{\mathrm{d}r} \tag{10.14}$$

and, by substituting for τ, above,

$$\frac{r^2}{2}\frac{\mathrm{d}}{\mathrm{d}x}(p+\rho gz) = r\mu\frac{\mathrm{d}u}{\mathrm{d}r} - C_1$$

and $du = \left[\dfrac{r}{2\mu}\dfrac{\mathrm{d}}{\mathrm{d}x}(p+\rho gz) + \dfrac{C_1}{r\mu}\right]\mathrm{d}r.$

Integrating with respect to r yields an expression for the velocity variation across the flow in terms of r and known system parameters:

$$u = \frac{r^2}{4\mu}\frac{\mathrm{d}}{\mathrm{d}x}(p+\rho gz) + \frac{C_1}{\mu}\log_e r + C_2. \tag{10.15}$$

Values of C_1 and C_2 may be evaluated from boundary conditions at $r = 0$ and $r = R$. At $r = 0$ it has been shown that $C_1 = 0$. At $r = R$, i.e. at the pipe wall, the local flow velocity u is zero; hence,

$$C_2 = -\frac{R^2}{4\mu}\frac{\mathrm{d}}{\mathrm{d}x}(p+\rho gz)$$

and $u = -\dfrac{(R^2-r^2)}{4\mu}\dfrac{\mathrm{d}}{\mathrm{d}x}(p+\rho gz).$ \hfill (10.16)

Equation (10.16) describes the variation of local fluid velocity u across the pipe and, from the form of the equation, this velocity profile may be seen to be parabolic. The negative sign is again present due to the fact that the pressure gradient will be negative in the flow direction.

The maximum velocity will occur on the pipe centreline, i.e. $r = 0$; hence

$$u_{\max} = -\frac{R^2}{4\mu}\frac{\mathrm{d}}{\mathrm{d}x}(p+\rho gz). \tag{10.17}$$

The volume flow rate through the pipe under these flow conditions may be calculated by integrating the incremental flow δQ through an annulus of radial width δr at radius r across the flow from $r = 0$ to $r = R$ (see Fig. 10.4):

$$\delta Q = u2\pi r\,\delta r,$$

$$Q = \int_0^R u2\pi r\,\mathrm{d}r. \tag{10.18}$$

Substitution for u at general radius r yields an expression

$$Q = -\frac{\pi}{2\mu}\frac{d}{dx}(p+\rho gz)\int_0^R (R^2 r - r^3)\,dr$$

$$= -\frac{\pi}{2\mu}\frac{d}{dx}(p+\rho gz)\left(R^2\frac{r^2}{2}-\frac{r^4}{4}\right)_0^R$$

$$= -\frac{\pi}{8\mu}\frac{d}{dx}(p+\rho gz)R^4$$

or, in terms of a pressure drop Δp over a length l of pipe of diameter d,

$$Q = \Delta p\pi d^4/128\mu l. \tag{10.19}$$

The mean flow velocity is given by Q/A, where A is the pipe cross-sectional area $\pi d^2/4$. Hence,

$$\bar{u} = -\frac{\pi}{8\mu}\frac{d}{dx}(p+\rho gz)R^2 = \tfrac{1}{2}u_{max}, \tag{10.20}$$

as shown in Fig. 10.5.

FIGURE 10.5

Velocity distribution in laminar flow in a circular pipe

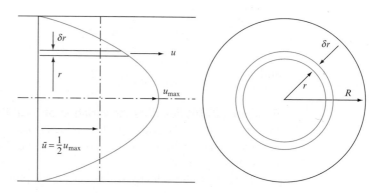

Equation (10.19) may be rearranged for the pressure loss giving the well-known Hagen–Poiseuille equation:

$$\Delta p = 128\mu l Q/\pi d^4. \tag{10.21}$$

Alternatively, substituting for $Q = (\pi d^2/4)\bar{u}$,

$$\Delta p = 32\mu l\bar{u}/d^2. \tag{10.22}$$

EXAMPLE 10.2

Glycerine of viscosity 0.9 N s m^{-2} and density 1260 kg m^{-3} is pumped along a horizontal pipe 6.5 m long of diameter $d = 0.01$ m at a flow rate of $Q = 1.8$ litres min^{-1}. Determine the flow Reynolds number and verify whether the flow is laminar or turbulent. Calculate the pressure loss in the pipe due to frictional effects and calculate the maximum flow rate for laminar flow conditions to prevail.

Solution

$$\text{Mean velocity, } \bar{u} = Q/A = \left(\frac{1.8}{60} \bigg/ \frac{\pi d^2}{4}\right) \times 10^{-3} \text{ m s}^{-1} = 0.382 \text{ m s}^{-1},$$

$$\text{Re} = \rho u d/\mu = 1260 \times 0.382 \times 0.01/0.9 = 0.535.$$

Therefore, flow is laminar as Re < 2000 (see Section 4.10).

Frictional losses may be calculated from the Hagen–Poiseuille equation (10.21):

$$\Delta p = 128 \mu l Q/\pi d^4$$
$$= 128 \times 0.9 \times 6.5 \times 3 \times 10^{-5}/(\pi \times 0.01^4) = \mathbf{715 \times 10^3 \text{ N m}^{-2}}.$$

Upper limit of laminar flow conditions is reached when

$$\text{Re}/\text{Re}_{\text{crit}} = Q/Q_{\text{crit}},$$
$$Q_{\text{crit}} = (Q/\text{Re})\text{Re}_{\text{crit}}$$
$$= (1.8/0.535) \times 2000 \text{ litres min}^{-1}.$$

Therefore, $Q_{\text{crit}} = \mathbf{112 \text{ litres s}^{-1}}$.

10.3 INCOMPRESSIBLE, STEADY AND UNIFORM TURBULENT FLOW IN BOUNDED CONDUITS

In the preceding sections expressions have been developed for the velocity distribution and pressure losses encountered during laminar flow. Reference to the Reynolds number of such flow, i.e. Re < 2000 in closed circular pipes, shows that such flow is restricted to relatively low flow rate conditions for all gases and those liquids that do not possess a high viscosity. Thus, in general, turbulent flow conditions are far more likely in most engineering situations. In this section, expressions will be developed for the losses incurred in turbulent flow in both closed and open conduits. However, it will be seen that completely analytical solutions are not available and that empirical relationships are needed in order to produce the necessary expressions.

Consider a small element of fluid within a conduit, as shown in Fig. 10.6. The flow is assumed to be uniform and steady so that the fluid acceleration in the flow direction is zero. Applying the momentum equation to the fluid element in the flow direction yields

$$p_1 A - p_2 A - \tau_0 l P + W \sin \theta = 0, \tag{10.23}$$

FIGURE 10.6

Turbulent flow in a bounded conduit

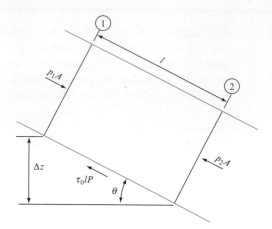

where P is the wetted perimeter of the element defined as that part of the conduit's circumference which is in contact with the fluid. It will be seen that including the area over which the shear stress τ_0 acts in the form of lP, as above, effectively renders the derivation applicable to both open or closed conduits. Putting $W = \rho g A l$ and $\sin\theta = -\Delta z/l$ yields

$$A(p_1 - p_2) - \tau_0 lP - \rho g A \,\Delta z = 0,$$

where p_1, p_2 are the static pressures in the flow at sections 1 and 2 (Fig. 10.6). Hence

$$\frac{1}{l}[(p_1 - p_2) - \rho g \,\Delta z] - \tau_0 \frac{P}{A} = 0$$

where the first term represents a drop in piezometric head over a length l of the conduit and the ratio A/P is known as the hydraulic mean depth, normally denoted by m; thus,

$$\tau_0 = m\frac{\mathrm{d}p^*}{\mathrm{d}x}, \tag{10.24}$$

where $\mathrm{d}p^*/\mathrm{d}x$ is the rate of loss of piezometric head along the conduit and τ_0 is the wall or boundary shear stress.

In order to express τ_0 in equation (10.24), the concept of a flow friction factor f is introduced, which is a non-dimensional, experimentally measured factor normally introduced in the form

$$\tau_0 = f\rho\bar{v}^2/2, \tag{10.25}$$

where $\bar{v}$ is the mean flow velocity. Hence,

$$\frac{\mathrm{d}p^*}{\mathrm{d}x} = f\rho\bar{v}^2/2m. \tag{10.26}$$

If the frictional head loss down a length l of the conduit is denoted by h_f, then the rate of loss of piezometric pressure may be expressed as

$$\frac{\mathrm{d}p^*}{\mathrm{d}x} = f\rho\bar{v}^2/2m = \rho g h_f/l$$

and

$$h_f = f l\bar{v}^2/2gm. \tag{10.27}$$

Now, as

$$\frac{\mathrm{d}p^*}{\mathrm{d}x} = \frac{\mathrm{d}}{\mathrm{d}x}(p + \rho g z),$$

where z is the elevation of the conduit above some datum, then for *open channels*, as the static pressure p may be assumed to remain constant along the channel, it follows that

$$\frac{\mathrm{d}p^*}{\mathrm{d}x} = \rho g \frac{\mathrm{d}z}{\mathrm{d}x} = \rho g \sin\theta$$

and, since for uniform flow the hydraulic gradient h_f/l is equal to the slope of the channel,

$$\frac{h_f}{l} = \sin\theta = i.$$

Then it follows, by equating equations (10.26) and (10.27), that

$$f\rho\bar{v}^2/2m = \rho g i$$

so that

$$\bar{v} = \sqrt{(2g/f)} \times \sqrt{(mi)}.$$

If, now

$$\sqrt{(2g/f)} = C \tag{10.28}$$

is substituted, the expression known as the Chezy formula is obtained:

$$\bar{v} = C\sqrt{(mi)}, \tag{10.29}$$

yielding the flow rate through a given channel of a given slope and roughness. Various values of C are employed in open-channel design.

For pipes running full of fluid, the wetted perimeter becomes the internal circumference of the pipeline; hence $A/P = m = \pi D^2/4\pi D = D/4$, so that equation (10.27) becomes for circular cross-sections

$$h_f = \frac{4fl}{d} \cdot \frac{\bar{v}^2}{2g}. \tag{10.30}$$

This expression is in every way equivalent to the Chezy equation above and follows directly from the study of the general condition illustrated in Fig. 10.5. It is known as the Darcy–Weisbach equation for head loss in circular pipes. The Darcy equation is also the equivalent of the Poiseuille equation derived for laminar flow, with one important exception, namely the inclusion of an empirical factor f to describe the friction loss in turbulent flow, which was not necessary in the case of laminar flow. This fundamental difference arises from the complexity of turbulent flow, resulting in the fact that the relationship $\tau = \mu(\mathrm{d}u/\mathrm{d}y)$ cannot be used and, therefore, an analytical solution is not possible.

10.4 INCOMPRESSIBLE, STEADY AND UNIFORM TURBULENT FLOW IN CIRCULAR CROSS-SECTION PIPES

The head loss in turbulent flow in a closed section pipe is given by the Darcy equation (10.30),

$$h_f = \frac{4fl}{d} \cdot \frac{\bar{v}^2}{2g}.$$

It will be seen from the above expression that all the parameters, with the exception of the friction factor f, are measurable. Results of extensive experimentation in this area led to the establishment of the following proportional relationships:

1. $h_f \propto l$;
2. $h_f \propto \bar{v}^2$;
3. $h_f \propto 1/d$;
4. h_f depends on the surface roughness of the pipe walls;
5. h_f depends on fluid density and viscosity;
6. h_f is independent of pressure.

The value of f must be selected so that the correct value of h_f will always be given by the Darcy equation and so cannot be a single-value constant. The value of f must depend on all the parameters listed above. Expressed in a form suitable for dimensional analysis this implies that

$$f = \phi(\bar{v}, d, \rho, \mu, k, k', \alpha), \tag{10.31}$$

where k is a measure of the size of the wall roughness, k' is a measure of the spacing of the roughness particles, both having dimensions of length, and α is a form factor, a dimensionless parameter whose value depends on the shape of the roughness particles. In the general rough pipe case, dimensional analysis yields an expression

$$f = \phi_2(\rho\bar{v}d/\mu, k/d, k'/d, \alpha)$$

or, in terms of Reynolds number,

$$f = \phi_2(\text{Re}, k/d, k'/d, \alpha). \tag{10.32}$$

Dimensional analysis can only indicate the best combination of parameters for an empirical solution; the actual algebraic format of the relation for friction factor in terms of the variables listed must be determined by experimentation.

Blasius, in 1913, was the first to propose an accurate empirical relation for the friction factor in turbulent flow in smooth pipes, namely

$$f = 0.079/\text{Re}^{1/4}. \tag{10.33}$$

This expression yields results for head loss to ±5 per cent for smooth pipes at Reynolds numbers up to 100 000.

At this point it may be useful to note that the value of friction factor quoted in many American texts is $4f$ in the notation employed in this text, and so the value of the constant in Blasius' equation will be changed. In this text the UK-recognized value of friction factor as defined by equation (10.30) will be used exclusively.

For rough pipes, Nikuradse, in 1933, proved the validity of the f dependence on the relative roughness ratio k/d by investigating the head loss in a number of pipes which had been treated internally with a coating of sand particles whose size could be varied. These tests in no way investigated the effect of particle spacing k'/d, or of particle shape factor α, on the friction factor, but did show that, for one type of roughness,

$$f = \phi_3(\text{Re}, k/d). \tag{10.34}$$

It may well be argued that experimental problems would make it virtually impossible to hold k'/d and α constant so that the effect of roughness size k/d might be investigated in isolation. However, the accuracy of the results obtained by basing the value of f simply on Reynolds number and k/d does suggest that the effects of particle spacing and shape are negligible compared with that of the relative roughness based solely on k/d.

Thus, the calculation of losses in turbulent pipe flow is dependent on the use of empirical results and the most common reference source is the Moody chart, which is a logarithmic plot of f vs. Re for a range of k/d values. This type of data presentation is commonly referred to as a Stanton diagram. A typical *Moody chart* is presented as Fig. 10.7 and a number of distinct regions may be identified and commented on.

1. The straight line labelled 'laminar flow', representing $f = 16/\text{Re}$, is a graphical representation of the Poiseuille equation (10.19), i.e.

$$Q = \Delta p \pi d^4 / 128 \mu l,$$

$$h_f = \Delta p / \rho g = 128 \mu l Q / \rho g \pi d^4.$$

Now, $Q = \pi(d^2/4)\bar{v}.$

Hence, from the Darcy equation,

$$h_f = 128 \mu l \pi d^2 \bar{v} / 4 \rho g \pi d^4 = 4 f l \bar{v}^2 / 2 g d$$

FIGURE 10.7

Variation of friction factor f with Reynolds number and pipe wall roughness for ducts of circular cross-section

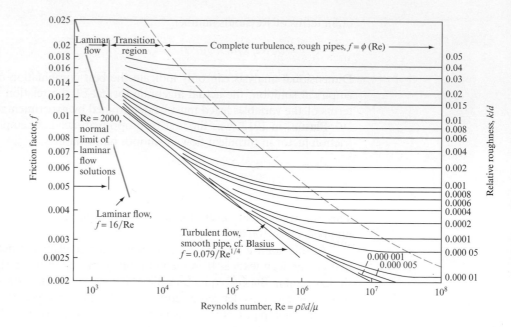

or $\qquad f = \dfrac{128}{8}\mu/\rho\bar{v}d$

$$f = 16/\text{Re}. \tag{10.35}$$

Equation (10.35) plots as a straight line of slope −1 on a log–log plot and is independent of pipe surface roughness. This relation also shows that the Darcy equation may be applied to the laminar flow regime provided that the correct f value is employed.

   2. For values of $k/d < 0.001$ the rough pipe curves of Fig. 10.7 approach the Blasius smooth pipe curve due to the presence of the laminar sublayer (discussed in Chapter 11), which develops in turbulent flow close to the pipe wall and whose thickness decreases with increasing Reynolds number. Thus, for certain combinations of surface roughness and Reynolds number, the thickness of the laminar sublayer is sufficient to cover the wall roughness and the flow behaves as if the pipe wall were smooth. For higher Reynolds numbers the roughness particles project above the now decreased thickness laminar sublayer and contribute to an increased head loss.

   3. At high Reynolds numbers, or for pipes having a high k/d value, all the roughness particles are exposed to the flow above the laminar sublayer. In this condition, the head loss is totally due to the generation of a wake of eddies by each particle making up the pipe roughness. This form of head loss is known as 'form drag' and is directly proportional to the square of the mean flow velocity; thus $h_f \propto \bar{v}^2$ and, hence, from the Darcy equation, f is a constant, depending only on the roughness particle size. This condition is represented on the Moody chart by portions of the f vs. Re curves which are parallel to the Re axis and which occur at high values of Re and k/d.

EXAMPLE 10.3

Calculate the loss of head due to friction and the power required to maintain flow in a horizontal circular pipe of 40 mm diameter and 750 m long when water (coefficient of dynamic viscosity 1.14×10^{-3} N s m^{-2}) flows at a rate: (a) 4.0 litres min^{-1}; (b) 30 litres min^{-1}. Assume that for the pipe the absolute roughness is 0.000 08 m.

Solution

(a) In order to establish whether the flow is turbulent or laminar it is first necessary to calculate the Reynolds number:

$$\mathrm{Re} = \rho \bar{v} d / \mu,$$

but $\quad Q = 4.0 \times 10^{-3}/60 = 66.7 \times 10^{-6} \,\mathrm{m^3\ s^{-1}}$

and $\quad$ Pipe area, $A = \pi d^2/4 = \pi(0.04)^2/4 = 1.26 \times 10^{-3}\,\mathrm{m^2}$,

so that the mean velocity in the pipe is given by

$$\bar{v} = Q/A = 66.7 \times 10^{-6}/1.26 \times 10^{-3} = 52.9 \times 10^{-3}\,\mathrm{m\ s^{-1}}.$$

Hence, $\quad \mathrm{Re} = \dfrac{10^3 \times 52.9 \times 10^{-3} \times 0.04}{1.14 \times 10^{-3}} = 1856.$

Therefore, flow is laminar, since Re < 2000. So, the loss due to friction may therefore be calculated by using either Poiseuille's equation (i) or the Darcy equation and $f = 16/\mathrm{Re}$ (ii):

(i) Poiseuille's equation:

$$\Delta p = \frac{128 \mu l Q}{\pi d^4} = \frac{128 \times 1.14 \times 10^{-3} \times 750 \times 66.7 \times 10^{-6}}{\pi(0.04)^4}$$

$$= 907.6\ \mathrm{N\ m^{-2}}.$$

Therefore, head lost due to friction is given by

$$h_f = \frac{\Delta p}{\rho g} = \frac{907.6}{10^3 \times 9.81} = \mathbf{92.4 \times 10^{-3}\,m\ of\ water}.$$

(ii) Darcy equation:

$$h_f = \frac{4fl}{d} \frac{\bar{v}^2}{2g}, \quad \text{but} \quad f = \frac{16}{\mathrm{Re}} = \frac{16}{1856} = 0.008\,62.$$

Hence, $\quad h_f = \dfrac{4 \times 0.008\,62 \times 750}{0.04} \times \dfrac{(52.9 \times 10^{-3})^2}{2 \times 9.81} = \mathbf{92.4 \times 10^{-3}\,m\ of\ water}.$

Power required to maintain flow,

$$P = \rho g h_f Q$$

$$= 10^3 \times 9.81 \times 92.4 \times 10^{-3} \times 66.7 \times 10^{-6} = \mathbf{0.0605\ W}.$$

(b) $Q = \dfrac{30 \times 10^{-3}}{60} = 0.5 \times 10^{-3}\,\mathrm{m^3\,s^{-1}}, \quad \bar{v} = \dfrac{0.5 \times 10^{-3}}{1.26 \times 10^{-3}} = 0.4\,\mathrm{m\,s^{-1}}.$

Therefore,

$$\mathrm{Re} = \frac{10^3 \times 0.4 \times 0.04}{1.14 \times 10^{-3}} = 14\,035$$

and the flow is turbulent, so the Darcy equation must be used. To determine the value of friction factor:

Relative roughness $= k/d = 0.000\,08/0.04 = 0.002.$

From the Moody chart, for $\mathrm{Re} = 1.4 \times 10^4$ and relative roughness of 0.002, $f = 0.008$. Therefore,

$$h_f = \frac{4fl}{d}\frac{\bar{v}^2}{2g} = \frac{4 \times 0.008 \times 750}{0.04} \times \frac{(0.4)^2}{2 \times 9.81} = \mathbf{4.89\ m\ of\ water}.$$

Power required, $P = \rho g h_f Q = 10^3 \times 9.81 \times 4.89 \times 0.5 \times 10^{-3}$

$$= \mathbf{24.0\ W}.$$

10.5 STEADY AND UNIFORM TURBULENT FLOW IN OPEN CHANNELS

It was shown in Section 10.3 that the general equation for head losses in turbulent flow could be derived concurrently for both open and closed section conduits. The general equation (10.27)

$$h_f = fl\bar{v}^2/2gm,$$

reduces to the Chezy equation (10.29)

$$\bar{v} = C\sqrt{(mi)},$$

when it is realized that, for open channels, provided the flow is steady and uniform, h_f/l is equal to the slope of the channel. This, however, is not the case in non-uniform flow discussed in Chapter 16.

Since $\bar{v}$ is the mean velocity in the channel of area A, it follows that the flow rate Q is given by:

$$Q = A\bar{v} = AC\sqrt{(mi)}. \tag{10.36}$$

Although C is referred to as the Chezy coefficient, implying a dimensionless constant, this is not so; since $C = 2g/f$, it has dimensions of $\mathrm{L^{1/2}T^{-1}}$.

It has been shown that, for pipe flow, the value of f depends both on the flow Reynolds number and on the surface roughness of the pipe material, so it would be reasonable to expect C to vary with Re and k/m, where m, the mean hydraulic depth, is employed as the characteristic length for the system. Generally, the dependence of C on Reynolds number is small and k/m is the predominant factor. For almost all open-channel work the flow may be assumed to be fully turbulent, with a high value of Reynolds number and, therefore, k/m may be taken as the sole factor affecting C values, provided the channel section shape remains simple.

Finally, it will be seen that, as values of C depend only on Re and k/m and not on the Froude number of the open-channel flow, the Chezy equation applies equally to rapid or tranquil flow, defined in Chapter 15. However, in cases of non-uniform flow, an important distinction between the slope of the channel and the hydraulic gradient has to be made. This will be discussed in Chapter 16.

EXAMPLE 10.4

A rectangular open channel has a width of 4.5 m and a slope of 1 vertical to 800 horizontal. Find the mean velocity of flow and the discharge when the depth of water is 1.2 m and if C in the Chezy formula is 49.

Solution

The mean velocity may be obtained using the Chezy formula:

$$\bar{v} = C\sqrt{(mi)}.$$

For the channel, $i = 1/800$ and $m = A/P$. Now,

$$A = 4.5 \times 1.2 = 5.4 \text{ m}^2,$$

$$P = 2 \times 1.2 + 4.5 = 6.9 \text{ m},$$

so that $m = 5.4/6.9 = 0.783$ m.

Substituting into the Chezy formula,

$$\bar{v} = 49\sqrt{(0.783/800)} = \mathbf{1.53 \text{ m s}^{-1}}.$$

The discharge is given by

$$Q = \bar{v}A = 1.53 \times 5.4 = \mathbf{8.27 \text{ m}^3 \text{ s}^{-1}}.$$

10.6   VELOCITY DISTRIBUTION IN TURBULENT, FULLY DEVELOPED PIPE FLOW

Owing to the nature of turbulent flow, there are difficulties in the derivation of expressions defining the distribution of velocity in pipe flow. The use of dimensional analysis, together with a series of assumptions based on the relative importance of the fluid viscosity and eddy viscosity terms in the laminar sublayer which is present in a fully developed turbulent boundary layer, do, however, allow the prediction of the form of the velocity distribution expressions. The algebraic format of these equations has been developed from experimental investigations and, although now well established and accepted, is empirical in nature.

For fully developed turbulent pipe flow in a circular cross-section pipe, it would be reasonable to suppose that the local velocity u at a distance y from the pipe wall would be given by a general function of the form

$$u = \phi(\rho, \mu, \tau_0, R, y, k), \tag{10.37}$$

where ρ, μ are the fluid density and viscosity, R is the pipe radius, k is the roughness particle size and τ_0 is the wall shear stress.

Dimensional analysis suggests an expression of the form

$$\frac{\bar{u}}{\sqrt{(\tau_0/\rho)}} = \phi_1\left[\left(\frac{\rho R}{\mu}\right)\sqrt{\left(\frac{\tau_0}{\rho}\right)}, \frac{y}{R}, \frac{k}{R}\right].$$

The term $\sqrt{(\tau_0/\rho)}$ has the dimensions of velocity and is referred to as the *shear stress velocity u^**.

$$\bar{u}/u^* = \phi_1(\rho u^* R/\mu, y/R, k/R), \tag{10.38}$$

where $\rho u^* R/\mu$ is a form of the Reynolds number.

To proceed beyond equation (10.38), it is necessary to make some assumptions about the importance of the various groups.

Surface roughness, represented by k/R, will affect the value of u^*, but will only be a significant factor in the flow zone close to the wall. Similarly, the fluid viscosity will only be of major importance in the laminar sublayer close to the pipe wall. Thus, the velocity in the central turbulent core of the flow will be assumed to depend only on the positional group y/R. It is customary to express this relationship in terms of the velocity defect, or the difference between the local velocity u at position y from the wall and the flow maximum velocity on the pipe centreline u_{max}. Hence

$$(u_{max} - u)/u^* = \phi_2(y/R). \tag{10.39}$$

This expression is referred to as the *velocity defect distribution* and is well supported by experimental work which shows that, for a wide range of flow Reynolds numbers, the velocity profiles only differ in the region close to the pipe wall. It is apparent from experimental results that, as the Reynolds number increases, the friction factor f and the shear stress τ_0 terms become smaller, and the velocity profile across the central core of the flow becomes progressively more uniform.

Prandtl proposed an empirical velocity distribution for this turbulent central core of the form

$$u/u_{max} = (y/R)^n, \tag{10.40}$$

where the value of $n = \frac{1}{7}$ for Re $< 10^7$ and decreases above this Reynolds number. This is well-supported experimentally, but does break down at $y = R$ as symmetry here demands that $du/dy = 0$, which cannot be justified by the expression.

If the case of the smooth pipe is considered, the k/R group becomes unimportant and so, close to the pipe wall, the effect of pipe radius R is negligible, so long as $y < R$; let $y = y_2$ be the limit for this assumption so that

$$u/u^* = \phi_3(\mathrm{Re}^*), \quad 0 < y < y_2, \tag{10.41}$$

where $\mathrm{Re}^* = \rho y u^*/\mu$, a group independent of pipe radius R.

If equation (10.39) is applicable from $y = y_1$ to R, and if experimental results which indicate that $y_2 > y_1$ are accepted, then it becomes apparent that there is a region in the flow, close to the pipe wall, where both equations (10.39) and (10.41) apply simultaneously.

For a smooth pipe, the relation

$$u/u^* = \phi_1(\mathrm{Re}, y/R) \tag{10.42}$$

applies for $y_1 < y < R$, where $\mathrm{Re} = \rho R u^*/\mu$ and, for $y = R$,

$$u_{\max}/u^* = \phi_4(\mathrm{Re}). \tag{10.43}$$

Adding equations (10.39) and (10.42) yields

$$u_{\max}/u^* = \phi_4(\mathrm{Re}) = \phi_1(\mathrm{Re}, y/R) + \phi_2(y/R) \tag{10.44}$$

for $y_1 < y < R$. Both equations (10.39) and (10.41) apply in the zone $y_1 < y < y_2$ and may be added to give

$$u_{\max}/u^* = \phi_2(y^*) + \phi_3(\mathrm{Re}^*) = \phi_4(\mathrm{Re}), \tag{10.45}$$

where $y^* = y/R$. Differentiating (10.45) with respect to Re yields

$$y^* \phi_3'(\mathrm{Re}^*) = \phi_4'(\mathrm{Re}), \tag{10.46}$$

as $\mathrm{Re}^* = \mathrm{Re}\, y^*$. Since $\phi_4'(\mathrm{Re})$ is independent of y^*, so then is $y^* \phi_3'(\mathrm{Re}^*)$, so that $\phi_3'(\mathrm{Re}^*)$ has the form

$$(1/y^*)\phi(\mathrm{Re}). \tag{10.47}$$

Similarly, differentiating equation (10.45) with respect to y^* yields

$$\phi_2'(y^*) + \mathrm{Re}\, \phi_3'(\mathrm{Re}^*) = 0, \tag{10.48}$$

and so $\phi_3'(\mathrm{Re}^*)$ has the form

$$(1/\mathrm{Re})\phi(y^*) \tag{10.49}$$

since $\phi_2'(y^*)$ is independent of Re.

In order for equations (10.49) and (10.47) to be satisfied simultaneously in the zone $y_1 < y < y_2$, it is necessary for $\phi(\mathrm{Re}) = \mathrm{constant}/\mathrm{Re}$ and $\phi(y^*) = \mathrm{constant}/y^*$; therefore, $\phi_3'(\mathrm{Re}^*) = A/\mathrm{Re}^*$, where A is a constant.

From (10.46),

$$\phi_4'(\mathrm{Re}) = y^* \phi_3'(\mathrm{Re}) = y^* A/\mathrm{Re}^* = A/\mathrm{Re}$$

or $\phi_4(\text{Re}) = A \log_e \text{Re} + \text{constant}.$ (10.50)

Similarly, from (10.48),

$$\phi_2'(y^*) = -\text{Re}\,\phi_3'(\text{Re}^*) = -\text{Re}\,A/\text{Re}^*$$
$$= -A/y^*$$

or, by integration,

$$\phi_2(y^*) = -A \log_e y^* + \text{constant}.$$ (10.51)

However, from equations (10.39), (10.45)

$$u/u^* = \phi_4(\text{Re}) - \phi_2(y^*).$$

Thus, $u/u^* = (A \log_e \text{Re} + \text{constant}) + (A \log_e y^* + \text{constant})$,

$$u/u^* = A \log_e \text{Re}^* + A_1,$$ (10.52)

where A and A_1 are constants to be determined experimentally.

Equation (10.52) is known as the *universal velocity distribution*. However, it is to be noted that, due to the restriction placed on equations (10.39) and (10.45), the expression only applies in the central turbulent core of the pipeline, as shown in Fig. 10.8.

FIGURE 10.8

Zones of application of empirical relations defining velocity distribution in turbulent pipe flow

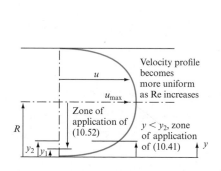
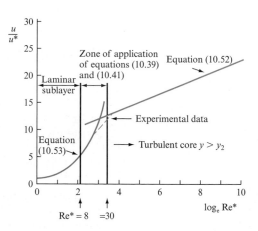

Close to the pipe wall, within the laminar sublayer, the effect of fluid viscosity is predominant and so the expression $\tau = \mu(du/dy)$ may be integrated to describe the velocity distribution in this zone:

$$u = \tau(y/\mu) + E,$$

where $E = 0$ when $u = 0$ at $y = 0$. Hence,

$$u = \tau y/\mu$$

or $u/\sqrt{(\tau/\rho)} = y\rho\tau^{1/2}/\mu\rho^{1/2},$

$$u/u^* = \text{Re}^*.$$ (10.53)

Equation (10.53) may be plotted, as shown in Fig. 10.8, and intersects the straight line relation of equation (10.52) at a point which, theoretically, defines the upper limit of the laminar sublayer. In practice, the upper limit of the laminar sublayer is ill defined, and experimental results tend to smooth the intersection (Fig. 10.8).

Nikuradse's results for smooth pipes show that the constants in equation (10.52) may be taken as

$$u/u^* = 2.5 \log_e \text{Re}^* + 5.5 = 5.75 \log_{10} \text{Re}^* + 5.5. \qquad (10.54)$$

As shown in Fig. 10.8, equation (10.53) applies accurately up to $\text{Re}^* \simeq 8$ and (10.52) applies from $\text{Re}^* \simeq 30$.

By employing equation (10.52), it is possible to relate the friction factor to mean flow velocity and flow Reynolds number. From equation (10.25),

$$f = 2\tau/\rho\bar{u}^2,$$

where $\bar{u}$ is the mean flow velocity. Mean velocity may be calculated by integration of equation (10.52) across the pipe, assuming the thickness of the laminar sublayer to be negligible, and dividing the result by the pipe cross-sectional area, as was done in the case of laminar flow (equation (10.20)). Substitution of friction factor for mean velocity through the relation above (equation (10.25)) yields an expression of the form

$$1/\sqrt{f} = F + G \log_e(\text{Re}\sqrt{f}), \qquad (10.55)$$

where $\text{Re} = \rho d\bar{u}/\mu$ and F, G are constants.

The constants in equation (10.55) may be obtained from experimental investigations, and Nikuradse's results for smooth pipes suggest an expression

$$1/\sqrt{f} = 4.07 \log_{10}(\text{Re}\sqrt{f}) - 0.6, \qquad (10.56)$$

although values of 4.0 and −0.4 for G and F do yield improved results. This expression has been verified for Reynolds numbers in the range $5000 < \text{Re} < 3 \times 10^6$. However, the Blasius expression for smooth pipes (equation (10.33)) gives reasonably accuracy for Re values up to 10^5.

Similar relations for velocity distribution and friction factor may be determined for rough pipes, the form of the expressions being deduced by dimensional analysis techniques and the algebraic format of the relations being obtained empirically by extensive testing.

The Moody chart (Fig. 10.7) showed that rough pipe turbulent flow falls into two regimes as far as friction factor calculation is concerned. First, a regime where friction factor is independent of Reynolds number and depends on surface roughness; second, a transitionary regime where the friction factor increases with Reynolds number for any particular pipe surface roughness, from the value appropriate for a smooth pipe up to the Reynolds-number-independent value mentioned. The mechanism responsible for this transition has already been explained in Section 10.4 in terms of the relation between roughness particle size and laminar sublayer thickness.

Nikuradse showed that, by describing the flow Reynolds number in terms of the roughness particle size then the three identifiable flow regimes could be described as follows:

1. smooth pipe f results apply for $\mathrm{Re} = \rho u^* k/\mu < 4$;
2. transition occurs for $4 < \mathrm{Re} < 70$;
3. f is independent of Re for $70 < \mathrm{Re}$.

For the f independent zone, the velocity profile may be expressed as

$$u/u^* = \phi(y/R, k/R) \tag{10.57}$$

and experimental results verify an equation of the form

$$u/u^* = 5.75 \log_{10}(y/k) + 8.48. \tag{10.58}$$

Integration to give mean velocity and the introduction of friction factor via equation (10.25) yields an expression similar to (10.55):

$$1/\sqrt{f} = 4 \log_{10}(d/k) + 2.28, \tag{10.59}$$

where d is the pipe diameter.

For the transition regime, $4 < \rho u^* k/\mu < 70$, an expression

$$1/\sqrt{f} = -4 \log_{10}(k/3.71d + 1.26/\mathrm{Re}\,\sqrt{f}), \tag{10.60}$$

where $\mathrm{Re} = \rho u d/\mu$, has been shown to be applicable and may be seen to converge to equation (10.59) or (10.56) for fully rough pipes or smooth pipes characterized by $\mathrm{Re} \to \infty$ or $k \to 0$. Equation (10.60) is known as the *Colebrook–White equation* and was employed by Moody in the preparation of the friction factor chart of Fig. 10.7 (see program CBW, Section 10.10).

As mentioned in the derivation of the laminar flow equations, the results only apply to fully developed pipe flow and so do not cover the entry length of a pipeline. However, as this is normally short in comparison to the pipe length, no appreciable error arises.

EXAMPLE 10.5

Assuming the following velocity distribution in a circular pipe

$$u = u_{\max}(1 - r/R)^{1/7},$$

where $u_{\max}$ is the maximum velocity, calculate (a) the ratio between the mean velocity and the maximum velocity, (b) the radius at which the actual velocity is equal to the mean velocity.

Solution

(a) The elementary discharge through an annulus dr is given by

$$dQ = 2\pi r u\, dr$$
$$= 2\pi u_{\max}(1 - r/R)^{1/7}\, dr,$$

and discharge through the pipe by

$$Q = 2\pi u_{max} \int_0^R r(1 - r/R)^{1/7} \, dr.$$

Let $1 - r/R = x$; then

$$\frac{dx}{dr} = -\frac{1}{R} \quad \text{and} \quad dr = -R \, dx$$

so that $R - r = xR$,   when $r = 0$, $x = 1$,

$r = R - xR = R(1 - x)$,   when $r = R$, $x = 0$.

Therefore, substituting,

$$Q = 2\pi u_{max} \int_1^0 R(1 - x)x^{1/7}(-R \, dx) = 2\pi R^2 u_{max} \int_0^1 (1 - x)x^{1/7} \, dx$$

$$= 2\pi R^2 u_{max} \left(\frac{7}{8}x^{8/7} - \frac{7}{15}x^{15/7} \right)_0^1$$

$$= 2\pi R^2 u_{max} \left(\frac{7}{8} - \frac{7}{15} \right) = 2\pi R^2 u_{max} \left(\frac{105 - 56}{120} \right) = \pi R^2 u_{max} \frac{49}{60},$$

and $\bar{u} = Q/\pi R^2 = \pi R^2 u_{max}\frac{49}{60}/\pi R^2 = \frac{49}{60}u_{max}$,

with the result that

$$\bar{u}/u_{max} = \mathbf{49/60}.$$

(b) $u = \bar{u} = 49u_{max}/60 = u_{max}(1 - r/R)^{1/7}$.

Therefore,

$$(49/60)^7 = 1 - r/R$$

and $r/R = 1 - (49/60)^7 = 1 - 0.242 = 0.758$.

Hence, $r = \mathbf{0.758}R$.

EXAMPLE 10.6

Assuming the universal velocity distribution for turbulent flow in a pipe,

$$\frac{u}{u^*} = 5.5 + 5.75 \log_{10} Re^*,$$

determine the radius at which the point velocity is equal to the mean velocity and the ratio of mean velocity to maximum velocity.

Solution

The point velocity is given by

$$u = u^*(5.5 + 5.75 \log_{10} \text{Re}^*),$$

but $u^* = \sqrt{(\tau_0/\rho)} = \text{constant and Re}^* = \rho y u^*/\mu.$

Let $\rho u^*/\mu = a$ and, for a pipe, let $y = r$.

Then, $u = u^*(5.5 + 5.75 \log_{10} ar).$

To obtain the mean velocity, it is necessary first to calculate the discharge which, when divided by the cross-sectional area of the pipe, will give the mean velocity. Thus, the elementary discharge through an annulus dr is given by

$$dQ = 2\pi r u \, dr = 2\pi u^*(5.5r + 5.75r \log ar) \, dr$$

and discharge through the pipe by

$$Q = 2\pi u^* \left(5.5 \int_0^R r \, dr + 5.75 \int_0^R r \log_{10} ar \, dr \right).$$

Now, $\displaystyle\int_0^R r \, dr = \left(\frac{r^2}{2} \right)_0^R = \frac{R^2}{2},$

and to obtain

$$\int_0^R r \log_{10} ar \, dr = \frac{1}{\log_e 10} \int_0^R r \log_e ar \, dr,$$

put $\log_e ar = y$. Hence,

$$dy = \frac{1}{ar} a \, dr = \frac{dr}{r}$$

and $r \, dr = dx.$

Therefore,

$$x = r^2/2.$$

Integrating now by parts,

$$\int y \, dr = yx - \int r \, dy.$$

Substituting

$$\int r \log_{10} ar \, dr = \frac{1}{2} r^2 \log_e ar - \frac{1}{2} \int r^2 \frac{1}{r} dr$$

$$= \tfrac{1}{2} r^2 \log_e ar - r^2/4$$

$$= (r^2/2)(\log_e ar - \tfrac{1}{2}).$$

Therefore,

$$\int_0^R r \log_{10} ar\, dr = \frac{1}{\log_e 10} \int_0^R r \log_e ar\, dr$$

$$= \frac{1}{\log_e 10} \left[\frac{r^2}{2} \left(\log_e ar - \frac{1}{2} \right) \right]_0^R$$

$$= \left[\frac{r^2}{2} \left(\log_{10} ar - \frac{1}{2 \log_e 10} \right) \right]_0^R$$

$$= \frac{R^2}{2} \left(\log_{10} aR - \frac{1}{2 \log_e 10} \right)$$

$$= \frac{R^2}{2} (\log_{10} aR - 0.217).$$

Substituting into the equation for Q,

$$Q = 2\pi u^* \left[5.5 \frac{R^2}{2} + 5.75 \frac{R^2}{2} (\log_{10} aR - 0.217) \right]$$

$$= \pi u^* R^2 (5.5 + 5.75 \log_{10} aR - 1.248)$$

$$= \pi u^* R^2 (4.252 + 5.75 \log_{10} aR).$$

Therefore, mean velocity in the pipe

$$\bar{u} = Q/\pi R^2 = u^* (4.252 + 5.75 \log_{10} aR).$$

The radius at which the point velocity is the same as the mean velocity is now obtained by equating the two expressions:

$$u^*(5.5 + 5.75 \log_{10} ar) = u^*(4.252 + 5.75 \log_{10} aR),$$

$$1.248 = 5.75(\log_{10} aR - \log_{10} ar).$$

$$0.217 = \log_{10} (R/r),$$

$$10^{0.217} = R/r,$$

so that, finally,

$$r = 0.607R.$$

The maximum velocity occurs at the centre of the pipe, where $r = 0$. Therefore,

$$u_{max} = 5.5u^*$$

and

$$\bar{u}/u_{max} = u^*(4.252 + 5.75 \log aR)/5.5u^*$$

$$= 0.773 + 1.045 \log_{10} aR$$

$$= \mathbf{0.773 + 1.045 \log_{10}(\rho u^*/\mu)R.}$$

10.7 VELOCITY DISTRIBUTION IN FULLY DEVELOPED, TURBULENT FLOW IN OPEN CHANNELS

In the use of the Chezy and related equations for open-channel flow, the assumption is made that the flow is uniform across the channel. This is, in practice, never achieved, and, further, due to the lack of symmetry in open-channel flow, the accepted central position of maximum flow velocity for pipe flow is also not reproduced. The actual velocity profiles across the flow are influenced by the presence of the channel solid boundaries and the free surface. Irregularities in the solid boundaries are generally so large and random that each channel has its own individual velocity distribution and there is no direct equivalent to the velocity distribution expressions derived for pipe flow. Figure 10.9 illustrates a typical velocity distribution, the maximum velocity occurring at some depth below the free surface, usually between 5 and 25 per cent of flow depth, and the mean velocity, which is usually some 80 to 85 per cent of the free surface velocity, occurs at about 60 per cent of the flow depth below the free surface.

FIGURE 10.9

Velocity distribution in a simple open channel under fully developed, turbulent flow conditions

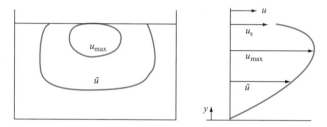

Normally, in open-channel calculations, the uncertainties involved in the flow parameters are so large as to render any variations of flow velocity away from the mean to be negligible and these are neglected.

10.8 SEPARATION LOSSES IN PIPE FLOW

Whenever the uniform cross-section of a pipeline is interrupted by the inclusion of a pipe fitting, such as a valve, bend, junction or flow measurement device, then a pressure loss will be incurred. The values of these losses, which are sometimes misleadingly referred to as 'minor losses', have to be included in a pipeline's total resistance if errors in pump and system matching or flow calculations for a given pressure differential are to be avoided. In this treatment, the term 'separation loss' has been chosen to define pressure losses across such fittings, as it is felt that this term describes well the physical phenomena which occur at such obstructions in the pipeline. Generally, the flow separates from the pipe walls as it passes through the obstructing pipe fitting, resulting in the generation of eddies in the flow, with consequent pressure loss, as shown in Fig. 10.10 for the case of a sudden enlargement. For small, complex pipe networks such as those found in some chemical process plants, aircraft fuel and hydraulic systems and in ventilation systems, the total effect

FIGURE 10.10

Separation loss in a sudden enlargement

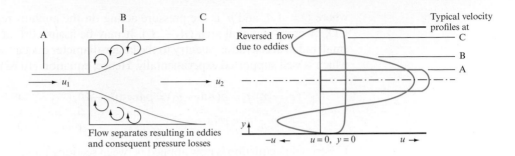

B   C

A

u_1 u_2

Reversed flow due to eddies

Typical velocity profiles at
C
B
A

Flow separates resulting in eddies and consequent pressure losses

$-u \leftarrow$ $u = 0, \ y = 0$ $u \rightarrow$

of separation losses may be the predominant factor in the system pressure loss calculation, exceeding the contribution of pipe friction at the design flow rate. Conversely, in large pipe systems, such as water distribution networks or overland oil pipelines, the losses due to pipe fitting may be negligible compared with the friction loss and may often be ignored.

10.8.1 Losses in sudden expansions and contractions

Generally, the losses due to pipe or duct fittings are determined experimentally. However, the case of a sudden expansion in a pipe or duct may be determined analytically. Figure 10.11 illustrates a sudden enlargement; consider a control volume ABCDEF as shown and let p_1 and p_2 be the pressures at sections 1 and 2, respectively, where the mean flow velocities are related by the continuity equation

FIGURE 10.11

Calculation of the loss coefficient for a sudden enlargement

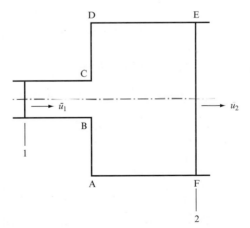

D   E

C

$\bar{u}_1$ u_2

B

1

A   F

2

$$A_1\bar{u}_1 = A_2\bar{u}_2 \qquad (10.61)$$

and represent the duct cross-sectional areas A_1, A_2.

By application of the momentum equation between sections 1 and 2 the following relation may be derived:

Resultant force in flow direction $=$ Rate of change of momentum in flow direction,

$$p_1A_1 + p'(A_2 - A_1) - p_2A_2 = \rho Q(\bar{u}_2 - \bar{u}_1), \qquad (10.62)$$

where $Q = A_2\bar{u}_2$ and p' is the pressure acting on the annulus represented by AB and CD, of cross-sectional area $(A_2 - A_1)$. It may be assumed that $p_1 = p'$, owing to the small radial acceleration at entry to the larger-diameter duct at section ABCD, a result which is well supported experimentally. Hence, equation (10.62) reduces to

$$(p_1 - p_2)A_2 = \rho Q(\bar{u}_2 - \bar{u}_1) = \rho\bar{u}_2 A_2(\bar{u}_2 - \bar{u}_1),$$

$$p_1 - p_2 = \rho\bar{u}_2(\bar{u}_2 - \bar{u}_1). \tag{10.63}$$

If Bernoulli's equation is now applied between sections 1 and 2, with a term h included to represent the separation loss, then an expression for the pressure differential $p_1 - p_2$ may be derived:

$$p_1/\rho g + \bar{u}_1^2/2g + Z_1 = p_2/\rho g + \bar{u}_2^2/2g + Z_2 + h,$$

where $Z_1 = Z_2$ if the enlargement is situated in a horizontal pipe or duct; thus,

$$h = (p_1 - p_2)/\rho g + (\bar{u}_1^2 - \bar{u}_2^2)/2g$$

and, substituting for $p_1 - p_2$ from equation (10.63)

$$h = \frac{\rho\bar{u}_2(\bar{u}_2 - \bar{u}_1)}{\rho g} + \frac{\bar{u}_1^2 - \bar{u}_2^2}{2g} = \frac{1}{2g}(2\bar{u}_2^2 - 2\bar{u}_2\bar{u}_1 + \bar{u}_1^2 - \bar{u}_2^2)$$

$$= \frac{1}{2g}(\bar{u}_1^2 - 2\bar{u}_1\bar{u}_2 + \bar{u}_2^2) = \frac{(\bar{u}_1 - \bar{u}_2)^2}{2g}.$$

Thus, the loss due to sudden enlargement is given by

$$h = (\bar{u}_1 - \bar{u}_2)^2/2g. \tag{10.64}$$

Alternatively, since, from equation (10.61)

$$\bar{u}_2 = \bar{u}_1(A_1/A_2),$$

$$h = (\bar{u}_1^2/2g)(1 - A_1/A_2)^2 = \frac{\bar{u}_2^2}{2g}\left(\frac{A_2}{A_1} - 1\right)^2. \tag{10.65}$$

This expression is sometimes referred to as the Borda–Carnot relationship and is usually within a few per cent of the experimental result for the separation loss incurred by sudden enlargement in coaxial pipelines.

The loss at exit from a pipe into a reservoir may be obtained by considering equation (10.65). It will be seen that as $A_2 \to \infty$, so $\bar{u}_2 \to 0$ and $h \to \bar{u}_1^2/2g$, i.e. the kinetic energy of the approaching flow. This case is obviously representative of a pipe discharging into a large tank or a duct discharging to atmosphere and is the accepted expression for conduit exit loss.

Sudden contractions in a duct or pipe may also be dealt with in this way, provided that there is little or no loss between the upstream large-section conduit and the vena contracta formed within the smaller conduit just downstream of the junction, as shown in Fig. 10.12.

FIGURE 10.12

Sudden contraction. Loss approximated by consideration of sudden enlargement between the vena contracta and section 2

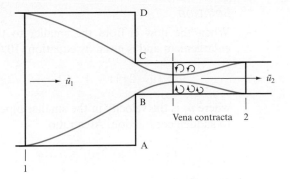

It is not possible to apply the momentum equation between sections 1 and 2 in Fig. 10.12, owing to the uncertain pressure distribution across the face ABCD. However, it has been shown experimentally that the majority of the pressure loss occurs as a result of the eddies formed as the flow area expands from the vena contracta area up to the full cross-section of the downstream pipe. If the area of the vena contracta is A_c then accurate results may be achieved by applying the sudden enlargement expression between A_c and A_2 at section 2; thus

$$h \simeq \frac{\bar{u}_2^2}{2g}\left(\frac{A_2}{A_c}-1\right)^2 = \frac{\bar{u}_2^2}{2g}\left(\frac{1}{C_c}-1\right)^2, \tag{10.66}$$

where C_c is the coefficient of contraction for the junction based on the smaller-pipe entry diameter BC. The above equation indicates, since the expression in brackets is constant for any given area ratio, that it may be generalized into the form

$$h = K\bar{u}_2^2/2g, \tag{10.67}$$

where K is known as the loss coefficient. Table 10.1 shows some experimental values of C_c and the corresponding values of K obtained with sharp pipe edges.

TABLE 10.1

Loss coefficients for sudden contraction

A_2/A_1	0.1	0.3	0.5	0.7	1.0
C_c	0.61	0.632	0.673	0.73	1.0
K	0.41	0.34	0.24	0.14	0

EXAMPLE 10.7

In a water pipeline there is an abrupt change in diameter from 140 mm to 250 mm. If the head lost due to separation when the flow is from the smaller to the larger pipe is 0.6 m greater than the head lost when the same flow is reversed, determine the flow rate.

Solution

When the flow is from the smaller to the larger pipe the loss is due to sudden enlargement and is given by equation (10.65):

$$h = (\bar{u}_1^2/2g)(1 - A_1/A_2)^2 = (\bar{u}_1^2/2g)(1 - 0.314)^2 = 0.47\,\bar{u}_1^2/2g,$$

where $\bar{u}_1$ is the velocity in the smaller pipe. When the flow is reversed, the loss is due to sudden contraction. Area ratio,

$$A_2/A_1 = (140/250)^2 = 0.314.$$

From Table 10.1,

$$K = 0.33 \text{ (say).}$$

Therefore, the loss,

$$h' = 0.33\,\bar{u}_2^2/2g,$$

where $\bar{u}_2$ is again the velocity in the smaller pipe. Since the flow rate is the same in both cases, then

$$\bar{u}_1 = \bar{u}_2 = \bar{u}$$

and $h - h' = 0.6,$

so that $(0.47 - 0.33)(\bar{u}^2/2g) = 0.6$

and $\bar{u} = 9.17 \text{ m s}^{-1}.$

10.8.2 Losses in pipe fittings, bends and at pipe entry

Losses in pipe fittings are usually expressed in the form already suggested for the loss at sudden contraction, namely

$$h = K(\bar{u}^2/2g),$$

where K is the fitting loss coefficient. It is a non-dimensional constant and its value is obtained experimentally for any pipe fitting. Table 10.2 sets out some typical values. The major advantage of expressing losses due to separation in the above form is that it can easily be incorporated into the steady flow energy equation, Section 6.4, as will be shown in Chapter 14.

Figure 10.13 illustrates the flow in a pipe bend, demonstrating the area of flow separation which results in the loss coefficients for bends listed in Table 10.2. As the bend becomes sharper, so the areas of separation become more extensive and the loss coefficient increases. Losses due to flow control devices are also illustrated in Figure 10.3.

Loss at entry to a pipe from a reservoir is a special case of sudden contraction, in which the velocity in the reservoir is considered to be zero. Owing to the fact that the fluid enters the pipe from all directions, a vena contracta is formed downstream of the pipe inlet and, consequently, the loss is associated with enlargement from the vena contracta to the full-bore pipe. This is the same situation as in the case of sudden contraction.

TABLE 10.2
Head loss coefficients
for a range of pipe
fittings

FITTING	LOSS COEFFICIENT K
Gate valve (open to 75 per cent shut)	$0.25 \rightarrow 25$
Globe valve	10
Spherical plug valve (fully open)	0.1
Pump foot valve	1.5
Return bend	2.2
90° elbow	0.9
45° elbow	0.4
Large-radius 90° bend	0.6
Tee junction	1.8
Sharp pipe entry	0.5
Radiused pipe entry	$\rightarrow 0.0$
Sharp pipe exit	0.5

FIGURE 10.13
Separation at bends and
valves

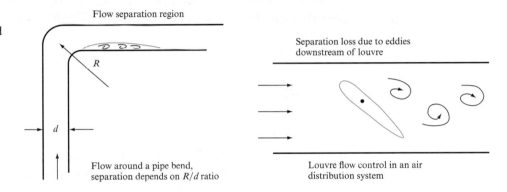

Flow separation region

R

d

Flow around a pipe bend,
separation depends on R/d ratio

Separation loss due to eddies
downstream of louvre

Louvre flow control in an air
distribution system

Figure 10.14 illustrates various types of pipe entry conditions. The results tabulated in Table 10.2 and indicated in Fig. 10.14 may be explained by reference to the flow separation at entry to the pipe explained above, so that the sharper the entry corner, the smaller is the vena contracta, and, hence, the greater the flow separation and the higher the value of K.

FIGURE 10.14
Pipe entry losses

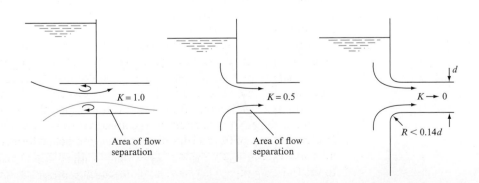

$K = 1.0$

Area of flow
separation

$K = 0.5$

Area of flow
separation

d

$K \rightarrow 0$

$R < 0.14d$

10.8.3 Equivalent length for pipe fitting loss calculations

Separation loss coefficients K may also be defined in terms of an equivalent length of straight pipe, of the same diameter as that including the fitting, that would result in the same frictional loss as that incurred by flow separation through the fitting. This is justified by consideration of the Darcy equation and equation (10.67),

$$h_f = 4fl_e\bar{u}^2/d2g = K\bar{u}^2/2g,$$

where l_e is the equivalent length of pipe, diameter d, that would yield a friction loss equivalent to the particular fitting. Thus,

$$l_e = Kd/4f \qquad\qquad (10.68)$$

and so l_e is normally calculated as a number of pipe diameters.

l_e may be the equivalent length for a single fitting or the summation of all the separation loss coefficients for a particular system. Hence, for the total pressure loss through a pipeline of length l and diameter d, the expressiom

$$h_f = 4f(l + l_e)\bar{u}^2/2dg \qquad\qquad (10.69)$$

may be employed.

10.8.4 Diffusers

In order to avoid the head losses incurred by the installation of sudden enlargements into pipe and duct flow, diffusers are commonly employed. Figure 10.15 illustrates a

FIGURE 10.15

Loss of pressure in a conical diffuser

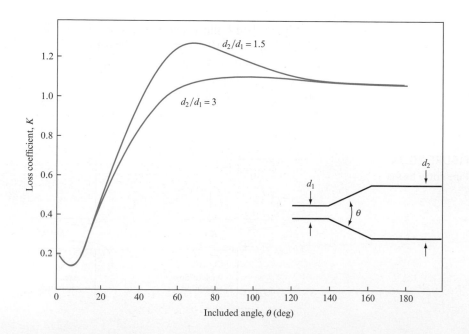

typical conical diffuser and the variation in pressure loss across it with diffuser-included angle.

The loss experienced depends on the area ratio between which the diffuser operates and the included angle of the diffuser θ. The total loss across the diffuser is made up of two components, the first due to fluid friction along the length of the diffuser, which, therefore, increases as θ decreases for a given area ratio, i.e. the diffuser length l increases as θ decreases and results in an increase in friction loss. The second contribution to the total loss is dependent on θ and is the separation loss, which increases with increasing included angle for a given area ratio, reaching a maximum when the diffuser approaches a sudden enlargement.

The minimum loss for any particular area ratio will, therefore, be a compromise, where the angle of the diffuser is sufficiently small to limit separation, or flow eddy, losses, but not so small as to increase the length of the diffuser to the point where the frictional losses become predominant. Normally 6° to 7° included angle is the minimum acceptable.

Diffusers are found in a wide range of applications where it is necessary to reduce the flow velocity by means of an area change, without undue pressure loss. For example, wind tunnel return circuits are at one end of the size spectrum and venturi meter discharge diffusers at the other.

10.9 SIGNIFICANCE OF THE COLEBROOK–WHITE EQUATION IN PIPE AND DUCT DESIGN

While the friction and separation loss equations already defined are essential to determine the values of these terms in a fluid network, the designer is often faced with the need to link delivered flow, whether liquid or gas, to the overall 'cost' in terms of the pressure loss to be overcome by a fan or pump, or by gravity if the terrain allows that option.

Thus an expression linking flow rate Q to pressure loss per unit length of system $\Delta p/L$ and incorporating the roughness of the pipe material is required. It will be appreciated that use of the equivalent length approach to represent separation losses is particularly attractive within this methodology.

By combining the Darcy equation for the pressure loss due to friction with the Colebrook–White friction factor relationship it is possible to eliminate the friction factor and arrive at a relationship linking, for any given fluid in a given duct or pipe, the flow rate to the conduit diameter and pressure loss per unit length. This technique was of considerable interest to network designers prior to the advent of readily available computer models. While the methodology remains of interest it should be seen as the basis for a computerized approach; clearly the design curves to be demonstrated could be made available within a program.

From the Darcy equation

$$\frac{\Delta p}{L} = \frac{\rho f v^2}{2m},$$

where $m = D/4$ for circular cross-section conduits and $v = Q/A$, A being the cross-sectional area of the conduit. Thus

$$\frac{\Delta p}{L} = \frac{4\rho f Q^2}{2D(\pi D^2/4)^2}$$

$$\frac{\Delta p}{L} = \frac{64\rho f Q^2}{2\pi^2 D^5}.$$

Solving for the friction factor f, yields

$$f = \frac{2D^5 \pi^2 \Delta p/L}{64\rho Q^2}.$$

From the definition of Reynolds number, for a circular cross-section conduit,

$$\mathrm{Re} = \rho V D/\mu = \rho Q D/\mu A = 4\rho Q/\mu \pi D.$$

(Note that a similar approach could be followed for any cross-sectional shape based upon the hydraulic mean depth m.)

Substituting for both f and Re in the Colebrook–White equation yields the required relationship between Q, $\Delta p/L$ and the conduit surface roughness:

$$\frac{1}{\sqrt{f}} = -4\log\left(\frac{k}{3.71D} + \frac{1.255}{\mathrm{Re}\sqrt{f}}\right)$$

$$\frac{Q}{\sqrt{[(2D^5\pi^2\Delta p/L)/64\rho]}}$$

$$= -4\log\left\{\frac{k}{3.71D} + \frac{1.255}{(4\rho Q/\mu\pi D)\sqrt{[(2D^5\pi^2\Delta p/L)/64\rho Q^2]}}\right\}. \qquad (10.70)$$

For a given fluid, i.e. density and viscosity constant and assuming a circular cross-section conduit, this expression reduces to

$$Q = -C_1\sqrt{[(\Delta p/L)D^5]}\log\left\{\frac{k}{3.71D} + C_2\frac{D}{\sqrt{[(\Delta p/L)D^5]}}\right\}. \qquad (10.71)$$

This expression may be solved for any circular cross-section conduit or may be presented in graphical form as illustrated in Fig. 10.16.

The curves illustrate a number of fundamental relationships that govern pressure loss in duct and pipe flow:

1. For a given flow rate the pressure loss rises as the conduit diameter reduces. The Darcy equation has already indicated that this is a fifth-power relationship.

FIGURE 10.16
Schematic of the circular duct sizing charts for air at 20 °C that may be constructed from the Colebrook–White loss equations

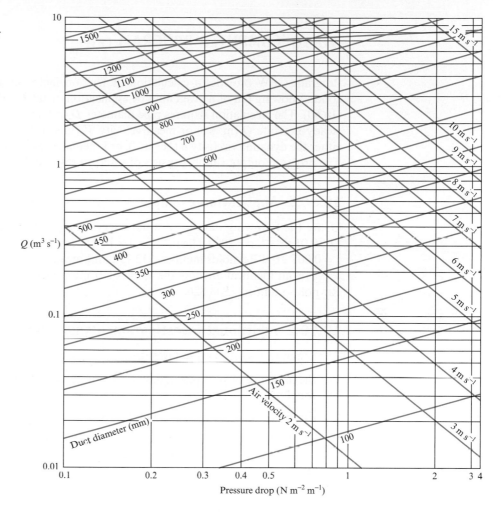

2. It is possible to retain the same flow velocity as flow rate rises by increasing the conduit diameter. The schematic chart in Fig. 10.16 illustrates this option, which might be important if acoustic considerations are important in the design of air ducting.

The inclusion of fittings in a pipe or duct network can be the main source of pressure loss. In order to represent such separation losses the addition of the fitting equivalent length to the length of the conduit allows the overall pressure loss to be calculated from the chart. It is stressed that this technique would in all probability now form the basis for computer-aided design calculations, but it does serve to illustrate some of the fundamental principles involved.

10.10 COMPUTER PROGRAM CBW

Program CBW provides a calculation of friction factor in a circular cross-section conduit flowing full of a given fluid by application of the Colebrook–White equation (10.60). The calculation requires pipe diameter D (m), pipe roughness k (mm), either steady flowrate Q (m³ s⁻¹) or mean velocity V (m s⁻¹), fluid density ρ (kg m⁻³) and either the dynamic viscosity μ (N s m⁻²), or kinematic viscosity υ (m² s⁻¹), for the fluid. The program output simply consists of a screen-displayed value for friction factor f.

10.10.1 Application example

Using the following data for a water carrying pipeline: $D = 0.1$ m; $k = 0.15$ mm; $Q = 0.01$ m³ s⁻¹; $\rho = 1000.0$ kg m⁻³; $\mu = 1 \times 10^{-3}$ N s m⁻², yields a value of friction factor $f = 0.007\ 91$ at Reynolds number $= 12\ 732$ and mean velocity $V = 1.273$ m s⁻¹.

For an identical air flow the value of friction factor becomes $f = 0.005\ 566$ at Reynolds number $= 848\ 826$.

10.10.2 Additional investigations using CBW

The program may be used to investigate:

1. the influence of pipe properties on friction factor, including diameter and roughness;
2. the effect of fluid properties on Reynolds number and friction factor;
3. the accuracy of Colebrook–White in the laminar region.

Concluding remarks

Chapter 10 has drawn upon earlier material both to differentiate between laminar and turbulent flows and to investigate the dependence of flow frictional losses on conduit and fluid properties. While the laminar flow study indicated that it was possible to develop theoretical expressions for both frictional losses and the velocity profiles within a bounded fluid stream, the equivalent expressions for turbulent flow require an empirical friction factor, defined either graphically by the Moody chart or by the Colebrook–White equation. The Darcy equation for frictional loss was developed and shown to be the equivalent of the Chezy expression for open channels. This understanding that the fundamental equations apply across the boundary between full-bore and free surface flow, and are independent of conduit cross-section so long as the flow remains uniform, is important and will be shown later to apply to both steady flows (Chapters 14 and 15) and unsteady flows (Chapters 20 and 21).

Frictional losses have been shown not to be the sole cause of resistance to flow in conduits. The concept of separation loss was introduced and shown to be dependent upon flow kinetic energy. The introduction of separation losses will enable a later treatment of flow balancing within networks (Chapter 14) and will also be important in the study of unsteady flow, particularly the effects of valve closure, in Chapter 20. The definition of separation losses in terms of the equivalent pipe length that would be necessary to generate the same effect via frictional loss was introduced and shown to be particularly helpful in system design via the Colebrook–White-based charts introduced in this chapter.

A treatment of velocity profiles within fully developed pipe flow, and similarly within free surface flows, introduced the concept of the boundary layer and further differentiated the flow regimes present that will be referred to in later material. The definitions of frictional and separation loss introduced in this chapter will be utilized throughout the remainder of the text.

Summary of important equations and concepts

1. This chapter emphasizes the differences between laminar and turbulent flow regimes, deriving the Hagen–Poiseuille equation (10.21) for laminar flows and contrasting this to the empirical approach necessary in turbulent flows, e.g. the Darcy and Chezy equations, (10.30) and (10.29), respectively, relying on the Colebrook–White expression for friction factor, equation (10.60).

2. The comparison of the open-channel Chezy and full-bore flow Darcy equations is important, relying upon the definition of hydraulic mean depth, i.e. (cross-sectional area A)/(wetted perimeter P).

3. While the velocity profile in laminar flow is derived, equations (10.5) and (10.20), velocity profiles in turbulent flow are dependent upon empirical results. Typical velocity distributions for full-bore flows and free surface channel flows are presented, in the latter case the observation that the free surface has a centre line velocity less than the maximum for the profile is important. The application of the 'no-slip' condition is stressed.

4. The expression of fitting separation losses in a form compatible with frictional losses, is essential to the easy application of the steady flow energy equation, Section 6.4.

Further reading

CIBSE (1986). *Guide to Current Practice*, Vol. C, Chartered Institution of Building Services Engineers.

Colebrook, C. F. (1939). Turbulent flow in pipes with particular reference to the transition region between smooth and rough pipe laws. *JICE*, **10**.

Jepson, R. W. (1976). *Analysis of Flow in Pipe Networks*. Ann Arbor Science, Ann Arbor.

Moody, L. F. (1944). Friction factors for pipe flow. *Trans. ASME*, **66**.

Ward-Smith, A. J. (1980). *International Fluid Flow: The Fluid Dynamics of Flow in Pipes and Ducts*. Oxford University Press.

Problems

10.1 Show that, for laminar flow between two, infinite, moving flate plates, distance z apart, the flow rate Q is given by an expression of the form

$$Q = -\frac{1}{\mu}\frac{dp}{dl}\frac{z^3}{12} + \frac{z}{2}(U+V),$$

where μ is the fluid viscosity, dp/dl is the pressure gradient in the flow direction and U and V are the absolute velocities of the two plates.

10.2 For the case set out in Problem 10.1 above derive the shear stress expressions for each plate surface:

$$\tau_0 = \mu\left(\frac{U'}{z} - \frac{z}{2\mu}\frac{dp}{dl}\right),$$

$$\tau_z = \mu\left(\frac{U'}{z} + \frac{z}{2\mu}\frac{dp}{dl}\right), \quad \text{where } U' = (U+V).$$

10.3 A thin film of oil, thickness z and viscosity μ, flows down an inclined plate. Show that the velocity profile is given by

$$u = \frac{\rho g}{2\mu}(z^2 - y^2)\sin\theta,$$

where u is the local velocity at a depth y below the free surface, θ is the plate inclination to the horizontal and ρ is the fluid density.

10.4 For the case in Problem 10.3 above calculate the flow rate per unit plate width if the fluid has a viscosity of 0.9 N s m^{-2}, a density of 1260 kg m^{-3}, the plate is inclined at 30° and the depth of flow is 10 mm.
[0.137 litres min^{-1} m^{-1}]

10.5 A film of fluid, density 2400 kg m^{-3}, flows down a vertical plate with a free surface velocity of 0.75 m s^{-1}. If the film is 20 mm thick determine the fluid viscosity.
[6.28 N s m^{-2}]

10.6 Fluid of density 1260 kg m^{-3} and viscosity 0.9 N s m^{-2} passes between two infinite parallel plates, 2 cm separation. If the flow rate is 0.5 litres s^{-1} per unit width calculate the pressure drop per unit length if both plates are stationary.
[0.68 kN m^{-2} m^{-1}]

10.7 The radial clearance between a hydraulic plunger and the cylinder wall is 0.15 mm, the length of the plunger 0.25 m and the diameter 150 mm. Calculate the leakage rate past the plunger at an instant when the pressure differential between the two ends of the plunger is 15 m of water. Viscosity of hydraulic fluids is 0.9 N s m^{-2}.
[5.2 × 10^{-3} litres min^{-1}]

10.8 For laminar flow in a tube calculate the position of the average cross-sectional velocity.
[0.293 × radius from tube wall]

10.9 An oil having a viscosity of 0.048 kg m^{-1} s^{-1} flows through a 50 mm diameter tube at an average velocity of 0.12 m s^{-1}. Calculate the pressure drop in 65 m of tube and the velocity 10 mm from the tube wall.
[4.8 kN m^{-2}, 0.154 m s^{-1}]

10.10 Oil of specific gravity 0.9 and kinematic viscosity 0.000 33 m^2 s^{-1} is pumped over a distance of 1.5 km through a 75 mm diameter tube at a rate of 25 × 10^3 kg h^{-1}. Determine whether the flow is laminar and calculate the pumping power required, assuming 70 per cent mechanical efficiency.
[Re = 397, laminar, 48.8 kW]

10.11 For the flow conditions set out in Problem 10.10 above calculate the shear stress at the tube walls.
[55.3 N m^{-2}]

10.12 Air at 20 °C is drawn through a 0.5 m diameter duct by a fan. If the volume flow rate is 4.5 m^3 s^{-1} and the duct is 12 m long, with a friction factor of 0.005, determine the fan shaft power necessary, assuming 80 per cent mechanical efficiency. Take air density as 1.2 kg m^{-3} and viscosity as 1.8 × 10^{-5} N s m^{-2}.
[0.85 kW]

10.13 Water at a density of 998 kg m^{-3} and kinematic viscosity 1 × 10^{-6} m^2 s^{-1} flows through smooth tubing at a mean velocity of 2 m s^{-1}. If the tube diameter is 30 mm calculate the pressure gradient per unit length necessary. Assume that the friction factor for a smooth pipe is given by 16/Re for laminar flow and 0.079/Re$^{1/4}$ for turbulent flow.
[1.34 kN m^{-2} m^{-1}]

10.14 In a laboratory the water supply is drawn from a roof storage tank 25 m above the water discharge point. If the friction factor is 0.008, the pipe diameter is 5 cm and the pipe is assumed vertical, calculate the maximum volume flow achievable, if separation losses are ignored.
[0.01 m^3 s^{-1}]

10.15 For the case set out in Problem 10.14 above calculate the relative roughness of the pipe used, if the water is at 0 °C.
[0.006]

10.16 The friction factor applicable to turbulent flow in a smooth glass pipe is given by $f = 0.079/\text{Re}^{1/4}$. Calculate the pressure loss per unit length necessary to maintain a flow of 0.02 m^3 s^{-1} of kerosene, specific gravity 0.82, viscosity 1.9 × 10^{-3} N s m^{-2}, in a glass pipe of 8 cm diameter. If the tube is replaced by a galvanized steel pipeline, wall roughness

0.15 mm, calculate the increase in pipe diameter to handle this flow with the same pressure gradient.

[1332 N m^{-2} m^{-1}, 6.75 per cent]

10.17   Define static regain along a diffuser and show that it may be calculated as

$$\text{Static regain} = \tfrac{1}{2}\rho(V_1^2 - V_2^2) - \tfrac{1}{2}K\rho V_1^2.$$

10.18   A 150 mm diameter pipe reduces in diameter abruptly to 100 mm. If the pipe carries water at 30 litres s^{-1} calculate the pressure loss across the contraction and express this as a percentage of the loss to be expected if the flow was reversed. Take the coefficient of contraction as 0.6.

[3.2 kN m^{-2}, 143 per cent]

10.19   An air duct, carrying a volume Q of air per second, is abruptly changed in section. Deduce the diameter ratio for the two duct sections if the pressure loss is to be independent of flow direction. Assume a value of 0.6 for the contraction coefficient.

[1.732]

11

Boundary Layer

AT THE INTERFACE BETWEEN A FLUID AND A SURFACE IN RELATIVE MOTION A CONDITION
known as 'no slip' dictates an equivalence between fluid and surface velocities. Away
from the surface the fluid velocity rapidly increases; the zone in which this occurs is
known as the boundary layer and its definition is fundamental to all calculations
of surface drag and viscous forces. This chapter will present both a qualitative
and quantitative treatment of the development of boundary layers, both laminar and
turbulent, including the laminar sublayer, and will introduce the velocity profiles
appropriate to each. The effect of the boundary layer on the velocity profiles already
discussed in terms of bounded flows, together with definitions of the boundary layer
in terms of its physical thickness and its effect upon the flow, quantified in terms of
displacement and momentum thickness, are presented. The dependence of boundary
layer effects on shear stress, Reynolds number and surface roughness will be discussed
and the application of the momentum equation in the determination of skin friction
demonstrated. The sensitivity of a boundary layer to pressure gradients imposed upon
the flow will be discussed, with particular reference to flow separation over aerofoil
sections and the loss of lift. ● ● ●

The drag on a body passing through a fluid may be considered to be made up of two components, the *form drag*, which is dependent on the pressure forces acting on the body, and the *skin friction drag*, which depends on the shearing forces acting between the body and the fluid. Form drag will be dealt with in detail in Chapter 12, while the mechanics of skin friction will be covered in this chapter.

In Chapter 10, it was shown that in both laminar and turbulent flow in pipes the fluid velocity is not uniform but varies from zero at the wall to a maximum at the pipe centre. It was further shown that, in general, the velocity distribution is dependent upon the Reynolds number, which defines the type of flow. This chapter will be concerned with the analysis of the effects the fluid viscosity has on the velocity gradient near a solid boundary and, hence, how it affects the skin friction. Such analysis is most conveniently carried out by the consideration of flow over a flat plate of infinite width.

The shear stress on a smooth plate is a direct function of the velocity gradient at the surface of the plate. That a velocity gradient should exist in a direction perpendicular to the surface is evident, because the particles of fluid adjacent to the surface are stationary whilst those some distance above the surface move with some velocity. The condition of zero fluid velocity at the solid surface is referred to as 'no slip' and the layer of fluid between the surface and the free stream fluid is termed the boundary layer. Thus, it will be appreciated that any calculations of surface resistance or skin friction forces will obviously involve the integration of the shear stress at the surface over the whole fluid immersed area and will be directly concerned with the patterns of flow within the boundary layer.

Within this context, the importance of Reynolds number becomes self-evident as, with the dramatic change in particle motion consequent upon a transition from a laminar to a turbulent type of flow, considerable changes in boundary layer flow patterns and velocity gradients must be expected that will materially affect any calculations of surface resistance.

11.1 QUALITATIVE DESCRIPTION OF THE BOUNDARY LAYER

As mentioned above, the boundary layer is taken as that region of fluid close to the surface immersed in the flowing fluid. Figure 11.1 illustrates such a flat plate in a free fluid stream. Only the top surface boundary layer is shown but there will, in practice, be symmetry between the upper and lower surface boundary layers, provided both surfaces are identical in nature. The fluid in contact with the plate surface has zero velocity, 'no slip' and a velocity gradient exists between the fluid in the free stream and the plate surface. Now, shear stress may be defined (equation (10.3)) as

$$\tau = \mu \frac{\partial u}{\partial y}, \tag{11.1}$$

where τ is the shear stress, μ the fluid viscosity and du/dy the velocity gradient.

This shear stress acting at the plate surface sets up a shear force which opposes the fluid motion and fluid close to the wall is decelerated. Further along the plate, the shear force is effectively increased owing to the increasing plate surface area affected,

FIGURE 11.1

Development of the boundary layer along a flat plate, illustrating variations in layer thickness and wall shear stress

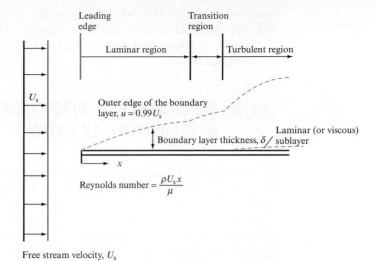

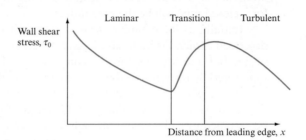

so that more and more of the fluid is retarded and the thickness of the fluid layer affected increases, as shown in Fig. 11.1. Returning to the Reynolds number concept, if the Reynolds number locally were based on the distance from the leading edge of the plate, then it will be appreciated that, initially, the value is low, so that the fluid flow close to the wall may be categorized as laminar. However, as the distance from the leading edge increases so does Reynolds number, until a point must be reached where the flow regime becomes turbulent.

For smooth, polished plates the transition may be delayed until Re equals 500 000. However, for rough plates or for turbulent approach flows, transition may occur at much lower values. Again, the transition does not occur in practice at one well-defined point but, rather, a transition zone is established between the two flow regimes, as shown in Fig. 11.1.

The random particle motion characterizing turbulent flow results in a far more rapid growth of the boundary layer in the turbulent region, so that the velocity gradient at the wall increases as does the corresponding shear force opposing motion.

Figure 11.1 also depicts the distribution of shear stress along the plate in the flow direction. At the leading edge, the velocity gradient is large, resulting in a high shear stress. However, as the laminar region progresses, so the velocity gradient and shear stress decrease with thickening of the boundary layer. Following transition the velocity gradient again increases and the shear stress rises.

Theoretically, for an infinite plate, the boundary layer goes on thickening indefinitely. However, in practice, the growth is curtailed by other surfaces in the vicinity. This is particularly the case for boundary layers within ducts, as will be described in Section 11.2, where the growth is terminated when the boundary layers from opposite duct surfaces meet on the duct centreline.

It must be appreciated that the thickness of the boundary layer, δ in Fig. 11.1, is much smaller than x.

11.2 DEPENDENCE OF PIPE FLOW ON BOUNDARY LAYER DEVELOPMENT AT ENTRY

The types of fluid flow considered in Chapter 10 have been confined to steady and uniform flow. The assumption of steady, uniform flow conditions led to the simplifying condition that the fluid elements were under no acceleration, either spatial or temporal, and allowed the development of the equations set out there. This assumption implies that the flow conditions are fully established and are not subject to any changes. This is, obviously, not the case in the initial length of a pipe where the boundary layer is still developing and growing in thickness up to its maximum, which for a closed pipe will be the pipe radius.

Initially, as the boundary layer develops, it will be laminar in form. However, as described earlier, the boundary layer will become turbulent, depending upon the ratio of inertial and viscous forces acting on the fluid, this condition being normally monitored by reference to the value of the flow Reynolds number.

For pipe flow, it is normal practice to base the Reynolds number, Re, on the mean flow velocity and the pipe diameter. Generally, for values of Re < 2000, the flow may be assumed to be laminar, although it has been shown possible to maintain laminar flow at higher values of Reynolds number under specialized laboratory conditions. Above Re = 2000 it is, however, reasonable to suppose that the flow will be turbulent and that the boundary layer development will include a transition and a turbulent region, as described for the flat plate. The only major difference is that, in the pipe flow case, there is a limit to the growth of the boundary layer thickness, namely the pipe radius.

If, therefore, this limit is reached before transition occurs, i.e. if laminar boundary layers meet at the pipe centre, the flow in the remainder of the pipe will be laminar. On the other hand, if transition within the boundary layer occurs before they fill the pipe, the flow in the rest of the pipe will be turbulent. These two cases are illustrated in Fig. 11.2.

Once the boundary layer, whether laminar or turbulent in nature, has grown to fill the whole pipe cross-section, the flow may be said to be fully developed and no further changes in velocity profile are to be expected downstream, provided that the pipeline characteristics (i.e. diameter, surface roughness) remain constant.

Theoretically, the entry length for a particular pipe (i.e. the distance from entry at which a laminar or turbulent boundary layer ceases to grow) is infinite. However, it is normally assumed that the flow has become fully developed when the maximum velocity, at the pipe centreline, becomes 0.99 of the theoretical maximum. Using this approximation, typical entry lengths for the establishment of fully developed laminar or turbulent flow may be taken as 120 and 60 pipe diameters, respectively. The entry length characteristic of turbulent flow is the shorter owing to the higher growth rate of the turbulent boundary layer.

Thus, the assumption of steady, uniform flow restricts the application of the equations derived for pipe flow to that part of a conduit beyond the entry length.

FIGURE 11.2

Development of fully
developed laminar
and turbulent flow
in a circular pipe.
(a) Laminar flow
conditions, Re < 2000.
(b) Turbulent flow
conditions, Re > 2000.

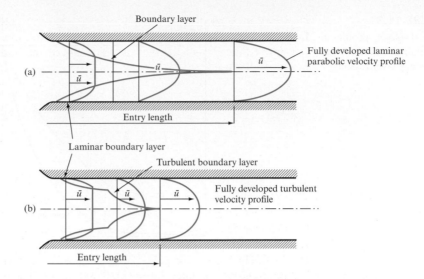

Normally, this is not a serious restriction as the entry length is usually small compared
with the total length of the pipeline.

Similarly, all the equations derived for laminar flow have depended on the shear
stress–viscosity relation of equation (11.1), $\tau = \mu \, du/dy$. In turbulent flow, owing to the
random nature of the motion of the fluid particles, the apparent shear stress may be
expressed as

$$\tau = (\mu + \varepsilon)\frac{du}{dy},$$ (11.2)

where ε is the *eddy viscosity* and is often much larger than μ. Since eddy viscosity is
difficult to determine, equations dealing with the calculations of pressure loss
associated with turbulent flow in pipes, established in Chapter 10, were developed,
introducing the concept of an empirical friction factor. However, it will be shown that
the friction factor is related to the skin friction coefficient defined later in this chapter.

11.3 FACTORS AFFECTING TRANSITION FROM LAMINAR TO TURBULENT FLOW REGIMES

As mentioned above, the transition from laminar to turbulent boundary layer
conditions may be considered as Reynolds number dependent, $\mathrm{Re} = \rho U_s x/\mu$, and a
figure of 5×10^5 is often quoted. However, this figure may be considerably reduced
if the surface is rough. For $\mathrm{Re} < 10^5$, the laminar layer is stable; however, at Re near
2×10^5 it is difficult to prevent transition.

The presence of a pressure gradient dp/dx can also be a major factor. Generally,
if dp/dx is positive, then transition Reynolds number is reduced, a negative dp/dx
increasing transition Reynolds number. This effect forms the basis of suction high-lift
devices designed for aircraft wings. Figure 11.3 illustrates typical velocity profiles
through the boundary layer in both the laminar and turbulent regions, the increased
velocity gradient du/dy being apparent. As mentioned, the growth of the boundary

FIGURE 11.3

Typical velocity profiles in the laminar and turbulent boundary layer regions

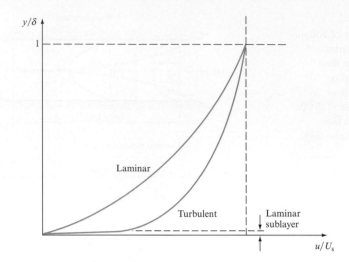

layer thickness is more rapid in the turbulent region, roughly varying as $x^{0.8}$ here compared with $x^{0.5}$ in the laminar region.

In calculations involving long plates, it is often reasonable to suppose that transition occurs close to the leading edge and, in such cases, the presence of the laminar section may be ignored.

The study of the turbulent boundary layer is the more important as in most engineering applications the flow Reynolds number is sufficiently high to ensure transition and the establishment of a turbulent boundary layer. However, it will be appreciated that the random motion of the fluid particles must die out very close to the surface to maintain the condition of 'no slip' at the plane/fluid interface. To accommodate this, the presence of a laminar sublayer in the turbulent region has been established, the thickness of this being small compared with the local boundary layer thickness, as shown in Fig. 11.1. The velocity profile across this sublayer is assumed linear and tangential to the velocity profile up through the turbulent boundary layer.

11.4 DISCUSSION OF FLOW PATTERNS AND REGIONS WITHIN THE TURBULENT BOUNDARY LAYER

Figure 11.4 illustrates the velocity distribution through one particular section in a turbulent boundary layer. As mentioned above, very close to the plane surface the flow remains laminar and a linear velocity profile may be assumed. In this region, the velocity gradient is governed by the fluid viscosity (equation (11.1))

$$\frac{du}{dy} = \tau_0/\mu.$$

Rearranging this expression yields, after integration,

$$u = (\tau_0/\mu)y \tag{11.3}$$

or
$$\frac{u}{\sqrt{(\tau_0/\rho)}} = \frac{\sqrt{(\tau_0/\rho)}}{v}y, \tag{11.4}$$

FIGURE 11.4

Eddy formation in the boundary layer

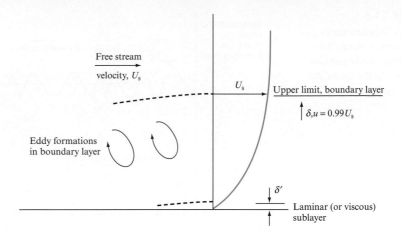

where v is the fluid kinematic viscosity μ/ρ. The term $\sqrt{(\tau_0/\rho)}$ is common in boundary layer theory and termed the shear stress velocity due to the units of velocity applicable to the combination (see Chapter 10) and is denoted by u^*. Thus,

$$\frac{u}{u^*} = \frac{y}{v/u^*}. \tag{11.5}$$

Experimentally, it has been shown that the laminar sublayer occurs in flows, where equation (11.5) has a value less than approximately 5, so that the thickness of the laminar sublayer $y = \delta'$ becomes

$$\delta' = 5v/u^*, \tag{11.6}$$

indicating that the sublayer thickness will be small for large shear stress flows, i.e. u^* large, and that it will increase in the downstream direction as shear stress decreases in this direction. Above the laminar sublayer, the flow regime is turbulent and equation (11.1) no longer adequately represents the shear forces acting. It is appropriate here to describe the mechanism of flow within this upper region.

Owing to the random motion of the fluid particles, eddy patterns are set up in the boundary layer which sweep small masses of fluid up and down through the boundary layer, moving in a direction perpendicular to the surface and the mean flow direction. Owing to these eddies, fluid from the upper higher-velocity areas is forced into the slower-moving stream above the laminar sublayer, having the effect of increasing the local velocity here relative to its value in the laminar sublayer. This increase in velocity of fluid close to the wall is shown in Fig. 11.3. Conversely, slow-moving fluid is lifted into the upper levels, slowing down the fluid stream and, by doing so, effectively thickening the boundary layer, explaining the more rapid growth of the turbulent boundary layer compared with the laminar one.

The process described is, effectively, a momentum transfer phenomenon. However, the effect is analogous to a shear stress applied to the fluid as the overall deceleration is increased as boundary layer thickness increases. In order to explain this process, and to retain the useful form of equation (11.1), a new viscosity term may be introduced, the eddy viscosity, ε, and equation (11.1) rewritten as the relationship mentioned earlier (equation (11.2)),

FIGURE 11.5

Velocity fluctuations in the mean flow direction and normal to the surface at a point in the turbulent boundary layer

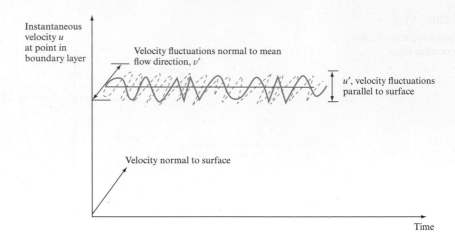

$$\tau = (\varepsilon + \mu)\frac{\mathrm{d}u}{\mathrm{d}y}.$$

Figure 11.5 illustrates the likely output from a velocity-measuring device positioned within the turbulent boundary layer. Here it will be seen that, although the mean velocity is in the flow direction, there are fluctuations in velocity corresponding to the random particle motion.

If the fluid velocity is made up of a mean value $\bar{u}$ and fluctuating components u' and v' in the flow direction and perpendicular to it, respectively, then it may be assumed that the apparent shear stress required to duplicate the eddy effects discussed above would be

$$\tau = -\rho\bar{u}'\bar{v}', \tag{11.7}$$

i.e. the shear stress opposing motion is given by the product of fluid density and the average product of the normal velocity fluctuations over an incremental time period.

11.5 PRANDTL MIXING LENGTH THEORY

In the form of equation (11.7), little further may be done. However, Prandtl (1875–1953) – who, in 1904, was responsible for stating the basics of boundary layer theory and proposing that all viscous effects are concentrated within it – developed the necessary theory to relate the apparent shear stress to mean velocity distribution through the boundary layer. This theory, summarized below, is known as the Prandtl mixing length theory.

Prandtl defined the mixing length as that distance l in which a particle loses its excess momentum and assumes the mean velocity of its surroundings, an idea in some respects similar to the mean free path. In practice, the loss or transfer of momentum would be gradual over the length l. Assuming that the changes of velocity u' and v' following from this particle motion would be equal – a not unreasonable assumption – it may be seen that $v' = u' = l\,\mathrm{d}u/\mathrm{d}y$ (Fig. 11.6) and, from equation (11.7),

FIGURE 11.6

Concept of mixing length
in Prandtl's theory

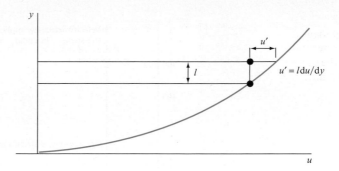

$$\tau = \rho l^2 \left(\frac{du}{dy}\right)^2.$$ (11.8)

Close to the surface, Prandtl assumed that l became dependent on the distance from the surface, or $l = ky$. This allows for l to have zero value at the boundary where $y = 0$. Hence

$$\tau = \rho k^2 y^2 \left(\frac{du}{dy}\right)^2,$$

where k was proposed as a universal constant having a value around 0.4. More recent work has shown distinct limitations in this approach and values varying from 0.4 have been recorded. However, the Prandtl mixing length theory was a major advance at the time and may still be of use in particular situations.

Close to the surface it may be assumed that the shear stress equals the surface value, so

$$\tau_0 = \rho k^2 y^2 \left(\frac{du}{dy}\right)^2$$ (11.9)

or $\qquad du = \dfrac{\sqrt{(\tau_0/\rho)}}{k} \dfrac{dy}{y},$

which, on integration, yields

$$u/u^* = (1/k) \log_e y + C.$$ (11.10)

Values of the integration constant C have been experimentally determined in the form

$$C = 5.56 - (1/k) \log_e(v/u^*),$$

so that a velocity distribution

$$u/u^* = (1/k) \log_e[y(u^*/v)] + 5.56$$ (11.11)

FIGURE 11.7

Overlap of velocity
distributions

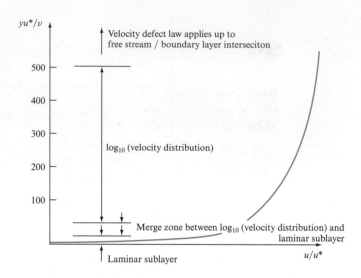

is obtained. Substituting 0.4 for k yields

$$u/u^* = 5.75 \log_{10}[y(u^*/v)] + 5.56 \tag{11.12}$$

in terms of log base 10.

Comparison of equation (11.12), which applies for $30 < yu^*/v < 500$, with equation (11.5) shows that there is a major change in velocity profile between the laminar sublayer and the turbulent boundary layer region. However, both profiles are related through the yu^*/v term.

Figure 11.7 illustrates these profiles. However, above $yu^*/v = 500$, experimental results indicate that a better fit is obtained by a velocity defect law of the form

$$(U_s - u)/u^* = f(y/\delta). \tag{11.13}$$

Thus, three zones of application of velocity distribution equations are apparent. These zones do not possess sharp boundaries; rather they merge into each other. In the intersection zones, experimental results straddle the predictions of each equation; however, the general boundaries of 30 and 500 for yu^*/v are adequate in this treatment.

In terms of experimental results, a simplified velocity profile, which applies to 90 per cent of the boundary layer thickness but not to the 10 per cent close to the plane surface, was proposed by Prandtl:

$$u/U_s = (y/\delta)^n. \tag{11.14}$$

The value of n for Reynolds numbers in the region $10^5 > \text{Re} > 10^7$ may be taken as $\frac{1}{7}$ and the expression (11.14) is known as the seventh-power law. In addition, it is usually assumed that the velocity profile through the laminar sublayer is linear and tangential to the seventh-power law.

11.6   DEFINITIONS OF BOUNDARY LAYER THICKNESSES

So far the boundary layer thickness has been referred to only in physical terms; namely, boundary layer thickness is defined as that distance from the surface where the local velocity equals 99 per cent of the free stream velocity:

$$\delta = y_{(u=0.99U_s)}, \tag{11.15}$$

where U_s is the free stream velocity. It is possible, however, to define boundary layer thickness in terms of the effect on the flow.

11.6.1 Displacement thickness δ^*

Owing to the presence of the boundary layer, the flow past a given point on the surface is reduced by a volume equivalent to the area ABC in Fig. 11.8. This volume reduction is given by the integral $\int(U_s - u)\,\mathrm{d}y$. If the area ABC is equated to an area ABDE, whose volume may be calculated as δ^*U_s, then the displacement thickness for the boundary layer may be defined as the distance the surface would have to move in the y direction to reduce the flow passing by a volume equivalent to the real effect of the boundary layer:

FIGURE 11.8
Displacement thickness

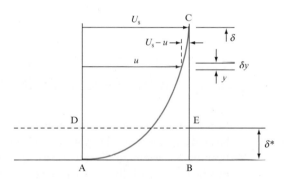

$$\delta^* = \int_0^\infty (1 - u/U_s)\,\mathrm{d}y. \tag{11.16}$$

11.6.2 Momentum thickness θ

The fluid passing through the element δy carries momentum at a rate $(\rho u \delta y)u$ per unit width, whereas, in the absence of the boundary layer, u would equal U_s, so that the total reduction in momentum flow is

$$\int_0^\infty \rho(U_s - u)u\,\mathrm{d}y,$$

FIGURE 11.9

Control volume applied to a general section of boundary layer over a flat plate

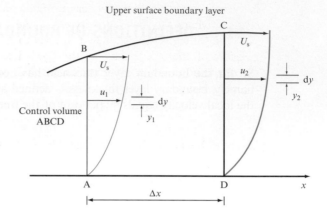

which may be equated to the momentum carried through a section θ deep per unit width at the free stream velocity (Fig. 11.9):

$$(\rho U_s \theta) U_s = \int_0^\infty \rho (U_s - u) u \, \mathrm{d}y,$$

$$\theta = \int_0^\infty \frac{u}{U_s}\left(1 - \frac{u}{U_s}\right)\mathrm{d}y. \tag{11.17}$$

11.7 APPLICATION OF THE MOMENTUM EQUATION TO A GENERAL SECTION OF BOUNDARY LAYER

Von Kármán first applied the momentum equation to a general section of a boundary layer. Regardless of the position of the section in either the laminar or turbulent boundary layer regions, it is possible to equate the skin friction drag force to the product of rate of change of momentum and mass of fluid affected by the boundary layer. Figure 11.9 illustrates a control volume ABCD around a general section of boundary layer. It will be assumed that the flow continues to be incompressible and that as $\mathrm{d}p/\mathrm{d}x$, the pressure gradient in the flow direction, is also zero so will be $\mathrm{d}U_s/\mathrm{d}x$, i.e. any change in free stream velocity. Flow enters the control volume through AB and BC as shown and leaves via CD. Assuming a unit width of surface, then the momentum equation applied to this control volume in the flow direction becomes

$$-\tau_0 \Delta x = \int_0^{\delta_2} \rho u_2^2 \, \mathrm{d}y - \int_0^{\delta_1} \rho u_1^2 \, \mathrm{d}y - \rho U_s^2 (\delta_2 - \delta_1). \tag{11.18}$$

Taking the terms in order: $\tau_0 \Delta x$ represents the shear force, opposing motion in the positive flow direction, over the immersed area Δx;

$$\int_0^{\delta_2} \rho u_2^2 \, \mathrm{d}y = \int_0^{\delta_2} \rho u_2 \, \mathrm{d}y \, u_2$$

is the rate of momentum transfer through a section dy on CD at a height y above the surface, where the local velocity is u_2; the third term is identical in form to the preceding integral except that it applies to a section dy on AB; the last term is a measure of the momentum, in the flow direction, carried into the control volume across the boundary layer upper surface BC. But

$$U_s(\delta_2 - \delta_1) = \int_0^{\delta_2} u_2 dy - \int_0^{\delta_1} u_1 dy, \tag{11.19}$$

i.e. the difference in flow rates past AB and CD. Hence,

$$-\tau_0 \Delta x = \int_0^{\delta_2} \rho u_2^2 \, dy - \int_0^{\delta_1} \rho u_1^2 \, dy - \rho U_s \left(\int_0^{\delta_2} u_2 \, dy - \int_0^{\delta_1} u_1 \, dy \right)$$

$$= \rho \left[\int_0^{\delta_2} (u_2^2 - U_s u_2) \, dy - \int_0^{\delta_1} (u_1^2 - U_s u_1) \, dy \right]$$

$$= \rho U_s^2 \left\{ \int_0^{\delta_2} \left[\left(\frac{u_2}{U_s} \right)^2 - \frac{u_2}{U_s} \right] dy - \int_0^{\delta_1} \left[\left(\frac{u_1}{U_s} \right)^2 - \frac{u_1}{U_s} \right] dy \right\}.$$

As Δx approaches zero in the limit, and multiplying both sides by -1, the equation above reduces to

$$\tau_0 = \rho U_s^2 \frac{d}{dx} \int_0^{\delta} \frac{u}{U_s} \left(1 - \frac{u}{U_s} \right) dy \tag{11.20}$$

or, by reference to the momentum thickness defined in equation (11.17),

$$\tau_0 = \rho U_s^2 \frac{d\theta}{dx}. \tag{11.21}$$

Equation (11.20) is the momentum equation applied to a general boundary layer section and is of general use in deriving further relations in the boundary layer. It is also of use in the area of heat transfer through boundary layers, although these applications are outside the scope of this text.

In the following sections, more detailed relations applying to the laminar and turbulent boundary layers individually will be presented.

11.8 PROPERTIES OF THE LAMINAR BOUNDARY LAYER FORMED OVER A FLAT PLATE IN THE ABSENCE OF A PRESSURE GRADIENT IN THE FLOW DIRECTION

In practice, the laminar section of the boundary layer formed as a result of flow over a surface is short; however, it will always exist, even in flows that are nominally turbulent. For example, consider the inlet section of a circular cross-section duct. As discussed in Chapter 10, the flow in the duct will be considered turbulent if the

Reynolds number based on mean fluid velocity and duct diameter $\mathrm{Re} = \rho \bar{u} d / \mu > 2000$. However, as far as the boundary layer is concerned, the transition from laminar to turbulent occurs at Reynolds numbers above 10^5 based on mean fluid velocity and distance measured from the entry to the duct, $\mathrm{Re} = \rho U_s x / \mu$, so that there will always be a finite length of laminar boundary layer.

Blasius developed a series of analytical solutions for the laminar boundary layer, which will be quoted in this section and compared with approximate results that may be derived from the assumptions already discussed; namely, a linear relation between shear stress and vertical distance to the surface and the absence of a pressure gradient across the flat surface.

In all laminar flow, equation (11.1) applies, i.e.

$$\tau_0 = \mu \left(\frac{\mathrm{d}u}{\mathrm{d}y} \right)_{y=0}, \tag{11.22}$$

and, from equation (11.20), the momentum equation,

$$\mu \left(\frac{\mathrm{d}u}{\mathrm{d}y} \right)_{y=0} = \rho \frac{\mathrm{d}}{\mathrm{d}x} \int_0^\delta u (U_s - u) \, \mathrm{d}y. \tag{11.23}$$

If the assumption is made that velocity profiles through the boundary layer are geometrically similar along the whole length of the laminar section, then this may be expressed as

$$u = U_s f(\eta), \tag{11.24}$$

where $\eta = y/\delta$, $u = 0$ at $y = 0$, or $\eta = 0$, $u = U_s$ at $y = \delta$, or $\eta = 1$. Substituting into (11.23) yields

$$\frac{\mu}{\delta} U_s \left[\frac{\mathrm{d}f(\eta)}{\mathrm{d}\eta} \right]_{\eta=0} = \rho \frac{\mathrm{d}}{\mathrm{d}x} \left\{ U_s^2 \delta \int_0^1 [1 - f(\eta)] f(\eta) \, \mathrm{d}\eta \right\}, \tag{11.25}$$

where the limits of integration have been changed as $\eta = 1$ at $y = \delta$.

Owing to the geometric similarity of velocity profiles, $f(\eta)$ is independent of position along the laminar section x, so that

$$\int_0^1 [1 - f(\eta)] f(\eta) \, \mathrm{d}\eta$$

is independent of x and may be regarded as a constant C_1 and $[\partial f(\eta)/\partial \eta]_0$ as a constant C_2. As a result, equation (11.25) becomes

$$\frac{\mu}{\delta} U_s C_2 = \rho U_s^2 C_1 \frac{\mathrm{d}\delta}{\mathrm{d}x}$$

or $\qquad \mu \dfrac{C_2}{C_1} = \rho U_s \delta \dfrac{\mathrm{d}\delta}{\mathrm{d}x}.$

Integrating,

$$\delta^2 / 2 = (\mu / \rho U_s)(C_2 / C_1) x + \text{constant},$$

so $\qquad \delta = \sqrt{[(2 C_2 / C_1)(\mu / \rho U_s)] x} = \sqrt{(2 C_2 / C_1)} x / \sqrt{(\mathrm{Re})} \tag{11.26}$

if $\delta = 0$ at $x = 0$, i.e. zero boundary layer thickness at the leading edge, and

$$\mathrm{Re}_x = \rho U_s x/\mu.$$

From (11.25),

$$\tau_0 = \rho U_s^2 C_1 \frac{\mathrm{d}\delta}{\mathrm{d}x} = \rho U_s^2 C_1 \sqrt{\left(\frac{2C_2}{C_1}\frac{\mu}{\rho U_s}\right)}\frac{\mathrm{d}(x^{1/2})}{\mathrm{d}x},$$

$$\tau_0 = \rho U_s^2 C_1 \sqrt{\left(\frac{2C_2\mu}{C_1\rho U_s}\right)}\frac{x^{-1/2}}{2} = \rho U_s^2 \sqrt{\left(\frac{C_1 C_2}{2\mathrm{Re}_x}\right)} \tag{11.27}$$

and, from the integral of shear stress at the wall over the length l of the laminar boundary layer, the total skin friction force per unit width of surface may be written as

$$F = \int_0^l \tau_0\,\mathrm{d}x = \int_0^l \rho U_s^2 C_1 \frac{\mathrm{d}\delta}{\mathrm{d}x}\,\mathrm{d}x = (\rho U_s^2 C_1 \delta)_0^l = \rho U_s^2 l \sqrt{\left(\frac{2C_1 C_2}{\mathrm{Re}_l}\right)},$$

when equations (11.27) and (11.26) are employed to substitute for τ_0 and δ. Simplifying yields

$$F = \rho U_s^2 \sqrt{(2C_1 C_2 \mu/\rho U_s l)}\,l$$
$$= \sqrt{(2C_1 C_2 \rho\mu\,U_s^3 l)}, \tag{11.28}$$

and the *skin friction coefficient* C_f may then be calculated as

$$C_f = F/\tfrac{1}{2}\rho U_s^2 l \quad \text{per unit width.} \tag{11.29}$$

However, equation (11.28) is of little value in the form shown owing to the presence of C_1 and C_2. If these constants could be evaluated, then the skin friction for a flat plane would be known. The values of C_1, C_2 depend on the assumptions made with respect to the variation of shear stress with distance above the plane, i.e. $\tau = f(y)$ or $f(\eta)$. Now, we have already mentioned that this function may be assumed to be linear and the boundary conditions at $y = 0$ and $y = \delta$ are known to be $\tau = \tau_0$ and $\tau = 0$, respectively. This may be expressed by a relation of the form

$$\tau = C_3(\delta - y), \tag{11.30}$$

$$\mu\frac{\mathrm{d}u}{\mathrm{d}y} = C_3(\delta - y)$$

$$\mu u = C_3(y\delta - y^2/2) + C_4.$$

Now $u = 0$ at $y = 0$. Therefore, $C_4 = 0$ and so

$$\mu u = C_3(y\delta - y^2/2).$$

As $u = U_s$ when $y = \delta$, $C_3 = 2\mu U_s/\delta^2$. Hence,

$$\mu u = 2\mu(U_s/\delta^2)(y\delta - y^2/2),$$
$$u/U_s = 2(y/\delta - y^2/2\delta^2) = 2(\eta - \eta^2/2),$$

and so $u/U_s = 2\eta - \eta^2$ \tag{11.31}

is the resulting velocity profile from the linear shear stress vs. distance above surface assumption. This allows C_1 and C_2 to be evaluated as

$$C_1 = \int_0^1 [1 - f(\eta)]f(\eta)\,d\eta = \int_0^1 [1 - (2\eta - \eta^2)](2\eta - \eta^2)\,d\eta$$

$$= 2/15$$

and $$C_2 = \left[\frac{\partial}{\partial\eta}f(\eta)\right]_{\eta=0} = \left[\frac{\partial}{\partial\eta}(2\eta - \eta^2)\right]_{\eta=0} = (2 - 2\eta)_{\eta=0}$$

$$= 2.$$

Substituting back, yields

$$\delta = \sqrt{(2 \times 2 \times \tfrac{15}{2})}\,[x/\sqrt{(Re_x)}] = 5.48x/\sqrt{(Re_x)} \tag{11.32}$$

and $$C_f = \frac{\sqrt{(2C_1 C_2 \rho\mu U_s^3 l)}}{\tfrac{1}{2}\rho U_s^2 l}$$

$$= \sqrt{\left(\frac{2 \times \frac{2}{15} \times 2 \times \rho\mu U_s^3 l}{\tfrac{1}{4}\rho^2 U_s^4 l^2}\right)}$$

$$= \sqrt{\frac{\left(\frac{32}{15}\right)}{Re_x}} \text{ per plate side}$$

$$C_f = 1.4\ Re_x^{-1/2}. \tag{11.33}$$

Blasius was able, by reference to the general equations of motion for boundary layers, to plot the velocity distribution up through the laminar boundary layer in a form

$$u/U_s = f(y\,Re_x^{-1/2}/x) \tag{11.34}$$

and, from this plot, again making the assumptions that $y = \delta$ when $u = 0.99U_s$ and that the velocity profiles are geometrically similar along the surface, Blasius was able to show that, approximately,

$$\delta = 5x\,Re_x^{-1/2}, \tag{11.35}$$

which is comparable with equation (11.32) above. Similarly, by taking the slope of the curve (11.34) at $y = 0$,

$$\left(\frac{du}{dy}\right)_{y=0} = 0.332\frac{U_s}{x}Re_x^{1/2}, \tag{11.36}$$

which, when substituted into

$$\tau_0 = \mu\left(\frac{du}{dy}\right)_{y=0}$$

and integrated from $x = 0$ to l, gives a skin friction coefficient,

$$C_f = 1.33 \mathrm{Re}_l^{-1/2}, \tag{11.37}$$

which is comparable with equation (11.33) above.

A number of alternative velocity profiles have been suggested to replace (11.31). However, the results do not differ substantially from those shown above.

Values for the displacement and momentum thicknesses of the boundary layer may also be calculated in terms of δ, initially from equation (11.31) substituted into equations (11.16) and (11.17), respectively. Hence,

$$\delta^* = \delta \int_0^1 [1 - f(\eta)] \, d\eta = \delta \int_0^1 (1 - 2\eta + \eta^2) \, d\eta,$$

$$\delta^* = \delta/3 = 1.86 x \, \mathrm{Re}_x^{-1/2} \tag{11.38}$$

and

$$\theta = \delta \int_0^1 f(\eta)[1 - f(\eta)] \, d\eta$$

$$= \delta \int_0^1 (2\eta - \eta^2)(1 - 2\eta + \eta^2) \, d\eta$$

$$\theta = \tfrac{2}{15}\delta = 0.73 x \, \mathrm{Re}_x^{-1/2}. \tag{11.39}$$

Experimental results verify the Blasius solution except close to the leading edge of the surface, where the assumption of a zero velocity component normal to the surface is not strictly valid. However, the results above are usable for the calculation of skin friction forces and boundary layer thicknesses.

Typical laminar boundary layer thicknesses are of the order of 0.75 mm in air at 100 m s^{-1}, Re $= 10^6$ and typical lengths, for a smooth flat plate, would be around 160 to 200 mm. Measurement of boundary layer velocity profiles is difficult and requires specialized information. The advent of the hot-wire anemometer has made life a lot easier here, but great care is still necessary to ensure that the results obtained are not a direct function of the experimental set-up. Again, it may be appreciated that the laminar section of the boundary layer is, generally, of secondary importance to the turbulent section, which will be dealt with in the next section.

EXAMPLE 11.1

Oil with a free stream velocity of 3.0 m s^{-1} flows over a thin plate 1.25 m wide and 2 m long. Determine the boundary layer thickness and the shear stress at mid-length and calculate the total, double-sided resistance of the plate ($\rho = 860$ kg m^{-3}, $v = 10^{-5}$ m^2 s^{-1}, $v = \mu/\rho$).

Solution

Calculate the Reynolds number at $x = 1$ m:

$$\mathrm{Re}_x = U_s x / \nu = 3x/10^{-5}.$$

Therefore

$$\mathrm{Re}_x^{1/2} = 5.48 \times 10^2.$$

Note that Re is low enough to allow the laminar boundary layer to survive over the whole plate. From equation (11.36):

$$\tau_0 = 0.332 \mu (U_s/x) \ \mathrm{Re}_x^{1/2}$$

$$= 0.332 \times 10^{-5} \times 860 \times \frac{3}{1} \times 5.48 \times 10^2 = 4.7 \ \mathrm{N/m^2}$$

The skin friction force is given by, double sided,

$$F = 2 \times \tfrac{1}{2} \rho U_s^2 l \times b \times C_f,$$

where l is plate length and b is plate width,

$$F = 2 \times \tfrac{1}{2} \times 860 \times 3^2 \times 2 \times 1.25 \times C_f,$$

where (from equation (11.37))

$$C_f = 1.33 \mathrm{Re}_l^{-1/2} = 1.33/(6 \times 10^5)^{1/2}.$$

Therefore,

$$F = 860 \times 18 \times 1.25 \times 1.33/[\sqrt{(60)} \times 10^2] = \mathbf{33.224 \ N}.$$

11.9 PROPERTIES OF THE TURBULENT BOUNDARY LAYER OVER A FLAT PLATE IN THE ABSENCE OF A PRESSURE GRADIENT IN THE FLOW DIRECTION

The majority of boundary layers met in engineering practice are turbulent over most of their length, and so the study of this section of the development of the boundary layer is usually regarded as of greater fundamental importance than that of the laminar section. In many cases, the laminar section of the boundary layer is short enough, compared with the total length of the surface, to be ignored in calculations of skin friction forces.

The momentum equation (11.20) may be applied to the turbulent boundary layer as no limiting assumptions were made in its derivation. However, as mentioned in Chapter 5, a new relation for the velocity profile up through the boundary layer will have to be found and the shear stress will no longer be obtained simply from the product of fluid viscosity and the gradient of the velocity profile.

Owing to the basic similarity between the development of boundary layers within circular cross-section pipes and over flat pipes, Prandtl suggested that the results from the pipe case be applied to the analysis of flat-plate turbulent boundary layers. As was mentioned in Section 11.2, the boundary layer growth in pipes is limited to the pipe radius R, so that $u = U_s$ at $y = R$, and the mean flow velocity in turbulent pipe flow is known to be about $0.8\, U_s$. The velocity distribution in such flow is adequately represented by the Prandtl power law,

$$u/U_s = (y/\delta)^n, \tag{11.40}$$

where $n = \frac{1}{7}$ for $\mathrm{Re}_x < 10^7$. Obviously, this profile breaks down at the wall, where $y = 0$. However, the presence of a laminar sublayer has already been discussed (Section 11.2) where the velocity decreases linearly to zero at the wall, this profile being tangential to the power law (Fig. 11.7).

To develop the analogy between flat plates and pipe flow, it is necessary to appreciate that $\delta = R$ in the fully developed region and to develop some relation for τ_0 to replace equation (11.1), which no longer applies.

Blasius proposed that, for smooth pipes, the shear stress at the wall could be expressed by

$$\tau_0 = f\tfrac{1}{2}\rho\bar{u}^2, \tag{11.41}$$

where $\bar{u}$ is the mean fluid velocity equal to $0.8 U_s$ and f is an empirical constant known as the friction factor, which is a function of flow Reynolds number ($\mathrm{Re} = \rho\bar{u}d/\mu$, d = pipe diameter) and the ratio of wall roughness to pipe diameter. Friction factors are covered in more detail in Chapter 10. Thus, $\tau_0 = \tfrac{1}{2}\rho(0.8 U_s^2)f$ and, as Blasius developed the expression

$$f = 0.079/\mathrm{Re}^{1/4} = 0.079/(\rho\bar{u}d/\mu)^{1/4}$$

to apply to smooth pipes, substitution yields an expression

$$\tau_0 = \tfrac{1}{2}\rho(0.8U_s)^2\, 0.079(\mu/\rho 0.8 U_s 2R)^{1/4}$$

and, if $\delta = R$, then,

$$\tau_0 = 0.0225\rho U_s^2\,(\mu/\rho U_s\delta)^{1/4}. \tag{11.42}$$

As the assumption of zero pressure gradient has been made, equation (11.20) can be applied. Thus,

$$\tau_0 = \rho U_s^2\frac{\mathrm{d}}{\mathrm{d}x}\int \frac{u}{U_s}\left(1 - \frac{u}{U_s}\right)\mathrm{d}y$$

$$= \rho U_s^2\frac{\mathrm{d}\delta}{\mathrm{d}x}\int_0^1 (1 - \eta^{1/7})\eta^{1/7}\,\mathrm{d}\eta, \tag{11.43}$$

where

$$u/U_s = (y/\delta)^{1/7} = \eta^{1/7}.$$

Therefore,

$$\tau_0 = \frac{7}{72}\rho U_s^2 \frac{d\delta}{dx}. \tag{11.44}$$

Equating these two expressions (11.42) and (11.44) for τ_0 yields

$$\delta^{1/4}\, d\delta = 0.234(\mu/\rho U_s)^{1/4}\, dx.$$

Integrating yields

$$\tfrac{4}{5}\delta^{5/4} = 0.234(\mu/\rho U_s)^{1/4}x + C_5.$$

Now, if the turbulent boundary layer is assumed to extend to the plate leading edge, which is reasonable if the plate is long compared with the length of the laminar layer, then $\delta = 0$ at $x = 0$ and $C_5 = 0$. Hence,

$$\delta^{5/4} = 0.292(\mu/\rho U_s)^{1/4}x,$$
$$\delta = 0.37x/(\rho U_s x/\mu)^{1/5}$$
$$= 0.37x\, \mathrm{Re}_x^{-1/5}. \tag{11.45}$$

Comparing equation (11.45) to (11.32), it may be seen that the turbulent boundary layer grows more rapidly than the laminar layer, the proportional to distance along the plate being to the power $x^{4/5}$ and $x^{1/2}$, respectively. The skin friction force on the flat surface may be determined by eliminating δ between equations (11.42) and (11.45). Hence,

$$\tau_0 = 0.029\rho U_s^2\,(\mu/\rho U_s x)^{1/5}$$

and

$$F = \int_0^l \tau_0\, dx \text{ per unit width,}$$

where l is plate length.

$$F = 0.036\rho U_s^2 l(\mu/\rho U_s l)^{1/5}$$
$$= 0.036\rho U_s^2 l\, \mathrm{Re}_l^{-1/5} \tag{11.46}$$

and the skin friction coefficient,

$$C_f = F/\tfrac{1}{2}\rho U_s^2 l \text{ per unit width}$$

$$C_f = 0.072\, \mathrm{Re}_l^{-1/5}. \tag{11.47}$$

The expression above is valid for Reynolds numbers up to 10^7, but experimental results indicate that a better approximation is given by

$$C_f = 0.074\, \mathrm{Re}_l^{-1/5}. \tag{11.48}$$

Prandtl has suggested subtracting the length of the laminar layer, resulting in an expression

$$C_f = 0.074\ \mathrm{Re}_l^{-1/5} - 1700\ \mathrm{Re}_l^{-1}$$

to apply from $\mathrm{Re}_l = 5 \times 10^5$ to 10^7.

To extend the Reynolds number range further, Schlichting employed the logarithmic velocity distribution for pipes under turbulent flow conditions, which have already been mentioned in Chapter 10, resulting in a semi-empirical relation,

$$C_f = 0.455(\log_{10}\ \mathrm{Re}_l)^{-2.58}, \tag{11.49}$$

applying from $10^6 < \mathrm{Re}_l < 10^9$.

Comparison of equation (11.47) with equation (11.37) shows that the skin friction is proportional to the $\frac{9}{5}$ power of velocity of the main stream and the $\frac{4}{5}$ power of plate length for the turbulent layer, compared with the $\frac{3}{2}$ and $\frac{1}{2}$ powers, respectively, for the laminar layer. Generally, then, it may be seen that retention of a laminar boundary as long as possible is desirable from a drag viewpoint. Figure 11.10, a plot of C_f vs. Re_l, illustrates the variations in skin friction coefficient.

FIGURE 11.10
Variation of skin friction coefficient with Reynolds number

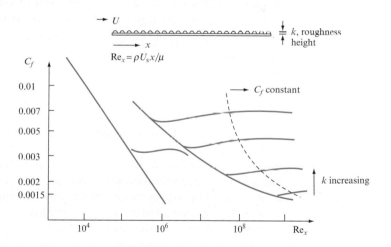

EXAMPLE 11.2

A smooth flat plate 3 m wide and 30 m long is towed through still water at 20 °C at a speed of 6 m s^{-1}. Determine the total drag on the plate and the drag on the first 3 m of the plate.

Solution

For the whole plate:

$$\mathrm{Re}_l = 1000 \times 6 \times 30/10^{-3} = 180 \times 10^6,$$

$$\rho = 1000\ \mathrm{kg\ m}^{-3}, \quad \mu = 10^{-3}\ \mathrm{N\ s\ m}^{-3},$$

$$C_f = 0.455/[\log_{10}(1.8 \times 10^8)]^{2.58} = 0.001\ 96.$$

Drag on both sides of the plate,

$$F = 2(\tfrac{1}{2}\rho U_s^2\, C_f) \times b \times l = 2 \times \tfrac{1}{2} \times 1000 \times 36 \times 0.001\,96 \times 90$$

$$= \mathbf{6.36\ kN}$$

Considering the point at which the boundary layer becomes turbulent, assume transition at $Re_l = 10^5$:

$$10^5 = \rho U_s l_t / \mu,$$

where l_t is the transition length,

$$l_t = 10^5 \mu / \rho U_s = 10^5 \times 10^{-3} / 10^3 \times 6 = 0.0167\ \text{m}.$$

Thus, it is reasonable to ignore the laminar layer compared with the 30 m plate length. Drag on the first 3 m is then calculated in the same way as shown above for the full plate length.

11.10 EFFECT OF SURFACE ROUGHNESS ON TURBULENT BOUNDARY LAYER DEVELOPMENT AND SKIN FRICTION COEFFICIENTS

Initially, the effect of surface roughness is to cause transition from laminar to turbulent conditions closer to the leading edge of the surface. Indeed, the method commonly used to trigger a turbulent boundary layer over model surfaces in wind tunnels is to fix a trip wire or a band of sandpaper or rough material along the surface leading edge to ensure correct drag readings from the models tested. Following transition, where the boundary layer is still thin, the value of k/δ may be significant and all the surface roughness protrudes above the boundary layer. In this case, all the drag is due to the eddies caused by the flow passing over the surface roughness and the drag is proportional to the square of the free stream velocity. As the boundary layer continues to develop so the layer depth increases, and the laminar sublayer eventually becomes thick enough to cover all the surface roughness, so that the eddy-related losses mentioned above do not occur. In this case, which occurs for high Reynolds numbers, the degree of roughness of the surface becomes unimportant, i.e. a change in roughness height k would not affect the drag force. Under these special conditions the surface is said to have become hydraulically smooth.

11.11 EFFECT OF PRESSURE GRADIENT ON BOUNDARY LAYER DEVELOPMENT

So far, the assumption made of zero pressure gradient in the flow direction across the flat surfaces considered has been unquestioned. The presence of a pressure gradient $\partial p/\partial x$ effectively means a $\partial u/\partial x$ term, i.e. the flow stream velocity changes across the

FIGURE 11.11

Variation of pressure
and velocity over a
curved surface

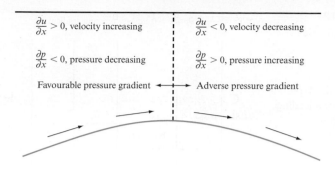

surface. If, for example, a curved surface is considered, then the velocity is seen to vary as shown in Fig. 11.11.

If the pressure *decreases* in the downstream direction, then the boundary layer tends to be reduced in thickness, and this case is termed a favourable pressure gradient.

If the pressure *increases* in the downstream direction, then the boundary layer thickens rapidly; this case is referred to as an adverse pressure gradient. This adverse pressure gradient, together with the action of the shear forces described in the boundary layer, if they act for a sufficient length, will bring the boundary layer to rest and the flow separates from the surface. This flow separation has serious consequences in the design of aerofoils, as, once the flow breaks away from the surface, all lift is lost. Owing to the continuing action of the adverse pressure gradient downstream of the separation point, reversed flow eddies are formed which act to increase drastically the drag force acting on the surface. Figure 11.11 illustrates the changes in boundary layer velocity profile under the conditions described above.

Generally, then, for the design of aerofoils or other lift-producing surfaces, such as pump and fan blades, the onset of separation should be avoided by design. In the particular case of aerofoil design, this has led to a number of ingenious lift-sustaining devices which act either to revitalize the slow-moving air layer by the introduction of a faster moving jet or to remove the surface layer prior to separation by sucking away this low-velocity layer. One of the earliest devices of the first type was the Handley Page leading edge slot, which passed high-velocity air from below the wing into the upper wing surface boundary layer prior to separation, thus preventing the change of shape of the velocity profile shown in Fig. 11.12. More recently, a French short takeoff and landing (STOL) transport relied on exhaust air from the turbines ducted and discharged along the wing leading edge to prevent separation and loss of lift at slow speed and high wing angles of attack.

The second method, sucking away the boundary layer, has been employed in the study of laminar flow wings for long-range transport aircraft where the marked reduction in skin friction drag that would follow from an entirely laminar boundary layer covering the wing would have obvious range and/or lifting capacity advantages.

The effects of wake formation are not solely concerned with aerofoils, but have resonant failure results in bridge design. Given certain wind speeds over and under a bridge span, the alternate breaking away of the flow from the upper and lower surfaces can impose cyclic loads which, under special conditions, can correspond to the structure's natural frequency.

The examples above have all dealt with the formation of the boundary layer external to a flat or curved surface. However, as mentioned for the pipe case, boundary

FIGURE 11.12
Effect of an adverse pressure gradient on boundary layer development. Flow separation of this type is illustrated for a leading edge slat study in Fig. 11.13

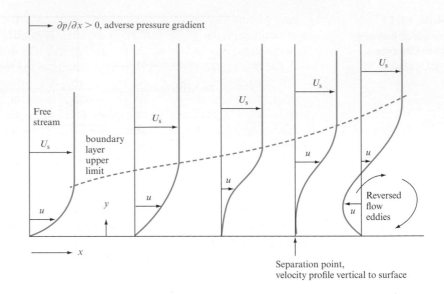

FIGURE 11.13
Surface flow patterns seen during an investigation into the effect of sealing slat tracks on a modern commercial aircraft, photo courtesy of Aerodynamics Laboratory, British Aerospace Systems, Filton, Bristol. Note air flow from bottom right to top left and the flow separation indicated in the surface oil layer due to the unsealed slat–leading edge join line

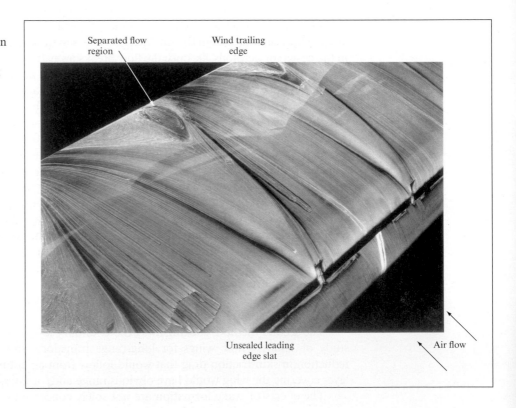

layers form within any duct and grow to fill the duct – imposing a velocity profile of some sort across the duct cross-section. Generally, this is of little concern. However, in the special case of aircraft engine intakes leading to the engine first-stage compressors, the development of the boundary layer can be adverse to the performance of the engine. For the best output, the velocity profile at entry to the compressor, normally now of an axial design, should be as uniform as possible, which cannot occur with a fully developed boundary layer. To prevent this it is now quite common to bleed or suck the boundary layer away down the length of the intake.

A particular complication arises in the case of an aircraft engine designed to operate at supersonic speed. (With two exceptions these are all military aircraft.) It is necessary to decelerate the air prior to entry to the compressors. However, this requires the generation of a shock wave pattern in the intake. If this shock wave pattern is represented as a step increase in pressure, then it will be seen as a concentrated adverse pressure gradient which could cause boundary layer separation in the intake. This, in turn, would cause the formation of an eddy wake which would be likely to stall the compressor – with obvious consequences of loss of engine power. To avoid this boundary layer, bleeding is again employed.

With the few examples given above, the study of the boundary layer can be seen to be of fundamental importance in the understanding of fluid flow phenomena.

Concluding remarks

This chapter has provided both a qualitative and quantitative view of the importance of the boundary layer formed at the interface between a fluid and a solid boundary. The treatment has included reference to the historically important development of this aspect of fluid flow analysis, including the work of Prandtl and von Kármán. The development of expressions defining the velocity profiles within both laminar and turbulent boundary layers has been presented, together with the importance of the laminar sublayer. The effect of surface roughness and applied pressure gradient upon boundary layer development has been explained.

The presence of the boundary layer and its reaction to an applied pressure gradient will be returned to in the treatment of aerofoil lift and stall conditions (Chapter 12) together with elements of the ideal flow theory developed in Chapter 7. The influence of boundary layer growth on the pipe or duct length necessary to produce fully developed, steady, uniform flow was discussed. This has a direct implication for flow monitoring by velocity measurement, as only when the flow has become fully developed can any local velocity measurement be used to calculate cross-sectional flow rate.

In fluid flow networks the boundary layer is a determinant of the flow development and resistance. It is essential in the possible heat transfer that can occur between a surface and a fluid. In the study of aerofoil lift and drag the boundary retaining the boundary layer in contact with the surface is essential to the continuance of lift; boundary layer separation leads to stall conditions. Therefore it may be seen that an understanding of the mechanism of boundary layer growth and retention will be indispensable to later sections of this text.

Summary of important equations and concepts

1. The condition of 'no slip' between a fluid and a surface or between two fluids is essential in the analysis of unseparated flow conditions.

2. The emphasis on the dependence of shear stress on both velocity gradient and viscosity, the identification of the random nature of fluid particle motion in turbulent flows and the introduction of eddy viscosity, equation (11.2), are all essential to the study of boundary layer development.

3. Sections 11.1 to 11.4 introduce and discuss the development of the boundary layer and represent important concepts to be drawn on later.

4. Section 11.5 introduces the Prandtl mixing length theory, leading to a definition of the velocity profile up through a boundary layer and boundary layer thickness.

5. Sections 11.8 and 11.9 apply the momentum concepts first introduced in Chapter 5 to both laminar and turbulent boundary layers, in each case leading to definitions of velocity profile and skin frictional drag.

6. Section 11.11 introduces the effect of pressure gradient on 'real' boundary layer development.

Further reading

Schlichting, H. (1979). *Boundary Layer Theory* (translated by J. Kestin). McGraw Hill, New York.

Problems

11.1 Air at 20 °C and with a free stream velocity of 40 m s^{-1} flows past a smooth thin plate which is 3 m wide and 10 m long in the flow direction. Assuming a turbulent boundary layer from the leading edge determine the shear stress, laminar sublayer thickness and the boundary layer thickness 6 m from the leading edge.

Take density = 1.2 kg m^{-3} and kinematic viscosity as 1.49×10^{-5} m^2 s^{-1}.   [2.015 N m^{-2}, 0.06 mm, 80 mm]

11.2 Determine the ratio of momentum and displacement thickness to the boundary layer thickness δ when the layer velocity profile is given by

$$u/U_s = (y/\delta)^{1/2},$$

where u is the velocity at a height y above the surface and the flow free stream velocity is U_s.   [0.166, 0.333]

11.3 Repeat Problem 11.2 above if the velocity profile is given by

$$\frac{u}{U_s} = \sin\left(\frac{\pi}{2}\frac{y}{\delta}\right).$$

[0.136, 0.36]

11.4 Oil with a free stream velocity of 2 m s^{-1} flows over a thin plate 2 m wide and 3 m long. Calculate the boundary layer thickness and the shear stress at the mid-length point and determine the total surface resistance of the plate.

Take density as 860 kg m^{-3}, kinematic viscosity as 10^{-5} m^2 s^{-1}.   [13.69 mm, 2.085 N m^{-2}, 35.44 N]

11.5 A flat plate is drawn submerged through still water at a velocity of 9 m s^{-1}. If the plate is 3 m wide and 20 m long determine the position of the laminar to turbulent transition and the total drag force acting on the plate. Take water temperature as 20 °C.   [0.01 m, 9.5 kN]

11.6 An open rectangular box section, sides 3 m × 20 m and 1.5 m × 20 m, is drawn submerged through still water, at 20 °C, at a velocity of 9 m s^{-1}. Determine the overall drag force, neglecting any edge effects.   [28.62 kN]

11.7 Estimate the skin friction drag on an airship 92 m long, average diameter 18 m, being propelled at 130 km h^{-1} through air at 90 kN m^{-2} absolute pressure and 27 °C.   [6.7 kN]

11.8 Assuming a velocity distribution defined by

$$u/U_s = \sin(\pi y/2\delta)$$

determine the general expressions for growth of the laminar boundary layer and for the surface shear stress for a smooth flat plate.   [$\delta = 4.8x/\mathrm{Re}_x^{1/2}$, $\tau_0 = 0.33\sqrt{(\mu U_s^3 \rho/x)}$]

11.9 Air at 20 °C and 760 mm Hg absolute pressure flows past a smooth wind tunnel wall, with a free stream velocity of 160 km h^{-1}. Determine the position along the wall, in the flow direction, at which the boundary layer becomes turbulent and the distance to a boundary layer thickness of 25 mm. All wall measurements may be assumed to be taken from the working section entrance and edge/corner effects may be ignored.   [33.6 mm, 1.4 m]

11.10 Show that, if a flat plate, sides a, b in length, is towed through a fluid so that the boundary layer is entirely laminar, the ratio of towing speeds so that the drag force remains constant regardless of whether a or b is in the flow direction is given by

$$U_a/U_b = \sqrt[3]{(a/b)},$$

where U_a is the free stream velocity if side a is in the flow direction and U_b is the corresponding fluid velocity if b is in the flow direction.

11.11 Repeat Problem 11.10 above if the boundary layer is considered fully turbulent.   [$U_a/U_b = \sqrt[9]{(a/b)}$]

Incompressible Flow around a Body

CHAPTERS 7 AND 11 CONTRIBUTE ELEMENTS TO THE TREATMENT OF INCOMPRESSIBLE flow around a body in that the likely flow patterns were established, the effect of pressure gradient identified and the importance of fluid viscosity in the formation of boundary layers established. This chapter extends this previous material to allow the determination of the forces acting upon a body moving in a fluid field. Drag, including pressure and skin friction effects, will be defined, together with lift. These effects will be treated for both fully and partially submerged bodies. Dimensional analysis and similarity will be utilized to determine the zones of dependence of these forces on Reynolds number, Froude number and Mach number. The flow over cylinders and spheres will be treated as a precursor to the consideration of flow over aerofoil sections, where the ideal flow treatment of Chapter 7 will be invoked in the discussion of lift generation. The importance of boundary layer separation will be stressed and the dependence of aerofoil lift on angle of incidence identified, together with the onset of stall conditions. The development of aerofoil trailing vortices and induced drag will be discussed and the momentum equation will be applied to wakes in order to determine pressure drag. A computer program to determine the drag on a body by means of wake traverse data is introduced. ● ● ●

12.1 REGIMES OF EXTERNAL FLOW

In Chapter 7, flow around a cylinder was discussed and expressions enabling the calculation of velocity and pressure in the flow field around the cylinder were derived. Clearly, such a flow may be described as external as it is concerned with the pattern of streamlines surrounding a solid body immersed in a moving fluid. However, the treatment of Chapter 7 excluded any effects which viscosity may have on the flow pattern because that chapter was concerned with ideal flow only.

Chapter 11 introduced the concept of a boundary layer and dealt with the effects viscosity has on a fluid adjacent to a solid surface and with the calculation of forces acting on the surface due to fluid friction. We are, therefore, now in a position to consider the external flows of real fluids, namely taking into account viscous effects. The knowledge of potential flow and of boundary layer theory makes it possible to treat an external flow problem as consisting broadly of two distinct regimes: that immediately adjacent to the body's surface, where viscosity is predominant and where frictional forces are generated, and that outside the boundary layer, where viscosity is neglected but velocities and pressures are affected by the physical presence of the body together with its associated boundary layer. In this outside zone, the theories of ideal flow may be used. In addition, there is the stagnation point at the front of the body (which may stretch into a stagnation region if the body is very blunt) and there is the flow region behind the body (which is known as the wake). These flow regimes are shown in Fig. 12.1.

FIGURE 12.1

Flow regimes around an immersed body

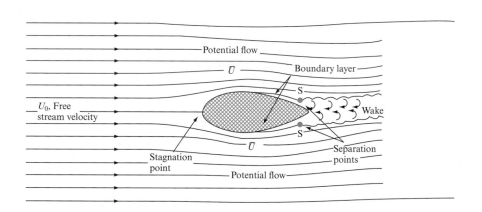

The wake, which starts from points S at which the boundary layer separation occurs, deserves a fuller description. It will be remembered from Chapter 11 that separation occurs due to adverse pressure gradient ($\partial p/\partial x > 0$), which, combined with the viscous forces on the surface, produces flow reversal, thus causing the stream to detach itself from the surface. The same situation exists at the rear edge of a body as it represents a physical discontinuity of the solid surface. In both cases the flow reversal produces a vortex, as shown in Fig. 12.2.

The flow in the wake is thus highly turbulent and consists of large-scale eddies. High-rate energy dissipation takes place there, with the result that the pressure in the wake is reduced. A situation is thus created whereby the pressure acting on the front

FIGURE 12.2

Formation of a vortex
in a wake

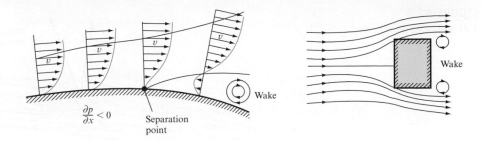

of the body (the stagnation pressure) is in excess of that acting on the rear of the body, so that a resultant force acting on the body in the direction of the relative fluid motion exists. This force, arising from the pressure difference, or more generally from the non-uniform pressure distribution on the body, is called *pressure drag*.

It is worth while to recollect the findings of Chapter 7 dealing with ideal flow. There, in the absence of viscosity, the flow pattern over the rear part of a body, such as a cylinder, was symmetrical with respect to that over the front half of the body. There were two stagnation points – at the front and at the rear – and the pressure at these points was the same. Thus, there was no resultant force due to pressure acting on the body in the direction of the relative motion (see Section 7.10).

A situation very similar to this exists in the case of flow of the real fluids around streamlined bodies, but only at very low Reynolds numbers. There is no wake then and the pressure at the rear stagnation point is nearly equal to that at the front stagnation point. The degree to which the rear stagnation pressure approaches the front stagnation pressure is sometimes called *pressure recovery*. In the ideal flow, the pressure recovery is complete. When the flow separates from the surface and the wake is formed, the pressure recovery is not complete. The larger the wake, the smaller is the pressure recovery and greater the pressure drag. The art of streamlining a body lies, therefore, in shaping its contour so that separation, and hence the wake, is eliminated, or at least in confining the separation to a small rear part of the body and, thus, keeping the wake as small as possible. Such bodies are known as *streamlined* bodies. Otherwise a body is referred to as *bluff* and a significant pressure drag is associated with it.

12.2 DRAG

Pressure drag was described in the preceding section, but asymmetry of pressure distribution, which is responsible for it, is not necessarily the only cause for the existence of a force acting on an immersed body in the direction of relative motion.

Thus, in general, when a body is immersed in a fluid and is in relative motion with respect to it, the *drag* is defined as that component of the resultant force acting on the body which is in the direction of the relative motion.

The force component perpendicular to the drag, i.e. acting in the direction normal to the relative motion, is called *lift* and was defined in Section 7.10. Both lift and drag components of the resultant force are shown in Fig. 12.3.

In Chapter 11, frictional drag was discussed in connection with the boundary layer theory. It is the force on the body acting in the direction of relative motion due to fluid shear stress at the surface. Thus, in external flow, the immersed body is

FIGURE 12.3
Lift and drag on a body

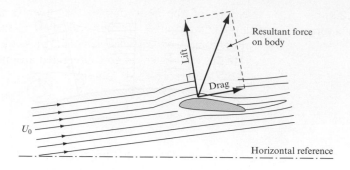

subjected to frictional drag over its entire surface. *Total drag* on the body, often called *profile drag*, is, therefore, made up of two contributions, namely the pressure (or form) drag and the *skin friction* drag. Thus,

$$\text{Profile drag} = \text{Pressure drag} + \text{Skin friction drag.} \qquad (12.1)$$

FIGURE 12.4

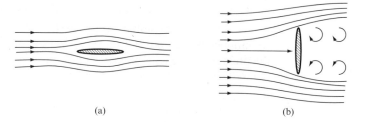

(a)   (b)

The relative contribution of pressure drag and friction drag to the profile drag depends upon the shape of the body and its orientation with respect to the flow. Take, for example, a small rectangular flat plate. If it is held in a fluid stream 'edge on', as shown in Fig. 12.4(a), the pressure drag will be negligible because even though the pressure recovery is incomplete, the resulting pressure difference will act on a very small frontal area (that perpendicular to the flow). The skin friction drag, however, will be substantial, owing to the formation of the boundary layer on both sides of the plate.

If, however, the plate is held perpendicularly to the flow (as in Fig. 12.4(b)) the drag will be almost entirely due to pressure difference, whereas the skin friction drag will be negligible.

The foregoing description of the two kinds of drag can now be formalized mathematically as follows. If (Fig. 12.5) p_s is the fluid pressure acting on the surface element ds, and it acts in the direction perpendicular to the surface, then the force on

FIGURE 12.5

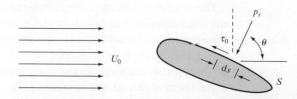

that part of the body due to the pressure is p_s ds. This may be resolved into components parallel and perpendicular to the relative direction of motion. The parallel component responsible for the pressure drag is $p_s \cos \theta$ ds.

If this component is now integrated around the whole contour of the body, the pressure drag is obtained. Thus, pressure drag,

$$D_p = \oint p_s \cos \theta \, \mathrm{d}s. \tag{12.2}$$

Similarly, the friction force on the body is manifested by the existence of the shear stress at the surface S. This also acts on the element ds and gives rise to a tangential force τ_0 ds, whose component in the direction of the motion is $\tau_0 \sin \theta \, \delta s$. Performing, again, the integration around the body's contour, the skin friction drag is obtained. Thus, skin friction drag,

$$D_f = \oint \tau_0 \sin \theta \, \mathrm{d}s. \tag{12.3}$$

Both contributions to the profile drag can, therefore, be theoretically calculated, but the first requires knowledge of the pressure distribution around the body and the other knowledge of the shear stress distribution on the surface. The determination of these could be very laborious and it is, therefore, usually simpler to measure the profile drag experimentally, as a force component in a wind tunnel. It is customary to relate the measured drag to the projected area of the body A, the fluid density ρ, and the free stream velocity U_0 by the expression

$$D = \tfrac{1}{2} C_D \rho U_0^2 A, \tag{12.4}$$

where C_D is known as the *drag coefficient* and A is the area of the body's projection on a plane perpendicular to the relative direction of motion.

A similar exercise of summation may be carried out for the force components normal to the direction of motion to give lift. This is also related to ρ, U_0 and A by an analogous expression,

$$L = \tfrac{1}{2} C_L \rho U_0^2 A. \tag{12.5}$$

The resultant force on the body is, of course, obtained by compounding lift and drag:

$$F = \sqrt{(L^2 + D^2)} = \tfrac{1}{2} \rho U_0^2 A \sqrt{(C_L^2 + C_D^2)}. \tag{12.6}$$

EXAMPLE 12.1

A kite, which may be assumed to be a flat plate of face area $1.2 \, \mathrm{m}^2$ and mass $1.0 \, \mathrm{kg}$, soars at an angle to the horizontal. The tension in the string holding the kite is $50 \, \mathrm{N}$ when the wind velocity is $40 \, \mathrm{km \, h^{-1}}$ horizontally and the angle of the string to the horizontal direction is $35°$. The density of air is $1.2 \, \mathrm{kg \, m^{-3}}$. Calculate the lift and the drag coefficients for the kite in the given position indicating the definitions adopted for these coefficients.

Solution

Since the wind is horizontal, the drag, by definition, will also be horizontal and the lift vertical. The kite is in equilibrium and, therefore, lift and drag must be balanced by the string tension and the weight of the kite. Resolving forces into horizontal and vertical components (Fig. 12.6),

FIGURE 12.6

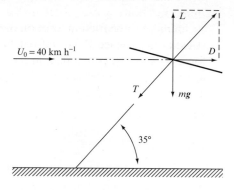

$$L = T \sin 35° + mg = 50 \sin 35° + 1.0 \times 9.81$$

$$= 38.49 \text{ N},$$

$$D = T \cos 35° = 50 \cos 35° = 40.95 \text{ N}.$$

But lift

$$L = \tfrac{1}{2} C_L \rho U_0^2 A$$

and, therefore,

$$C_L = \frac{2L}{\rho U_0^2 A} = \frac{2 \times 38.49}{1.2(40 \times 1000/3600)^2 1.2}$$

$$= \textbf{0.432}.$$

Similarly, the drag coefficient,

$$C_D = \frac{2D}{\rho U_0^2 A} = \frac{2 \times 40.95}{1.2(40 \times 1000/3600)^2 1.2}$$

$$= \textbf{0.460}.$$

Both coefficients have been based on the full area of the kite, because the projected area varies with incidence. This is also the accepted practice in the case of aerofoils.

12.3   DRAG COEFFICIENT AND SIMILARITY CONSIDERATIONS

In order to obtain some idea of the nature of the drag coefficient, it is informative to carry out a dimensional analysis exercise (see Chapter 8) in which the drag on an immersed body is considered to be the dependent variable while the following are included as independent variables: the fluid density ρ, its viscosity μ, free stream velocity U_0, a linear dimension of the body l, the weight of the fluid per unit mass $\mathbf{g}$ (acceleration due to gravity), surface tension σ, and bulk modulus K. Thus,

$$D = f(\rho, \mu, U_0, l, \sigma, \mathbf{g}, K)$$

or $$D = \rho^a \mu^b U_0^c l^d \mathbf{g}^e \sigma^f K^h,$$

and, substituting the dimensions,

$$\frac{ML}{T^2} = \left[\frac{M}{L^3}\right]^a \left[\frac{M}{LT}\right]^b \left[\frac{L}{T}\right]^c [L]^d \left[\frac{L}{T^2}\right]^e \left[\frac{M}{T^2}\right]^f \left[\frac{M}{T^2 L}\right]^h.$$

Equating indices:

[M] $1 = a + b + f + h;$

therefore,

$$a = 1 - b - f - h;$$

[T] $-2 = -b - c - 2e - 2f - 2h.$

Thus,

$$c = 2 - b - 2e - 2f - 2h;$$

[L] $1 = -3a - b + c + d + e - h,$

from which,

$$d = 1 + 3a + b - c - e + h$$
$$= 1 + 3(1 - b - f - h) + b - 2 + b + 2e + 2f$$
$$+ 2h - e + h = 2 - b - f + e.$$

Therefore,

$$D = \rho^{(1-b-f-h)} \mu^b U_0^{(2-b-2e-2f-2h)} l^{(2-b-f+e)} \mathbf{g}^e \sigma^f K^h$$

$$= \rho U_0^2 l^2 \left(\frac{\mu}{\rho U_0 l}\right)^b \left(\frac{\sigma}{\rho U_0^2 l}\right)^f \left(\frac{gl}{U_0^2}\right)^e \left(\frac{K}{\rho U_0^2}\right)^h.$$

But $\rho U_0 l / \mu = \mathrm{Re}$ (Reynolds number), $U_0^2 / gl = \mathrm{Fr}^2$ (Froude number),

$\rho l U_0^2 / \sigma = \mathrm{We}^2$ (Weber number), $U_0^2 / (K/\rho) = \mathrm{Ma}^2$ (Mach number),

so that

$$D = \rho U_0^2 l^2 \phi(\text{Re, Fr, We, Ma}). \tag{12.7}$$

Comparing this expression for drag with that of equation (12.4),

$$\tfrac{1}{2} C_D \rho U_0^2 A = \rho U_0^2 l^2 \phi(\text{Re, Fr, We, Ma}).$$

Since, for a body of a fixed shape,

$$A = \lambda l^2,$$

where λ is a numerical constant, and incorporating the constant $\tfrac{1}{2}$ as well as λ into the function ϕ', such that

$$\phi'(\ ) = \phi(\ )/\tfrac{1}{2}\lambda,$$

we obtain

$$\tfrac{1}{2}\lambda C_D \rho U_0^2 l^2 = \rho U_0^2 l^2 \phi(\text{Re, Fr, We, Ma})$$

and, finally,

$$C_D = \phi'(\text{Re, Fr, We, Ma}). \tag{12.8}$$

Equation (12.8) demonstrates that the drag coefficient is not a numerical constant, but a proportionality coefficient whose numerical value is a function of a series of dimensionless groups. These groups, and also others which, for simplicity, were not incorporated into the analysis (such as relative roughness, free stream turbulence level, cavitation number), come into play if the kind of forces represented by them are of significance. For example, Re will predominate in cases where viscous forces are dominant, Fr will only be significant in the presence of gravity waves (wave-making drag), Ma will dominate at high compressibility rates associated with high-speed gas flow, cavitation number will not be important unless cavitation occurs, etc.

It may, therefore, be said that, in general, the drag coefficients (and lift coefficients as well, since an analogous expression may be derived for them) for two geometrically similar situations will be the same if the other parameters are the same. For example, the drag on a smooth sphere in an incompressible fluid without cavitation will be such that

$$C_D = f(\text{Re}),$$

which means that so long as Re is the same the drag coefficient for any sphere of any size in any fluid will be the same provided other parameters are insignificant or irrelevant. For instance, if the free stream level of turbulence is not the same C_D will vary. This is why the value of C_D for a sphere falling in a stationary liquid (zero turbulence) may be different from that obtained when the sphere is stationary and the fluid is moving past it. Similarly, the boundary layer transition and separation affect both lift and drag and, hence, their values for two situations may not be the same unless all the parameters involved are the same.

It must also be remembered that if one effect is absent from (say) the model situation of a dynamically simpler system, it must also be absent from the prototype situation. For example, aerofoils tested in water must not cavitate if their performance in air is required, or submarine hulls when tested in a wind tunnel should not be subjected to high velocities to avoid Mach number effects.

Although the values of drag coefficient vary with Re and other parameters described, they depend primarily upon the shape of the body and its orientation with respect to the fluid flow. Appendix 2 gives values of C_D at specified Re for a variety of commonly encountered shapes.

12.4 RESISTANCE OF SHIPS

So far in this chapter, it was assumed that the body which is in relative motion with respect to the fluid is totally immersed in it. An important case, however, exists when the body is partly immersed in a liquid, an example being a ship. When it travels on the surface of water, two main sets of waves are produced, one originating at the bow and the other at the stern of the ship, both diverging from each side of the hull. Energy is required to generate these waves, and this energy originates from the propulsion system of the ship, which must therefore overcome not only the skin friction drag and the form drag but also the additional resistance in generating the waves. This additional resistance is known as *wave-making drag* or wave drag. (Note, however, that the term 'wave drag' is also used to describe the compressibility effects at supersonic velocities. See Section 13.1.)

It is not possible to measure the wave resistance directly. It is, therefore, normally obtained by measurement of the total drag and subtracting from it the calculated value of the skin friction drag:

$$\text{Wave-making resistance} = \text{Total drag} - \text{Skin friction drag.} \tag{12.9}$$

In this equation the form drag (due to the wake at the stern) is included in the wave-making resistance.

The application of dimensional analysis to the problem carried out in Section 12.3 indicated that the friction drag is dependent upon Reynolds number and the wave-making resistance upon Froude number. The latter is the ratio of inertia forces to the gravity forces and, in the present context, is defined as

$$\text{Fr} = v/\sqrt{(gl)}, \tag{12.10}$$

where v is the velocity of the ship and l is its length.

It may also be shown by dimensional analysis that the velocity of propagation of surface waves, sometimes called *celerity*, is given by

$$c = \sqrt{(gL)}\phi(d/L, h/L), \tag{12.11}$$

where L is the wavelength, h is the height of the waves, d is depth of water. If the ratio h/L is small, the celerity is not affected by it and is given by

$$c = \sqrt{[(gL/2\pi) \tan h(2\pi d/L)]},$$

which, for deep-water waves, where $d \gg L$, reduces to

$$c = \sqrt{(gL/2\pi)}. \tag{12.12}$$

Experiments which involve towing model ships indicate that the bow and stern waves produced by them travel at the same speed as the ship. This may be demonstrated by suddenly stopping the ship and measuring the wave velocity. Thus,

$$v = C = \sqrt{(gL/2\pi)}.$$

But, from (12.10),

$$v = \mathrm{Fr}\sqrt{(gl)},$$

so that, equating the two expressions, the Froude number may be written as

$$\mathrm{Fr} = 0.4\sqrt{(L/l)}. \tag{12.13}$$

This expression is important because it indicates two things. First, it shows that the Froude number describes completely the interrelation between the ship's length and the wavelength produced by it and, hence, determines the wave-making flow pattern at a given speed; it, therefore, also demonstrates that, for dynamic similarity of the wave-making resistance, the Froude number must be the same for the model and for the prototype. Second, it indicates how many wavelengths there are in a ship's length and, hence, describes the interaction between the bow and the stern systems of waves, which may be beneficial or detrimental to the ship's resistance. For example, at certain speeds the waves superpose in such a way that a travelling mound of water is built at the stern. The hydrostatic pressure of this mound acts on the ship pushing it forward and, hence, diminishing its wave-making resistance. However, at other speeds the superposition may produce a travelling trough at the stern, thus increasing the resistance. This is very pronounced at Fr of about 0.6, when the ship rides on the back of its first bow wave crest with the stern in the trough. This 'uphill' ride means a very much increased wave-making resistance. The most dramatic situation occurs, however, at Fr = 1, attained by planing speed boats, at which the boat rides on top of the wave crest and the wave-making resistance is then reduced very considerably. Figure 12.7

FIGURE 12.7

Wave-making resistance of a ship

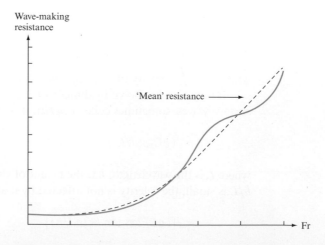

shows the 'ups and downs' of the wave-making resistance due to the interaction of the wave systems as the Froude number is increased. A 'mean' resistance curve is also shown. A more extensive 'mean' wave-making resistance curve is drawn, together with the frictional resistance curve, in Fig. 12.8. The two curves add up to the total resistance of a ship. Note the drop of the wave-making resistance at Fr = 1 and the consequent effect upon the total resistance.

FIGURE 12.8

Ship's resistance

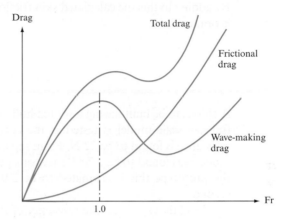

The procedure for predicting the total resistance of a ship during its design stage is based on towing model tests which are aimed at determination of the wave-making resistance (including form drag) and on calculations of frictional drag related to the mean wetted area of the ship. Model towing tests are carried out at a corresponding speed based on Froude number as the criterion. Thus, if suffix 'm' refers to the model and suffix 'p' to the prototype,

$$(\mathrm{Fr})_\mathrm{m} = (\mathrm{Fr})_\mathrm{p}$$

or $\quad v_\mathrm{m}/\sqrt{(gl_\mathrm{m})} = v_\mathrm{p}/\sqrt{(gl_\mathrm{p})},$

from which the corresponding speed

$$v_\mathrm{m} = v_\mathrm{p}\sqrt{(l_\mathrm{m}/l_\mathrm{p})}. \tag{12.14}$$

The total drag of the model D_m is measured at this speed. The frictional drag of the model $D_{f\mathrm{m}}$ is calculated using the boundary layer theory or empirical formulae determined by towing thin plates. Hence, the model wave-making resistance R_m is obtained:

$$R_\mathrm{m} = D_\mathrm{m} - D_{f\mathrm{m}}.$$

This is then scaled up to predict the wave-making resistance of the prototype R_p using the general drag relationship (equation (12.7)), but neglecting all parameters except Froude number, so that

$$R_\mathrm{p} = \rho_\mathrm{p} v_\mathrm{p}^2 l_\mathrm{p}^2\, \phi(\mathrm{Fr})_\mathrm{p}$$

and

$$R_m = \rho_m v_m^2 l_m^2 \, \phi(\text{Fr})_m.$$

Dividing one equation by the other, and since $(\text{Fr})_m = (\text{Fr})_p$ by design of the towing tests (corresponding speed), it follows that

$$R_p = R_m(\rho_p/\rho_m)(v_p/v_m)^2(l_p/l_m)^2. \tag{12.15}$$

By adding to this the calculated skin friction drag for the prototype D_{fp}, the total drag is obtained:

$$D_p = R_p + D_{fp}. \tag{12.16}$$

EXAMPLE 12.2

A ship is to be built having a wetted hull area of 2500 m² to cruise at 12 m s⁻¹. A 1/40 full-size scale model is tested at the corresponding speed and the measured total resistance is found to be 32 N. From separate tests, the skin friction resistance for the model was found to be $3.7v^{1.95}$ (newtons per square metre of wetted area) whereas for the prototype this is estimated to be $2.9v^{1.8}$, where v is the velocity in metres per second.

   Find the expected total resistance of the full-size ship if it operates in sea water of density 1025 kg m⁻³ whereas the model is tested in fresh water.

Solution

The corresponding speed at which the model must be tested is given by equation (12.14), obtained by equating the Froude number for the model and for the prototype:

$$v_m = v_p \sqrt{(l_m/l_p)} = 12\sqrt{(1/40)} = 1.90 \, \text{m s}^{-1}.$$

Now, the skin friction drag of the model at this test speed will be

$$D_{fm} = 3.7v_m^{1.95} A_m = 3.7(1.90)^{1.95}(2500/40^2) = 20.2 \, \text{N}$$

and, hence, the model's wave-making resistance will be

$$R_m = D_m - D_{fm} = 32 - 20.2 = 11.8 \, \text{N}.$$

Now, using equation (12.15), the wave-making resistance of the ship is obtained:

$$R_p = R_m \frac{\rho_p}{\rho_m}\left(\frac{l_p}{l_m}\right)^2\left(\frac{v_p}{v_m}\right)^2 = 11.8\frac{1025}{1000}(40)^2\left(\frac{12}{1.90}\right)^2 = 771.9 \, \text{kN}.$$

Now the skin friction drag for the ship is calculated from

$$D_{fp} = 2.9v_p^{1.8} \times A_p = 2.9(12)^{1.8}2500 \times 10^{-3} = 635.13 \, \text{kN}$$

and, hence, the total drag of the ship,

$$D_p = R_p + D_{fp} = 771.9 + 635.13 = \mathbf{1407.06 \, kN}.$$

12.5 FLOW PAST A CYLINDER

In this section a thin circular cylinder of infinite length, placed transversely in a fluid stream, will be used to discuss in greater detail the changes in flow pattern and in the drag coefficient which accompany the variation of Reynolds number. It must be remembered that for a given cylinder of a given diameter immersed in a given fluid the Reynolds number is directly proportional to the velocity and, therefore, the variation with Reynolds number could be imagined as the variation with velocity for a given cylinder. We assume also that there are no end effects and therefore that the flow is two dimensional.

At very small values of Re, say below 0.5, the inertia effects are negligible and the flow pattern is very similar to that for ideal flow, the pressure recovery being nearly complete. Thus, pressure drag is negligible and the profile drag is nearly all due to skin friction. Figure 12.9(a) shows the flow pattern and associated pressure distribution for such a case. Figure 12.10 indicates a straight line relationship between C_D and Re in this range, from which we conclude that the drag D is directly proportional to velocity U_0.

At increased Re, say between 2 and 30, separation of the boundary layer occurs at points S as indicated in Fig. 12.9(b). Two symmetrical eddies, rotating in opposition to one another, are formed. They remain fixed in position and the main flow closes behind them. The separation of the boundary layer is reflected in the variation of C_D graph by the curvature of the line indicating that the drag is now proportional to U_0^n, where $n \to 2$.

Further increase of Re tends to elongate the fixed eddies, which then begin to oscillate until at about Re = 90, depending upon the free stream turbulence level, they break away from the cylinder as shown in Fig. 12.9(c). The breaking away occurs alternately from one and then the other side of the cylinder, the eddies being washed away by the main stream. This process is intensified by a further increase of Re, whereby the shedding of eddies from alternate sides of the cylinder is continuous, thus forming in the wake two discrete rows of vortices, as shown in Fig. 12.9(c). This is known as a *vortex street* or von Kármán vortex street. At this stage the contribution of pressure drag to the profile drag is about three-quarters. Von Kármán showed analytically, and confirmed experimentally, that the pattern of vortices in a vortex street follows a mathematical relationship, namely

$$h/l = (1/\pi) \sinh^{-1}(l) - 0.281, \tag{12.17}$$

where h and l are indicated in Fig. 12.11.

It will be seen that shedding of each vortex produces circulation and, hence, gives rise to a lateral force on the cylinder. Since these forces are periodic following the frequency of vortex shedding, the cylinder may be subjected to a forced vibration. The familiar 'singing' of telephone wires is due to this phenomenon, caused by a lateral wind, whereas the collapse of suspension bridges and the 'flutter' of aerofoils are the result of a resonance between the natural frequency of the body and the frequency of forced vibration due to vortex shedding.

The frequency of such forced vibration, sometimes called self-induced vibration, may be calculated from an empirical formula due to Vincent Strouhal:

$$fd/U_0 = 0.198(1 - 19.7/\text{Re}), \tag{12.18}$$

FIGURE 12.9

Flow past a cylinder

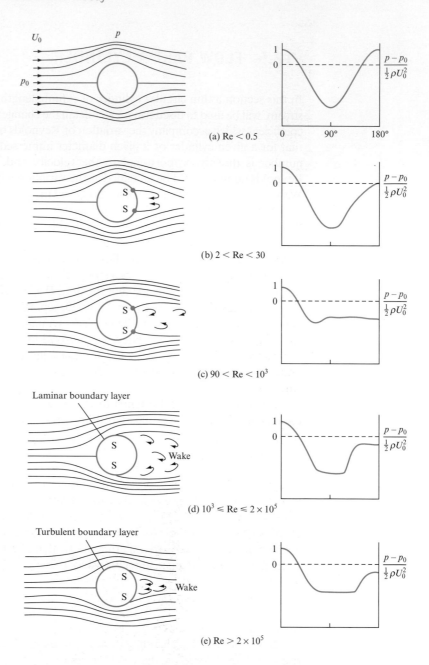

(a) Re < 0.5

(b) 2 < Re < 30

(c) 90 < Re < 10³

(d) 10³ ≤ Re ≤ 2 × 10⁵

(e) Re > 2 × 10⁵

in which

$$fd/U_0 = \text{Str} \tag{12.19}$$

and is known as *Strouhal number*. The formula applies to $250 < \text{Re} < 2 \times 10^5$.

It is fortunate that at higher values of Reynolds number the vortices disappear because of high rates of shear and are then replaced by a highly turbulent wake. This produces an increase in the value of C_D at about $\text{Re} = 3 \times 10^4$. Pressure drag is now responsible for nearly all the drag.

FIGURE 12.10
Drag coefficient for a sphere and a cylinder

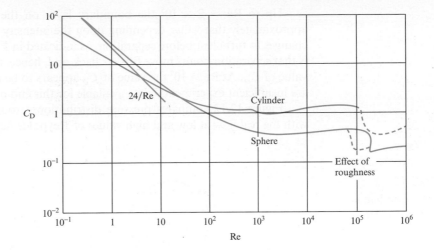

FIGURE 12.11
(a) Von Kármán vortex street.
(b) Smoke visualization showing von Kármán vortex street

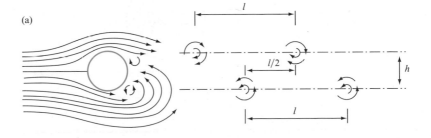

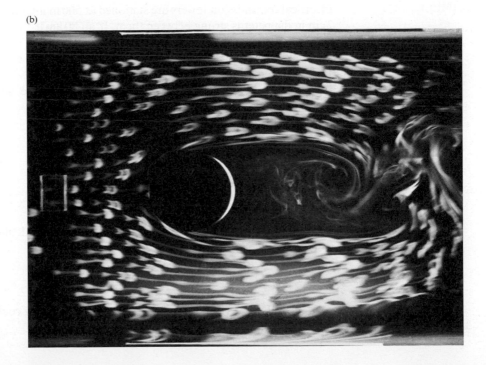

Up to Re $\simeq 2 \times 10^5$ the boundary layer on the cylinder is laminar, but at approximately that value, depending upon the intensity of free stream turbulence, it changes to turbulent before separation, as indicated in Fig. 12.9(d). The effect of this is that separation points move further back and, hence, there is a marked drop in the value of C_D. At Re $> 10^7$ the value of C_D appears to be independent of Re, but there are insufficient experimental data available for this end of the range.

Figure 12.12 compares pressure distributions around a cylinder for ideal flow with the real flow at low and high values of Reynolds number.

FIGURE 12.12

Pressure distribution around a cylinder

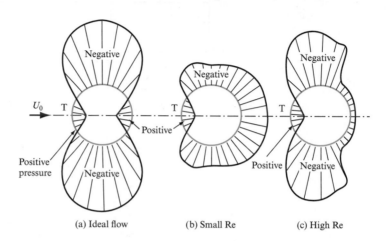

(a) Ideal flow   (b) Small Re   (c) High Re

EXAMPLE 12.3

Electrical transmission towers are stationed at 500 m intervals and a conducting cable 2 cm in diameter is strung between them. If an 80 km h^{-1} wind is blowing transversely across the wires, calculate the total force each tower carrying 20 such cables is subjected to. Assume there is no interference between the wires and take air density as 1.2 kg m^{-3} and viscosity 1.7×10^{-5} N s m^{-2}.

Also establish whether the wires are likely to be subjected to self-induced vibrations and if so what would be the frequency?

Solution

Drag on one wire,

$$D = \tfrac{1}{2}\rho C_D U_0^2 A.$$

In order to establish the value of C_D, it is necessary to calculate the value of Re first.

$$\mathrm{Re} = \frac{\rho U_0 d}{\mu} = \frac{1.2 \times 80 \times 1000 \times 0.02}{3600 \times 1.7 \times 10^{-5}} = 3.14 \times 10^4.$$

Now, from Fig. 12.10, for the above value of Re the drag coefficient is

$$C_D = 1.2.$$

The projected area of a single wire between towers,

$$A = 0.02 \times 500 = 10 \text{ m}^2.$$

Hence, drag on wire,

$$D = \tfrac{1}{2} \times 1.2 \times 1.2 \left(\frac{80 \times 1000}{3600} \right)^2 \times 10 = 3556 \text{ N}.$$

Therefore, the force on each tower due to 20 cables is

$$F = 20 \times 3556 = \textbf{71.11 kN}.$$

Since $250 < \text{Re} < 10^5$, 'singing' may occur. Using equation (12.18),

$$f = 0.198 \frac{U_0}{d} \left(1 - \frac{19.7}{\text{Re}} \right)$$

$$= 0.198 \frac{80 \times 1000}{3600 \times 0.02} \left(1 - \frac{19.7}{3.14 \times 10^4} \right)$$

$$= \textbf{219.9 Hz}.$$

12.6   FLOW PAST A SPHERE

So far our discussion of drag has been confined to two-dimensional flow. We will now examine the flow past the simplest of all three-dimensional bodies, the sphere. There is a great similarity in the development of drag at increasing Re between the sphere and the cylinder, except that the vortex street associated with the latter and two-dimensional bodies such as aerofoils is not formed in the case of three-dimensional bodies. Instead, a vortex ring occurs, which for a sphere is formed at about Re = 10 and becomes unstable at $200 < \text{Re} < 2000$ when it tends to move downstream of the body, to be immediately replaced by a new ring. This process is not periodic, however, and does not give rise to vibrations of the sphere.

The study of flow past a sphere is of great practical importance because it is the foundation of a branch of fluid mechanics, namely *particle mechanics*. This subject concerns itself with all problems associated with the flow of solid particles in a fluid or liquid particles in a gas and encompasses practical problems such as pneumatic conveying, particle separation, sedimentation, filtration, etc. In practice, most particles are not spheres and there are ways of classifying them in accordance with shape, but in the end they are always related to the sphere as the simplest theoretical shape and, therefore, most amenable to both analytical as well as experimental investigations.

At very low values of Re, during so-called 'creeping' flow around a sphere, it may be assumed that the inertial effects are negligible and, hence, the steady flow Navier–Stokes equations may be greatly simplified by omission of the inertia term, thus enabling the calculation of viscous drag.

Stokes obtained the solution for drag by expressing the simplified Navier–Stokes equation together with the continuity equation in polar coordinates and using the boundary conditions that all velocity components are zero at the surface of the sphere. His solution is the well-known equation,

$$D = 6\pi\mu R U_0, \tag{12.20}$$

in which R is the radius of the sphere, U_0 is the free stream velocity of the fluid and μ is its dynamic viscosity. This relationship holds true for Re < 0.1, but may be used with negligible error up to Re = 0.2. In this range, often referred to as *Stokes flow*, the drag coefficient may be calculated by equating the general drag equation (12.4) to the Stokes solution,

$$\tfrac{1}{2}C_D \rho U_0^2 A = 6\pi\mu R U_0,$$

but $A = \pi R^2,$

so that

$$C_D = 12\mu/\rho U_0 R = 24\mu/\rho U_0 d,$$

where $d = 2R$ is the diameter of the sphere. But $\rho U_0 d/\mu = $ Re, based on the sphere diameter, and, therefore,

$$C_D = 24/\text{Re} \quad \text{for Re} < 0.2. \tag{12.21}$$

At larger values of Re, separation of the boundary layer occurs and the Navier–Stokes equations cannot be used. It is, therefore, necessary to rely on empirical expressions. One such formula extends Stokes' law to Re < 100 and is as follows:

$$C_D = (24/\text{Re})(1 + \tfrac{3}{16}\text{Re})^{1/2}. \tag{12.22}$$

Beyond Re = 100 it is necessary to use values of C_D as a function of Re from the graph, as in Fig. 12.13.

Stokes' formula forms the basis for the determination of viscosity of oils, which consists in allowing a sphere of known diameter to fall freely in the oil. After initial

FIGURE 12.13

Drag coefficient for a sphere

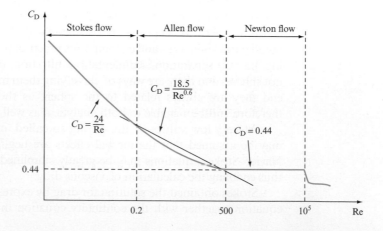

acceleration, the sphere attains a constant velocity known as *terminal velocity* which is reached when the external drag on the surface and buoyancy, both acting upwards and in opposition to the motion, become equal to the downward force due to gravity. At this equilibrium condition,

$$6\pi\mu U_t R + \tfrac{4}{3}\pi R^3 \rho g = \tfrac{4}{3}\pi R^3 \rho_p g,$$

$$(\text{Drag}) + (\text{Buoyancy}) = (\text{Gravity})$$

where ρ = density of the fluid, ρ_p = density of the sphere material, U_t = terminal velocity. Thus,

$$6\mu U_t = \tfrac{4}{3}R^2(\rho_p - \rho)g$$

and $$U_t = (2R^2/9\mu)(\rho_p - \rho)g$$

or, in terms of the sphere's diameter d,

$$U_t = (d^2/18\mu)(\rho_p - \rho)g. \tag{12.23}$$

By timing the rate of fall of the sphere, the terminal velocity is measured and, hence, the viscosity of the fluid may be determined using the above equation. Alternatively, the method may also be used to determine the mean diameter of spherical particles by allowing them to settle freely in a liquid of known viscosity.

At large values of Reynolds number, the flow over the front half of a sphere may be divided into a thin boundary layer region, where viscosity effects are dominant, and an outer region, in which the flow corresponds to that of an inviscid fluid. The pressure is decreasing over the front half of the sphere from the stagnation point onwards, thus having a stabilizing effect on the boundary layer, which remains laminar up to about $\mathrm{Re} = 5 \times 10^5$. Beyond the minimum pressure point on the sphere (at about 80°) the boundary layer is subjected to an adverse pressure gradient and separation occurs. At low Re it begins at the rear stagnation point and with increasing Re it moves forward, reaching the 80° point from the front stagnation at a Reynolds number of about 1000. Pressure drag begins to dominate and C_D becomes independent of Re until, at about $\mathrm{Re} = 5 \times 10^5$, transition in the boundary layer occurs, becoming turbulent before separation. This moves the separation point to the rear, making the wake smaller and abruptly reducing the value of C_D from about 0.5 to 0.2.

The experimental determination of C_D for a sphere is difficult because, first, the method of supporting the sphere in the wind tunnel affects the results and, second, because the results depend upon the free stream turbulence level, which is difficult to control, and upon the roughness of the sphere, which is difficult to reproduce. It is not surprising, therefore, that the early experimenters produced conflicting results. Any data which do not specify the method of support and free stream turbulence level should be viewed with caution.

However, for the purposes of particle mechanics the flow past a sphere is subdivided into three regimes as follows:

1. Stokes flow, $\mathrm{Re} < 0.2,$ $C_D = 24/\mathrm{Re};$

2. Allen flow, $0.2 < \mathrm{Re} < 500,$ $C_D = f(\mathrm{Re});$

3. Newton flow, $500 < \mathrm{Re} < 10^5,$ $C_D = \text{constant} = 0.44.$

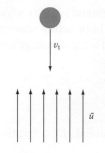

FIGURE 12.14

Spherical particle falling into a vertical fluid stream

These regimes are shown in Fig. 12.13.

The calculation of terminal velocity is of great importance in particle mechanics because it forms the basis of such operations as settling or sorting. In sorting, for example, solid particles are introduced into a vertical stream of fluid, as shown in Fig. 12.14. If the fluid were stationary, that is $\bar{u} = 0$, the particle would attain a constant terminal descending velocity v_t. If, however, the fluid is moving vertically up with a velocity $\bar{u}$ there are three distinct possibilities:

1. when $\bar{u} < v_t$ the particle will be falling down with an absolute velocity $v = v_t - \bar{u}$, where v_t is now the relative velocity between the fluid and the particle and governs the drag on the particle;

2. when $\bar{u} = v_t$ the particle will be suspended, having an absolute velocity of $v = 0$;

3. when $\bar{u} > v_t$ the particle will move upwards with the fluid with an absolute velocity $v = \bar{u} - v_t$.

Thus, if particles of different size or weight are introduced into a vertical fluid stream, some will be carried over and some will descend, thus enabling *sorting* to be carried out.

The difficulty in deciding the correct sorting velocity $\bar{u}$ lies in the fact that it usually corresponds to the particle terminal velocity in the Allen flow regime, where C_D is a function of Reynolds number. This cannot be calculated until the velocity is known, which is precisely the variable we are trying to establish. To demonstrate the difficulty let us examine a general case for terminal velocity. It occurs when

$$\text{Drag} + \text{Buoyancy} = \text{Gravitational force,}$$

but drag

$$D = \tfrac{1}{2}\rho C_D A v^2.$$

For a spherical particle,

$$A = \pi d^2/4$$

and

$$D = \tfrac{1}{8}\rho C_D \pi d^2 v^2,$$

so that

$$\tfrac{1}{8}\rho C_D \pi d^2 v^2 + (\pi d^3/6)\rho g = (\pi d^3/6)\rho_p g.$$

Simplifying and rearranging,

$$\tfrac{1}{8}\rho C_D v^2 = (d/6)(\rho_p - \rho)g$$

and

$$v = v_t = \sqrt{[\tfrac{4}{3}dg(\rho_p - \rho)/C_D\rho]}. \tag{12.24}$$

Thus,

$$v = f(C_D) = f_1(\text{Re}) = f_2(v).$$

For Stokes flow, $C_D = 24/\text{Re}$, and the substitution gives

$$v_t = d^2(\rho_p - \rho)g/18\mu$$

which is the equation (12.23) already derived.

For Allen flow two alternatives are possible.

1. The $C_D = f(\text{Re})$ curve may be approximated to a straight line, giving

$$C_D = 18.5/\text{Re}^{0.6}. \tag{12.25}$$

This yields a cumbersome and inaccurate equation for v_t.

2. A more satisfactory procedure is to eliminate v_t from equation (12.24) and to replace it by Re in the following manner. From (12.24),

$$C_D = 4d(\rho_p - \rho)g/3v_t^2 \rho;$$

multiplying both sides by

$$\text{Re}^2 = (v_t \, d\rho/\mu)^2,$$
$$C_D \, \text{Re}^2 = 4d^3(\rho_p - \rho)\rho g/3\mu^2. \tag{12.26}$$

The right-hand side of this equation can be calculated for any given fluid and particle combination, since all relevant parameters are known. Thus, the value of $C_D \, \text{Re}^2$ becomes known. This is then referred to a graph relating $C_D \, \text{Re}^2$ to Re, shown in Fig. 12.15. This graph is simply a replot of the Allen part of the $C_D = f(\text{Re})$ graph.

FIGURE 12.15

The $C_D \, \text{Re}^2$ vs. Re graph

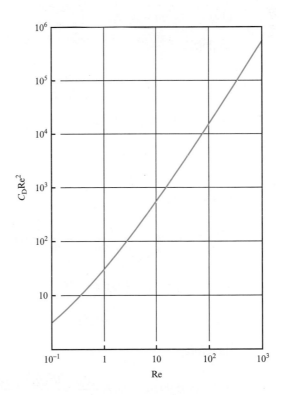

EXAMPLE 12.4

A particle of 1 mm diameter and density $1.1 \times 10^3 \, \text{kg m}^{-3}$ is falling freely from rest in an oil of $0.9 \times 10^3 \, \text{kg m}^{-3}$ density and $0.03 \, \text{N s m}^{-2}$ viscosity. Assuming that Stokes' law applies, how long will the particle take to reach 99 per cent of its terminal velocity? What is the Reynolds number corresponding to this velocity?

Solution

The equation of motion for the particle is

$$\text{Mass} \times \text{Acceleration} = \text{Resultant force on the body in the direction of motion}$$

$$= \text{Gravity} - \text{Buoyancy} - \text{Drag},$$

$$m\frac{\mathrm{d}v}{\mathrm{d}t} = mg - m_0 g - D,$$

where m = mass of particle, m_0 = mass of oil displaced by the particle, D = drag on the particle. Thus

$$\frac{\mathrm{d}v}{\mathrm{d}t} = g - \frac{m_0}{m}g - \frac{D}{m},$$

but $D = 3\pi\mu dv$ and $m = \frac{1}{6}\pi d^3 \rho_\mathrm{p}.$

Therefore, $D/m = 18\mu v/d^2\rho_\mathrm{p}$;

also $m_0/m = \rho_0/\rho_\mathrm{p},$

so that

$$\frac{\mathrm{d}v}{\mathrm{d}t} = g\left(1 - \frac{\rho_0}{\rho_\mathrm{p}}\right) - \frac{18\mu v}{d^2\rho_\mathrm{p}}.$$

To facilitate integration, let $A = g(1 - \rho_0/\rho_\mathrm{p})$ and $B = 18\mu/d^2\rho_\mathrm{p}$, so that

$$\frac{\mathrm{d}v}{\mathrm{d}t} = A - Bv.$$

Hence,

$$t = \int_0^{0.99v_\mathrm{t}} \frac{\mathrm{d}v}{A - Bv} = \left[-\frac{1}{B}\log_\mathrm{e}(A - Bv)\right]_0^{0.99v_\mathrm{t}}$$

$$= -\frac{1}{B}\log_\mathrm{e}(A - 0.99Bv_\mathrm{t}) + \frac{1}{B}\log_\mathrm{e} A = \frac{1}{B}\log_\mathrm{e}\left(\frac{A}{A - 0.99Bv_\mathrm{t}}\right).$$

But $v_\mathrm{t} = d^2(\rho_\mathrm{p} - \rho)g/18\mu$;

therefore,

$$0.99Bv_\mathrm{t} = 0.99\frac{18\mu}{d^2\rho_\mathrm{p}} \times \frac{d^2(\rho_\mathrm{p} - \rho)g}{18\mu} = 0.99\frac{(\rho_\mathrm{p} - \rho)g}{\rho_\mathrm{p}}$$

$$= 0.99(1 - \rho/\rho_\mathrm{p})g = 0.99A.$$

Hence,

$$t = \frac{1}{B}\log_\mathrm{e}\left(\frac{A}{A - 0.99A}\right) = \frac{1}{B}\log_\mathrm{e}\left(\frac{1}{1 - 0.99}\right) = \frac{1}{B}\log_\mathrm{e} 100 = \frac{4.60}{B}.$$

But

$$B = 18\mu/d^2\rho_p = 18 \times 0.03/[(0.1 \times 10^{-2})^2 \times 1.1 \times 10^3] = 490$$

and $\quad t = 4.60/490 = \textbf{0.0094 s}.$

Terminal velocity,

$$v_t = d^2(\rho_p - \rho_0)g/18\mu = 10^{-6}(1.1 - 0.9)10^3 \times 9.81/(18 \times 0.03)$$
$$= 3.63 \times 10^{-3}\,\text{m s}^{-1}.$$

Reynolds number at this velocity,

$$\text{Re} = \rho v_t d/\mu = 0.9 \times 10^3 \times 3.63 \times 10^{-3} \times 10^{-3}/0.03$$
$$= \textbf{0.1089}.$$

EXAMPLE 12.5

A solid particle of specific gravity 2.4, when settling in oil of specific gravity 0.9 and viscosity 0.027 P, attains a terminal velocity of 3×10^{-3} m s^{-1}. What should be the velocity of an air stream (density 1.3 kg m^{-3}), blowing vertically up, in order to carry the particle at a velocity of 0.5 m s^{-1}? Viscosity of air may be taken as 1.7×10^{-5} N s m^{-2}.

Solution

It is first necessary to determine the diameter of the particle. This may be done from the settling data in oil, assuming Stokes flow. This assumption will have to be checked. From equation (12.23),

$$v_t = d^2(\rho_p - \rho)g/18\mu.$$

Therefore,

$$d = \sqrt{\left[\frac{18\mu v_t}{(\rho_p - \rho)g}\right]} = \sqrt{\left[\frac{18 \times 0.027 \times 10^{-1} \times 3 \times 10^{-3}}{(2.4 - 0.9)10^3 \times 9.81}\right]}$$

$$= 0.0995 \times 10^{-3}\,\text{m}.$$

To check the flow regime,

$$\text{Re} = \frac{vd\rho}{\mu} = \frac{3 \times 10^{-3} \times 0.0995 \times 10^{-3} \times 0.9 \times 10^3}{0.0027}$$

$$= 0.0995.$$

Therefore Re < 0.1 and so the assumption of Stokes flow was correct.

Now, for the particle to move vertically upwards with absolute velocity v in an air stream of absolute velocity u, the relative velocity v_t is

$$v_t = u - v.$$

It is this relative velocity which is responsible for the drag on the particle. Therefore,

$$v_t = \sqrt{[4dg(\rho_p - \rho_{air})/3C_D\rho_{air}]}.$$

Since we do not know C_D, we calculate

$$
\begin{aligned}
C_D \, Re^2 &= 4d^3(\rho_p - \rho_{air})g\rho_{air}/3\mu^2 \\
&= 4(0.0995 \times 10^{-3})^3(2.4 \times 10^3 - 1.3)9.81 \times 1.3/[3 \times (1.7 \times 10^{-5})^2] \\
&= 139
\end{aligned}
$$

and from Fig. 12.15 we read that $Re = 5$ and hence the relative velocity v_t may be obtained from the expression for Re:

$$Re = \rho v d/\mu;$$

therefore,

$$
\begin{aligned}
v_t &= \mu \, Re/\rho d = 1.7 \times 10^{-5} \times 5/[1.3 \times 0.0995 \times 10^{-3}] \\
&= 0.657 \text{ m s}^{-1}.
\end{aligned}
$$

The upward air velocity required:

$$u = v_t + v = 0.657 + 0.5 = \mathbf{1.157 \text{ m s}^{-1}}.$$

12.7   FLOW PAST AN INFINITELY LONG AEROFOIL

An aerofoil may be defined as a streamlined body designed to produce lift. There are other lift-producing surfaces such as hydrofoils or circular arcs. In general the following elementary aerofoil theory also applies to these surfaces. There is an accepted terminology concerning aerofoils and familiarization with it is necessary in order to understand the discussion of flow past aerofoils.

Figure 12.16 shows an aerofoil section and the following are some of the most important terms relating to it:

FIGURE 12.16
An aerofoil

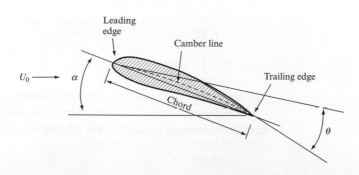

Leading edge	the front, or upstream, edge, facing the direction of flow;
Trailing edge	the rear, or downstream, edge;
Chord line	a straight line joining the centres of curvature of the leading and trailing edges;
Chord, c	the length of chord line between the leading and trailing edges;
Camber line	the centreline of the aerofoil section;
Camber, δ	the maximum distance between the camber line and the chord line;
Percentage camber	$= 100\delta/c$ per cent is a measure of aerofoil curvature;
Span, b	the length of the aerofoil in the direction perpendicular to the cross-section;
Plan area, A	the area of the projection of the aerofoil on the plane containing the chord line. If the aerofoil is of constant cross-section, $A = c \times b$;
Mean chord, $\bar{c}$	$= A/b$;
Aspect ratio, AR	$= (\text{Span})/(\text{Mean chord}) = b/c - b^2/A$;
Deviation, θ	angle between the tangent to camber line at trailing edge and the tangent to camber line at leading edge;
Angle of attack (incidence)	the angle between the direction of the relative motion and the chord line;
Pressure coefficient, C_p	$= (p - p_0)/\frac{1}{2}\rho U_0$, where p is the local pressure and p_0 is the pressure far upstream of the aerofoil where velocity is V_0.

The primary purpose of an aerofoil is to produce lift when placed in a fluid stream. It will, of course, experience drag at the same time. In order to minimize drag, an aerofoil is a streamlined body. A measure of its usefulness as a wing section of an aircraft or as a blade section of a pump or turbine is the ratio of lift to drag. The higher this ratio is, the better the aerofoil, in the sense that it is capable of producing high lift at a small drag penalty. In an aircraft it is the lift on the wing surfaces which maintains the plane in the air. At the same time it is the drag which absorbs all the engine power necessary for the craft's forward motion. Similarly, in pumps, the head generated is due to the lift produced by the impeller blades, whereas the torque necessary to rotate the blades overcomes the drag on them. Thus, the lift/drag ratio,

$$\frac{\text{Lift}}{\text{Drag}} = \frac{\frac{1}{2}\rho C_L U_0^2 A}{\frac{1}{2}\rho C_D U_0^2 A} = \frac{C_L}{C_D}. \tag{12.27}$$

The creation of lift is, therefore, of primary importance. How does an aerofoil produce lift? How does it start when the motion of the aerofoil begins and how is it maintained during the motion? We will attempt to answer these questions by reference to potential flow theory, expounded in Chapter 7, and by reference to the boundary layer theory.

The Kutta–Joukowsky law, derived for a cylinder with circulation, relates lift to circulation. It is not limited to cylinders, but may be shown to apply to any two-dimensional section. The important point, however, is that lift exists only if there is circulation around the section.

A rotating cylinder placed in a real fluid produces circulation by viscous action of its rotating surface on the fluid. Aerofoils, however, do not rotate and, hence, there must be a different mechanism of producing and maintaining circulation. Let us first consider how the circulation starts. It was shown earlier how a vortex is formed, either due to separation of a stream or at the rear of a blunt body. Similarly, if two parallel

FIGURE 12.17
Starting vortex

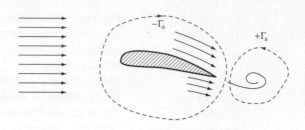

streams of unequal velocity meet, there is a discontinuity of velocity at their interface and that produces a vortex. In the same manner, when a slightly inclined aerofoil starts motion it splits the flow into two streams: one over the upper surface and one over the lower surface. The velocities in these streams are not equal due to the inclination of the aerofoil and, therefore, when they meet at the trailing edge a starting vortex is produced as shown in Fig. 12.17. This vortex is cast off soon after the beginning of the motion. It does, however, give rise to circulation around the aerofoil, which is equal in strength (but opposite in sign) to the circulation of the starting vortex.

Let us now consider the situation during the motion of the aerofoil. For the circulation to exist there must be vorticity in the stream which cuts across the circulation contour. The flow around the aerofoil may be considered as potential outside the boundary layer and, therefore, it is irrotational there. Hence, there is no vorticity and there cannot be any circulation associated with it. Within the boundary layer, however, the flow is viscous and owing to the velocity gradient vorticity exists there.

FIGURE 12.18
Circulation in the
boundary layer

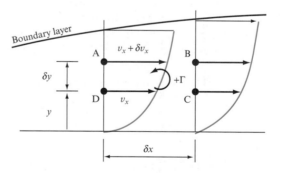

Consider an element of fluid ABCD in the boundary layer as shown in Fig. 12.18. Taking δx as small, the change of velocity in the x direction ($(\partial v_x / \partial x)\,\mathrm{d}x$) may be neglected. The circulation for the element becomes

$$\Gamma_{\text{ABCD}} = -(v_x + \delta v_x)\delta x + v_x \delta x = -\delta v_x \delta x$$

and vorticity,

$$\zeta_{\text{ABCD}} = \delta v_x \delta x / \delta x \delta y = \delta v_x / \delta y.$$

FIGURE 12.19

(a) Flow around an aerofoil, with circulation.
(b) Helium bubble in air flow visualization around an aerofoil

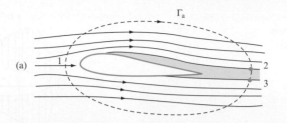

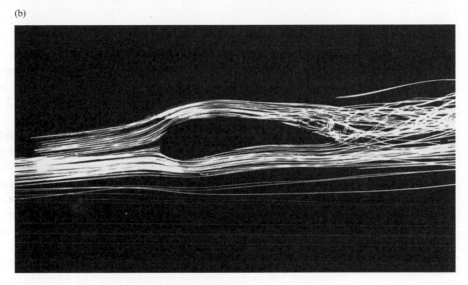

Thus, there is vorticity in the boundary layer and its value depends upon the velocity gradient.

Figure 12.19 shows the flow around an aerofoil with an exaggerated boundary layer and vorticity within it. It will be noticed that, because the velocity gradients are of opposite sign on the top and bottom surfaces, the vorticity in the upper boundary layer is clockwise whereas the vorticity in the lower boundary layer is anticlockwise. If these two vorticities are equal in strength, they cancel each other and the resultant circulation around the contour is zero. This occurs, for example, in the case of a symmetrical aerofoil without camber placed in a fluid stream at zero angle of incidence. Because of complete flow symmetry, the growth of the boundary layer at the top and bottom is identical and hence vorticities are the same in strength and of opposite rotation. However, if the vorticity over the top surface exceeds that over the bottom, the resultant circulation around the aerofoil will be clockwise, as shown in Fig. 12.19(a). The circulation contour may be drawn arbitrarily around the aerofoil and, provided it contains the whole of the boundary layer, the value of circulation will not be affected because the irrotational flow outside the boundary layer makes no contribution to it.

We therefore deduce that the circulation around the aerofoil, $\Gamma_a = \oint v\,\mathrm{d}s$, will be clockwise if the velocities over the upper surface are greater than the velocities over the lower surface, or, more exactly, if

$$\oint_1^2 v\,\mathrm{d}s > \oint_3^1 v\,\mathrm{d}s.$$

FIGURE 12.20
Pressure distribution
around an aerofoil

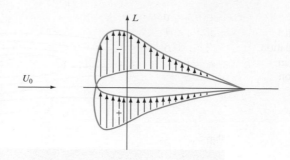

Such velocity distributions must, in accordance with Bernoulli's equation, be accompanied by higher pressures on the bottom surface and lower pressures on the top surface. Such a pressure distribution, as shown in Fig. 12.20, gives rise to a resultant upward force, namely the lift.

Summarizing, at a small angle of incidence the fluid flowing over the bottom surface of an aerofoil is slowed down, thus increasing the pressure, which means that the pressure gradient there is favourable, the boundary layer thickness is small and the anticlockwise vorticity in it is also small. Over the upper surface the vorticity is greater, the pressure gradient adverse, the boundary layer thicker and the clockwise vorticity in it greater. Thus, the resulting pressure difference gives rise to lift, which may be related to the circulation around the aerofoil. By the same argument, a negative lift (downward force) may exist for negative values of angles of incidence.

The foregoing discussion indicates a strong dependence of lift upon the incidence angle. Let us consider this in greater detail by referring to the potential flow around a cylinder as our model. It will be remembered from our discussion of this topic in Chapter 7 that the increase of circulation around the cylinder alters the position of stagnation points, as shown in Fig. 7.29. Since lift (by Kutta–Joukowsky, $L = \rho U_0 \Gamma$) depends upon circulation, it may therefore be related to the position of stagnation points. The analogy between the cylinder and the aerofoil is illustrated in Fig. 12.21.

The stream function for the flow around a cylinder with circulation is given by equation (7.64):

$$\Psi = U_0(r - a^2/r) \sin \theta + (\Gamma/2\pi) \log_e r,$$

FIGURE 12.21
Relationship between
the zero-lift line on the
aerofoil and the position
of stagnation points on a
cylinder with circulation

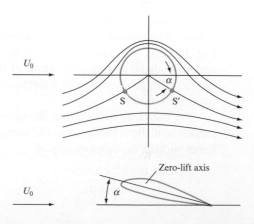

where U_0 = upstream velocity; a = radius of the cylinder; Γ = circulation around the cylinder; r, θ = cylindrical coordinates.

The velocity on the cylinder surface is the tangential velocity v_θ (since there is no velocity into or out of the cylinder) at $r = a$. Thus, since (from equation (7.13))

$$v_\theta = -\frac{\partial \Psi}{\partial r},$$

it follows that

$$v_\theta = -U_0(1 + a^2/r^2) \sin \theta - \Gamma/2\pi r$$

and, making $r = a$,

$$v_\theta = -2U_0 \sin \theta - \Gamma/2\pi a. \tag{12.28}$$

At stagnation points, $v_\theta = 0$, and, therefore, $\Gamma/2\pi a = -2U_0 \sin \theta$, so that the location of the stagnation points is given by

$$\theta = \sin^{-1}(-\Gamma/4\pi U_0 a).$$

This equation results in two solutions, namely

$$\theta_1 = -\alpha \quad \text{and} \quad \theta_2 = -(180° - \alpha).$$

The corresponding angle of incidence α for the aerofoil is, therefore, measured from the position of the aerofoil in the stream such that there is zero lift. The axis of the aerofoil parallel to the direction of flow under this condition and drawn through the trailing edge is known as the zero-lift axis and is shown in Fig. 12.22. Thus, α_0 is the negative angle of incidence corresponding to no lift.

FIGURE 12.22
The zero-lift axis of an aerofoil

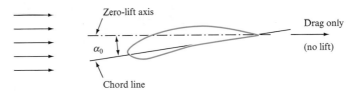

Returning now to our analogy with the cylinder, it is possible to express the circulation in terms of α (equation (7.65)):

$$\Gamma = -4\pi U_0 a \sin \alpha$$

and, using Kutta–Joukowsky's expression, which is for clockwise circulation defined as negative, the lift becomes

$$L = \rho U_0 \Gamma = 4\pi a \rho \sin \alpha U_0^2. \tag{12.29}$$

Comparing this with equation (12.5) for lift, namely

$$L = \tfrac{1}{2} C_L \rho U_0^2 A,$$

the following relationship is obtained:

$$4\pi a \rho U_0^2 \sin \alpha = \tfrac{1}{2} C_L \rho U_0^2 A,$$

so that

$$C_L = (8\pi a / A) \sin \alpha, \tag{12.30}$$

indicating that the coefficient of lift is directly proportional to sin α, which, for small angles of incidence, means that it is proportional to the angle of incidence. This theory is in good agreement with experimental results. Figure 12.23 shows the calculated values of C_L for a given aerofoil together with the measured values, as functions of the angle of attack. The drag coefficient is also shown.

FIGURE 12.23

Calculated and experimental values of the coefficient of lift for an aerofoil

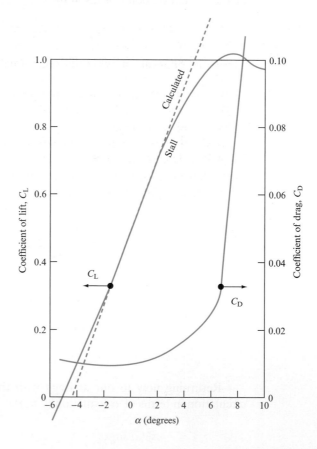

The good agreement at small angles of attack is related to the fact that there is no separation of the boundary layer at these small angles. As the angle of incidence is increased, however, separation occurs at the top surface near the trailing edge, thus reducing slightly the rate of increase of the lift with the angle of attack. As the

FIGURE 12.24

Separation due to increased angle of incidence of an aerofoil

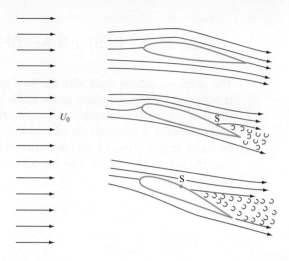

incidence is increased, the point of separation moves forward, as shown in Fig. 12.24. It will be seen that the wake widens and, hence, the drag increases until at some stage the separation point moves to a position such that any further increase of incidence no longer produces an increase of lift. This position is called *stall* and constitutes a critical angle of attack above which the lift drops rapidly, as indicated in Figs 12.23 and 12.25. The stall is accompanied by a rapidly increasing drag, which is mainly due to the increasing wake.

Typical aerofoil characteristics are shown in Fig. 12.25. In addition to curves of lift and drag coefficients the diagram also shows the lift/drag ratio and pressure coefficient C_p.

FIGURE 12.25

Typical characteristics of an aerofoil

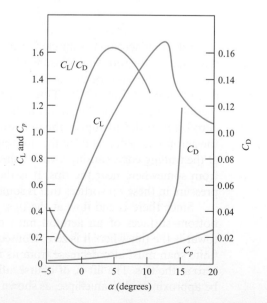

12.8 FLOW PAST AN AEROFOIL OF FINITE LENGTH

The previous section dealt with the flow past an infinitely long aerofoil or one which is bounded by parallel plates at the ends. Such conditions or assumptions mean that the flow is truly two-dimensional and there is no spanwise variation of flow patterns and forces for a constant chord aerofoil. In three-dimensional flow, the aerofoil is of finite length (span) b and without walls at the ends, so that they extend freely into the surrounding fluid. This has a considerable effect on the spanwise distribution of lift.

FIGURE 12.26

End effects on an aerofoil of finite span

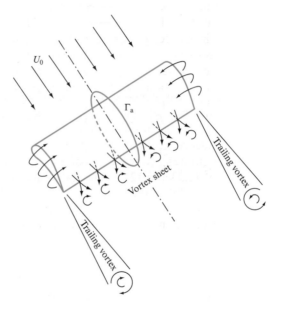

When an aerofoil is subjected to lift force, the pressure on its underside is greater than that on the top. This pressure difference between the upper and the lower surface causes flow around the tips of the aerofoil from the underside to the upper surface, as indicated in Fig. 12.26. This end flow affects the rest of the flow pattern in the following manner. The flow on the underside is deflected towards the tips of the aerofoil in order to supply the necessary end flow, whereas the flow at the top of the aerofoil is deflected from the tips towards the centre. This produces unstable flow at the trailing edge, causing a vortex sheet which rolls up into two vortices emanating from somewhere near the tips. It is the condensation of water vapour due to low pressure in these tip vortices that is sometimes seen at the tips of aircraft wings.

Since there is end flow at the tips, the pressure difference between the top and bottom surfaces of an aerofoil must decrease from a maximum at the mid-span towards the tips where it is zero. Consequently, the circulation around the aerofoil of finite span must also decrease from its maximum value Γ_a at the centreline down to zero at the tips. The lift is, of course, affected in the same way. The distribution may be approximated to an ellipse, as shown in Fig. 12.27.

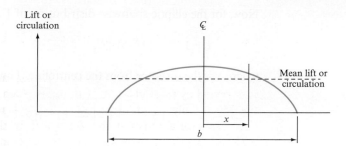

A further consequence of the tip vortices is that they induce a downward velocity component which is known as *downwash velocity* $\bar{v}_i$. Its presence means that the relative velocity of motion between the fluid and the aerofoil is no longer the free stream velocity U_0 but velocity U, deflected from U_0 by an angle ε known as the induced angle of incidence. The resulting geometry is shown in Fig. 12.28. What follows is that, in accordance with the definition of lift, which stipulates that it is perpendicular to the relative direction of motion, the true lift is normal to U. However, since it is more convenient and customary to relate lift and drag to the direction of the free stream relative to the aerofoil, the true lift L_0 is resolved into L, the component perpendicular to U_0, and D_i, the component parallel to U_0. This latter component, which is in the same direction as drag, is known as *induced drag* and is added to pressure drag and the skin friction drag to give the total drag on an aerofoil. The expression for induced drag is derived as follows. The true lift per unit length of span is given by

$$L_0 = \rho U \Gamma;$$

hence, the induced drag per unit span,

$$D_i' = L_0 \sin \varepsilon = \rho U \Gamma \sin \varepsilon.$$

But $\sin \varepsilon = v_i/U$ and, using Prandtl's approximation for elliptical spanwise lift distribution that $v_i = \Gamma_0/2b$, where Γ_0 is the maximum circulation at the centreline,

$$\sin \varepsilon = \Gamma_0/2bU$$

and

$$D_i' = \rho U \Gamma (\Gamma_0/2bU) = \rho \Gamma (\Gamma_0/2b).$$

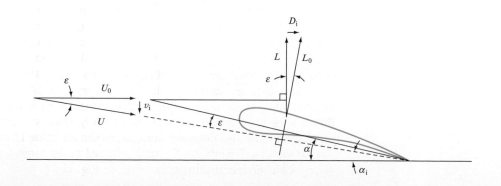

Now, for the elliptic spanwise distribution of Γ,

$$\Gamma = \Gamma_0[1 - (2x/b)^2]^{1/2},$$

where x is the distance from the centreline. Thus, the induced drag for the total span,

$$D_i = \frac{\rho}{2b}\Gamma_0^2 \int_{-b/2}^{+b/2} \left[1 - \left(\frac{2x}{b}\right)^2\right]^{1/2} dx = \frac{\rho}{2b}\Gamma_0^2 \frac{b\pi}{4}$$

$$= \rho\pi\Gamma_0^2/8, \tag{12.31}$$

is obtained by substituting $2x/b = \sin\theta$. But

$$L_0 = \int_{-b/2}^{+b/2} \rho U \Gamma \, dx = \rho U \Gamma_0 \int_{-b/2}^{+b/2} \left[1 - \left(\frac{2x}{b}\right)^2\right]^{1/2} dx = \rho U \Gamma_0 b \frac{\pi}{4},$$

from which

$$\Gamma_0 = 4L_0/\rho U b\pi$$

and, substituting into (12.31),

$$D_i = (\rho\pi/8)(4L_0/\rho U b\pi)^2 = 2L_0^2/\rho\pi U^2 b^2.$$

However, from similar triangles,

$$L_0/U = L/U_0$$

and, hence,

$$D_i = (2/\rho\pi b^2)(L/U_0)^2. \tag{12.32}$$

If the coefficient of induced drag is defined as

$$C_{D_i} = D_i/\tfrac{1}{2}\rho U^2 A$$

and, since $C_L = L/\tfrac{1}{2}\rho U^2 A$, by substitution

$$C_{D_i} = \frac{D_i}{L/C_L} = \frac{C_L 2L^2}{L\rho\pi b^2 U^2} = 2C_L\frac{L}{\rho\pi b^2 U^2}$$

$$= 2C_L\frac{\tfrac{1}{2}C_L\rho U^2 A}{\rho\pi b^2 U^2} = C_L^2\frac{A}{\pi b^2} = C_L^2\frac{cb}{\pi b^2}$$

$$= \frac{C_L^2}{\pi} \times \frac{c}{b}.$$

But

$$\frac{c}{b} = \frac{1}{(\text{Aspect ratio})},$$

so that

$$C_{D_i} = \frac{C_L^2}{\pi(\text{Aspect ratio})}. \tag{12.33}$$

This equation shows that a large aspect ratio minimizes the induced drag, as would be expected.

EXAMPLE 12.6

A wing of an aircraft of 10 m span and 2 m mean chord is designed to develop a lift of 45 kN at a speed of 400 km h^{-1}. A 1/20 scale model of the wing section is tested in a wind tunnel at 500 m s^{-1} and $\rho = 5.33$ kg m^{-3}. The total drag measured is 400 N. Assuming that the wind tunnel data refer to a section of infinite span, calculate the total drag for the full-size wing. Assume an elliptical lift distribution and take air density as 1.2 kg m^{-3}.

Solution

Wing area,

$$A = 2 \times 10 = 20 \text{ m}^2.$$

Coefficient of drag from the model data,

$$C_D = \frac{D}{\frac{1}{2}\rho U^2 A} = \frac{400}{\frac{1}{2} \times 5.3(500)^2 20/(20)^2} = 0.012.$$

For the prototype,

$$U = 400 \text{ km h}^{-1} = 111.1 \text{ m s}^{-1}$$

and the lift coefficient,

$$C_L = \frac{L}{\frac{1}{2}\rho U^2 A} = \frac{45\,000}{\frac{1}{2} \times 1.2(111.1)^2 20} = 0.304.$$

Now, assuming an elliptical distribution, the coefficient of induced drag,

$$C_{D_i} = C_L^2/\pi(\text{AR}) = (0.304)^2/\pi(\tfrac{10}{2}) = 0.0059.$$

Hence, the total drag coefficient,

$$C_{D_t} = C_D + C_{D_i} = 0.012 + 0.0059 = 0.0179$$

and the total drag on the wing,

$$D = \tfrac{1}{2}C_{D_t}\rho U^2 A = \tfrac{1}{2} \times 0.0179 \times 1.2(111.1)^2 \times 20 = 2648.9 \text{ N}.$$

Therefore,

$$D = 2.65 \text{ kN}.$$

12.9   WAKES AND DRAG

It was explained in Section 12.1 that pressure drag is closely related to boundary layer separation and the formation of a wake at the rear of the body. The size of the wake and the pressure within it are the two factors which determine the magnitude of the pressure drag. The wider the wake the greater is the area over which the pressure difference between the front and the rear of the body acts and hence the greater is the drag. Equally, the lower the pressure within the wake, the greater is the pressure difference acting on the body and hence the drag. The two effects are in fact related: as the width of the wake is reduced the pressure within it increases and approaches the free stream pressure. This interdependence was first shown theoretically by Helmholtz, who assumed a stagnant wake region behind the body. For such theoretical flow (known as Helmholtz flow) the separation points move towards the rear of the body and the wake is reduced as the pressure within the wake is increased. Thus, when pressure recovery is assumed to be complete, that is when the pressure within the wake is the same as the free stream pressure, the wake disappears completely and there is no pressure drag. Empirical evidence of flow past cylinders, spheres and other bodies supports the above principles. In particular, the work of Eisenberg and Reichardt provided strong evidence of linear correlation between the drag of various bodies and the pressure coefficient of the cavity behind them.

It follows, therefore, that in order to minimize pressure drag it is important to reduce the width of the wake as much as possible. This is achieved by preventing or delaying boundary layer separation from the surface of the body. In Chapter 10 it was shown that the turbulent boundary layer separates less easily than the laminar boundary layer and therefore in the former case the separation points are always further to the rear of the body, the wake is narrower and the drag coefficient is considerably smaller than for the laminar boundary layer. For example, for a cylinder and for a specific Re, laminar separation before transition occurs at $\theta = \pm 98°$ (measured from the rear stagnation point) and the drag coefficient $C_D = 1.2$, whereas turbulent separation after transition occurs at $\theta = \pm 60°$ and $C_D = 0.3$. Similarly for a sphere: for laminar separation $\theta = \pm 100°$ and $C_D = 0.44$, but for turbulent separation $\theta = \pm 60°$ and $C_D = 0.22$. (Figure 12.10 shows that C_D varies by up to 10 per cent with Re over small Re ranges.) The level of free stream turbulence has little effect on the value of the drag coefficient as such, but it does affect the Reynolds number at which transition from a laminar to a turbulent boundary layer takes place (see Section 12.5). Generally higher levels of turbulence cause earlier transition.

The fluid velocity in the wake is greatly reduced compared with that upstream of the body. It is in general not uniform, unsteady and sometimes oscillatory. Much work has been done on the nature of this very complicated unsteady flow which can give rise to significant body forces, particularly on bluff bodies at subcritical Reynolds number. For most flows, however, it is sufficient to use time-averaged velocities, as measured by a Pitot–static tube for example.

The drag force, and hence the drag coefficient of a body, due to the relative motion of a fluid over it, may be determined either by direct force measurement or by calculation from detailed velocity and pressure distributions in the wake. The latter method is based on the application of Newton's second law of linear motion to an immersed body which causes fluid deceleration in the wake and hence a change of fluid linear momentum, but also a difference of pressure between the front and rear of the body.

FIGURE 12.29

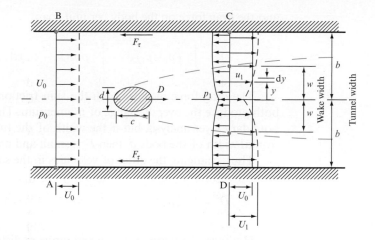

Consider flow past a two-dimensional body mounted in a wind tunnel as shown in Fig. 12.29. Let the free stream velocity and static pressure in front of the body be U_0 and p_0, respectively, and the drag force on the body per unit span be D'. Let also the velocity and pressure profiles downstream of the body be as shown, so that at any distance y from the centreline the velocity is $u_1 = f(y)$ and the pressure is $p_1 = f'(y)$. Now, if the half-width of the wake w is defined as that distance from the centreline at which the velocity becomes constant, then u_1 varies within the band $-w < y < +w$ and becomes equal to U_1 outside the wake where it is constant.

Taking the control volume ABCD such that AB is far upstream of the body where both the velocity and the pressure are constant, CD is downstream of the body and cuts through the wake, and BC and AD are coincident with the tunnel walls, the forces acting on the fluid in the x direction are: pressure forces due to pressure difference, the negative drag force D with which the body acts on the fluid, and forces due to shear stresses F_τ along the tunnel walls resulting from the boundary layer there. The sum of all these forces must be equal to the rate of change of linear momentum in the x direction.

Now,

Pressure force on fluid element dy, per unit span $= (p_0 - p_1)\,dy$,

Resultant force on the fluid within the control volume, per unit span

$$= \int_{-b}^{+b} (p_0 - p_1)\,dy - D' - F_\tau,$$

Mass flow rate through fluid element dy, per unit span $= \rho u_1\,dy$,

Change of velocity $= (u_1 - U_0)$.

Therefore,

Rate of change of fluid momentum $= \displaystyle\int_{-b}^{+b} \rho u_1(u_1 - U_0)\,dy,$

and equating the forces to the rate of change of momentum:

$$\int_{-b}^{+b} \rho u_1(u_1 - U_0)\,dy = \int_{-b}^{+b} (p_0 - p_1)\,dy - D' - F_\tau,$$

and the drag force on the body per unit span:

$$D' = \int_{-b}^{+b} (p_0 - p_1)\,\mathrm{d}y - \int_{-b}^{+b} \rho u_1 (u_1 - U_0)\,\mathrm{d}y - F_\tau. \tag{12.34}$$

The drag force includes both the skin friction drag and pressure drag because both produce the overall change of momentum. The force F_τ may be calculated using boundary layer analysis, but if the width of the tunnel b is large compared with the frontal width of the body d, then F_τ is small and may be neglected.

Thus, changing the order of velocities in the second integral:

$$D' = \int_{-b}^{+b} (p_0 - p_1)\,\mathrm{d}y + \int_{-b}^{+b} \rho u_1 (U_0 - u_1)\,\mathrm{d}y. \tag{12.35}$$

However, the typical velocity and pressure distributions indicate that at distances greater than w from the centreline the fluid is unaffected by the body and the velocity and pressure there are constant. If the wake is enclosed, that is within the walls of a tunnel, the constant velocity U_1 outside the wake is not equal to the free stream velocity, because continuity demands that the velocity defect within the wake is made up outside it.

From the continuity equations, therefore,

$$U_0 b = U_1 (b - w) + \int_{-w}^{+w} u_1\,\mathrm{d}y,$$

from which

$$U_1 = \frac{U_0 b}{(b - w)} - \frac{1}{(b - w)} \int_{-w}^{+w} u_1\,\mathrm{d}y. \tag{12.36}$$

Similarly, the constant pressure p_1' outside the wake may not be equal to p_0, so that in such a case the integration must include the whole tunnel width. So, to evaluate drag both the velocity and the pressure distributions are necessary if the traversing is carried out close behind the body. However, some simplifications and approximations are possible in appropriate cases. First, for a free wake, such as that forming behind an aircraft wing during flight, there is no restriction of x direction mass flow continuity imposed by the tunnel walls and therefore the velocity outside the wake U_1 is equal to U_0. Similarly the pressure outside the wake p_1' is equal to the free stream pressure p_0. Thus, both the integrals of equation (12.35) may have their limits changed to $\pm w$ because they become equal to zero outside these limits. Therefore, equation (12.35) for a free wake becomes

$$D' = \int_{-w}^{+w} (p_0 - p_1)\,\mathrm{d}y + \int_{-w}^{+w} \rho u_1 (U_0 - u_1)\,\mathrm{d}y. \tag{12.37}$$

Now, since the drag is obviously independent of any traverse which may be carried out at any distance downstream of the body, it follows that the sum of the two integrals of equation (12.37) must remain constant and independent of x. So, as the wake diffuses with increasing distance from the body, its width will increase and the velocity defect $(U_0 - u)$ will decrease, as shown in Fig. 12.30. At some distance,

FIGURE 12.30

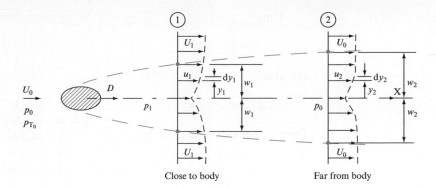

Close to body   Far from body

sufficiently far away downstream of the body, the static pressure across the wake and flow will be constant and equal to the upstream free stream static pressure p_0. Thus the pressure integral in equation (12.37) becomes equal to zero, and the drag becomes

$$D' = \int_{-w_2}^{+w_2} \rho u_2 (U_0 - u_2)\, dy_2,$$

(12.38)

where suffix 2 refers to the faraway downstream section.

Using the definition of the coefficient of drag of equation (12.4),

$$D = \tfrac{1}{2} \rho C_D U_0^2 A,$$

and remembering that $D' = D/L$, we obtain

$$C_D = \frac{D}{\tfrac{1}{2}\rho U_0^2 Lc} = \frac{D'}{\tfrac{1}{2}\rho U_0^2 c} = \frac{2}{c} \int_{-w_2}^{+w_2} \frac{u_2}{U_0}\left(1 - \frac{u_2}{U_0}\right) dy_2.$$

(12.39)

Note that in the above equation the value of C_D would be based on the span area, since it was taken that $A = cL$, which is usual for lifting surfaces such as plates or aerofoils. However, for bluff bodies the frontal area $A = dL$ is commonly used in which case equation (12.39) would take the form

$$C_D = \frac{2}{d} \int_{-w_2}^{+w_2} \frac{u_2}{U_0}\left(1 - \frac{u_2}{U_0}\right) dy_2.$$

(12.39a)

To be consistent, the numerical values of Reynolds number at which values of C_D are determined and quoted are customarily also based on c for lifting surfaces and on d for bluff bodies.

Equations (12.38) and (12.39) are convenient to use provided it is practicable to carry out the traversing at such a distance that $p_2 = p_0$. If this is not possible equation (12.37) must be used unless an approximation is acceptable. One such approximation was developed by B. M. Jones and has been used extensively together with a Pitot rake behind a body. The method assumes that within the wake along any given streamtube between sections 1 and 2 the total pressure remains constant. It is then possible to modify equation (12.38) so that it refers to section 1 using also continuity to account for the larger wake width at section 2.

Thus:

$$p_T = p_1 + \tfrac{1}{2}\rho u_1^2,$$

and also

$$p_T = p_2 + \tfrac{1}{2}\rho u_2^2.$$

Thus the velocities are given by

$$u_1 = \sqrt{\left[\frac{2}{\rho}(p_T - p_1)\right]} \quad \text{and} \quad u_2 = \sqrt{\left[\frac{2}{\rho}(p_T - p_2)\right]},$$

but since $p_2 = p_0$,

$$u_2 = \sqrt{\left[\frac{2}{\rho}(p_T - p_0)\right]}.$$

Similarly the free stream velocity U_0 may be expressed in terms of the total pressure there:

$$U_0 = \sqrt{\left[\frac{2}{\rho}(p_{T_0} - p_0)\right]}.$$

Continuity:

$$u_1 \, \mathrm{d}y_1 = u_2 \, \mathrm{d}y_2,$$

from which

$$\mathrm{d}y_2 = \frac{u_1}{u_2}\mathrm{d}y_1.$$

Substituting into equation (12.38),

$$D' = \int_{-w_1}^{+w_1} \rho u_2 (U_0 - u_2)\frac{u_1}{u_2}\mathrm{d}y_1 = \int_{-w_1}^{+w_2} \rho u_1 (U_0 - u_2)\,\mathrm{d}y_1,$$

and replacing the velocities,

$$D' = 2\int_{-w_1}^{+w_1} \sqrt{(p_T - p_1)}[\sqrt{(p_{T_0} - p_0)} - \sqrt{(p_T - p_0)}]\,\mathrm{d}y_1. \tag{12.40}$$

Also, equation (12.39) becomes

$$C_D = \frac{2}{c}\int_{-w_1}^{+w_1}\left(\frac{p_T - p_1}{p_{T_0} - p_0}\right)^{1/2}\left[1 - \left(\frac{p_T - p_0}{p_{T_0} - p_0}\right)^{1/2}\right]\mathrm{d}y_1. \tag{12.41}$$

In order to evaluate the drag of a body and its drag coefficient using the above equations a traverse of total pressures and static pressures at some section down-stream of the body is required together with the reading of total and static pressures upstream of the body. The integration may be performed graphically – the method was devised prior to the widespread availability of computers. It must be remembered, however, that the method is approximate and therefore it is preferable to use equation (12.37) in conjunction with a suitable numerical integration method and a computer.

12.10 COMPUTER PROGRAM WAKE

Program WAKE calculates the drag per unit span for a body in an airstream and the associated drag coefficient from a static and stagnation pressure traverse carried out at right angles to the wake downstream of the body. The calculation assumes a two-dimensional incompressible flow and requires data detailing the flow conditions, air temperature and traverse experimental results. Equations (12.37) and (12.39) are invoked, together with the equation of state (1.13) and Bernoulli's equation, with the integration of equation (12.37) being undertaken by use of the trapezoidal rule.

12.10.1 Application example

The calculation requires the following data:

1. test site barometric pressure, B mm of Hg;
2. static pressure upstream of the body, p_{Su} in mm of H_2O gauge;
3. temperature upstream of the body, T °C;
4. body characteristic dimension, i.e. frontal width or chord, C mm;
5. upstream free stream velocity, U_0 m s^{-1}, or flow stagnation pressure, p_{T_0} in mm of H_2O gauge;
6. number of traverse readings, N, maximum 50, and their spacing, H mm (it is assumed that readings are taken at equispaced locations across the wake);
7. up to 50 static pressure readings, p_s, and a stagnation pressure reading, p_T, all in mm of H_2O gauge, or the value of the constant static pressure, p_s, at the traverse section and up to 50 values of the stagnation pressure, p_T, as recorded at each traverse point, all in mm of H_2O gauge. Note that static pressure will always be less than the stagnation pressure due to the kinetic pressure term.

For the following data: $B = 770$ mm Hg; $p_{Su} = 78.5$ mm H_2O; $T = 20$ °C; $C = 20$ mm; $U_0 = 45.0$ m s^{-1}; $N = 15$; $H = 5$ mm; for a constant traverse section static pressure $p_s = 48.8$ mm H_2O.

The 15 stagnation pressure values p_T in mm H_2O are

176,   170,   150,   136,   110,   85,   70,   82,   111,   135,   173,   175,   176,

illustrating the expected velocity distribution across the wake.

As a result of the program calculation the drag per unit span was 41.08 kN m^{-1} and the drag coefficient C_D was 1.65, with a reference free stream velocity of 45 m s^{-1}, air flow density of 1.23 kg m^{-3} and a body characteristic dimension of 20 mm.

12.10.2 Additional investigations using WAKE

The computer program calculation may also be used to investigate

1. the influence of traverse intervals on the predicted drag;
2. the effect of local air conditions, pressure and temperature on the drag coefficient.

Concluding remarks

The application of earlier material defining flow regimes, together with the dimensional analysis and similarity techniques, were utilized in the treatment of drag forces acting on fully and partially submerged bodies and the definition of zones of influence for each of the major dimensionless groupings, Reynolds, Froude and Mach numbers. The earlier treatment of ideal flow was seen to have some application in the flow external to the body. Similarly the effect of an adverse pressure gradient on the boundary layer at the fluid/surface interface was referred to.

The importance of the momentum equation was again emphasized by its use in developing drag forces from wake traverses. The flow over a cylinder was utilized as a means of illustrating separation effects, together with vortex generation, and the dependence of both these effects upon flow conditions, while the flow over a sphere was utilized to discuss the measurement of fluid viscosity, Stokes method. The detail flow experienced over an aerofoil was considered, with particular attention being paid to the generation of lift and the minimization of drag.

The material covered in this chapter will be utilized in the treatment of rotodynamic machinery where the mechanisms of energy transfer are dependent upon the forces acting between the moving blades and the fluid passing through the machine.

Summary of important equations and concepts

1. This chapter introduces the concepts of lift and drag forces and their respective non-dimensional coefficients, equations (12.4) and (12.5). In the case of drag, also differentiating between skin friction and pressure drag, equation (12.1).

2. The dependence of drag forces on a range of non-dimensional groups is emphasized and the zone of influence of each identified. In particular Section 12.4 emphasizes the role of Froude number in ship resistance.

3. With reference to flow past cylinders and spheres the applicable flow regimes are identified based on Reynolds number, in particular the Stokes law is identified, together with Allen and Newton flow conditions for a sphere and the concept of terminal velocity highlighted, equation (12.23).

4. Lift and drag on an aerofoil section is introduced, together with the definition of angle of attack and stall conditions.

5. Flow conditions in the wake downstream of a body in a fluid stream is considered in Section 12.9 and a computer calculation is presented in Section 12.10.

Problems

12.1 A wing of a small aircraft is rectangular in plan having a span of 10 m and a chord of 1.2 m. In straight and level flight at 240 km h^{-1} the total aerodynamic force acting on the wing is 20 kN. If the lift/drag ratio is 10 calculate the coefficient of lift and the total weight the aircraft can carry. Assume air density to be 1.2 kg m^{-3}.

[0.622, 1990 kg]

12.2 A screen across a pipe of rectangular cross-section 2 m by 1.2 m consists of well-streamlined bars of 25 mm

maximum width and at 100 mm centres, their coefficient of total drag being 0.30. A water stream of 5.5 m^3 s^{-1} passes through the pipe. What is the total drag on the screen? If a rectangular block of wood 1 m by 0.3 m and about 25 mm thick is held by the screen, making suitable assumptions, estimate the increase of the drag.   [449 N, 3759 N]

12.3 A parachute of 10 m diameter when carrying a load W descends at a constant velocity of 5.5 m s^{-1} in atmospheric air at a temperature of 18 °C and pressure of

1.0×10^5 N m^{-2}. Determine the load W if the drag coefficient for the parachute is 1.4.   [1.992 kN]

12.4   Prove that the viscous resistance F of a sphere of diameter d moving at constant speed v through a fluid of density ρ and viscosity μ may be expressed as

$$F = k \frac{\mu^2}{\rho} f\left(\frac{\rho v d}{\mu}\right), \quad \text{where } k \text{ is a constant.}$$

Two balls made of steel and aluminium are allowed to sink freely in an oil of specific gravity 0.9. Determine the ratio of their diameters if dynamic similarity must be obtained when the balls attain their terminal sinking velocities. The specific gravities of steel and aluminium are 7.8 and 2.7, respectively.   [0.639]

12.5   The drag and bending moment on a structure in a 40 km h^{-1} wind is to be studied using a 1/20 scale model in a pressurized wind tunnel. If the tunnel and ambient temperatures are the same but the air density in the tunnel is eight times that of the ambient air, calculate the air speed in the tunnel and the bending moment for the structure if that measured on the model is 30 N m.   [100 km h^{-1}, 4800 N m]

12.6   A submarine periscope is 0.15 m in diameter and is travelling at 15 km h^{-1}. What is the frequency of the alternating vortex shedding and the force per unit length of the periscope? Take the density of water as 1.03×10^3 kg m^{-3} and kinematic viscosity as 1.25 mm^2 s^{-1}.   [5.5 Hz, 805 N m^{-1}]

12.7   A 1/20 model of a cargo ship 120 m long is towed in fresh water at a velocity of 2.5 m s^{-1}. The measured total drag is 105 N. The skin friction drag coefficient is 0.002 72 and the wetted area is 6.5 m^2. The estimated skin friction drag coefficient for the prototype is 0.0018. Determine: (a) the wave drag for the model, (b) the wave drag coefficient for the model, (c) the wave drag and the total drag for the ship, and (d) the power required to tow the model and to propel the ship at its design cruising speed.
[(a) 49.75 N, (b) 0.002 45, (c) 408 kN, 435 kN, (d) 262.5 W, 4863 kW]

12.8   A spherical weather balloon of 2 m diameter is filled with hydrogen. The total mass of the balloon skin and the instruments it carries is 29.682 kg. If at a certain altitude the density of air is 1.0 kg m^{-3} and is ten times the density of hydrogen in the balloon, determine the steady upward velocity of the balloon. Take viscosity of air to be 1.8×10^{-5} N s m^{-2}.   [0.932 m s^{-1}]

12.9   A small spherical water droplet falls freely at a constant speed in air. An air bubble of the same diameter rises freely at a constant rate in water. Derive an expression for the ratio of the distances travelled by the droplet and the air bubble during the same time and calculate this ratio if

$$\rho_{air} = 1.2 \text{ kg m}^{-3}, \mu_{air} = 1.7 \times 10^{-5}, \mu_{water} = 10^{-3} \text{ N s m}^{-2}.$$

[58.8 for $d < 0.0446$ mm; depends on d for $0.0446 < d < 2.04$ mm; 28.9 for $d > 2.04$ mm]

12.10   (a) A submarine is deeply submerged and moving along a straight course. Describe the physical phenomena that give rise to resistance to its motion. The submarine now comes to the surface and continues on course. What changes occur in the resistance phenomena?

(b) The following data refer to a 1/20 scale model of a cargo vessel under test in a model basin:

Model speed	1.75 m s^{-1}
Total resistance	34.25 N
Model length	6.20 m
Wetted surface area	5.91 m^2
Basin water density	998 kg m^{-3}
Kinematic viscosity	0.1010×10^{-5} m^2 s^{-1}

The ITTC coefficients may be calculated from

$$C_F = 0.075/(\log_{10} R - 2)^2,$$

where R is the Reynolds number.
What will be the total resistance for the smooth ship at the corresponding speed in sea water of kinematic viscosity 0.1188×10^{-5} m^2 s^{-1}?   [244.5 N]

12.11   When a slender body held transversely is tested in a wind tunnel it is found that the decrease in velocity in the wake is approximately linear. It decreases from the undisturbed velocity u_0 at double the solid width to $0.2U_0$ at the axis, the pressure in the wake being constant throughout and the same as that in the undisturbed stream. If such a body of 1.5 m width moves, under dynamically similar conditions, through still air at 150 m s^{-1}, calculate the drag on the solid per unit length and the drag coefficient. The air is at 5 °C and a pressure of 510 mm of mercury. Take the density of mercury as 13.6×10^3 kg m^{-3} and the gas constant for air as $R = 287$ J kg^{-1} K^{-1}.   [21.49 kN, 1.493]

13

Compressible Flow around a Body

THE DIMENSIONAL ANALYSIS INTRODUCED IN CHAPTER 12 INDICATED THAT THE EXTERNAL flow Mach number becomes a determinant of the forces acting on a body in supersonic flows. The formation of shock waves, effectively reducing fluid velocity to less than the speed of sound, leads to sudden changes in pressure, generating wave drag. This becomes the predominant factor in defining supersonic body profiles. This chapter introduces these concepts, defining the changes in pressure and temperature across shock waves by reference to both the energy equation (Chapter 5) and the equation of state for gases (Chapter 1). Normal and oblique shock waves are treated, together with a discussion of the supersonic flow expansion and compression to be found at changes in surface orientation, leading to the propagation of Mach waves that can coalesce into a shock wave at concave corners. The calculation of flow conditions across a shock is included in the form of a computer program. ● ● ●

13.1 EFFECTS OF COMPRESSIBILITY

In the previous chapter, the discussion of drag in external flow was limited to the influence of Reynolds number and Froude number. The former involves the relative influence of two fluid properties, namely the density and the viscosity, whereas the latter is concerned with the effects of gravity. However, in Section 12.3 it was shown by dimensional analysis that Mach number, which is a measure of the importance of elastic forces in the fluid, may be of significance. This occurs when changes of density are appreciable, and the flow is then called compressible. Mach number is also the ratio of the free stream velocity and the velocity of propagation of pressure waves, called the velocity of sound (see Section 5.13):

$$\mathrm{Ma} = U_0/c. \tag{13.1}$$

Since very significant changes occur at $\mathrm{Ma} = 1$, the flows are classified into subsonic for $\mathrm{Ma} < 1$ and supersonic for $\mathrm{Ma} > 1$.

In subsonic flow, at relatively low velocities the viscous forces and, hence, Re are of predominant importance. The density changes are small, Ma is also small and its influence is negligible. As the velocity is increased we know from previous paragraphs that at some value of Re the drag coefficient becomes independent of it. This is, however, accompanied by a simultaneous increase of Mach number, whose influence becomes more and more pronounced, and cannot be neglected.

In supersonic flow, shock waves are formed. They not only affect the boundary layer and, hence, the skin friction drag and the position of separation, which controls the form drag, but also produce an abrupt change of pressure. This gives rise to additional drag known as *wave drag*. Since the wave drag is not related to viscosity, but to pressure change across the shock wave, it would be present in an ideal fluid at supersonic flow.

At supersonic flow, the wave drag constitutes the largest contribution to total drag and, therefore, streamlining the rear part of the body, which is so important in subsonic flow, has little effect. As will be shown later, in order to reduce the wave drag in supersonic flow the nose of the body must be sharp and pointed. This confines the shock wave to only a small region. Thus, the streamlining requirements for supersonic flow are completely the reverse of those for subsonic flow. Whereas the latter requires a rounded nose and long, gradually pointed tail, the former requires a sharp, pointed nose and rounded, blunt tail.

The effect of Mach number on the coefficient of drag for projectiles is shown in Fig. 13.1, which also indicates the considerable reduction in C_D achieved by a pointed nose.

The effects of compressibility in external flow are not confined to drag. Since, fundamentally, they take into account the variations of fluid density, all parameters are affected. As an illustration, let us consider the very important conditions at the front stagnation point on a body. Let the pressure, temperature and density at the stagnation point be denoted by suffix T and those in the free stream some distance upstream of the body by a suffix 0, as indicated in Fig. 13.2. The conditions at the stagnation point may be expressed in terms of those upstream by the application of Bernoulli's equation and by remembering that the velocity at the stagnation point is zero. First, assuming incompressible flow, we obtain

FIGURE 13.1

Effect of Mach number on the coefficient of drag for projectiles

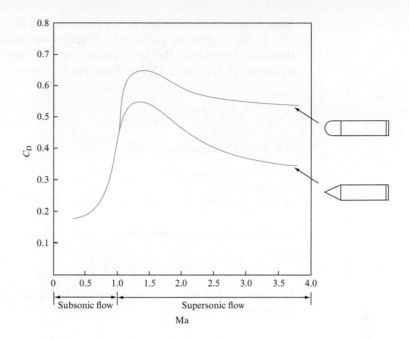

$$p_T = p_0 + \tfrac{1}{2}\rho U_0^2 \tag{13.2}$$

and, since $\rho_0 = \rho_T = \rho$, it follows from Boyle's law ($p/\rho = f(T)$) that

$$T_T/T_0 = p_T/p_0.$$

Hence, the stagnation temperature is obtained:

$$T_T = T_0(p_T/p_0) = T_0[(p_0 + \tfrac{1}{2}\rho U_0^2)/p_0]$$

$$T_T = T_0[1 + \tfrac{1}{2}(\rho/p_0)U_0^2]. \tag{13.3}$$

But the equation of state (equation (1.13)) gives

$$p_0 = \rho R T_0, \tag{13.4}$$

from which $\rho/p_0 = 1/RT_0$, which, on substitution into (13.3), gives

$$T_T = T_0(1 + \tfrac{1}{2}U_0^2/RT_0) = T_0 + \tfrac{1}{2}U_0^2/R. \tag{13.5}$$

FIGURE 13.2

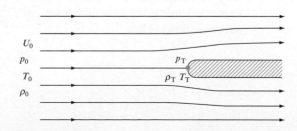

Now, assuming that the flow is compressible and the process by which it is brought to rest at the stagnation point is frictionless and adiabatic (no heat exchange) and, therefore, isentropic, the appropriate form of Bernoulli's equation, derived from equations (5.21) and (1.17) (see Example 13.1) gives

$$\frac{\gamma}{\gamma-1}\frac{p_T}{\rho_T} = \frac{\gamma}{\gamma-1}\frac{p_0}{\rho_0} + \frac{U_0^2}{2}. \tag{13.6}$$

But, from the equation of state,

$$p/\rho = RT,$$

so that $[\gamma/(\gamma-1)]RT_T = [\gamma/(\gamma-1)]RT_0 + U_0^2/2$

and the stagnation temperature,

$$T_T = T_0 + [(\gamma-1)/\gamma R](U_0^2/2). \tag{13.7}$$

However, it was also shown in Section 5.13 that the velocity of sound is given by equation (5.31),

$$c = \sqrt{(\gamma RT)} = \sqrt{[(\gamma-1)c_p T]} = \sqrt{(\gamma p/\rho)}, \tag{13.8}$$

from which $\gamma R = c^2/T_0$ may be substituted into equation (13.7), giving

$$T_T = T_0 + [(\gamma-1)/2](U_0^2/c^2)T_0.$$

But $U_0/c = \mathrm{Ma}_0,$

so that, finally,

$$T_T = T_0[1 + \tfrac{1}{2}(\gamma-1)\mathrm{Ma}_0^2]. \tag{13.9}$$

Now, since, for isentropic processes,

$$T_T/T_0 = (p_T/p_0)^{(\gamma-1)/\gamma},$$

the stagnation pressure may be obtained by

$$p_T = p_0(T_T/T_0)^{\gamma/(\gamma-1)} = p_0[1 + \tfrac{1}{2}(\gamma-1)\,\mathrm{Ma}_0^2]^{\gamma/(\gamma-1)}. \tag{13.10}$$

In order to compare this expression with equation (13.2) for incompressible flow, it may be rearranged as a pressure ratio and then the expression in square brackets expanded using the binomial theorem (justified because $\tfrac{1}{2}(\gamma-1)\,\mathrm{Ma}_0^2 < 1$ for subsonic flow). Thus,

$$\frac{p_T}{p_0} = \left(1 + \frac{\gamma-1}{2}\mathrm{Ma}_0^2\right)^{\gamma/(\gamma-1)}$$

$$= 1 + \frac{\gamma}{2}\mathrm{Ma}_0^2 + \frac{\gamma}{8}\mathrm{Ma}_0^4 + \frac{\gamma(2-\gamma)}{48}\mathrm{Ma}_0^6 + \cdots.$$

Rearranging and taking $(\gamma/2)\,\mathrm{Ma}_0^2$ outside the bracket,

$$\frac{p_T}{p_0} - 1 = \frac{\gamma}{2}\mathrm{Ma}_0^2\left[1 + \frac{\mathrm{Ma}_0^2}{4} + \frac{(2-\gamma)}{24}\mathrm{Ma}_0^4 + \cdots\right]$$

and $\quad p_T - p_0 = \dfrac{\gamma}{2}p_0\,\mathrm{Ma}_0^2\left[1 + \dfrac{\mathrm{Ma}_0^2}{4} + \dfrac{(2-\gamma)}{24}\mathrm{Ma}_0^4 + \cdots\right].$

But $\quad \dfrac{\gamma}{2}p_0\,\mathrm{Ma}_0^2 = \dfrac{\gamma}{2}p_0\dfrac{U_0^2}{c^2} = \dfrac{\gamma}{2}p_0\dfrac{U_0^2}{\gamma p_0/\rho_0} = \dfrac{1}{2}p_0U_0^2,$

so that, finally,

$$p_T - p_0 = \tfrac{1}{2}\rho_0 U_0^2\left[1 + \frac{\mathrm{Ma}_0^2}{4} + \frac{(2-\gamma)}{24}\mathrm{Ma}_0^4 + \cdots\right]. \tag{13.11}$$

Now, if the flow is considered incompressible, the corresponding pressure difference $(p_T - p_0)_{\rho=\text{constant}}$ given by equation (13.2) is less than the correct pressure difference given above. The ratio between the two is called the *compressibility factor*. Thus

$$\text{Compressibility factor} = \frac{p_T - p_0}{(p_T - p_0)_{\rho=\text{constant}}} = \frac{p_T - p_0}{\frac{1}{2}\rho_0 U_0^2}$$

$$\text{Compressibility factor} = \left[1 + \frac{\mathrm{Ma}_0^2}{4} + \frac{(2-\gamma)}{24}\mathrm{Ma}_0^4 + \cdots\right]. \tag{13.12}$$

Figure 13.3 shows the variation of compressibility factor with Ma, indicating that for Ma < 0.2 the error in assuming the flow to be incompressible amounts to less than 1 per cent, but for Ma > 0.5 exceeds 5 per cent and becomes 27.6 per cent at Ma = 1.

FIGURE 13.3

Variation of compressibility factor with Mach number

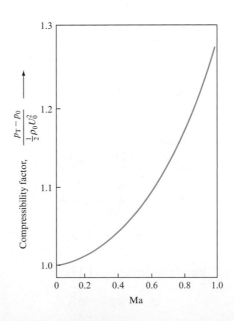

EXAMPLE 13.1   Show that for horizontal isentropic flow Bernoulli's equation takes the form

$$\frac{\gamma}{\gamma-1}\frac{p}{\rho}+\frac{\bar{v}^2}{2} = \text{constant.}$$

Calculate, working from the above equation, the stagnation pressure, temperature and density for an airstream at Ma = 0.7 and density $\rho = 1.8$ kg m^{-3} and temperature of 75 °C. Take $R = 287$ J kg^{-1} K^{-1} and $\gamma = 1.4$.

Solution

Euler's equation (5.21) states

$$\frac{1}{\rho}\frac{dp}{ds}+\bar{v}\frac{d\bar{v}}{ds}+g\frac{dz}{ds} = 0,$$

which, upon integration, becomes

$$\int\frac{dp}{\rho}+\frac{\bar{v}^2}{2}+gz = \text{constant.}$$

Now, for horizontal flow, $z = 0$, and for isentropic flow, $p/\rho^\gamma = $ constant (equation (1.17)). Therefore,

$$\rho = \left(\frac{p}{\text{constant}}\right)^{1/\gamma} = \frac{p^{1/\gamma}}{G}$$

and $$\int\frac{dp}{\rho} = G\int p^{-1/\gamma}\,dp = \frac{G}{1-1/\gamma}p^{(\gamma-1)/\gamma}+C.$$

But $G = p^{1/\gamma}/\rho$ and so, substituting,

$$\int\frac{dp}{\rho} = \frac{G\gamma}{\gamma-1}p^{(\gamma-1)/\gamma}+C = \frac{\gamma}{\gamma-1}\frac{p^{1/\gamma}}{\rho}p^{(\gamma-1)/\gamma}+C = \frac{\gamma}{\gamma-1}\frac{p}{\rho}+C.$$

Therefore, Euler's equation, after integration for isentropic conditions, becomes

$$\frac{\gamma}{\gamma-1}\frac{p}{\rho}+\frac{\bar{v}^2}{2} = \text{constant.}$$

At stagnation point, $v = 0$; therefore, applying Bernoulli's equation to a point in the free stream and the stagnation point (suffix T),

$$\frac{\gamma}{\gamma-1}\frac{p_T}{\rho_T} = \frac{\gamma}{\gamma-1}\frac{p_0}{\rho_0}+\frac{\bar{v}_0^2}{2},$$

$$\frac{p_T}{\rho_T} = \frac{p_0}{\rho_0}+\frac{\gamma-1}{\gamma}\frac{\bar{v}_0^2}{2}.$$

But, from equation (1.17),

$$\rho_T = \rho_0(p_T/p_0)^{1/\gamma}. \tag{I}$$

Therefore

$$\frac{p_T}{\rho_0}\left(\frac{p_0}{p_T}\right)^{1/\gamma} = \frac{p_0}{\rho_0} + \frac{\gamma-1}{\gamma}\frac{\bar{v}_0^2}{2},$$

$$p_T^{(\gamma-1)/\gamma} = p_0^{(\gamma-1)/\gamma} + \rho_0\left(\frac{\gamma-1}{\gamma}\right)\frac{\bar{v}_0^2}{2p_0^{1/\gamma}} \qquad\qquad (II)$$

Now, from the equation of state,

$$p_0 = \rho_0 RT = 1.8 \times 287(273 + 75) = 179.8 \text{ kN m}^{-2};$$

velocity of sound,

$$c = (\gamma RT)^{1/2} = \sqrt{[1.4 \times 287(273 + 75)]} = 373.9 \text{ m s}^{-1};$$

stream velocity,

$$\bar{v}_0 = \text{Ma } c = 0.7 \times 373.9 = 261.7 \text{ m s}^{-1}.$$

Substituting into (II),

$$p_T^{0.4/1.4} = (179.8 \times 10^3)^{0.4/1.4} + 1.8\left(\frac{0.4}{1.4}\right) \times \frac{(261.7)^2}{2(179.8 \times 10^3)^{0.714}}$$

$$= 31.72 + 3.12 = 34.84.$$

Hence, $p_T = (34.84)^{3.5} = \textbf{249.6 kN m}^{-2}$.

Now, from (I),

$$\rho_T = 1.8\left(\frac{249.6}{179.8}\right)^{0.714} = \textbf{2.275 kg m}^{-3}$$

and, therefore,

$$T_T = \frac{p_T}{\rho_T R} = \frac{249.6 \times 10^3}{2.275 \times 287} = \textbf{382.3 K}$$

$$= \textbf{109.3 °C}.$$

13.2 SHOCK WAVES

Weak pressure change in the fluid is propagated through the fluid continuum with the velocity of sound, which is a function of the elastic properties of the fluid. Thus, if a periodic pressure disturbance occurs at a point S in Fig. 13.4 in a stationary fluid, the resulting pressure waves will travel radially outwards from point S as concentric spheres. If the period of the disturbance is Δt, then the distance travelled by a wave

FIGURE 13.4

Wave propagation in a
stationary fluid

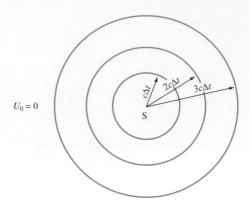

between the first and second disturbance will be $c\Delta t$. By the time the second wave
covered the distance $c\Delta t$, the first wave, being $c\Delta t$ ahead of the second, would have
travelled a total distance of $2c\Delta t$. Thus, all the successive waves are equidistant from
each other in all directions, the distance being $c\Delta t$.

FIGURE 13.5

Wave propagation in a
fluid moving with a
velocity smaller than that
of sound

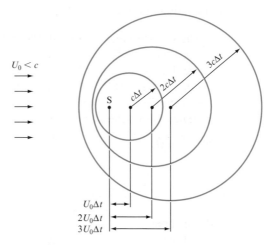

Now, consider a situation (as shown in Fig. 13.5) in which the source of periodic
disturbance S is placed in a moving fluid whose velocity U_0 is less than the velocity of
sound. The waves are still concentric spheres, but are being swept away by the moving
fluid. The lateral distance over which each sphere moves during the periodic time Δt
is $U_0\Delta t$. Thus, the absolute velocity with which the disturbance is now propagated
depends upon the direction, being $(U_0 + c)$ in the direction of fluid motion but only
$(U_0 - c)$ in the opposite direction. The source remains within the spheres, but the
distance between the consecutive waves will be large downstream of S and small
upstream of it. This concentration of spheres' surfaces upstream will increase as the
velocity U_0 approaches the velocity of sound c, until, when $U_0 = c$ and Ma = 1, all the
spherical waves become tangential to each other at S. If the fluid velocity is increased
further so that $U_0 > c$ and Ma > 1, the spheres are swept away faster than they are generated, the distance $U_0\Delta t$ being greater than $c\Delta t$. Such a situation is shown in Fig. 13.6.

FIGURE 13.6

Wave propagation in
a fluid moving with a
velocity greater than
that of sound

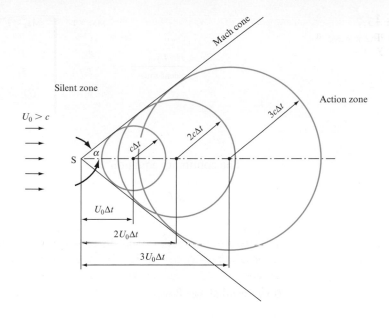

The surface tangential to all the spherical waves is, of course, a cone, known as
the *Mach cone*, which contains within itself the subsonic region called the *action zone*.
Outside the cone, in the *silent zone*, the flow is supersonic and hence the disturbance
generated at S is not 'communicated' to any part of the zone. This is the reason for it
to be called silent. It follows from the geometry of the situation that the greater is U_0
the greater will be the distances travelled by the spheres ($U_0\Delta t$) and, since $c\Delta t$ remains
constant, the Mach angle α, defined as

$$\sin \alpha = c\Delta t/U_0\Delta t = c/U_0 = 1/\mathrm{Ma}$$

or $$\alpha = \sin^{-1}(1/\mathrm{Ma}),$$ (13.13)

will decrease.

If the source of disturbance S is replaced by a thin wedge, as shown later in
Fig. 13.11, every point on the body becomes a source of disturbance and generates
weak Mach waves. The pattern resulting from the superimposition of these waves
yields a *shock wave* across which finite changes of flow parameters occur. If the plane
of the shock wave is perpendicular to the direction of flow the shock wave is known
as a *normal shock* wave. Consider such a shock wave: let the parameters upstream of
the shock wave, in the silent zone where Ma > 1, be denoted by a suffix 1 and those
downstream of the shock, in the action zone where Ma < 1, be denoted by suffix 2,
as indicated in Fig. 13.7. The flow is considered to be adiabatic and frictionless, but
not isentropic. This is because there is dissipation of mechanical energy across the
shock which results in an increase of entropy. The process is, thus, an irreversible one.
A perfect gas is also stipulated. The derivation of the relationship between the
upstream and downstream Mach numbers, and, hence, between the remaining
parameters, is based on the four fundamental relationships: the continuity equation,
the steady flow energy equation, the momentum equation and the equation of state.
They are applied to a horizontal streamtube of constant cross-section.

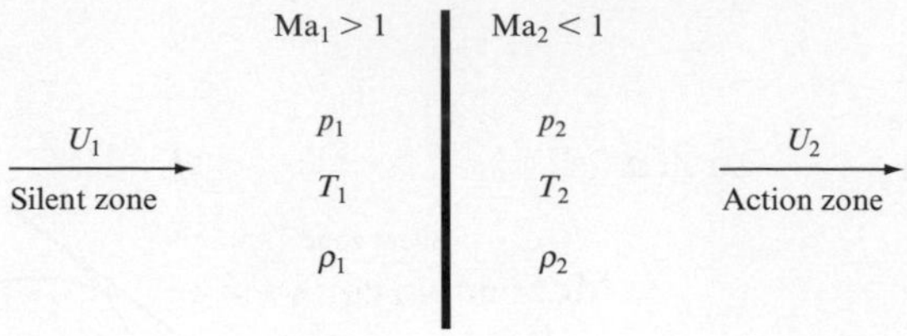

FIGURE 13.7

Normal shock wave

1. The continuity equation

$$\rho_1 U_1 A = \rho_2 U_2 A;$$

therefore,

$$\rho_1 U_1 = \rho_2 U_2.$$

But, for adiabatic flow,

$$\text{Ma} = U/c = U/\sqrt{(\gamma RT)},$$

so that $U = \text{Ma}\sqrt{(\gamma R T)},$

which gives

$$\rho_1 \text{Ma}_1 \sqrt{(\gamma R T_1)} = \rho_2 \text{Ma}_2 \sqrt{(\gamma R T_2)}.$$

Therefore,

$$\rho_1 \text{Ma}_1 \sqrt{T_1} = \rho_2 \text{Ma}_2 \sqrt{T_2}. \tag{13.14}$$

However, from the equation of state,

$$p_1/\rho_1 T_1 = p_2/\rho_2 T_2,$$

so that $\rho_1/\rho_2 = (p_1/p_2)(T_2/T_1).$

Substituting into (13.14) in order to eliminate the density ratio:

$$(p_1/p_2)(T_2/T_1)\,\text{Ma}_1 \sqrt{T_1} = \text{Ma}_2 \sqrt{T_2},$$

which gives

$$p_1\,\text{Ma}_1/\sqrt{T_1} = p_2\,\text{Ma}_2/\sqrt{T_2}. \tag{13.15}$$

2. Steady flow energy equation (6.10):

$$U_1^2/2 + H_1 = U_2^2/2 + H_2 = H_T,$$

where H_1 and H_2 are the enthalpies and H_T is the stagnation enthalpy, which remains constant across the shock waves. But $H_T = c_p T_T$ and, by equation (13.9),

$$T_T = T_0 \left(1 + \frac{\gamma - 1}{2} \mathrm{Ma}_0^2 \right),$$

so that
$$T_1 \left(1 + \frac{\gamma - 1}{2} \mathrm{Ma}_1^2 \right) = T_2 \left(1 + \frac{\gamma - 1}{2} \mathrm{Ma}_2^2 \right). \tag{13.16}$$

3. Momentum equations:

$$p_1 A - p_2 A = \rho_1 A U_1 (U_2 - U_1),$$
$$p_1 - p_2 = \rho_1 U_1 U_2 - \rho_1 U_1^2.$$

But $\rho_1 U_1 = \rho_2 U_2$, so that

$$p_1 - p_2 = \rho_2 U_2^2 - \rho_1 U_1^2$$

or
$$p_1 + \rho_1 U_1^2 = p_2 + \rho_2 U_2^2.$$

However,

$$U^2 = \gamma R T \mathrm{Ma}^2,$$

so that
$$p_1 + \gamma R T_1 \mathrm{Ma}_1^2 \rho_1 = p_2 + \gamma R T_2 \mathrm{Ma}_2^2 \rho_2$$

and
$$p_1 (1 + \gamma \mathrm{Ma}_1^2 R T_1 \rho_1 / p_1) = p_2 (1 + \gamma \mathrm{Ma}_2^2 R T_2 \rho_2 / p_2).$$

In addition

$$R T \rho / p = 1,$$

which gives

$$p_1 (1 + \gamma \mathrm{Ma}_1^2) = p_2 (1 + \gamma \mathrm{Ma}_2^2). \tag{13.17}$$

In order to obtain the relationship between Ma_1 and Ma_2, it is necessary to eliminate pressures and temperatures from equations (13.15), (13.16) and (13.17). The former objective is realized by dividing equation (13.17) by equation (13.15), which gives

$$\left(\frac{1 + \gamma \mathrm{Ma}_1^2}{\mathrm{Ma}_1} \right) \sqrt{T_1} = \left(\frac{1 + \gamma \mathrm{Ma}_2^2}{\mathrm{Ma}_2} \right) \sqrt{T_2}.$$

This equation is now divided by the square root of equation (13.16). The result is

$$\frac{1 + \gamma \mathrm{Ma}_1^2}{\mathrm{Ma}_1 [1 + \frac{1}{2}(\gamma - 1) \mathrm{Ma}_1^2]^{1/2}} = \frac{1 + \gamma \mathrm{Ma}_2^2}{\mathrm{Ma}_2 [1 + \frac{1}{2}(\gamma - 1) \mathrm{Ma}_2^2]^{1/2}} = f(\mathrm{Ma}, \gamma). \tag{13.18}$$

It shows that the above particular function of Mach number and γ is constant across the shock and determines the relationship between the upstream and downstream values of the Mach number. It is plotted in Fig. 13.8 for air ($\gamma = 1.4$).

By solving equation (13.18), the expression for Ma_2 is obtained:

$$\mathrm{Ma}_2 = \left\{ \frac{\mathrm{Ma}_1^2 + 2/(\gamma - 1)}{[2\gamma/(\gamma - 1)] \mathrm{Ma}_1^2 - 1} \right\}^{1/2}. \tag{13.19}$$

FIGURE 13.8

Change of Mach number across a shock wave

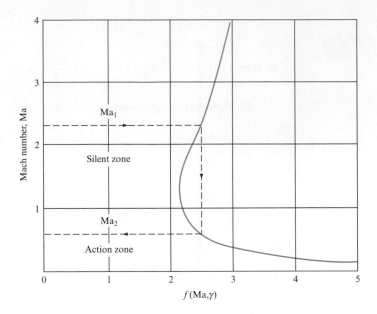

This equation can now be substituted into equations (13.15), (13.16) and (13.17) to give the following ratios across the shock wave:

$$\frac{T_2}{T_1} = \frac{\gamma(\gamma-1)}{(\gamma+1)^2\,\mathrm{Ma}_1^2}\left(1 + \frac{\gamma-1}{2}\,\mathrm{Ma}_1^2\right)\left(\frac{2\gamma}{\gamma-1}\,\mathrm{Ma}_1^2 - 1\right), \tag{13.20}$$

$$\frac{p_2}{p_1} = \frac{2\gamma}{\gamma+1}\,\mathrm{Ma}_1^2 - \frac{\gamma-1}{\gamma+1}, \tag{13.21}$$

$$\frac{\rho_2}{\rho_1} = \frac{U_1}{U_2} = \frac{\gamma+1}{2}\,\frac{\mathrm{Ma}_1^2}{1 + [(\gamma-1)/2]\,\mathrm{Ma}_1^2}. \tag{13.22}$$

The above three ratios are functions of Ma and γ only and are plotted in Fig. 13.9 for $\gamma = 1.4$.

The *strength* of a shock wave is defined as the ratio of the pressure rise across the shock to the upstream pressure. Thus,

$$\text{Shock strength} = (p_2 - p_1)/p_1 = p_2/p_1 - 1. \tag{13.23}$$

By substitution from equation (13.21),

$$\text{Shock strength} = 2\gamma(\mathrm{Ma}_1^2 - 1)/(\gamma + 1). \tag{13.24}$$

FIGURE 13.9

Changes of parameters of state across a shock wave

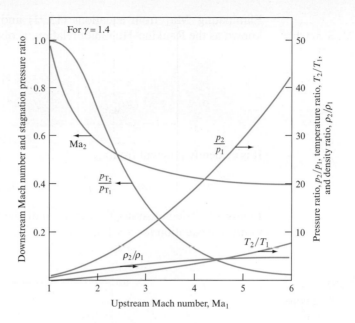

It is also useful to have an expression for the ratio of the stagnation pressures, which may be obtained using equation (13.10),

$$p_T = p_0\{1 + [(\gamma - 1)/2]\mathrm{Ma}_0^2\}^{\gamma/(\gamma-1)},$$

together with (13.21). This procedure, although using an isentropic equation, is justified because the stagnation pressure is defined as resulting from a reversible adiabatic and, hence, isentropic process and, in any case, would take place either upstream or downstream of the shock. Thus, from (13.10),

$$\frac{p_{T_1}}{p_{T_2}} = \frac{p_1}{p_2}\left\{\frac{1 + [(\gamma - 1)/2]\,\mathrm{Ma}_1^2}{1 + [(\gamma - 1)/2]\,\mathrm{Ma}_2^2}\right\}^{\gamma/(\gamma-1)}.$$

Substituting for p_1/p_2 from (13.21),

$$\frac{p_{T_1}}{p_{T_2}} = \left(\frac{2\gamma}{\gamma + 1}\mathrm{Ma}_1^2 - \frac{\gamma - 1}{\gamma + 1}\right)^{-1}\left\{\frac{1 + [(\gamma - 1)/2]\,\mathrm{Ma}_1^2}{1 + [(\gamma - 1)/2]\,\mathrm{Ma}_2^2}\right\}^{\gamma/(\gamma-1)}.$$

Now, eliminating Ma_2^2 using equation (13.19) and simplifying gives

$$\frac{p_{T_1}}{p_{T_2}} = \left[\frac{(\gamma + 1)\,\mathrm{Ma}_1^2}{(\gamma - 1)\,\mathrm{Ma}_1^2 + 2}\right]^{\gamma/(\gamma-1)}\left[\frac{\gamma + 1}{2\gamma\mathrm{Ma}_1^2 - (\gamma - 1)}\right]^{1/(\gamma-1)}. \tag{13.25}$$

It is now possible to show that the flow across the shock is irreversible, and (hence) accompanied by an increase of entropy, by obtaining the relationship between pressure and density and comparing it with the isentropic relationship (equation (1.17)), namely

$$p/\rho^\gamma = \text{constant}.$$

Eliminating Ma_1^2 from equations (13.21) and (13.22), the following relationship, known as the Rankine–Hugoniot relation, is obtained:

$$\frac{\rho_2}{\rho_1} = \left[\left(\frac{\gamma+1}{\gamma-1}\right)\frac{p_2}{p_1} + 1 \right] \Big/ \left[\frac{p_2}{p_1} + \left(\frac{\gamma+1}{\gamma-1}\right) \right]. \tag{13.26}$$

It is evidently different from

$$p/\rho^\gamma = \text{constant} \quad \text{or} \quad \rho_2/\rho_1 = (p_2/p_1)^{1/\gamma}.$$

Figure 13.10 demonstrates the deviation of the Rankine–Hugoniot relation from the isentropic equation for $\gamma = 1.4$.

FIGURE 13.10

Comparison of Rankine–Hugoniot and isentropic curves for $\gamma = 1.4$

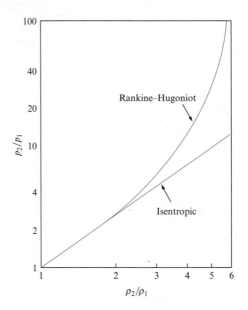

The increase of entropy across a shock is obtained from:

$$S_2 - S_1 = \int_1^2 \frac{dQ}{T} = c_v \int_1^2 \frac{dT}{T} + \int_1^2 \frac{p}{T} d\left(\frac{1}{\rho}\right)$$

$$= c_v \log_e\left(\frac{T_2}{T_1}\right) + \int_1^2 R\rho \, d\left(\frac{1}{\rho}\right)$$

$$= c_v \log_e\left(\frac{T_2}{T_1}\right) - R \log_e\left(\frac{\rho_2}{\rho_1}\right). \tag{13.27}$$

An alternative expression for the specific entropy in terms of pressure ratio may be obtained as follows:

$$\frac{S_2 - S_1}{c_v} = \log_e\left(\frac{T_2}{T_1}\right) - \frac{R}{c_v}\log_e\left(\frac{p_2}{p_1}\right),$$

but $\dfrac{R}{c_v} = \dfrac{c_p - c_v}{c_v} = (\gamma - 1)$ and $\dfrac{p_2}{p_1} = \dfrac{T_1}{T_2} \times \dfrac{\rho_2}{\rho_1},$

so that $\dfrac{S_2 - S_1}{c_v} = \log_e\left(\dfrac{T_2}{T_1}\right) - (\gamma - 1)\left[\log_e\left(\dfrac{T_1}{T_2}\right) + \log_e\left(\dfrac{\rho_2}{\rho_1}\right)\right]$

$$= \log_e\left(\frac{T_2}{T_1}\right) - \gamma\log_e\left(\frac{T_1}{T_2}\right) + \log_e\left(\frac{T_1}{T_2}\right) - (\gamma - 1)\log_e\left(\frac{\rho_2}{\rho_1}\right)$$

$$= \log_e\left(\frac{T_2}{T_1}\right) + \gamma\log_e\left(\frac{T_2}{T_1}\right) - \log_e\left(\frac{T_2}{T_1}\right) - (\gamma - 1)\log_e\left(\frac{\rho_2}{\rho_1}\right).$$

Finally,

$$\frac{S_2 - S_1}{c_v} = \gamma\log_e\left(\frac{T_2}{T_1}\right) - (\gamma - 1)\log_e\left(\frac{\rho_2}{\rho_1}\right). \tag{13.27a}$$

EXAMPLE 13.2

A Pitot–static tube is inserted into an airstream of velocity U_0, pressure 1.02×10^5 N m^{-2} and temperature 28 °C. It is connected differentially to a mercury U-tube manometer. Calculate the difference of mercury levels in the two limbs of the manometer if the velocity U_0 is (a) 50 m s^{-1}, (b) 250 m s^{-1} and (c) 420 m s^{-1}. Take the specific gravity of mercury as 13.6 and for air $\gamma = 1.4$ and $R = 287$ J kg^{-1} K^{-1}.

Solution

The two limbs of the manometer are connected one to the total (or stagnation) connection of the Pitot–static tube and the other to the static connection. Thus, the manometer 'reads' the difference between the two so that

$$p_T - p = \rho_{Hg}gh,$$

where ρ_{Hg} is the density of mercury and h is the difference between the mercury levels. Thus,

$$h = (p_T - p)/\rho_{Hg}g.$$

It is, therefore, necessary to obtain $(p_T - p)$ for the three cases. First, the value of the Mach number must be calculated in order to establish the type of flow taking place, which will govern the choice of appropriate equations.

(a) $\mathrm{Ma} = \dfrac{U_0}{c} = \dfrac{U_0}{\sqrt{(\gamma R T)}} = \dfrac{50}{\sqrt{(1.4 \times 287 \times 301)}} = \dfrac{50}{347.77} = 0.14.$

Therefore the flow may be considered as incompressible and equation (13.2) may be used:

$$p_T - p_0 = \tfrac{1}{2}\rho U_0^2.$$

But, $p/\rho = RT,$

from which

$$\rho = p/RT = 1.02 \times 10^5/(287 \times 301) = 1.18 \text{ kg m}^{-3}$$

and $p_T - p = \tfrac{1}{2}\rho U_0^2 = \tfrac{1}{2} \times 1.18(50)^2 = 1475 \text{ N m}^{-2},$

$$h = 1475/(13.6 \times 10^3 \times 9.81) = 11.06 \times 10^{-3} \text{ m of mercury}$$

$$= \textbf{11.06 mm of mercury.}$$

(b) $\text{Ma} = 250/347.77 = 0.719.$

Compressibility effects must be taken into account and, therefore, either (i) equation (13.10) is used or (ii) the value of the compressibility factor is obtained from Fig. 13.3.

(i) $p_T = p_0\{1 + [(\gamma - 1)/2]\text{Ma}_0^2\}^{\gamma/(\gamma-1)}$

$$= 1.02 \times 10^5[1 + (0.4/2)(0.719)^2]^{1.4/0.4} = 1.44 \times 10^5 \text{ N m}^{-2}.$$

Therefore,

$$p_T - p_0 = (1.44 - 1.02)10^5 = 0.42 \times 10^5 \text{ N m}^{-2}.$$

(ii) From Fig. 13.3, for $\text{Ma} = 0.719$, $(p_T - p_0)/\tfrac{1}{2}\rho_0 U_0^2 = 1.135$. Therefore,

$$p_T - p_0 = 1.135 \times \frac{1.18}{2}(250)^2 = 0.418 \times 10^5 \text{ N m}^{-2}.$$

Taking 0.42×10^5 as more accurate,

$$h = \frac{0.42 \times 10^5}{13.6 \times 10^3 \times 9.81} = 315 \times 10^{-3} \text{ m of mercury}$$

$$= \textbf{315 mm of mercury.}$$

(c) $\text{Ma} = 420/347.77 = 1.208.$

The flow is supersonic and, therefore, a shock wave will be formed owing to the disturbance created by the Pitot–static tube. As the nose of the tube is rounded, it is reasonable to assume that the shock will be detached and a section of it just upstream of the Pitot–static tube will be normal to it. Thus, the pressure downstream of the shock and upstream of the tube will be given by equation (13.21):

$$p_2 = p_1\left(\frac{2\gamma}{\gamma+1}\text{Ma}_1^2 - \frac{\gamma-1}{\gamma+1}\right) = 1.02 \times 10^5\left[\frac{2.8}{2.4}(1.208)^2 - \frac{0.4}{2.4}\right]$$

$$= 1.567 \times 10^5 \text{ N m}^{-2}.$$

Now, in order to calculate the stagnation pressure there, using equation (13.10), it is necessary first to determine the Mach number in the action zone between the shock wave and the Pitot–static tube. This may be obtained from equation (13.19):

$$Ma_2 = \left\{ \frac{Ma_1^2 + 2/(\gamma - 1)}{[2\gamma/(\gamma - 1)]Ma_1^2 - 1} \right\}^{1/2}$$

or

$$Ma_2^2 = \frac{(1.208)^2 + 2/0.4}{(2.8/0.4)(1.208)^2 - 1} = 0.70.$$

Hence,

$$p_{T_2} = p_2\left(1 + \frac{\gamma - 1}{2}Ma_2^2\right)^{\gamma/(\gamma - 1)} = 1.567 \times 10^5\left(1 + \frac{0.4}{2} \times 0.70\right)^{1.4/0.4}$$

$$= 2.479 \times 10^5 \text{ N m}^{-2}.$$

Therefore,

$$h = \frac{(2.479 - 1.567)10^5}{13.6 \times 10^3 \times 9.81} = 683.4 \times 10^{-3} \text{ m} = \textbf{683 mm of mercury}.$$

13.3 OBLIQUE SHOCK WAVES

When a shock wave is not perpendicular to the direction of flow, it is called an oblique shock wave (Fig. 13.11). It occurs during flow past a wedge or sharp object or when the supersonic flow is forced to change direction by a solid boundary, as shown in Fig. 13.12.

One way of treating an oblique shock wave is to consider its normal and tangential components. The normal component undergoes changes associated with the normal shock wave, whereas the tangential component remains unchanged. Thus, only the normal velocity component is reduced causing the deflection of the flow.

It is important to note that although u_{2n} must be subsonic, being downstream of the normal shock, the resultant velocity downstream of the oblique shock, namely

$$U_2 = \sqrt{(u_{2n}^2 + u_{2t}^2)},$$

may be supersonic, provided u_{2t} is large enough. Thus, Ma_2 is always smaller than Ma_1, but it may be greater than one.

The equations derived for the normal shock wave are valid provided they are applied to the normal velocity components. Since

$$u_{1n} = U_1 \sin\beta$$

and

$$u_{2n} = U_2 \sin(\beta - \theta),$$

where β = shock angle (with respect to upstream flow direction) and θ = deflection angle, it is sufficient to substitute these expression – as well as $Ma_1 \sin\beta$ for Ma_1 and $Ma_2 \sin(\beta - \theta)$ for Ma_2 – in the normal shock equations. The angles β and θ are related by

FIGURE 13.11

Oblique shock wave

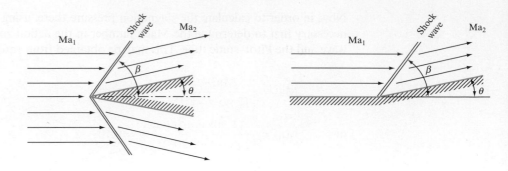

FIGURE 13.12

Flow deflection due to an
oblique shock wave

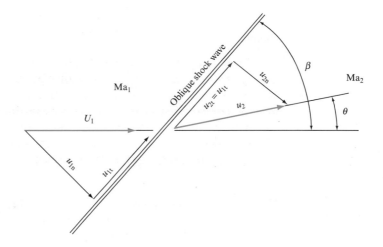

$$\frac{u_{2n}}{u_{1n}} = \frac{\tan(\beta - \theta)}{\tan \beta} = \frac{\rho_1}{\rho_2} = \frac{p_1 T_1}{p_2 T_2}. \tag{13.28}$$

Using equations previously derived, it may be shown that

$$\frac{\tan(\beta - \theta)}{\tan \beta} = \frac{2 + (\gamma - 1) \, \text{Ma}_1^2 \sin^2 \beta}{(\gamma + 1) \, \text{Ma}_1^2 \sin^2 \beta} \tag{13.29}$$

and $$\tan \theta = \frac{2 \cot \beta (\text{Ma}_1^2 \sin \beta - 1)}{\text{Ma}_1^2 (\gamma + \cos 2\beta) + 2}, \tag{13.30}$$

from which the deflection angle may be determined. This equation has two real roots,
giving two values of β for each value of θ and Ma_1 as shown by the plot in Fig. 13.13.
The two values correspond to a strong and a weak wave, respectively. For the strong
shock wave, the downstream flow is always subsonic and the shock angle is large; for
the weak wave, the downstream flow is usually supersonic and the shock angle is
smaller. The chain curve in Fig. 13.13 separates the region of $\text{Ma}_2 < 1$ from that in
which $\text{Ma}_2 > 1$. The heavy line, however, joins the maximum values of $\theta \, (= \theta_{max})$ and
thus separates the weak shock from the strong.

FIGURE 13.13

Oblique shock angles for $\gamma = 1.4$

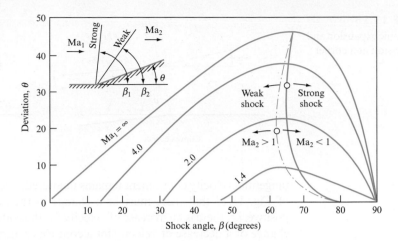

The plot also indicates that, for the shock to occur, θ must be smaller than θ_{max} for a given value of the upstream Mach number. If the physical situation is such that this condition is not satisfied – for example, if the wedge angle is greater than θ_{max} for the flow Mach number – the shock will detach itself from the wedge, thus creating a subsonic space just in front of the wedge. Such a shock wave is always curved, as shown in Fig. 13.14. It, thus, extends further and significantly increases the wave drag on the body. Hence a sharp, pointed nose is a better shape for supersonic flow.

FIGURE 13.14

Detached shock wave

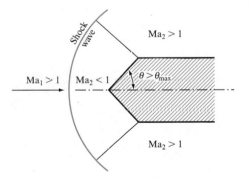

13.4 SUPERSONIC EXPANSION AND COMPRESSION

Consider supersonic flow round an infinitesimal corner, which may be convex or concave as shown, with the angle $\delta\theta$ greatly exaggerated (Fig. 13.15). The corner constitutes a disturbance and, since $\delta\theta$ is very small, the disturbance is small, thus generating a very weak shock wave. Such a wave of infinitesimal strength is called a *Mach wave*. It may be represented by a Mach line whose angle μ is given by

$$\sin \mu = 1/\text{Ma}. \tag{13.31}$$

When the supersonic flow is forced to negotiate a corner, the flow remains parallel to both the upstream and the downstream solid boundary surfaces. Since the

FIGURE 13.15
Supersonic expansion and
compression at a corner

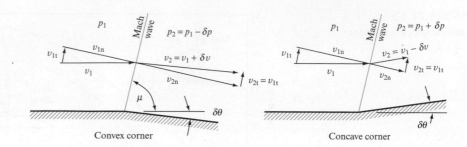

Convex corner   Concave corner

tangential velocity component remains unaltered, it follows from the velocity triangles
of Fig. 13.15 that there must be a change in the resultant velocity. This change is
positive, i.e. there is an increase of velocity for the convex corner and there is a negative
change or a decrease of velocity for a concave corner. These changes must be accompanied by the corresponding changes in pressure. Thus, there is a pressure drop or
expansion at a convex corner and a pressure rise or compression at a concave corner.

Any convex corner of finite deflection θ may be regarded as a series of consecutive
infinitesimal corners of deflections $\delta\theta$. This gives rise to a fan of Mach waves (or
characteristics), as shown in Fig. 13.16, through which smooth and isentropic
expansion takes place. This is known as Prandtl–Mayer flow. The evaluation of
changes of pressure and Mach number is carried out in steps through each successive
Mach line. Such a step-by-step method is called the method of characteristics and is
beyond the scope of this book.

FIGURE 13.16
Prandtl–Mayer expansion

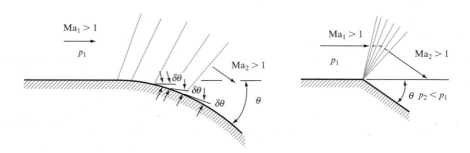

A concave corner of finite deflection gives rise to a series of Mach lines which converge into an envelope and thus form a shock wave as shown in Fig. 13.17. Such a compression process is, therefore, not isentropic, since the changes occur across a shock wave.

FIGURE 13.17
Shock wave at a
concave corner

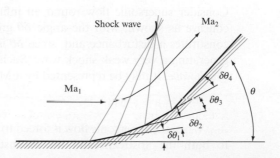

13.5 COMPUTER PROGRAM NORSH

Program NORSH calculates the Mach number, celerity, gas velocity and the parameters of state downstream of a normal shock, together with the entropy change across the shock. The program also determines the upstream parameters not input as data and presents the output in tabular form. The calculation invokes equations (1.3), (5.31), (13.1), (13.10), (13.19), (13.21), (13.22) and (13.27).

13.5.1 Application example

The calculation requires the following data:

1. the values of gas constant, R J kg^{-1} K^{-1}, and the ratio of specific heats, γ;
2. the values of any two of the following parameters of state upstream of the shock: static pressure, p_1 kN m^{-2}, temperature, T_1 K, and density, ρ kg m^{-3};
3. the value of either the Mach number or the gas velocity, u_1 m s^{-1}, upstream of the shock.

For the following data $R = 287$ J kg^{-1} K^{-1}; $\gamma = 1.4$; $p_1 = 102$ kN m^{-2}; $T_1 = 301$ K; $u_1 = 420$ m s^{-1}, the output table details both the downstream conditions and the unstated upstream conditions:

		UPSTREAM	DOWNSTREAM
Mach number		1.21	0.84
Static pressure	kN m^{-2}	102	156.57
Stagnation pressure	kN m^{-2}	249.85	247.86
Temperature	K	301	340.98
Density	kg m^{-3}	1.181	1.6
Celerity	m s^{-1}	347.77	370.14
Flow velocity	m s^{-1}	420	309.96
Entropy change		2.29 J kg^{-1} K^{-1}	

13.5.2 Additional investigations using NORSH

The computer program may be used to investigate the dependence of downstream conditions on systematic changes in one upstream parameter for the range of input data possible.

Concluding remarks

The effects of fluid compressibility have been discussed in this chapter, with particular reference to the development of shocks within a flow external to a body. The effect of shock waves on the drag forces acting on the body were discussed and solutions aimed at minimizing these forces introduced. The equation of state and the energy equation, introduced in Chapters 1 and 5, were essential in the development of relationships linking flow parameters across the shock. The discussion featured both normal and oblique shocks, together with the generation of a detached shock ahead of a body.

Summary of important equations and concepts

1. The effects of compressibility are emphasized and a compressibility factor, equation (13.12) introduced following the development of expressions for stagnation temperature and pressure, equations (13.3) and (13.10).

2. The concept of a shock wave is introduced, Section 13.2, and equations defining temperature, pressure, and density ratios across normal shocks derived, equations (13.20), (13.21) and (13.22). Entropy change across the shock is also derived as equation (13.27). These equations are utilized in a computer program NORSH, presented in Section 13.5.

3. The application of normal shock equations to an oblique shock is emphasized in Section 13.3.

Problems

13.1 A Pitot–static tube is inserted into an airstream and the mercury manometer connected differentially to it shows a difference in levels of 300 mm. The free stream temperature and pressure are 40 °C and 150 kN m^{-2} absolute. Calculate the air velocity and the percentage error which would have been committed if the flow was considered as incompressible. (Specific gravity of mercury = 13.6).

[219 m s^{-1}, 4.6 per cent]

13.2 If the difference between static and stagnation pressure in standard air ($p = 101.3$ kN m^{-2}, $T = 288$ K) is 600 mm of mercury, compute the air velocity assuming (a) the air is incompressible, (b) the air is compressible, and hence calculate the compressibility factor.

[(a) 361 m s^{-1}, (b) 316 m s^{-1}, 1.31]

13.3 An airstream issues from a nozzle into the atmosphere where the barometric pressure is 750 mm of mercury and the temperature is 20 °C. Assuming that for air the difference between the stagnation temperature and the free stream temperature is given by

$$T_\mathrm{T} - T_0 = \left(\frac{V_0}{45}\right)^2 \quad °C,$$

where $V_0 = 250$ m s^{-1} is the free stream velocity, calculate the stagnation temperature, pressure density and the Mach number of the flow. For air $R = 287$ J kg^{-1} K^{-1} and $\gamma = 1.4$.

[324 K, 144.1 kN m^{-2}, 1.55 kg m^{-3}, 0.729]

13.4 A Pitot–static tube is inserted into the test section of a subsonic wind tunnel. It indicates a static pressure of 80 kN m^{-2}, while the difference between stagnation and static pressure is shown as 120 mm of mercury. The barometric pressure is 760 mm of mercury and the stagnation temperature is 40 °C. Calculate the Mach number and the air velocity.

[0.35, 123 m s^{-1}]

13.5 Starting from the differential form of Euler's equation

$$\frac{1}{\rho}\frac{\mathrm{d}p}{\mathrm{d}x} + \bar{v}\frac{\mathrm{d}v}{\mathrm{d}x} + g\frac{\mathrm{d}z}{\mathrm{d}x} = 0,$$

show that for air ($R = 287$ J kg^{-1} K^{-1}; $\gamma = 1.4$), assuming horizontal, isentropic flow, the difference between stagnation temperature and free stream temperature is approximately given by

$$T_\mathrm{T} - T_0 = \left(\frac{V_0}{45}\right)^2 \quad °C,$$

where V_0 is the free stream velocity in metres per second. Calculate also the percentage error involved in the above approximation.

[0.8 per cent]

13.6 An airstream with velocity 500 m s^{-1}, static pressure 60 kN m^{-2} and temperature −18 °C undergoes a normal shock. Determine the air velocity and the static and stagnation conditions after the wave.

[255 m s^{-1}, 160.8 kN m^{-2}, 255 kN m^{-2}]

13.7 Given that the Mach number downstream of a normal shock is expressed in terms of the Mach number upstream of the shock as follows:

$$\mathrm{Ma}_2^2 = \frac{\mathrm{Ma}_1^2 + 2/(\gamma - 1)}{[2\gamma/(\gamma - 1)]\mathrm{Ma}_1^2 - 1}$$

derive an expression for the pressure ratio across the shock wave and hence an expression for the density ratio in terms of pressure ratio and Ma_1.

13.8 A normal shock moves into still air with a velocity of 1500 m s^{-1}. The still air is at 10 °C and 80 kN m^{-2}. Calculate the stagnation pressure and temperature behind the wave.

$$[12.8 \text{ kN m}^{-2}, 700 \text{ °C}]$$

13.9 The Mach angle as measured from a Schlieren photograph of a bullet has a magnitude of 30°. Estimate the speed of the bullet if the temperature and pressure of the atmosphere were 5 °C and 90 kN m^{-2}, respectively. What Mach angle would indicate the same velocity in air at 15 °C and 101 kN m^{-2}?

$$[668.4 \text{ m s}^{-1}, 30.6°]$$

13.10 Show that for a normal shock wave

$$p_1(1 + \gamma Ma_1^2) = p_2(1 + \gamma Ma_2^2),$$

where p_1 and p_2 are pressures upstream and downstream of a shock wave, respectively, Ma_1 and Ma_2 are the Mach numbers upstream and downstream of the shock, respectively, and γ is the ratio of specific heats.

A projectile with a rounded nose moves through still air at Ma = 5. The air pressure is 60 kN m^{-2} and the temperature is −10 °C. Assuming that the shock wave formed at the nose of the projectile is detached and normal to it determine the stagnation pressure and temperature at the nose. Take $\gamma = 1.4$ and use the relationship given in Problem 13.8.

$$[1959 \text{ kN m}^{-2}, 1525 \text{ K}]$$

13.11 A two-dimensional wedge is used to measure the Mach number of the flow in a supersonic wind tunnel using air. If the total wedge angle is 20° and the shock wave angle is 60° calculate the Mach number in the tunnel and downstream of the shock.

$$[1.36, 1.11]$$

13.12 (a) Starting from the momentum considerations and given that the Mach number downstream of a normal shock Ma_2 is related to the Mach number upstream of the shock Ma_1 by the equation

$$Ma_2^2 = \frac{Ma_1^2 + 2/(\gamma - 1)}{[2\gamma/(\gamma - 1)]Ma_1^2 - 1},$$

show that for air the shock strength is given by

$$\frac{p_2 - p_1}{p_1} = 1.167(Ma_1^2 - 1).$$

(b) A supersonic aircraft flies horizontally overhead at 3000 m through still air. The time interval between the instant the aircraft is directly overhead an observer on the ground and the instant the shock wave is detected by him is 7.0 s. If the velocity of sound in air is 335 m s^{-1} calculate the velocity of the aircraft and the stagnation pressure on its nose. Take the atmospheric pressure at 3000 m to be 70 kN m^{-2}. Note that for normal shock in air

$$Ma_2^2 = \frac{Ma_1^2 + 5}{7Ma_1^2 - 1}.$$

Take $\gamma = 1.4$ for air.

$$[538.4 \text{ m s}^{-1}; 188.8 \text{ kN m}^{-2}]$$

Steady Flow in Pipes, Ducts and Open Channels

Supply to a hydroelectric power station in the Scottish Highlands, photo courtesy of Scottish and Southern Energy plc

IN THE PREVIOUS PART, THE BEHAVIOUR OF REAL FLUIDS has been examined and, in particular, the energy losses which occur due to friction and other causes. In the following chapters, consideration is given to the practical design of pipelines and channels. It is usual to treat liquids under steady flow conditions as if they were incompressible, since the changes of pressure are not large enough to produce significant changes of density. This permits the use of the simple constant density form of the continuity and energy equations, as shown in Chapter 14, which also covers the analysis of pipe networks under such conditions.

Pipes and ducts can have a number of different functions; the most common is to convey fluids from point to point, in which case almost the whole of the head available to produce flow is used in overcoming resistance in the pipeline. Power from a pump, pressure vessel or high-level reservoir may also be transmitted along a pipeline if the fluid travelling through the pipeline arrives at the point of use under pressure or at high velocity.

The flow of liquids through open channels is dealt with in two parts. In Chapter 15 we consider uniform flow and the design of channel cross-sections for optimum performance, while Chapter 16 is concerned with non-uniform flow phenomena and the water surface profiles which can occur under these conditions.

When gases flow through pipelines it is, usually, necessary to take changes of density and temperature along the length of the pipe into account. In Chapter 17, the basic equations of compressible flow are considered and first applied to frictionless flow through orifices, venturi contractions and nozzles. The formation of a normal shock wave in a diffuser is discussed. For pipelines of constant cross-section with frictional resistance, an analysis is made for both adiabatic and isothermal conditions.

Internal views of the London ring main, photo courtesy of Thames Water

Steady Incompressible Flow in Pipe and Duct Systems

THE CONCEPTS OF CONTINUITY OF MASS FLOW AND ENERGY ARE UTILIZED IN THIS chapter to develop the steady flow energy equation and to demonstrate its application to both pipe and duct flows and flows possessing a free surface. A computer program designed to illustrate the application of the steady flow energy equation to flow in pipes and ducts is discussed. The definitions of frictional and separation losses introduced in Chapter 10 are included to allow the determination of system losses and the dependence of the flow in networks on the relative resistance of the alternative flow paths available. Network analysis fundamentally based on Kirchhoff's laws is applied to introduce the Hardy–Cross technique for the prediction of system flow distribution and a computer program to analyze network flow distributions is introduced. ● ● ●

14.1   GENERAL APPROACH

This section is concerned with the analysis of the steady flow of a fluid in closed or open conduits. A *closed conduit* is a pipe or duct through which the fluid flows while completely filling the cross-section. Since the fluid has no free surface, it can be either a liquid or a gas, its pressure may be above or below atmospheric pressure and this pressure may vary from cross-section to cross-section along its length. An *open conduit* is a duct or open channel along which a liquid flows with a free surface. At all points along its length the pressure at the free surface will be the same, usually atmospheric. An open conduit may be covered providing that it is not running full and the liquid retains a free surface; a partly filled pipe would, for example, be treated as an open channel.

In either case, as the fluid flows over the solid boundary a shear stress will be developed at the surface of contact (as discussed in Chapter 11) which will oppose fluid motion. This so-called frictional resistance results in an energy transfer within the system, experienced as a 'loss', measurable in a fluid flow by changes in fluid pressure or head. In addition to the losses attributable to friction, separation losses due to the flow disruption at changes in section, direction or around valves and other flow obstructions also contribute to the overall energy transfers to be accounted for. The first approach to the analysis of bounded systems is therefore to consider the energy balance between two chosen locations along the flow. In Fig. 14.1, for flow across the control volume boundaries represented by the conditions at A and B, the energy audit may be expressed, in terms of energy per unit volume, as

FIGURE 14.1
Energy change

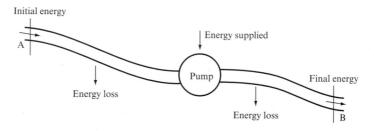

$$p_A + \tfrac{1}{2}\rho v_A^2 + \rho g z_A + \Delta p_{\text{pump}} = p_B + \tfrac{1}{2}\rho v_B^2 + \rho g z_B + \tfrac{1}{2}\rho K u^2,$$

where all terms are defined in the dimensions of pressure and hence are amenable to direct experimental measurement for any particular flow condition.

The pressure loss experienced as a result of friction and separation of the flow from the walls of the conduit has been shown to be defined by a term of the form $\tfrac{1}{2}\rho K u^2$, where u is the local flow velocity and K is a constant dependent upon the conduit parameters, i.e. length, diameter, roughness or fitting type, utilized here to represent both frictional and separation losses.

This form of the steady flow energy equation is particularly suited to the study of steady flow conditions in air duct systems as the constituent terms are all amenable to measurement by pressure transducers, or, more simplistically, by manometers. Traditionally in the study of water conduits the steady flow energy equation has been cast in its form of energy per unit weight, resulting in all the terms having the dimensions of head:

$$h_A + v_A^2/2g + z_A + \Delta h_{pump} = h_B + v_B^2/2g + z_B + Ku^2/2g.$$

While this format is correct and accepted for water-carrying systems, care should be taken in its application in general as all too often it is forgotten that the 'head' term is measured in 'metres of flowing fluid'. Hence pump characteristic data in metres of water will, for example, not apply without modification if the fluid is oil of a given specific gravity. In general the head form of the equation will only be used for water examples; the pressure form is generally applicable for all systems and is recommended.

Also, for steady flow to be maintained it is necessary that

$$\frac{\text{Mass per unit time entering}}{\text{the control volume at A}} = \frac{\text{Mass per unit time leaving}}{\text{the control volume at B.}}$$

For incompressible flow the density remains constant and hence the continuity of mass flow equation above reduces to

$$\frac{\text{Volume per unit time entering}}{\text{the control volume at A}} = \frac{\text{Volume per unit time leaving}}{\text{the control volume at B.}}$$

Analysis of all steady flow problems in pipes and channels is based on the application of the steady flow energy equation and the continuity of volumetric flow equation, applied between suitable points in the system.

14.2   INCOMPRESSIBLE FLOW THROUGH DUCTS AND PIPES

For incompressible flow, since there is no change of density with pressure, the steady flow energy equation reduces to a form of Bernoulli's equation with the addition of terms for the energy losses due to friction and separation, for work done by the fluid in driving turbines or for work done on the fluid by the introduction of a pump or fan. All these terms represent energy per unit volume, measured in pressure units, or energy per unit weight, measured in terms of the head of the fluid concerned.

The pressure loss, Δp, or energy lost per unit volume due to friction, may be conveniently expressed via the Darcy equation

$$\Delta p = 4fL\rho v^2/2D \tag{14.1}$$

for a circular cross-section conduit flowing full. In terms of head this expression becomes

$$\Delta h = 4fLv^2/2gD.$$

Both forms of the Darcy equation may be applied to either laminar or turbulent flow provided the correct form of friction factor, f, is introduced. It should be noted that in laminar flow $f = 16/\text{Re}$ and hence depends only on flow velocity, v. This second form of the Darcy equation may also be utilized in the study of steady, uniform, free surface flow, provided that the conduit diameter hydraulic mean depth, $m = D/4$, is replaced by an appropriate value of m.

Separation losses may be expressed either as a pressure term, $K\frac{1}{2}\rho v^2$, or as a head term, $Kv^2/2g$, where the value of K depends on the type of fitting encountered. Alternatively, fitting losses may be included by the addition of an equivalent length of pipe or duct that would generate the same friction loss as the separation of flow around the fitting generates; this extra equivalent length is simply added to the conduit length and is normally expressed as so many conduit diameters. Tabular values exist for a wide range of fittings and partial valve-opening settings.

Often the engineer is more concerned with the flow deliverable rather than the flow velocity in the conduit, although this too can be of prime interest acoustically or where scouring is a concern. An alternative form of the Darcy equation may be obtained by writing

$$v = \frac{Q}{\text{Pipe cross-section area}} = \frac{Q}{\pi D^2/4}.$$

Substituting in the Darcy equation yields

$$\Delta p = \frac{64fL\rho Q^2}{2D(\pi D^2)^2}, \quad \Delta h = \frac{64fLQ^2}{2gD(\pi D^2)^2}.$$

Both these expressions indicate that the dependence of frictional loss on conduit diameter is a fifth-power relationship, making the reduction of pipe diameter a potentially costly exercise.

In SI units, $g = 9.81\,\text{m\,s}^{-2}$, these expressions reduce to

$$\Delta p = 3.24fL\rho Q^2/D^5, \quad \Delta h = fLQ^2/3.03D^5$$

or, within an error of 1 per cent for the head definition,

$$\Delta h = fLQ^2/3D^5. \tag{14.2}$$

In general, for all pipes, ducts and fittings, the loss of pressure or head may be expressed as either Δp or $\Delta h = KQ^2$, where K is a resistance coefficient.

EXAMPLE 14.1

Water discharges from a reservoir A (Fig. 14.2) through a 100 mm pipe 15 m long which rises to its highest point at B, 1.5 m above the free surface of the reservoir, and discharges direct to the atmosphere at C, 4 m below the free surface at A. The length of pipe l_1 from A to B is 5 m and the length of pipe l_2 from B to C is 10 m. Both the entrance and exit of the pipe are sharp and the value of f is 0.08. Calculate (a) the mean velocity of the water leaving the pipe at C and (b) the pressure in the pipe at B.

FIGURE 14.2

Flow through a siphon

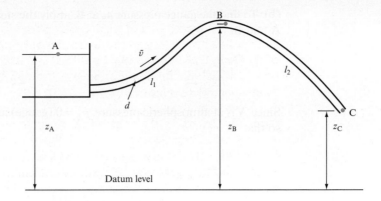

Solution

(a) To determine the velocity $\bar{v}$, first apply the steady flow energy equation between point A on the free surface and point C at the exit from the pipe, since the pressure and elevation of these points are known:

$$\frac{\text{Total energy per unit}}{\text{weight at A}} = \frac{\text{Total energy per unit}}{\text{weight at C}} + \text{Losses.} \qquad \text{(I)}$$

Since the entrance to the pipe is sharp, there will be a loss of $0.5\bar{v}^2/2g$ (see Section 10.8.2). The loss due to friction in the length of pipe AC is given by the Darcy formula as

$$\frac{4f(l_1 + l_2)}{d} \frac{\bar{v}^2}{2g}.$$

There will be no loss of energy at the exit because, although the pipe exit is sharp, the water emerges into the atmosphere without any change of the cross-section of the stream.

At both A and C the pressure is atmospheric, so that $p_A = p_C = $ zero gauge pressure. Also, if the area of the free surface of the reservoir is large, the velocity at A is negligible. Thus,

$$z_A = z_C + \bar{v}^2/2g + \Sigma \text{Losses.}$$

Substituting in (I),

$$z_A = \left(z_C + \frac{\bar{v}^2}{2g} \right) + 0.5\frac{\bar{v}^2}{2g} + \frac{4f(l_1 + l_2)}{d} \frac{\bar{v}^2}{2g},$$

$$z_A - z_C = \frac{\bar{v}^2}{2g} \left[1 + 0.5 + \frac{4f(l_1 + l_2)}{d} \right].$$

Putting $z_A - z_C = 4\,\text{m}$, $l_1 = 5\,\text{m}$, $l_2 = 10\,\text{m}$, $d = 100\,\text{mm} = 0.1\,\text{m}$, $f = 0.08$,

$$4 = \frac{\bar{v}^2}{2 \times 9.81} \left[1 + 0.5 + \frac{4 \times 0.08 \times 15}{0.1} \right] \text{m},$$

$$\bar{v}^2 = \frac{4 \times 2 \times 9.81}{49.5} = 1.585,$$

$$\bar{v} = 1.26\ \text{m s}^{-1}.$$

(b) To find the gauge pressure p_B at B, apply the steady flow energy equation between A and B:

$$\left(\frac{p_A}{\rho g}+\frac{\bar{v}_A^2}{2g}+z_A\right)=\left(\frac{p_B}{\rho g}+\frac{\bar{v}^2}{2g}+z_B\right)+0.5\frac{\bar{v}^2}{2g}+\frac{4fl_1}{d}\frac{\bar{v}^2}{2g}.$$

Since A is at atmospheric pressure, $p_A = 0$ (gauge) and, if the reservoir is large, $\bar{v}_A = 0$, so that

$$z_A=\frac{p_B}{\rho g}+z_B+\frac{\bar{v}^2}{2g}\left(1+0.5+\frac{4fl_1}{d}\right),$$

$$p_B=\rho g(z_A-z_B)-\rho\frac{\bar{v}^2}{2}\left(1.5+\frac{4fl_1}{d}\right).$$

Substituting $(z_A - z_B) = -1.5\,\text{m}$, $\bar{v} = 1.26\,\text{m s}^{-1}$, $f = 0.08$, $l_1 = 5\,\text{m}$, $d = 100\,\text{mm} = 0.1\,\text{m}$, $\rho = 10^3\,\text{kg m}^{-3}$,

$$p_B=10^3\times9.81\times(-1.5)-\frac{10^3\times1.26^2}{2}\left(1.5+\frac{4\times0.08\times5}{0.1}\right)$$

$$=-14.71\times10^3-13.87\times10^3\,\text{N m}^{-2}$$

$$=-28.58\times10^3\,\text{N m}^{-2}$$

$$=\textbf{28.58 kN m}^{-2}\textbf{ below atmospheric pressure.}$$

14.3   COMPUTER PROGRAM SIPHON

This program uses the steady flow energy equation, along with the separation and frictional loss equations already introduced, to investigate the flow of a fluid between two reservoirs or pressurized tanks, via a series pipe network of up to five pipes. The simulation presents pressure and flow data along the series pipe system and is capable of dealing with the possibility of a high point in the pipeline profile, similar to that addressed in Example 14.1.

The program calculates the maximum flow rate between the supply and collection reservoirs/tanks and calculates the local pressure profile along the system to check against a violation of the vapour pressure limit. A maximum flow is identified to avoid cavitation.

The program accepts data on the absolute pressure level 'above' the fluid in the two reservoirs or tanks. Fluid surface level in both reservoirs is required, together with the entry/exit depth of the pipeline connection to both. General data are also required for fluid density, vapour pressure and the number of pipes in the series.

For each pipe in the line, data defining entry level above a datum (synonymous with the exit from the upstream pipe), pipe length, diameter and friction factor are required and data concerning separation losses per pipe are also requested with a location expressed as a percentage of pipe length from entry. The discharge level for the final system pipe is also required.

14.3.1 Application example

Consider a two-pipe system leading from one tank to an open reservoir. The pressure in the upstream tank is 300 kN m^{-2} absolute and the fluid in the downstream reservoir is open to atmosphere, i.e. 100 kN m^{-2}. The fluid level in the upstream tank is 10 m above a datum, that in the reservoir 6 m. Fluid density is 1000 kg m^{-3} and its vapour pressure is 20 N m^{-2}.

Pipe data are as follows:

	ENTRY ELEVATION (m)	PIPE LENGTH (m)	DIAMETER (m)	FRICTION FACTOR	SEPARATION LOSS FACTOR	LOCATION (% LENGTH)
Pipe 1	9	5	0.1	0.08	0.5	0.0
Pipe 2	11.5	10	0.1	0.08	1.0	100.0

Pipe 2 discharges into the downstream reservoir at an elevation of 5 m above datum.

The simulation indicates that there is no violation of the vapour pressure lower limit, the flow velocity in each pipe was 1.75 m s^{-1} and the flow rate was 0.14 m^3 s^{-1}.

14.3.2 Additional investigations using SIPHON

The simulation may be used to investigate:

1. the effect of changes in the pressure 'above' the fluid in each tank or reservoir;
2. the effect of variations in pipe diameter or friction factor;
3. the influence of separation losses, up to five per pipe, on the flow conditions, analogous to the introduction of valves along the pipeline;
4. following Example 14.1 the influence of pipeline profile may also be considered, the simulation visual graphical output illustrating both this and the associated pressure profile.

14.4   INCOMPRESSIBLE FLOW THROUGH PIPES IN SERIES

When pipes of different diameters are connected end-to-end to form a pipeline, so that the fluid flows through each in turn, the pipes are said to be in series. The total loss of energy, or pressure loss, over the whole pipeline will be the sum of the losses for each pipe together with any separation losses such as might occur at the junctions, entrance or exit.

EXAMPLE 14.2

Two reservoirs A and B (Fig. 14.3) have a difference level of 9 m and are connected by a pipeline 200 mm in diameter over the first part AC, which is 15 m long, and then 250 mm diameter for CB, the remaining 45 m length. The entrance to and exit from the pipes are sharp and the change of section at C is sudden. The friction coefficient f is 0.01 for both pipes.

FIGURE 14.3

Pipes in series, showing head losses and the total energy line and hydraulic gradient

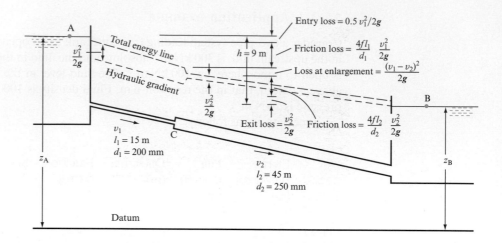

(a) List the losses of head (energy per unit weight) which occur, giving an expression for each. (b) Use program SIPHON (Section 14.3) to calculate the system flow rate and hydraulic gradient.

Solution

(a) The losses of head which will occur are as follows:

    (i) Loss at entrance to pipe AC. This is a separation loss and, since the entrance is described as sharp and is below the free surface of the reservoir (from Section 10.8.2), the value of k will be 0.5:

        Loss of head at entry, $h_1 = 0.5\bar{v}_1^2/2g$.

    (ii) Friction loss in AC. Using the Darcy formula, we have

$$\text{Loss of head in friction in AC,} = h_{f_1} = \frac{4fl_1}{d_1}\frac{\bar{v}_1^2}{2g}.$$

    (iii) Loss at change of section at C. There will be a separation loss at the sudden change of section. From Section 10.8.1, the loss at a sudden enlargement will be

$$\text{Loss of head at sudden enlargement,} \; h_2 = (\bar{v}_f - \bar{v}_2)^2/2g.$$

    (iv) Friction loss in CB. Using the Darcy formula,

$$\text{Loss of head in friction in CB,} = h_{f_2} = \frac{4fl_2}{d_2}\frac{\bar{v}_2^2}{2g}.$$

    (v) Loss of head at exit. Since the exit is described as sharp and is beneath the surface of the reservoir B, there will be a separation loss as explained in Section 10.8:

        Loss of head at exit, $h_3 = \bar{v}_2^2/2g$.

(b) Volume flow rate, $Q = 0.158 \text{ m}^3 \text{ s}^{-1}$.

14.5 INCOMPRESSIBLE FLOW THROUGH PIPES IN PARALLEL

When two reservoirs are connected by two or more pipes in parallel, as shown in Fig. 14.4, the fluid can flow from one to the other by a number of alternative routes. The difference of head h available to produce flow will be the same for each pipe. Thus, each pipe can be considered separately, entirely independently of any other pipes running in parallel. For incompressible flow, the steady flow energy equation can be applied for flow by each route and the total volume rate of flow will be the sum of the volume rates of flow in each pipe.

FIGURE 14.4
Pipes in parallel

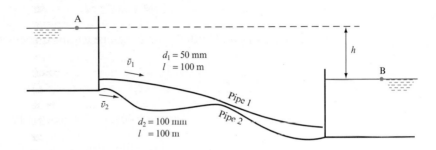

EXAMPLE 14.3

Two sharp-ended pipes of diameter $d_1 = 50$ mm, and $d_2 = 100$ mm, each of length $l = 100$ m, are connected in parallel between two reservoirs which have a difference of level $h = 10$ m, as in Fig. 14.4. If the Darcy coefficient $f = 0.008$ for each pipe, calculate: (a) the rate of flow for each pipe, (b) the diameter D of a single pipe 100 m long which would give the same flow if it was substituted for the original two pipes.

Solution

(a) Since the two pipes are in parallel, we can deal with each pipe independently and apply the steady flow energy equation between points A and B on the free surfaces of the upper and lower reservoirs, respectively.
    For flow by way of pipe 1,

$$\left(\frac{p_A}{\rho g} + \frac{\bar{v}_A^2}{2g} + z_A\right) = \left(\frac{p_B}{\rho g} + \frac{\bar{v}_B^2}{2g} + z_B\right) + \left(0.5\frac{\bar{v}_1^2}{2g} + \frac{4fl}{d_1}\frac{\bar{v}_1^2}{2g} + \frac{\bar{v}_1^2}{2g}\right).$$

Since $p_A = p_B =$ atmospheric pressure and, if the reservoirs are large, $\bar{v}_A$ and $\bar{v}_B$ will be negligible,

$$z_A - z_B = \left(1.5 + \frac{4fl}{d_1}\right)\frac{\bar{v}_1^2}{2g}.$$

Putting $z_A - z_B = h = 10\,\text{m}$, $f = 0.008$, $l = 100\,\text{m}$, $d_1 = 50\,\text{mm} = 0.05\,\text{m}$,

$$10 = \left(1.5 + \frac{4 \times 0.008 \times 100}{0.05}\right)\frac{\bar{v}_1^2}{2g},$$

$$\bar{v}_1^2 = 2g \times 10/(1.5 + 64),$$

$$\bar{v}_1 = 1.731\,\text{m s}^{-1}.$$

Volume rate of flow through pipe 1, $Q_1 = (\pi/4)d_1^2\,\bar{v}_1$

$$= (\pi/4) \times 0.05^2 \times 1.731 = \mathbf{0.0034\,m^3\,s^{-1}}.$$

For flow by way of pipe 2,

$$\left(\frac{p_A}{\rho g} + \frac{\bar{v}_A^2}{2g} + z_A\right) = \left(\frac{p_B}{\rho g} + \frac{\bar{v}_B^2}{2g} + z_B\right) + \left(0.5\frac{\bar{v}_2^2}{2g} + \frac{4fl}{d_2}\frac{\bar{v}_2^2}{2g} + \frac{\bar{v}_2^2}{2g}\right).$$

Since $p_A = p_B$ and both $\bar{v}_A$ and $\bar{v}_B$ can be assumed negligible,

$$z_A - z_B = \left(1.5 + \frac{4fl}{d_2}\right)\frac{\bar{v}_2^2}{2g}.$$

Putting $z_A - z_B = h = 10\,\text{m}$, $f = 0.008$, $l = 100\,\text{m}$, $d_2 = 100\,\text{mm} = 0.10\,\text{m}$,

$$10 = \left(1.5 + \frac{4 \times 0.008 \times 100}{0.10}\right)\frac{\bar{v}_2^2}{2g},$$

$$\bar{v}_2^2 = 2g \times 10/(1.5 + 32),$$

$$\bar{v}_2 = 2.42\,\text{m s}^{-1}.$$

Volume rate of flow through pipe 2,

$$Q_2 = (\pi/4)d_2^2\,\bar{v}_2 = (\pi/4) \times 0.10^2 \times 2.42 = \mathbf{0.0190\,m^3\,s^{-1}}.$$

(b) Replacing the two pipes by the equivalent single pipe which will convey the same total flow,

Volume rate of flow through single pipe,

$$Q = Q_1 + Q_2 = 0.0034 + 0.0190 = 0.0224\,\text{m}^3\,\text{s}^{-1}.$$

If $\bar{v}$ is the velocity in the single pipe, $Q = (\pi/4)D^2\bar{v}$. Therefore,

$$\bar{v} = \frac{4Q}{\pi D^2} = \frac{4 \times 0.0224}{\pi D^2} = \frac{0.028\,52}{D^2}.$$

Applying the steady flow energy equation between A and B,

$$\left(\frac{p_A}{\rho g} + \frac{\bar{v}_A^2}{2g} + z_A\right) = \left(\frac{p_B}{\rho g} + \frac{\bar{v}_B^2}{2g} + z_B\right) + \left(0.5\frac{\bar{v}^2}{2g} + \frac{4fl}{D}\frac{\bar{v}^2}{2g} + \frac{\bar{v}^2}{2g}\right).$$

Making the same assumptions as before,

$$z_A - z_B = \left(1.5 + \frac{4fl}{D}\right)\frac{\bar{v}^2}{2g}.$$

Putting $z_A - z_B = h = 10\,\text{m}$, $f = 0.008$, $l = 100\,\text{m}$, $\bar{v} = 0.028\,52/D^2$,

$$10 = \left(1.5 + \frac{4 \times 0.008 \times 100}{D}\right) \times \frac{(0.028\,52)^2}{2gD^4}$$

$$= (1.5D + 3.2)(0.028\,52)^2/2gD^5.$$

Therefore,

$$241\,212D^5 - 1.5D - 3.2 = 0. \tag{I}$$

This equation can be solved graphically or by successive approximations. An approximate answer can be obtained by omitting the second term; then,

$$241\,212D^5 = 3.2 \quad \text{and} \quad D = 0.1058\,\text{m}.$$

To obtain a more precise answer, let the left-hand side of (I) be called $f(D)$; then, if $D = 0.1058\,\text{m}$,

$$f(D) = 3.198 - 0.159 - 3.2 = -0.161.$$

The negative value of $f(D)$ suggests that the value chosen for D was too small. If $D = 0.1070\,\text{m}$

$$f(D) = 3.383 - 0.161 - 3.2 = +0.022.$$

Comparing these two results, the correct value of D will be a little less than $0.107\,\text{m}$. This result is sufficiently accurate for practical purposes.

$$\text{Diameter of equivalent single pipe} = 0.107\,\text{m} = \textbf{107\,mm}.$$

14.6 INCOMPRESSIBLE FLOW THROUGH BRANCHING PIPES. THE THREE-RESERVOIR PROBLEM

If the flow from the upper reservoir passes through a single pipe which then divides and the two branch pipes lead to two separate reservoirs with different surface levels, as shown in Fig. 14.5, the problem is more complex, particularly as it is sometimes difficult to decide the direction of flow in one of the pipes. Thus, in Fig. 14.5, if we draw the hydraulic gradient lines as shown, flow will be from D to B if the level of the hydraulic gradient at D is above the level of the free surface at B, but if it is below the level of B then flow will be in the reverse direction from B to D. Unfortunately, the hydraulic gradient cannot be drawn until the problem has been solved and so its value,

FIGURE 14.5

The three-reservoir
problem

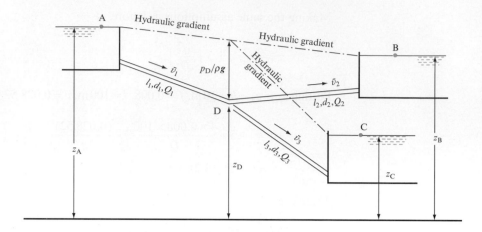

$(z_D + p_D/\rho g)$, at D cannot be determined initially. In many cases, the direction of flow is reasonably obvious, but if it is doubtful, e.g. in DB, imagine that this branch is closed and calculate the value of $(z_D + p_D/\rho g)$ when there is flow from A to C only. If $(z_D + p_D/\rho g)$ is greater than z_B for this condition, flow will initially be from D to B when branch DB is opened. In some cases, conditions at D might then change sufficiently for the flow to reverse, but, if the correct assumption has been made, the continuity requirement that the sum of the flows into the junction is equal to the sum of the flows leaving the junction will be satisfied. If this is not the case, the assumed direction of flow must be reversed and a new solution calculated.

EXAMPLE 14.4

Water flows from a reservoir A (Fig. 14.5) through a pipe of diameter $d_1 = 120\,\text{mm}$ and length $l_1 = 120\,\text{m}$ to a junction at D, from which a pipe of diameter $d_2 = 75\,\text{mm}$ and length $l_2 = 60\,\text{m}$ leads to reservoir B in which the water level is 16 m below that in reservoir A. A third pipe, of diameter $d_3 = 60\,\text{mm}$ and length $l_3 = 40\,\text{m}$, leads from D to reservoir C, in which the water level is 24 m below that in reservoir A. Taking $f = 0.01$ for all the pipes and neglecting all losses other than those due to friction, determine the volume rates of flow in each pipe.

Solution

In this case, the levels of reservoirs B and C are such that flow is obviously from D to B and D to C, as indicated in Fig. 14.5. There are three unknowns, $\bar{v}_1$, $\bar{v}_2$ and $\bar{v}_3$, and the necessary three equations are obtained by applying the steady flow energy equation, first for flow from A to B, then for flow from A to C, and finally writing the continuity of flow equation for the junction D.

For flow from A to B,

$$\left(\frac{p_A}{\rho g} + \frac{\bar{v}_A^2}{2g} + z_A\right) = \left(\frac{p_B}{\rho g} + \frac{\bar{v}_B^2}{2g} + z_B\right) + \frac{4fl_1}{d_1}\frac{\bar{v}_1^2}{2g} + \frac{4fl_2}{d_2}\frac{\bar{v}_2^2}{2g}.$$

Putting $p_A = p_B$ and treating $\bar{v}_A$ and $\bar{v}_B$ as negligibly small,

$$z_A - z_B = \frac{4fl_1}{d_1}\frac{\bar{v}_1^2}{2g} + \frac{4fl_2}{d_2}\frac{\bar{v}_2^2}{2g}.$$

Substituting $z_A - z_B = 16\,\text{m}, f = 0.01, l_1 = 120\,\text{m}, d_1 = 0.120\,\text{m}, l_2 = 60\,\text{m}, d_2 = 0.075\,\text{m}$,

$$16 = \frac{4 \times 0.01 \times 120\bar{v}_1^2}{0.120 \times 2g} + \frac{4 \times 0.01 \times 60\bar{v}_2^2}{0.075 \times 2g} = 2.0387\bar{v}_1^2 + 1.6310\bar{v}_2^2. \tag{I}$$

For flow from A to C,

$$\left(\frac{p_A}{\rho g} + \frac{\bar{v}_A^2}{2g} + z_A\right) = \left(\frac{p_C}{\rho g} + \frac{\bar{v}_C^2}{2g} + z_C\right) + \frac{4fl_1}{d_1}\frac{\bar{v}_1^2}{2g} + \frac{4fl_3}{d_3}\frac{\bar{v}_3^2}{2g},$$

giving $$z_A - z_C = \frac{4fl_1}{d_1}\frac{\bar{v}_1^2}{2g} + \frac{4fl_3}{d_3}\frac{\bar{v}_3^2}{2g}.$$

Putting $z_A - z_C = 24\,\text{m}, f = 0.01, l_1 = 120\,\text{m}, d_1 = 0.120\,\text{m}, l_3 = 40\,\text{m}, d_3 = 0.060\,\text{m}$,

$$24 = \frac{4 \times 0.01 \times 120\bar{v}_1^2}{0.120 \times 2g} + \frac{4 \times 0.01 \times 40\bar{v}_3^2}{0.060 \times 2g} = 2.0387\bar{v}_1^2 + 1.3592\bar{v}_3^2. \tag{II}$$

For continuity of flow at D,

Flow through AD = Flow through DB + Flow through DC,

$$Q_1 = Q_2 + Q_3,$$

$$(\pi/4)d_1^2\,\bar{v}_1 = (\pi/4)d_2^2\,\bar{v}_2 + (\pi/4)d_3^2\,\bar{v}_3,$$

$$\bar{v}_1 = (d_2/d_1)^2\bar{v}_2 + (d_3/d_1)^2\bar{v}_3.$$

Substituting numerical values,

$$\bar{v}_1 = (0.075/0.120)^2\bar{v}_2 + (0.060/0.120)^2\bar{v}_3$$

$$\bar{v}_1 - 0.3906\bar{v}_2 - 0.2500\bar{v}_3 = 0. \tag{III}$$

Values of $\bar{v}_1, \bar{v}_2$ and $\bar{v}_3$ are found by solution of the simultaneous equations (I), (II) and (III). From (I),

$$\bar{v}_2 = \sqrt{(9.81 - 1.25\bar{v}_1^2)}. \tag{IV}$$

From (II)

$$\bar{v}_3 = \sqrt{(17.657 - 1.5\bar{v}_1^2)}. \tag{V}$$

Substituting in equation (III),

$$\bar{v}_1 - 0.3906\sqrt{(9.81 - 1.25\bar{v}_1^2)} - 0.25\sqrt{(17.657 - 1.5\bar{v}_1^2)} = 0. \tag{VI}$$

Equation (VI) can be solved graphically or by successive approximations. In the latter case, if the square roots are to be real, the value of $\bar{v}_1$ cannot exceed the lowest value that will make one of the terms under the square root signs equal to zero; this will be given by $\bar{v}_1^2 = 9.81/1.25 = 7.848$, so that $\bar{v}_1$ must be less than $\sqrt{(7.848)} = 2.80\,\text{m s}^{-1}$. Calling the left-hand side of equation (VI) $f(\bar{v}_1)$ and choosing, as a first approximation, a value of $\bar{v}_1$ less than $2.80\,\mathrm{m\,s^{-1}}$ which, by inspection, will make $f(\bar{v}_1)$ approximately zero, calculate $f(\bar{v}_1)$. If this is not zero, choose further values of $\bar{v}_1$ until a value is found that makes $f(\bar{v}_1)$ sufficiently close to zero to be acceptable. Thus, if

$$\bar{v}_1 = 1.9\,\mathrm{m\,s^{-1}}, \qquad f(\bar{v}_1) = 1.9 - 0.8990 - 0.8747 = +0.1263;$$

$$\bar{v}_1 = 1.8\,\mathrm{m\,s^{-1}}, \qquad f(\bar{v}_1) = 1.8 - 0.9374 - 0.8943 = -0.0317;$$

$$\bar{v}_1 = 1.82\,\mathrm{m\,s^{-1}}, \qquad f(\bar{v}_1) = 1.82 - 0.9300 - 0.8905 = -0.0005.$$

Taking $\bar{v}_1 = 1.82\,\mathrm{m\,s^{-1}}$ as a sufficiently accurate result,

$$\text{Volume rate of flow in AD, } Q_1 = (\pi/4)d_1^2\,\bar{v}_1$$
$$= (\pi/4)(0.120)^2 \times 1.82 = \mathbf{0.0206\,m^3\,s^{-1}}.$$

From equation (IV),

$$\bar{v}_2 = \sqrt{(9.81 - 1.25 \times 1.82^2)} = 2.381\,\mathrm{m\,s^{-1}},$$

$$\text{Volume rate of flow in DB, } Q_2 = (\pi/4)d_2^2\,\bar{v}_2 = (\pi/4)(0.075)^2 \times 2.381$$
$$= \mathbf{0.0105\,m^3\,s^{-1}}.$$

From equation (V),

$$\bar{v}_3 = \sqrt{(17.657 - 1.5 \times 1.82^2)} = 3.562\,\mathrm{m\,s^{-1}}.$$

$$\text{Volume rate of flow in DC, } Q_3 = (\pi/4)d_3^2\,\bar{v}_3$$
$$= (\pi/4)(0.060)^2 \times 3.562 = \mathbf{0.0101\,m^3\,s^{-1}}.$$

Checking for continuity at D,

$$Q_2 + Q_3 = 0.0105 + 0.0101 = 0.0206 = Q_1.$$

14.7　INCOMPRESSIBLE STEADY FLOW IN DUCT NETWORKS

The steady flow energy equation may be used to calculate the pressure at any point along a pipe or duct and may be seen to represent an overall system pressure loss or balance relationship. It is necessary to reinforce two points that will be met again in fan or pump and system matching.

Series pipes/ducts

The pressure loss along a series of pipes is the sum of the pressure loss along each.

Referring to Fig. 14.6, the pressure loss from A to D via the three series pipes is given by the sum of the individual pressure losses in each pipe at their common flow rate Q. (Note that while the flow rate Q in each of the three series pipes is constant the

FIGURE 14.6

Flow in a series pipe/duct system

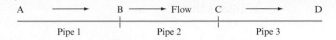

velocity of the fluid in each pipe may well be different as the flow velocity $V = Q/A$, where A is the individual pipe's cross-sectional area.)

Thus

$$\Delta p_{A-D} = \sum \left[\frac{4\rho f (L + L_e) Q^2}{2DA^2} \right]_{\text{pipes 1-3}},$$

where L_e is the sum of the equivalent lengths for all the separation losses in that particular pipe.

Parallel pipes/ducts

The pressure loss between two points in a system is the same regardless of the route taken. While this may seem self-evident it is the basis for all system balancing, as the proportion of the flow that travels in any one particular flow path between two points depends on the relative resistance of each of the paths. Thus if two points are connected by two pipes, one of large diameter and the other of much smaller diameter, then it follows that the majority of the flow will arrive via the large-diameter pipe. However, the insertion of a valve in the large-diameter pipe can alter this ratio as the separation loss caused by this valve can be arranged so that the flow resistance is greater than that for the smaller-diameter pipe.

A natural result of this is the statement, akin in form to Kirchhoff's laws of d.c. electrics, that the total pressure loss around a loop in a pipe circuit is zero.

In Fig. 14.7, the pressure loss from A to F is identical via route BCE and BDE, but the flow will divide inversely as to the resistance of either path.

FIGURE 14.7

Flow in a looped pipe/duct system

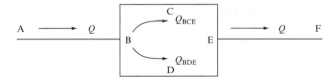

$$\Delta p_{BCE} = \left[\frac{4\rho f (L + L_e) Q^2}{2DA^2} \right]_{\text{route BCE}},$$

$$\Delta p_{BDE} = \left[\frac{4\rho f (L + L_e) Q^2}{2DA^2} \right]_{\text{route BDE}}.$$

However, as

$$\Delta p_{BCE} = \Delta p_{BDE}$$

it follows that

$$\frac{Q_{BCE}}{Q_{BDE}} = \sqrt{\left\{\left[\frac{4\rho f(L+L_e)}{2DA^2}\right]_{\text{route BDE}} \bigg/ \left[\frac{4\rho f(L+L_e)}{2DA^2}\right]_{\text{route BCE}}\right\}},$$

where L_e is the equivalent length added owing to the presence of any separation losses along either route. It should be noted that the separation losses associated with valves change as the valve setting changes and thus Fig. 14.7 illustrates the basis for any system flow balancing.

Continuity of volumetric flow across the control volume represented by the network between A and F implies that

$$Q = Q_{BCE} + Q_{BDE},$$

and therefore the flow in either of the looped paths may be determined as

$$\frac{Q_{BCE}}{Q - Q_{BCE}} = \sqrt{\left\{\left[\frac{4\rho f(L+L_e)}{2DA^2}\right]_{\text{route BDE}} \bigg/ \left[\frac{4\rho f(L+L_e)}{2DA^2}\right]_{\text{route BCE}}\right\}}.$$

EXAMPLE 14.5

A horizontal duct system (Fig. 14.8) draws atmospheric air into a 0.3 m diameter duct, 10 m long, through an entry grille with a separation loss coefficient equivalent to $4 \times$ the flow kinetic pressure. Following passage through an axial flow fan the air passes along a further 30 m of rectangular duct, 0.2 m by 0.4 m in cross-section. The transition may be represented by a separation loss equivalent length of $20 \times$ the approach duct diameter. The air is discharged to the ventilated space via a ceiling grille with a separation loss factor equivalent to $10 \times$ the flow kinetic pressure. Determine the fan pressure input necessary if the required air flow is $0.8\,\text{m}^3\,\text{s}^{-1}$. Assume that the friction factor for both ducts is 0.008 and that the density of air is $1.2\,\text{kg m}^{-3}$.

FIGURE 14.8
Duct layout in Example 14.5

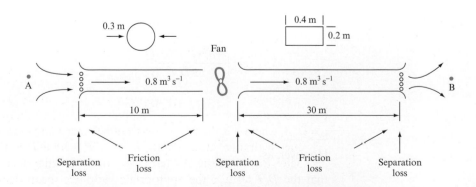

Solution

Application of the steady flow energy equation between points A and B in Fig. 14.8, where the local flow velocity may be taken as zero, the local pressure may be assumed atmospheric and any difference in the potential energy terms may be ignored for air and a horizontal duct, reduces the full form of the steady flow energy equation

$$p_A + \tfrac{1}{2}\rho v_A^2 + \rho g z_A + \Delta p_{fan} = p_B + \tfrac{1}{2}\rho v_B^2 + \rho g z_B + \Sigma(\tfrac{1}{2}\rho K u^2)_{\text{ducts } 1-2}$$

to $\quad \Delta p_{fan} = \Sigma(\tfrac{1}{2}\rho K u^2)_{\text{ducts } 1-2},$

where u is the local velocity in each duct and K is the combined friction and separation loss coefficient for that duct.

For duct 1, the loss terms from entry to the transition into duct 2 may be expressed as

Entry loss at grille + Friction loss along duct 1 + Loss at transition.

These terms may be expressed in terms of the air flow, Q, through the ductwork as Q is a constant for both ducts. The flow velocity in either duct may be calculated as $u = Q/A$, where A is the duct cross-sectional area.

$$\tfrac{1}{2}\rho K u^2_{\text{duct 1}} = \frac{1}{2}\frac{\rho K_{\text{entry}} Q^2}{A^2} + \frac{4 f L \rho Q^2}{2 D A^2} + \frac{4 f L_e \rho Q^2}{2 D A^2},$$

where L_e is the equivalent length of duct necessary to generate an equivalent frictional loss to the separation loss representing the transition. This type of expression for duct total loss is conveniently reduced to

$$\tfrac{1}{2}\rho K u^2_{\text{duct 1}} = \frac{1}{2}\frac{\rho Q^2}{A^2}\left[K_{\text{entry}} + \frac{4 f (L + L_e)}{D}\right].$$

Substitution of the appropriate values yields

$$\Delta p_{\text{duct 1}} = 0.5 \times 1.2 \frac{0.8^2}{(\pi 0.3^2/4)^2}\left[4 + \frac{4 \times 0.008}{0.3}(10 + 20 \times 0.3)\right]$$

$$= \frac{0.384}{0.0707^2}(4 + 1.71) = 76.83 \times 5.71 = 438 \, \text{N m}^{-2}.$$

For duct 2, the loss terms from the transition into duct 2 to the exit to the room may be expressed as

Friction loss along duct 2 + Loss at exit grille.

These terms may be expressed in terms of the air flow, Q, through the ductwork as Q is a constant for both ducts. The flow velocity in either duct may be calculated as $u = Q/A$, where A is the duct cross-sectional area. Care must now be taken to introduce the duct hydraulic mean depth, $m = A/P$, where P is the duct internal 'wetted' perimeter, to replace the $D/4$ term used for the circular duct cross-section. It is stressed that the $D/4$ term is the appropriate hydraulic mean depth for circular ducts and therefore the form of the Darcy equation used should arguably always feature m rather than $D/4$. However, circular cross-section pipes and ducts are more common and hence it is customary to refer to this special case of the Darcy equation

$$\tfrac{1}{2}\rho K u_{\text{duct 2}}^2 = \frac{fL\rho Q^2}{2mA^2} + \frac{1}{2}\frac{\rho K_{\text{exit}} Q^2}{A^2}.$$

This expression for duct total loss may be conveniently reduced to

$$\tfrac{1}{2}\rho K u_{\text{duct 2}}^2 = \frac{1}{2}\frac{\rho Q^2}{A^2}\left(\frac{fL}{m} + K_{\text{exit}}\right).$$

The appropriate value of the hydraulic mean depth, $m = A/P$, for the rectangular section duct is $m = (0.2 \times 0.4)/[(0.2 + 0.4)2]$.

Substitution of the appropriate values yields

$$\Delta p_{\text{duct 1}} = 0.5 \times 1.2 \frac{0.8^2}{(0.2\times 0.4)^2}\left\{\frac{0.008 \times 30.0}{[(0.2\times 0.4)/1.2]} + 10\right\}$$

$$= \frac{0.384}{0.08^2}(3.6 + 10)$$

$$= 60.0 \times 13.6 = 816\,\text{N m}^{-2}.$$

The pressure rise to be generated by the fan is therefore the sum of the losses incurred in both ducts:

$$\Delta_{\text{fan}} = 438 + 816 = \mathbf{1254\,N\,m^{-2}}.$$

EXAMPLE 14.6

A network of fume cupboard extract ducts (Fig. 14.9) may be represented by four, horizontal, 150 mm diameter pipes of 10 m length joining to form a 20 m long 300 mm duct that rises vertically to discharge to atmosphere. The separation losses at each 150 mm diameter duct entry may be taken as 4 × the local flow kinetic pressure and

FIGURE 14.9
Extract duct layout in Example 14.6

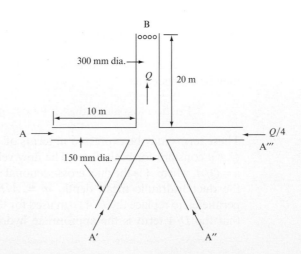

the equivalent length separation loss at exit from the 300 mm diameter duct as equal to 20 × the duct diameter. Ignore any loss at the duct junction. Determine the pressure rise necessary across a fan in the 300 m diameter duct if the flow extracted from each fume cupboard is 0.2 m³ s⁻¹. Take the friction factor for all ducts as 0.008 and the density of air as 1.2 kg m⁻³.

Solution

The point to be realized in this example is that the four horizontal ducts each carry an equal flow and are parallel paths as discussed in the preceding section. Thus the pressure drop from the entry to the junction of the ducts is the same irrespective of which path is followed. Thus the fan will need to meet the pressure drop requirements calculated on the basis of one duct at the stated flow rate. In addition the fan must meet the pressure loss in the larger-diameter duct up to the roof discharge.

Application of the steady flow energy equation between points A and B in Fig. 14.9, where the local flow velocity may be taken as zero, the local pressure may be assumed atmospheric and any difference in the potential energy terms may be ignored for air and a horizontal duct, reduces the full form of the steady flow energy equation

$$p_A + \tfrac{1}{2}\rho v_A^2 + \rho g z_A + \Delta p_{\text{fan}} = p_B + \tfrac{1}{2}\rho v_B^2 + \rho g z_B + \Sigma(\tfrac{1}{2}\rho K u^2)_{\text{ducts 1-2}}$$

to $\Delta p_{\text{fan}} = \Sigma(\tfrac{1}{2}\rho K Q^2/A^2)_{\text{ducts 1-2}},$

where Q is the local flow rate in each duct and K is the combined friction and separation loss coefficient for that duct.

Employing the techniques discussed in more detail in the previous example it follows that the fan pressure rise may be expressed as

$$\Delta p_{\text{fan}} = \frac{1}{2}\rho\left(\frac{Q^2}{A^2}\right)_{\text{duct 1}}\left(K_{\text{entry}} + \frac{4fL}{D}\right)_{\text{duct 1}} + \frac{1}{2}\rho\left(\frac{Q^2}{A^2}\right)_{\text{duct 2}}\left[\frac{4f(L+L_e)}{D}\right]_{\text{duct 2}},$$

which may be reduced further if the substitution $Q_{\text{duct 2}} = 4 \times Q_{\text{duct 1}}$ is made:

$$\Delta p_{\text{fan}} = \frac{1}{2}\rho Q_{\text{duct 1}}^2\left\{\frac{1}{A_{\text{duct 1}}^2}\left(K_{\text{entry}} + \frac{4fL}{D}\right)_{\text{duct 1}} + \frac{16}{A_{\text{duct 2}}^2}\left[\frac{4f(L+L_e)}{D}\right]_{\text{duct 2}}\right\}.$$

Substitution of the appropriate values yields

$$\Delta p_{\text{fan}} = 0.5 \times 1.2 \times 0.2^2\left[\frac{1}{(\pi \times 0.15^2/4)^2}\left(4 + \frac{4 \times 0.008 \times 10}{0.15}\right)\right]$$

$$+ 0.5 \times 1.2 \times 0.2^2\left\{\frac{16}{(\pi \times 0.3^2/4)^2}\left[\frac{4 \times 0.008}{0.3}(20 + 20 \times 0.3)\right]\right\}$$

$$= 0.024\left\{\frac{1}{0.0177^2}(4 + 2.13) + \frac{16}{0.0707^2}[0.107(20 + 6)]\right\}$$

$$= 469.6 + 213.72 = \mathbf{683.30\,N\,m^{-2}}.$$

14.7.1 Fan and system pressure relationships

While the frictional relationships detailed above and earlier in Chapter 6 give a means of determining the overall 'loss' of pressure across a system, it is also useful to look in detail at the variations in total, static and kinetic energy terms along the ducts making up a network. This has to some extent already been introduced in Chapter 6, in the discussion of the steady flow energy equation, and demonstrated for water transfer between reservoirs by Fig. 6.8.

At any location it remains true that the available air flow total pressure, p_t, is equal to the sum of the static, p_s, and kinetic pressure. Thus

$$p_t = p_s + \tfrac{1}{2}\rho V^2,$$

where V is the local section mean velocity. Frictional and separation losses reduce the static pressure in the flow direction. Figure 14.10 illustrates the resulting variation in air pressure along a duct incorporating a ventilation fan and a filter unit. As will be shown later (Chapter 25), the fan output will match the requirements of the network at the operating point. It will be seen that the local static pressure drops along the duct from entry to the fan location. The kinetic pressure, shown here as the difference at any location between the total and static pressures, increases between points a and c as the flow passes through the duct entry reducer. The flow velocity and hence kinetic pressure remains constant along the constant diameter duct, c to f; however, the total and static pressure rise across the fan owing to the energy input that this represents. Within the enlarged filter section the flow velocity decreases and hence the difference between the total and static pressures falls. Note, however, that friction and separation losses in this section act to reduce the static pressure.

The flow is then accelerated through the reducer section linking the filter section to the downstream duct, with a consequent reduction in static pressure as the flow kinetic pressure term rises with the square of velocity. Note that it is possible for the

FIGURE 14.10

Total, static and kinetic pressure variations along a fan and duct system (Woods Air Movement Ltd)

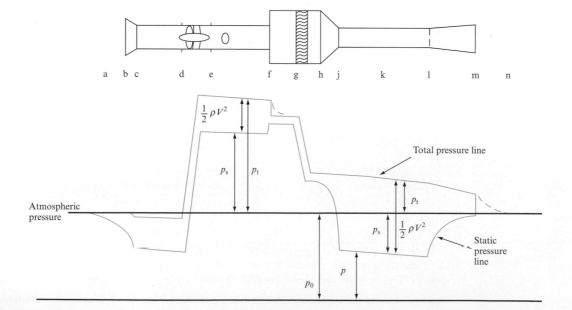

FIGURE 14.11

Illustration of static
regain at a duct junction.
Note slope of static
pressure line decreases
downstream of junction

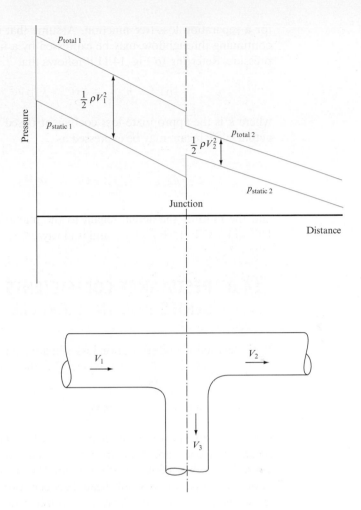

static pressure to be below atmosphere along this duct section. At exit the flow kinetic energy is lost, relative to a position, n, remote from the duct exit diffuser, as explained in the treatment of the steady flow energy equation in Chapter 6. While this process is identical in nature to that demonstrated for gravity or pumped water or other liquid transfer between storage tanks, it is useful to appreciate the interrelationship of total, static and kinetic pressures in a regime where subatmospheric pressures are encountered without the complications of cavitation experienced in liquid systems.

The interrelationship of the total, static and kinetic pressure terms also leads to the introduction of 'static regain' as a technique in the design of air distribution networks, particularly where a fan-supplied duct distributes air through a series of offtakes positioned in series along its length. Figure 14.11 illustrates such a duct layout. The underlying principle is that the duct static pressure is held 'constant' along its length by resizing the duct after each offtake so that the flow velocity is reduced and hence the difference between the flow total and static pressures is decreased. The velocity reduction in the main duct after each offtake is chosen so that the 'regain' balances the separation loss incurred at the junction plus the duct friction for that section. The 'regain' can never be 100 per cent owing to the separation loss incurred at the junction; the shortfall may be expressed as a percentage of the ideal condition for a separation loss-free junction. Assume that the junction loss, as applied to the continuing throughflow, may be expressed by a factor times the downstream kinetic pressure. Referring to Fig. 14.11 it follows that

$$p_{static\,1} + \tfrac{1}{2}\rho V_1^2 = p_{static\,2} + \tfrac{1}{2}\rho V_2^2 + k\tfrac{1}{2}\rho V_2^2,$$

where k is the appropriate loss coefficient based on the downstream flow. Thus the actual static regain may be expressed as

$$p_{static\,2} - p_{static\,1} = \tfrac{1}{2}\rho V_1^2 - (1+k)\tfrac{1}{2}\rho V_2^2,$$

and the ratio of the actual regain to the maximum possible, when $k = 0$, becomes $[V_1^2 - (1+k)V_2^2]/(V_1^2 - V_2^2)$, and is always < 1.

14.8 RESISTANCE COEFFICIENTS FOR PIPELINES IN SERIES AND IN PARALLEL

Both the friction and separation losses in a pipeline are functions of the mean velocity of flow $\bar{v}$ and, since $\bar{v} = Q/A$, where Q is the volume rate of flow and A the cross-sectional area of the pipe,

$$\text{Total head loss, } h = KQ^n, \tag{14.3}$$

where n is some power which depends on the type of flow. For turbulent flow, n will be equal to two and, if separation losses are negligible, equation (14.3) becomes $h = KQ^2$, which is identical with equation (14.2). The form $h = KQ^2$ gives a misleading impression that the loss of head between two points in a pipeline is the same irrespective of the sign of Q, i.e. that h is independent of the direction of flow, which is of course absurd. It would be preferable to write $h = KQ|Q|$, where $|Q|$ means the numerical value of Q without regard to sign. Similarly, equation (14.3) could be written $h = KQ(|Q|)^{n-1}$.

For pipes in series (Fig. 14.12), Q is the same for each pipe and the losses of head $h_1, h_2, \ldots$, in each pipe are additive:

FIGURE 14.12
Resistances in series

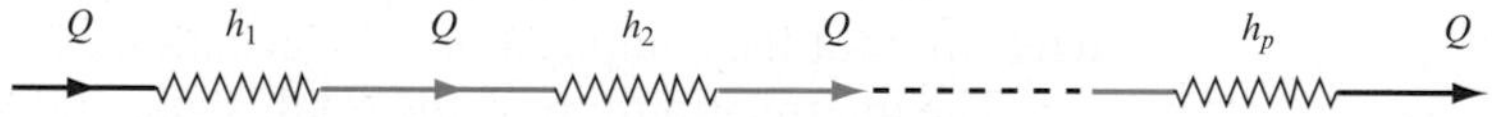

$$\text{Total loss of head} = h_1 + h_2 + \cdots + h_p$$
$$= K_1 Q^n + K_2 Q^n + \cdots + K_p Q^n$$
$$= (K_1 + K_2 + \cdots + K_p)Q^n. \tag{14.4}$$

If several pipes are connected in parallel, as shown in Fig. 14.13, the loss of head between A and B must be the same for each pipe. Thus,

$$h = K_1 Q_1^n = K_2 Q_2^n = \cdots = K_p Q_p^n. \tag{14.5}$$

FIGURE 14.13
Resistances in parallel

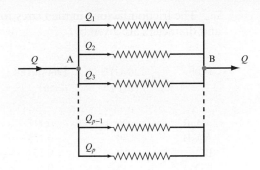

Also, for continuous flow,

Total flow, Q = Sum of the flows through each pipe

$$= Q_1 + Q_2 + \cdots + Q_p.$$

Substituting for $Q_1, Q_2, \ldots, Q_p$ from equation (14.5),

$$Q = \sqrt[n]{(h/K_1)} + \sqrt[n]{(h/K_2)} + \cdots + \sqrt[n]{(h/K_p)}. \tag{14.6}$$

The set of parallel pipes can be considered as equivalent to a single pipe with a resistance coefficient K carrying the total flow Q for which

$$h = KQ^n \quad \text{or} \quad Q = \sqrt[n]{(h/K)}.$$

Substituting in equation (14.6),

$$\sqrt[n]{(h/K)} = \sqrt[n]{(h/K_1)} + \sqrt[n]{(h/K_2)} + \cdots + \sqrt[n]{(h/K_p)}$$

or, assuming that $n = 2$,

$$1/\sqrt{K} = 1/\sqrt{K_1} + 1/\sqrt{K_2} + \cdots + 1/\sqrt{K_p}. \tag{14.7}$$

EXAMPLE 14.7

A system of pipes conveying water is connected in parallel and in series, as shown in Fig. 14.14. The section DE represents the resistance of a valve for controlling the flow, which has a resistance coefficient $K_{DE} = (4000/n)^2$, where n is the percentage valve opening.

FIGURE 14.14
Resistance network

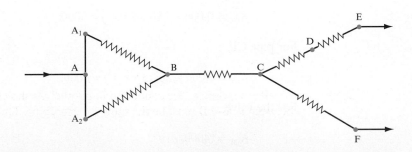

The friction factor f in the Darcy formula is 0.006 for all pipes, and their lengths and diameters are given by

Pipe	Length l (m)	Diameter d (m)
AA_1B	30	0.1
AA_2B	30	0.125
BC	60	0.15
CD	15	0.1
CF	30	0.1

The head at A is 100 m, at E is 40 m and at F is 60 m. If the valve is adjusted to give equal discharge rates at E and F, calculate the head at C, the total volume rate of flow through the system and the percentage valve opening. Neglect all losses except those due to friction.

Solution

For any pipe, neglecting separation losses and working in SI units,

$$\text{Head lost in friction, } h = \frac{4fl}{d}\frac{\bar{v}^2}{2g} = \frac{flQ^2}{3d^5} = KQ^2.$$

Therefore,

$$K = fl/3d^5.$$

For pipe AA_1B,

$$K_{AA_1B} = 0.006 \times 30/(3 \times 0.1^5) = 6000.$$

For pipe AA_2B,

$$K_{AA_2B} = 0.006 \times 30/(3 \times 0.125^5) = 1966.$$

For pipe BC,

$$K_{BC} = 0.006 \times 60/(3 \times 0.15^5) = 1580.$$

For pipe CD,

$$K_{CD} = 0.006 \times 15/(3 \times 0.1^5) = 3000.$$

For pipe CF,

$$K_{CF} = 0.006 \times 30/(3 \times 0.1^5) = 6000.$$

For the valve,

$$K_{DE} = (4000/n)^2.$$

First, combine the resistance of pipes AA_1B and AA_2B, which are in parallel, using equation (14.7):

$$1/\sqrt{K_{AB}} = 1/\sqrt{K_{AA_1B}} + 1/\sqrt{K_{AA_2B}} = (\sqrt{K_{AA_1B}} + \sqrt{K_{AA_2B}})/\sqrt{(K_{AA_1B} \times K_{AA_2B})},$$

$$K_{AB} = (K_{AA_1B} \times K_{AA_2B})/(\sqrt{K_{AA_1B}} + \sqrt{K_{AA_2B}})^2$$

$$= (6000 \times 1966)/(\sqrt{6000} + \sqrt{1966})^2 = 795.$$

Now combine K_{AB} with K_{BC} to find the equivalent resistance when BC is assessed in series with the pipes between A and B (Fig. 14.15). From equation (14.4)

FIGURE 14.15
Equivalent pipeline

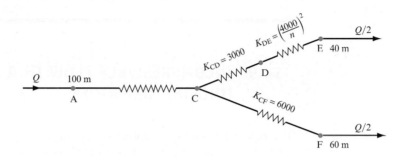

$$K_{AC} = K_{AB} + K_{BC} = 795 + 1580 = 2375.$$

If Q is the total volume rate of flow entering at A,

$$\text{Loss of head between A and C} = H_A - H_C = K_{AC}Q^2 = 2375Q^2. \tag{I}$$

Since it is required that the discharges at E and F should be equal, the flows through CE and CF will each be $\frac{1}{2}Q$:

$$\text{Loss of head between C and F} = H_C - H_F = K_{CF}(\tfrac{1}{2}Q)^2$$

$$= 6000Q^2/4 = 1500Q^2. \tag{II}$$

From equations (I) and (II),

$$(H_A - H_C)/2375 = Q^2 = (H_C - H_F)/1500,$$

$$1500H_A - 1500H_C = 2375H_C - 2375H_F,$$

$$3875H_C = 1500H_A + 2375H_F.$$

Putting $H_A = 100$ m and $H_F = 60$ m,

$$3875H_C = 1500 \times 100 + 2375 \times 60$$

$$H_C = \textbf{75.48 m}.$$

From equation (I),

$$2375Q^2 = H_A - H_C = 100 - 75.48 = 24.52 \text{ m},$$

$$Q = \textbf{0.1016 m}^3 \textbf{ s}^{-1}.$$

Since pipe CD and valve DE are in series,

Loss of head between C and E $= H_C - H_E = (K_{CD} + K_{DE})(\frac{1}{2} Q)^2.$

Substituting numerical values,

$$75.48 - 40 = [3000 + (4000/n)^2](0.0508)^2,$$
$$35.48 = 7.74 + 41\ 290/n^2,$$
$$27.74n^2 = 41\ 290,$$
$$n = \textbf{38.58 per cent}.$$

14.9   INCOMPRESSIBLE FLOW IN A PIPELINE WITH UNIFORM DRAW-OFF

If a pipeline has a large number of tappings along its length from which the fluid is discharged, as in the case of a perforated pipe used as a sprinkler, the problem can be treated as if fluid was being drawn off at a uniform rate per unit length. Under these circumstances, the volume rate of flow across successive cross-sections will decrease as the distance from the point of input increases. If the pipe is of constant diameter, the velocity and, therefore, the frictional loss of head per unit length will also decrease. Such problems are solved by considering a short length of pipe and then integrating to obtain the result for the whole pipe.

14.10   INCOMPRESSIBLE FLOW THROUGH A PIPE NETWORK

A pipe network is a set of pipes which are interconnected so that the flow from a given input or to a given outlet may come through several different routes. Thus, in Fig. 14.16, the input at a may be divided between pipes ab, ad, and af; part of this may leave at c and part at h, each combining with part of the input Q_2. An attempt to apply Bernoulli's equation and the continuity of flow equation to the various elements in the network would lead to a very large number of simultaneous equations which would be cumbersome to solve. The alternative approach is to use a method of successive

FIGURE 14.16
Pipe network

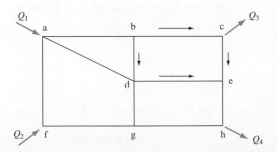

approximations, assuming values for the flow in each pipe, or the heads at each junction, and checking whether the values chosen satisfy the requirements that:

1. the loss of head between any two junctions must be the same for all routes between these junctions (e.g. in loop bced (Fig. 14.16)),

$$\frac{\text{Loss of head}}{\text{in pipe bc}} + \frac{\text{Loss of head}}{\text{in pipe de}} = \frac{\text{Loss of head}}{\text{in pipe bd}} + \frac{\text{Loss of head}}{\text{in pipe ce;}}$$

2. the inflow to each junction must equal the outflow from that junction.

If the values chosen do not satisfy these conditions throughout the network, they must be corrected by successive approximations until they do so within the required degree of accuracy. The Hardy–Cross method provides a system for calculating the value of the correction to be made, each loop or junction being considered in turn and corrected assuming that conditions in the remainder of the network remain unaltered. Obviously, corrections to one element will affect conditions elsewhere and the required balance of heads and flows will not be reached as a result of the first correction. However, each successive repetition of the process will bring the system nearer to the final balanced condition.

14.11 HEAD BALANCE METHOD FOR PIPE NETWORKS

The head balance method is used when the total volume rate of flow through the network is known, but the heads or pressures at junctions within the network are unknown. For each pipe, an assumption must first be made of the direction and volume rate of flow so as to satisfy condition (2) (that the inflow to each junction must equal the outflow from that junction). In the loop bced in Fig. 14.16, the directions of flow might be as indicated by the arrows; thus the flow in bc and ce is clockwise round this loop and the flow in bd and de is anticlockwise. To satisfy condition (1), the loss of head between b and e must be the same by either the clockwise or anticlockwise route. Neglecting losses other than friction for any pipe,

$$\text{Head lost, } h = KQ^n,$$

where Q = volume rate of flow in the pipe, K = resistance coefficient which, in SI units, would be $fl/3D^5$ and n is a constant which, for turbulent flow, would be 2.

If Σ_c and Σ_{cc} represent summations of quantities in the clockwise and anticlockwise (counterclockwise) directions, respectively,

$$\frac{\text{Loss of head in pipes in}}{\text{which flow is clockwise}} = \Sigma_c h = \Sigma_c KQ^n,$$

$$\frac{\text{Loss of head in pipes in}}{\text{which flow is anticlockwise}} = \Sigma_{cc} h = \Sigma_{cc} KQ^n.$$

The values initially chosen for the volume rate of flow in each pipe Q are unlikely to meet the requirement that

$$\Sigma_c h = \Sigma_{cc} h.$$

If it is assumed that $\Sigma_c h > \Sigma_{cc} h$,

$$\text{Out of balance head} = \Sigma_c h - \Sigma_{cc} h = \Sigma_c KQ^n - \Sigma_{cc} KQ^n.$$

To remove this out of balance head, while keeping the total flow through the loop constant, the clockwise flow must be *reduced* by an amount δQ and the anticlockwise flow increased by δQ, so that

$$\Sigma_c h - \Sigma_{cc} h = \Sigma_c K(Q - \delta Q)^n - \Sigma_{cc} K(Q + \delta Q)^n = 0.$$

Expanding the terms in brackets and neglecting all terms involving the second or higher orders of δQ, which is a small quantity compared with Q,

$$\Sigma_c K(Q^n - nQ^{n-1}\delta Q) = \Sigma_{cc} K(Q^n + nQ^{n-1}\delta Q),$$

from which

$$\delta Q = \frac{\Sigma_c KQ^n - \Sigma_{cc} KQ^n}{n(\Sigma_c KQ^{n-1} + \Sigma_{cc} KQ^{n-1})}.$$

Now $KQ^n = h$ and $KQ^{n-1} = h/Q$; therefore,

$$\delta Q = \frac{\Sigma_c h - \Sigma_{cc} h}{n[\Sigma_c (h/Q) + \Sigma_{cc} (h/Q)]}.$$

Adopting a sign convention that values of h and Q are to be regarded as positive in pipes in which the flow is clockwise *with regard to the loop under consideration* and negative if anticlockwise,

$$\delta Q = -\frac{\Sigma h}{n \Sigma (h/Q)}.$$

The negative sign indicates that the positive (clockwise) values of Q are to be reduced and the negative (anticlockwise) values of Q are to be increased. When a system has a number of loops, corrections to one loop will unbalance adjoining loops which will require further correction. Also pipes common to two loops will receive corrections for each loop. The process is therefore iterative and must be continued until the desired degree of accuracy is achieved.

14.12 COMPUTER PROGRAM HARDYC

HARDYC calculates the distribution of flow in a network using the Hardy–Cross method, allowing for frictional and separation losses. The program provides for a network design of up to 10 nodes, each node having the capacity to represent inflow or outflow to the network. Loops are identified with one pipe in common and each pipe is defined in terms of its length, diameter, friction factor and separation loses.

Once a network has been designed and analyzed the simulation allows changes to any of the pipeline properties to allow an evaluation of the change, for example the effect of valve setting, represented by its separation loss, on redirection of flow within a network.

In response to screen prompts the network may be designed and external applied inflow/outflow values at each node defined. For each pipe the following data are required: pipe length, diameter, friction factor and separation loss coefficient. An annotated sketch showing all this data will aid use of the program.

14.12.1 Application example

Consider the 10-pipe system illustrated in Fig. 14.17. The individual pipe data are shown on the figure, including applied inflow at nodes 0 and 3 and extracted outflows at nodes 6 and 7. All pipes are initially assumed to be of 1 m diameter and none have a defined separation loss. The fluid density is taken as $1000 \, \text{kg} \, \text{m}^{-3}$.

FIGURE 14.17
Network example for
HARDYC

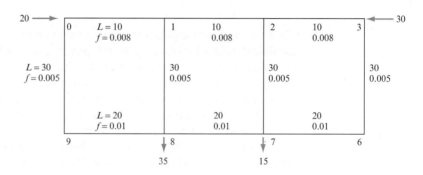

The output from HARDYC indicated that the flow distribution is as follows after nine iterations, with the sign convention imposed that flow from a lower node number to a higher is deemed positive:

PIPE FROM NODE TO NODE		FLOW DIRECTION	FLOW (m³ s⁻¹)
0	1	+ ve	8.15
1	2	− ve	7.5
2	3	− ve	17.51
0	9	+ ve	11.85
1	8	+ ve	15.66
2	7	+ ve	10.01
3	6	+ ve	12.49
6	7	+ ve	12.49
7	8	+ ve	7.49
8	9	− ve	11.84

14.12.2 Additional investigations using HARDYC

The simulation may be used to investigate:

1. the effect of changes to pipe length, diameter or friction factor on the distribution of a particular inflow under constant extraction conditions;

2. the effect of changes in separation loss values to represent a valve action in any network pipe.

Returning to the network illustrated in Fig. 14.17, the flow in the pipe linking node 3 to node 6 is reduced to $0.09\,\text{m}^3\,\text{s}^{-1}$ if the diameter of the pipe is reduced to $0.1\,\text{m}$. Similarly for the original network the flow in the pipe linking node 2 to node 7 is reduced to $3.04\,\text{m}^3\,\text{s}^{-1}$ if a valve with a 25 loss coefficient is introduced into the line. Further examples may be run with ease.

14.13 THE QUANTITY BALANCE METHOD FOR PIPE NETWORKS

When the heads at various points in a pipe network are known and it is necessary to calculate the quantities flowing in each pipe, the quantity balance method can be used. An estimate is made of the head at each junction (or node) in the network and the volume rate of flow Q is calculated for each pipe from the difference of head h between the junctions at each end of the pipe and its resistance coefficient K, using the formula $h = KQ^n$. If the resulting inflow does not equal the outflow at each junction, the original estimates of head must be corrected.

If, in Fig. 14.18, the head at b has been overestimated by an amount δh relative to the head at a, c and d, the values of Q_{ab} and Q_{cb} will be too small and the value of Q_{bd} will be too great. Differentiating the equation $h = KQ^n$, we have

FIGURE 14.18
Pipe network – quantity balance network

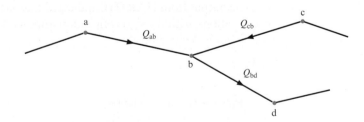

$$\delta h = KnQ^{n-1}\,\delta Q$$

or, since $K = h/Q^n$,

$$\delta Q = (Q/nh)\delta h.$$

If, therefore, the original estimate of the head at b is reduced by an amount δh, the flows in ab and cb will be increased to $Q_{ab} + \delta Q_{ab}$ and $Q_{cb} + \delta Q_{cb}$. If the flows are now correct, inflow and outflow at b will balance:

$$\text{Flow in ab} \quad + \text{Flow in cb} = \text{Flow in bd},$$

$$(Q_{ab} + \delta Q_{ab}) + (Q_{cb} + \delta Q_{cb}) = (Q_{bd} - \delta Q_{bd}). \tag{14.8}$$

If h_{ab}, h_{cb} and h_{bd} are the assumed losses of head in pipes ab, cb and bd used to calculate Q_{ab}, Q_{cb} and Q_{bd},

$$\delta Q_{ab} = \frac{Q_{ab}}{nh_{ab}} \delta h, \quad \delta Q_{cb} = \frac{Q_{cb}}{nh_{cb}} \delta h \quad \text{and} \quad \delta Q_{bd} = \frac{Q_{bd}}{nh_{bd}} \delta h.$$

Substituting in equation (14.8),

$$\left(Q_{ab} + \frac{Q_{ab}}{nh_{ab}} \delta h \right) + \left(Q_{cb} + \frac{Q_{cb}}{nh_{cb}} \delta h \right) = \left(Q_{bd} - \frac{Q_{bd}}{nh_{bd}} \delta h \right),$$

$$\delta h = -\frac{(Q_{ab} + Q_{cb} - Q_{bd})}{Q_{ab}/nh_{ab} + Q_{cb}/nh_{cb} + Q_{bd}/nh_{bd}}.$$

Using the sign convention that for flow towards the junction b both Q and h are positive,

$$\delta h = -\frac{\Sigma Q}{\Sigma(Q/nh)},$$

where $\Sigma Q = Q_{ab} + Q_{cb} - Q_{bd} = $ Algebraic sum of the flows towards the junction.

When δh has been calculated, the new value of the head at b can be used to determine revised values of Q_{ab}, Q_{cb} and Q_{bd}. The process is repeated until ΣQ has been reduced to a negligible quantity. The following simple example shows the application of this method to the three-reservoir problem.

EXAMPLE 14.8

A reservoir A (Fig. 14.19) with its surface 60 m above datum supplies water to a junction D through a 300 mm diameter pipe, 1500 m long. From the junction, a 250 mm diameter pipe, 800 m long, feeds reservoir B, in which the surface level is 30 m

FIGURE 14.19
The three-reservoir problem

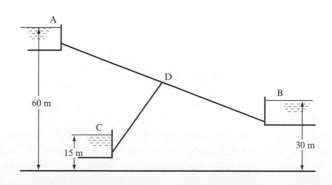

above datum, while a 200 mm diameter pipe, 400 m long, feeds reservoir C, in which the surface level is 15 m above datum. Calculate the volume rate of flow to each reservoir. Assume that the loss of head due to friction is given by $h = flQ^2/3d^5$ for each pipe and that $f = 0.01$.

Solution

First choose, by inspection, a value for the head at D. Since the elevation of A is much greater than that of B or C, flow is likely to be from D to B and C (as indicated in the question) and the head at D would, therefore, be greater than that at B. Assume a trial value for the head at D of 35 m. Then, initially,

$$\text{Head loss in AD, } h_{AD} = 60 - 35 = 25\,\text{m,}$$

$$\text{Head loss in DB, } h_{DB} = 35 - 30 = 5\,\text{m,}$$

$$\text{Head loss in DC, } h_{DC} = 35 - 15 = 20\,\text{m.}$$

For each pipe, $h = KQ^2$ where $K = fl/3d^5$. Therefore, the flow Q for any value of h is $Q = \sqrt{(h/K)}$.

$$\text{For pipe AD, } K_{AD} = \frac{0.01 \times 1500}{3 \times 0.300^5} = 2058.$$

$$\text{For pipe DB, } K_{DB} = \frac{0.01 \times 800}{3 \times 0.250^5} = 2730.$$

$$\text{For pipe DC, } K_{DC} = \frac{0.01 \times 400}{3 \times 0.200^5} = 4167.$$

TABLE 14.1

The calculations for junction D

					FIRST CORRECTION			
PIPE	K	ASSUMED VALUE OF h (m)	$Q = \sqrt{(h/K)}$ (m³ s⁻¹)	$\dfrac{Q}{2h}$	$\delta h = \dfrac{-\Sigma Q}{\Sigma(Q/2h)}$	$h_1 = h + \delta h$	$Q = \sqrt{(h_1/K)}$ (m³ s⁻¹)
AD	2058	25	+0.1102	0.002 20	+0.23	+25.33	+0.1107
DB	2730	−5	−0.0428	0.004 28	+0.23	−4.77	−0.0418
DC	4167	−20	−0.0693	0.001 73	+0.23	−19.77	−0.0689
			$\Sigma -0.0019$	$\Sigma -0.008\,21$			0

The calculations for junction D can now be set out as shown in Table 14.1. After the first correction it can be seen from Table 14.1 that ΣQ is zero to four places of decimals and no further correction is necessary. The flows in the pipes are

$Q_{AD} = 0.1107\ \mathrm{m^3\,s^{-1}}$ from reservoir A,

$Q_{DB} = 0.0418\ \mathrm{m^3\,s^{-1}}$ to reservoir B,

$Q_{DC} = 0.0689\ \mathrm{m^3\,s^{-1}}$ to reservoir C.

Concluding remarks

The principle of conservation of mass flow in Chapter 4 and the steady flow energy equation developed in Chapter 6, together with the concepts of frictional and separation losses developed in Chapter 10, have been utilized in this chapter to address the prediction of steady, incompressible liquid or gas flow conditions in both pipes and ducts. The most important element of this treatment concerns the choice in the application of the steady flow energy equation as to the control volume entry and exit locations to simplify the problem. A wide range of system applications were referred to, some of which will be returned to later in Chapter 25 in a treatment of machine and network matching.

The use and control of separation losses to balance effectively flow in networks was introduced and may be seen to be the basis for even the most complex flow-balancing applications. The basic Kirchhoff laws applying to potential change and flow rate around closed loops were harnessed to determine the flow distribution in complex networks via the Hardy–Cross technique, a previously tedious methodology that is well suited to a computer.

Summary of important equations and concepts

1. Chapter 14 reinforces the fundamental concepts of incompressible fluid flow in pipe systems. In particular it defines frictional losses, equation (14.1), emphasizing the utility of the hydraulic mean depth approach to the equation. Separation losses are highlighted, Section 14.2, and the combination of friction and separation loss reinforced. A computer program SIPHON is included to demonstrate series pipeline application of the steady flow energy equation, Section 14.3.

2. Series and parallel pipe systems are addressed, the importance of losses being cumulative in the flow direction in series systems, equation (14.4) and the equality of parallel path pressure losses emphasized as an essential concept in flow balancing.

3. The combination of parallel losses to give an overall loss coefficient is presented in equation (14.7), an identical approach to that found in direct current applications and Kirchoff's laws.

4. Network analysis is addressed, and in particular the Hardy–Cross method is used as a basis for discussion. The application of Kirchoff's laws is again apparent as flow summation at a junction must be zero, pressure loss around a loop must be zero and pressure loss in a link must be accounted for through the frictional and separation loss expressions. A computer program HARDYC illustrates this section.

Further reading

Jeppson, R. W. (1979). *Analysis of Flow in Pipe Networks*. Ann Arbor Science, Michigan.

Problems

14.1 Two vessels in which the difference of surface levels is maintained constant at 2.4 m are connected by a 75 mm diameter pipeline 15 m long. If the frictional coefficient f may be taken as 0.008, determine the volume rate of flow through the pipe. [11.9 litre s^{-1}]

14.2 The difference in surface levels in two reservoirs connected by a syphon is 7.5 m. The diameter of the syphon is 300 mm and its length 750 m. The friction coefficient f is 0.0064. If air is liberated from solution when the absolute pressure is less than 1.2 m of water, what will be the maximum length of the inlet leg of the syphon to run full, if the highest point is 5.4 m above the surface level in the upper reservoir? What will be the discharge?

[350 m, 107 dm^3 s^{-1}]

14.3 Two reservoirs whose difference of level is 15 m are connected by a pipe ABC whose highest point B is 2 m below the level in the upper reservoir A. The portion AB has a diameter of 200 mm and the portion BC a diameter of 150 mm, the friction coefficient being the same for both portions. The total length of the pipe is 3 km.

Find the maximum allowable length of the portion AB if the pressure head at B is not to be more than 2 m below atmospheric pressure. Neglect the secondary losses.

[1815 m]

14.4 A pipeline 30 m long connects two tanks which have a difference of water level of 12 m. The first 10 m of pipeline from the upper tank is of 40 mm diameter and the next 20 m is of 60 mm diameter. At the change in section a valve is fitted. Calculate the rate of flow when the valve is fully opened assuming that its resistance is negligible and that f for both pipes is 0.0054. In order to restrict the flow the valve is then partially closed. If k for the valve is now 5.6, find the percentage reduction in flow.

[0.007 39 m^3 s^{-1}, 25.9 per cent]

14.5 A smooth walled tube is used in a 3000 m long pipeline carrying water at 15 °C between two reservoirs whose surface elevations are 6 m apart. Entry is sharp edged and the outlet is also abrupt to the downstream reservoir. The pipeline contains six 45° bends and two globe valves. Determine the necessary pipe diameter so that the discharge should be 28 litre s^{-1} to the lower reservoir.

Take the equivalent length of each bend as 26.5 diameters, the valves as 75 diameters and the entry as 30 diameters.

[222 mm]

14.6 A horizontal duct system draws atmospheric air into a circular duct of 0.3 m diameter, 20 m long, then through a centrifugal fan and discharges it to atmosphere through a rectangular duct 0.25 m by 0.20 m, 50 m long. Assuming that the friction factor for each duct is 0.01 and accounting for an inlet loss of one-half of the velocity head and also for the kinetic energy at outlet, find the total pressure rise across the fan to produce a flow of 0.5 m^3 s^{-1}.

Sketch also the total energy and hydraulic gradient lines putting in the most important values. Assume the density of air to be 1.2 kg m^{-3}. [695 N m^{-2}]

14.7 For flow through pipes at high Reynolds number, the coefficient of friction is given by the following relation,

$$\frac{1}{\sqrt{f}} - 4\log_{10}\left(\frac{r}{\varepsilon}\right) = 3.48,$$

where r = pipe radius and ε = mean height of roughness projections. A pipe of internal diameter 0.15 m is formed of a material for which ε is 0.000 38 m. The pipe is 1524 m long and it connects two water reservoirs whose surface levels are maintained at the same height. Water may be pumped along the pipe and the maximum pumping power available is 82 kW. Calculate the maximum rate of flow in the pipe.

[0.06 m^3 s^{-1}]

14.8 A pipeline conveying water between reservoirs A and B is of 30.5 cm diameter and 366 m long. The difference of head between the two surfaces is 4.12 m. Determine the flow rate if $f = 0.005$.

It is required to increase the flow by 50 per cent by duplicating a portion of the pipe. If the head and friction factor are unchanged and minor losses are ignored, find the length of the second pipe which is of the same diameter as the first. [0.134 m^3 s^{-1}; 270 m]

14.9 There is a pressure loss of 300 kN m^{-2} when water is pumped through pipeline A at a rate of 2 m^3 s^{-1} and there is a pressure loss of 250 kN m^{-2} when water is pumped at a rate of 1.4 m^3 s^{-1} through pipeline B. Calculate the pressure loss which will occur when 1.5 m^3 s^{-1} of water are pumped through pipes A and B jointly if they are connected (a) in series, (b) in parallel, assuming that junction losses may be neglected. In the latter case calculate the volume rate of flow through each pipe. [(a) 456 kN m^{-2}, (b) 54.1 kN m^{-2}, 0.849 m^3 s^{-1}, 0.651 m^3 s^{-1}]

14.10 A complex ventilation system for a coal mine may be reduced to the system shown in Fig. 14.20, where R_1, R_2 and R_3 represent the equivalent resistances of the three main sections of the mine. Assuming an air density of 1.17 kg m^{-3} these resistances are:

R_1 = 49 mm of water total pressure at 100 m^3 s^{-1}

R_2 = 73 mm of water total pressure at 100 m^3 s^{-1}

R_3 = 10 mm of water total pressure at 200 m^3 s^{-1}.

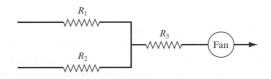

FIGURE 14.20

The fan characteristic at a density of 1.2 kg m^{-3} is:

Discharge Q (m^3 s^{-1})	0	100	150	200
Fan total pressure (mm of water)	175	180	175	160

Discharge Q (m^3 s^{-1})	250	300	350
Fan total pressure (mm of water)	135	100	60

(*a*) Determine the volume rate of flow handled by the fan and the fan total pressure. (*b*) If, owing to the increased length of workings, the resistance of the whole system changes and is found to be 150 mm of water total at $200\,m^3\,s^{-1}$ and density $1.17\,kg\,m^{-3}$, determine the percentage increase of fan speed required to maintain the same flow through the fan.

[(*a*) $265\,m^3\,s^{-1}$, $122\,mm$ of water; (*b*) 30.5 per cent]

14.11 Water flows in the parallel pipe system shown in Fig. 14.21 for which the following data are available:

PIPE	DIAMETER (m)	LENGTH (m)	f
AaB	0.10	300	0.0060
AbB	0.15	250	0.0055
AcB	0.20	500	0.0050

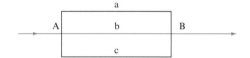

FIGURE 14.21

The supply pipe to point A is of 0.30 m diameter and the mean velocity of water in it is $3\,m\,s^{-1}$. If the elevation of point A is 100 m and the elevation of point B is 30 m above datum, calculate the pressure at point B if that at A is $200\,kN\,m^{-2}$. What is the discharge in each pipe? Neglect all minor losses.

[$559.7\,kN\,m^{-2}$, $0.024\,m^3\,s^{-1}$, $0.075\,m^3\,s^{-1}$, $0.114\,m^3\,s^{-1}$]

14.12 Water is handled by a system of pipes as shown in Fig. 14.22, the details being as follows:

PIPE	LENGTH (m)	DIAMETER (m)	f
$A_1B = A_2B$	100	0.50	0.0055
BC	300	0.75	0.0050
CD	500	0.30	0.0060
CF	400	0.25	0.0060
CF	500	0.30	0.0060

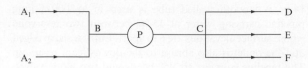

FIGURE 14.22

The elevation of outlets D, E and F is 100 m above the elevation of inlets A_1 and A_2. All outlets and inlets are open to atmosphere. If the mean velocity in the pipes A_1B and A_2B is $2.5\,m\,s^{-1}$, calculate the flow rate through the pump P, the pressure difference across the pump and the power consumed. Take the pump efficiency as 76 per cent.

[$0.98\,m^3\,s^{-1}$, $1538\,kN\,m^{-2}$, $1983\,kW$]

14.13 A horizontal water main comprises 1500 m of 150 mm diameter pipe followed by 900 m of 100 mm diameter pipe, the friction factor f for each pipe being 0.007. All the water is drawn off at a uniform rate per unit length along the pipe. If the total input to the system is $25\,dm^3\,s^{-1}$, find the total pressure drop along the main, neglecting all losses other than pipe friction. Also draw the hydraulic gradient taking the pressure head at inlet as 54 m.   [20.50 m]

14.14 A 675 mm water main runs horizontally for 1500 m and then branches into two 450 mm mains each 3000 m long. In one of these branches the whole of the water entering is drawn off at a uniform rate along the length of the pipe. In the other branch one-half of the quantity entering is drawn off at a uniform rate along the length of the pipe. If $f = 0.006$ throughout, calculate the total difference of head between inlet and outlet when the inflow to the system is $0.28\,m^3\,s^{-1}$. Consider only frictional losses and assume atmospheric pressure at the end of each branch.

[4.41 m]

14.15 Water entering a 150 mm diameter pipe 1300 m long is all drawn off at a uniform rate per metre of length along the pipe. Neglecting all losses other than pipe friction, find the volume rate of flow entering the pipe when the pressure drop along the pipe is 2.55 bar. Take $f = 0.008$. Draw the hydraulic gradient for the system if the pressure at entry to the pipe is 2.8 bar.   [$0.0415\,m^3\,s^{-1}$]

14.16   The head loss for flow in a duct can be written as $h_f = rQ^n$, where r is the pipe resistance and Q is the volume rate of flow. The fuel gallery for a small gas turbine is shown in Fig. 14.23. Each injection nozzle passes 5 litres of kerosene per minute.

The relationships between the pipe resistances are as follows:

$$r_{BC} = r_{CD} = r_{AO} = r_{OE}, \quad r_{AB} = r_{DE},$$

$$r_{AB} = 2r_{BC}, \quad r_{OC} = 3r_{BC}.$$

If the pipe OC is 2 m long, 0.01 m in diameter and the friction factor f is 0.010 in the formula

$$h_f = \frac{4fL}{d}\frac{v^2}{2g},$$

find the pressure drop between O and C.   [3360 N m^{-2}]

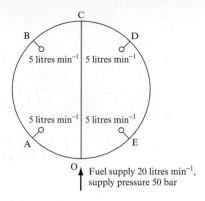

FIGURE 14.23

Uniform Flow in Open Channels

THE TREATMENT OF STEADY UNIFORM FREE SURFACE FLOWS FOLLOWS DIRECTLY FROM the principles of conservation of mass and energy outlined in Chapter 10 and utilized for fully bounded flows in Chapter 14. It will be shown that the frictional representations already developed in terms of the Darcy and Colebrook–White equations may be used to provide a better understanding of open-channel or partially filled pipe flows. Alternative loss representations, particularly the Chezy and Manning coefficients will be introduced. The calcuation of fully developed steady uniform flow depth based on flow rate, channel cross-section slope and roughness, known as the flow normal depth, will be introduced. The dependence of free surface flow rate on channel cross-section and depth will be discussed and the influence of velocity distribution within a fully developed flow considered. ● ● ●

15.1 FLOW WITH A FREE SURFACE IN PIPES AND OPEN CHANNELS

As explained in Section 14.1, flow in an open channel or a duct in which the liquid has a free surface differs from flow in pipes in so far as the pressure at the free surface is constant (normally atmospheric) and does not vary from point to point in the direction of flow, as the pressure can do in a pipeline. A further difference is that the area of cross-section is not controlled by the fixed boundaries, since the depth can vary from section to section without restraint.

The types of flow occurring in open channels can be classified as steady if conditions do not vary with time and uniform if they do not vary from cross-section to cross-section. Thus *steady uniform flow* will occur in long channels of constant cross-section and slope over that portion which is far enough from entry or exit for the flow to have reached its terminal velocity. Such a situation will occur when the energy loss due to friction is exactly supplied by the reduction in potential energy which occurs owing to the fall in bed level. Under these conditions the depth is constant and known as the *normal depth*.

At entry and exit where the depth is varying and wherever the cross-section is changing, as would be the case in most rivers or natural channels, *steady non-uniform flow* will occur if conditions do not change with time. This is also referred to as *varied flow*. Since the flow in an open channel has a free surface, gravity waves can be formed which are an example of *unsteady non-uniform flow*, a result of the fact that conditions of depth and velocity change with time relative to a fixed point on the bed of the channel.

Both laminar and turbulent flow can occur depending on the value of the Reynolds number. In a pipe, laminar flow occurs when the value of the Reynolds number $\rho \bar{v} d / \mu < 2000$, ρ and μ being the mass density and dynamic viscosity of the fluid, $\bar{v}$ the mean velocity and d the pipe diameter. This relation can also be applied to a channel if the diameter d is replaced by the *hydraulic mean depth m*, which is defined as the ratio A/P of the cross-sectional area A of the liquid flowing to the *wetted perimeter P* (the length of the line of contact between the liquid and the channel boundary at that section). Thus, for a rectangular channel of width B in which the depth of the liquid is D,

$$\text{Cross-sectional area} = BD,$$

$$\text{Wetted perimeter} = B + 2D,$$

$$\text{Hydraulic mean depth} = BD/(B + 2D).$$

For a pipe of diameter d running full, $A = (\pi/4)d^2$ and $P = \pi d$, so that $m = d/4$. Replacing m by $d/4$ in the Reynolds number, the criterion for the type of flow in channels will be

$$\text{Laminar flow, } \rho \bar{v}(4m)/\mu < 2000 \quad \text{or} \quad \rho \bar{v} m / \mu < 500.$$

For values of $\rho \bar{v} m / \mu$ between 500 and 2000, flow will be transitional and, if $\rho \bar{v} m / \mu > 2000$, flow is generally turbulent.

In practice, laminar flow is rare in channels and will only occur if the kinematic viscosity μ/ρ is very high or m is very small, as, for example, in the flow of a thin film of liquid over an inclined surface. Normally, flow is turbulent and in this section this will be assumed to be the case.

The continuity, momentum and energy equations can be applied to channel flow in the same way as for pipe flow. Thus, in Fig. 15.1, since there is no change of density between sections 1 and 2, for continuity of steady flow the volume rate of flow Q must be the same at both sections:

$$Q = B_1 d_1 \bar{v}_1 = B_2 d_2 \bar{v}_2, \tag{15.1}$$

FIGURE 15.1
Channel flow

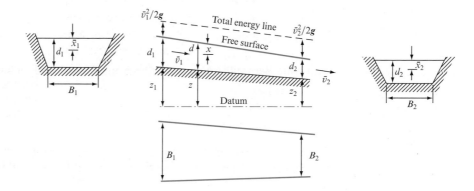

where $\bar{v}_1$ and $\bar{v}_2$ are the mean velocities at the two sections. For wide channels of approximately rectangular section it is sometimes convenient to consider the flow per unit width q, so that

$$q_1 = Q/B_1 = \bar{v}_1 d_1 \quad \text{and} \quad q_2 = Q/B_2 = \bar{v}_2 d_2.$$

In travelling from section 1 to section 2 there will be a change of momentum per second of the liquid corresponding to the change of velocity:

$$\text{Rate of change of momentum} = \text{Mass per second} \times \text{Change of velocity}$$

$$= \rho Q(\bar{v}_2 - \bar{v}_1).$$

This change is produced by the difference in the hydrostatic forces at sections 1 and 2. From equation (3.2),

$$\text{Force in direction of motion at section } 1 = \rho g A_1 \bar{x}_1,$$

$$\text{Force opposing motion at section } 2 = \rho g A_2 \bar{x}_2,$$

where $\bar{x}_1$ and $\bar{x}_2$ are the depths from the free surface to the centroids of the cross-sections.

The resultant force in the direction of motion is $\rho g (A_1 \bar{x}_1 - A_2 \bar{x}_2)$. By Newton's second law

$$\text{Force} = \text{Rate of change of momentum}$$

$$\rho g (A_1 \bar{x}_1 - A_2 \bar{x}_2) = \rho Q(\bar{v}_2 - \bar{v}_1),$$

$$(A_1 \bar{x}_1 - A_2 \bar{x}_2) = Q(\bar{v}_2 - \bar{v}_1)/g. \tag{15.2}$$

The steady flow energy equation, i.e. Bernoulli's equation with a term for loss of energy, can be used, since the fluid flowing in the channel can be assumed to be incompressible. Considering conditions at a point on any streamline at a depth x below the free surface (Fig. 15.1),

$$\text{Total energy per unit weight, } H = \frac{p}{\rho g} + \frac{\bar{v}^2}{2g} + (z + d - x).$$

Now p is the hydrostatic pressure at a depth x below the free surface; therefore $p/\rho g = x$ and

$$\begin{array}{l}\text{Total energy at any point}\\ \text{per unit weight, } H\end{array} = z + d + \frac{v^2}{2g}.$$

Applying Bernoulli's equation to sections 1 and 2 and including the head loss h,

$$z_1 + d_1 + \frac{v_1^2}{2g} = z_2 + d_2 + \frac{v_2^2}{2g} + h. \tag{15.3}$$

In the special case of steady uniform flow $\bar{v}_1 = \bar{v}_2$ and $d_1 = d_2$. Therefore, from equation (15.3), the head loss h must be equal to the difference of bed level, $h = z_1 - z_2$, and equation (15.3) reduces to

$$d_1 + \bar{v}_1^2/2g = d_2 + \bar{v}_2^2/2g,$$

which corresponds to Bernoulli's equation for the frictionless flow of a liquid in a channel with a horizontal bed.

The total energy per unit weight measured above bed level, $(d + \bar{v}^2/2g)$, is termed the *specific energy*. For problems involving steady flow it is often easier to analyze the situation using specific energy instead of total energy and omitting losses of energy due to friction, which correspond to the difference between specific energy and total energy.

The hydraulic gradient line for steady, uniform or non-uniform flow will coincide with the free surface (as shown in Fig. 15.1), while the total energy line will lie $\bar{v}^2/2g$ above the free surface. The gradients of the total energy line, the water surface and the bed will normally differ although they are interrelated. They will only be the same for steady uniform flow in open channels.

15.2   RESISTANCE FORMULAE FOR STEADY UNIFORM FLOW IN OPEN CHANNELS

The analysis of the resistance to the flow of a liquid in a channel is the same as that for flow in a pipe, as shown in Section 10.5. For steady flow, the resistance due to the shear force on the channel boundaries is exactly equal and opposite to the component of the force due to gravity acting in the direction of flow. The resistance laws for pipes can be applied to channels if they are written in terms of the hydraulic mean depth m. Thus, the Darcy formula for the loss of head h_f in length l can be written

$$h_f = \frac{fl}{m}\frac{\bar{v}^2}{2g},$$

(15.4)

since, for a pipe running full, $m = A/P = (\pi/4)d^2/\pi d = d/4$ and, in the form of equation (15.4), the Darcy formula can be applied directly to open channels of any form.

Many problems involve steady flow at uniform depth and constant cross-sectional area. Under these conditions, the bed slope s is equal to the slope of the total energy line i, which is the loss of energy per unit weight per unit length h_f/l. It is, therefore, more convenient for resistance formulae for channels to be written in terms of i. From equation (15.4),

$$\bar{v}^2 = \frac{2g}{f} m \frac{h_f}{l},$$

so that

$$\bar{v} = \sqrt{(2g/f)}\sqrt{(mi)}$$

or, putting $\sqrt{(2g/f)} = C$,

$$\bar{v} = C\sqrt{(mi)}.$$

(15.5)

This is the *Chezy formula*. Unlike the Darcy coefficient f, which is dimensionless, the Chezy coefficient C has dimensions $L^{1/2}T^{-1}$ and, therefore, its numerical value will vary with the system of units employed. In Section 10.3 it was seen that, for pipes, the Darcy friction coefficient f and, therefore, the Chezy coefficient C varied with the Reynolds number and relative roughness of the boundary. The same relationship will hold for open channels, but the Reynolds number will be calculated as $\bar{v}m/\nu$, so that C will depend on the mean velocity $\bar{v}$, the hydraulic mean depth m, the kinematic viscosity ν and the relative roughness. There is experimental evidence that the value of the resistance coefficient does vary with the shape of the channel and therefore with m and possibly also with the bed slope s, which for uniform flow will be equal to i, the relationship for velocity being of the form $\bar{v} = Km^x i^y$, where K, x and y are constants.

While the frictional losses in full-bore flow pipes have been fully investigated, e.g. by Colebrook and White, a similar complete investigation of the Chezy coefficient C has not been completed. This is due not only to the extra variables involved in free surface flows but also to the wide range of surface roughnesses met in practice and the difficulty in achieving steady uniform conditions outside the laboratory. Defects in the setting of channel or sewer slopes, obstructions due to faulty junctions and the inherently unsteady nature of much free surface flow combine to make such a study difficult. The American Society of Civil Engineers' 1963 study concluded that the behaviour of Chezy C could be inferred directly from the full-bore flow friction factor as

$$C = \sqrt{(2g/f)},$$

(15.6)

where the appropriate value of the friction factor f was to be determined from the Colebrook–White equation with the characteristic length being taken as the hydraulic mean depth, $m = A/P$. Normally the Colebrook–White equation is recognized in its full-bore flow form, i.e. equation (10.60), where the appropriate value of m is $D/4$. Hence the coefficients will change in the equation as

$$\frac{1}{\sqrt{f}} = -4\log_{10}\left(\frac{k}{14.8m} + \frac{0.315}{\mathrm{Re}\sqrt{f}}\right), \tag{15.7}$$

where Re is defined as $\rho Vm/\mu$, V being the local cross-section mean flow velocity and k being the local roughness. Figure 15.2 illustrates this dependence, while Table 15.1 presents appropriate values of the roughness k for use in the Colebrook–White equation.

FIGURE 15.2

Dependence of the Chezy coefficient C on Reynolds number and channel surface roughness

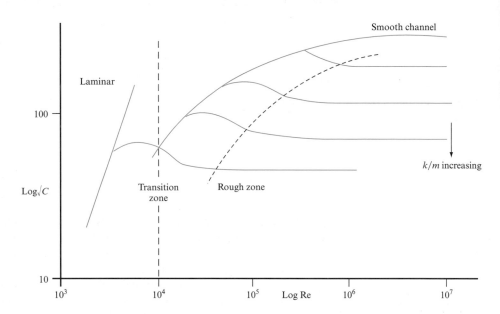

TABLE 15.1

Values of k appropriate to the Colebrook–White equation

	k (mm)
Cast iron (coated)	0.15
Cast iron (uncoated)	0.30
Concrete	0.15
Glazed paper	0.06
UPVC	0.06
Glass	0.03

Experimental investigations have shown that for partially filled pipe flow, where the pipe diameter is less than a metre, the Colebrook–White-based resistance is more accurate than the predictions of the Manning expression. At constant channel slope Manning's n was found to vary with flow depth in partially filled pipe flows where the maximum depth was restricted to less than a metre.

A number of empirical formulae have been proposed for the determination of the value of C in the Chezy formula. In 1869 Ganguillet and Kutter proposed the following formula based on an analysis of the behaviour of rivers and open channels, which, stated in SI units, is

$$C = \frac{23 + 0.001\,55/s + 1/n}{1 + (23 + 0.001\,55/s)n/\sqrt{m}}, \tag{15.8}$$

where s is the bed slope and n is a roughness coefficient which increases with increasing roughness of the channel boundary. Typical values of n are given in Table 15.2. Equation (15.8) is usually referred to as the *Kutter formula*. It is not very convenient to use unless presented in the form of tables giving values of C for values of m, s and n. There is some doubt whether the terms involving s are justified.

TABLE 15.2

Values of n in Manning's formula for flow in open channels

SURFACE OF CHANNEL	CONDITION	
	GOOD	POOR
Neat cement	0.010	0.013
Cement mortar	0.011	0.015
Concrete, *in situ*	0.012	0.018
Concrete, precast	0.011	0.013
Cement rubble	0.017	0.030
Dry rubble	0.025	0.035
Brick with cement mortar	0.012	0.017
Plank flumes, planed	0.010	0.014
unplaned	0.011	0.015
Metal flumes, semicircular, smooth	0.011	0.015
corrugated	0.022	0.030
Cast iron	0.013	0.017
Steel, riveted	0.017	0.020
Canals, earth, straight and uniform	0.017	0.025
dredged earth	0.025	0.033
rock cuts, smooth	0.025	0.035
rock cuts, jagged	0.035	0.045
rough beds with weeds on sides	0.025	0.040
Natural streams, clean, smooth and straight	0.025	0.035
rough	0.045	0.060
very weedy	0.075	0.150

The simplest formula, and one very widely used, is that published in 1890 by Robert Manning, who found from the experimental data then available that C varied as $m^{1/6}$ and was dependent on the roughness coefficient n of the channel boundaries. The *Manning formula*, obtained by putting Manning's value of C in the Chezy formula, stated in SI units, is usually written in the form

$$\bar{v} = (1/n)m^{2/3}i^{1/2}, \tag{15.9}$$

where n has the same value as in the Kutter formula. Equation (15.9) is also known as the *Strickler formula* and $1/n$ as the Strickler coefficient. The dimensions of n in the Manning formula are $L^{-1/3}T$.

The *Bazin formula*, published in 1897, does not relate C to the bed slope s. Stated in SI units,

$$C = \frac{86.9}{1 + k/\sqrt{m}}, \tag{15.10}$$

where k depends on the surface roughness. Typical values of k are given in Table 15.3.

TABLE 15.3
Values of k in the Bazin formula (SI units)

SURFACE OF CHANNEL	k
Smooth cement or planed wood	0.060
Planks, ashlar and brick	0.160
Rubble masonry	0.460
Earth channels of very regular surface	0.850
Ordinary earth channels	1.303
Exceptionally rough channels	1.750

A number of other formulae have been put forward, but the experimental investigation of open-channel flow is complicated by the effects of the free surface and of cross-currents, as well as the variety of bed conditions which can occur. It should also be remembered that such formulae apply, strictly, only to uniform flow and, if applied to non-uniform flow, it must be assumed that the loss of energy per unit weight at a given section is the same as for uniform flow at the same depth. In practice, if the flow is diverging, this loss will be greater because of increased turbulence, while for converging flow the loss will be decreased. The results obtained provide a means of estimating flow in channels, but their relation to the actual flow will depend upon the experience of the user in selecting the most suitable formula and appropriate resistance coefficient (see Section 16.3).

EXAMPLE 15.1

An open channel has a cross-section in the form of a trapezium (Fig. 15.3) with a bottom width B of 4 m and side slopes of 1 vertical to $1\frac{1}{2}$ horizontal. Assuming that the roughness coefficient n is 0.025, the bed slope is 1/1800 and the depth of the water is 1.2 m, find the volume rate of flow Q using (a) the Chezy formula with C determined from the Kutter formula, and (b) the Manning formula.

FIGURE 15.3

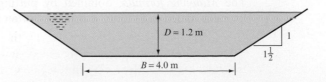

$D = 1.2$ m

$B = 4.0$ m

$1\frac{1}{2}$ 1

Solution

$$\text{Width of water surface} = B + 2 \times 1.5D$$

$$= 4 + 3 \times 1.2 = 7.6 \text{ m},$$

$$\text{Area of cross-section, } A = [(7.6 + 4)/2] \times 1.2 = 6.96 \text{ m}^2,$$

$$\text{Wetted perimeter, } P = B + 2\sqrt{(1.2^2 + 1.8^2)}$$

$$= 4 + 2 \times 1.2\sqrt{(1.44 + 2.25)} = 8.33 \text{ m},$$

$$\text{Hydraulic mean depth, } m = A/P = 6.96/8.33 = 0.836 \text{ m}.$$

For uniform steady flow,

$$\text{Total energy gradient } i = \text{Bed slope } s = 1/1800.$$

(a) From the Kutter formula,

$$C = \frac{23 + 0.001\,55/i + 1/n}{1 + (23 + 0.001\,55/i)n/\sqrt{m}} \quad \text{in SI units}$$

$$= \frac{23 + 0.001\,55 \times 1800 + 1/0.025}{1 + (23 + 0.001\,55 \times 1800)0.025/\sqrt{(0.836)}} = 38.6.$$

$$\text{Volume rate of flow, } Q = CA\sqrt{(mi)}$$

$$= 38.6 \times 6.96\sqrt{(0.836/1800)} = \mathbf{5.79 \text{ m}^3 \text{ s}^{-1}}.$$

(b) Using the Manning formula,

$$\text{Volume rate of flow, } Q = A(1/n)m^{2/3}i^{1/2}$$

$$= 6.96 \times 0.836^{2/3}/0.025 \times (1/1800)^{1/2} = \mathbf{5.82 \text{ m}^3 \text{ s}^{-1}}.$$

In addition to energy losses due to friction, there will be separation losses wherever the flow is disturbed, as, for example, at a reservoir entrance or exit, a change of section or a bend. These are similar to those occurring in pipe flow and can be expressed in the form $k(\bar{v}^2/2g)$, where $\bar{v}$ is the mean velocity and k is a coefficient of the same order as for pipes. Such separation losses are normally small compared with overall friction losses, but can be of local importance, since the loss of head will appear as a change in level of the free surface.

15.3 OPTIMUM SHAPE OF CROSS-SECTION FOR UNIFORM FLOW IN OPEN CHANNELS

The shape of a channel or, more precisely, the cross-section of flow, will affect the ratio of the area of flow A to the wetted perimeter P and, therefore, the value of the hydraulic mean depth m. For uniform flow with a given bed slope, the mean velocity and discharge depend on m and so the shape of a channel will affect its hydraulic effectiveness. Given complete freedom of choice of cross-section, the optimum shape, hydraulically, would be that producing a maximum discharge for a given area, bed slope and surface roughness, which would be that with the smallest wetted perimeter. Such a channel would also have the smallest cross-section of flow for a required discharge and, since its wetted perimeter is a minimum, would require the least amount of lining material or surface finishing. The optimum cross-section would, therefore, also tend to be the cheapest.

In practice, the choice of cross-section may be dictated by other factors. Of all sections with an open surface, a semicircle has the smallest wetted perimeter for a given area, but semicircular channels are not easy to construct in many materials. Channels excavated in the ground are usually trapezoidal in cross-section, but although the optimum side slopes can be calculated, the nature of the ground will determine the slope that can be used. Unlined earth banks will normally not stand at slopes steeper than $1\frac{1}{2}$ horizontal to 1 vertical and, in sandy soil, the side slopes may be as flat as 3 to 1. Channels cut in rock, lined with concrete or constructed from timber or metal can be built with vertical sides if required. The following example shows how the optimum shape of trapezoidal channels can be calculated.

EXAMPLE 15.2

Find the proportions of a trapezoidal channel (Fig. 15.4) which will make the discharge a maximum for a given cross-sectional area of flow and given side slopes. Show also that if the side slopes can be varied the most efficient of all trapezoidal sections is a half-hexagon.

FIGURE 15.4

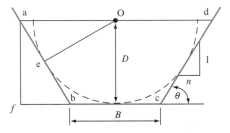

A trapezoidal channel has side slopes of 3 horizontal to 4 vertical and the slope of its bed is 1 in 2000. Determine the optimum dimensions of the channel if it is to carry water at $0.5 \text{ m}^3 \text{ s}^{-1}$. Use the Chezy formula, assuming that $C = 80 \text{ m}^{1/2} \text{ s}^{-1}$.

Solution

Using the Chezy formula,

$$Q = ACm^{1/2}i^{1/2} = AC(A/P)^{1/2}i^{1/2}.$$

Maximum discharge for given values of A, C and i will, therefore, occur when P is a minimum.

In Fig. 15.4, base width $= B$, depth $= D$ and the side slope is 1 vertical to n horizontal.

$$\text{Area of flow, } A = (B + nD)D, \tag{I}$$

from which

$$B = A/D - nD. \tag{II}$$

$$\text{Wetted perimeter, } P = bc + 2cd$$

$$= B + 2D\sqrt{(n^2 + 1)}.$$

Substituting from equation (II),

$$P = A/D + [2\sqrt{(n^2 + 1)} - n]D. \tag{III}$$

If A and n are fixed, P will be a minimum when $dP/dD = 0$. Differentiating equation (III),

$$\frac{dP}{dD} = -A/D^2 + 2\sqrt{(n^2 + 1)} - n = 0.$$

$$A = D^2[2\sqrt{(n^2 + 1)} - n].$$

Substituting for A from equation (I),

$$BD + nD^2 = D^2[2\sqrt{(n^2 + 1)} - n].$$

For maximum discharge,

$$\boldsymbol{B = 2D[\sqrt{(n^2 + 1)} - n].} \tag{IV}$$

(Note special case of rectangular sections, i.e. $n = 0$.)

We now find the side slopes for the section which will have the greatest possible efficiency. The value of B will be that for optimum efficiency, given by equation (IV),

$$\text{Area, } A = BD + nD^2 = 2D^2[\sqrt{(n^2 + 1)} - n] + nD^2$$

$$= D^2[2\sqrt{(n^2 + 1)} - n].$$

So, for maximum efficiency,

$$D = A^{1/2}/[2\sqrt{(n^2 + 1)} - n]^{1/2}.$$

Substituting in equation (III),

$$P = 2A^{1/2}[2\sqrt{(n^2 + 1)} - n]^{1/2}. \tag{V}$$

Since P^2 will also be a minimum when P is a minimum, it is convenient to square equation (V):

$$P^2 = 4A[2\sqrt{(n^2 + 1)} - n].$$

Differentiating and equating to zero,

$$\frac{d(P^2)}{dn} = 4A\left[\frac{2n}{\sqrt{(n^2 + 1)}} - 1\right] = 0,$$

$$2n = \sqrt{(n^2 + 1)}.$$

Squaring,

$$4n^2 = n^2 + 1 \quad \text{and} \quad n = 1/\sqrt{3}.$$

If θ is the angle of the side of the horizontal, $\tan \theta = 1/n = \sqrt{3}$ and $\theta = 60°$. Thus, the cross-section of flow of greatest possible efficiency will be a half-hexagon.

For the given cross-section in this problem, $n = \frac{3}{4}$; therefore, substituting in equation (IV) for maximum discharge,

$$B = 2D\{\sqrt{[(\tfrac{3}{4})^2 + 1]} - \tfrac{3}{4}\} = 2D(\tfrac{5}{4} - \tfrac{3}{4}) = D,$$

$$\text{Area of cross-section, } A = BD + \tfrac{3}{4}D^2$$

$$= D^2 + \tfrac{3}{4}D^2 = \tfrac{7}{4}D^2,$$

$$\text{Wetted perimeter, } P = B + 2 \times \tfrac{5}{4}D = \tfrac{7}{2}D,$$

$$\text{Hydraulic mean depth, } m = A/P = \tfrac{1}{2}D.$$

Substituting in the Chezy formula,

$$Q = ACm^{1/2}i^{1/2} = \tfrac{7}{4}D^2 \times C(\tfrac{1}{2}D)^{1/2}i^{1/2}.$$

Putting $Q = 0.5 \text{ m}^3 \text{ s}^{-1}$, $C = 80 \text{ m}^{1/2} \text{ s}^{-1}$ and $i = 1/2000$,

$$0.5 = \tfrac{7}{4} \times \frac{80D^{5/2}}{(2 \times 2000)^{1/2}}$$

$$\text{Depth, } D = \left(\frac{4 \times 0.5 \times 63.25}{7 \times 80}\right)^{2/5} = \mathbf{0.552 \text{ m}},$$

$$\text{Base width, } B = D = \mathbf{0.552 \text{ m}}.$$

It can also be shown that for a channel of optimum proportions the sides and base are tangential to a semicircle with its centre O on the free surface. Drawing Oe perpendicular to ab from the midpoint of the free surface (Fig. 15.4):

$$\sin \widehat{Oae} = \frac{Oe}{Oa} = \frac{Oe}{\tfrac{1}{2}(B + 2nD)},$$

$$\sin \widehat{abf} = \frac{af}{ab} = \frac{D}{D\sqrt{(n^2 + 1)}}.$$

But $\widehat{Oae} = \widehat{abf}$ and so

$$\frac{Oe}{\tfrac{1}{2}(B + 2nD)} = \frac{D}{D\sqrt{(n^2 + 1)}},$$

$$Oe = \frac{B + 2nD}{2\sqrt{(n^2 + 1)}}.$$

From equation (IV) in Example 15.2,

$$B = 2D[\sqrt{(n^2 + 1)} - n] = 2D\sqrt{(n^2 + 1)} - 2nD, \tag{15.11}$$

and so

$$\text{Oe} = \frac{2D\sqrt{(n^2 + 1)} - 2nD + 2nD}{2\sqrt{(n^2 + 1)}} = D.$$

As ab is perpendicular to Oe it will be a tangent to a semicircle of centre O to which bc and cd will also be tangential.

The hydraulic mean depth for an optimum trapezoidal section will be

$$m = \frac{BD + nD^2}{B + 2D\sqrt{(n^2 + 1)}} = \frac{B + nD}{B/D + 2\sqrt{(n^2 + 1)}}.$$

Substituting for B from equation (15.11),

$$m = \frac{2D\sqrt{(n^2 + 1)} - nD}{2\sqrt{(n^2 + 1)} - 2n + 2\sqrt{(n^2 + 1)}} = \frac{D}{2}.$$

15.4   OPTIMUM DEPTH FOR FLOW WITH A FREE SURFACE IN COVERED CHANNELS

For covered channels which are not flowing full, there will be optimum depths of flow for maximum velocity and for maximum discharge. This arises because, as the level of the free surface rises, a point is reached beyond which the wetted perimeter increases very rapidly in comparison with the area of flow, causing the hydraulic mean depth and, therefore, the velocity to decrease. Since the discharge is the product of the area and velocity, there will also come a point at which the discharge will start to diminish as the depth continues to increase, because the increase in area is more than offset by the reduction in mean velocity. The method of determining these optimum conditions is similar to that used in Section 15.3 but, since the shape of the conduit is fixed, the area A can no longer be treated as constant.

For the steady uniform flow in the circular conduit shown in Fig. 15.5 the Chezy formula gives

$$\text{Mean velocity, } \bar{v} = Cm^{1/2}i^{1/2} = C(A/P)^{1/2}i^{1/2},$$

and so, for constant values of C and i, the maximum value of $\bar{v}$ will occur when A/P is a maximum.

If the free surface subtends an angle 2θ at the centre O for any depth Z,

$$\text{Area of flow, } A = \text{Sector OSTU} - \text{Triangle OSU}$$

$$= \tfrac{1}{2}r^2 \times 2\theta - r^2 \sin\theta \cos\theta$$

$$= r^2(\theta - \tfrac{1}{2}\sin 2\theta),$$

$$\text{Wetted perimeter, } P = 2r\theta.$$

FIGURE 15.5

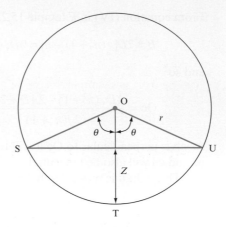

For maximum velocity,

$$\frac{\mathrm{d}(A/P)}{\mathrm{d}\theta} = \frac{1}{P^2}\left(P\frac{\mathrm{d}A}{\mathrm{d}\theta} - A\frac{\mathrm{d}P}{\mathrm{d}\theta}\right) = 0$$

or $\qquad P\dfrac{\mathrm{d}A}{\mathrm{d}\theta} = A\dfrac{\mathrm{d}P}{\mathrm{d}\theta}.$

Substituting for P, A, $\mathrm{d}A/\mathrm{d}\theta$ and $\mathrm{d}P/\mathrm{d}\theta$,

$$2r\theta \times r^2(1 - \cos 2\theta) = r^2(\theta - \tfrac{1}{2}\sin 2\theta) \times 2r,$$

$$\theta(1 - \cos 2\theta) = \theta - \tfrac{1}{2}\sin 2\theta,$$

$$2\theta = \tan 2\theta,$$

giving $\quad 2\theta = 257.5°.$

Depth of flow, $Z = r - r\cos\theta$

$$= r(1 + 0.62) = 1.62r$$

$$= \mathbf{0.81 \times Pipe\ diameter}.$$

For maximum discharge, the result will depend on the choice of resistance formula. Using the Chezy formula,

Discharge, $Q = ACm^{1/2}i^{1/2} = AC(A/P)^{1/2}i^{1/2} = C(A^3/P)^{1/2}i^{1/2}.$

For given values of C and i, the discharge Q will be a maximum when (A^3/P) is a maximum. Differentiating with respect to θ and equating to zero,

$$\frac{\mathrm{d}(A^3/P)}{\mathrm{d}\theta} = \frac{1}{P^2}\left(3PA^2\frac{\mathrm{d}A}{\mathrm{d}\theta} - A^3\frac{\mathrm{d}P}{\mathrm{d}\theta}\right) = 0,$$

$$3P\frac{\mathrm{d}A}{\mathrm{d}\theta} - A\frac{\mathrm{d}P}{\mathrm{d}\theta} = 0.$$

Substituting for P, A, $\mathrm{d}A/\mathrm{d}\theta$ and $\mathrm{d}P/\mathrm{d}\theta$,

$$3 \times 2r\theta \times r^2(1 - \cos 2\theta) - r^2(\theta - \tfrac{1}{2}\sin 2\theta) \times 2r = 0.$$

Dividing by r^3 and simplifying,

$$4\theta - 6\theta \cos 2\theta + \sin 2\theta = 0,$$

from which

$$2\theta = 308°,$$
$$\theta = 154° = 2.68 \text{ rad}.$$

Depth for maximum discharge,

$$Z = r(1 - \cos \theta)$$
$$= r(1 + 0.90)$$

Depth for maximum discharge $= \mathbf{0.95 \times Pipe\ diameter}$.

Since any slight increase in depth will cause a reduction in the volume rate of flow that the channel can carry, it is usual to design on the assumption that the section will run full.

$$\text{Hydraulic mean depth for maximum discharge, } m_1 = \frac{r^2(2.68 - \tfrac{1}{2}\sin\ 308°)}{2r \times 2.68}$$
$$= 0.574r.$$

$$\text{Hydraulic mean depth running full, } m_2 = 0.5r.$$

$$\frac{\text{Discharge running full}}{\text{Maximum discharge}} = \left(\frac{m_2}{m_1}\right)^{1/2} = \left(\frac{0.5}{0.574}\right)^{1/2}$$

$$\text{Discharge running full} = \mathbf{0.933 \times Maximum\ discharge}.$$

Concluding remarks

The treatment of steady uniform flow in open channels presented in this chapter has drawn heavily on the earlier material, particularly the principle of conservation of mass and the steady flow energy equation. The fundamental differences between full-bore and free surface flow conditions have been stressed, particularly the dependence of flow depth on both flow rate and the channel properties, i.e. slope, roughness and cross-sectional shape. The result that there is an optimum depth of flow for a maximum discharge was investigated for partially filled pipe flows.

While steady, uniform, free surface flows rarely exist, the treatment in this chapter is essential to the development of gradually varied flow theory in Chapter 16. The assumption of flow normal depth, i.e. the depth that would be attained in fully developed, uniform, free surface flow, is useful in providing either upstream or downstream control conditions for a gradually varied flow profile. The differentiation of free surface flows into subcritical and supercritical regimes will also be seen to be essential, both to the gradually varied flow described in Chapter 16 and the unsteady flow treated later in Chapter 20.

Figure 15.6 illustrates the variation in flow velocity and flow rate as depth increases at constant channel slope for a partially filled pipe flow. It must be stressed that the maximum discharge depends on the choice of resistance formula. It is interesting to note that the maximum velocity and maximum flow rate do not occur at the same depth.

FIGURE 15.6

Variation of flow velocity and discharge with depth at constant slope

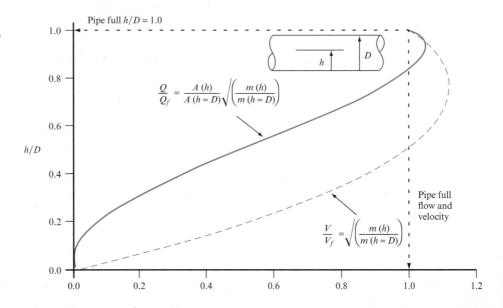

Summary of important equations and concepts

1. This chapter emphasizes the equations governing steady free surface flow and in particular draws parallels between the Darcy (full-bore flow) and Chezy (free surface flow) equations (15.4) and (15.5), and emphasizes further the applicability of the Colebrook–White expression for frictional loss, comparing equation (10.60) to (15.6).

2. Through equation (15.3) the term specific energy of a free surface flow is introduced, a concept linking depth and flow kinetic energy that will be returned to in later chapters.

Further reading

Anon. (March 1963). ASCE Report of task force on friction factors in open channels. *J. Hydraul. Div. ASCE*, **89** (HY2), 97.

Chow, V. T. (1959). *Open Channel Hydraulics*. McGraw-Hill, New York.

French, R. H. (1985). *Open Channel Hydraulics*. McGraw-Hill, New York.

Henderson, F. M. (1966). *Open Channel Flow*. Macmillan, New York.

Rouse, H. (June 1965). Critical analysis of open channel roughness. *J. Hydraul. Div. ASCE*, **91**, 1–15.

Swaffield, J. A. and Galowin, L. S. (1992). *The Engineered Design of Building Drainage Systems*. Ashgate, Aldershot.

Problems

15.1 A rectangular channel is 2.5 m wide and has a uniform bed slope of 1 in 500. If the depth of flow is constant at 1.7 m calculate (*a*) the hydraulic mean depth, (*b*) the velocity of flow, (*c*) the volume rate of flow. Assume that the value of the coefficient C in the Chezy formula is 50 in SI units.   [(*a*) 0.72 m, (*b*) 1.9 m s^{-1}, (*c*) 8.1 m^3 s^{-1}]

15.2 An open channel has a vee-shaped cross-section with sides inclined at an angle of 60° to the vertical. If the rate of flow is 80 dm^3 s^{-1} when the depth at the centre is 0.25 m, what must be the slope of the channel assuming $C = 45$ in SI units?   [1 in 401]

15.3 A channel 5 m wide at the top and 2 m deep has sides sloping 2 vertically in 1 horizontally. The slope of the channel is 1 in 1000. Find the volume rate of flow when the depth of water is constant at 1 m. Take C as 53 in SI units.

What would be the depth of water if the flow were to be doubled?   [4.79 m^3 s^{-1}, 1.6 m]

15.4 Water is conveyed in a channel of semicircular cross-section with a slope of 1 in 2500. The Chezy coefficient C has a value of 56 in SI units. If the radius of the channel is 0.55 m, what will be the volume in cubic decimetres per second flowing when the depth is equal to the radius?

If the channel had been rectangular in form with the same width of 1.1 m and depth of flow of 0.55 m, what would be the discharge for the same slope and value of C?   [279 dm^3 s^{-1}, 355 dm^3 s^{-1}]

15.5   A 900 mm diameter conduit 3600 m long is laid at a uniform slope of 1 in 1500 and connects two reservoirs. When the levels in the reservoirs are low the conduit runs partly full and it is found that a normal depth of 600 mm gives a rate of flow of 0.322 m^3 s^{-1}.

The Chezy coefficient C is given by Km^n, where K is a constant, m is the hydraulic mean depth and $n = \frac{1}{6}$. Neglecting losses of head at entry and exit obtain (a) the value of K, (b) the discharge when the conduit is flowing full and the difference in level between the two reservoirs is 4.5 m.

[(a) 67.6, (b) 0.562 m^3 s^{-1}]

15.6   An earth channel is trapezoidal in cross-section with a bottom width of 1.8 m and side slopes of 1 vertical to 2 horizontal. Taking the friction coefficient k in the Bazin formula as 1.3 and the slope of the bed as 0.57 m per kilometre, find the discharge in cubic metres per second when the depth of water is 1.5 m.   [5.69 m^3 s^{-1}]

15.7   The water supply for a turbine passes through a conduit which for convenience has its cross-section in the form of a square with one diagonal vertical. If the conduit is required to convey 8.5 dm^3 s^{-1} under conditions of maximum discharge at atmospheric pressure when the slope of the bed is 1 in 4900, determine its size assuming that the velocity of flow is given by

$$v = 80i^{1/2}m^{2/3}.$$

[0.222 m side]

15.8   A trapezoidal channel is to be designed to carry 280 m^3 per minute of water. Determine the cross-sectional dimensions of the channel if the slope is 1 in 1600, side slopes 45° and the cross-section is to be a minimum. Take $C = 50$ in SI units.   [$D = 1.53$ m, $B = 1.27$ m]

15.9   A circular section open conduit conveys liquid under maximum velocity conditions. Show that the depth of liquid is 81 per cent of the diameter. Show, without complete

calculation, that this will not be the maximum discharge condition.

Such a conduit having a diameter of 0.8 m is to discharge 0.6 m^3 s^{-1} at maximum velocity. Find the required channel slope if the Chezy constant is 90 in SI units.

[1 in 1050]

15.10   It is required to excavate a canal out of rock. It is to be of rectangular cross-section and to bring 14.2 m^3 of water from a distance of 6.5 km with a velocity of 2.25 m s^{-1}. Determine the gradient and the most suitable section.

[1 in 186, $D = 1.78$ m, $B = 3.56$ m]

15.11   An egg-shaped sewer has a section formed by circular arcs, the top being a semicircle of radius R. The area and wetted perimeter of the section below the horizontal diameter of the semicircle are $3R^2$ and $4.82R$, respectively. Prove that, if C in the Chezy formula is constant, the maximum flow will occur when the water surface subtends an angle of approximately 55° at the centre of curvature of the semicircle.

15.12   The upper portion of the cross-section of an open channel is a semicircle of radius a, the lower portion is a semiellipse of width $2a$, depth $2a$ and perimeter $4.847a$, whose minor axis coincides with the horizontal diameter of the semicircle. The channel is required to convey 14 m^3 s^{-1} when running three-quarters full (i.e. with three-quarters of the vertical axis of symmetry immersed), the slope of the bed i being 0.001. Assuming that the mean velocity of flow is given by Manning's formula $v = 80i^{1/2}m^{2/3}$, determine the dimensions of the section and the depth under maximum flow conditions.   [$a = 1.29$ m, 3.68 m]

15.13   The cross-section of a closed channel is a square with one diagonal vertical, s is the side of the square and y is the depth of the waterline below the apex. Show that for maximum discharge $y = 0.127s$ and that for maximum velocity $y = 0.414s$.

15.14 Find an expression for the theoretical depth for maximum velocity in a closed circular channel in terms of the diameter d.

Compare the discharge at maximum velocity with that when the channel is running full, assuming that the Chezy coefficient is unaltered, and that the pressure remains atmospheric.   [$0.81d$, 0.964]

15.15 An open channel of economic trapezoidal cross-section with sides inclined at $60°$ to the horizontal is required to give a discharge of 10 m^3 s^{-1} when the slope of the bed is 1 in 1600. Calculate the dimensions of the cross-section assuming $v = 74i^{1/2}m^{2/3}$.

[$D = 2.34$ m, $B = 0.72$ m]

16

Non-uniform Flow in Open Channels

FREE SURFACE FLOWS RARELY ATTAIN FULL UNIFORM FLOW CONDITIONS OWING TO local changes in channel, sewer or pipe slope. This chapter introduces the flow definitions necessary to allow these effects to be predicted, including the classification of free surface flows into subcritical and supercritical based upon the local flow Froude number. The flow critical depth corresponding to this classification will be introduced, together with a computer program to calculate both critical and normal flow depths in an open channel or partially filled pipe. Transition from one flow classification to the other requires the traversing of a flow discontinuity, a hydraulic jump, akin in many respects to a shock wave, with consequent changes in flow depth and local mean velocity. The momentum equation is utilized to determine the depth changes inherent across a jump, while the conservation of mass and energy principles will be invoked to determine the flow/depth profile as the flow approaches boundary conditions, e.g. obstructions in the flow, free discharges or changes in channel cross-section or roughness properties. The determination of the location of hydraulic jumps, positioned upstream of discontinuities sufficient to force a local change of flow classification in supercritical flows, will also be demonstrated. The determination of gradually varied flow/depth profiles along rectangular section open channels or partially filled pipes will be illustrated by means of the computer program described in this chapter. Free surface annular downflow is also introduced and the rate of entrained airflow to be expected illustrated, based upon a dimensional analysis presented in Chapter 9. ● ● ●

16.1 SPECIFIC ENERGY AND ALTERNATIVE DEPTHS OF FLOW

As explained in Section 15.1, specific energy E is defined as the energy per unit weight of the liquid at a cross-section measured above bed level at that point. If D is the depth and $\bar{v}$ is the mean velocity

$$E = D + \frac{\bar{v}^2}{2g}.$$

(16.1)

An examination of this equation shows that, for a given specific energy, the possible depths of flow are limited. Considering a wide rectangular channel, width B, cross-sectional area A, through which there is a volume rate of flow Q,

$$\bar{v} = Q/A = Q/BD.$$

Substituting in equation (16.1),

$$E = D + \frac{1}{2g}\left(\frac{Q}{BD}\right)^2$$

or, putting $Q/B = q =$ volume rate of flow per unit width,

$$E = D + q^2/2gD^2$$

(16.2)

$$D^3 - ED^2 + q^2/2g = 0.$$

(16.3)

Equation (16.3) has three roots, of which two are positive and real and the other is negative and unreal. For a constant value of specific energy E, there are normally two, and only two, alternative depths for a given discharge q, as can be seen from Fig. 16.1(a). Similarly, as shown in Fig. 16.1(b), for a constant value of the discharge per unit width q, there will normally be two, and only two, alternative depths for a given value of specific energy.

FIGURE 16.1
Alternative depths of flow

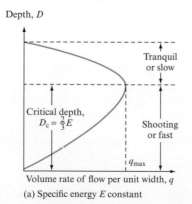

(a) Specific energy E constant

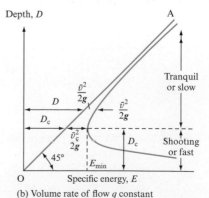

(b) Volume rate of flow q constant

The larger of these two values corresponds to the condition of deep slow flow, which is known as *subcritical, tranquil* or *streaming* flow. The smaller value is that for shallow fast flow, which is known as *supercritical* or *shooting* flow.

From Fig. 16.1(a) and (b) it can be seen that there is a *critical depth* D_c at which the two roots coincide, when the discharge for a given specific energy is a maximum and the energy required for a given discharge is a minimum. To find this value, differentiate equation (16.2) assuming that q is constant:

$$\frac{dE}{dD} = 1 - 2q^2/2gD^3.$$

When dE/dD is zero, flow will be at the critical depth D_c. Thus,

$$\text{Critical depth, } D_c = (q^2/g)^{1/3} = (Q^2/gB^2)^{1/3}. \tag{16.4}$$

The corresponding value of the specific energy will be E_{min} and is obtained by substituting the value of $q^2 = gD_c^3$ from equation (16.4) in equation (16.2) giving

$$E = D_c + gD_c^3/2gD_c^2 = \tfrac{3}{2}D_c.$$

Thus, the critical depth of flow D_c in a rectangular channel will be $\tfrac{2}{3}E$.

The same result could have been obtained by differentiating equation (16.3), assuming that E is constant, since

$$q = D[2g(E - D)]^{1/2}, \tag{16.5}$$

$$\frac{dq}{dD} = \sqrt{(2g)}[(E - D)^{1/2} - \tfrac{1}{2}D/(E - D)^{1/2}].$$

For maximum discharge, when $D = D_c$ at the critical depth, $dq/dD = 0$ and so $(E - D_c) - \tfrac{1}{2}D_c = 0$, from which $D_c = \tfrac{2}{3}E$.

The maximum discharge per unit width for a given value of E is found by substituting this value in equation (16.5):

$$q_{max} = \tfrac{2}{3}E[2g(E - \tfrac{2}{3}E)]^{1/2} = g^{1/2}(\tfrac{2}{3}E)^{3/2}$$

or

$$q_{max} = \sqrt{(gD_c^3)}. \tag{16.6}$$

The velocity of flow corresponding to critical depth is known as the *critical velocity* v_c. From equation (16.1), putting $E = \tfrac{3}{2}D_c$ and $D = D_c$ for critical flow conditions,

$$\tfrac{3}{2}D_c = D_c + v_c^2/2g,$$

$$v_c = \sqrt{(gD_c)}.$$

Referring to Section 5.14, it will be seen that this is the velocity of propagation of a surface wave. Thus, for critical flow conditions, the Froude number $\bar{v}/\sqrt{(gD)} = v_c/\sqrt{(gD_c)} = 1$.

For tranquil flow, the velocity will be less than the critical velocity v_c and this is sometimes termed *subcritical* flow. Similarly, for shooting flow, the velocity will be greater than v_c and may be termed *supercritical* flow. An important difference between them is that for tranquil flow the mean velocity $\bar{v}$ is less than the velocity of propagation of a disturbance relative to the stream and, therefore, disturbances can be propagated both up and downstream, so enabling downstream conditions to determine the behaviour of the flow. For shooting flow, the velocity of the stream exceeds the velocity of propagation and, therefore, disturbances cannot travel upstream and downstream conditions cannot control the behaviour of the flow.

Examination of Fig. 16.1 shows that, when flow is in the region of the critical depth, small changes of energy or flow rate are associated with relatively large changes of depth. Small surface waves are, therefore, easily formed but, since the velocity of propagation is equal to the critical velocity, these waves will be stationary or *standing waves*, and their presence is an indication of critical flow conditions.

In Fig. 16.1(b) a line OA can be drawn at 45° through the origin. Assuming that the scales for E and D are the same, horizontal distances from the vertical axis to this line will be equal to the depth D and, since $E = D + \bar{v}^2/2g$, the distance from OA to the specific energy curve will represent $\bar{v}^2/2g$. For tranquil flow, $\bar{v}$ will decrease as D increases if q is constant so that the specific energy curve is asymptotic to OA. Similarly, as D decreases $\bar{v}$ increases and the specific energy curve will be asymptotic to the E axis.

Although at a cross-section there can be two alternative depths for a given specific energy, the maintenance of uniform flow at one or other of these depths is dependent on the slope of the channel. Energy losses are a function of velocity. For shooting flow, therefore, the slope must be greater than for tranquil flow, since energy losses will be greater. The slope of the channel which will just maintain flow at the critical depth is known as the *critical slope*. For uniform tranquil flow the slope is said to be *mild* and for uniform shooting flow it is termed *steep*.

16.2 CRITICAL DEPTH IN NON-RECTANGULAR CHANNELS

For a channel of any shape and cross-sectional area A (Fig. 16.2) the specific energy for any depth D is

$$E = D + \bar{v}^2/2g$$

FIGURE 16.2

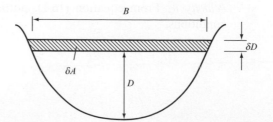

or, since $\bar{v} = Q/A$,

$$E = D + Q^2/2A^2g. \tag{16.7}$$

For flow at critical depth and velocity the specific energy is a minimum for a given value of Q and $\mathrm{d}E/\mathrm{d}D$ is zero. Differentiating equation (16.7),

$$1 - \frac{2Q^2A^{-3}}{2g}\frac{\mathrm{d}A}{\mathrm{d}D} = 0. \tag{16.8}$$

Referring to Fig. 16.2, a change of depth δD will produce a change in cross-sectional area of $\delta A = B\delta D$; therefore, $\mathrm{d}A/\mathrm{d}D = B$. Substituting in equation (16.8) under critical flow conditions,

$$Q^2B/A^3g = 1, \tag{16.9}$$

where B and A are the breadth and area of the flow under these conditions.

Critical velocity, $v_c = Q/A = (Ag/B)^{1/2}$

and from equation (16.9)

$$v_c = (Ag/B)^{1/2} \quad \text{or} \quad v_c = (g\bar{D})^{1/2} \tag{16.10}$$

where $\bar{D}$ is the average depth, defined as A/B for critical flow conditions.

EXAMPLE 16.1

Find, in terms of the specific energy E, the critical velocity and the critical depth in a trapezoidal channel with bottom width B and side slopes $1:n$.

Solution

The cross-section most frequently used for large open channels is trapezoidal. Thus, in Fig. 16.3

Area of section, $A = (B + nD)D$,

Specific energy, $E = D + \bar{v}^2/2g$,

giving $\bar{v} = \sqrt{[2g(E - D)]}$.

FIGURE 16.3

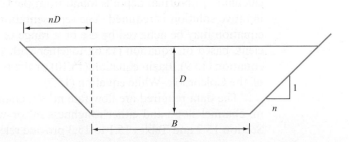

Volume rate of flow, $Q = A\bar{v} = D(B + nD)\sqrt{[2g(E - D)]}.$ (I)

Under critical flow conditions for a constant value of E, the discharge Q is a maximum and $dQ/dD = 0$. From (I),

$$\log_e Q = \log_e D + \log_e(B + nD) + \tfrac{1}{2}\log_e 2g + \tfrac{1}{2}\log_e(E - D).$$

Differentiating with respect to D,

$$\frac{1}{Q}\frac{dQ}{dD} = \frac{1}{D} + \frac{n}{B + nD} + \frac{-1}{2(E - D)}.$$

Since $dQ/dD = 0$ when $D = $ critical depth D_c,

$$\frac{1}{D_c} + \frac{n}{B + nD_c} - \frac{1}{2(E - D_c)} = 0,$$

$$5nD_c^2 + (3B - 4nE)D_c - 2BE = 0.$$ (II)

If n is not zero

$$D_c = \frac{-(3B - 4nE) \pm \sqrt{(9B^2 - 24BnE + 16n^2E^2 + 40BnE)}}{10n}$$ (III)

$$= \frac{4nE - 3B + \sqrt{(16n^2E^2 + 16nEB + 9B^2)}}{10n}.$$

This is a general solution. If $B = 0$, the channel is triangular and (III) gives $D_c = \tfrac{4}{5}E$. If $n = 0$, (III) does not apply, but (II) gives $D_c = \tfrac{2}{3}E$ for a rectangular section.

In practice, tables and curves are available for the determination of the critical depth in trapezoidal channels having any of a number of bottom widths and side slopes.

16.3 COMPUTER PROGRAM CRITNOR

Program CRITNOR calculates the critical and normal depths for free surface flows in partially filled circular cross-section drains or rectangular section channels. Critical depth is determined by an iterative solution to the general form of the critical depth relationship, equation (16.9) (Note that the rectangular case may be found by direct calculation.) Normal depth is found from the Chezy equation (15.5), where again an iterative solution is required. The representation of channel roughness in the Chezy equation may be achieved by use of a range of representations of the Chezy coefficient, based on equation (15.6), together with the Kutter equation (15.8), Manning equation (15.9), Bazin equation (15.10) or the now more acceptable free surface form of the Colebrook–White equation (15.7).

The data required are flow rate m^3 s^{-1}, channel depth and width m, or diameter m, channel slope and either roughness m, or values of Manning loss coefficient n. Section 15.2 and Tables 15.1 to 15.3 provide relevant data.

16.3.1 Application example

Using the following data for a circular cross-section drain, 0.1 m in diameter, Manning n 0.009, at a 0.01 slope, carrying 0.0002 m³ s⁻¹ of water, the flow critical and normal depths were found to be 13.86 mm and 11.13 mm. (Note that these values are also generated as part of the screen display of initial data by FM4WAVE, Chapter 21, for the same drain initial flow conditions.)

For a rectangular channel with the same full flow area and maximum depth, and hence a width of 0.0786 m, the critical and normal flow depths become 8.69 mm and 6.9 mm, respectively, at 0.01 slope, $n = 0.009$ and 0.0002 m³ s⁻¹ throughflow.

16.3.2 Additional investigations using CRITNOR

The computer program calculation may also be used to investigate:

1. the dependence of critical flow depth on channel slope and cross-section;
2. the channel slope required to move from a subcritical to supercritical flow regime for a particular flow rate;
3. the relative differences in critical and normal flow depth predictions at any flow rate caused by the Chezy coefficient representation chosen.

16.4   NON-DIMENSIONAL SPECIFIC ENERGY CURVES

The curve in Fig. 16.1(a) is drawn for a single value of E and is, therefore, one of a family of curves for other values of E which would all be similar in shape. These can be reduced to a single curve in non-dimensional form applying to all values of E by dividing equation (16.2) by q_{max}^2, the square of the maximum discharge occurring at critical depth:

$$\frac{E}{q_{max}^2} = \frac{D}{q_{max}^2} + \frac{1}{2gD^2}\left(\frac{q}{q_{max}}\right)^2.$$

Now, from equation (16.6), $q_{max} = (gD_c^3)^{1/2}$. Hence,

$$\frac{E}{gD_c^3} = \frac{D}{gD_c^3} + \frac{1}{2gD^2}\left(\frac{q}{q_{max}}\right)^2,$$

$$\left(\frac{q}{q_{max}}\right)^2 = \frac{2E}{D_c}\left(\frac{D}{D_c}\right)^2 - 2\left(\frac{D}{D_c}\right)^3.$$

For a rectangular channel, $E = \frac{3}{2}D_c$; therefore,

$$\left(\frac{q}{q_{max}}\right)^2 = 3\left(\frac{D}{D_c}\right)^2 - 2\left(\frac{D}{D_c}\right)^3.$$

Similarly, Fig. 16.1(b) can be presented in a non-dimensional form applicable to all values of q by dividing equation (16.2) by D_c:

$$\frac{E}{D_c} = \frac{D}{D_c} + \frac{1}{2gD_c}\left(\frac{q}{D}\right)^2$$

or, since for minimum energy $q^2 = gD_c^3$,

$$\frac{E}{D_c} = \frac{D}{D_c} + \frac{1}{2}\left(\frac{D_c}{D}\right)^2.$$

16.5 OCCURRENCE OF CRITICAL FLOW CONDITIONS

Since, at the critical depth, the volume rate of flow is a maximum for a given specific energy, cross-sections at which the flow passes through the critical depth are known as *control sections*. Such sections are a limiting factor in the design of a channel and can be expected to occur under the following circumstances:

FIGURE 16.4

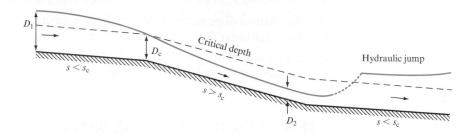

1. *Transition from tranquil to shooting flow.* This may occur as shown in Fig. 16.4 where there is a change of bed slope *s*. Upstream the slope is mild and *s* is less than the critical slope s_c. Over a considerable distance the depth will change smoothly from D_1 to D_2 and at the break in the slope the depth will pass through the critical depth forming a control section which regulates the depth upstream. The reverse transition from shooting to tranquil flow occurs abruptly by means of a hydraulic jump (see Section 16.11).

FIGURE 16.5

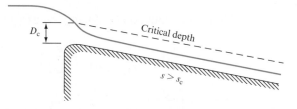

2. *Entrance to a channel of steep slope from a reservoir.* If the depth of flow in the channel is less than the critical depth for the channel the water surface must pass through the critical depth in the vicinity of the entrance (Fig. 16.5), since conditions in the reservoir correspond to tranquil flow.

3. *Free outfall from a channel with a mild slope.* In Fig. 16.6 if the slope *s* of the channel is less than s_c the upstream flow will be tranquil. At the outfall there is no resistance to flow so that, theoretically, it will be a maximum and the depth should be critical. In practice, the gravitational acceleration creates a curvature of the streamlines and an increase of velocity at the brink so that the depth is less than critical.

FIGURE 16.6

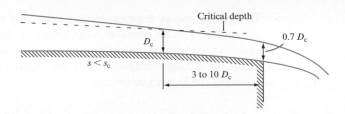

Experiments indicate that critical depth occurs at a distance of $3D_c$ to $10D_c$ from the brink and that the depth at the brink is approximately $0.7D_c$. If the slope of the channel is steep, s being greater than s_c, the upstream flow will be shooting and the depth will everywhere be less than the critical depth.

4. *Change of bed level or channel width.* Under certain circumstances flow will occur at critical depth if a hump is formed in the bed of the channel or the width of the channel is reduced. These cases are discussed in Sections 16.6 and 16.7.

16.6   FLOW OVER A BROAD-CRESTED WEIR

A broad-crested weir consists of an obstruction in the form of a raised portion of the bed extending across the full width of the channel with a flat upper surface or crest sufficiently broad in the direction of flow for the surface of the liquid to become parallel to the crest. The upstream edge is rounded to avoid the separation losses which would occur at a sharp edge.

FIGURE 16.7

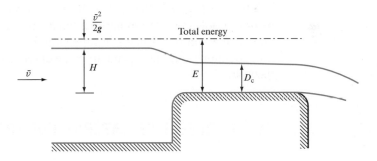

In Fig. 16.7 the flow upstream is tranquil and the conditions downstream allow a free fall over the weir. Since there is no restraining force on the liquid, the discharge over the weir will be the maximum possible and flow over the weir will take place at the critical depth. For a rectangular channel, from equation (16.4),

$$D_c = (Q^2/gB^2)^{1/3}$$

so that $Q = B(g D_c^3)^{1/2}.$

Since $D_c = \tfrac{2}{3}E,$

$$Q = B(g \times \tfrac{8}{27} E^3)^{1/2} = 1.705BE^{3/2} \quad \text{in SI units.} \tag{16.11}$$

The specific energy E measured above the crest of the weir will, assuming no losses, be equal to $H + \bar{v}^2/2g$, where H is the height of the upstream water level above the crest and $\bar{v}$ is the mean velocity at a point upstream where the flow is uniform. If the depth upstream is large compared with the depth over the weir, $\bar{v}^2/2g$ is negligible and equation (16.11) can be written

$$Q = 1.705BH^{3/2} \quad \text{in SI units.} \tag{16.12}$$

A single measurement of the head H above the crest of the weir would then be sufficient to determine the discharge Q.

Since the critical depth $D_c = (Q^2/gB^2)^{1/3}$, the depth over the crest of the weir is fixed, irrespective of its height. Any increase in the height of the weir will not alter D_c but will cause an increase in the depth of flow upstream.

FIGURE 16.8

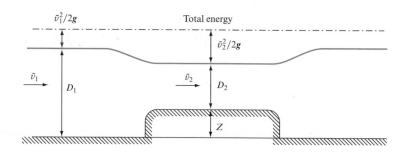

If, as in Fig. 16.8, the level of the flow downstream is raised, the surface level will be drawn down over the hump, but the depth may not fall to the critical depth. The rate of flow can be calculated by applying Bernoulli's equation and the continuity of flow equations and will depend upon the difference in surface level upstream and over the weir.

16.7 EFFECT OF LATERAL CONTRACTION OF A CHANNEL

When the width of a channel is reduced while the bed remains flat (Fig. 16.9), the discharge per unit width increases. If losses are neglected, the specific energy remains constant and so, from Fig. 16.1(a), for tranquil flow the depth will decrease while for shooting flow the depth will increase as the channel narrows.

FIGURE 16.9

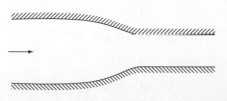

FIGURE 16.10

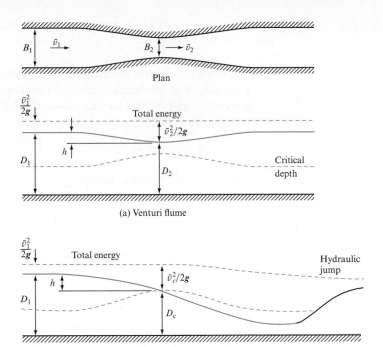

(a) Venturi flume

(b) Standing wave flume

A lateral contraction followed by an expansion can be used for flow measurement as an alternative to the broad-crested weir. If the conditions are such that the free surface does not pass through the critical depth, the arrangement forms a *venturi flume* analogous to the venturi meter used for flow in pipes. Referring to Fig. 16.10(a), for continuity of flow,

$$B_1 D_1 \bar{v}_1 = B_2 D_2 \bar{v}_2. \tag{16.13}$$

Applying Bernoulli's equation to the upstream and throat sections and ignoring losses,

$$D_1 + \bar{v}_1^2/2g = D_2 + \bar{v}_2^2/2g.$$

Substituting for $\bar{v}_1$ from equation (16.13),

$$\frac{\bar{v}_2^2}{2g}\left(1 - \frac{B_2^2 D_2^2}{B_1^2 D_1^2}\right) = D_1 - D_2 = h,$$

$$\bar{v}_2 = \sqrt{\left[\frac{2gh}{1 - (B_2 D_2/B_1 D_1)^2}\right]},$$

Volume rate of flow, $Q = B_2 D_2 \bar{v}_2$

$$= B_2 D_2 \sqrt{\left[\frac{2gh}{1 - (B_2 D_2/B_1 D_1)^2}\right]}. \tag{16.14}$$

Owing to energy losses, the actual discharge in practice will be slightly less than this value and is given by

$$Q = C_\mathrm{d} B_2 D_2 \sqrt{\left[\frac{2gh}{1 - (B_2 D_2 / B_1 D_1)^2}\right]},$$

where C_d is a coefficient of discharge of the order of 0.95 to 0.99.

If the degree of contraction and the flow conditions are such that the upstream flow is tranquil, and the free surface passes through the critical depth in the throat as shown in Fig. 16.10(b),

$$Q = B_2 D_\mathrm{c} v_\mathrm{c} = B_2 D_\mathrm{c} \sqrt{[2g(E - D_\mathrm{c})]},$$

where E is the specific energy measured above the bed level at the throat and, since the critical depth $D_\mathrm{c} = \frac{2}{3} E$,

$$Q = B_2 \times \tfrac{2}{3} E \sqrt{(2g \times \tfrac{1}{3} E)} = 1.705 B E^{3/2} \quad \text{in SI units}.$$

Assuming, as for the broad-crested weir, that the upstream velocity head is negligible,

$$Q = 1.705 B H^{3/2} \quad \text{in SI units}, \tag{16.15}$$

where H is the height of the upstream free surface above bed level at the throat. In some cases, in addition to the lateral contraction, a hump is formed in the bed (as shown in Fig. 16.11), in which case $H = D_1 - Z$.

FIGURE 16.11

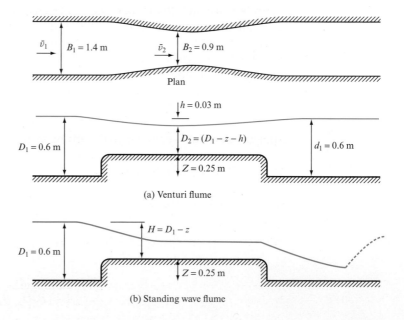

(a) Venturi flume

(b) Standing wave flume

If the upstream conditions are tranquil and the bed slope is the same downstream as upstream, it will not be possible for shooting flow to be maintained for any great distance from the throat. It will revert to tranquil flow downstream by means of an hydraulic jump or standing wave. A venturi flume operating in this mode is known as a *standing wave flume*.

EXAMPLE 16.2

A venturi flume is formed in a horizontal channel of rectangular cross-section 1.4 m wide by constricting the width to 0.9 m and raising the floor level in the constricted section by 0.25 m above that of the channel. If the difference in levels of the free surface between the throat and upstream is 30 mm and both upstream and downstream depths are 0.6 m, calculate the volume rate of flow.

If the downstream conditions are changed so that a standing wave forms clear of the constriction, what will be the volume rate of flow if the upstream depth is maintained at 0.6 m?

Solution

The venturi flume is shown in Fig. 16.11 and, from equation (16.14),

$$Q = B_2 D_2 \sqrt{\left[\frac{2gh}{1 - (B_2 D_2 / B_1 D_1)^2}\right]}$$

$$= 0.9 \times 0.32 \sqrt{\left[\frac{2 \times 9.81 \times 0.03}{1 - (0.9 \times 0.32 / 1.4 \times 0.6)^2}\right]}$$

$$= \mathbf{0.2352 \ m^3 \ s^{-1}}.$$

The standing wave flume is shown in Fig. 16.10(b) and, from equation (16.15),

$$Q = 1.705 B_2 E^{3/2} \quad \text{in SI units.}$$

If it is assumed that $E = H = D_1 - Z = 0.35$ m, then

$$Q = 1.705 \times 0.9 \times 0.35^{3/2} = 0.3177 \ m^3 \ s^{-1}.$$

Checking to determine the effect of neglecting the upstream velocity $\bar{v}_1$, since $Q = B_1 \bar{v}_1 D_1$,

$$\bar{v}_1 = Q / B_1 D_1 = 0.3177 / (1.4 \times 0.6) = 0.3782 \ m \ s^{-1},$$

$$\bar{v}_1^2 / 2g = (0.3782)^2 / (2 \times 9.81) = 0.0073 \ m.$$

This is very small compared with H.

16.8 NON-UNIFORM STEADY FLOW IN CHANNELS

In non-uniform flow, the depth and cross-sectional area of the actual flow may vary from point to point along the length of the channel in contrast to uniform flow, in which the slope, shape and area of cross-section are constant and flow takes place at the *normal* depth. True uniform flow is found only in artificial channels and, even then, flow will be non-uniform near the entrance and exit. In natural streams, the slope of the bed, and the shape and size of the cross-section vary considerably and true uniform flow does not exist. Nevertheless, this type of flow can be analyzed using the equations for uniform flow, such as the Chezy or Manning formulae, by dividing its length into sections, or *reaches*, within which conditions are approximately constant.

Non-uniform flow either may be *gradually varied flow*, in which the change in conditions extends over a long distance, or it can be local non-uniform flow, known as *rapidly varied flow*, in which changes take place suddenly, as at an hydraulic jump, or over a short distance, as at the entrance to a channel or at an obstruction such as a bridge pier or weir. For the purpose of analysis, the distinction between these types is that, for gradually varied flow, it is assumed that changes take place sufficiently slowly for the effects of acceleration to be negligible. In the analysis of rapidly varied flow, the acceleration of the fluid and the resulting rate of change of momentum cannot be overlooked.

It was seen in Section 15.1 that, for uniform flow, the total energy line, the water surface and the bed are all parallel, so that the bed slope s is equal to the slope of the total energy line i and the depth is constant. In non-uniform flow, the depth is variable, s may be greater or less than i for any particular reach and the flow may be accelerating or decelerating. In practice, the variation in depth may be of considerable importance, e.g. in determining the possibility of the occurrence of flooding.

16.9 EQUATIONS FOR GRADUALLY VARIED FLOW

For channels of regular cross-section it is possible to derive analytically an expression for the variation of depth from point to point and so to determine, theoretically, the profile of the free surface if it is assumed that:

1. the channel is rectangular, straight and of constant roughness;

2. the bed slope is small, so that the depth measured normal to the bed can be assumed equal to the vertical depth;

3. flow is steady and streamlines are approximately parallel, so that pressure distribution is hydrostatic.

Consider a length of channel δl (Fig. 16.12) of rectangular cross-section and bed slope s. At section A the depth is D and the velocity v, while at section B the depth is $D + \delta D$ and the velocity $v + \delta v$. The loss of energy between A and B will be $i\delta l$, where i is the slope of the total energy line, and the fall in the bed level will be $s\delta l$. Applying Bernoulli's equation to A and B,

FIGURE 16.12

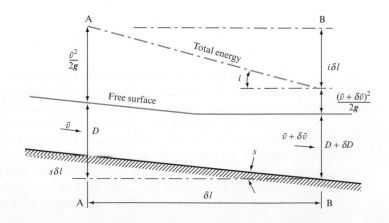

$$s\delta l + D + v^2/2g = (D + \delta D) + (v + \delta v)^2/2g + i\delta l,$$

$$s\delta l = \delta D + (v\delta v/g) + i\delta l$$

(neglecting the second order of small quantities),

$$\delta D/\delta l = s - (v\delta v/g\delta l) - i. \tag{16.16}$$

Assuming a constant width of channel, for continuity of flow the discharge per unit width is constant from section to section and so

$$vD = (v + \delta v)(D + \delta D),$$

$$vD = vD + v\delta D + D\delta v + \text{second-order terms,}$$

$$\delta v = -v\delta D/D.$$

Substituting in equation (16.16)

$$\frac{\delta D}{\delta l} = s - i + \frac{v^2}{gD}\frac{\delta D}{\delta l},$$

$$\frac{\delta D}{\delta l} = \frac{s - i}{1 - v^2/gD}. \tag{16.17}$$

This is the basic equation for non-uniform flow and has a number of alternative forms. Since $v/\sqrt{(gD)}$ is the Froude number Fr, equation (16.17) becomes

$$\delta D/\delta l = (s - i)/(1 - \text{Fr}^2). \tag{16.18}$$

Also, since $v = q/D$,

$$v^2/gD = q^2/gD^3 = (D_c/D)^3,$$

where D_c is the critical depth. Substituting in equation (16.17),

$$\frac{\delta D}{\delta l} = \frac{s - i}{1 - (D_c/D)^3}. \tag{16.19}$$

Moreover, since i is the slope of the total energy gradient and, therefore, is equal to the bed slope required to maintain the given flow at a normal depth D, if a resistance formula of the form $v = Km^a i^b$ is assumed,

$$\text{Discharge per unit width, } q = DKm^a i^b,$$

where m is the hydraulic mean depth at depth D. If D_n is the normal depth for the given flow q on the actual bed slope s,

$$q = D_n K m_n^a s^b$$

and so $Dm^a i^b = D_n m_n^a s^b.$

For a wide channel, $m = D$ and $m_n = D_n$. Hence,

$$(i/s)^b = (D_n/D)^{1+a}$$

or $i/s = (D_n/D)^c,$

where $c = (1 + a)/b$. Substituting in equation (16.19),

$$\frac{\delta D}{\delta l} = s\,\frac{1 - (D_n/D)^c}{1 - (D_c/D)^3}.$$

(16.20)

If the width of the channel is not constant, but changes from B to $B + \delta B$, the continuity of flow equation will be

$$Q = B\bar{v}D = (B + \delta B)(\bar{v} + \delta\bar{v})(D + \delta D)$$

and, neglecting second-order terms,

$$\delta\bar{v} = \frac{\bar{v}}{D}\delta D - \frac{\bar{v}}{B}\delta B.$$

Substituting in equation (16.16),

$$\frac{\delta D}{\delta l} = s - i + \frac{\bar{v}^2}{gD}\frac{\delta D}{\delta l} + \frac{\bar{v}^2}{gB}\frac{\delta B}{\delta l}$$

$$\frac{\delta D}{\delta l} = \frac{s - i + (\bar{v}^2/gB)(\delta B/\delta l)}{1 - \bar{v}^2/gD}.$$

(16.21)

From this equation, it can be seen that, if $s = i$, and $\delta B = 0$, dD/dl is zero and so flow is uniform. Also, if $\bar{v}^2/gD = 1$ or Fr = 1, dD/dl is infinite, so that, theoretically, the slope of the water surface is vertical. For any other conditions, these equations give the slope of the water surface at any point and so, using a step-by-step method of integration, the profile of the free surface can be constructed.

16.10 CLASSIFICATION OF WATER SURFACE PROFILES

It can be seen from equation (16.20) that the gradient of the free surface dD/dl may be positive or negative, depending on the signs of the numerator and denominator, which in turn will depend on the relative values of the actual depth D, the critical depth D_c, and the normal depth D_n. Whether the normal depth is above or below the critical depth will depend on the classification of the bed slope. Surface profiles are classified by a letter and a number. The letter refers to the bed slope, which may be one of the following categories:

M	Mild slope	$D_n > D_c$;
C	Critical slope	$D_n = D_c$;
S	Steep slope	$D_n < D_c$;
H	Horizontal	$s = 0$;
A	Adverse	bed slope s is negative. (Note positive is 'downhill'.)

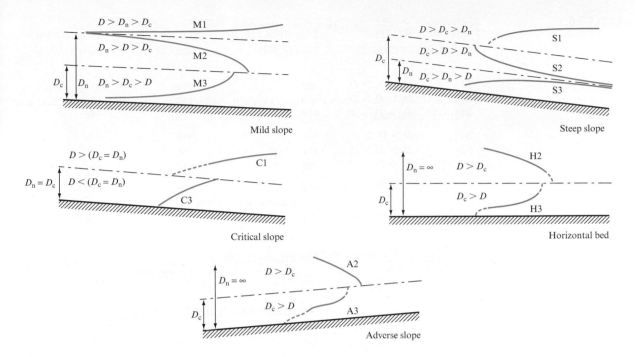

FIGURE 16.13
Water surface profiles

The number refers to the relation between the actual depth of flow D, the normal depth D_n and the critical depth D_c as follows:

1.   free surface of stream lies above both normal and critical depth lines;
2.   free surface of stream lies between normal and critical depth lines;
3.   free surface of stream lies below both normal and critical depth lines.

Although there are five categories of slope and three categories of depth only 12 combinations are possible. For horizontal and adverse slopes, uniform flow is impossible, thus eliminating the H1 and A1 curves; while for critical slope, critical depth and normal depth coincide, thus eliminating the C2 curve. The water profiles corresponding to the various categories are shown in Fig. 16.13 with an indication of how they may occur. Note that the vertical scale has been greatly exaggerated and that even a steep slope, in this context, would amount to only a few degrees inclination to the horizontal. Water surface curves in which the depth increases in the direction of flow are known as *backwater* curves. If the depth decreases in the direction of flow they are called *drop down* or *draw down* curves. Backwater curves have a positive value of dD/dl and draw down curves have a negative value of dD/dl. The type of curve corresponding to each of the classifications can, therefore, be determined by examining whether the numerator and denominator are positive or negative in equation (16.20) for the M, S and C curves, or in equation (16.19) for the H and A curves, as shown in Table 16.1.

For all type 1 curves, the surface will approach the horizontal asymptotically since, as the depth increases, the velocity decreases and tends to zero. All curves approach the normal depth line asymptotically since the steady state is one of uniform flow at a distance remote from the disturbance. In theory, curves cross the critical depth line vertically, since the denominator in equation (16.20) becomes zero, making

TABLE 16.1

Class	Relative Depths	Equation (16.20) $1 - (D_n/D)^c$	$1 - (D_c/D)^3$	Free surface slope dD/dl
M1	$D > D_n > D_c$	+ve	+ve	+ve
M2	$D_n > D > D_c$	−ve	+ve	−ve
M3	$D_n > D_c > D$	−ve	−ve	+ve
S1	$D > D_c > D_n$	+ve	+ve	+ve
S2	$D_c > D > D_n$	+ve	−ve	−ve
S3	$D_c > D_n > D$	−ve	−ve	+ve
C1	$D > (D_n = D_c)$	+ve	+ve	+ve
C2		Not feasible since D_n and D_c coincide		
C3	$D < (D_n = D_c)$	−ve	−ve	+ve

		Equation (16.19) $s - i$	$1 - (D_c/D)^3$	
H1		Not feasible since D_n is indeterminate		
H2	$D > D_c$	−ve	+ve	−ve
H3	$D < D_c$	−ve	−ve	+ve
A1		Not feasible since D_n is indeterminate		
A2	$D > D_c$	−ve	+ve	−ve
A3	$D < D_c$	−ve	−ve	+ve

dD/dl infinite. However, in this region the theory does not apply, since it conflicts with the assumption (that changes will be gradual) on which it is based.

EXAMPLE 16.3

A wide canal has a bed slope of 1 in 1000 and conveys water at a normal depth of 1.2 m. A weir is to be constructed at one point to increase the depth of flow to 2.4 m. How far upstream of the weir will the depth be 1.35 m?

Take C in the Chezy formula as 55 in SI units.

Solution

From equation (16.20), using the Chezy formula and putting $a = \frac{1}{2}$, $b = \frac{1}{2}$ so that $c = (1 + \frac{1}{2})/\frac{1}{2} = 3$,

$$\delta l = \delta D \,[1 - (D_c/D)^3]/s[1 - (D_n/D)^3], \tag{I}$$

normal depth, $D_n = 1.2$ m, bed slope, $s = 1/1000$. The discharge per unit width Q for uniform flow is found from the Chezy formula, taking the hydraulic mean depth m as equal to the depth, since the channel is wide.

$$Q/B = q = D_n C(D_n s)^{1/2} = 1.2 \times 55(1.2/1000)^{1/2} = 2.286 \text{ m}^2 \text{ s}^{-1},$$

Critical depth, $D_c = (q^2/g)^{1/3} = 0.811$ m.

Using a step-by-step method, the distance δl corresponding to a change in depth δD can be calculated working from the known conditions at the weir back upstream. The general rule is to work from the cause of the change towards the point at which flow at normal depth occurs, i.e. upstream for tranquil flow and downstream for shooting flow.

Dividing the range of depths from 2.4 to 1.35 m into seven steps of 0.15 m, the length for each step can be calculated from (I) using the mean depth $\bar{D}$ for each step. Putting $\delta D = 0.15$ m and $s = 1/1000$,

$$\delta l = 150 \frac{[1 - (0.811/\bar{D})^3]}{[1 - (1.2/\bar{D})^3]} \text{ m.}$$

The work is best carried out in tabular form:

DEPTH (m)	MEAN DEPTH D_m	$1 - (0.811/\bar{D})^3$	$1 - (1.2/\bar{D})^3$	L (m)
2.4				
	2.325	0.9576	0.8625	166.54
2.25				
	2.175	0.9482	0.8321	170.93
2.10				
	2.025	0.9358	0.7919	177.26
1.95				
	1.875	0.9191	0.7379	186.83
1.80				
	1.725	0.8961	0.6634	202.62
1.65				
	1.575	0.8635	0.5577	232.25
1.50				
	1.425	0.8157	0.4028	303.76
1.35				
			Total	1440.19

Distance upstream at which depth is 1.35 m = **1440 m approx**.

16.11 THE HYDRAULIC JUMP

The hydraulic jump is an important example of local non-uniform flow. As can be seen from the water surface profiles in Fig. 16.14, there is no possibility of a smooth transition from shooting to tranquil flow since, theoretically, the slope of the water surface should be vertical as it passes through the critical depth. In practice this cannot occur, and the transition takes the form of the hydraulic jump with a steep upward-sloping water surface and violently turbulent conditions accompanied by a substantial loss of energy. As for steady flow, the mass per unit time flowing upstream and downstream of the jump (Fig. 16.14) will be equal and, since the velocity upstream is greater than the velocity downstream, there will be a change in the momentum of the stream per unit time as it passes through the jump. For a channel with a moderate bed slope, the force slowing down the stream and producing this rate of change of momentum is due to the difference in the resultant forces caused by the hydrostatic pressure at the downstream and the upstream cross-sections.

FIGURE 16.14

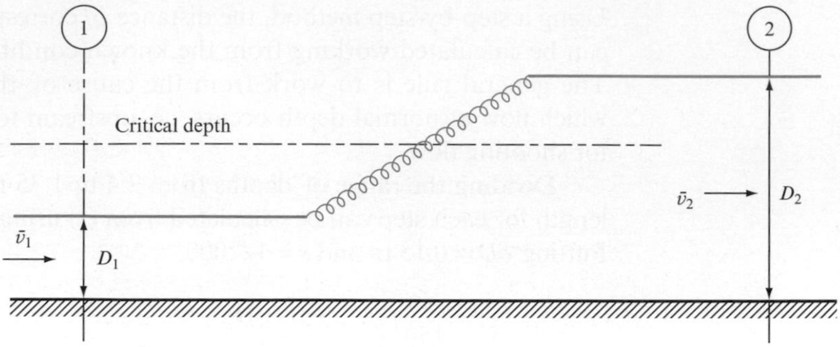

If Q is the volume rate of flow, B the width of the channel, assumed to be rectangular, D_1 and D_2, $\bar{v}_1$ and $\bar{v}_2$ the depths and mean velocities at the upstream and downstream sections, respectively, for continuity of flow,

$$Q = B\bar{v}_1 D_1 = B\bar{v}_2 D_2 \quad \text{and so} \quad \bar{v}_2 = \bar{v}_1 (D_1/D_2),$$

$$\begin{array}{l}\text{Rate of change of momentum} \\ \text{between sections 1 and 2}\end{array} = \rho Q(\bar{v}_1 - \bar{v}_2)$$

$$= \rho B\bar{v}_1^2 D_1(1 - D_1/D_2),$$

$$\begin{array}{l}\text{Hydrostatic force acting downstream} \\ \text{at section 1}\end{array} = \tfrac{1}{2}\rho g D_1^2 B,$$

$$\begin{array}{l}\text{Hydrostatic force acting upstream} \\ \text{at section 2}\end{array} = \tfrac{1}{2}\rho g D_2^2 B,$$

$$\text{Resultant force acting upstream} = \tfrac{1}{2}\rho g B(D_2^2 - D_1^2).$$

By Newton's second law,

$$\rho B\bar{v}_1^2 D_1(1 - D_1/D_2) = \tfrac{1}{2}\rho g B(D_2^2 - D_1^2),$$

$$D_2^2 - D_1^2 = (2\bar{v}_1^2 D_1/gD_2)(D_2 - D_1),$$

$$D_2 + D_1 = 2\bar{v}_1^2 D_1/gD_2,$$

$$D_2^2 + D_1 D_2 - 2\bar{v}_1^2 D_1/g = 0,$$

$$D_2 = \tfrac{1}{2}D_1[-1 + \sqrt{(1 + 8\bar{v}_1^2/gD_1)}]. \tag{16.22}$$

From equation (16.22), the conjugate depths D_1 and D_2 before and after the jump can be determined, Since $\bar{v}_1/\sqrt{(gD_1)} = \text{Fr}_1$, the Froude number at the upstream section,

$$D_2 = \tfrac{1}{2}D_1[-1 + \sqrt{(1 + 8\text{Fr}_1^2)}].$$

The loss of energy in the jump will be equal to the difference of specific energies at the upstream and downstream sections.

$$\text{Loss of energy} = (D_1 + \bar{v}_1^2/2g) - (D_2 + \bar{v}_2^2/2g).$$

16.12 LOCATION OF AN HYDRAULIC JUMP

It is often desirable to be ale to determine the position at which an hydraulic jump will occur. For example, in the design of a spillway over a dam (Fig. 16.15), the energy of the fast-flowing stream must be partially dissipated to prevent erosion of the bed downstream. This can be done by arranging for the formation of an hydraulic jump, but to prevent damage this must occur on the apron. Shooting flow down the face of the dam is retarded by the flatter slope of the apron, which is insufficient to maintain its original high velocity, and a jump will occur. An obstruction can be introduced on the apron to force the jump to form at the desired position but, if this is not done, the position at which the jump will occur naturally can be estimated using equation (16.22) to determine the possible conjugate upstream and downstream depths for the known upstream and downstream conditions.

FIGURE 16.15

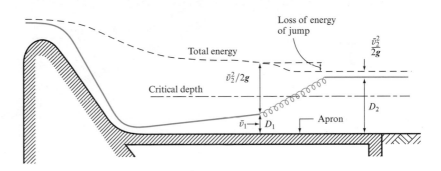

If the discharge and velocity at the foot of the spillway are known and the downstream depth D_2 is fixed, the conjugate depth D_1 can be determined. From equation (16.22), if q is the discharge per unit width,

$$D_1 = \tfrac{1}{2} D_2[-1 + \sqrt{(1 + 8q^2 D_2/g)}]. \tag{16.23}$$

Starting from the known conditions at the foot of the dam, the distance along the apron at which the depth of flow will have increased to D_1 can be calculated as explained in Section 16.10.

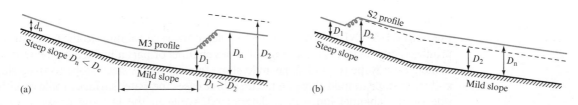

(a)

(b)

FIGURE 16.16

If the downstream conditions are not fixed, the position of the jump can still be determined by applying equation (16.22) to find the conjugate depths which are compatible with the surface profiles upstream and downstream. For example, if the slope of the channel changes from steep to mild (as in Fig. 16.16), the jump may occur either upstream or downstream of the break in the slope. To decide which of the two alternatives is possible, first determine the normal depth for the upstream and downstream slopes. Considering the possibility of the jump occurring upstream of the break, calculate the conjugate depth corresponding to the normal depth on the upstream slope. If this conjugate depth is less than the normal depth on the downstream slope, the jump will form on the upstream slope and be followed by an S2 curve leading to the normal depth downstream. On the other hand, if it is greater than the normal depth on the downstream slope, the jump cannot occur upstream of the break. The jump must, therefore, occur downstream of the break (as in Fig. 16.16(a)), the depth after the jump being normal depth on the downstream slope and the corresponding conjugate depth D_1 occurring immediately before the jump. An M3 curve is formed upstream of the jump.

FIGURE 16.17

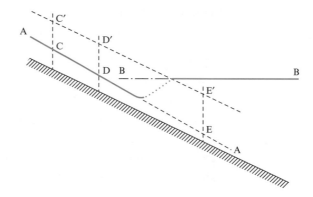

The position of the jump can also be determined graphically by plotting the water surface profiles for the upstream and downstream flow AA and BB, respectively (Fig. 16.17); in this case an accelerating flow upstream and a backwater curve starting from an obstacle downstream which is backing up the flow. If, for a number of points such as C, D and E on the upstream profile, the conjugate depths C′, D′ and E′ are plotted, the position of the hydraulic jump will be the intersection of the line joining C′, D′ and E′ with the downstream backwater curve.

16.13 COMPUTER PROGRAM CHANNEL

Program CHANNEL provides a step-by-step numerical solution to the water surface profile equation (16.17) for a range of series channel or partially filled pipe flow cases, namely mild slope to mild slope, mild slope to steep slope, steep slope to steep slope and steep slope to mild slope. In the first three cases the water surface profile on either side of the channel junction is determined, while in the last the presence of an hydraulic jump is identified and the surface profiles upstream and downstream of the jump determined. In addition, the backwater profile upstream from an exit-critical depth control is determined, e.g. as would be the case for a free discharge. The program is supported by screen graphics as well as a text data file defining flow depth along each channel. Circular and rectangular cross-section channels are included.

16.13.1 Application example

Determine the location of an hydraulic jump formed in a mild slope channel downstream of a transition from a steep approach channel as the slope of the upstream channel is altered. The downstream channel slope is 0.0001, the partially filled pipe has a diameter of 1.0 m and carries a flow rate of 1.0 m³ s⁻¹, both channels have a Manning *n* value of 0.009.

Operating the program and extracting the jump position from the screen displayed data file yields the following result:

Upstream slope	0.4	0.3	0.2	0.1	0.05	0.02
Distance to jump (m)	87.2	83.9	72.8	54.7	35.3	3.1

At upstream slopes below 0.0 the jump moves into the upstream pipe.

16.13.2 Additional investigations using CHANNEL

The computer program may be used to investigate the following areas:

1. backwater surface profiles upstream of transitions from mild to steep channel slopes;

2. the influence of Manning *n* on the backwater profiles and on normal and critical depths;

3. changes in Manning *n* between series channels nominally at the same slope.

16.14 ANNULAR WATER FLOW CONSIDERATIONS

There are a number of practical examples of gravity-driven free surface flows that are amenable to analysis by the techniques of steady non-uniform flow developed in this chapter. Possibly the most readily recognizable are downflows associated with rainwater or domestic drainage. In these cases water entering a vertical pipe, or stack, adheres to the pipe walls and falls as a film. The falling water film entrains an air core and this air movement is accompanied by a reduction in the air pressure within the stack, as would be expected from an application of the steady flow energy equation to the air movement from atmospheric conditions outside the stack.

Flows entering a stack, normally from a side entry branch, tend to arch across the vertical to impinge on the opposite pipe wall (Fig. 16.18). The subsequent change in flow direction imparts a swirl component to the flow that allows the annular flow regime to be established. As the flow moves downwards it accelerates under gravity and the swirl, or circumferential, velocity component dies away. Early researchers were surprised to discover that the annular water flow velocity, for a given flow rate, was independent of stack height. However, this result is easily explained as the annular water flow attains a terminal velocity, a condition where the wall frictional forces are balanced by the gravitational mass force.

Figure 16.19 illustrates the development of the annular film thickness and the attainment of terminal velocity under steady uniform flow conditions, where it is also assumed that the flow swirl velocity component is zero.

Application of Newton's second law yields

FIGURE 16.18

Branch discharge flow enters the vertical stack, effectively obstructing air flow passages

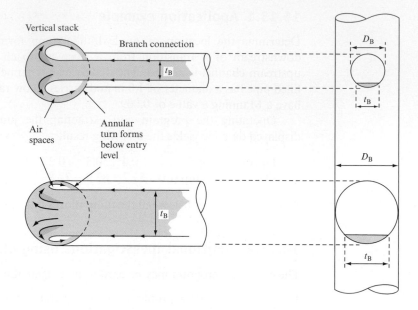

FIGURE 16.19

Annular water flow development and the establishment of terminal conditions

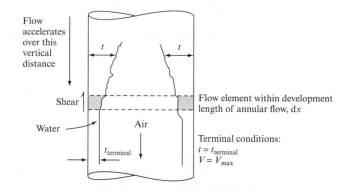

$$\rho g(\pi D t\, \mathrm{d}x) - \tau_0 \pi D\, \mathrm{d}x = \rho \pi D t\, \mathrm{d}x \frac{\mathrm{d}V_{\mathrm{w}}}{\mathrm{d}t'}$$

if the water to air shear stress is neglected, so

$$\rho g t - \tau_0 = \rho t \frac{\mathrm{d}V_{\mathrm{w}}}{\mathrm{d}t'}.$$

At terminal velocity the flow ceases to accelerate; hence

$$\tau_0 = \rho g t. \tag{16.24}$$

The shear stress may be expressed as

$$\tau_0 = \tfrac{1}{2}\rho f V_{\mathrm{w}}^2,$$

where f is the appropriate friction factor. Thus terminal velocity and annular thickness are linked as

$$t = (f/2g)V_w^2$$
$$V_w = \sqrt{(2gt/f)}.$$

Hence

$$1/\sqrt{f} = V_w/\sqrt{(2gt)} = Q_w/\pi Dt\sqrt{(2gt)}. \tag{16.25}$$

It has been shown that the Colebrook–White equation for the friction factor may be applied to free surface flows provided that the hydraulic mean depth is used as the characteristic length term in both the Reynolds number and relative roughness groups. Hence the appropriate form of the Colebrook–White equation becomes

$$1/\sqrt{f} = -4 \log_{10}[(k/14.84m) + (0.313\,75/Re\sqrt{f})],$$

where $Re = \rho(Vm/\mu)$.

For an annular flow where the thickness is small compared with the pipe circumference, i.e. $t \ll \pi D$, the hydraulic mean depth may be expressed as

$$m = \pi Dt/\pi D = t,$$

and $$Re = \rho\frac{V}{\mu}t = \frac{\rho}{\mu}\frac{Q_w}{\pi D}.$$

Substituting in the Colebrook–White equation above yields

$$\frac{Q_w}{\pi Dt\sqrt{(2gt)}} = -4\log_{10}\left\{\frac{k}{14.84t} + \frac{0.313\,75}{(\rho Q_w/\mu\pi D)[\pi Dt\sqrt{(2gt)}]/Q_w}\right\}$$

$$\frac{Q_w}{\pi Dt\sqrt{(2gt)}} = -4\log_{10}\left(\frac{k}{14.84t} + \frac{\mu\times 0.313\,75}{\rho t\sqrt{2gt}}\right) \tag{16.26}$$

For annular water flow in stacks of unknown roughness, it follows that this equation may be used to calculate the terminal water film thickness and hence terminal velocity as

$$V_t = Q_w/\pi Dt.$$

Table 16.2 illustrates representative results for a smooth stack, i.e. $k = 0$, and identifies the maximum stack flow in each case based on annular flow area less than one-quarter of the stack cross-section, or $t \le \frac{1}{16}D$, which is a current design limit for this type of flow in building drainage applications. It will be seen that the stack capacities are highly dependent on the stack diameter.

In the analysis of annular downflows of this type it would be useful to deduce from the preceding analysis an expression linking terminal velocity directly to stack diameter and flow rate. Application of Newton's second law to the falling water film identified

$$\frac{dV_w}{dt'} = g - \frac{\tau_0}{\rho t},$$

TABLE 16.2
Terminal velocities, V_t, water film thicknesses, t, and terminal distances, L_t, calculated for a smooth stack

WATER DOWNFLOW (litres s^{-1})	STACK DIAMETER (mm)								
	75			100			150		
	t (mm)	V_t (m s^{-1})	L_t (m)	t (mm)	V_t (m s^{-1})	L_t (m)	t (mm)	V_t (m s^{-1})	L_t (m)
1.0	1.84	2.3	0.84	1.56	2.04	0.66	1.23	1.73	0.48
2.0	2.76	3.08	1.51	2.33	2.73	1.19	1.84	2.31	0.84
3.0	3.51	3.62	2.08	2.96	3.23	1.66	2.33	2.73	1.19
4.0	4.16	4.08	2.65	3.51	3.63	2.10	2.76	3.08	1.51
4.80	4.68	4.39	3.06	3.91	3.91	2.43	3.07	3.32	1.75
10.60	–	–	–	6.25	5.39	4.62	4.92	4.57	3.32
31.20	–	–	–	–	–	–	9.38	7.05	7.90

where

$$\tau_0 = f \tfrac{1}{2} \rho V_w^2, \quad t = \frac{Q_w}{\pi D V_w} \quad \text{and} \quad t' = \text{Time}.$$

Hence

$$\frac{dV_w}{dt'} = g - f \tfrac{1}{2} \rho \frac{\pi D V_w^3}{\rho Q_w}$$

$$\frac{dV_w}{dt'} = g - \tfrac{1}{2} f \frac{\pi D}{Q_w} V_w^3.$$

Thus the terminal velocity, i.e. V_w at $t = t_{\text{terminal}}$, becomes

$$V_t = \left(\frac{2g}{f} \frac{Q_w}{\pi D} \right)^{1/3}$$

as $dV_w/dt' = 0$ once terminal conditions are reached.

The $(2g/f)$ term will be recognized as related to the Chezy coefficient C:

$$C = \sqrt{(2g/f)} = m^{1/6}/n,$$

where m is the hydraulic mean depth in a free surface channel flow where the channel roughness is expressed by the Manning coefficient, n.

As the annular flow thickness is limited to one-sixteenth of the stack diameter, it is an appropriate approximation to consider the annular flow as if it were in a rectangular channel where the ratio of flow width to depth was 16. The appropriate hydraulic mean depth m has already been shown to be equivalent to the flow thickness, t; thus as $t = Q_w/\pi D V_w$ it follows that

$$V_t = \left[\frac{1}{n^2} \left(\frac{Q_w}{\pi D V_t} \right)^{1/3} \frac{Q_w}{\pi D} \right]^{1/3}, \tag{16.27}$$

i.e. substituting for C^2. Thus

$$V_t V_t^{1/9} = \left(\frac{1}{n^2} \frac{1}{\pi^{4/3}} \frac{Q_w^{4/3}}{D^{4/3}} \right)^{1/3}$$

$$V_t^{10/9} = \left(\frac{1}{n^2} \frac{1}{\pi^{4/3}} \right)^{1/3} \left(\frac{Q_w}{D} \right)^{4/9}$$

$$V_t = K \left(\frac{Q_w}{D} \right)^{0.4}, \tag{16.28}$$

where $K = 0.2173^{0.3}/n^{0.6} = 0.632/n^{0.6}$, $\tag{16.29}$

yielding a value of K for a typical smooth UPVC pipe of 12.4. This value compares closely with empirical results quoted by researchers in this area, thus confirming the use of the Chezy approximation for the annular frictional resistance and the treatment of this annular flow as an example of steady, non-uniform, free surface flow.

The distance necessary to achieve terminal velocity may also be determined from the application of Newton's second law. Rearranging as follows:

$$\frac{dV}{dt'} = \frac{dV}{dz} \frac{dz}{dt'} = V \frac{dV}{dz}$$

$$\frac{dV}{dz} = \frac{1}{V} \frac{dV}{dt'}$$

$$= \frac{1}{V} g - \tfrac{1}{2} f \frac{\pi D}{Q_w} V^3,$$

where t' is again time;

$$dz = \frac{V \, dV}{g - \tfrac{1}{2} f \pi D (V^3/Q_w)}$$

$$= \frac{1}{g} \frac{V \, dV}{[1 - (f/2g)(\pi D/Q_w) V^3]}.$$

Substituting for terminal velocity yields

$$dz = \frac{1}{g} V_t^2 \frac{(V/V_t) \, d(V/V_t)}{1 - (V/V_t)^3}$$

$$dz = \frac{V_t^2}{g} \frac{\theta \, d\theta}{(1 - \theta^3)}, \tag{16.30}$$

where $\theta = V/V_t$, an expression that may be integrated to yield the distance to terminal velocity and thickness, z_t:

$$\int_{z=0}^{z=z_t} dz = \frac{V_t^2}{g} \int_{\theta=0}^{\theta=1} \frac{\theta \, d\theta}{1 - \theta^3}. \tag{16.31}$$

However, the result will be infinite as the flow will approach the limiting condition asymptotically. Normal procedure is to calculate the distance to 99 per cent of terminal conditions, i.e. $\theta = 0.99$ as an upper integration limit.

The vertical fall required to attain terminal velocity is thus

$$Z_t = 0.159\, V_t^2 \tag{16.32}$$

in units of metres and metres per second, respectively.

From Table 16.2 it will be seen that the terminal lengths required rarely exceed one or two storey heights, a result that supports the use of terminal velocity over the greater length of a vertical stack in a multi-storey building.

The equations determining terminal flow velocity and thickness, and the distance of fall required to attain these conditions, confirm the expected relationships, e.g. the dominance of stack diameter in these results. Clearly the terminal velocity is lower in rougher pipes, the distance required also decreasing. The terminal thickness will increase as the final velocity achieved decreases.

The falling annular water film entrains an air core down the stack. The mechanism that develops this entrained air flow down the vertical stack is the condition of 'no slip' that exists between the annular water film and the air core. To sustain this air movement it follows that outside air must be drawn into the stack and that this incurs a pressure drop due to friction in its passage through the dry upper levels of the stack and past active branch connections. (The passage of the air through water films, generated by side branch water flows into the stack, closing, or partially closing, the central air path results in local pressure reductions analogous to separation losses in full-bore flow.)

The frictional loss in the dry stack may be determined from an application of Darcy's equation for the entrained air flow and the stack parameters of pipe diameter and roughness. Separation losses at entry and due to passing through the discharging branch flow may also be determined so long as the appropriate loss coefficients are known. A 'back pressure' is also experienced at the base of the vertical stack as the entrained air flow is forced through the water curtain formed as the annular stack flow transforms itself into partially filled horizontal pipe flow downstream. Thus these pressure losses, which are returned to later in Chapter 21, may be combined as

$$\Delta p_{\text{total}} = \Delta p_{\text{entry}} + \Delta p_{\text{dry pipe friction}} + \Delta p_{\text{branch junction}} + \Delta p_{\text{back pressure}} \tag{16.33}$$

The 'motive force' to entrain this air flow and compensate for these 'pressure losses' is derived from the shear force between the annular terminal velocity water layer and the air in the wet portion of the stack. Hence a 'negative' friction factor may be postulated that generates an equal pressure rise to that determined from equation (16.33) – the equivalent to a fan drawing air through the stack. Ongoing research has identified the format and relationships governing this shear force representation, and allows the prediction of the transient response of the stack network to variations in applied water downflows.

It is normal practice to refer to the annular water film discussed above attaining a terminal velocity; however, it must be appreciated that this is calculated as a mean velocity within the water film. Figure 16.20 illustrates the likely velocity profiles in both the water film and the air core. The interface velocities are equal, the water velocity reduces to zero at the pipe wall, i.e. a no-slip condition, while the air core velocity

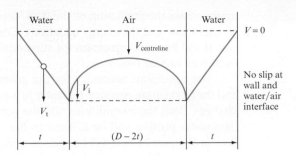

FIGURE 16.20
Velocity profiles
assumed within the
annular water film and
entrained air core

ideally reduces to some minimum at the stack centreline. It must be remembered that such profiles are unlikely to be so simplistic as the air flow will have passed through a number of active branch connections which will have disrupted the flow, and the flow itself will in any case not be steady. If a linear velocity profile is assumed through the water film then it follows that the interface velocity, V_i, is given by

$$V_i = 2V_t. \tag{16.34}$$

However, there are no published data to support this assumption.

The air flow rate, Q_a, may be calculated as

$$Q_a = \frac{\pi}{4}(D - 2t)^2 V_a, \tag{16.35}$$

where V_a is the mean air velocity, which will be less than the common interface velocity V_i. The ratio of air to water flow may be as high as 100:1, with air flow rates in the stack up to $0.2\,\mathrm{m^3\,s^{-1}}$. Ample evidence exists to support the annular mechanism, including experimental work that established that only a small fraction of the water flow falls as water droplets in the air core, and measurements of both the annular water film velocity and thickness using a laser anemometer. The air flow is thus primarily caused by the shear between the water annulus and the initially stationary air core.

The relationship between the annular water downflow rate, the associated pressure reduction in the stack and the entrained air flow rate is of interest to designers. A dimensional analysis and similarity treatment of this condition was included in Chapter 9 and utilized the terminal velocity dependence on Manning's n as a basis for the development of the defining groups. While the annular thickness remains small compared with the pipe diameter and perimeter, this result remains valid. However, once the annular film thickness increases there is a probability of instability in the flow, as the film will close across the pipe and plug flow will result, generating large swings in air core pressure.

Concluding remarks

Changes of channel slope, roughness or cross-section, or the presence of outfalls, obstructions or junctions, inevitably dictate that a free surface flow, whether in an open channel or a partially filled pipe, will display the variable flow/depth profiles described in this chapter. It will be seen that the assumption of a gradually varied condition, i.e. one where the locally steady flow condition results may be assumed to apply, allows the prediction of the flow/depth profile by application of the continuity of mass flow and the energy equations. Control conditions imposed downstream for subcritical flow and upstream for supercritical flow then allow the necessary integration to be carried out.

The conditions necessary for the generation of hydraulic jumps were discussed and the momentum equation (Chapter 5) was utilized to determine the consequent depth changes. Both these conditions and the general development of the gradually varied backwater profiles will be utilized in the later treatment of unsteady free surface flow in Chapter 21 as it will be necessary to generate base conditions for a steady flow throughout a free surface network prior to the initiation of transient conditions.

The definition of free surface flows was extended to include annular water downflow with an entrained air core. Under quasi-steady conditions the annular film terminal velocity was developed and the mechanism of air entrainment discussed. This section drew on the definitions of 'no slip' set out earlier in the text, together with the momentum equation applied to the terminal annular flow equilibrium condition. This presentation will be referred to in the treatment of unsteady annular downflow in Chapter 20.

Summary of important equations and concepts

1. The identification of a free surface flow regime depends upon the flow Froude number, Fr < 1 subcritical, Fr > 1 supercritical. Transition from supercritical to subcritical in a channel requires the formation of an hydraulic jump.

2. The flow-critical depth is independent of channel slope and corresponds to a minimum flow specific energy.

3. Flow normal depth is dependent upon both channel cross-sectional considerations and slope and roughness.

4. Numerical solution of the water surface equation (16.17) allows the transition from one flow type to another to be determined and identifies the location and severity of any hydraulic jumps formed in the flow.

5. The basic concepts of shear force and gravity balance, together with the frictional resistance expressions, may be extended from the expected channel applications of free surface flow to include annular flows in vertical pipes with a consequent necessity to consider air entrainment.

Further reading

Henderson, F. M. (1996). *Open Channel Flow.* Macmillan, New York.

Stephenson, D. (1981). *Stormwater Hydrology and Drainage.* Elsevier, Amsterdam.

Problems

16.1 A channel has a trapezoidal cross-section with a base width of 0.6 m and sides sloping at 45°. When the flow along the channel is 20 m³ min⁻¹ determine the critical depth.

[0.27 m]

16.2 Contrast and relate the Chezy and Manning formulae for the mean velocity of flow in open channels. A rectangular channel is 6 m wide and will carry a discharge

of 22.5 m³ s⁻¹ of water. Determine the necessary slopes to achieve uniform flow at (*a*) a depth of 3 m, (*b*) a depth of 0.6 m, (*c*) the critical depth. Assume that for this channel $n = 0.012$ in the Manning equation.

[(*a*) 1/7631, (*b*) 1/70, (*c*) 1/482]

16.3 Water flows across a broad-crested weir in a rectangular channel 400 mm wide. The depth of the water just upstream of the weir is 70 mm and the crest of weir is 40 mm above the channel bed. Calculate the fall of the surface level and the corresponding discharge assuming that the velocity of approach is negligible. [10 mm, 3.54×10^{-3} m^3 s^{-1}]

16.4   A venturi flume with a level bed is 12 m wide and 1.5 m deep upstream with a throat width of 6 m. Assuming that a standing wave forms downstream calculate the rate of flow of water if the discharge coefficient is 0.94. Correct for the velocity of approach. [18.54 m^3 s^{-1}]

16.5   Show that the equation

$$Q = a_1 \left[\frac{2g(h_1 - h_2)}{r^2 - 1} \right]^{1/2},$$

where $r = a_1/a_2$, commonly derived for the frictionless flow of water with a rate of discharge Q through a venturi meter, is also applicable to the frictionless flow of water over zero bed slope (a) through a venturi flume, (b) over a broad-crested weir, (c) under a sluice gate, (d) over a rounded crest of a spillway to the horizontal bed at the toe of a dam. Define a_1 and a_2 in each of the above cases and state for which of these cases and for what special conditions the equation $Q = 1.704BE^{3/2}$ is valid. Derive this equation for the relevant conditions, carefully specifying E.

16.6   A venturi flume is placed in an open channel 2 m wide in which the throat width is 1.2 m. The upstream depth is 1 m and the floor is effectively horizontal. Calculate the flow when (a) the depth at the throat is 0.9 m and (b) a standing wave is produced beyond the throat.
[(a) 1.8 m^3 s^{-1}, (b) 2.05 m^3 s^{-1}]

16.7   A venturi flume in a rectangular channel of width B has a throat width b. The depth of liquid at entry is H and at the throat is h. Derive an expression for the theoretical volume flow rate of the liquid in terms of H, b and the ratio h/H. Develop also a relationship between the ratios h/H and b/B. State what assumptions you make regarding the downstream flow.

$$\left[Q = 3.13bH^{3/2} \left(\frac{h}{H} \right)^{3/2} ; \frac{b}{B} = \sqrt{3} \left(\frac{H}{h} \right) - \sqrt{2} \left(\frac{H}{h} \right)^{3/2} \right]$$

16.8   A rectangular prismatic channel 1.2 m wide has a uniform slope of 1 in 1600 and a normal depth of 0.6 m when the flow rate is 0.72 m^3 s^{-1}. When a sluice is lowered the upstream depth is increased to 1 m. Determine the distance upstream from the sluice where the depth of water

is 0.8 m. Using a step-by-step method to solve the problem, divide the range of depth into two equal parts. [472 m]

16.9   A rectangular channel of slope 0.001 carries 40 m^3 s^{-1} of water and is 5 m wide. If an overflow weir is installed across the channel which raises the water level at the weir to a depth of 6 m, (a) compute the normal depth of flow, (b) compute in two steps the distance upstream to the point where the depth of water is 5.8 m and (c) classify the surface profile. Take the value of n in the Manning formula as 0.02.
[(a) 3.84 m, (b) 297 m, (c) M1]

16.10   A sluice across a channel 6 m wide discharges a stream 1.2 m deep. What will be the flow if the upstream depth is 6 m?
The conditions downstream cause an hydraulic jump to occur at a place where concrete blocks have been placed in the bed. What will be the force on the blocks if the downstream depth is 3.06 m? [71.3 m^3 s^{-1}, 196 kN]

16.11   The stream issuing from beneath a vertical sluice gate is 0.3 m deep at the vena contracta. Its mean velocity is 6 m s^{-1}. A standing wave is created on the level bed below the sluice gate. Find the height of the jump, the loss of head and the power dissipated per unit width of sluice.
[1.04 m, 0.7 m, 12.36 kW]

16.12   Water issuing from a sluice enters a horizontal rectangular channel with uniform velocity v and depth y. Show that if y is less than the critical depth an hydraulic jump will be formed.
If the velocity when the water enters the channel is 4 m s^{-1} and the Froude number is 1.4, obtain (a) the depth of flow after the jump, (b) the loss of specific energy due to the formation of the jump. [(a) 1.28 m, (b) 0.02 m]

16.13   In a rectangular channel 0.6 m wide a jump occurs where the Froude number is 3. The depth after the jump is 0.6 m. Estimate the total loss of head and the power dissipated at the jump. [0.225 m, 0.79 kW]

16.14   A wide channel with uniform rectangular section has a change of slope from 1 in 95 to 1 in 1420 and the flow is 3.75 m^3 s^{-1} per metre width. Determine the normal depth of flow corresponding to each slope and show that an hydraulic jump will occur in the region of the junction. Calculate the height of the jump and sketch the surface profiles between the upstream and downstream regions of uniform flow. Manning's coefficient $n = 0.013$ and it may be assumed that the channel is wide in comparison with the depth of flow, so that the hydraulic mean depth is approximately equal to the depth of flow. [0.639, 1.44, 0.58 m]

Compressible Flow
in Pipes

THE ANALYSIS OF COMPRESSIBLE FLOW IN DUCTS WILL BE BASED UPON MATERIAL presented earlier that detailed the general energy relationships, via Euler's equation (Chapter 5) and the equation of state for gases (Chapter 1). The variation of density segregates this work from that covered in Chapters 10 and 12, where it was ignored. The influence of changes in duct cross-section on supersonic flows is introduced, leading to a treatment of nozzle flows and the prediction of maximum discharge rates under supersonic flow conditions. The use of venturi meters to determine mass flow is discussed. The momentum equation is again used to determine conditions across shock waves. Duct flow under adiabatic and isothermal conditions is introduced, together with fluid frictional effects. ● ● ●

17.1   COMPRESSIBLE FLOW. THE BASIC EQUATIONS

When considering flow in ducts and pipes, in Chapters 10 and 14, it was assumed that the fluid could be treated as if it were incompressible and, therefore, of constant density. For a wide range of fluids employed in engineering this assumption is valid because the pressure changes which occur are normally too small to cause an appreciable change in density. For gases, however, this assumption cannot be made, since large variations of density can be produced as a result of the changes of pressure which occur in normal engineering applications: this compressibility must be taken into account except where such pressure changes are very small. Thus, in considering the continuous flow of a compressible fluid, the relationship between density and the other factors affecting fluid flow must be considered. This will be the equation of state relating the absolute pressure p, absolute temperature T and the mass density ρ which was given in Chapter 1, equation (1.13), for a perfect gas as $p = \rho RT$, where R is the gas constant for the particular gas concerned; hence, we have

$$\rho = p/RT. \tag{17.1}$$

The continuity of flow equation, which arises from the principle of conservation of mass as discussed in Section 4.12, must also be used, in the form

$$\dot{m} = \rho A \bar{v} = \text{constant}, \tag{17.2}$$

where $\dot{m}$ is the mass flow rate through a cross-section A at which the velocity and the density of the fluid are $\bar{v}$ and ρ, respectively.

The steady flow energy equation was discussed in Section 6.2. For compressible flow in a horizontal plane, equation (6.10) becomes

$$\tfrac{1}{2}v_1^2 + H_1 + q - w = \tfrac{1}{2}v_2^2 + H_2, \tag{17.3}$$

where H is the enthalpy, q the heat added per unit mass and w the work done per unit mass. If q and w are zero, equation (17.3) reduces to

$$\tfrac{1}{2}v^2 + H = \text{constant} = H_0, \tag{17.4}$$

where $H_0 = $ total or stagnation enthalpy $= c_p T_0$, where T_0 is the stagnation temperature.

For frictionless flow, the Euler equation, as derived in Section 5.12, is also applicable. Equation (5.21), which states that along a streamline

$$\frac{\mathrm{d}p}{\rho} + v\,\mathrm{d}v + g\,\mathrm{d}z = 0, \tag{17.5}$$

can be integrated when the relationship between p and ρ is known. If friction and other forces act, the momentum equation can be applied in its basic form, as given in equation (5.5).

17.2   STEADY ISENTROPIC FLOW IN NON-PARALLEL-SIDED DUCTS NEGLECTING FRICTION

Although isentropic flow, which is frictionless flow under adiabatic conditions, is an ideal which cannot be fully realized in practice, the assumption of isentropic conditions gives a satisfactory approximation for the analysis of flow through short transitions, orifices, venturi meters and nozzles in which friction and heat transfer are minor effects which can be neglected.

Since, for an incompressible fluid, ρ was constant, it was possible to write equation (17.2) as $A\bar{v} = $ constant, indicating that for steady flow the velocity must increase if the area of the stream decreases. For compressible flow, this need not be the case since ρ is also variable. Considering a horizontal stream, from equation (17.5),

$$\frac{\mathrm{d}p}{\rho} + \bar{v}\,\mathrm{d}\bar{v} = 0, \tag{17.6}$$

but, from equation (5.31), $\mathrm{d}p/\mathrm{d}\rho = c^2$, where c is the velocity of sound, so that

$$\frac{\mathrm{d}p}{\rho} = c^2 \frac{\mathrm{d}\rho}{\rho}.$$

Substituting in equation (17.6),

$$c^2 \frac{\mathrm{d}\rho}{\rho} + \bar{v}\,\mathrm{d}\bar{v} = 0. \tag{17.7}$$

Differentiating equation (17.2) and dividing by $\rho A \bar{v}$,

$$\frac{\mathrm{d}\rho}{\rho} + \frac{\mathrm{d}\bar{v}}{\bar{v}} + \frac{\mathrm{d}A}{A} = 0. \tag{17.8}$$

Eliminating $\mathrm{d}\rho/\rho$ between equations (17.7) and (17.8),

$$\frac{\bar{v}\,\mathrm{d}\bar{v}}{c^2} - \frac{\mathrm{d}\bar{v}}{\bar{v}} - \frac{\mathrm{d}A}{A} = 0.$$

Dividing through by $\mathrm{d}\bar{v}/A$,

$$\frac{\mathrm{d}A}{\mathrm{d}\bar{v}} = \frac{A}{\bar{v}}\left(\frac{\bar{v}^2}{c^2} - 1\right)$$

or, since v/c is the Mach number Ma,

$$\frac{\mathrm{d}A}{\mathrm{d}\bar{v}} = \frac{A}{\bar{v}}(\mathrm{Ma}^2 - 1). \tag{17.9}$$

From equation (17.9), it can be seen that for steady frictionless flow with no restriction on heat transfer:

1. If $Ma < 1$ (subsonic flow), $dA/d\bar{v}$ is always negative, indicating that the velocity must increase as the cross-sectional area of the duct decreases.

2. If $Ma > 1$ (supersonic flow), $dA/d\bar{v}$ is always positive, indicating that for the velocity to increase the area of the duct must also increase.

3. If $Ma = 1$ (sonic velocity), $dA/d\bar{v}$ is zero. Since

$$\frac{dA}{d\bar{v}} = \frac{dA}{dx} \Big/ \frac{d\bar{v}}{dx},$$

and $d\bar{v}/dx$ cannot be infinite, the value of dA/dx must be zero, indicating that the cross-sectional area must be a minimum, since the second derivative is positive, when the velocity reaches the velocity of sound, as in Fig. 17.1.

FIGURE 17.1
Convergent–divergent
nozzle

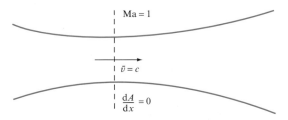

The effect of a convergent–divergent nozzle on the flow of a compressible fluid will therefore depend upon the Mach number. On the upstream side and initially, supersonic flow will be decelerated towards $Ma = 1$ by the convergent section, while subsonic flow will be accelerated towards $Ma = 1$. Once the throat (or minimum area of cross-section) is passed, the flow will accelerate for the supersonic case. For the subsonic case, if $Ma = 1$ is not attained in the throat, the flow will decelerate in the divergent section. It is only at a throat or minimum area of cross-section that the velocity can be sonic and the Mach number unity. To obtain supersonic steady flow of a compressible fluid flowing initially at subsonic velocity or contained at rest in a reservoir, the fluid must pass through a convergent–divergent nozzle.

17.3   MASS FLOW THROUGH A VENTURI METER

When a gas flows through a venturi meter (Fig. 17.2), the mass flow rate can be determined using the method explained in Section 6.10, but the form of Bernoulli's equation will be that obtained by integrating equation (17.5), which is

FIGURE 17.2

Flow through a
venturi meter

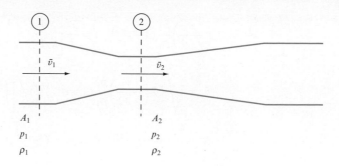

$$\int \frac{\mathrm{d}p}{\rho} + \frac{\bar{v}^2}{2} + gz = \text{constant}. \tag{17.10}$$

For the short distance between the full-bore and throat sections, conditions can be considered as adiabatic and the relationship between pressure and density will be $p/\rho^\gamma = \text{constant} = k$. Putting $\rho = (p/k)^{1/\gamma}$ in equation (17.10),

$$k^{1/\gamma} \int p^{-1/\gamma}\,\mathrm{d}p + \tfrac{1}{2}\bar{v}^2 + gz = \text{constant}.$$

Integrating and putting $k = p/\rho^\gamma$,

$$\left(\frac{\gamma}{\gamma-1}\right)\frac{p}{\rho} + \tfrac{1}{2}\bar{v}^2 + gz = \text{constant},$$

or, for two points on a horizontal streamline corresponding to sections 1 and 2 in Fig. 17.2,

$$\left(\frac{\gamma}{\gamma-1}\right)\left(\frac{p_1}{\rho_1} - \frac{p_2}{\rho_2}\right) + \tfrac{1}{2}(\bar{v}_1^2 - \bar{v}_2^2) = 0. \tag{17.11}$$

Also, since for adiabatic flow

$$p_1/\rho_1^\gamma = p_2/\rho_2^\gamma,$$

$$\rho_2 = \left(\frac{p_2}{p_1}\right)^{1/\gamma}\rho_1 \quad \text{and} \quad \frac{p_2}{\rho_2} = \frac{p_1}{\rho_1}\left(\frac{p_2}{p_1}\right)^{(\gamma-1)/\gamma}$$

or, putting $p_2/p_1 = r$,

$$\frac{p_2}{\rho_2} = \frac{p_1 r^{(\gamma-1)/\gamma}}{\rho_1}. \tag{17.12}$$

For continuity of flow by mass, $\rho_1 A_1 \bar{v}_1 = \rho_2 A_2 \bar{v}_2$,

$$\bar{v}_2 = \frac{A_1}{A_2}\left(\frac{\rho_1}{\rho_2}\right)\bar{v}_1 = \frac{A_1}{A_2}\left(\frac{1}{r}\right)^{1/\gamma}\bar{v}_1. \tag{17.13}$$

Substituting from equations (17.12) and (17.13) in equation (17.11),

$$\left(\frac{\gamma}{\gamma-1}\right)\frac{p_1}{\rho_1}(1-r^{(\gamma-1)/\gamma}) = \frac{\bar{v}_1^2}{2}\left[\left(\frac{A_1}{A_2}\right)^2\left(\frac{1}{r}\right)^{2/\gamma}-1\right],$$

$$\bar{v}_1 = \sqrt{\left\{2\left(\frac{\gamma}{\gamma-1}\right)\frac{p_1}{\rho_1}(1-r^{(\gamma-1)/\gamma})\Big/\left[\left(\frac{A_1}{A_2}\right)^2\left(\frac{1}{r}\right)^{2/\gamma}-1\right]\right\}}.$$

Mass flow per unit time, $\dot{m} = C_d A_1 \bar{v}_1 \rho_1$,

where C_d is a coefficient of discharge; therefore,

$$\dot{m} = C_d A_1 \rho_1 \sqrt{\left\{2\left(\frac{\gamma}{\gamma-1}\right)\frac{p_1}{\rho_1}(1-r^{(\gamma-1)/\gamma})\Big/\left[\left(\frac{A_1}{A_2}\right)^2\left(\frac{1}{r}\right)^{2/\gamma}-1\right]\right\}}.$$

EXAMPLE 17.1

A venturi meter having an inlet diameter of 75 mm and a throat diameter of 25 mm is used for measuring the rate of flow of air through a pipe. Mercury U-tube gauges register pressures at the inlet and throat equivalent to 250 mm and 150 mm of mercury, respectively. Determine the volume of air flowing through the pipe per unit time in cubic metres per second. Assume adiabatic conditions ($\gamma = 1.4$). The density of air at the inlet is 1.6 kg m^{-3} and the barometric pressure is 760 mm of mercury.

Solution

$$p_1 = \frac{760+250}{1000}\times13.6\times10^3\times9.81 = 134\,750 \text{ N m}^{-2},$$

$$p_2 = \frac{760+150}{1000}\times13.6\times10^3\times9.81 = 121\,408 \text{ N m}^{-2},$$

$$\rho_1 = 1.6\,\text{kg m}^{-3}, \quad \rho_2 = \rho_1\left(\frac{p_2}{p_1}\right)^{1/\gamma} = 1.6\left(\frac{121\,408}{134\,750}\right)^{1/1.4} = 1.485\,\text{kg m}^{-3}.$$

For continuous flow, $A_1\bar{v}_1\rho_1 = A_2\bar{v}_2\rho_2$. Therefore,

$$\bar{v}_2 = \frac{A_1}{A_2}\frac{\rho_1}{\rho_2}\bar{v}_1 = \frac{d_1^2}{d_2^2}\frac{\rho_1}{\rho_2}\bar{v}_1 = \left(\frac{75}{25}\right)^2\left(\frac{1.6}{1.485}\right)\bar{v}_1 = 9.697\bar{v}_1.$$

Applying Bernoulli's equation for adiabatic conditions,

$$\left(\frac{\gamma}{\gamma-1}\right)\left(\frac{p_1}{\rho_1}-\frac{p_2}{\rho_2}\right) = \frac{\bar{v}_2^2-\bar{v}_1^2}{2},$$

$$\frac{1.4}{0.4}\left(\frac{134\,750}{1.6}-\frac{121\,408}{1.485}\right) = \frac{\bar{v}_1^2}{2}(9.697^2-1),$$

$$\bar{v}_1 = 13.6 \text{ m s}^{-1}.$$

Volume of flow $= A_1\bar{v}_1 = (\pi/4)(0.075)^2 \times 13.6 = \mathbf{0.060 \text{ m}^3 \text{ s}^{-1}}.$

17.4 MASS FLOW FROM A RESERVOIR THROUGH AN ORIFICE OR CONVERGENT–DIVERGENT NOZZLE

Conditions for flow through an orifice or a nozzle, as shown in Fig. 17.3(a) and (b), can be taken as adiabatic and Bernoulli's equation in the form of equation (17.11) will apply. But, if the reservoir is large, $\bar{v}_1 = 0$ and $\bar{v}_2 = \bar{v}$ so that equation (17.11) reduces to

FIGURE 17.3

Mass flow from a large reservoir

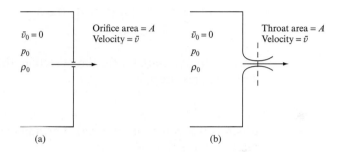

$$\left(\frac{\gamma}{\gamma-1}\right)\left(\frac{p_0}{\rho_0} - \frac{p}{\rho}\right) = \frac{\bar{v}^2}{2}. \tag{17.14}$$

Since, for adiabatic conditions, $p_0/\rho_0^\gamma = p/\rho^\gamma$,

$$\rho = \rho_0 r^{1/\gamma},$$

where $r = (p/p_0)$ and $p/\rho = p_0 r^{(\gamma-1)/\gamma}/\rho_0$. Substituting in equation (17.14),

$$\left(\frac{\gamma}{\gamma-1}\right)\frac{p_0}{\rho_0}(1 - r^{(\gamma-1)/\gamma}) = \frac{\bar{v}^2}{2}$$

$$\bar{v} = \sqrt{\left[2\left(\frac{\gamma}{\gamma-1}\right)\frac{p_0}{\rho_0}(1 - r^{(\gamma-1)/\gamma})\right]}. \tag{17.15}$$

Mass flow per unit time, $\dot{m} = A\bar{v}\rho = A\rho_0 r^{1/\gamma}\bar{v}$

$$\dot{m} = A\rho_0\sqrt{\left[2\left(\frac{\gamma}{\gamma-1}\right)\frac{p_0}{\rho_0}r^{2/\gamma}(1 - r^{(\gamma-1)/\gamma})\right]}. \tag{17.16}$$

In practice, the actual discharge will be $C_d\dot{m}$, where C_d is a coefficient of discharge.

17.5 CONDITIONS FOR MAXIMUM DISCHARGE FROM A RESERVOIR THROUGH A CONVERGENT–DIVERGENT DUCT OR ORIFICE

For the throat section, where $p = p_t$ and $\rho = \rho_t$, it can be seen from equation (17.16) that for maximum discharge under given initial conditions p_0 and ρ_0, the quantity $r^{2/\gamma}(1 - r^{(\gamma-1)/\gamma})$ must be a maximum. This will occur for the value of r which makes

$$\frac{d}{dr}[r^{2/\gamma}(1 - r^{(\gamma-1)/\gamma})] = 0$$

$$(2/\gamma)r^{(2-\gamma)/\gamma} - [(\gamma+1)/\gamma]r^{1/\gamma} = 0$$

$$r^{(\gamma-1)/\gamma} = 2/(\gamma+1),$$

where $r = p_t/p_0$. Thus, for the throat section, the ratio of the throat pressure p_t to the upstream pressure p_0 is

$$p_t/p_0 = [2/(\gamma+1)]^{\gamma/(\gamma-1)},$$

which is 0.528 for air ($\gamma = 1.4$). For adiabatic conditions, $p/\rho^\gamma = $ constant and $\rho_t/\rho_0 = (p_t/p_0)^{1/\gamma}$. Thus,

$$\rho_t/\rho_0 = [2/(\gamma+1)]^{1/(\gamma-1)},$$

which is 0.634 for air ($\gamma = 1.4$). Also,

$$\frac{T_t}{T_0} = \frac{p_t}{p_0}\frac{\rho_0}{\rho_t} = \left(\frac{p_t}{p_0}\right)^{(\gamma-1)/\gamma},$$

so that $T_t/T_0 = 2/(\gamma+1)$, which is 0.833 for air ($\gamma = 1.4$).

The throat velocity for maximum discharge is obtained by putting $r = p_t/p_0 = [2/(\gamma+1)]^{\gamma/(\gamma-1)}$ in equation (17.15) and

$$\frac{p_0}{\rho_0} = \frac{p_t}{\rho_t}\left(\frac{1}{r}\right)^{(\gamma-1)/\gamma}.$$

Then, for maximum discharge,

$$\text{Throat velocity, } \bar{v}_t = \sqrt{\left\{2\left(\frac{\gamma}{\gamma-1}\right)\frac{p_t}{\rho_t}\left[\left(\frac{1}{r}\right)^{(\gamma-1)/\gamma} - 1\right]\right\}}$$

$$= \sqrt{\left[2\left(\frac{\gamma}{\gamma-1}\right)\frac{p_t}{\rho_t}\left(\frac{\gamma+1}{2} - 1\right)\right]}$$

$$\bar{v}_t = \sqrt{\left(\frac{\gamma p_t}{\rho_t}\right)} = c_t,$$

where c_t is the local velocity of sound in the throat or orifice.

17.6 THE LAVAL NOZZLE

Named after its Swedish inventor, de Laval (1845–1913), this nozzle is designed to produce supersonic flow. It takes the form of a convergent–divergent nozzle with subsonic flow in the converging section, critical or transonic conditions in the throat and supersonic flow in the diverging section.

If ρ, $\bar{v}$ and A are the density, velocity and cross-sectional area at any section of the nozzle and ρ_t, $\bar{v}_t$, A_t are the critical values at the throat, then, since the mass flow rate is the same at each cross-section,

$$\rho \bar{v} A = \rho_t \bar{v}_t A_t,$$
$$A/A_t = \rho_t \bar{v}_t / \rho \bar{v}. \tag{17.17}$$

The velocity at any point can be expressed in terms of the Mach number at that point and the local speed of sound,

$$\bar{v} = \mathrm{Ma}\, c = \mathrm{Ma}\sqrt{(\gamma RT)}$$

for adiabatic conditions.

At the throat, $\mathrm{Ma} = 1$ and $T = T_t$, so that $\bar{v}_t = \sqrt{(\gamma R T_t)}$. Substituting in equation (17.17),

$$\frac{A}{A_t} = \frac{\rho_t}{\rho}\left(\frac{T_t}{T}\right)\frac{1}{\mathrm{Ma}}. \tag{17.18}$$

Now, for isentropic flow from a large reservoir in which the conditions are given by p_0, ρ_0 and T_0 and $\bar{v}_0$ is zero, from Bernoulli's equation at any section of the nozzle,

$$\frac{\bar{v}^2}{2} = \left(\frac{\gamma}{\gamma-1}\right) R(T_0 - T).$$

Dividing by c^2, where $c = \sqrt{(\gamma RT)}$, the local velocity of sound, and rearranging,

$$\frac{\bar{v}^2}{c^2} = \mathrm{Ma}^2 = \frac{2}{\gamma-1}\left(\frac{T_0}{T}-1\right),$$
$$\frac{T_0}{T} = 1 + \left(\frac{\gamma-1}{2}\right)\mathrm{Ma}^2 \tag{17.19}$$

and since, for isentropic flow,

$$\frac{T_0}{T} = \left(\frac{p_0}{p}\right)^{(\gamma-1)/\gamma} = \left(\frac{\rho_0}{\rho}\right)^{(\gamma-1)},$$
$$\frac{p_0}{p} = \left(1 + \frac{\gamma-1}{2}\,\mathrm{Ma}^2\right)^{\gamma/(\gamma-1)} \tag{17.20}$$

and $\quad \dfrac{\rho_0}{\rho} = \left(1 + \dfrac{\gamma-1}{2}\,\mathrm{Ma}^2\right)^{1/(\gamma-1)}, \tag{17.21}$

we have,

$$\frac{\rho_t}{\rho} = \frac{\rho_t}{\rho_0} \times \frac{\rho_0}{\rho} = \left\{ \frac{1 + [(\gamma-1)/2]Ma^2}{(\gamma+1)/2} \right\}^{1/(\gamma-1)}$$

and

$$\frac{T_t}{T} = \frac{T_t}{T_0} \times \frac{T_0}{T} = \left\{ \frac{1 + [(\gamma-1)/2]Ma^2}{(\gamma+1)/2} \right\}.$$

Substituting these values in equation (17.18),

$$\frac{A}{A_t} = \frac{1}{Ma} \left\{ \frac{1 + [(\gamma-1)/2]Ma^2}{(\gamma+1)/2} \right\}^{(\gamma+1)/2(\gamma-1)}, \tag{17.22}$$

in which A is the area at the section at which the Mach number is Ma and A_t is the area at the throat. The value of A/A_t will never be less than unity and, for any given value of A/A_t, there will be two values of the Mach number, one less than unity and the other greater than unity.

The maximum mass flow rate $\dot{m}_{max}$ will be given by

$$\dot{m}_{max} = \rho_t A_t \bar{v}_t,$$

which can be expressed in terms of the reservoir conditions ρ_0, p_0 and T_0 since $\bar{v}_t = \sqrt{(\gamma R T_t)}$ and, from Section 17.5, $\rho_t/\rho_0 = [2/(\gamma+1)]^{1/(\gamma-1)}$ and $T_t/T_0 = 2/(\gamma+1)$, giving

$$\dot{m}_{max} = \rho_0 \left(\frac{2}{\gamma+1} \right)^{1/(\gamma-1)} A_t \sqrt{\left(\frac{2\gamma R T_0}{\gamma+1} \right)}.$$

Putting $\rho_0 = p_0/R T_0$,

$$\dot{m}_{max} = \frac{A_t p_0}{\sqrt{T_0}} \sqrt{\left[\frac{\gamma}{R} \left(\frac{2}{\gamma+1} \right)^{(\gamma+1)/(\gamma-1)} \right]}. \tag{17.23}$$

If $\gamma = 1.4$, this becomes

$$\dot{m}_{max} = 0.686 A_t p_0 / \sqrt{(R T_0)},$$

indicating that the mass flow varies linearly with A_t and p_0, but inversely as the square root of the absolute temperature.

EXAMPLE 17.2   A supersonic wind tunnel consists of a large reservoir containing gas under high pressure which is discharged through a convergent–divergent nozzle to a test section of constant cross-sectional area. The cross-sectional area of the throat of the nozzle is 500 mm² and the Mach number in the test section is 4. Calculate the cross-sectional area of the test section assuming $\gamma = 1.4$.

Solution

From equation (17.22), putting $\gamma = 1.4$ and Ma $= 4$,

$$\frac{A}{A_t} = \frac{1}{4}\left(\frac{1 + 0.2 \times 4^2}{1.2}\right)^{2.4/0.8} = 10.72,$$

Area of test section $= 10.72 \times 500 = \mathbf{5360 \ mm^2}$.

Equation (17.23) shows that the maximum mass flow is a function only of the reservoir conditions and the throat area, and cannot be affected by reducing the outlet pressure. Such a change could only be propagated upstream at the velocity of sound and, therefore, could not pass through the throat where the fluid velocity is sonic. Under these conditions, the nozzle is said to be choked.

When the mass flow rate is a maximum, the flow downstream of the throat can be either supersonic or subsonic depending on the downstream pressure. From equations (17.16) and (17.23) at any section of area A,

$$\dot{m} = A\rho_0 \sqrt{\left\{2\frac{\gamma}{(\gamma - 1)}\frac{p_0}{\rho_0}\left(\frac{p}{p_0}\right)^{2/\gamma}\left[1 - \left(\frac{p}{p_0}\right)^{(\gamma-1)/\gamma}\right]\right\}}$$

$$= \frac{A_t p_0}{\sqrt{T_0}}\sqrt{\left[\frac{\gamma}{R}\left(\frac{2}{\gamma + 1}\right)^{(\gamma+1)/(\gamma-1)}\right]}.$$

Eliminating $\dot{m}$,

$$\left(\frac{p}{p_0}\right)^{2/\gamma}\left[1 - \left(\frac{p}{p_0}\right)^{(\gamma-1)/\gamma}\right] = \frac{\gamma - 1}{2}\left(\frac{2}{\gamma + 1}\right)^{(\gamma+1)/(\gamma-1)}\left(\frac{A_t}{A}\right)^2.$$

Thus, for a given value of A_t/A, which must be less than unity, in the diverging duct there will be two possible values of p/p_0 between zero and unity, the upper value corresponding to subsonic flow and the lower to supersonic flow. For all other values of p/p_0 less than the upper value, isentropic flow is impossible and shock waves occur.

Flow through a nozzle can be classified by reference to the exit conditions. Figure 17.4 shows the variations of pressure and Mach number through a Laval nozzle. In the converging section, the pressure falls from the stagnation pressure p_0 at the entry, where the Mach number is small, to the value corresponding to the critical pressure ratio $p_t/p_0 = [2/(\gamma + 1)]^{\gamma/(\gamma-1)}$ at the throat, where Ma $= 1$. The pressure then continues to decrease until it reaches the exit and the Mach number increases correspondingly as shown. The exit pressure, p_e, of the fluid issuing from the nozzle will not necessarily be the same as the back pressure of the fluid outside into which the nozzle is discharging. If the exit pressure is higher than the back pressure, p_b, the nozzle is *under-expanded*, since the flow could have expanded further, and, therefore, expansion waves form at the nozzle exit (Fig. 17.5(a)).

FIGURE 17.4
Variation of static
pressure and Mach
number in a Laval nozzle

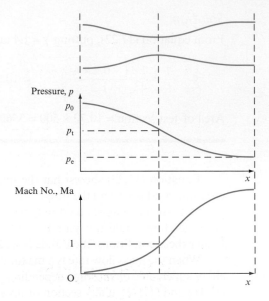

FIGURE 17.5
Flow through a nozzle

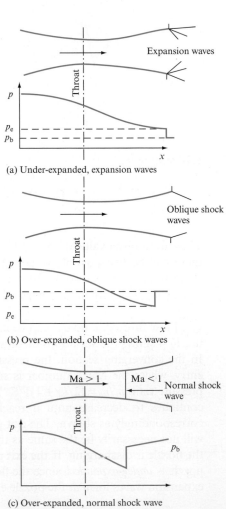

(a) Under-expanded, expansion waves

(b) Over-expanded, oblique shock waves

(c) Over-expanded, normal shock wave

FIGURE 17.6

Variation of pressure ratio
and Mach number
through a nozzle

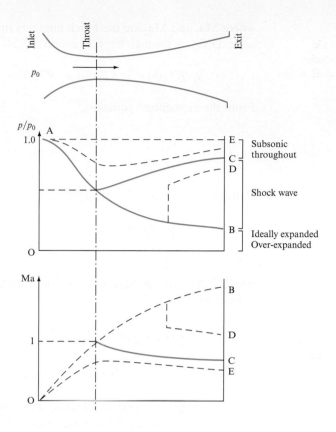

If the exit pressure is less than the back pressure, shock waves occur and the nozzle is said to be *over-expanded*. If the difference is small, oblique shock waves form at the exit (Fig. 17.5(b)), but for larger differences of pressure, a normal shock wave will form in the nozzle (Fig. 17.5(c)). Figure 17.6 summarizes the variations of pressure ratio and Mach number.

17.7 NORMAL SHOCK WAVE IN A DIFFUSER

When a normal shock wave forms in a diffuser, a supersonic flow is decelerated to a subsonic flow with a consequent increase in stagnation temperature, pressure and density. An analogy can be drawn to the hydraulic jump discussed in Chapter 16.

Taking a control volume enclosing the wave (Fig. 17.7) of cross-sectional area A, for steady flow

$$\rho_1 \bar{v}_1 A = \rho_2 \bar{v}_2 A. \tag{17.24}$$

Putting $\rho_1 = p_1/RT_1$,

$$\bar{v}_1 = \text{Ma}_1\sqrt{(\gamma RT_1)},$$

and if $\rho_2 = p_2/RT_2$,

$$\bar{v}_2 = \text{Ma}_2\sqrt{(\gamma RT_2)},$$

where Ma_1 and Ma_2 are the Mach numbers upstream and downstream of the shock wave. Dividing through by A, equation (17.24) becomes

$$(p_1/RT_1)Ma_1\sqrt{(\gamma RT_1)} = (p_2/RT_2)Ma_2\sqrt{(\gamma RT_2)}. \tag{17.25}$$

From the momentum equation,

$$(p_1 - p_2)\,A = \dot{m}(\bar{v}_2 - \bar{v}_1)$$
$$p_1 - p_2 = \rho_2\bar{v}_2^2 - \rho_1\bar{v}_1^2. \tag{17.26}$$

Putting $\rho = p/RT$ and $\bar{v} = Ma\sqrt{(\gamma RT)}$,

$$p_1 + (p_1/RT_1)\bar{v}_1^2 = p_2 + (p_2/RT_2)\bar{v}_2^2,$$
$$p_1(1 + \gamma Ma_1^2) = p_2(1 + \gamma Ma_2^2), \tag{17.27}$$

which is the same as equation (13.17). Thus, the static pressure ratio across a shock wave is given by

$$p_2/p_1 = (1 + \gamma Ma_1^2)/(1 + \gamma Ma_2^2), \tag{17.28}$$

and, since $Ma_1 > 1$ and $Ma_2 < 1$, the static pressure increases across the shock wave ($p_2 > p_1$).

Assuming adiabatic conditions, there will be no change in the stagnation temperature across the shock wave, so that $(T_0)_1 = (T_0)_2 = T_0$. From equation (17.19),

$$\frac{T_2}{T_1} = \frac{T_2}{T_0} \times \frac{T_0}{T_1} = \frac{1 + [(\gamma-1)/2]\,Ma_1^2}{1 + [(\gamma-1)/2]\,Ma_2^2}, \tag{17.29}$$

which corresponds to equation (13.16). Substituting in equation (17.25) from equations (17.28) and (17.29),

$$\frac{Ma_1}{1 + \gamma Ma_1^2}\left(1 + \frac{\gamma-1}{2}\,Ma_1^2\right)^{1/2} = \frac{Ma_2}{1 + \gamma Ma_2^2}\left(1 + \frac{\gamma-1}{2}\,Ma_2^2\right)^{1/2}.$$

If this is solved for Ma_2 in terms of Ma_1, there are two solutions. The first is $Ma_1 = Ma_2$, which is the case for no shock wave. The second is

$$Ma_2^2 = \frac{(\gamma-1)Ma_1^2 + 2}{2\gamma Ma_1^2 - (\gamma-1)}, \tag{17.30}$$

which corresponds to equation (13.19).

EXAMPLE 17.3

Air is flowing through a duct and a normal shock wave is formed at a cross-section at which the Mach number is 2.0. If the upstream pressure and temperature are 105 bar and 15 °C, respectively, find the Mach number, pressure and temperature immediately downstream of the shock waves. Take $\gamma = 1.4$.

Solution

From equation (17.30),

$$Ma_2^2 = \frac{(1.4-1)2^2+2}{2\times 1.4 \times 2^2 - (1.4-1)} = 0.333,$$

$$Ma_2 = \mathbf{0.577}.$$

From equation (17.28),

$$p_2 = p_1 \frac{1+1.4\times 2^2}{1+1.4\times 0.577^2} = 4.5 p_1 = 4.5 \times 105$$

$$= \mathbf{473\ bar}.$$

From equation (17.29),

$$T_2 = T_1 \frac{1+0.2\times 2^2}{1+0.2\times 0.577^2} = 1.687 T_1 = 1.687 \times 288$$

$$= \mathbf{486\ K\ or\ 213\ °C}.$$

FIGURE 17.7

Normal shock wave

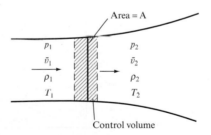

An insight into the nature of the changes in flow conditions which occur across a shock wave, where the area can be considered to be constant, can be obtained by examining the relationship graphically. If the upstream conditions (Fig. 17.7) are taken as fixed, curves can be drawn showing all the corresponding possible conditions downstream of the shock wave. It is possible to draw one set of curves, known as *Fanno lines*, in which each curve represents conditions which, for a particular mass flow, satisfy the continuity and energy equations, which are

$$\text{Mass flow per unit area, } G = \dot{m}/A = \rho \bar{v} = \text{constant} \qquad (17.31)$$

and

$$\text{Stagnation enthalpy, } H_0 = H + \bar{v}^2/2 = \text{constant}. \qquad (17.32)$$

It is instructive to plot the Fanno lines as a graph of enthalpy H against entropy S. The entropy equation for a perfect gas is

$$S - S_1 = c_v \log_e \left[\frac{p}{p_1} \left(\frac{\rho_1}{\rho} \right)^{\gamma} \right] \qquad (17.33)$$

and

$$H = c_p T = c_p p / R\rho. \qquad (17.34)$$

Combining equations (17.31) to (17.34), we have

$$S = S_1 + c_v \log_e [H(H_0 - H)^{(\gamma-1)/2}] + \text{constant}, \qquad (17.35)$$

where the constant is determined by the mass flow per unit area G and the upstream conditions.

This is shown plotted for a given mass flow per unit area in Fig. 17.8. Maximum entropy occurs at P, the conditions being found by differentiating equation (17.35) with respect to H and putting $\mathrm{d}S/\mathrm{d}H = 0$ for $H = H_p$, the value at point P.

FIGURE 17.8

Fanno and Rayleigh lines

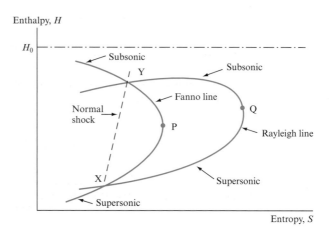

$$\frac{\mathrm{d}S}{\mathrm{d}H} = \frac{1}{H_P} - \frac{(\gamma-1)}{2}\frac{1}{H_0 - H_P} = 0,$$

$$H_P = [2/(\gamma+1)]H_0,$$

$$H_0 = [(\gamma+1)/2]H_P = H_P + v_P^2/2,$$

$$v_P^2 = (\gamma-1)H_P = (\gamma-1)c_p T_P = (\gamma-1)[\gamma R/(\gamma-1)]T_P,$$

$$v_P = \sqrt{(\gamma R T_P)} = \text{Velocity of sound.}$$

Thus, the Fanno line shows that maximum entropy occurs at the point P for which the Mach number is 1 and conditions are sonic. If $H > H_P$, the flow is subsonic and, if $H < H_P$, the flow is supersonic. For a shock wave, the conditions before and after the shock must both lie on the Fanno line for the mass flow and area of the section at which the shock occurs.

To determine the position of these points, we now consider the requirement that both the continuity and momentum equations must be satisfied for a given mass flow. A curve known as the *Rayleigh line* can be drawn showing all the points on the *H–S* diagram which satisfy these requirements. The momentum equation is

$$p_1 - p = (m/A)(\bar{v} - \bar{v}_1)$$

and, combining this with equation (17.31),

$$p + \rho\bar{v}^2 = p_1 + \rho_1\bar{v}_1^2 = \text{constant}$$

or $p + G^2/\rho = \text{constant} = B.$

Substituting for p in equation (17.33),

$$S = S_1 + c_v \log_e \left[\frac{(B - G^2/\rho)}{\rho^\gamma} \right] + \text{constant}, \tag{17.36}$$

where the constant depends upon the upstream conditions. Now

$$H = c_p T = \frac{c_p}{R} \frac{1}{\rho} \left(B - \frac{G^2}{\rho} \right). \tag{17.37}$$

From equations (17.36) and (17.37), the Rayleigh line can be plotted (Fig. 17.8) for the given mass flow.

The conditions corresponding to the point Q for maximum entropy are found by differentiating equations (17.36) and (17.37) to give dS/dρ and dH/dρ; then, dividing and equating to zero,

$$\frac{\mathrm{d}S}{\mathrm{d}H} = \frac{\mathrm{d}S}{\mathrm{d}\rho}\frac{\mathrm{d}\rho}{\mathrm{d}H} = \frac{c_v}{c_p} R\rho_Q \frac{\{G^2/[\rho_Q(B - G^2/\rho_Q)]\} - \gamma}{2G^2/\rho_Q - B} = 0.$$

If the denominator is not zero, this gives

$$\frac{G^2}{\rho_Q(B - G^2/\rho_Q)} - \gamma = 0$$

or, substituting for G in terms of $\bar{v}$ from equation (17.31),

$$\bar{v}_Q = \sqrt{(\gamma p_Q/\rho_Q)} = \text{Velocity of sound.}$$

Thus, sonic conditions occur at the point of maximum entropy. The upper limb of the curve corresponds to subsonic flow and the lower limb to supersonic flow.

For a shock wave the continuity, energy and momentum equations must be satisfied and it is, therefore, clear that conditions before and after the shock wave must lie on both the appropriate Fanno and Rayleigh lines, namely points X and Y in Fig. 17.8. The value of the entropy S will be greater at Y than at X since the shock occurs from supersonic to subsonic conditions.

Note that an alternative plot for Fanno and Rayleigh lines is a *T–S* diagram, but, since $H = c_p T$, the diagrams are similar.

17.8 COMPRESSIBLE FLOW IN A DUCT WITH FRICTION UNDER ADIABATIC CONDITIONS. FANNO FLOW

The flow of a liquid through a duct against resistance due to friction was discussed in Chapter 10. The analysis of the flow of a compressible fluid under similar circumstances is, fundamentally, the same, but is complicated by the interdependence of density, pressure and temperature, all of which will change from point to point along the length of the duct. It is necessary to make some assumptions that will simplify the problem.

One such assumption is that the duct or pipe is perfectly insulated and that conditions in the fluid are adiabatic. This is known as Fanno flow. For steady flow in a duct of constant cross-sectional area, the continuity equation can be written as $\rho\bar{v} =$ constant, where ρ is the density and $\bar{v}$ the velocity at any cross-section. Differentiating,

$$\frac{d\bar{v}}{v} + \frac{d\rho}{\rho} = 0. \tag{17.38}$$

The energy equation will be $H + \bar{v}^2/2 =$ constant, where H is the enthalpy and conditions are adiabatic. Now, $H = c_p T$ and, for a perfect gas, it can be shown that $c_p = \gamma R/(\gamma - 1)$, so that the energy equation reduces to

$$\frac{\gamma R}{\gamma - 1} T + \frac{\bar{v}^2}{2} = \text{constant}$$

or differentiating,

$$\frac{\gamma R}{\gamma - 1} dT + \bar{v}\, d\bar{v} = 0. \tag{17.39}$$

The momentum equation is derived from consideration of the control volume shown in Fig. 17.9. Neglecting gravitational forces, which are small, and assuming a shear stress τ_0 at the wall of the pipe, the forces in the direction of motion are equated to the rate of change of momentum in that direction across the system boundaries:

$$A[p - (p + \delta p)] - \tau_0 P\, dx = \rho\bar{v}A[(\bar{v} + \delta\bar{v}) - \bar{v}], \tag{17.40}$$

where P is the perimeter of the duct cross-section.

FIGURE 17.9

Friction in a duct

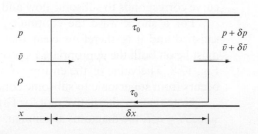

From equation (10.25), $\tau_0 = f\rho\bar{v}^2/2$, where f is the resistance coefficient in the Darcy equation

$$h_f = \frac{fl}{m}\frac{\bar{v}^2}{2g}.$$

Simplifying, equation (17.40) becomes

$$-\mathrm{d}p - f\rho\frac{\bar{v}^2}{2}\frac{P}{A}\,\mathrm{d}x = \rho\bar{v}\,\mathrm{d}\bar{v} \tag{17.41}$$

or, putting $(A/P) = m$,

$$\rho\bar{v}\,\mathrm{d}\bar{v} + \mathrm{d}p + \frac{f\rho}{m}\frac{\bar{v}^2}{2}\,\mathrm{d}x = 0, \tag{17.42}$$

which is the general equation for flow with friction in ducts.

Since both velocity and temperature will be changing as the gas flows along the duct, the value of the Mach number will also vary from point to point. The amount of change can be found by combining the equation of state with the momentum equation. From equation (17.1), $p/\rho = RT = c^2/\gamma$, where c is the sonic velocity, which will be $\sqrt{(\gamma RT)}$ for adiabatic conditions. Putting $\rho = \gamma p/c^2$ and dividing by p, equation (17.42) becomes

$$\gamma\frac{\bar{v}^2}{c^2}\frac{\mathrm{d}\bar{v}}{\bar{v}} + \frac{\mathrm{d}p}{p} + \frac{f}{m}\frac{\gamma}{c^2}\frac{\bar{v}^2}{2}\,\mathrm{d}x = 0$$

or $\qquad \gamma\,\mathrm{Ma}^2\frac{\mathrm{d}\bar{v}}{\bar{v}} + \frac{\mathrm{d}p}{p} + \gamma\frac{f}{m}\frac{\mathrm{Ma}^2}{2}\,\mathrm{d}x = 0, \tag{17.43}$

where Ma is the Mach number $(\bar{v}/c)$.

Differentiating equation (17.1), which is the equation of state,

$$\frac{\mathrm{d}p}{p} = \frac{\mathrm{d}\rho}{\rho} + \frac{\mathrm{d}T}{T}. \tag{17.44}$$

Substituting in equation (17.44) for ρ in terms of $\bar{v}$ from equation (17.38) and for T in terms of $\bar{v}$ from equation (17.39), we have

$$\frac{\mathrm{d}p}{p} = -\frac{\mathrm{d}\bar{v}}{\bar{v}} - (\gamma - 1)\,\mathrm{Ma}^2\frac{\mathrm{d}\bar{v}}{\bar{v}}. \tag{17.45}$$

Putting this value of $\mathrm{d}p/p$ in equation (17.43),

$$(\mathrm{Ma}^2 - 1)\frac{\mathrm{d}\bar{v}}{\bar{v}} + \frac{\gamma f}{m}\frac{\mathrm{Ma}^2}{2}\,\mathrm{d}x = 0. \tag{17.46}$$

Now, the Mach number is defined as $\mathrm{Ma} = \bar{v}/c = \bar{v}/\sqrt{(\gamma RT)}$, which, on differentiation, gives

$$\frac{\mathrm{dMa}}{\mathrm{Ma}} = \frac{\mathrm{d}\bar{v}}{\bar{v}} - \frac{1}{2}\frac{\mathrm{d}T}{T}. \tag{17.47}$$

From equation (17.39),

$$\frac{\mathrm{d}T}{T} = -\frac{\bar{v}\,\mathrm{d}\bar{v}(\gamma-1)}{\gamma R T} = -(\gamma-1)\,\mathrm{Ma}^2\,\frac{\mathrm{d}\bar{v}}{\bar{v}},$$

so that equation (17.47) becomes

$$\frac{\mathrm{dMa}}{\mathrm{Ma}} = \frac{\mathrm{d}\bar{v}}{\bar{v}}\left[1 + \frac{(\gamma-1)}{2}\,\mathrm{Ma}^2\right], \qquad (17.48)$$

from which $\mathrm{d}\bar{v}/\bar{v}$ can be eliminated from equation (17.46) to obtain

$$\frac{(1-\mathrm{Ma}^2)\,\mathrm{dMa}}{\mathrm{Ma}^3\{1+[(\gamma-1)/2]\mathrm{Ma}^2\}} = \frac{\gamma f}{2m}\,\mathrm{d}x. \qquad (17.49)$$

From equation (17.49), it can be seen that if flow is subsonic and $\mathrm{Ma} < 1$, then $(\mathrm{dMa}/\mathrm{d}x) > 0$ and the Mach number increases with distance along the duct.

If the flow is supersonic, $\mathrm{Ma} > 1$ and $(\mathrm{dMa}/\mathrm{d}x) < 0$, the Mach number will decrease along the duct. Thus, the effect of pipe friction is always to cause the Mach number to approach unity. It is impossible for the Mach number of a compressible flow to change from subsonic to supersonic in a duct of constant cross-section and, consequently, the maximum Mach number that can be attained by an initially subsonic flow is unity, reached at the exit from the duct. Conversely, supersonic flow may only become subsonic owing to the occurrence of shock waves in the duct.

In order to integrate equation (17.49) to obtain values of Ma against length, some reasonable assumption must be made about the variation of friction factor f along the duct. The continuity equation states that $\rho\bar{v}$ is constant along the length of the duct; therefore, any variation of Reynolds number, which governs the friction factor f, can only occur as a result of a change in the viscosity of the fluid. Viscosity is dependent on temperature. For example, a change in temperature of 20 per cent, which can frequently occur in compressible flow, would produce a 10 per cent change in viscosity in the case of air. However, the resulting 10 per cent change in Reynolds number would normally give rise to a much smaller change in the friction factor f and, so, it is reasonable to assume a constant value for f when integrating equation (17.49). This value is equal to the average value of f along the duct. Reducing the left-hand side of equation (17.49) to partial fractions yields

$$\left[\!\left[\frac{1}{\mathrm{Ma}^3} + \frac{\gamma+1}{2\mathrm{Ma}} + \frac{(\gamma+1)(\gamma-1)\mathrm{Ma}}{4\{1+[(\gamma-1)/2]\mathrm{Ma}^2\}}\right]\!\right]\mathrm{dMa} = \frac{\gamma f}{2m}\,\mathrm{d}x. \qquad (17.50)$$

Integrating both sides,

$$-\frac{1}{2\mathrm{Ma}^2} - \frac{\gamma+1}{2}\log_e\mathrm{Ma} + \frac{\gamma+1}{4}\log_e\left[1 + \frac{(\gamma-1)}{2}\mathrm{Ma}^2\right] = \frac{\gamma f x}{2m} + C. \quad (17.51)$$

To determine the constant of integration C, let x_1 be the distance along the pipe at which the Mach number becomes unity. Then,

$$C = -\frac{\gamma f x_1}{2m} - \frac{1}{2} + \frac{(\gamma+1)}{4}\log_e\left(\frac{\gamma+1}{2}\right). \qquad (17.52)$$

Substitution of C into equation (17.51) yields an expression linking Mach number to distance along the duct:

$$\frac{1-\mathrm{Ma}^2}{\gamma\mathrm{Ma}^2} + \frac{\gamma+1}{2\gamma}\log_e\left[\frac{(\gamma+1)\mathrm{Ma}^2}{2+(\gamma-1)\mathrm{Ma}^2}\right] = \frac{f(x_1-x)}{m}. \tag{17.53}$$

EXAMPLE 17.4

Air (for which $\gamma = 1.4$) flows along a circular pipe with a diameter d of 50 mm. Assuming that conditions are adiabatic and that the Mach number at the entrance to the pipe is 0.2, calculate the distance from the entrance of the pipe to the section at which the Mach number will be (a) 1.0, (b) 0.6. Take $f = 0.003\,75$.

Solution

(a) The distance x_1 at which the Mach number is unity can be found from equation (17.53), since $\gamma = 1.4$, $m = d/4$, $f = 0.003\,75$, and when $x = 0$, Ma $= 0.2$. Substituting these values,

$$\frac{1-0.2^2}{1.4\times0.2^2} + \frac{2.4}{2.8}\log_e\left(\frac{2.4\times0.04}{2+0.4\times0.04}\right) = \frac{4\times0.003\,75(x_1-0)}{d},$$

$$x_1 = 968.9d. \tag{17.54}$$

Putting $d = 50$ mm $= 0.05$ m,

$$x_1 = \mathbf{48.44\ m}.$$

(b) The distance from the entrance, where Ma $= 0.2$ and $x = x_{0.2} = 0$, to the section at which Ma $= 0.6$ and $x = x_{0.6}$ cannot be found directly. First, find the distance from $x_{0.6}$ to x_1 from equation (17.53):

$$\frac{1-0.6^2}{1.4\times0.6^2} + \frac{2.4}{2.8}\log_e\left(\frac{2.4\times0.36}{2+0.4\times0.36}\right) = \frac{4\times0.003\,75(x_1-x_{0.6})}{d},$$

$$x_1 - x_{0.6} = 32.7d.$$

Substituting for x_1 from equation (17.54),

$$x_{0.6} = (968.9 - 32.7)d = 936.2d.$$

Putting $d = 0.05$ m,

$$x_{0.6} = \mathbf{46.81\ m}.$$

The variation of pressure along the length of the duct can also be obtained from equation (17.45):

$$\frac{\mathrm{d}p}{p} = -\frac{\mathrm{d}\bar{v}}{\bar{v}}[1+(\gamma-1)\mathrm{Ma}^2].$$

Substituting this value in equation (17.48) and rearranging,

$$\frac{\mathrm{d}p}{p} = -\frac{\mathrm{dMa}}{\mathrm{Ma}}\left\{\frac{1+(\gamma-1)\mathrm{Ma}^2}{1+[(\gamma-1)/2]\mathrm{Ma}^2}\right\}, \tag{17.55}$$

from which it can be seen that $\mathrm{d}p/\mathrm{dMa}$ is negative, indicating that the pressure decreases with increasing Mach number. Thus, the observed pressure decrease for subsonic flow along a duct corresponds to an increase of Mach number. From equation (17.55)

$$\frac{\mathrm{d}p}{p} = \left\{-\frac{1}{\mathrm{Ma}} - \frac{[(\gamma-1)/2]\mathrm{Ma}}{1+[(\gamma-1)/2]\mathrm{Ma}^2}\right\}\mathrm{dMa}.$$

Integrating,

$$\log_e p = \log_e \mathrm{Ma} - \tfrac{1}{2}\log_e\left[1+\left(\frac{\gamma-1}{2}\right)\mathrm{Ma}^2\right] + C. \tag{17.56}$$

The constant of integration C can be evaluated in terms of the pressure p_1 corresponding to $\mathrm{Ma} = 1$, giving

$$\log_e p_1 = -\tfrac{1}{2}\log_e\left(\frac{\gamma+1}{2}\right) + C$$

and so, from equation (17.56),

$$\frac{p}{p_1} = \frac{1}{\mathrm{Ma}}\left[\frac{\gamma+1}{2+(\gamma-1)\mathrm{Ma}^2}\right], \tag{17.57}$$

where p is the pressure corresponding to Mach number Ma.

17.9 ISOTHERMAL FLOW OF A COMPRESSIBLE FLUID IN A PIPELINE

When a gas flows at low velocities in a long duct through which heat transfer can occur readily, conditions may be approximately isothermal so that the temperature T can be considered constant. From the equation of state, $p/\rho = \text{constant} = p_1/\rho_1$, where p_1 and ρ_1 are the values of pressure and density at a given point and p and ρ the corresponding values at any other point, or

$$\rho = (p/p_1)\rho_1. \tag{17.58}$$

For a duct of constant cross-sectional area A, the continuity equation $\rho A \bar{v} = \rho_1 A_1 \bar{v}_1 = \text{constant}$, so that $\rho \bar{v} = \rho_1 \bar{v}_1$ and, hence,

$$\bar{v} = \bar{v}_1(\rho_1/\rho).$$

Substituting for ρ from equation (17.58), the velocity at any section is

$$\bar{v} = \bar{v}_1(p_1/p).\tag{17.59}$$

For flow with frictional resistance, from equation (17.42),

$$\bar{v}\,\mathrm{d}\bar{v} + \frac{\mathrm{d}p}{\rho} + \frac{f}{m}\frac{\bar{v}^2}{2}\,\mathrm{d}x = 0.\tag{17.60}$$

By integration, the pressure drop along the duct can be determined as follows (Sections 17.9.1 and 17.9.2).

17.9.1 Approximate solution neglecting velocity change

Under these conditions, $\mathrm{d}v = 0$ and equation (17.60) becomes

$$\frac{\mathrm{d}p}{\rho} + \frac{f}{m}\frac{\bar{v}^2}{2}\,\mathrm{d}x = 0.$$

Substituting for ρ and $\bar{v}$ from equations (17.58) and (17.59),

$$\frac{\mathrm{d}p}{p}\frac{p_1}{\rho_1} + \frac{f}{2m}\bar{v}_1^2\left(\frac{p_1}{p}\right)^2\mathrm{d}x = 0,$$

$$p\,\mathrm{d}p = -\frac{f}{2m}\rho_1 p_1 \bar{v}_1^2\,\mathrm{d}x.\tag{17.61}$$

Integrating and putting $\rho_1 = p_1/RT$,

$$p^2 - p_1^2 = -fp_1^2\bar{v}_1^2(x - x_1)/mRT,$$

$$p = p_1\sqrt{[1 - f(x - x_1)\bar{v}_1^2/mRT]},\tag{17.62}$$

where p is the pressure at a distance x downstream from the point at which the pressure is p_1 and m is the hydraulic mean depth.

17.9.2 Solution allowing for velocity changes

Since for continuity of flow $\rho\bar{v} = $ constant and for a perfect gas $p = \rho \times$ constant, for isothermal conditions,

$$\frac{\mathrm{d}\bar{v}}{\bar{v}} = -\frac{\mathrm{d}\rho}{\rho} = -\frac{\mathrm{d}p}{p}.\tag{17.63}$$

Substituting for $\mathrm{d}p$ in equation (17.60) and dividing by v^2,

$$\frac{f\,\mathrm{d}x}{2m} = \frac{p}{\rho}\frac{\mathrm{d}\bar{v}}{\bar{v}^3} - \frac{\mathrm{d}\bar{v}}{\bar{v}}.\tag{17.64}$$

Now, as T is constant and as viscosity may be assumed to be a function only of T at normal pressures, Reynolds number is constant and, for uniform roughness along the duct, the friction factor f may also be treated as constant.

As $p/\rho = RT =$ constant, equation (17.64) may be integrated directly:

$$\frac{f}{2m}(x_1 - x) = \frac{RT}{2}\left(\frac{1}{\bar{v}_1^2} - \frac{1}{\bar{v}^2}\right) - \log_e\left(\frac{\bar{v}}{\bar{v}_1}\right),$$

$$\frac{f}{2m}(x_1 - x) = \frac{1}{2\gamma}\left(\frac{1}{\mathrm{Ma}_1^2} - \frac{1}{\mathrm{Ma}^2}\right) - \log_e\left(\frac{\mathrm{Ma}}{\mathrm{Ma}_1}\right), \tag{17.65}$$

where the suffix 1 refers to conditions at a known point in the duct.

There is a limitation arising from these results in respect of the maximum attainable Mach number for a subsonic isothermal flow. Substituting equation (17.63) into equation (17.42),

$$\frac{\mathrm{d}p}{\mathrm{d}x} = \left(\frac{f\rho v^2}{2m}\right)\bigg/\left(\frac{\rho\bar{v}^2}{p} - 1\right) = \frac{f\rho\bar{v}^2}{2m(\gamma\mathrm{Ma}^2 - 1)}. \tag{17.66}$$

Equation (17.66) shows that for $\mathrm{Ma} < (1/\gamma)^{1/2}$, the value of $\mathrm{d}p/\mathrm{d}x$ is negative but, when $\mathrm{Ma} = (1/\gamma)^{1/2}$, the value of $\mathrm{d}p/\mathrm{d}x$ becomes infinite and discontinuities arise in both pressure and velocity variations. Thus, there is a maximum flow length for isothermal flow, obtained by putting $\mathrm{Ma} = (1/\gamma)^{1/2}$ in equation (17.65), comparable with the limiting flow length for adiabatic flow which was shown to be limited by $\mathrm{Ma} = 1$. In practice, $\mathrm{Ma} = (1/\gamma)^{1/2}$ is never achieved, as $\mathrm{d}p/\mathrm{d}x$ would have to be infinite. If the actual length of the pipe exceeds the limiting flow length, choking would occur and the rate of flow would adjust until conditions were such that the value $\mathrm{Ma} = (1/\gamma)^{1/2}$ would not be reached until the end of the actual pipe.

From equation (17.65), substituting for $\bar{v}$ from equation (17.59),

$$\frac{\mathrm{d}p}{\mathrm{d}x} = \left(\frac{f}{2m}\frac{p}{p_1}\frac{\rho_1\bar{v}_1^2 p_1^2}{p_2}\right)\bigg/\left(\frac{p}{p_1}\frac{\rho_1\bar{v}_1^2 p_1^2}{p^3} - 1\right)$$

$$= \frac{f}{2m}\rho_1\bar{v}_1^2 p_1^2 \frac{p}{(\rho_1\bar{v}_1^2 p_1 - p^2)},$$

$$p\,\mathrm{d}p - \rho_1\bar{v}_1^2 p_1\frac{\mathrm{d}p}{p} = -\frac{f}{2m}\rho_1\bar{v}_1^2 p_1^2\mathrm{d}x. \tag{17.67}$$

This equation can be compared with the approximate solution of equation (17.61), to which it will reduce if $\rho_1\bar{v}_1^2 p_1(\mathrm{d}p/p)$ is small. Integrating equation (17.67) and putting $\rho_1 = p_1/RT$,

$$\frac{p^2 - p_1^2}{2} - \bar{v}_1^2\frac{p_1^2}{RT}\log_e\left(\frac{p}{p_1}\right) = -\frac{fp_1^2\bar{v}_1^2(x - x_1)}{2mRT}. \tag{17.68}$$

EXAMPLE 17.5

Air flows along a pipe 100 mm in diameter under isothermal conditions. At the entrance the pressure is 200 kN m^{-2}, the volume rate of flow is 28 m^3 min^{-1} and the temperature is constant at 15 °C. If the pipe is 60 m long and the value of f is 0.004, calculated the pressure at the outlet assuming $R = 287$ J kg^{-1} K^{-1} (a) neglecting changes of velocity, (b) allowing for velocity changes.

Solution

(a) At entrance,

$$\bar{v}_1 = \frac{Q}{\pi/4 d^2} = \frac{28}{(\pi/4) \times 0.01 \times 60} = 59.40 \text{ m s}^{-1}.$$

Substituting in equation (17.62), $p_1 = 200 \times 10^3$ N m^{-2}, $f = 0.004$, $(x - x_1) = 60$ m, $m = d/4 = 0.025$ m, $R = 287$ J kg^{-1} K^{-1}, $T = 15\,°C = 288$ K,

$$p = 200 \times 10^3 \sqrt{\left[1 - \frac{0.004 \times 60 \times (59.4)^2}{0.025 \times 287 \times 288}\right]} \text{ N m}^{-2}$$

$$= 153.6 \text{ kN m}^{-2}.$$

(b) In this case use equation (17.68), which can be solved by trial. Substituting the numerical values,

$$\frac{p^2}{2} - \frac{(200 \times 10^3)^2}{2} - 59.4^2 \frac{(200 \times 10^3)^2}{287 \times 288} \log_e \left(\frac{p}{200 \times 10^3}\right)$$

$$= -\frac{0.004(200 \times 10^3)^2 \times 59.4^2 \times 60}{2 \times 0.025 \times 287 \times 288}$$

$$\frac{1}{2}\left(\frac{p}{10^3}\right)^2 - 20\,000 - 1707 \log_e \left(\frac{p}{200 \times 10^3}\right) = -8196.$$

Try $p = 152$ kN m^{-2}:

Left-hand side $= 11\,552 - 20\,000 + 468 = -7980.$

Try $p = 150$ kN m^{-2}:

Left-hand side $= 11\,250 - 20\,000 + 491 = -8259.$

Try $p = 150.5$ kN m^{-2}:

Left-hand side $= 11\,325 - 20\,000 + 485 = -8190,$

which is approximately equal to the right-hand side. Therefore,

Pressure at outlet $= \mathbf{150.5}$ **kN m^{-2}**.

Concluding remarks

The objective of the treatment of compressible pipe and duct flow presented was to provide the basis for a comparison of effects of compressibility on the pipe flow relationships developed earlier for incompressible pipe flow (Chapter 14). The necessity of introducing the equation of state to provide a linkage between flow pressure and density was established, although this then led to the need to define the flow further in terms of its adiabatic index. The relationship between the changes of duct area and supersonic flow velocity was emphasized and the shock equations developed to deal with diffuser flow. Consideration was given to nozzle flow, including the Laval nozzle. The techniques introduced have their own application as discussed in this chapter. However, in addition the introduction of the equation of state to define the pressure–density relationship will also be utilized in the later work on low-amplitude air pressure transients in Chapter 21 and the treatment of compressors in Chapter 22.

Summary of important equations and concepts

1. The steady flow energy equation introduced for incompressible flow in Chapter 6 is recast in a form suitable for compressible flow conditions, equations (17.2) to (17.5).

2. The expressions previously derived for incompressible flow through venturi meters and orifices are reworked to include the effects of compressibility. The maximum discharge is identified, Section 17.5, and the expressions governing the operation of a Laval nozzle developed, Section 17.6 and equation (17.23).

3. The formation of a shock during flow deceleration in a diffuser is considered in Section 17.7, and an analogy to hydraulic jump formation in free surface flows converting from supercritical to subcritical is drawn.

4. Adiabatic and isothermal flow conditions for compressible pipeline flow are developed in Sections 17.8 and 17.9, together with a discussion of frictional effects, equation (17.49) indicating that friction will act to force the Mach number towards unity.

Problems

17.1   Air at 5 bar and 560 K is expanded in steady flow in a horizontal convergent–divergent duct to an exit velocity of 640 m s^{-1}. The walls of the duct are heated so as to keep the temperature drop to one-half the value of an isentropic expansion to the same velocity and pressure from the same initial conditions. Calculate, assuming negligible initial velocity, (a) the heat supplied, (b) the final temperature, (c) the mass flow rate per square metre of exit area.
[(a) 102.4 kJ kg^{-1}, (b) 458 K, (c) 499 kg m^{-2} s^{-1}]

17.2   A venturi meter which has a throat diameter of 25 mm is installed in a horizontal pipeline of 75 mm diameter conveying air. The pressure at the inlet to the meter is 133.3 kN m^{-2} and that at the throat is 100 kN m^{-2}, both pressures being absolute. The temperature of the air at the inlet is 15 °C. Assuming isentropic flow, determine the mass flow rate in kilograms per second. For air $\gamma = 1.4$ and $R = 287$ J kg^{-1} K^{-1}.
[0.138 kg s^{-1}]

17.3   A sharp-edged circular orifice of 45 mm diameter is used to measure the flow of air from the atmosphere into a large tank. Barometric pressure is 735 mm of mercury, temperature is 17 °C and the difference of pressure between

the atmosphere and the inside of the tank is equivalent to a head of 20 mm of water. Determine the mass of air in kilograms per minute passing into the tank if the coefficient of discharge for the orifice is 0.6. Take $R = 287$ J kg^{-1} K^{-1}.
[0.949 kg min^{-1}]

17.4   Calculate the maximum mass flow possible through a frictionless, heat-insulated, convergent nozzle if the entry or stagnation conditions are 5 bar and 15 °C and the throat area is 6.5 cm^2. Also calculate the temperature of the air at the throat. Take $c_p = 1.00$ kJ kg^{-1} K^{-1} and $\gamma = 1.4$.
[0.775 kg s^{-1}, 240 K]

17.5   A convergent–divergent nozzle is supplied with compressed air from a reservoir at a pressure of 1 MN m^{-2} absolute. The throat area is 8 cm^2 and the nozzle expands to a parallel section of area 13.5 cm^2 before discharging into a region where the absolute pressure is 100 kN m^{-2}. Calculate the exit Mach number.
[2.16]

17.6   Air initially at standard temperature and pressure flows into an evacuated tank through a convergent nozzle contracting to a diameter of 4 cm. What pressure must be maintained in the tank to produce a sonic jet? What is the mass flow?   [53.6 kN m^{-2}, 0.3 kg s^{-1}]

17.7 Show that if the Mach number upstream of a normal shock wave in air is large the density ratio across the shock wave is 6 and the downstream Mach number is 0.378.

17.8 A shock wave occurs in a duct carrying air where the upstream Mach number is 2 and the upstream temperature and pressure are 15 °C and 20 kN m^{-2} absolute. Calculate the Mach number, pressure, temperature and velocity after the shock wave.   [0.577, 90 kN m^{-2}, 214 °C, 255 m s^{-1}]

17.9 Air flows through a parallel passage in which a shock wave is formed. If the suffixes 1 and 2 refer to conditions just before and just after the wave, show that

$$\frac{u_1^2}{2} + \frac{\gamma}{\gamma-1}\frac{p_1}{\rho_1} = \frac{u_2^2}{2} + \frac{\gamma}{\gamma-1}\left(\frac{u_2}{u_1}\right)\left(u_1^2 - u_1 u_2 + \frac{p_1}{\rho_1}\right).$$

If $p_1 = 690$ kN m^{-2} absolute, $\rho_1 = 5.45$ kg m^{-3} and $u_1 = 450$ m s^{-1}, calculate the values of p_2 and u_2 immediately after the shock wave given that $\gamma = 1.4$. Also calculate the Mach number before and after the wave.
[400 m s^{-1}, 813 kN m^{-2}, 1.07, 0.936]

17.10 A normal shock wave forms in an airstream at a static temperature of 22 K, the total temperature being 400 K. Estimate the Mach number and static temperature behind the shock wave.   [0.539, 105 °C]

17.11 Fanno flow (adiabatic flow with friction) prevails as air moves through a pipe of 50 mm diameter. At a certain point along the pipe the Mach number is 0.2. Find the maximum distance from this point to the exit from the pipe if choking is avoided. (Assume that the friction factor f is 0.006. Start from the energy equation and the momentum equation in differential form.)   [30.27 m]

17.12 Air flows adiabatically at the rate of 2.7 kg s^{-1} through a horizontal 100 mm diameter pipe for which a mean value of f is 0.006. If the initial pressure and temperature are 1.8 bar absolute and 50 °C, what is the maximum length of the pipe for which choking will not occur? What are then the temperature and pressure at the exit end and half-way along the pipe?
[4.74 m; 9.1 °C, 82.7 kN m^{-2}; 44.2 °C, 168.9 kN m^{-2}]

17.13 Air flows through a pipe isothermally which is 50 mm in diameter and 1200 m long. Calculate the flow, in cubic metres per minute of free air at 15 °C and 101.3 kN m^{-2} absolute, if the initial pressure is 1 MN m^{-2} absolute and the final pressure 0.7 MN m^{-2} absolute. The temperature is constant at 5 °C and the friction coefficient $f = 0.004$.
[12.4 m^3 min^{-1}]

17.14 Air passes steadily through a horizontal duct of diameter 15 cm and length 300 m. The mass flow rate is 4.5 kg s^{-1}, the pressure at entry is 5 bar absolute and at exit is 1.25 bar absolute. Assuming the flow to be shock free and isothermal at 60 °C determine (*a*) the friction factor for the duct which may be assumed to be constant, (*b*) the heat transfer rate in watts to the air in the duct, (*c*) the Mach number at exit.

What percentage error would have occurred in your answer to (*a*) had the velocity term been neglected? Take $R = 287$ J kg^{-1} K^{-1} and $\gamma = 1.4$.
[0.005, 80 kW, 0.532, 4 per cent]

Unsteady Flow in Bounded Systems

Pressure surge affects all fluid carrying systems, examples range from in-flight refuelling to the need for surge shaft pressure relief in hydroelectric schemes and water supply systems illustrated by the Thames Water Tower surge relief stack that incorporates a solar powered barometer. Photos © Crown Copyright/MOD and courtesy of Thames Water

UNSTEADY FLOW MAY BE DEFINED AS A STATE IN WHICH the flow parameters are time dependent, governed by partial differential equations requiring, in their complete form, numerical methods of solution using computers. By considering the rates of change of the various flow parameters it is possible to place most unsteady flow phenomena into one of three categories:

1. *Quasi-steady flows* in which the rate of change of mass flow is continuous with time, but the fluid acceleration and the forces responsible for acceleration are negligible. In such cases the steady flow equations may be applied with reasonable accuracy (e.g. continuous filling and emptying of reservoirs and tanks).

2. *Mass oscillation* in which the rate of change of fluid velocity is sufficient for the forces causing fluid acceleration to be important, but still so slow as to permit the compressibility of the fluid to be ignored. The pressures generated within the affected system are often termed *surge pressures*. Examples include reciprocating machinery and oscillatory fluid motion such as that found in pipe systems with more than one free surface.

3. Flows in which the time taken to change fluid velocity is comparable with the period of the system based on the wave propagation velocity through the fluid, modified by the pipe properties, and the pipe length. If these times are comparable then the compressibility of the fluid becomes significant and the solution requires graphical or computer-based numerical techniques. These unsteady flow conditions, historically referred to as *waterhammer*, may result from rapid valve operation, pump shutdown or turbine load rejection and are commonly termed *pressure* or *fluid transients*.

Unsteady flow conditions also occur in free surface liquid flows, either open-channel or partially filled pipe flow, and in gaseous flows, either entrained or forced. All of these examples may be treated by the same numerical approach. In the free surface case the wave propagation speed, dependent on flow depth and channel geometry, and the fluid velocities are of the same order, while for low-amplitude transients in gas flows the conduit properties may be ignored.

This part of the text will present these flow conditions in order of transient severity and will culminate in the presentation of the finite difference numerical methods used to generate the simulations available with the text for each constituent of this family of flow conditions.

18

Quasi-steady Flow

THIS CHAPTER REPRESENTS A DIVERGENCE FROM THE MATERIAL PRESENTED UP TO THIS point in the text, in that it introduces the concept of time dependency. The principles of conservation of mass (Chapter 4), the steady flow energy equation (Chapters 6 and 12) and the finite difference approach to numerical integration (Chapter 5) are utilized to analyze flow conditions while the controlling parameters change slowly with time. This chapter should be regarded as an introduction to the wider spectrum of unsteady flow conditions that extends from mass oscillation to transient propagation. ● ● ●

18.1 DEPTH-DEPENDENT DISCHARGE

Let us consider the case of a tank emptying through an orifice into the atmosphere. It may be assumed that the rate of change of fluid discharge as the fluid level in the tank drops is sufficiently slow to permit application of the steady-state relationship at any instant during the discharge.

Hence, application of the quasi-steady approximation results in the following relationship between the instantaneous rate of change of the head H above the orifice and the outflow through the orifice:

$$A_s\left(\frac{-dH}{dt}\right) = C_d A_o \sqrt{(2gH)},\tag{18.1}$$

where the surface area A_s is much greater than the cross-sectional area A_o of the orifice. Note that the surface area A_s will be a function $f(H)$ of the head and that a negative sign is included in the rate of change of head since the fluid level in the tank is falling.

Rearrangement and incorporation of the fact that $A_s = f(H)$ yields an expression which may be integrated to give the time taken by the fluid level to fall from $H = H_1$ to $H = H_2$ (Fig. 18.1):

FIGURE 18.1

Discharge from a tank through an orifice of area A_s and discharge coefficient C_d

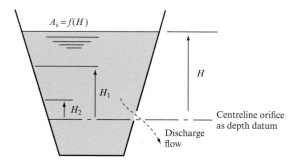

$$t = -\int_{H_1}^{H_2} [f(H)H^{-1/2}/C_d A_o \sqrt{(2g)}]\,dH.\tag{18.2}$$

The method of integration of equation (18.2) will be dependent upon the form of $f(H)$. Normally, for this type of example, it may be assumed that the coefficient of discharge at the orifice C_d is a constant.

If the tank or reservoir has a constant inflow Q_{in} during the discharge considered above, then equation (18.1) becomes

$$A_s\left(-\frac{dH}{dt}\right) = C_d A_o\sqrt{(2gH)} - Q_{in},\tag{18.3}$$

and equation (18.2) may be re-expressed as

$$t = -\int_{H_1}^{H_2}\{f(H)/[C_d A_o\sqrt{(2gH)} - Q_{in}]\}\,dH.\tag{18.4}$$

Let us now consider a system in which the orifice has been replaced by a pipe of length l and diameter d. Then, for quasi-steady conditions, it follows that the instantaneous head H balances the total losses in the discharge pipeline. Hence,

$$H = (4fl/d + k)(\bar{u}^2/2g),\tag{18.5}$$

where k is the sum of all the separation loss coefficients for the discharge pipeline and could include the loss through a partially shut valve and $\bar{u}$ is the mean flow velocity in the pipe.

Consideration of the continuity of flow between the fluid surface and the pipe discharge yields

$$A_s\left(-\frac{dH}{dt}\right) = A_o\bar{u} = C\sqrt{H},\tag{18.6}$$

where $\quad C = A_o\sqrt{[2g/(4fl/d + k)]}.\tag{18.7}$

Rearrangement again results in an expression for the time taken for the surface to fall from H_1 to H_2:

$$t = -\int_{H_1}^{H_2}[f(H)/C\sqrt{H}]\,dH.\tag{18.8}$$

Integration of equation (18.8) will depend upon $f(H)$, but, if the tank is assumed to have a constant cross-section over the depth range under consideration, then the solution for t is given by

$$t = 2A_s(H_2^{1/2} - H_1^{1/2})/C.\tag{18.9}$$

Strictly, the friction factor f which appears in equation (18.7) is a function of both the discharge flow Reynolds number and the discharge pipe roughness. However, it is usually sufficient to assume a constant value for f based on the initial flow rate. If it is required to introduce friction factor variation into the solution, this may be done by relating the friction factor to the instantaneous head through the Reynolds number relations and equation (18.5), and integrating the resulting expression by an incremental method.

18.2 FINITE DIFFERENCE REPRESENTATION OF DEPTH-DEPENDENT DISCHARGE

There may be circumstances in which the direct integration for discharge time demonstrated in equation (18.9) becomes impossible, for example if the dependence of cross-sectional area on depth is complex or if periodic inflows to the reservoir occur during part but not all of the discharge period. Under these conditions it may be necessary to resort to a finite difference approach to the numerical integration of discharge time.

Referring to equations (18.1) or (18.6) it may be seen that both may be represented by

$$\frac{\Delta h}{\Delta t} = \frac{Q}{A_{\text{fluid surface}}}. \tag{18.10}$$

Assume that the time increment for the fluid depth to fall from h^{n+1} to h^n is required, i.e. $\Delta h = h^{n+1} - h^n$, then this may be defined in terms of equation (18.10), where the flow rate Q is based on that appropriate at the mid-depth increment value of h. Hence if $Q = A_{\text{pipe}}(kh)^{0.5}$, as is the case in both equations (18.1) and (18.6), it follows that:

$$\frac{h^{n+1} - h^n}{\Delta t} = [0.5k(h^{n+1} + h^n)]^{0.5}\left(\frac{A_{\text{pipe cross-section}}}{A_{\text{fluid surface}}}\right). \tag{18.11}$$

The total time to fall from h_1 to h_2 is then the summation of the constituent Δt values from h_1 to h_2. However, the accuracy of this approach will depend entirely upon the choice of Δh, too large a value will lead to errors, too small a value wastes computing time, suggesting that any such technique requires an initial test to determine appropriate incremental depth values.

EXAMPLE 18.1

A rectangular cross-section tank of 6 m² surface area is filled to a depth of 1.5 m with water. Calculate the time to reduce the depth in the tank by 1.25 m if it is drained through a 1.5 m long, 0.02 m diameter pipe discharging 2 m below the base of the tank and having a friction factor of 0.01 and separation losses due to bends etc. of 0.9. Compare the integral for discharge time with a range of Δh incremental values to demonstrate the accuracy of a finite difference approach.

Solution

Let A_s be the tank surface area $= 6\,\text{m}^2$, A_p be the pipe cross-sectional area, $0.3142 \times 10^{-3}\,\text{m}^2$. From equation (18.7), the flow rate from the tank at any depth h is given by:

$$Q = A_p(2gh)^{0.5}/(4fl/d + k + 1.0)^{0.5}$$

$$= A_p\, h^{0.5}(19.62/(4 \times 0.01 \times 1.5/0.02 + 1.9)^{0.5}) = 2A_p\, h^{0.5}.$$

Thus equation (18.11) becomes

$$\Delta t = \Delta h\, A_s/[\![(2A_p)\, \{0.5[h + (h - \Delta h)]\}^{0.5}]\!],$$

where h is measured vertically from the discharge level, i.e. (depth in tank $+\ 2$) m, allowing the time taken for the level to drop Δh to be calculated. Summing Δt until the depth falls to the lower limit yields the emptying time of the tank. (These calculations are best attempted by use of a spreadsheet such as Excel.)

For the values quoted in this example the direct integral time from 1.5 m to 0.25 m tank depth and the numerical integration totals, dependent upon Δh values are given below.

Direct integration:	equation (18.9), $h_2 = 3.5$, $h_1 = 1.25$	Time to drain $= 7081.39$ s
Numerical integration, (initial Δh choice $= 0.25$ m):	$\Delta h(1) = 0.250$ m	$= 8717.09$ s
	$\Delta h(2) = 0.125$ m	$= 7887.90$ s
	$\Delta h(3) = 0.0083$ m	$= 7616.62$ s
	$\Delta h(4) = 0.0063$ m	$= 7481.90$ s
	$\Delta h(5) = 0.0050$ m	$= 7401.37$ s
	$\Delta h(6) = 0.0042$ m	$= 7081.34$ s

where in this case for $i = 2$ to 6, $\Delta h(i) = \Delta h(1)/(i)$, and the trials cease when the difference between the summation of Δt and the direct integration result falls below 1 in 10^5.

EXAMPLE 18.2

If the tank described in Example 18.1 was subject to a constant or intermittent steady inflow during the emptying period discuss how this would affect the numerical integration considered in Example 18.1 and outline how a stabilization depth could be determined.

Solution

If the inflow rate is Q_{in} then the level in the tank will fall if $Q_{in} < Q_{out} = (kh)^{0.5}$ and rise if $Q_{in} > Q_{out} = (kh)^{0.5}$. Thus the effect of any inflow depends on the depth in the tank at the time the inflow occurs and the discharge resistance of the tank as described by equation (18.7). Care must be taken to recognize that the value of Δt will tend to infinity as the inflow rate approaches the outflow, as demonstrated by

$$\Delta t = \Delta h\, A_{\rm s}/[\![(2A_{\rm p})\{0.5[h+(h-\Delta h)]\}^{0.5}-Q_{\rm in}]\!] = \Delta h\, A_{\rm s}/(Q_{\rm out}-Q_{\rm in}).$$

This will lead to computational failure.

Similarly if the inflow is initiated at some point during the tank discharge it may be that the fluid level should rise in the tank. Hence any numerical solution should check the inflow against the mean outflow across any Δh increment to determine whether the integration should continue or whether the Δh value should become negative, i.e. surface level rises until stabilization occurs when $Q_{\rm out} = Q_{\rm in}$.

For any system the stabilization depth may be determined from an expression of the form

$$Q_{\rm in} = (kh_{\rm stabilization\ depth})^{0.5}.$$

EXAMPLE 18.3   Indicate the advantages of a numerical integration scheme.

Solution

1. The case of intermittent inflow has been addressed in Example 18.2 and will be included below.

2. The main advantage of a numerical solution lies in the treatment of tank cross-sectional area $A_{\rm s}$ in the equations in Section 18.1 and Examples 18.1 and 18.2. As an example, replace the rectangular tank in Example 18.1 with a tank where the surface area at any depth is given by tabular data. In this case it would be necessary to rewrite

$$\Delta t = \Delta h\, A_{\rm s}/[\![(2A_{\rm p})\{0.5[h+(h-\Delta h)]\}^{0.5}-Q_{\rm in}]\!]$$

    as

$$\Delta t = \Delta h\, 0.5(A_{{\rm s}(h)}+A_{{\rm s}(h-\Delta h)})/[\![(2A_{\rm p})\{0.5[h+(h-\Delta h)]\}^{0.5}-Q_{\rm in}]\!],$$

    with the values of $A_{\rm s}$ being interpolated between the nearest bracketing depth values. The comments made concerning stabilization level and the effect of inflow rate continue to apply.

3. Numerical integration would also allow a variable friction factor to be introduced, dependent upon the average fluid velocity across any depth increment.

18.3 TWO-RESERVOIR PROBLEM

Quasi-steady flow may also occur when a tank or reservoir is connected to a higher-level tank and fluid transfer occurs under gravity (Fig. 18.2). The major difference between this case and those previously described is the presence of two free surfaces.

FIGURE 18.2
Unsteady flow transfer
between two reservoirs

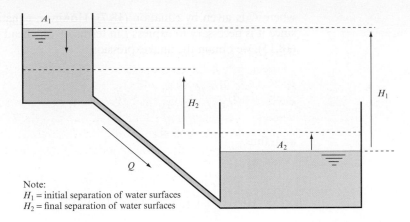

Note:
H_1 = initial separation of water surfaces
H_2 = final separation of water surfaces

Consider the case of two reservoirs connected by a single pipeline of length l and diameter d.

The instantaneous head difference h for quasi-steady conditions balances the total losses in the connecting pipeline. Hence,

$$h = (4fl/d + k)(\bar{u}^2/2g), \tag{18.12}$$

where, again, k represents the sum of all the separation losses in the connecting pipeline.

Now, from the application of continuity of flow between the two reservoirs at any instant, it follows that the rate of change of the fluid surface levels is linked, by the volume flow rate Q:

$$A_1\left(-\frac{dh_1}{dt}\right) = Q = A_2\left(\frac{dh_2}{dt}\right). \tag{18.13}$$

Note the difference in sign as one surface level rises while the other falls.

The rate of change of h, the difference in the two levels, is given by the sum of the rate of change of the fluid levels in the individual reservoirs. Hence,

$$-\frac{dh}{dt} = -\frac{dh_1}{dt} + \frac{dh_2}{dt}. \tag{18.14}$$

Substituting from equation (18.13),

$$\frac{dh_1}{dt} = \frac{dh}{dt}\Bigg/\left(1 + \frac{A_1}{A_2}\right). \tag{18.15}$$

From equations (18.12) and (18.13),

$$-A_1\frac{dh_1}{dt} = A_o\bar{u} = C\sqrt{h}, \tag{18.16}$$

where C is given by equation (18.7). However, equation (18.16) is still not solvable since it is necessary to express t in terms of one head only. Therefore, from equation (18.15), we obtain the final expression

$$-A_1\frac{\mathrm{d}h}{\mathrm{d}t}\bigg/\left(1+\frac{A_1}{A_2}\right) = C\sqrt{h} \tag{18.17}$$

and, thus,

$$t = -\int_{H_1}^{H_2} [A_1/C\sqrt{h}(1 + A_1/A_2)]\,\mathrm{d}h. \tag{18.18}$$

It has, of course, been assumed that the surface areas of the two reservoirs remain constant over the depth range H_1 to H_2. If this is not the case, then the solution of equation (18.18) may become difficult and it may be necessary to employ numerical or graphical methods of integration.

It must be stressed that the H_1 and H_2 limits in equation (18.18) refer to the initial and final vertical separation of the fluid surfaces in the two reservoirs, as shown in Fig. 18.2.

A finite difference approach to numerical integration may also be used for the two-reservoir case. From equation (18.17) and using the format introduced in equation (18.11) it follows that

$$\frac{h^{n+1} - h^n}{\Delta t} = [0.5k(h^{n+1} + h^n)]^{0.5}\left(\frac{A_\text{pipe cross-section}}{A_\text{surface 1}}\right)\left(1 + \frac{A_\text{surface 1}}{A_\text{surface 2}}\right). \tag{18.19}$$

The total time for the surface levels in the two tanks to change from the initial to final required conditions may then be summed as before.

EXAMPLE 18.4

For the tank and discharge pipe described in Example 18.1, with an initial tank depth of 1.5 m, determine the time taken for the difference in depth in the two tanks to fall from 3.0 m to 2.0 m depth if the discharge pipe is connected, at its previous discharge level of 2 m below the base of the first tank, into a second tank, surface area 4.5 m². The initial depth in the second tank is 0.5 m above the discharge pipe entry and the entry loss coefficient may be assumed to be 1.0.

Solution

From Example 18.1, $A_{\text{surface 1}} = 6\,\text{m}^2$, $A_{\text{surface 2}} = 4.5\,\text{m}^2$, $A_{\text{pipe}} = 0.3142 \times 10^{-3}\,\text{m}^2$ and the discharge flow rate is given by $Q = 2.0\, A_{\text{pipe}}\, h^{0.5}$, where h is the differential height between the fluid surfaces in each tank.

Thus equation (18.19) becomes

$$\Delta t = \Delta h\, A_{\text{s1}} / [\![2 A_{\text{p}} \{ 0.5[h + (h - \Delta h)] \}^{0.5} (1 + A_{\text{s1}}/A_{\text{s2}})]\!],$$

where h is the differential fluid surface separation height, which varies from 3 m (i.e. 1.5 m in first tank + 2.0 m drop to pipe discharge − 0.5 m fluid surface in second tank above pipe entry) to 2.0 m.

For the values quoted in this example the direct integral time from 3.0 m to 2.0 m differential tank depth and the numerical integration totals, dependent upon Δh values are given below.

Direct integration:	equation (18.9), $h_1 = 3.0,\, h_2 = 2.0$	Time to drain $= 2601.20\,\text{s}$
Numerical integration, (initial Δh choice $= 0.25$ m):	$\Delta h(1) = 0.250\,\text{m}$	$= 3347.44\,\text{s}$
	$\Delta h(2) = 0.125\,\text{m}$	$= 2968.46\,\text{s}$
	$\Delta h(3) = 0.0083\,\text{m}$	$= 2844.78\,\text{s}$
	$\Delta h(4) = 0.0063\,\text{m}$	$= 2783.42\,\text{s}$
	$\Delta h(5) = 0.0050\,\text{m}$	$= 2746.75\,\text{s}$
	$\Delta h(6) = 0.0042\,\text{m}$	$= 2601.17\,\text{s}$

The total exit flow summed over each Δt within the discharge period is $= 2.57\,\text{m}^3$, indicating a drop in the first tank surface level of $2.57/6.0 = 0.42\,\text{m}$ and a rise in the second tank surface level of $2.57/4.5 = 0.58\,\text{m}$, i.e. a change of 1 m in differential height.

Concluding remarks

The objective of this short treatment was to provide a starting point for a more detailed analysis of time-dependent flows. The assumption that conditions, while time dependent, changed sufficiently slowly for the equivalent steady flow relationships to be applied allowed the cases discussed to be solved, either by direct integration or by the application of finite difference methods. The quasi-steady assumptions will be met again later where it is often accepted that while the steady relationships, for example for frictional loss in transient flow, are inadequate, there is no suitably verified alternative. Similarly the finite difference approach, introduced in Chapter 5, will be found applicable to the modelling of transient behaviour later.

Summary of important equations and concepts

1. The analysis presented is predicated upon the acceptance of steady state relationships to apply at particular time instants in a time-dependent flow, hence Darcy's equation for frictional and separation loss in pipe flow may be applied to determine discharge rates.

2. The depth terms used in the direct integration or as limits within the finite difference solutions relate the fluid surface to the discharge point. In the case of a tank with a discharge pipe, then it is the vertical distance between interfaces to atmosphere that determine the depth term. Similarly where fluid flows between two or more reservoirs, it is the vertical separation of the surfaces that determines the depth term.

Problems

18.1 A vertical cylindrical tank, 0.4 m in diameter and 3 m high, is used as part of a flow calibration diverter unit. If the water collected in the tank, up to a depth of 2.5 m, is discharged through an orifice and valve in the tank base, that may be represented by a 50 mm diameter orifice of discharge coefficient 0.6, calculate the time to empty half the collected volume and express as a percentage of the time to empty fully.   [22.3 s, 29.3 per cent]

18.2 A vertical axis tank is conical in shape, the diameter increasing uniformly from 1 m at the base to 1.75 m diameter at a height of 3 m. The tank is to be emptied by means of a 50 mm orifice in the base having a discharge coefficient of 0.6. Calculate the time to reduce the water level from 2 m to 1 m above the base.   [233.8 s]

18.3 For the case set out in Problem 18.2 above, calculate the inflow necessary to hold the liquid level at 1.5 m above the base.   [383 litres min⁻¹]

18.4 For the tank in Problem 18.2, calculate the time to discharge the full contents of the tank if the discharge is carried away by a 25 mm diameter pipe, length 4 m, friction factor 0.005. Assume that the effect of the orifice to pipe connection can be represented by a separation loss having a k value of 2, and that final discharge is at tank base level.   [5249 s]

18.5 A rectangular cross-section tank, 2 m × 3 m, is filled with water up to a depth of 2 m. Calculate the time to reduce the volume in the tank by 50 per cent if the discharge is via a 40 mm diameter pipe, 6 m long, for which a friction factor of 0.005 may be assumed and the separation losses may be

represented by a k value of 0.9. Assume final discharge 2 m below tank base level.   [1276 s]

18.6 A 1.2 m deep rectangular tank is 2 m × 1 m in area and has a vee notch in one side. The lowest point of the vee notch is 770 mm above the base of the tank. A water supply to the tank of 1036 litres min⁻¹ establishes a steady depth of 1 m above the base of the tank. If the water inflow ceases, calculate the time needed for the level to fall to 150 mm.   [68.9 s]

18.7 A cylindrical tank is 1.8 m in diameter and 3 m long and is mounted horizontally. Oil of specific gravity 0.87 stored in the tank is drawn off through an orifice, 20 mm diameter, 0.6 discharge coefficient, at the tank's lowest point. Calculate the time taken to reduce the level in the talk from 0.9 m to 0.8 m above the orifice.   [699.4 s]

18.8 A 2 m deep tank is 2 m × 3 m in area and is divided into two equal halves by a vertical separation plate. Flow from one tank to the other takes place through a square orifice, 1 cm side, having a discharge coefficient of 0.6. If water is initially at 1.5 m depth on one side of the plate and 0.5 m depth on the other, calculate the time taken for the depths in both tanks to be equal.   [11290 s]

18.9 A rectangular cross-section tank of 12 m² surface area is filled to a depth of 2.5 m with water. Calculate the time to reduce the depth in the tank by 2.0 m if it is drained through a 3 m long, 0.04 m diameter pipe discharging 2 m below the base of the tank and having a friction factor of 0.01 and separation losses due to bends etc. of 0.9. Compare the integral for discharge time with a range of Δh incremental values to demonstrate the accuracy of a finite difference approach.   [5144.59 seconds]

18.10   For the tank and discharge pipe described in Problem 18.9, with an initial tank depth of 4.5 m, determine the time taken for the difference in depth in the two tanks to fall from 6.0 m to 5.0 m depth if the discharge pipe is connected, at its previous discharge level of 2 m below the base of the first tank, into a second tank, surface area 4.5 m². The initial depth in the second tank is 0.5 m above the discharge pipe entry and the entry loss coefficient may be assumed to be 1.0.
                    [554.4 seconds]

19

Unsteady Flow in Closed Pipeline Systems

THIS CHAPTER CONTINUES THE DEVELOPMENT OF THE STUDY OF TIME-DEPENDENT FLOWS BY considering mass oscillations. In particular surge conditions, such as load adjustment within hydroelectric power stations, that may be controlled by the installation of surge shafts, are investigated and illustrated by the available computer program, SHAFT. The equation of momentum is utilized to determine the surge pressure arising from gradual changes in the flow conditions. The limitations of the analysis as the rate of change of flow conditions increases are defined. ● ● ●

Whenever the steady-state operating conditions of a fluid system are changed, either intentionally by a planned valve or pump operation or inadvertently due to some system failure, then the change in the prevailing equilibrium flow conditions will be communicated to the system as a whole by pressure waves, travelling at the appropriate sonic velocity and propagating away from the point in the system where the change in the steady flow conditions was imposed. After a short time the system will attain a new equilibrium condition, assuming, of course, that the pressures generated during the change in steady conditions have not been sufficient to damage the system. For example, if the flow into a pipe network is increased during operation, then, after some time has elapsed, the flow will again become steady, with a new flow distribution throughout the pipe network. The transition from one set of steady conditions to the next will always be accompanied by the propagation of pressure and velocity waves throughout the system and a consequent variation of pressure at all points in the pipe network. The severity of these transient pressures, which may be of destructive proportions, will, for any particular system, depend on the rate of change of the flow velocity imposed locally at the item of equipment whose operation introduced the fluid acceleration. The potentially destructive nature of the pressure waves that are necessary to convey a change in operating conditions through a pipe system, although they may have a very short duration, explains the considerable practical importance of unsteady flow analysis.

The rate of change of conditions imposed on the system is of prime importance and governs, for any particular system, the method to be employed in calculating the effects of the pressure wave propagation. If the rate of change of the flow velocity is slow in comparison with the time taken for a pressure wave to pass through the system, then it is possible to consider the fluid as incompressible, i.e. a change in flow conditions is instantaneously transmitted through the system as the fluid is implied to have an infinite sonic velocity. The analysis of unsteady flow conditions from this assumption is known as rigid column theory and is considerably simpler and cheaper than a full solution involving fluid compressibility.

In recent years, the trend in engineering fluid systems has been towards higher flow rates with systems incorporating ever larger pumps and turbines, this being particularly the case in the power generation industry. The need to control such systems within a reasonable time has meant that the propagation of severe pressure transients as a result of system operation has become more common, with the majority of cases falling within the third unsteady flow category, where fluid compressibility and system elasticity can no longer be neglected. Historically, this category has been referred to as waterhammer, a title which in no way reflects the occurrence of phenomena in all fluid systems, whether the moving fluid is liquid ammonia, water, crude oil or natural gas. The historic method of solution was the graphical method proposed in the 1930s by Schnyder and Bergeron. However, since the early 1960s, the advent of the digital computer has meant that solutions by numerical methods can be attempted, particularly using the method of characteristics. Developments in these applications of numerical analysis will be presented in Chapters 20 and 21.

It is worth mentioning here that, although the rigid column theory may only be applied to transient conditions displaying slow rates of flow acceleration, the analysis based on the wave theory, incorporating the fluid compressibility, may be universally applied. Effectively, the rigid column theory is a limiting case of the wave theory. However, considerations of time and cost have in the past dictated that rigid column theory be applied whenever possible.

Rigid column theory will be outlined first, followed by an introduction to pressure transient wave theory and the numerical solution based on the method of characteristics.

19.1 RIGID COLUMN THEORY

Consider a simple pipeline from an open reservoir discharging into the atmosphere through a valve (Fig. 19.1). Friction is neglected, so that all points along the pipe are at a head H. The valve is now shut, with the result that after time t the velocity in the pipe has been reduced by $\Delta \bar{v}$ and the head at the valve has risen by ΔH. The total force accelerating the flow is $-\rho g A \Delta H$ and the acceleration of the flow is $-d\bar{v}/dt$. Note the negative signs here as the flow is retarded. Hence, from Newton's second law,

FIGURE 19.1
Simple pipeline of constant cross-section A terminated by a valve

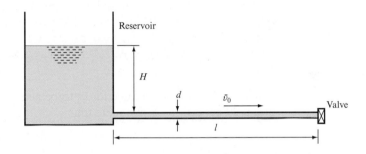

$$-\rho g A \Delta H = -\frac{d\bar{v}}{dt} \rho A l,$$

$$\Delta H = \frac{l}{g} \frac{d\bar{v}}{dt}. \tag{19.1}$$

ΔH in equation (19.1) is referred to as the surge or inertia pressure resulting from the closure of the valve.

Equation (19.1) may also be employed to describe the establishment of flow on valve opening in the pipeline illustrated in Fig. 19.1, provided that the flow acceleration involved is low. The flow is accelerated from rest, following valve opening, by the difference in head between the reservoir surface and the valve discharge level. Initially, the full reservoir head H is available to accelerate the flow. However, as the flow builds up, so this accelerating head is decreased because of the increasing frictional and separation losses along the pipeline. The final velocity $\bar{v}_0$ attained by the flow is given by the energy balance expression

$$H = 4fl_e\bar{v}_0^2/2dg, \tag{19.2}$$

where f is the friction coefficient, l_e is the equivalent length of pipe (and so represents all the separation losses in the system) and d is the pipe diameter. Thus, at any instant during acceleration, the accelerating head ΔH is

$$\Delta H = H - 4fl_e\bar{v}^2/2dg,$$

where $\bar{v}$ is the mean instantaneous fluid velocity at that time. From equation (19.1) it follows that

$$H - \frac{4fl_e\bar{v}^2}{2dg} = \frac{l}{g}\frac{\mathrm{d}\bar{v}}{\mathrm{d}t} \qquad (19.3)$$

and, by substitution from equation (19.2),

$$H(1 - \bar{v}^2/\bar{v}_0^2) = \frac{l}{g}\frac{\mathrm{d}\bar{v}}{\mathrm{d}t}.$$

Therefore,

$$t = \frac{l\bar{v}_0^2}{gH}\int_0^{\bar{v}_0} \frac{1}{\bar{v}_0^2 - \bar{v}^2}\,\mathrm{d}\bar{v} \qquad (19.4)$$

or

$$t = \tfrac{1}{2}(l\bar{v}_0^2/gH)\log_e[(\bar{v}_0 + \bar{v})/(\bar{v}_0 - \bar{v})], \qquad (19.5)$$

which implies an infinite time for $\bar{v}_0$ to be achieved. However, the flow velocity does attain 99 per cent of $\bar{v}_0$ in a finite time:

$$t_{0.99\bar{v}_0} = \tfrac{1}{2}(l\bar{v}_0^2/gH)\log_e(1.99/0.01)$$

$$t_{0.99\bar{v}_0} = 2.646\,l\bar{v}_0^2/gH.$$

The apparent anomaly suggested by the infinite time required to establish flow, as indicated by equation (19.5), arises from the assumption of fluid incompressibility. In practice, the propagation of pressure waves through a fluid results in equilibrium conditions being established in a shorter time than indicated by equation (19.5).

19.2 SURGE TANKS AND SHAFTS

It has been mentioned that rigid column theory may only be applied to unsteady flow conditions where the rate of change of the flow boundary conditions is slow compared with the transient pipe periods within the system. There is one very important area of unsteady flow where rigid column theory may be applied successfully, and this is to the analysis of waterhammer in hydroelectric installations and, particularly, the study of surge tanks for pressure surge control.

Figure 19.2 illustrates a typical hydroelectric scheme, where both the tunnels supplying water to the turbines and the surge shafts designed to limit surge pressures are cut into the rock.

If the electrical output taken from the generators varies, there is a tendency for the turbines to change speed; the turbine governors then operate control valves in the penstock approach to the turbines to stabilize the turbine speed. In the limiting case of load rejection, the penstock valves close, resulting in a pressure rise in the penstock that propagates towards the supply reservoir. If no surge control devices are provided,

FIGURE 19.2(a)

View down one of the Thames Water London Ring Main shafts that provide surge protection to a section of the London Ring Main – a £250 million project forming an 80 kilometre loop around London, at depths up to 75 m, with an average diameter of 2.5 m and with a capacity of 1300 million litres (Photo courtesy of Thames Water)

FIGURE 19.2(b)

Schematic of a surge shaft layout in a hydroelectric scheme. (Note that in a pump storage scheme, water is returned to the upper lake during low electricity demand periods.)

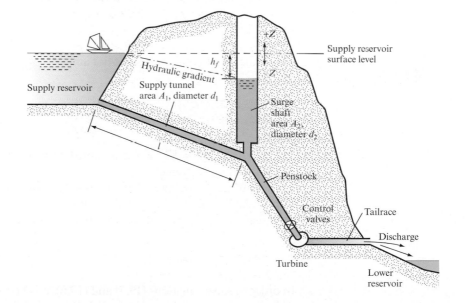

the system may then suffer severe damage, so the rate of deceleration of the water supply is reduced by, effectively, using the pressure surge in the penstock to divert flow into the surge tank, as shown.

Conversely, as the generators come back on load, the surge tank water re-enters the penstock, effectively increasing the flow acceleration and minimizing any negative surges that would otherwise occur. Apart from preventing damage to the installation, the provision of a surge tank also aids in the design of the turbine governors, as the presence of periodic pressure waves in the system following every load change would cause the governors to hunt. It is worth noting that the surge tank should be as close as possible to the turbines, as the penstock will suffer the unreduced surge pressure and an adjacent surge tank will reduce the penstock pipe period and the consequent surges following load change. The penstock has, therefore, to be designed to take the full surge pressure, a point sometimes not fully stressed.

Returning to Fig. 19.2, it will be seen that the surge tank or shaft extends up above the reservoir surface to prevent overflow during operation, although this has been known to happen inadvertently in some installations owing to bad design or an increase in system flow rate since the shaft was designed. Under normal conditions, the level in the tank will be below the reservoir level by an amount h_f equal to the friction and separation losses in the tunnels from the reservoir to the surge shaft.

As the penstock valves shut, so flow is diverted up the surge shaft and the water level rises to above the reservoir surface level. At any time t after the initiation of surge by load change, the level in the surge shaft has reached a level Z above the reservoir surface; then the head opposing h is

$$h = Z \pm h_f, \tag{19.6}$$

where h_f is positive if flow is into the shaft and vice versa.

The force opposing the motion of water in the supply tunnel is $\rho g A_1(Z \pm h_f)$ and the rate of change of momentum in the supply tunnel is $\rho A_1 l \, d\bar{v}/dt$. Applying rigid column theory to the mass oscillation that is now established between the reservoir and the surge shaft yields

$$\rho g A_1(Z \pm h_f) = -\rho A_1 l \frac{d\bar{v}}{dt},$$

where A_1 is the supply tunnel cross-sectional area

$$\frac{l}{g} \frac{d\bar{v}}{dt} + Z \pm h_f = 0, \tag{19.7}$$

where $h_f = 4fl\bar{v}^2/2d_1 g = K\bar{v}^2$ if l is assumed to include, as equivalent lengths, any separation losses in the tunnel.

At the entrance to the surge shaft, continuity of flow must apply, i.e. the flow along the supply tunnel at time t must equal the flow Q to the turbines plus the flow up into the surge shaft. Thus, as the surface velocity in the surge shaft is dZ/dt, it follows that

$$A_1 \bar{v} = A_2 \frac{dZ}{dt} + Q. \tag{19.8}$$

In order to solve equations (19.7) and (19.8), it is necessary to make assumptions about the variation of friction factor and the rate of flow reduction to the turbine. Even under total load rejection, when $Q = 0$, the variable friction factor still prevents an analytical solution and so it is now usual to use a step-by-step numerical integration technique, preferably utilizing a digital computer. If friction is ignored, then an approximation to maximum surge level and its period can be made. Neglecting f, the equations become

$$\frac{l}{g}\frac{d\bar{v}}{dt} + Z = 0 \qquad (19.9)$$

and, if $Q = 0$ for full load rejection, then

$$A_2\frac{dZ}{dt} = A_1\bar{v}.$$

When $t = 0$, $Z = 0$ in the frictionless case as $h_f = 0$ and $dZ/dt = Q_0/A_2$, where Q_0 is the steady flow before load rejection.

A solution to equations (19.7) and (19.8) under these conditions is

$$Z = (Q_0/A_2)\sqrt{(A_2l/A_1g)}\sin[\sqrt{(A_1g/A_2l)}]t,$$

so that the maximum level is

$$Z_{\text{max}} = (Q_0/A_2)\sqrt{(A_2l/A_1g)} \qquad (19.10)$$

and the period of the mass oscillation is

$$T = 2\pi\sqrt{(A_2l/A_1g)}. \qquad (19.11)$$

If friction is included, then the maximum level reached is lower and the water-level oscillation damps at a calculable frequency about the steady-state value of $-h_f$. Care should be taken to ensure that the turbine governors cannot respond to this frequency or that it does not correspond to a system resonant frequency. One important point here is that the minimum extreme of the oscillation, represented by $-Z_{\text{max}}$ in the frictionless case, should not be greater than the distance to the penstock entry, as this would result in air entrainment into the penstock.

In order to solve for the friction case, equations (19.7) and (19.8) can be treated numerically as shown by Example 19.1 below. The various designs of surge shaft will be discussed later, in Chapter 20.

Equation (19.7) is, however, amenable to finite difference numerical integration in a manner similar to that introduced in Sections 5.16 and 18.2. Rewriting equation (19.7) as

$$\Delta v = \frac{g}{l}\left(-Z \pm \frac{4flv|v|}{2d}\right)\Delta t. \qquad (19.12)$$

Assigning superscript $n + 1$ to conditions at one time-step in the future and n to the current time, then equation (19.12) becomes

$$v^{n+1} - v^n = \frac{g}{l}\left(-Z^n \pm \frac{4flv^n|v^n|}{2d}\right)\Delta t, \qquad (19.13)$$

and by continuity, assuming that the turbine has shut down,

$$Z_{\text{shaft}}^{n+1} = Z_{\text{shaft}}^n + \frac{0.5(v^{n+1} + v^n)A_{\text{tunnel}}\Delta t}{A_{\text{shaft}}}. \qquad (19.14)$$

The result of such a numerical integration clearly depends on experience in the choice of Δt and may be investigated further through the computer program SHAFT available with this text.

19.3 COMPUTER PROGRAM SHAFT

Program SHAFT calculates the water level position within a vertical constant cross-section surge shaft following total load rejection at the turbines. In addition the program displays shaft water levels at each time-step and the theoretical frictionless surge amplitude and period. The program utilizes rigid column surge theory and, in particular, the momentum and continuity equations for a simple system, equations (19.7) to (19.9).

The data required include tunnel length L (m), tunnel diameter d_1 (m), tunnel friction factor f, to include any separation losses, see Section 19.2, initial steady flow to the turbines, Q ($\mathrm{m^3\,s^{-1}}$), surge shaft diameter d_2 (m), selected time step Δt (s), and overall simulation time T (s).

19.3.1 Application example

Example 19.1 and Fig. 19.3 present the data output for a typical application of SHAFT.

EXAMPLE 19.1

In a hydroelectric scheme the supply tunnel is 1.25 m in diameter and has a friction factor of 0.01. At 200 m along the tunnel there is an open surge shaft of 4 m diameter. The steady flow to the turbines is 2 m³ s⁻¹. Determine the period of the mass oscillation and the peak water level in the surge shaft for both frictionless and friction loss cases, where the friction term is assumed to have incorporated any separation losses through an equivalent tunnel length extension of 20 m.

Solution
If f = 0.0 then

$$Z_{max} = (Q/A_{shaft})\,(A_{shaft}\,L/A_{tunnel}\,g)^{0.5},$$

where $A_{tunnel} = 1.227\ \mathrm{m^2}$, $A_{shaft} = 12.57\ \mathrm{m^2}$, $Q = 2\ \mathrm{m^3\,s^{-1}}$ and $L = 200$ m.
Then

$$Z_{max} = (2.0/12.57) \times (12.57 \times 200.0/1.227 \times 9.81)^{0.5} = 2.29\ \mathrm{m}.$$

The period of the mass oscillation in a frictionless case is given by equation (19.11),

$$T = 2\pi\,(A_{shaft}\,L/A_{tunnel}\,g)^{0.5} = 2 \times (22/7) \times (12.57 \times 200/1.227 \times 9.81)^{0.5}$$

$$= 90.8\ \mathrm{s}.$$

The numerical solution requires the use of equations (19.13) and (19.14), either within a spreadsheet or as the basis for a simple program. The separation losses due along the tunnel and due to the entry to the surge shaft have been incorporated for simplicity in this example by including an equivalent length in the calculation of

$h_f = 4fLV^2/(2gD_{\text{tunnel}})$; however, many texts differentiate between the tunnel losses and the shaft entry loss. Also the equivalent length is excluded from the frictionless case calculations of peak surge and mass oscillation period.

Program SHAFT available with this text may be used to investigate the interaction between the variables discussed. Note that there is no explicit data entry for separation losses, these have to be represented within SHAFT by factoring the assumed friction factor, e.g. if the equivalent length as used in this example is denoted by L_e then

$$f_{(\text{friction+separation})} = f_{(\text{friction only})} \times (L + L_e)/L,$$

hence for this example f becomes $0.01 \times 220/200 = 0.011$.

The SHAFT program requires the following data to be input in response to screen prompts: tunnel length and diameter (m), tunnel friction factor (note points above), initial turbine steady flow ($m^3\ s^{-1}$), surge shaft diameter (m), calculation time-step (s), and simulation duration time (s).

Figure 19.3 illustrates the surge shaft surface vertical velocity and elevation relative to the reservoir surface for this example, with a time-step of 0.5 s. The frictionless maximum surge is 2.29 m and the frictionless mass oscillation period is 90.8 s.

The figure illustrates that with friction and separation losses included, the peak surge becomes 2.65 m (i.e. $-0.867 + 1.780$) above the reservoir surface.

Program SHAFT also stores surge shaft water surface elevation and vertical velocity at each time-step in a file SHAFT.TXT, which may be used with a standard spreadsheet to generate output similar to that illustrated by Fig. 19.3.

FIGURE 19.3

Example of surge shaft surface oscillation and velocity

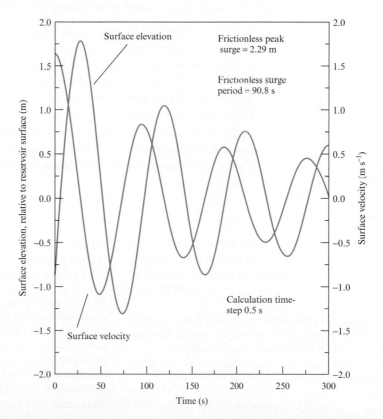

19.3.2 Additional investigations using SHAFT

The simulation may be used to investigate:

1. for a constant set of tunnel and turbine flow data the effect of surge shaft diameter;
2. the importance of frictional and separation losses to the maximum surge predicted;
3. the importance of time-step choice.

Concluding remarks

The increasing severity of the time dependency represented by the material presented in this chapter allows the analysis of mass oscillations, i.e. those cases where the surge condition is still predictable without recourse to a full transient analysis. The determination of surge shaft dimensions presented in this chapter offers a practical example of the scope of this level of simulation. The treatment of time-dependent flows will continue in Chapters 20 and 21 with a classical waterhammer, or pressure transient, analysis, together with a discussion of the associated cases of free surface wave propagation and low-amplitude gas flow transients.

Summary of important equations and concepts

1. Mass oscillation and the inertia approach to waterhammer surge pressure prediction can only be effective in cases where the rate of change of the boundary condition is slow or where the elasticity of the conduit and the compressibility of the fluid may be ignored. In addressing this area this chapter lays the basis for later more comprehensive analysis.
2. Finite difference solutions depend on the experience of the user. Time-step size in this case and incremental depth in Chapter 18 are both examples that illustrate that it is necessary to investigate the sensitivity of the solution to such variable choice to prevent instability, as discussed in Section 5.16.

Problems

19.1   The outlet valve on a 25 m pipeline is closed in 2 seconds. If the initial velocity of flow was 8 m s⁻¹ calculate the pressure rise on valve closure if the pipe is assumed to be rigid.   [100 kN m⁻²]

19.2   Show that the time taken to establish flow in a pipeline from a large reservoir by opening a valve on the pipe discharge is infinite if rigid column theory is applied.

Further show that the velocity–time curve following valve operation is given by an equation of the form

$$\text{Time to } nV_0 = \frac{0.102}{2} \, lV_0 \log_e[(1 + n)/(1 - n)],$$

where V_0 is final flow velocity after an infinite time, l is the ratio of pipe length to head available in the reservoir, and n is the percentage of final velocity attained at any time.

19.3   A valve positioned at the discharge end of a 30 m pipe is opened at a time when the level in the tank supplying the pipe is 6 m above the pipe inlet. Calculate the time taken to accelerate the flow to 50 per cent of its final value and plot the curve of discharge velocity against time following valve operation. (Assume the friction factor for the pipe is 0.01, its diameter is 0.05 m and any separation losses may be represented by 50 diameters of equivalent length.)

                                                                                      [0.6 s]

19.4 The pressure rise on closure of a valve on the discharge from a 100 m long pipe supplied by a constant head reservoir whose fluid surface level is 6 m above the valve is to be limited to 190 kN m^{-2}, this figure being calculated by rigid column theory. If the pipe diameter is 25 mm, the pipe friction factor applicable is 0.01, and if the separation losses may be represented by the k value of 10 for the valve fully open, calculate the minimum valve closure time to comply with the design pressure limitation.   [0.6 s]

19.5 For the case set out in Problem 19.4 above, determine the time taken to re-establish flow if the valve is reopened at some later time. Assume 99 per cent flow time to be sufficiently accurate.   [5.13 s]

19.6 In a small hydroelectric plant the supply tunnel is 1.2 m in diameter and the friction factor is estimated as 0.01. A simple surge shaft of 3.5 m diameter is positioned 150 m downstream of the supply reservoir. The initial flow to the turbines is 2.25 m^{3}s^{-1}. Calculate the maximum upsurge in the shaft relative to the initial steady-state level.
                    [1.7 m, value dependent on time-step chosen]

19.7 In a hydroelectric scheme the water flow in the low-pressure supply tunnel under full load is 12 m^{3}s^{-1}. The low-pressure tunnel is 980 m long, of 2.133 m diameter, and is protected by a simple surge shaft of 6 m diameter. The friction factor applicable to the supply tunnel is 0.005 and the entrance to the surge shaft from the supply tunnel is at a level 21.5 m below the water surface in the supply reservoir.

On start-up the penstock valves are suddenly opened to full load condition and the surge shaft level falls to supply water to the penstock. Calculate by step-by-step integration the minimum surge shaft water-level.   [9.7 m]

19.8 Show that, if friction loss is proportional to the square of flow velocity, the mass oscillation in a simple surge shaft following a sudden load rejection on flow shutdown is given by an expression of the form

$$\frac{\mathrm{d}^2 z}{\mathrm{d}t^2} \pm 2f\frac{d^2}{D^3}\left(\frac{\mathrm{d}z}{\mathrm{d}t}\right)^2 + \frac{Ag}{a\mathrm{L}}z = 0,$$

where D, A, d, a are the diameter and cross-sectional areas of the supply tunnel and shaft, respectively: f is the friction factor for the supply tunnel; L is the length of the supply tunnel and z is the shaft water surface level relative to the supply reservoir.

19.9 In a hydroelectric scheme the supply tunnel is 2.25 m in diameter and has a friction factor of 0.01. At 500 m along the tunnel there is an open surge shaft of 8 m diameter. The steady flow to the turbines is 20 m^{3}s^{-1}. Determine the period of the mass oscillation and the peak water level in the surge shaft for both frictionless and a friction loss case, where the friction term is assumed to have incorporated any separation losses through an equivalent tunnel length extension of 100 m.   [159.6 s, 10.09 m, 16.56 m]

Pressure Transient Theory and Surge Control

THIS CHAPTER PRESENTS A TREATMENT OF TIME-DEPENDENT FLOW WHERE THE RATE OF change of the controlling parameters is sufficient to propagate pressure transients throughout the system. The analysis presented will concentrate on the classical treatment of waterhammer, detailing the pressure oscillations that would accompany rapid changes in flow condition in a simple pipeline and introducing the Joukowsky pressure rise following instantaneous flow stoppage. The simplified wave equations necessary to illustrate the development of the reflection and transmission coefficients appropriate for a range of boundary conditions will be introduced and linked back to earlier derivations within the treatment of the finite difference solution of the Navier–Stokes equations. The central importance of wave speed will be emphasized and the dependence of this transient propagation velocity on both pipe and fluid properties, and particularly on free gas, will be defined. A computer program to facilitate wave speed calculation, including all these variables, will be introduced and demonstrated.

The objective of a pressure transient analysis will generally be to define the likely surge pressures to be encountered as a result of any system operation and hence to identify the necessity for the introduction of surge control devices. The objective of surge analysis is almost exclusively control rather than eradication of the surge and this chapter will introduce a range of commonly found surge limiting devices or operating procedures, including surge shafts, air chambers, relief valves or the specification of single or multiple phase valve closure rates. ● ● ●

Pressure surge, transient propagation, waterhammer and unsteady flow are all synonyms for the frequent and entirely normal flow regime in any fluid-carrying system when the running condition is subject to change – either intentional, for example by selection of an increased pump speed, or due to some system failure, for example a power loss or some form of fluid/system interaction. Thus pressure surge or transient conditions are a normal occurrence and as such might be expected to be well understood. While transient propagation may be found in all fluid systems, from aircraft in-flight refuelling to building drainage vent systems, the subject is still shrouded in misunderstanding and its 'difficult' mythology, despite the fact that several of the largest civil engineering consultants spend inordinate time evaluating surge as a design problem. Unchecked and misunderstood pressure transient propagation can lead to catastrophic system failure.

Changes in system operating conditions lead to transient propagation as this is the only mechanism available to transmit information through the system. Information is transmitted at the appropriate acoustic velocity, which depends not only on the fluid but also on both the cross-sectional dimensions and elasticity of the carrying conduit and the levels of free gas in the fluid. For any particular system/fluid combination, the magnitude of the surge depends on the rate of change of the flow conditions generating the transient. Thus the study of transients is really a study of control rather than eradication. Pressure transients arise due to changes in flow condition and increase in severity as the rate of change of the flow conditions increases, thus a basic premise of all transient control strategies is to reduce the rate of change of the flow condition, achievable by a whole series of techniques having in common the maintenance of flow velocity and/or its gradual change to the new desired or imposed condition. Examples that come readily to mind in the wider area of civil engineering hydraulics include air chambers to sustain flow on pump failure or surge shafts to divert flow on turbine speed change. In drainage and vent system applications the air admittance valve is a ready example. The simplest way to limit positive transients is to blow off fluid through a relief valve: for negative transients the simplest solution is to allow inflow of gas or liquid to maintain flow conditions, for example in aircraft fuel systems.

Transient theory has a long developmental history which is truly international and truly interdisciplinary, making this area of fluid mechanics an exciting and stimulating area of study and research. The literature relating to pressure transient, or waterhammer as the subject was known, is extensive and covers the last hundred years. Contributions to the literature range from the purely theoretical, such as early attempts to use Laplace transforms, through discussions of analysis methods to reports of particular transient conditions in individual plant or fluid systems and the development of computer simulations.

It would be inappropriate to include much of this work, so the scope of this review will be restricted to the development of the basic theory of transient propagation, the development of modern analysis techniques, concentrating on the use of finite difference-based computing models, and the treatment of some specialist areas applicable to utility services.

While there was associated acoustics research in the 1800s, the first major contribution to transient theory was due to Joukowsky[1], working at the Moscow Water Works in 1900, who published the results of a comprehensive experimental programme that laid the foundations of all subsequent waterhammer analysis. Joukowsky's work is notable as it derived for the first time the relationship between surge pressure rise Δp, wave speed c, fluid density ρ and the change in fluid velocity ΔV. For an instantaneous flow stoppage Joukowsky developed the expression,

$$\Delta p = \rho c V_0.$$

This form of the Joukowsky expression only applies to instantaneous flow stoppages, or more accurately those completed in less than one pipe period and is often misquoted leading to gross overestimates of surge pressure levels. More accurately the change in pressure may be linked to a change in the flow velocity, but again it would be necessary to track wave reflections to obtain a pressure prediction:

$$\Delta p = \rho c \Delta V.$$

Thus pressure change depends on density, velocity change and wave speed. Wave speed is the velocity at which a pressure wave would move through the fluid, propagating information on the change in system conditions. As such wave speed is not a constant but depends upon the fluid and the enclosing conduit properties, as well as being heavily dependent on any free gas content. The more elastic the conduit or the more compressible the fluid/gas mixture, the lower the wave speed.

Once wave speed has been determined it is possible to determine the wave travel times to the system boundaries, or to any intermediate change in conduit properties, that would develop a reflected wave. The round-trip time to a reflector at distance L from the source of the transient is thus $2L/c$ – known as the pipe period. A range of reflecting boundaries may be identified and the reflection and possible transmission coefficients calculated.

The essential point is that if the reflection arrives 'back' at the source of the transient before the change in flow condition is complete then the reflected wave must be superimposed on the generated wave, altering the transient pressure generated. Thus 'slow' valve closures – defined as closure taking longer than one pipe period – do not generate the full Joukowsky instantaneous pressure rise. The simplest way to limit a transient is therefore to slow down the rate of change of the flow condition, either by slowing a valve action or by providing make up flow or an alternative flow path.

Joukowsky realized the importance of wave speed and for the first time established a rational explanation for the complex pressure variations experienced within pipe networks by introducing the concept of reflection and transmission at pipe boundaries, recognizing the importance of pipe period and developing an expression linking pressure rise to fluid density, wave speed and velocity change. The results presented in Simin's 1904 translation show agreement within ±15 per cent on pressure measurement and ±2 per cent on wave speed, remarkable achievements for the instrumentation employed.

While Joukowsky is generally credited with the foundation of waterhammer theory, there is evidence[2] that an American researcher, J. P. Frizell, developed similar expressions linking pressure variations to wave speed and flow velocity for sudden valve closure in 1897.

Allievi[3] later established the Joukowsky relationships by making the same assumptions, namely a frictionless pipe, uniform pipe dimensions, homogenous wall material and uniform velocity distribution, extending Joukowsky's results to 'slow' valve closures, i.e. valve closures that take longer than one pipe period, and to the prediction of pressure variation at points along a pipeline by introducing wave superposition, establishing definition of 'time' in terms of pipe periods as central to the understanding of pressure transient propagation.

It was 1925 before an English translation of Allievi's work was generally available. In the interim a wide range of approximations were used without an understanding of their limitations, resulting in a range of predicted surge pressures for any case, e.g. from 65 to 535 per cent over-prediction for a simple valve closure.[4]

The expansion of large-scale water distribution networks and hydroelectric plants, particularly in the USA, made the provision of an accurate design and analysis method vital, a need filled by the Schnyder and Bergeron[5,6,7] graphical methods. The major advantage of graphical techniques was the inclusion of friction by means of discrete pressure drops, referred to as friction joints, along the length of the pipeline analyzed. The disadvantages included the possibility of cumulative graphical errors, the inherent complexity accompanying accurate friction modelling and the inability to use the technique as an iterative design tool prior to the computerization of the technique in the early 1960s.[8] Many of the complex pipe network graphical solutions generated in this period were almost works of art. The 1930s did, however, see an enormous increase in interest in waterhammer as an engineering problem, particularly in the USA where ASME/ASCE held symposia in 1933 and 1937.

In parallel to these developments a number of analytical solutions were proposed, involving the linearization of the friction term. These were poor approximations in the turbulent flow regime normally associated with surge and the periodic flow reversals that accompany transient reflections, resulting in complex mathematical solutions for even the simplest surge case, reinforcing the superiority of graphical techniques up to the 1950s when the next developmental stage began, namely the use of digital computing allied to the finite difference-based method of characteristics.

The method of characteristics is a general mathematical method particularly suited to the solution of pairs of quasi-linear hyperbolic partial differential equations linking two dependent and two independent variables. The equations defining pressure transient propagation within fluid piping systems and wave propagation in open-channel or partially filled pipe flow belong to this family of problems. The method was first proposed by Rieman in 1860 while he was studying finite amplitude sound wave propagation in air. In 1900 Massau[9] employed the method in a study of unsteady flow in open channels.

The first application to pressure transient analysis was due to Lamoen[10], while two later contributions by Gray[11,12] received wider attention, although neither author considered the use of a computer. Following Gray's work, a considerable number of papers were presented in the USA employing this technique, among the first being Ezekial and Paynter[13] and most significantly Streeter with a range of co-authors.[14,15,16,17] Much of the published work rested on the presentation of the method of characteristics by Mary Lister[18], which remains the basis for the analysis techniques in this presentation.

In the UK and Europe interest in the use of the method of characteristics developed rapidly[19,20,21,22,23,24], covering both the traditional civil engineering applications and other new problems, including transients in nuclear power station systems, aircraft fuel systems, in-flight refuelling and offshore oil rig networks – both undersea and rig based. By the early 1970s this technique was established as the industry standard for transient analysis, a fact clearly evident in the wide range of international papers presented at the BHRA Pressure Surge conferences held regularly since 1972.[25]

Despite claims[14] that the method of characteristics treated frictional loss as equally distributed along the whole pipeline, in fact the method extends and improves upon the graphical technique's friction joint approach by utilizing the computer's ability to introduce as many friction joints as the analyst would wish, thus allowing the model to approach the continuously distributed friction loss case.[26] The graphical technique and its computerized version are now of historic interest only.

The introduction and acceptance of the method of characteristics as the most suitable technique for the analysis of pressure transient phenomena opened up the whole subject, allowing previously difficult transient conditions to be treated and designs for systems modified to prevent potentially damaging transient propagation. Areas that have particularly developed since the introduction of computing methods have included problems involving gas release as line pressure falls in response to negative transient pressure wave propagation and problems involving cavity formation modelling during column separation, both with and without gas release.[23] In this later case the advantages of the computing methods over their earlier graphical counterparts is particularly clear as the accurate monitoring of cavity growth and collapse is essential in determining the final interface fluid velocity, which in turn determines, via Joukowsky's relationship, the resulting pressure surge on cavity collapse. The analysis of trapped air as a boundary condition[27] has wide-ranging implications, from accurate transient pressure measurement to the fast priming of fire-fighting networks[28,29] and the prevention of potentially hazardous transients.[30]

Similarly, the introduction of computing methods based around the method of characteristics has allowed progress in the study of the interface between fluid-borne transients and the vibration of structures, a subject of obvious importance in the study of structurally transmitted sound waves. The interaction of machine characteristics and pipe flows can also be studied utilizing these methods, improving understanding within the design of hydroelectric power stations and surge alleviation devices. Accurate prediction of transient response in a wide range of systems has also led to further development of the interpolation techniques inherent in the method of characteristics' dependence on both an initial steady flow condition and base values at each time-step that may not be located at calculation nodes.[31,32]

The establishment of the method of characteristics as the industry standard for waterhammer prediction encouraged its application in the study of other unsteady flow conditions belonging to the same family of transient phenomena defined by the St Venant equations of continuity and momentum. Recent applications to free surface flows[33,34,35,36] and low amplitude air pressure transients[37,38,39] demonstrate the flexibility of the methodology. Examples of all these applications are included in Chapter 21, following the development of the underlying waterhammer case in this chapter.

The introduction of rapid computing techniques has implications for the use of pressure transient analysis as a design tool rather than as a problem-solving technique only utilized following the demonstration of a surge-generated problem within a fluid system. This has far reaching consequences and illustrates the correct use of computing power within the engineering system design process.

Before embarking on the analysis of pressure transient phenomena and the derivation of the appropriate wave equations, it will be useful to describe the general mechanism of pressure propagation by reference to the events following the instantaneous closure of a valve positioned at the mid-length point of a frictionless pipeline carrying fluid between two reservoirs. The two pipeline sections upstream and downstream of the valve are identical in all respects. Transient pressure waves will be propagated in both pipes by valve operation and it will be assumed that the rate of valve closure precludes the use of rigid column theory.

As the valve is closed, so the fluid approaching its upstream face is retarded with a consequent compression of the fluid and an expansion of the pipe cross-section. The increase in pressure at the valve results in a pressure wave being propagated upstream, which conveys the retardation of flow to the column of fluid approaching the valve along the upstream pipeline. This pressure wave travels through the fluid at the appropriate sonic velocity, which will be shown to depend on the properties of the fluid and the pipe material.

Similarly, on the downstream side of the valve the retardation of flow results in a reduction in pressure at the valve, with the result that a negative pressure wave is propagated along the downstream pipe which, in turn, retards the fluid flow. It will be assumed that this pressure drop in the downstream pipe is insufficient to reduce the fluid pressure to either its vapour pressure or its dissolved gas release pressure, which may be considerably different.

Thus, closure of the valve results in the propagation of pressure waves along both pipes and, although these waves are of different sign relative to the steady pressure in the pipe prior to valve operation, the effect is to retard the flow in both pipe sections. The pipe itself is affected by the wave propagation as the upstream pipe swells as the pressure rise wave passes along it, while the downstream pipe contracts owing to the passage of the pressure-reducing wave. The magnitude of the deformation of the pipe cross-section depends on the pipe material and can be well demonstrated if, for example, thin-walled rubber tubing is employed. The passage of the pressure wave through the fluid is preceded, in practice, by a strain wave propagating along the pipe wall at a velocity close to the sonic velocity in the pipe material. However, this is a secondary effect and, while knowledge of its existence can explain some parts of a pressure–time trace following valve closure, it has little effect on the pressure levels generated in practical transient situations.

Following valve closure, the subsequent pressure–time history will depend on the conditions prevailing at the boundaries of the system. In order to describe the events following valve closure in the simple pipe system outlined above, it will be easier to refer to a series of diagrams illustrating conditions in the pipe at a number of time-steps (Fig. 20.1).

Assuming that valve closure was instantaneous, the fluid adjacent to the valve in each pipe would have been brought to rest and pressure waves conveying this information would have been propagated at each pipe at the appropriate sonic velocity c. At a later time t, the situation is as shown in Fig. 20.1(a), the wavefronts having moved a distance $l' = ct$ in each pipe. The deformation of the pipe cross-section will also have travelled a distance l' as shown.

The pressure waves reach the reservoirs terminating the pipes at a time $t = l/c$ following valve closure (Fig. 20.1(b)). At this instant, an unbalanced situation arises at the pipe–reservoir junction, as it is clearly impossible for the layer of fluid adjacent to the reservoir inlet to maintain a pressure different from that prevailing at that depth in the reservoir. Hence, a restoring pressure wave having a magnitude sufficient to bring the pipeline pressure back to its value prior to valve closure is transmitted from each reservoir at a time l/c. For the upstream pipe, this means that a pressure wave is propagated towards the closed valve, reducing the pipe pressure to its original value and restoring the pipe cross-section. The propagation of this wave also produces a fluid flow from the pipe into the reservoir as the pipe ahead of the moving wave is at a higher pressure than the reservoir. Now, as the system is assumed to be frictionless, the magnitude of this reserved flow will be the exact opposite of the original flow velocity, as shown in Fig. 20.1(c).

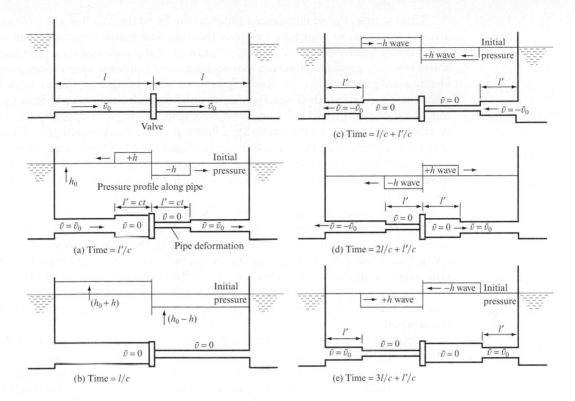

FIGURE 20.1

Pressure and pipe diameter deformation at a number of instants following an instantaneous valve closure. Frictional effects have been neglected

At the downstream reservoir, the converse occurs, resulting in the propagation of a pressure rise wave towards the valve and the establishment of a flow from the downstream reservoir towards the valve (Fig. 20.1(c)).

For the simple pipe considered here, the restoring pressure waves in both pipes reach the valve at a time $2l/c$. The whole of the upstream pipe has, thus, been returned to its original pressure and flow has been established out of the upstream pipe. At time $2l/c$, as the wave has reached the valve, there remains no fluid ahead of the wave to support the reversed flow. A low-pressure region, therefore, forms at the valve, destroying the flow and giving rise to a pressure-reducing wave which is transmitted upstream from the valve, once again bringing the flow to rest along the pipe and reducing the pressure within the pipe (as shown in Fig. 20.1(d)). It is assumed that the pressure drop at the valve is insufficient to reduce the pressure to the fluid vapour pressure. As the system has been assumed to be frictionless, all the waves will have the same absolute magnitude and will be equal to the pressure increment, above steady running pressure, generated by the closure of the valve. If this pressure increment is h, then all the waves propagating will be $\pm h$, as shown in Fig. 20.1. Thus, the wave propagating upstream from the valve at time $2l/c$ has a value $-h$, and reduces all points along the pipe to $-h$ below the initial pressure by the time it reaches the upstream reservoir at time $3l/c$.

Similarly, the restoring wave from the downstream reservoir that reached the valve at time $2l/c$ had established a reversed flow along the downstream pipe towards the closed valve. This is brought to rest at the valve, with a consequent rise in pressure which is transmitted downstream as a $+h$ wave arriving at the downstream reservoir at $3l/c$, at which time the whole of the downstream pipe is at pressure $+h$ above the initial pressure with the fluid at rest.

Thus, at time $3l/c$ an unbalanced situation similar to the situation at $t = l/c$ again arises at the reservoir–pipe junctions with the difference that it is the upstream pipe which is at a pressure below the reservoir pressure and the downstream pipe that is above reservoir pressure. However, the mechanism of restoring wave propagation is identical with that at $t = l/c$, resulting in a $-h$ wave being transmitted from the upstream reservoir, which effectively restores conditions along the pipe to their initial state (as shown in Fig. 20.1(e)), and a $+h$ wave being propagated upstream from the downstream reservoir, which establishes a flow out of the downstream pipe. Thus, at time $t = 4l/c$ when these waves reach the closed valve, the conditions along both pipes are identical to the conditions at $t = 0$, i.e. the instant of valve closure. However, as the valve is still shut, the established flow cannot be maintained and the cycle described above repeats.

The pipe system chosen to illustrate the cycle of transient propagation was a special case as, for convenience, the pipes upstream and downstream of the valve were identical. In practice, this would be unusual. However, the cycle described would still apply, except that the pressure variations in the two pipes would no longer show the same phase relationship. The period of each individual pressure cycle would be $4l/c$, where l and c took the appropriate values for each pipe. It is important to note that once the valve is closed the two pipes will respond separately to any further transient propagation.

The period of the pressure cycle described is $4l/c$. However, a term often met in transient analysis is 'pipe period'; this is defined as the time taken for a restoring reflection to arrive at the source of the initial transient propagation and, thus, has a value $2l/c$. In the case described, the pipe period for both pipes was the same and was the time taken for the reflection of the transient wave propagated by valve closure to arrive at the closed valve from the reservoirs.

From the description of the transient cycle above (Fig. 20.1), it is possible to draw the pressure–time records at points along the pipeline (as shown in Fig. 20.2). These variations are arrived at simply by calculating the time at which any one of the $\pm h$ waves reaches a point in the system assuming a constant propagation velocity c. The major interest in pressure transients lies in methods of limiting excessive pressure rises and one obvious method is to reduce valve speeds. However, reference to Fig. 20.2 illustrates an important point: no reduction in generated pressure will occur until the valve closing time exceeds one pipe period. The reduction in peak pressure achieved by slowing the valve closure arises as a result of the arrival of negative waves from the upstream reservoir at the valve prior to valve closure and, as no reflection can return to the valve before a time $2l/c$ from the start of valve motion, no beneficial pressure relief can be achieved if the valve is not open beyond this time. Generally, valve closures in less than a pipe period are referred to as *rapid* and those taking longer than $2l/c$ are *slow*.

In the absence of friction, the cycle would continue indefinitely. However, in practice, friction damps the pressure oscillations within a short period of time. In systems where the frictional losses are high, the neglect of frictional effects can result in a serious underestimate of the pressure rise following valve closure. In these cases, the head at the valve is considerably lower than the reservoir head. However, as the flow is retarded, so the frictional head loss is reduced along the pipe and the head at the valve increases towards the reservoir value. As each layer of fluid between the valve and the reservoir is brought to rest by the passage of the initial $+h$ wave, so a series of secondary positive waves, each of a magnitude corresponding to the friction head

FIGURE 20.2

Pressure variation following an instantaneous valve closure at points along the two identical pipes linking the reservoirs. Note that the closer to the reservoir the recording point, the shorter is the duration of the pressure change

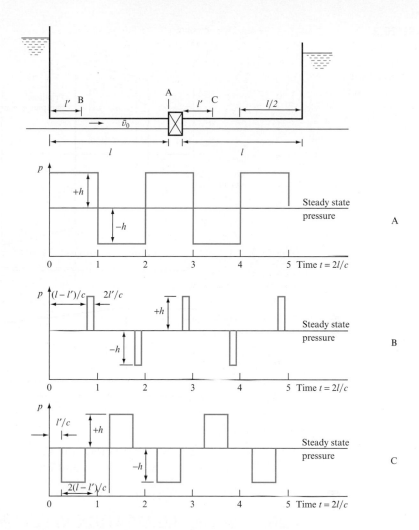

recovered, is transmitted towards the valve, resulting in the full effect being felt at time $2l/c$ (as shown in Fig. 20.3). As the flow reverses in the pipe during time $2l/c$ to $4l/c$, the opposite effect is recorded at the valve because of the re-establishment of a high friction loss, these variations being shown by lines AB and CD (Fig. 20.3). In certain cases, such as long-distance oil pipelines, this effect may contribute the larger part of the pressure rise following valve closure, known as 'line packing'.

In addition to the assumptions made with regard to friction in the cycle description, mention was also made of the condition that the pressure drop waves at no time reduced the pressure in the system to the fluid vapour pressure. If this had occurred, then the fluid column would have separated and the simple cycle described would have been disrupted by the formation of a vapour cavity at the position where the pressure was reduced to vapour level. In the system described, this could happen on the valve's downstream face at time 0 or on the upstream face at time $2l/c$. The formation of such a cavity is followed by a period of time when the fluid column moves under the influence of the pressure gradients between the cavity and the system boundaries. This period is normally terminated by the generation of excessive pressure on the final collapse of the cavity. This phenomenon is generally referred to as *column*

FIGURE 20.3

Effect of friction on the pressure variation recorded at the valve following instantaneous valve closure

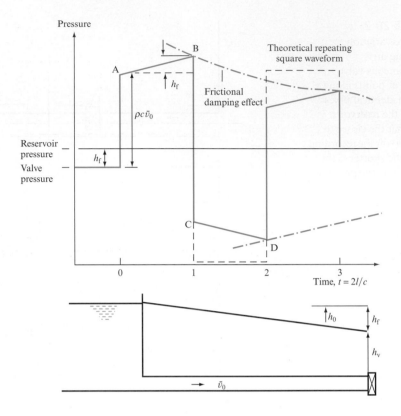

separation and is frequently made more complex by the release of dissolved gas in the vicinity of the cavity (see Section 21.3.1 and Examples 21.4 and 21.5).

20.1   WAVE PROPAGATION VELOCITY AND ITS DEPENDENCE ON PIPE AND FLUID PARAMETERS AND FREE GAS

Wave propagation velocity is the most important parameter within transient theory and analysis. It determines both the magnitude of the Joukowsky pressure differential (note that the Joukowsky pressure change can be positive or negative) and determines whether a particular system operation is defined as 'rapid' or 'slow' due to its controlling presence in the calculation of pipe period. Changes in wave speed brought about by changes in pipe material even if the flow cross-section remains the same will generate reflection and transmission coefficients that will affect the pressure time history at any point within a system. It is essential that the determination of wave speed is fully understood together with the effect of pipe and fluid parameters, and the influence of free gas.

If even a small quantity of free gas is present in a liquid flow, the change in mixture compressibility can be large enough to cause considerable reductions in the wave speed.

FIGURE 20.4

Effect of a pressure wave
in a gas/fluid pipeline

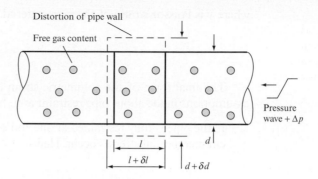

Consider a flow of a liquid of density ρ_f and bulk modulus K_f along a pipe of diameter d and wall thickness e, where the Young's modulus of the pipe wall is E and the Poisson's ratio is v (Fig. 20.4). The mean density $\bar{\rho}$ of the fluid/gas mixture is given by

$$\bar{\rho} = y\rho_g + (1 - y)\rho_f, \tag{20.1}$$

where y is the proportion of free gas by volume and ρ_g is the gas density.

Now, detailing the volumetric changes that occur if a pressure wave compresses a volume V_f of fluid by dV_f yields

$$dV_f = -(V_f/K_f)\,dp \quad \text{for the fluid,}$$

$$dV_g = -(V_g/K_g)\,dp \quad \text{for the gas,}$$

where V_g is the gas volume present, K_g is the gas bulk modulus and $V_f + V_g =$ total element volume, V_t. The pipe section containing the initial volume V_f can be distorted both radially and longitudinally.

If the original volume $V_t = \pi d^2 l/4$, where l is the length of pipe element chosen in the direction of flow, then the distorted volume increase due to the passage of a pressure wave dp is given by

$$dV_p = \pi d(\delta d/2)l \quad \text{(Volume increase due to radial distortion } \delta d\text{),} \tag{20.2}$$

where products of small quantities (e.g. $\delta d\,\delta l$) are neglected.

From thin-walled pipe theory, the longitudinal stress F_L and the circumferential stress F_C are given by

$$F_L = (d/4e)\,dp \quad \text{and} \quad F_C = (d/2e)\,dp,$$

while the longitudinal strain is given by

$$\delta l/l = (F_L - vF_C)/E$$

and the circumferential strain by

$$\delta d/d = (F_C - vF_L)/E,$$

where v is Poisson's ratio of the pipe material. Thus,

$$dV_p = \pi d^2 l (F_C - v F_L)/2E.$$

The final form of the volumetric strain expression above will depend upon the assumptions made about pipe restraint and, hence, longitudinal stress and strain:

1. If the pipe is fully restrained at one end only, then both longitudinal and circumferential stresses occur. Hence,

    $$dV_p/V = [(d/Ee)(1 - v/2)]\, dp.$$

2. If the pipe is fully restrained against axial movement along its whole length, then longitudinal strain $\delta l/l = 0$ and $F_L = v F_C$. Hence,

    $$dV_p/V = [(d/Ee)(1 - v^2)]\, dp.$$

3. The pipe is supplied with expansion joints at regular intervals along its length, which may be necessary to take up thermal expansion. These are standard in many systems, including aircraft fuel systems. In this case, the longitudinal stress $F_L = 0$ and

    $$dV_p/V = (d/Ee)\, dp.$$

A general form of the expression for pipe distortion is, therefore,

$$dV_p/V = (d/Ee)C'\, dp. \tag{20.3}$$

The total change in volume for the fluid, gas and pipe section is then

$$dV_t = dV_p - dV_f - dV_g$$
$$= [(d/Ee)C'V_t + V_f/K_f + V_g/K_g]\, dp,$$

where

$$V_f = (1 - y)V_t,$$
$$V_g = yV_t$$
$$dV_t = [dC'/Ee + (1 - y)/K_f + y/K_g]V_t\, dp.$$

As a result, an overall effective bulk modulus for the pipe, fluid and gas combination can be written

$$K_{eff} = [(1 - y)/K_f + y/K_g + dC'/Ee]^{-1}$$

and an expression for the wave speed c may be deduced as

$$c = \sqrt{(K_{eff}/\bar{\rho})}$$
$$= \sqrt{\{[y\rho_g + (1 - y)\rho_f]^{-1}[(1 - y)/K_f + y/K_g + dC'/Ee]^{-1}\}}. \tag{20.4}$$

At low temperatures and pressures, the gas bulk modulus K_g can be approximated by the initial absolute pressure of the gas p_a. This is an approximation, as the passage of the pressure wave compresses the gas and changes its pressure; however, this is acceptable as a first approximation.

In the absence of free gas the wave speed expression reduces to

$$c = \sqrt{\left[\frac{K}{\rho}\left(1 + \frac{\mathrm{d}K_f}{Ee}C'\right)\right]},\qquad(20.5)$$

and Fig. 20.5 illustrates the effect of the pipe bore to wall thickness ratio d/e on wave speed through water in steel, cast iron and glass pipelines. It will be seen that the effect of increasing d/e, effectively increasing the elasticity of the pipe wall, reduces the wave speed through the fluid.

FIGURE 20.5

Influence of the pipe diameter/pipe wall thickness ratio on the wave speed in water

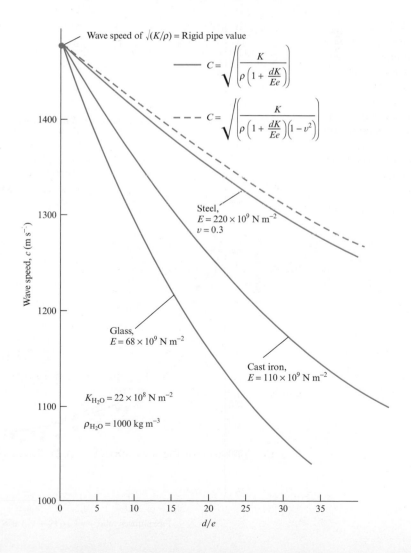

Wave speed of $\sqrt{(K/\rho)}$ = Rigid pipe value

$$C = \sqrt{\left(\frac{K}{\rho\left(1 + \frac{\mathrm{d}K}{Ee}\right)}\right)}$$

$$C = \sqrt{\left(\frac{K}{\rho\left(1 + \frac{\mathrm{d}K}{Ee}\right)\left(1 - v^2\right)}\right)}$$

Steel, $E = 220 \times 10^9$ N m^{-2}
$v = 0.3$

Glass, $E = 68 \times 10^9$ N m^{-2}

Cast iron, $E = 110 \times 10^9$ N m^{-2}

$K_{H_2O} = 22 \times 10^8$ N m^{-2}

$\rho_{H_2O} = 1000$ kg m^{-3}

Wave speed, c (m s^{-1})

d/e

It will also be seen that incorporating the effects of longitudinal strain has little effect over the normal working range; indeed, for glass and cast iron, as the value of Poisson's ratio is around 0.25, the value of C' tends to unity anyway. If the pipe wall is assumed rigid, i.e. $E \to \infty$, then the wave speed expression reduces to

$$c = \sqrt{(K_f/\rho)}$$

and has the value corresponding to the acoustic velocity in an expanse of fluid. Therefore, the value is the same irrespective of pipe material (as shown in Fig. 20.5). Similarly, for the rigid column theory the fluid bulk modulus tends to infinity as the fluid is assumed incompressible and pressure wave propagation may therefore be assumed to be instantaneous.

The presence of even a small quantity of free gas becomes important in reducing the wave speed through the mixture. Figure 20.6 illustrates the predicted reduction in wave speed through a kerosene and nitrogen mixture flowing in a polythene pipeline.

FIGURE 20.6

Influence of the free gas content on wave speed through a nitrogen–kerosene mixture in a polythene pipeline of 50 mm diameter, 6 mm wall thickness and 15 m length. The free gas pressure has been taken as 175 kN m^{-2}

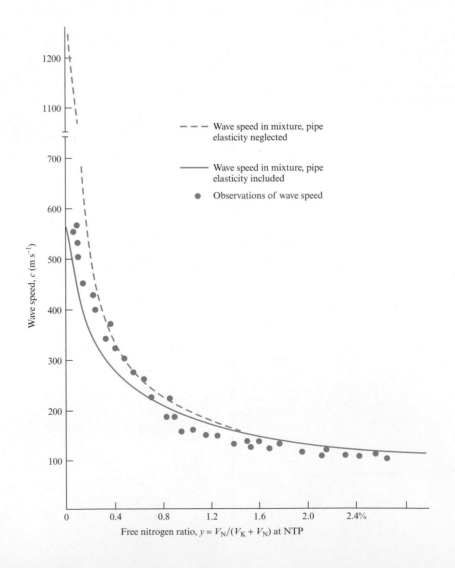

The results of a series of wave speed measurements are also included as a comparison with the predicted figures. It will be seen that the effect of neglecting pipe elasticity is only important at low gas contents. As the percentage of free gas increases, so the wave speed reducing effect of the pipe elasticity is numerically swamped by the effect of the gas terms; see Section 20.3 and program WAVESPD, Section 20.2.

It is sometimes stated that reduction in wave speed can be beneficial in reducing transient pressures. However, this is not automatically true, as will be seen later. Reducing wave speed increases the pipe period for any system, so that any valve operation taking a fixed time becomes effectively faster in terms of pipe periods, which is the only time measurement having any relevance in transient pressure prediction.

Table 20.1 presents values of Young's modulus and Poisson's ratio for some common pipe materials, while Table 20.2 details some values of common fluid bulk modulus and density.

TABLE 20.1

Values of Young's modulus of elasticity, Poisson's ratio and shear modulus for a range of common pipe wall materials

Material	Young's Modulus (10^{-9} N m^{-2})	Poisson's Ratio	Shear Modulus (10^{-9} N m^{-2})
Aluminium	70.0	0.3	27.6
Cast iron	80.0–110.0	0.25	40.0–80.0
Concrete	20.0–30.0	0.1–0.3	–
Copper	107.0–130.0	0.34	–
Glass	68.0	0.24	–
GRP	50.0	0.35	3.92
Polythene	3.1	–	1.10
PTFE plastic	0.35	–	–
PVC plastic	2.4–2.8	–	–
Reinforced concrete	30.0–60.0	0.15	–
Rubber	0.7–7.0	0.46–0.49	–
Steel	200.0–214.0	0.3	82.7
Titanium	103.4	0.34	44.8

TABLE 20.2

Values of bulk modulus and density for a range of common fluids

Fluid		Bulk Modulus (10^{-8} N m^{-2})	Density (kg m^{-3})
Carbon tetrachloride	at 15 °C	11.2	1590.0
Ethyl alcohol	at 15 °C	12.0	779.0
Kerosene		13.6	814.0
Oil, s.a.e. 10		16.7	880.0–940.0
s.a.e. 30		18.6	880.0–940.0
Sea water	at 10 °C	22.0	1026.0
Water	at 0 °C	20.5	1000.0
	at 20 °C	20.5	998.0
	at 80 °C	20.5	972.0

20.2 COMPUTER PROGRAM WAVESPD

The preceding section has identified the importance of the pipe, fluid and free gas parameters in the determination of wave speed. In order to calculate wave speed, and hence determine the category of any system operation into 'rapid' – less than one pipe period, or 'slow' – many pipe periods, it is preferable to use a simple program to undertake the calculation so that the effect of wall elasticity, fluid properties and free gas content can be included. The program WAVESPD uses equations (20.4) and (20.5) and relates to Figs 20.5 and 20.6. Data from Tables 20.1 and 20.2 may also be used.

The required data include pipe material Young's modulus of elasticity, fluid bulk modulus and density, free gas density, absolute pressure and free gas content as a ratio to the total volume of gas and fluid, and the pipeline length if pipe period is requested.

20.2.1 Application example

WAVESPD may be used to determine the required transient propagation velocity in Example 20.1 below.

EXAMPLE 20.1

Determine the wave speed and pipe period for a simple circular cross-section pipeline where the pipe Young's modulus, diameter, wall thickness, Poisson's ratio are known, along with the fluid bulk modulus and density and free gas pressure and volume as a percentage of the total fluid plus gas volume.

Solution

The pipe properties are Young's modulus $E = 70.0 \ 10^9 \, \mathrm{N\,m^{-2}}$, pipe diameter $d = 0.05$ m, $e = 0.006$ m, Poisson's ratio 0.3, expansion joint code 3, i.e. $C = 1.0$.

The fluid properties are bulk modulus $K = 13.6 \ 10^8 \, \mathrm{N\,m^{-2}}$, density $800 \, \mathrm{kg\,m^{-3}}$.

The gas properties are density $1.2 \, \mathrm{kg\,m^{-3}}$, pressure $175 \, \mathrm{kN\,m^{-2}}$, content by volume 0.004.

The pipe length in the pipe period calculation is 23 m.

The calculated wave speed is **230.66 m s^{-1}** and the pipe period is **0.2 s** (WAVESPD).

20.2.2 Additional investigations using WAVESPD

WAVESPD may be used to investigate:

1. the effect of both pipe properties and wall thickness ratio, or the influence of fluid density or bulk modulus on wave speed;
2. the influence of the expansion joint code on wave speed (this can be quantified for a particular fluid or pipe material);
3. the influence of free gas volume and pressure on the wave speed.

EXAMPLE 20.2

Wave propagation velocity is a parameter within all forms of transient propagation, from the waterhammer case developed above to free surface waves in channels and partially filled pipe flows and low-amplitude air pressure transient propagation. Drawing on this and earlier chapters develop a tabular format defining the relevant equations for wave speed and the likely range of values to be expected.

Solution

The relevant equations for waterhammer wave propagation velocity have been developed in this chapter. The acoustic velocity in a gas and the surface wave velocity in free surface flows are developed in Chapter 5, equations (5.30) and (5.35).

TABLE 20.3

Wave propagation velocity calculations

Table 20.3 below presents these wave speed calculations for a range of flow conditions, together with the pressure levels generated where appropriate for a $1\,\mathrm{m\,s^{-1}}$ instantaneous reduction in flow velocity.

APPLICATION	WAVE SPEED RANGE ($\mathrm{m\,s^{-1}}$)	DEFINING EQUATION
Rigid pipe	$c \to 1500$	$c = \sqrt{\left(\dfrac{K_f}{\rho_f}\right)}$
Elastic pipe	$200 < c < 1500$	$c = \sqrt{\dfrac{K_f}{\rho_f\{1 + [(DK_f/Ee)C_1]\}}}$
Gas/fluid pipe flow	$150 < c < 600$	$c = \sqrt{\left(\dfrac{K_{eff}}{\rho_{eff}}\right)}$
Low-amplitude gas	$c \to 350$	$c = \sqrt{\left(\gamma\dfrac{p}{\rho}\right)}$
Free surface wave	$0.25 < c < 5.0$	$c = \sqrt{\left(g\dfrac{A}{T}\right)}$
where	$K_{eff} = \left\{\dfrac{1}{(1 - y/K_f) + (y/K_g) + [(D/Ee)C_1]}\right\},$	$\rho_{eff} = (1 - y)\rho_f + y\rho_g$

Pressure surge values for a $1.0\,\mathrm{m\,s^{-1}}$ reduction in flow velocity vary from $1500\,\mathrm{kN\,m^{-2}}$ for water in a rigid pipe to 40 mm water gauge for a $1.0\,\mathrm{m\,s^{-1}}$ reduction in entrained air flow velocity in a building drainage vent stack. For partially filled circular cross-section pipe flow the half-diameter depth wave speed reduces to $2.776D$. For air density ρ may be taken as 1.3 at 0 °C, atmospheric pressure p as $101\,\mathrm{kN\,m^{-2}}$ and γ as 1.39 at 1 atm and 0 °C, so that wave speed $c = 329\,\mathrm{m\,s^{-1}}$.

20.3   SIMPLIFICATION OF THE BASIC PRESSURE TRANSIENT EQUATIONS

In order to describe some basic aspects of transient propagation it is necessary to rewrite the equations of continuity (5.59) and motion (5.60) in the simplified form,

$$\text{Continuity,} \quad \frac{\partial p}{\partial t} + \rho c^2 \frac{\partial \bar{v}}{\partial x} = 0; \tag{20.6}$$

$$\text{Motion,} \quad \frac{\partial p}{\partial x} + \rho \frac{\partial \bar{v}}{\partial t} = 0. \tag{20.7}$$

In this form, the equations apply to a frictionless horizontal pipeline, where the convective terms $\bar{v}\,\partial\bar{v}/\partial x$ and $\bar{v}\,\partial p/\partial x$ may be ignored in comparison with $\partial v/\partial t$ and $\partial p/\partial t$. The general solution of these two equations, due to D'Alembert, is

$$p - p_0 = F(t + x/c) + f(t - x/c), \tag{20.8}$$

$$\bar{v} - \bar{v}_0 = -(1/\rho c)[F(t + x/c) - f(t - x/c)]. \tag{20.9}$$

The f and F functions are entirely arbitrary and may be selected to satisfy the conditions imposed at the system boundaries. Referring to the simple pipe of Fig. 20.1, the F function may be interpreted as a pressure wave moving in the $-x$ direction, i.e. upstream, as x must decrease with time. From (20.8), the dimensions of both functions are those of pressure. Similarly, the f function may be interpreted as a pressure wave moving in the $+x$ direction, i.e. downstream. Both waves propagate at the sonic velocity c. The significance of equation (20.8) now becomes clear: it implies that at any time t following the initial disturbance, the pressure at any point in the pipe may be found from a summation of the F and f waves that have passed that point in the time t. It may be assumed that the pressure waves travel at a uniform velocity c and do not attenuate or change their shape as they propagate along the pipeline.

20.4   APPLICATION OF THE SIMPLIFIED EQUATIONS TO EXPLAIN PRESSURE TRANSIENT OSCILLATIONS

Referring again to Fig. 20.1, equation (20.8) may be used to calculate the pressure variations on either side of the valve following an instantaneous closure. At $t = 0$, the valve will be fully closed and an F wave will be propagated upstream while an f wave will be transmitted along the downstream pipe. On the upstream side of the valve at $t = 0$, therefore, the f function in equation (20.8) will be zero, so that

$$p - p_0 = F(t + x/c),$$

$$\bar{v} - \bar{v}_0 = (1/\rho c)F(t + x/c).$$

Eliminating $F(\ )$ yields

$$\Delta p = \rho c\,(\bar{v}_0 - \bar{v}) = \rho c\bar{v}_0 \tag{20.10}$$

as $\bar{v} = 0$ at time $t = 0$ because the closure was instantaneous. This is the maximum pressure rise possible due to valve closure and the expression (20.10) is commonly referred to as the Joukowsky pressure rise, Joukowsky having first demonstrated its validity in 1897. Equations (20.8) and (20.9) apply equally on the downstream side of the valve. However, at $t = 0$ here, it is the $F(\ )$ function that has zero value and so the magnitude of the initial $f(\ )$ wave propagated downstream will be given by

$$p - p_0 = f(t - x/c),$$
$$\bar{v} - \bar{v}_0 = -(1/\rho c)[-f(t - x/c)].$$

Hence,

$$\Delta p = -\rho c \bar{v}_0. \tag{20.11}$$

Thus, for the closure case described previously for Fig. 20.1, it follows that the pressure waves referred to were of a magnitude $\pm \rho c \bar{v}_0$.

Equations (20.8) and (20.9) can also be employed to calculate the reflections produced when an incident wave reaches the reservoirs terminating the pipeline in Fig. 20.1. Consider the upstream reservoir–pipe junction: at time $t = l/c$, the first $F(\ )$ wave arrives. Now, the overall pressure change at the reservoir–pipe junction must be zero. Hence, from equation (20.8),

$$\Delta p = 0 = F(t + x/c) + f(t - x/c),$$

and so

$$f(t - x/c) = -F(t + x/c). \tag{20.12}$$

Hence, the reflected wave that is transmitted downstream from the reservoir is equal in magnitude but of opposite sign to the incident wave. If the above analysis is applied to the downstream reservoir at time $t = l/c$, then the reflected $F(\ )$ wave produced will be equal to $-f(\ )$, the initial wave propagating downstream from the valve. If the concept of a reflection coefficient is introduced at this stage, defined as the ratio of the reflected wave to the incident wave, i.e. $C_R = f(\ )/F(\ )$ for the upstream reservoir, then it will be seen that the reservoir has a reflection coefficient of -1.0. This is true of all constant pressure boundary conditions and, for example, applies equally to vapour cavities formed during column separation as long as the cavity pressure remains a constant at vapour level.

Similarly, it is possible to calculate the appropriate reflection coefficient for the closed valve. At time $t = 2l/c$, the $f(\ )$ and $F(\ )$ waves in the two pipes of Fig. 20.1 arrive at the closed valve. However, so long as the pressure remains above vapour level, there can be no change in the velocity of the fluid adjacent to the valve, i.e. it must remain at rest. Hence, from equation (20.9)

$$\Delta \bar{v} = 0 = -(1/\rho c)[F(t + x/c) - f(t - x/c)],$$

and so

$$F(t + x/c) = f(t - x/c) \tag{20.13}$$

and the reflection coefficient may be shown to be $+1.0$.

The same analysis applies to the case of a transient arriving at the end of a closed pipe, and it is to be noted that this implies that the pressure recorded at the dead end will be twice the value of the incident pressure wave – a very important consideration if the dead-ended pipe happens to be a transducer connection.

The use of equations (20.8) and (20.9), together with known boundary conditions, allows reflection and transmission coefficients to be calculated for a range of cases likely to be met in any reasonably complex pipe network. However, this approach is limited to the propagation of fairly sharp transients, which may be approximated by step functions. In order to consider more gradual pressure changes it is necessary to employ the graphical method of solution of equations (20.8) and (20.9).

EXAMPLE 20.3

An outlet control valve on a long water distribution main is partially shut in 20 ms to restrict delivery. Calculate the length of the wavefront so propagated if the wave propagation speed in the pipe is $1250\,\mathrm{m\,s^{-1}}$.

If the change in velocity produced by the valve action is $0.5\,\mathrm{m\,s^{-1}}$, show that the convective terms $\bar{v}\,\partial\bar{v}/\partial x$ and $\bar{v}\,\partial p/\partial x$ may be ignored in equations (5.59) and (5.60). Under what circumstances would this simplification be unacceptable?

Solution

Small pressure waves may be imagined to propagate from the closing valve throughout its closing motion. Thus, pressure waves leave the valve for a time of 20 ms and so the length of the wavefront will be given by

$$c\Delta t = 1250 \times 0.02 = 25\,\mathrm{m}.$$

If the change in flow velocity is $0.5\,\mathrm{m\,s^{-1}}$, then the associated pressure rise is given by equation (20.10):

$$\Delta p = \rho c(\bar{v}_0 - \bar{v}) = 1000 \times 1250 \times \Delta\bar{v},$$

where $\Delta\bar{v} = 0.5$. Therefore, $\Delta p = 625\,\mathrm{kN\,m^{-2}}$. Now,

$$\frac{\partial\bar{v}}{\partial t} = 0.5/0.02 = 25, \quad \bar{v}\frac{\partial\bar{v}}{\partial x} = 0.5 \times 0.5/25 = 0.01,$$

where ∂x is the length of the wavefront, i.e. $\partial x = 25\,\mathrm{m}$. Thus, $\bar{v}\,\partial\bar{v}/\partial x$ may be ignored with respect to $\partial\bar{v}/\partial t$. Similarly,

$$\frac{\partial p}{\partial t} = 625/0.02 = 31\,250, \quad \bar{v}\frac{\partial p}{\partial x} = 0.5 \times 625/25 = 12.5,$$

and so $\bar{v}\,\partial p/\partial x$ may also be ignored in this case.

As stated, this case is typical of the values met in most transient examples. However, if the wave speed becomes very low, then the values obtained for the convective terms approach the $\partial\bar{v}/\partial t$ and $\partial p/\partial t$ values because of the reduction in wavefront length and this convenient approximation will no longer be valid.

EXAMPLE 20.4

Derive an expression for the reflection and transmission coefficients that describe the response of a three-pipe junction to the arrival of a transient along one of the pipelines (Fig. 20.7).

A small-bore branch pipe of cross-sectional area A_2, length l and wave speed c is connected to a main pipeline of area $A_1 = 25A_2$ and wave speed c. Draw the pressure variations at the mid-length point on the branch pipe for four branch pipe periods following the arrival of a step, ΔP, transient at the junction, if the branch is terminated by a closed valve. It may be assumed that no further transients arrive at the junction during the period of time under consideration. Compare this approach to that followed in Examples 5.11 and 5.12.

FIGURE 20.7
(a) Pipe layout.
(b) Pressure variation at mid-length point X of closed-ended branch pipe. In (b), ΔP in pipe 1 reaches junction at $t = 0$; a = partial transmission of ΔP into branch; b = total positive reflection at pipe end of pipe 3 at dead end; c = partial negative reflection at end of pipe 3 at junction; d = total positive reflection at pipe end of pipe 3 at dead end

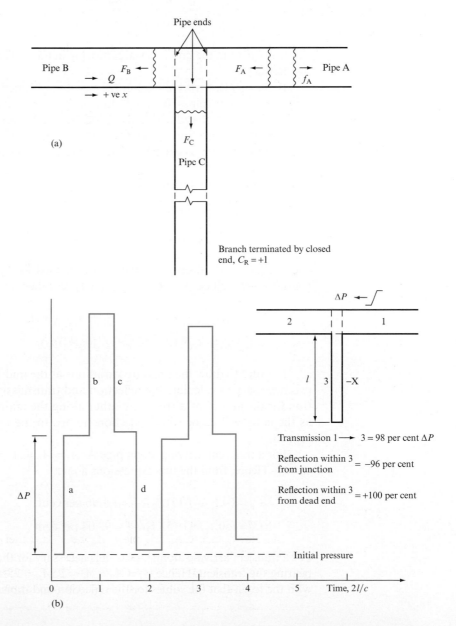

Solution

Assuming a frictionless system, then, from considerations of continuity at the junction, represented by pressure and flow conditions on the pipe ends,

$$p_A = p_B = p_C \quad \text{and} \quad \Delta p_A = \Delta p_B = \Delta p_C, \tag{I}$$

$$Q_A = Q_B + Q_C \quad \text{and} \quad \Delta Q_A = \Delta Q_B + \Delta Q_C. \tag{II}$$

From equations (20.8) and (20.9),

$$\Delta p_A = F_A + f_A, \quad \Delta Q_A = -(A_A/\rho c_A)(F_A - f_A), \tag{III}$$

$$\Delta p_B = F_B + f_B, \quad \Delta Q_B = -(A_B/\rho c_B)(F_B - f_B), \tag{IV}$$

$$\Delta p_C = F_C + f_C, \quad \Delta Q_C = -(A_C/\rho c_C)(F_C - f_C), \tag{V}$$

where F, f refer to the pressure waves moving in the negative and positive x directions, respectively.

If the incident wave moves in the negative x direction in pipe A, then this will generate waves in pipes B and C moving in the negative x direction, i.e. F-type transmission waves, and it will also produce a reflection of itself in pipe A moving in the positive x direction, i.e. an f-type wave. Thus, the f_B, f_C pressure waves included in the above equations are zero as there are no waves present in pipes B and C of this type.

Substituting into (I) and (II)

$$f_A + F_A = F_B = F_C \tag{VI}$$

$$F_A - f_A = (c_A/A_A)[(A_B/c_B)F_B + (A_C/c_C)F_C]. \tag{VII}$$

Let the junction reflection coefficient be defined as $C_R = f_A/F_A$ and the junction transmission coefficient by $C_T = F_B/F_A = F_C/F_A$; then, from (VI) and (VII),

$$C_R = (A_A/c_A - A_B/c_B - A_C/c_C)/(A_A/c_A + A_B/c_B + A_C/c_C),$$

$$C_T = (2A_A/c_A)/(A_A/c_A + A_B/c_B + A_C/c_C).$$

In order to draw the pressure variations at the mid-length point of the branch, it is necessary to calculate the reflection and transmission coefficients for the junction for the arrival of a transient either along the main pipe or along the branch, as the latter will occur when reflections return to the junction from the end of the branch.

For a transient arriving along pipe A, $A_1 = A_A = A_B = 25A_C = 25A_2$, $c_1 = c_2 = c_A = c_B = c_C$. Hence, from the two expressions above,

$$C_R = (-A_2/c_2)/(51A_2/c_2) = -1.96 \text{ per cent},$$

$$C_T = (50A_2/c_2)/(51A_2/c_2) = 98.04 \text{ per cent}.$$

For a transient arriving along pipe C, it is necessary for the suffix A to refer to the pipe bearing the transient. Hence $A_2 = A_B = A_C = 25A_A = 25A_1$ and $c_1 = c_2 = c_A = c_B = c_C$, with the result that the values for the reflection and transmission coefficients are

$$C_R = -49/51 = -96.08 \text{ per cent},$$

$$C_T = 2/51 = 3.92 \text{ per cent}.$$

The pressure variations over four branch pipe periods may now be drawn by considering the individual waves travelling within the branch (Fig. 20.7).

As the branch is terminated by a closed valve, the reflection coefficient at the end of the branch is +1. This means that any wave arriving at the closed end is reflected as an equal magnitude wave with no change of sign.

20.5 SURGE CONTROL

Pressure transient problems within pipe and duct systems may be characterized as due to either changes in the flow conditions at the boundaries of the system (i.e. changes that generate transients as a means of propagating the information on the change throughout the network) or local interactions between these transients and local pipeline features. The first category obviously includes any change in pump demand or valve adjustment while the latter encompasses such problems as column separation at high points in the local pipeline profile.

In general the severity of any transient is dependent upon the rate of change of the conditions as its source; for example, the rate of valve closure, or opening, determines for any given pipeline the magnitude of the resulting transient. A well-understood analysis of transient propagation may allow potentially damaging transients to be avoided by good design. However, this is not always possible for a range of reasons and in these cases it may be necessary to incorporate surge control and suppression devices.

The overall generic objective of surge control devices must therefore be a reduction in the rate of change of boundary flow conditions; for example, air chambers allow the maintenance of flow conditions while pumps run down or valves close, and inwards relief valves maintain pressure levels downstream of either pumps or closing valves. To achieve this end surge, control devices must be positioned as close as possible to the source of the transient.

It would be tempting to propose that all pressure surge problems can be suppressed by control devices. Unfortunately this is not the case for a variety of reasons, including cost, probability of transient generation, inadequate design information and system operating requirements. In large pumping main systems surge relief devices may be practical, but in other systems, equally prone to pressure surge, operating conditions and design criteria preclude the more common surge control devices. An example of the latter case is in aircraft fuel systems where system weight and the need to provide rapid changes in engine fuel supply combine to present particular design problems.

The following sections will introduce a range of devices and techniques to achieve these objectives, in each case developing a suitable boundary model to allow the effects of the control device to be incorporated into the method of characteristics analysis presented in previous chapters.

A wide range of surge suppression devices exist, all having the common factor that they tend to slow down the rate of change of flow in the system:

1. *Valve closure control.* By introducing controls onto the valve closure mechanism, particularly to slow down the final stages of closure, worthwhile reductions in surge pressure can be achieved. Alternatively, closing a valve in a system concurrently with some negative surge pressure-generating phenomenon can propagate a positive transient that will maintain the system pressure above vapour or gas release levels.

2. *Increasing pump inertia.* As mentioned, one of the major problems with pumped systems is the prevention of column separation following power failure. If the rate of change of flow can be decreased by keeping the pump turning longer, then the danger of column separation is reduced. This can be achieved by incorporating a flywheel into the pump design – not a good solution as it has to be driven while the pump starts up.

3. *Surge shafts or air vessels.* Positive surge pressures can be alleviated by allowing fluid to leave the pipeline and enter vertical surge shafts, thus absorbing some of the excess energy. If the pressures are very high, air chambers or accumulators where gas or air is compressed may be used.

4. *Air or fluid admission valves.* During flow stoppage it is possible for low-pressure regions to form and column separation to occur in the fluid in the pipeline. Examples of this are downstream of a valve or pump on flow stoppage. An inflow valve placed near to the likely separation site can allow air or fluid to enter the system, thereby filling the cavity, restoring system pressure and alleviating the surge pressure rise likely on cavity collapse. In many cases, introducing air into the system by such valves is troublesome when the system is restarted, so that fluid inflow valves should be used if practical.

5. *Relief valves.* An outflow valve can be arranged to blow off excess fluid as the pressure rises. This is a cheap but rather dangerous solution, as the jamming open of such valves could cause a substantial loss of fluid from the system.

6. *Bypass systems.* These are, effectively, extensions to the inflow fluid relief valve devices and merely allow water from a sump to bypass the pump and enter the downstream pipe section, if the pressure here, following pump shutdown, falls below sump pressure.

In a study of available surge suppression devices, it is probably best to consider the range of solutions available for a particular problem. Here it is best to take, as an example, the negative pressure surge problems caused by flow stoppage, which can lead to column separation and large resurge pressures on cavity collapse, as this type of surge problem is common.

Consider the case of a mains pipeline pumping fluid from one reservoir to another over some intervening high ground. The potential pressure surge problems arise in the following manner:

1. excessive positive pressure rise on terminal valve closure;
2. column separation at the pump discharge during emergency shutdown;
3. column separation at a high point in the system following pump shutdown.

Figure 20.8 illustrates the pipe network involved.

FIGURE 20.8

Layout of a pump supply
system prone to surge on
pump shutdown

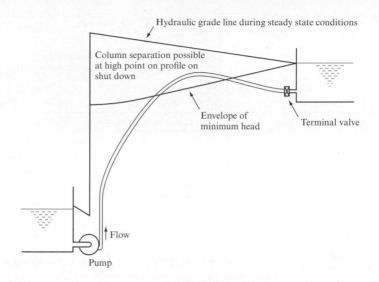

20.6 CONTROL OF SURGE FOLLOWING VALVE CLOSURE, WITH PUMP RUNNING AND SURGE TANK APPLICATIONS

Valve closure will produce a positive pressure surge that will propagate through the system. The simplest solution is to decrease the valve closure rate. However, merely increasing the operation time of the valve is not the most efficient technique. The majority of the pressure surge generated will be produced by the final stages of valve closure (perhaps the last 15 per cent of the operation accounting for 80 per cent of the flow reduction), so that the most efficient method of reducing surge generation is to introduce a two-speed valve closure operation, the last 15 per cent of travel taking proportionately longer than the first 85 per cent (see Fig. 20.9).

The boundary equation at the closing valve is supplied by the valve characteristic:

$$\tau = (\bar{v}/\bar{v}_0)\sqrt{(\Delta p_0/\Delta p)}, \tag{20.14}$$

FIGURE 20.9

Flow–time relation for
one- and two-stage valve
closure

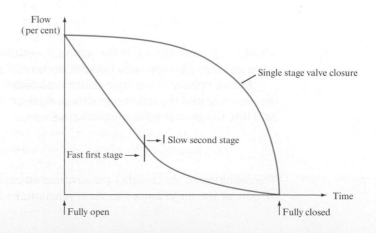

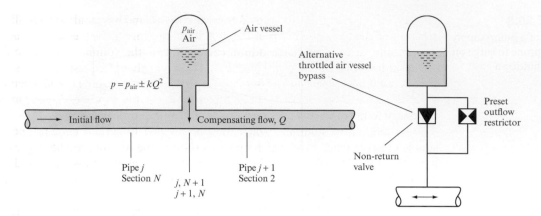

FIGURE 20.10

Air vessel installed upstream of a closing valve, or downstream of a pump, to reduce surge pressures

which varies from $1 \geqslant \tau \geqslant 0$ during valve closure. Now τ is known as a function of valve position, say $\tau = f(\theta)$, where θ is the angle of closure of the valve or a linear distance moved by a gate-type valve. Thus, as position may be monitored with time and may be expressed for each stage of closure as $\theta = f(t)$, then τ is known as a function of time.

An alternative solution here is to provide a bypass valve adjacent to the closing valve which will open to pass a small quantity of flow as the main valve closes and then shuts itself. However, this is merely an adaptation of the method described above.

A common solution to this surge generation case, and to the negative surge situation, is to install an air chamber on the pipeline (see Fig. 20.10). The purpose of this device is to allow fluid to be diverted out of the pipeline under high-pressure conditions. In many large civil engineering projects, open surge tanks may be used, but often the pressures generated are too high for these to be practical. Figure 20.10 illustrates a typical air vessel–pipeline junction. Initially, the air in the vessel is at the same pressure as the fluid in the pipeline. As the valve on the pipeline closes, so excess pressure in the main forces fluid into the air vessel, compressing the gas and effectively delaying and reducing the surge pressure wave propagated through the pipeline fluid. Alternatively, where the pressure in the pipeline is falling due to an upstream valve closure or pump trip, fluid is forced out of the air chamber to maintain flow and again reduce the rate of change of the flow conditions, possibly preventing column separation, see Section 21.8. The gas compression may be assumed to follow the relation

$$pV^n = \text{constant},$$

where p is gas pressure, V is the volume it occupies and n is an exponent (normally taken as 1.2 as a compromise between isothermal and adiabatic gas compression).

The air volume at any time t can be calculated from the air volume at the start of the time-step and the inflow, or outflow, through the tank orifice over that time-step Δt. Thus, the general volume equation is

$$V_t = V_{t-\Delta t} + Q \times \Delta t.$$

The definition of air chamber pressure and an expression representing the continuity of flow at the air chamber to main junction, including the air chamber entry loss coefficient, k in Fig. 20.10, provides a suitable boundary condition to allow the solution of the wave equations at the junction. Solution of these equations, achieved numerically, at each time-step allows calculation of the volume–time curve for the vessel under a particular surge condition. The vessel should be resized to ensure that it does not empty when the pressure in the pipe falls or fails to be useful when the air volume becomes too small as the pressure rises in the pipeline. A compressor is necessary to make up any air losses due to absorption or leakage.

An alternative solution, which may be applied in certain circumstances, is a simple pressure relief valve which opens as soon as the pressure reaches a preset level and allows fluid to blow out of the system. As soon as the pressure falls, the valve reseats. The obvious problem is the relief valve, which does not close after use and allows fluid to drain out of the pipeline.

As mentioned, in large-scale projects, such as hydroelectric power stations, surge tanks or shafts may be included in the station tunnelling.

Surge tanks are examples of fluid mass oscillation and, as described earlier (Fig. 19.2), may be analyzed by means of rigid column theory. This provides considerable savings in computer time and cost and, in addition, it is possible to use a frictionless approximation to gain a first approximation to the period and amplitude of the mass oscillation, namely

$$T = 2\pi \sqrt{(lA_2/gA_1)} \tag{20.15}$$

and

$$Z = (Q_0/A_2)\sqrt{(lA_2/gA_1)}. \tag{20.16}$$

In general, surge tanks or shafts, as they are commonly cut out of rock in hydroelectric schemes, may be classified as simple, orifice or differential. Figure 20.11 illustrates some common designs. The simple surge shaft has an unrestricted entry from the penstock–supply tunnel junction and must be so designed that it will not overflow on an upsurge, unless overspill provision is made, nor empty during downsurge, caused by turbine acceleration or by fluid oscillation. If the shaft empties, then air will be drawn into the tunnel–penstock system, and this should be avoided wherever possible. The period of mass oscillation is relatively long for this type of shaft.

The orifice surge tank has a restricted opening from the tunnel system. The orifice losses aid in dissipating the upsurge pressures caused by turbine load rejection. Also, during the downsurge cycle, the orifice damps the oscillation and helps to prevent the shaft emptying and reduces the air influx problem.

Conical or variable section surge tanks may be of either the simple or orifice type. These types of shaft again aid in reducing the probability of air entry on downsurge.

The differential surge tank is a combination of the orifice and simple designs. For rapid valve closures, on load rejection, the benefits of the orifice shaft are available, the central tank being designed to overspill into the larger outer shaft. In the valve-opening case, on load acceptance water is directly available in the central shaft with no orifice restriction to assist in accelerating the penstock flow. This is, later, supplemented by flow from the outer shaft via the base orifices.

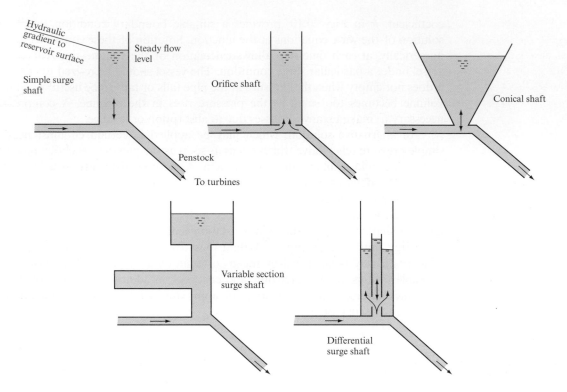

FIGURE 20.11

Schematics of some
common surge shaft
designs

The final design of surge tank for a given situation depends on the frequency of
any particular surge-producing system operating and in each case an analysis of the
likely advantages of each design should be investigated, preferably with the aid of a
computer simulation.

20.7   CONTROL OF SURGE FOLLOWING PUMP SHUTDOWN

Referring to Fig. 20.8, negative surge problems arise if the pump fails, thus generating
a negative pressure wave relative to the steady state pressure of the system. The effect
of this is to reduce the pressure of the system and, if this falls to gas release or vapour
pressure, then column separation may occur with potentially dangerous resurge
pressures being generated on collapse of the vapour cavity. In order to limit these
pressure rises, the principle followed is to increase the fluid pressure by passing fluid
from some control device into the pipeline as soon as the pressure in the pipeline falls
low enough to indicate possible column separation. There are a number of possible
methods here. Obviously, the air vessel discussed in Section 20.6 can be employed if it
is positioned closely to the potential low-pressure region; normally, such a vessel is
mounted at the pump discharge. The equations outlined for the application of the air
vessel to the relief of a positive surge apply without modification.

Figure 20.12 illustrates the installation of an air vessel to protect a water pumping
main against the effect of column separation following a pump failure. The initial
steady flow through the system prior to pump trip is 3.52 m³ s⁻¹. Figure 20.13
illustrates the effect of varying vessel capacity on the pressure surge recorded at the

FIGURE 20.12
Pumping main protected
by an air vessel, flow
$3.52\,\mathrm{m^3\,s^{-1}}$ (Swaffield and
Boldy 1993)

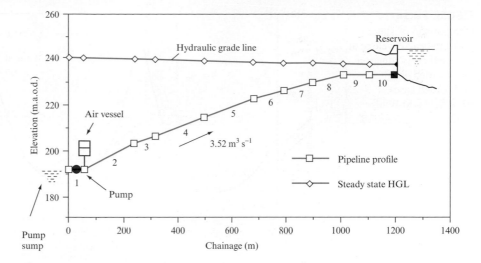

FIGURE 20.13
Variation of static
pressure head against
time for air vessel with
initial air volume of $28\,\mathrm{m^3}$
(Swaffield and Boldy
1993)

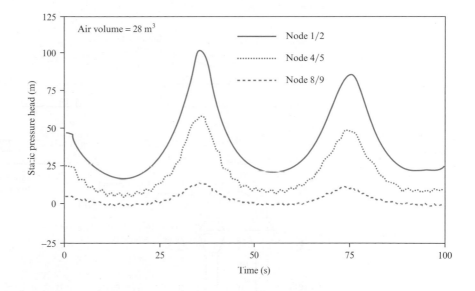

pump–vessel–pipeline junction. As shown in Fig. 20.10 air vessels are often fitted with
a bypass to restrict the inflow to the vessel as the pressure in the pipeline rises. Outflow
from the air vessel, designed to provide a continuation of the pump delivery as the
pump slows down, and therefore a reduction in the rate of change of the system
boundary conditions, is not affected. However, restricting inflow into the chamber is
desirable as it damps the pressure fluctuations following flow reversal and limits the
maximum head generated. Figure 20.14 illustrates this effect for the $28\,\mathrm{m^3}$ air vessel of
Fig. 20.13.

An alternative solution is to introduce an inwards relief valve, which could pass
either air or fluid into the pipeline as soon as pipeline pressure falls. Figure 20.15
illustrates such a valve employed on an aircraft refuelling system to pass relief air or
fuel into the pipeline following the emergency shutdown of the main refuelling valve,
which could occur under aircraft power failure conditions. The necessary equations
are similar to those for the air vessel, the main divergence, or simplification, being that

FIGURE 20.14

Comparison of static pressure head envelope at air vessel entry junction for throttled and unthrottled cases (Swaffield and Boldy 1993)

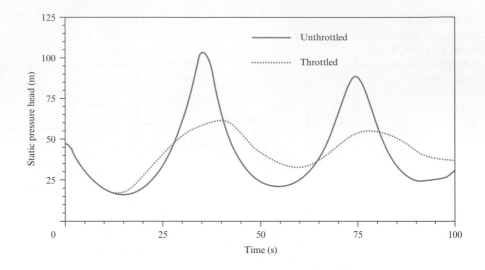

FIGURE 20.15

Comparison between the twin-hose supply to the Concorde RCU and the layout assumed in the analysis

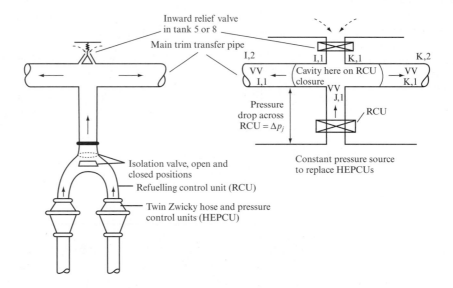

the pressure outside the relief valve can usually be taken as a constant, normally atmospheric pressure. Figure 20.16 illustrates a comparison between measured pressure traces for such a valve and an analysis based on the techniques described in Chapter 21, including the effects of air release. In this application such a valve is acceptable because there is no danger of contamination, as there would be if such a system were used to introduce air into a water supply main.

Column separation, caused by main pipeline pressure falling to gas release or fluid vapour pressure, can occur at any point in the pipe system. It is more likely, however, to be sited close to the pump or closing valve. One prime site for column separation is a high point in the pipe profile (Fig. 20.8), e.g. where the pipe goes over the top of a hill, thus reducing the fluid static pressure. If an analysis of the system indicates that separation may occur at such positions, then an air vessel or inwards

FIGURE 20.16

Predicted and observed pressure variations downstream of the Concorde RCU during and after its closure in an all-tanks-refuelling case

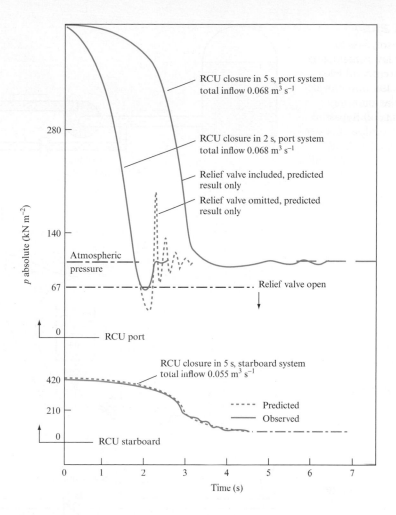

relief valve may be positioned at these locations. The introduction of air through such a relief valve, or from a badly sized air vessel, can lead to system priming problems when the pumps are restarted and so, as a general rule, the introduction of air should be avoided.

To prevent line pressures falling too low downstream of a closing valve, or a tripped pump, it may be possible to introduce some form of flow continuation that effectively reduces the rate of change of the flow conditions at this boundary. Figure 20.17 illustrates two simple techniques that may be employed, namely a bypass valve or a feed tank kept full by the system initial flow condition. Both of these solutions prevent the ingress of air into the pipeline and are particularly suitable in low-head systems.

Figure 20.17(a) illustrates a bypass valve network. The pump is protected against backflow by the non-return valve in the main. A parallel, smaller-bore, bypass pipe includes a control valve that opens as the pump trips and is subsequently allowed to close slowly. (Note that 'slow' should be thought of in pipe periods.) The bypass valve must not open too early.

Figure 20.17(b) illustrates a feed tank installed to provide an inflow into the main following pump trip. This is particularly useful if air ingress to the pipeline is to be

FIGURE 20.17

(a) Pump bypass to control downstream surge on pump trip. (b) Feed tank locations to provide protection following pump trip. (c) Bypass to cater for localized booster pump failure in long pipelines. (Swaffield and Boldy 1993)

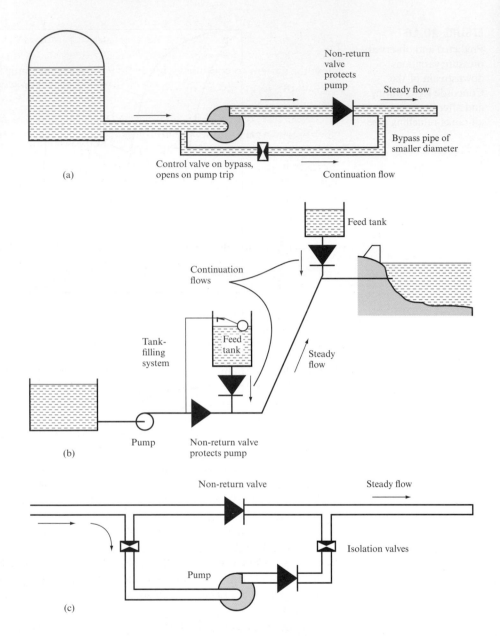

avoided. Fluid flows into the pipeline as the pressure falls downstream of the pump, thus preventing cavity formation and providing protection against both subatmospheric pressures and the subsequent high-pressure surge following cavity collapse. This is effectively the same as the solution proposed in the aircraft ground refuelling simulation (Fig. 20.15). Feed tanks may also be used at high points in a pipeline profile to prevent column separation.

The volume of fluid required in the tank may be determined by simulations similar to those already presented for the modelling of column separation on valve closure and pump trip. Piping arrangements must be made to ensure that the feed tank is kept filled prior to the next transient event.

In the case of long pipelines it may be necessary to introduce booster pumps. Local booster pump trip may be the cause of transient problems as the pump becomes a major flow obstruction. Bypass networks, as illustrated in Fig. 20.17(c), allow the system to continue at lower flow rates and minimize the surge implications of local pump failure. In this case the bypass piping diameter should be similar to the main.

While this section has been concerned with surge control and suppression, it must be appreciated that an essential first stage in any discussion of control or alleviation device selection is an estimate of the severity of the problem. The method of characteristics, to be introduced fully in Chapter 21, has been established as the most effective methodology upon which to base such simulations. In particular the method's ability to incorporate specialized system boundary conditions, while retaining the general pipe and more common boundary equations unaltered, allows a wide range of system simulations to be undertaken with minimal rewriting of the basic model. The development of computing power now makes this approach to simulation readily accessible.

Concluding remarks

Chapter 20 has attempted to put the study of waterhammer and pressure transient propagation in both an historical and engineering context. The treatment of the basic waterhammer or pressure surge phenomena has been traditional in that it has concentrated on the Joukowsky pressure rise on instantaneous flow stoppage and the modifications to this when valve closures are slowed. The concept of measuring time in terms of pipe period was also introduced, along with the importance of the reflection and transmission coefficients at pipe terminations or changes in cross-sectional area, wall thickness or material. The importance of wave speed was emphasized, and equations and a computer program, WAVESPD, provided to allow its calculation. The importance of free gas was also emphasized and its effect demonstrated.

Any surge analysis will inevitably be aimed at the control rather than the eradication of transient propagation. Chapter 20 also presented the basic approaches available for surge control, while recognizing that in some design cases control by traditional means becomes impractical, for example in any weight-conscious aircraft application or particularly in in-flight refuelling where emergency disconnection can lead to extremely fast valve closures.

Chapter 20 therefore lays the foundation for the numerical analysis of a whole range of transient pressure applications in Chapter 21, drawing upon the finite difference discussions already presented in Chapter 5. Chapter 21 will demonstrate that the fundamental concepts derived from a study of waterhammer phenomena apply equally to free surface wave propagation and to low-amplitude air pressure transient propagation.

Summary of important equations and concepts

1. Valve closures in less than one pipe period, calculated as the return time for a transient generated by initial valve motion, will generate the full Joukowsky pressure rise, equation (20.11).

2. Increases in valve closure time that remain within a pipe period have no effect.

3. Wave speed may be calculated by reference to the fluid effective bulk modulus and density. The first term is governed by both fluid and pipe wall properties and may be 'swamped' by free gas. The second term can be dominated by free gas density. Equation (20.4) presents the full calculation, also including the effect of pipe restraint.

4. All boundaries may be represented by reflection and transmission coefficients that generate transients. Changes in pipe wall material even if not accompanied by changes in wall thickness or cross-sectional area act as a boundary and will generate reflection and transmission coefficients; Example 20.3 demonstrates this result, it is also developed in Example 5.12.

5. Dead ends have a +1 reflection coefficient and transients propagating from a small diameter to much larger diameter conduit are almost entirely reflected, hence care must be taken in the choice of transducer connections, Example 20.4. (This case will be returned to in Chapter 21, Example 21.3.)

6. Surge analysis aims to control transients. The most basic approach is to reduce the rate of change of the flow conditions and this may be achieved by a variety of techniques, from variable speed valve closures, through air chambers to inwards or outwards pressure relief valves.

7. Transients may propagate as positive or negative pressure waves, leading to vapour cavity formation, gas release from solution and, in extreme cases, pipe implosion. Negative transients are extremely dangerous and are often forgotten as waterhammer is widely associated with pressure rise.

References

The development of pressure transient analysis has over the past 100 years been a truly international and interdisciplinary undertaking. The following references are offered as a source for the reader to investigate further.

1. Joukowsky, N. (1900). Uber den hydraulisher Stoss in Wasser – lietungsrohren. *Memoirs de l'Academie Imperiale des Sciences de St Petersburgh 1900*, trans. Simin, O. (1904). Waterhammer. *Proc. AWWA*, **24**, 341–424.

2. Wood, F. M. (1970). History of waterhammer. *Research Report No. 65*, Dept of Civil Eng., Queens University, Kingston, Ontario.

3. Allievi, L. (1904). *Notes I–IV* translated as *Theory of Waterhammer* by E. E. Halmos, Ricardo-Garoni, Rome, 1925.

4. Kerr, S. L. (1968). Surge problems in pipelines – oil and water. *Trans. ASME*, **98**, 113–21.

5. Schnyder, O. (1929). Waterhammer in pump discharge lines. *Schweizerische Bauxeitung*, **94**, (22), 23.

6. Bergeron, L. (1932). Variations in flow in water conduits. *Comptes Rendres des Travaux de la Soc. Hyd. de France*, Paris.

7. Bergeron, L. (1957). *Waterhammer in Hydraulics and Wave Surges in Electricity*. Wiley, New York.

8.   Thorley, A. R. D. and Enever, K. J. (1979). *Control and Suppression of Pressure Surges in Pipelines and Tunnels.* Construction Industry Research and Information Association, London.

9.   Massau, J. (1900). Memoirs sur l'integration graphique des equations aux derivees partialles. *Ann. Ass. Ingrs. Sortiis des Ecoles Speciales de Gand*, **23**, 95–215. Translated as *Unsteady Flow*, H. J. Putnam, Rocky Mountain Hydraulic Laboratory, Colorado, 1948.

10. Lamoen, J. (1947). *Le coup de bellier d'Allievi, compte tenu des pertes de charge continues.* Bull. Centre des Etudes, de Recherches et 'Essais Scientifiques des Constructions du Gerrie Civil et d'Hydraulique Fluviale, Tome II, Dosoer, Liege.

11. Gray, C. A. M. (1953). The analysis of the dissipation of energy in waterhammer. *Australian Journal of Applied Sciences*, **5** (2), 125–31.

12. Gray, C. A. M. (1954). Analysis of waterhammer by characteristics. *Proc. ASCE*, **119**, 106–12.

13. Ezekial, F. D. and Paynter, H. M. (1957). Computer representation of engineering systems involving fluid transients. *Trans. ASME*, **79**, 1840–50.

14. Streeter, V. L. and Lai, C. (1962). Waterhammer analysis including fluid friction. *J. Hydraul. Div. ASCE*, **88**, 151–72.

15. Streeter, V. L. (1966). Computer solutions of surge problems. *Proc. IMechE*, **180** (3E), 11–23.

16. Streeter, V. L. (1969). Waterhammer analysis. *J. Hydraul. Div. ASCE*, **95** (HY6), 1959–72.

17. Wylie, E. B. and Streeter, V. L. with Lisheng Suo (1993). *Fluid Transients in Systems.* Prentice Hall, Englewood Cliffs, NJ.

18. Lister, M. (1960). The numerical solution of hyperbolic partial differential equations by the method of characteristics. *Mathematical Methods for Digital Computers.* Wiley, New York.

19. Fox, J. A. (1968). The use of digital computers in the solution of waterhammer problems. *Proc. ICE*, **39**, 127–33.

20. Fox, J. A. (1989). *Transient Flow in Pipes, Open Channels and Sewers.* Ellis Horwood, Chichester, p. 283.

21. Evangelisti, G. (1969). Waterhammer analysis by the method of characteristics. *L'Energia Elletrica*, **10** (10–12), 673–9, 759–72, 839–58.

22. Swaffield, J. A. (1970). A study of column separation in a pipeline carrying aviation kerosene. *Proc. IMechE*, **186**, 693–703.

23. Doyle, T. J. and Swaffield, J. A. (1972). Evaluation of the method of characteristics applied to a pressure transient analysis of the BAC/SNAIS Concorde refuelling system. *Proc. IMechE*, **186** (40/72), 509–18.

24. Thorley, A. R. D. and Twyman, J. W. R. (1976). Propagation of transient pressure waves in a sodium cooled fast reactor. In H. S. Stephens, A. C. King and C. A. Stapleton (Eds) *Proc. 2nd Int. Conf. on Pressure Surge*, BHRA, London pp. A2–15.

25. BHRA. *British Hydromechanics Research Association Pressure Surge Conferences*, 1972, 1976, 1980, 1983, 1986, 1989. Cranfield.

26. Swaffield, J. A. and Boldy, A. P. (1993). *Pressure Surge in Pipe and Duct Systems*, Avebury Technical, Gower Press, Aldershot, p. 360.

27. Martin, C. S. (1976). Entrapped air in pipelines. In H. S. Stephens, A. C. King and C. A. Stapleton (Eds) *Proc. 2nd Int. Conf. on Pressure Surge*, BHRA, Bedford, September, pp. F2–115.

28. Lawson, J. D., O'Neill, I. C. and Graze, H. R. (1963). Pressure surge in fire services in tall buildings. In *Proc. 1st Australasian Conf. on Hydraulics and Fluid Mechanics*, Pergamon Press, Sydney, pp. 353–68.

29. Hope, P. and Papworth, M. U. (1980). Fire main failure due to rapid priming of dry lines. In H. S. Stephens and J. A. Hanson (Eds) *Proc. 3rd Int. Conf. on Pressure Surge*, BHRA, Cranfield.

30. Ballanco, J. A. (1998). Investigation and analysis of violently fracturing water closets. In S. Wolfson (Ed.) *Technical Proceedings ASPE 1998 Convention*, Indianapolis, October, pp. 109–50.e.

31. Swaffield, J. A. and Maxwell-Standing, K. (1986). Improvements in the application of the numerical method of characteristics to predict attenuation in unsteady partially filled pipe flow. *NBS Journal of Research*, **91** (3), 389–93.

32. Goldberg, D. E. and Wylie, E. B. (1983). Characteristics method using time-line interpolations. *J. Hydraul. Div. ASCE*, **109** (5), 670–83.

33. Swaffield, J. A. and Galowin, L. S. (1992). *The Engineered Design of Building Drainage Systems*. Ashgate, Aldershot, p. 405.

34. Swaffield, J. A., Escarameia, M. and Campbell, D. P. (1999). Unsteady gutter flow: development and application of simulation. *Proc. Chartered Institution of Building Services Engineers, Series A: BSER&T*, **20** (1), 29–40.

35. Swaffield, J. A., McDougall, J. A. and Campbell, D. P. C. (1999). Drainage flow and solid transport simulation in defective building drainage networks. *Proc. Chartered Institution of Building Services Engineers, Series A: BESR&T*, **20** (2), 73–82.

36. Arthur, S. and Swaffield, J. A. (1999). Numerical modelling of the priming of a siphonic rainwater drainage system. *Proc. Chartered Institution of Building Services Engineers, Series A: BSER&T*, **20** (2), 83–92.

37. Swaffield, J. A. and Campbell, D. P. (1992). Numerical modelling of air pressure transient propagation in building drainage systems, including the influence of mechanical boundary conditions. *Building and Environment*, **27** (4), 455–67.

38. Swaffield, J. A. and Wright, G. B. (1998). Drainage ventilation for underground structures I: Transient analysis of operation. *Proc. Chartered Institution of Building Services Engineers, Series A: BSER&T*, **19** (4), 187–94.

39. Swaffield, J. A. and Wright, G. B. (1998). Drainage ventilation for underground structures II: Simulation of transient response. *Proc. Chartered Institution of Building Services Engineers, Series A: BSER&T,* **19** (4), 195–202

Further reading

Bergeron, L. (1961). *Waterhammer in Hydraulics and Wave Surges in Electricity.* John Wiley, New York.

BHRA. *International Conferences on Pressure Surge,* Nos 1 to 4 (1972, 1976, 1980, 1983, 1986, 1989, 1992). British Hydromechanics Research Association, Cranfield.

Chow, V. T. (1959). *Open Channel Hydraulics.* McGraw-Hill, New York.

Finite Elements in Water Resources, Proc. 4th International Conference, Hanover (June 1982).

Fox, J. A. (1977). *Hydraulic Analysis of Unsteady Flow in Pipe Networks.* Macmillan, London.

Fox, J. A. (1989). *Transient Flow in Pipes, Open Channels and Sewers,* Ellis Horwood, Chichester.

Henderson, F. M. (1966). *Open Channel Flow.* Macmillan, New York.

Lister, M. (1960). The numerical solution of hyperbolic partial differential equations by the method of characteristics. In A. Ralston and H. S. Wilf (Eds) *Mathematical Methods for Digital Computers.* Wiley, New York, pp. 163–79.

Parmakian, J. (1963). *Waterhammer Analysis.* Dover, New York.

Streeter, V. L. and Wylie, E. B. (1984). *Fluid Mechanics.* McGraw-Hill, New York.

Swaffield, J. A. and Boldy, A. P. (1993). *Pressure Surge in Pipe and Duct Systems.* Avebury Technical, Aldershot.

Swaffield, J. A. and Galowin, L. S. (1992). *The Engineered Design of Building Drainage Systems.* Ashgate, Aldershot.

Symposium on Surges in Pipelines. Institute of Mechanical Engineers, London (1965).

Thorley, A. R. D. and Enever, K. J. (1979). *Control and Suppression of Pressure Surges in Pipelines and Tunnels.* CIRIA Report 84, London.

Towson, J. M. (1991). *Free-surface Hydraulics.* Unwin Hyman, London.

Urban Drainage Systems, Proc. 1st International Conference, Southampton (1982). Pitman Advanced Publishing Program, London.

Watters, G. Z. (1979). *Modern Analysis and Control of Unsteady Flow in Pipelines.* Ann Arbor Science, Michigan.

Wylie, E. B. and Streeter, V. L. (1993). *Fluid Transients in Systems,* Prentice Hall, New Jersey.

Problems

20.1   Water at a temperature of 20 °C flows through a pipe system made up, over a considerable length, of a range of different pipe sizes and materials. For each of the sections detailed below calculate the appropriate wave propagation velocity. Assume that the effects of longitudinal strain, as represented by the inclusion of Poisson's ratio, may be neglected.

(1)   50 mm diameter steel, 5 mm wall,
      Young's modulus $E = 204 \times 10^9 \, N \, m^{-2}$

(2)   50 mm diameter aluminium, 3 mm wall,
      $E = 70 \times 10^9 \, N \, m^{-2}$

(3)   25 mm diameter glass, 5 mm wall,
      $E = 1 \times 10^9 \, N \, m^{-2}$

(4)   75 mm diameter copper, 3 mm wall,
      $E = 2.5 \times 10^9 \, N \, m^{-2}$

(5)   900 mm diameter cast iron, 15 mm wall,
      $E = 2.0 \times 10^9 \, N \, m^{-2}$

(6)   1.2 m diameter concrete, 75 mm wall,
      $E = 0.5 \times 10^9 \, N \, m^{-2}$

Assume density of water as $998 \, kg \, m^{-3}$ and bulk modulus as $2 \times 10^9 \, N \, m^{-2}$.
$$[1351 \, m \, s^{-1}, \, 1165 \, m \, s^{-1}, \, 427 \, m \, s^{-1}, \, 309 \, m \, s^{-1},$$
$$181 \, m \, s^{-1}, \, 176 \, m \, s^{-1}]$$

20.2   For the aluminium pipe in Problem 20.1 above, determine the ratio of pipe diameter to wall thickness above which the effects of longitudinal strain, as expressed through the inclusion of Poisson's ratio $\mu = 0.3$, should be included in wave speed calculations. Take 10 per cent error as the boundary. $[D/e > 30]$

20.3   For each of the pipelines described in Problem 20.1 determine the physical length of the wavefront propagated by a valve closure in 0.8 seconds. Assume each pipe is infinitely long.
$$[1081 \, m, \, 932 \, m, \, 342 \, m, \, 247 \, m, \, 145 \, m, \, 141 \, m]$$

20.4   If the flow velocity in each case referred to in Problem 20.3 was $2.58 \, m \, s^{-1}$ calculate the pressure rise recorded at the closing valve.
$$[3479 \, kN \, m^{-2}, \, 3000 \, kN \, m^{-2}, \, 1100 \, kN \, m^{-2},$$
$$796 \, kN \, m^{-2}, \, 466 \, kN \, m^{-2}, \, 453 \, kN \, m^{-2}]$$

20.5   A 20 m long, 75 mm diameter, steel pipeline, wall thickness 6 mm, carries water from a large reservoir tank, held at a constant head of 6 m. Discharge is $0.022 \, m^2 \, s^{-1}$ through a variable speed valve positioned 10 m from the supply tank. Discharge is to a second constant head tank held at 2 m head.

If the valve closure is instantaneous determine the theoretical magnitudes of the pressure waves propagated away from the valve, under frictionless conditions.

Comment on the downstream pressure variation and draw pressure–time curves at points 5 m, 2.5 m and 0.5 m from the upstream tank.
$$[\pm 677 \, m, \text{ flow separates downstream of valve}]$$

20.6   Explain why, in cases where the valve closure time is fixed and longer than one pipe period, the belief that reducing the wave speed, either by free air or by changing the pipe material, will reduce peak pressures is not necessarily correct.

20.7   Show that the transient reflection and transmission coefficient for a junction of n pipes is given by an expression

$$C_R = \left(\frac{2A_j}{c_j} - \sum_{i=1}^{n} \frac{A_i}{c_i} \right) \Big/ \sum_{i=1}^{n} A_i c_i,$$

$$C_T = \left(\frac{2A_j}{c_j} \right) \Big/ \left(\sum_{i=1}^{n} \frac{A_i}{c_i} \right),$$

where the transient approaches the junction along pipe j.

20.8   Show that the reflection coefficient at a closed end is $+1$ and at a constant pressure boundary is -1. Fluid flows between two reservoir tanks held at constant pressures p_1 and p_2 absolute. A valve is placed mid-way between the tanks. Assuming an instantaneous valve closure plot the theoretical pressure variations at the downstream face of the valve and mid-way between the valve and the downstream reservoir. Assume frictionless flow with all the pressure loss $p_1 - p_2$ occurring at the valve. If the pressure drop on valve closure, $\rho c V_0$, is greater than (p_2 − fluid vapour pressure) sketch the resultant pressure variation on the valve downstream face if the first and second cavities last for six and four pipe periods, respectively. Show that during the existence of the first vapour cavity at the valve, the pressure variations recorded at the mid-downstream pipe position oscillate between fluid vapour pressure and downstream reservoir pressure at a frequency of c/L, where L is the pipe length and c is the wave speed.

20.9   A pressure transducer is attached to an infinite pipe, area A, wave speed C, by means of a connection tube, area a, wave speed c. Show that if $A \gg a$ and $c \to C$ and if the pressure front approaching the transducer junction is short in comparison with the length of the transducer connection, the error in recording pipeline pressure will approach 100 per cent.

If the pressure front is linear and has a length equal to the $6 \times$ length of the transducer connection tube, show that the error in recording pipeline pressure is of the order of 50 per cent.

20.10 Write a short computer program that will allow the relation between transducer connection tubing length and diameter and incoming wavefront length to be investigated for the type of installation described in Problem 20.9.

21

Simulation of Unsteady Flow Phenomena in Pipe, Channel and Duct Systems

THE INTRODUCTION OF COMPUTING METHODS HAS TRANSFORMED THE ANALYSIS AND simulation of fluid flow conditions. The brief notes included earlier in Chapter 5 demonstrated the development of the general Navier–Stokes equations and the application of finite difference methods, which although known and available for a considerable time, were impractical to apply due to the number of calculations necessary if the time-step and internodal lengths used were to give any realistic model of the flow.

Applications of computer assisted solutions have been spread throughout this text. In general these may be seen to fall naturally into a number of categories: first, simple computerized versions of calculations to determine such constants as friction factor or normal or critical flow depths; second, programs were introduced which predicted steady flow conditions for a range of free surface or duct flow conditions, or analyzed the steady flow in networks or determined the operating point of a particular fan or pump within a network under steady flow conditions. In Part VI of the text consideration has been given to a range of time-dependent flow conditions, ranging from mass oscillation to transient propagation, defined by the St Venant equations of continuity and momentum. These cases are suitable for numerical analysis and this chapter presents a specific computer-based methodology capable of simulating a range of unsteady flow conditions, from the 'slow' to 'rapid', in full or partially filled ducts with air or liquid as the working fluid.

The propagation of pressure transients has been introduced, together with the concept of wave speed and boundary reflection/transmission coefficients. These effects are well established, however, the numerical solution of the underlying equations may be seen to apply to a whole family of flow conditions subjected to changes in operating condition, ranging from the traditional waterhammer examples, with or without gas or vapour complications, through to the effects of fan speed or damper setting alteration in ventilation or air conditioning systems, and including a whole range of free surface flow applications. It will be noted that in the limit each of the cases developed below will revert to the steady state conditions already developed elsewhere in this text.

As shown in Chapter 5 computational fluid dynamics offers finite difference approaches to the solution of the defining Navier–Stokes equations. In this chapter a particular subset of quasi-linear hyperbolic partial differential equations of continuity and momentum will be addressed and a particular solution technique, long established as the industry standard for pressure surge simulation, the method of characteristics, will be shown to have application across the whole family of flow conditions considered. ● ● ●

21.1 DEVELOPMENT OF THE ST VENANT EQUATIONS OF CONTINUITY AND MOTION

Figure 21.1 represents the unsteady flow conditions for an element of one-dimensional flow, where F is the net force in the flow direction and HGL = hydraulic grade line (static pressure line), so that:

$$F = pA + \left(p + \frac{\partial p}{\partial x}dx\right)\left(A + \frac{\partial A}{\partial x}dx\right) + \left(p + \frac{1}{2}\frac{\partial p}{\partial x}dx\right)\frac{\partial A}{\partial x}dx$$
$$- \tau_0 P dx + mg \sin \alpha. \tag{21.1}$$

Noting that there are two entry routes for fluid into the control volume it follows that

$$F = \rho A dx \left(\frac{\partial V}{\partial t} + \frac{\partial V}{\partial x}\frac{dx}{dt}\right) + \rho q dx V, \tag{21.2}$$

where the lateral inflow term is assumed normal to the flow direction and does not contribute to the initial momentum in the flow direction.

Neglecting second order terms the equation of motion becomes:

$$-A\frac{\partial p}{\partial x}dx - \tau_0 P dx + \rho A dx g \sin \alpha = \rho A dx \left(\frac{\partial V}{\partial t} + \frac{\partial V}{\partial t}\frac{dx}{dt}\right) + \rho q dx V, \tag{21.3}$$

$$\frac{1}{\rho}\frac{\partial p}{\partial x} + \left(\frac{\partial V}{\partial t} + V\frac{\partial V}{\partial x}\right) - g \sin \alpha + \frac{\tau_0 P}{\rho A} + \frac{qV}{A} = 0, \tag{21.4}$$

expressed as

$$\text{Term I}^{\text{m}} + \text{Term II}^{\text{m}} - \text{Term III}^{\text{m}} + \text{Term IV}^{\text{m}} + \text{Term V}^{\text{m}} = 0.$$

Each of these terms has its own significance, e.g. Term III^{m} represents gravitational forces, while Term IV^{m} represents the frictional resistance acting to oppose the local flow, and Term V^{m} represents the acceleration of the lateral inflow.

The continuity equation may also be derived from Fig. 21.1. The conduit is assumed to be linearly elastic, only subjected to small deformations. The fluid is assumed to undergo small changes in density compared with the magnitude of its density. Figure 21.1 illustrates the mass flow through the element between two fixed sections δs apart. As these two sections are fixed in space, not relative to the conduit wall, it follows that $\partial s/\partial t = 0$. The conduit and fluid do not move together as a rigid body so that the fluid is not directly affected by the change in conduit wall length. Any axial movement of the conduit wall affects the axial stress and hence, through the material Poisson's ratio, the pipe diameter.

$$\text{Mass (inflow} - \text{outflow)} = \text{Rate of change of mass storage} \tag{21.5}$$

$$\rho A V - \left[\rho A V + \frac{\partial}{\partial x}(\rho A V)\delta s\right] + q\rho\delta s = \frac{d}{\partial t}(\rho A \delta s), \tag{21.6}$$

where q is the lateral inflow per unit length of conduit.

FIGURE 21.1
(a) Forces acting on a
small element of fluid in
a conduit. (b) Mass
flow through an element
of fluid in a conduit,
note lateral inflow.
(c) Relationship between
flow area and depth
change for a general
section free surface flow

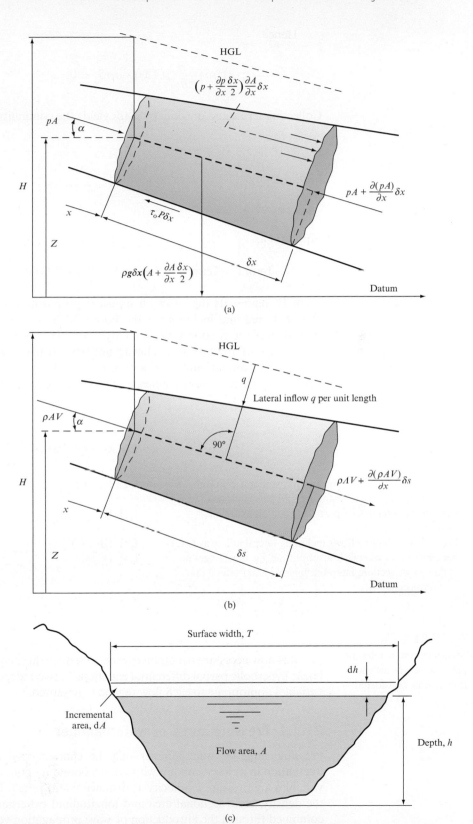

Hence

$$-\frac{\partial}{\partial x}(\rho A V)\delta s - \frac{\partial}{\partial t}(\rho A \delta s) + q\rho\delta s = 0. \tag{21.7}$$

Collecting terms and dividing by $\rho A \delta s$ yields the continuity equation:

$$\frac{\partial V}{\partial x} + \frac{1}{A}\left[\frac{\partial A}{\partial t} + V\frac{\partial A}{\partial x}\right] + \frac{1}{\rho}\left[\frac{\partial \rho}{\partial t} + V\frac{\partial \rho}{\partial x}\right] - \frac{q}{A} = 0, \tag{21.8}$$

expressed as

$$\textbf{Term I}^c + \textbf{Term II}^c + \textbf{Term III}^c - \textbf{Term IV}^c = \textbf{0.}$$

Term II^c represents the effect on transient propagation of a change in flow cross-sectional area and includes via the Poisson's ratio term the effect of longitudinal extension of the conduit wall. Similarly Term III^c represents density changes due to the passage of the transient. Having understood the function of each term in the equations of motion and continuity it is possible to identify which are relevant for each of the unsteady flow regimes to be considered, see Table 21.1.

FLOW REGIME	EQUATION OF MOTION	CONTINUITY EQUATION
Waterhammer (closed conduit flow, siphonic rainwater systems, all terms relevant except lateral inflows)	I, II, III, IV	I, II, III
Free surface channel or partially filled pipe flow, i.e. no density changes	I, II, III, IV	I, II only
Free surface channel flows including lateral inflow as $f(x, t)$	I, II, III, IV, V	I, II, IV only
Air pressure transients, low amplitude, i.e. no changes in flow cross-section, lateral inflows or longitudinal duct extensions	I, II, IV only	I, III only

TABLE 21.1
Identification of relevant terms for each unsteady flow regime considered

It is now necessary to combine terms to reduce these equations to a pair of quasi-linear hyperbolic partial differential equations in two independent and two dependent variables appropriate to each flow regime represented.

21.1.1 Pressure surge or waterhammer

Pressure surge or waterhammer may be characterized as a large-scale pressure fluctuation in a closed conduit with known boundary conditions, see Chapter 20. The variables are pressure p, velocity V, distance x and time t. In the continuity equation the density, cross-sectional area and longitudinal extension variation terms may be combined through the introduction of wave propagation velocity, c.

The density change Term III^c may be expressed as

$$\text{Term III}^c = \frac{1}{K}\left[\frac{\partial p}{\partial t} + \frac{\partial p}{\partial x}\frac{dx}{dt}\right], \tag{21.9}$$

where K is the fluid bulk modulus defined as $dp/K = d\rho/\rho$.

The cross-sectional area Term II^c may be expressed, Section 20.1, in terms of the axial and lateral stresses imposed on the conduit wall by the passing pressure transient, the conduit material Young's modulus of elasticity, the conduit wall thickness and the degree of restraint applied to the conduit through its supports. (The effect of conduit restraint is normally included by introducing a constant C_1 whose value depends on the conditions imposed, e.g. $C_1 = 1$ if the conduit is fully restrained along its whole length.)

It may be shown that Term II^c becomes

$$\text{Term II}^c = \frac{D}{Ee}C_1\left[\frac{\partial p}{\partial t} + \frac{\partial p}{\partial x}\frac{dx}{dt}\right], \tag{21.10}$$

allowing a combination of Terms II^c and III^c and an identification of this term as representing wave propagation velocity in the fluid as modified by the presence of an elastic pipe wall.

It is convenient to write an equivalent bulk modulus for the fluid/pipe wall combination as

$$\frac{1}{K_{\text{eff}}} = \frac{1}{K} + \frac{D}{Ee}C_1, \tag{21.11}$$

and the wave propagation velocity in a fluid as

$$c = \sqrt{\frac{K_{\text{eff}}}{\rho_{\text{eff}}}}, \tag{21.12}$$

where the bulk modulus and density terms are seen to incorporate the effect of wall distortion.

The possible presence of free gas or air in the fluid may be included by adding both a gas equivalent bulk modulus and modifying the density term to include gas density and a free gas content, normally a percentage by volume at NTP. The full expression for wave speed in a gas/fluid mixture in an elastic pipe may thus be written as

$$c = \sqrt{\frac{K_{\text{eff}}}{\rho_{\text{eff}}}} = \sqrt{\left(\frac{(1-y)/K_f + y/K_g + DC_1/Ee}{(1-y)\rho_f + y\rho_g}\right)}, \tag{21.13}$$

where suffixes g and f refer to the free gas and liquid, respectively, Section 20.1. This expression is relevant to a wide range of systems where free gas or gas released from solution may be present, for example sections of conduit where the line pressure falls below atmosphere, such as siphonic rainwater systems.

Thus from equations (21.8) and (21.11) the continuity equation may be written as

$$\rho c^2 \frac{\partial V}{\partial x} + \frac{\partial p}{\partial t} + V \frac{\partial p}{\partial x} = 0. \qquad (21.14)$$

Referring to the equation of motion, equation (21.4), it may be seen that for a full-bore flow case the frictional resistance Term IV^m may be expressed in terms of the Darcy equation and a suitable friction factor, preferably determined from the Colebrook–White expression, equation 10.60,

$$\text{Term } IV^m = \frac{fV|V|}{2m}, \qquad (21.15)$$

where m is the hydraulic mean depth or hydraulic radius, equal to $D/4$ for full-bore flow in circular cross-section conduits.

The equation of motion, equation (21.4), thus becomes

$$\frac{1}{\rho}\frac{\partial p}{\partial x} + \left[\frac{\partial V}{\partial t} + V\frac{\partial V}{\partial x}\right] - g\sin\alpha + \frac{fV|V|}{2m} = 0. \qquad (21.16)$$

21.1.2 Free surface flows with or without a lateral inflow

The variables in this case are flow depth h, related to pressure as $p = \rho g h$, velocity V, distance x and time t. The density change Term III^c may be ignored. For partially filled conduit flow without a lateral inflow Term IV^c may be ignored. The area change Term II^c may be expressed in terms of flow depth h and surface width T:

$$dA = T\,dh. \qquad (21.17)$$

The surface wave propagation velocity in a partially filled conduit flow may be shown to be, Section 5.14,

$$c = \sqrt{\frac{gA}{T}}, \qquad (21.18)$$

and the continuity equation for this flow regime, with or without lateral inflow becomes

$$c^2\frac{\partial V}{\partial x} + g\left[\frac{\partial h}{\partial t} + V\frac{\partial h}{\partial x}\right] - \frac{c^2 q}{A} = 0. \qquad (21.19)$$

Referring to the equation of motion, equation (21.4), it may be seen that for a partially filled conduit flow the frictional resistance Term IV^m may be expressed in terms of the

Chezy equation and a suitable friction factor, again preferably determined from the Colebrook–White expression, Section 15.2,

$$\text{Term IV}^m = \left[\frac{1}{2}\rho f V^2 \frac{P}{\rho A}\right] = \left[\frac{1}{2}\frac{f}{m}\right]\left[\frac{2g}{f}Sm\right] = gS,$$
(21.20)

where m is the hydraulic mean depth or hydraulic radius A/P, and S is the friction slope.

The equation of motion, equation (21.4), thus becomes

$$g\frac{\partial h}{\partial x} + \left[\frac{\partial V}{\partial t} + V\frac{\partial V}{\partial x}\right] + g(S - S_0) + \frac{Vq}{A} = 0,$$
(21.21)

where S_0 is the channel slope.

21.1.3 Low-amplitude air pressure transient propagation

The dependence of air density on pressure dictates that the variables become air flow velocity u and wave propagation velocity c, linking density to pressure, distance x and time t.

In the continuity equation only Terms I^c and III^c apply if it may be assumed that air pressure transient propagation will be at low amplitude and insufficient to cause any change in conduit cross-sectional area.

The relationship between fluid density and pressure for a low-amplitude transient may be expressed in terms of wave speed c, equation (5.31),

$$c = \sqrt{\frac{\gamma p}{\rho}},$$
(21.22)

and under the assumed isentropic process conditions

$$\frac{p}{\rho^\gamma} = \text{constant}.$$
(21.23)

Substitution from equations (21.22) and (21.23) yields

$$\partial\rho = \left[\frac{2}{\gamma - 1}\right]\frac{\rho}{c}\partial c$$
(21.24)

and

$$\partial p = \left[\frac{2}{\gamma - 1}\right]\rho c\,\partial c.$$
(21.25)

Substitution and exclusion of non-relevant terms then enables the equations of continuity and momentum for the low-amplitude air pressure transient case to be written as:

$$c^2 \frac{\partial V}{\partial x} + \left[\frac{2}{\gamma - 1} \right] c \left[\frac{\partial c}{\partial t} + V \frac{\partial c}{\partial x} \right] = 0 \qquad (21.26)$$

$$\left[\frac{2}{\gamma - 1} \right] c \frac{\partial c}{\partial x} + \left[\frac{\partial u}{\partial t} + u \frac{\partial u}{\partial x} \right] + \frac{fu|u|}{2m} = 0. \qquad (21.27)$$

As these equations link wave speed to fluid velocity it will also be necessary to determine the pressure at each node at each time-step as:

$$P_p = \left[(p_o / \rho_o^\gamma)(\gamma / c^2)^\gamma \right]^{1/(1-\gamma)}. \qquad (21.28)$$

21.2   THE METHOD OF CHARACTERISTICS

In each of the cases dealt with above the equations of continuity and motion may be recognized as a pair of quasi-linear hyperbolic partial differential equations that may be solved numerically provided a scheme to transform the partial into total derivatives can be identified. The method of characteristics provides this linkage. The equations of continuity and motion in two dependent variables, u_1 and u_2, and two independent variables, x and t, may be written as:

$$L_1 = a_c \frac{\partial u_1}{\partial x} + b_c \left[\frac{\partial u_2}{\partial t} + u_1 \frac{\partial u_2}{\partial x} \right] + c_c = 0, \qquad (21.29)$$

$$L_2 = a_m \frac{\partial u_2}{\partial x} + b_m \left[\frac{\partial u_1}{\partial t} + u_1 \frac{\partial u_1}{\partial x} \right] + c_m = 0, \qquad (21.30)$$

where the values of the coefficients a, b, c, for each case are listed in Table 21.2.

Combining equations (21.28) and (21.29) as $L_1 + \lambda L_2$ yields:

$$\lambda b_m \left[\frac{\partial u_1}{\partial t} + \frac{\partial u_1}{\partial x} \left(u_1 + \frac{a_c}{\lambda b_m} \right) \right] + b_c \left[\frac{\partial u_2}{\partial t} + \frac{\partial u_2}{\partial x} \left(u_1 + \frac{\lambda a_m}{b_c} \right) \right] + c_c + \lambda c_m = 0. \qquad (21.31)$$

By inspection it follows that the combined expression, equation (21.31), may be expressed as a total differential equation of the form

$$b_c \frac{du_2}{dt} + \lambda b_m \frac{du_1}{dt} + (c_c + \lambda c_m) = 0, \qquad (21.32)$$

provided that

$$\frac{dx}{dt} = u_1 + \frac{a_c}{\lambda b_m} = u_1 + \frac{\lambda a_m}{b_c}; \qquad (21.33)$$

	u_1	u_2	a_c	b_c	c_c	a_m	b_m	c_m	$\lambda=(a_cb_c/a_mb_m)^{0.5}$	dx/dt
			CONTINUITY EQUATION			**EQUATION OF MOTION**				
Waterhammer (inc. siphonic rainwater)	V	p	ρc^2	1	0	$1/\rho$	1	$(-g\sin\alpha+fV|V|/2m)$	$\pm\rho c$	$V\pm c$
Free surface	V	h	c^2	g	0	g	1	$[g(S-S_0)]$	$\pm c$	$V\pm c$
Free surface with lateral inflow	V	h	c^2	g	$(-c^2q/A)$	g	1	$[g(S-S_0)+qV/A]$	$\pm c$	$V\pm c$
Air pressure transients	u	c	c^2	$2c/(\gamma-1)$	0	$2c/(\gamma-1)$	1	$(fu|u|/2m)$	$\pm c$	$u\pm c$

TABLE 21.2

Identification of dependent variables and coefficients in the equations of continuity and motion developed for a range of unsteady flow conditions

hence

$$\lambda = \pm\sqrt{\frac{a_cb_c}{a_mb_m}} \tag{21.34}$$

and

$$\frac{\mathrm{d}x}{\mathrm{d}t} = u_1 \pm \sqrt{\frac{a_ca_m}{b_cb_m}}. \tag{21.35}$$

The combined total differential equation may thus be written as

$$\frac{\mathrm{d}u_1}{\mathrm{d}t} \pm C_2\frac{\mathrm{d}u_2}{\mathrm{d}t} + C_3 = 0, \tag{21.36}$$

provided that

$$\frac{\mathrm{d}x}{\mathrm{d}t} = u_1 \pm c. \tag{21.37}$$

This relationship between time-step, internodal length and the flow and wave propagation velocities is central to the method of characteristics solution of the St Venant equations. It is known as the Courant criterion and applies to all the cases of transient propagation to be covered in this chapter. Adherence to the Courant criterion also dictates the time-step for any network, as all constituent pipes must have the same time-step to allow continuity at junction boundaries. It follows, therefore, that the time-step depends upon the largest combined value of flow and wave propagation velocities found within the network and this in turn leads to some of the computational difficulties to be discussed later, in particular the possibility of rounding errors or backwater profile collapse due to excessive interpolation brought about by very different wave and flow velocities within the same network.

FIGURE 21.2

C^+, C^- characteristics in an x–t plane, illustrating the effect of wave speed, relative to flow velocity, on the slope of these characteristics for a free surface flow application

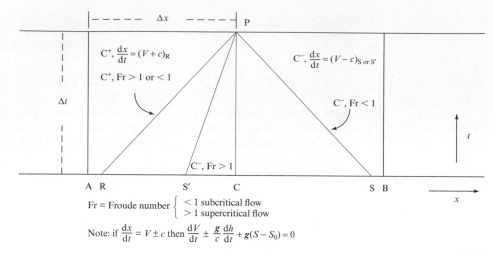

Referring to Fig. 21.2 it will be seen that the St Venant equations may be represented by two first order finite difference equations based on equations (21.38) and (21.40) below, known respectively as C^+ and C^- characteristics linking known conditions at time t to conditions at P at one time-step in the future:

$$u_1^P - u_1^R + C_2[u_2^P - u_2^R] + C_3\Delta t = 0, \tag{21.38}$$

provided that

$$x^P - x^R = [u_1^R + c^R]\,\Delta t; \tag{21.39}$$

and

$$u_1^P - u_1^S - C_2[u_2^P - u_2^S] + C_3\Delta t = 0, \tag{21.40}$$

provided that

$$x^P - x^S = [u_1^S - c^S]\,\Delta t; \tag{21.41}$$

where the coefficients C_2 and C_3 are given by Table 21.3 and superscripts P, R and S refer to Fig. 21.2, which illustrates a number of points inherent in the application of

TABLE 21.3

Identification of the coefficients in the finite difference equations applicable to the unsteady flow applications considered

	$C_2 = b_c/\lambda b_m$	$C_3 = (c_c + \lambda c_m)/\lambda b_m$	WAVE SPEED c
Waterhammer (inc. siphonic rainwater systems)	$1/\rho c$	$-g\sin\alpha + fV\lvert V\rvert/2m$	equation (21.13)
Free surface	g/c	$g(S - S_0)$	equation (21.18)
Free surface with lateral inflow	g/c	$g(S - S_0) + q(V \pm c)/A$	equation (21.18)
Air pressure transients	$2c/(\gamma - 1)$	$fu\lvert u\rvert/2m$	equation (21.22)

the method of characteristics to this family of equations, considered in more detail below.

21.2.1 Initial conditions

Initial conditions at time zero must be established along the whole conduit length to give values of the variables u_1, u_2 and x at each node. For full-bore flow cases, e.g. waterhammer and entrained or driven air flows, these conditions may be zero flow velocity and some system pressure, possibly atmospheric. In the case of free surface fluid flows it is necessary to define a 'trickle' flow, with flow depth and velocity calculated at each node. Care must be taken in cases where the free surface flow is not supercritical or where there may be initial hydraulic jumps providing a transition from subcritical to supercritical flow regimes.

21.2.2 Interpolation techniques

Figure 21.2 illustrates a form of the solution technique known as specified time intervals, where the time-step is set by reference to the largest combination of velocity and wave speed

$$\Delta t = \frac{\Delta x}{u_{1_{max}} + c_{max}}. \tag{21.42}$$

The rectangular grid is based on fixed and chosen internodal lengths and a time-step calculation that ensures that points R and S always fall within an internodal length upstream or downstream of the calculation node, defined at P. It will be appreciated that for a network the time-step must be constant for all the interlinked pipes and this may lead to the necessity to vary the internodal length between pipes. As Fig. 21.2 presents the specified time interval approach, interpolations to determine conditions at R and S are required along the distance axis at time t. This may lead to rounding errors in some cases as discussed later.

The time-step depends upon the internodal length chosen and the maximum values of velocity and wave speed. In waterhammer calculations the wave speed dictates the time-step. Wave propagation velocity in a full-bore metal water main may be as high as 1200–1500 m s^{-1}. This value falls dramatically if there is even a small percentage of free air but will still exceed the fluid flow velocity by several orders of magnitude.

In air pressure transient calculations the wave speed of 320 m s^{-1} exceeds the entrained air flow velocity to the extent that it is acceptable to disregard air velocity in the time-step calculation. In free surface water flow situations the opposite is true as the surface wave speed, dependent on flow depth, is of the same magnitude as, or less than, the flow velocity. In partially filled pipe or free surface flows the fluid velocity may exceed the wave speed, i.e. supercritical flow conditions. For a 1 metre internodal length the time-step can therefore vary from milliseconds for the waterhammer example to tenths of a second for free surface cases.

It will be seen from Fig. 21.2 that halving the internodal length will quadruple the number of calculation steps to complete any given simulation time, a consideration when the computing capacity available was limited.

Variable wave speed along the length of a pipe, duct or channel leads to different slopes for the C$^+$ and C$^-$ characteristics. Assuming that initial conditions are known at all nodes at time t, then it is necessary to determine the conditions at R and S at time t in order to apply the characteristics solution already developed. The most common solution to this situation is to interpolate linearly between conditions at A and C to obtain conditions at R and similarly between C and B to obtain conditions at S. (Note here that if $c < V$, i.e. the supercritical free surface or supersonic flow condition, then both R and S lie between A and C in Fig. 21.2.)

It is worth noting that interpolation implies that a pressure transient or surface wave arriving at A or B at time t determines conditions at R or S at that time. This effectively increases the speed of propagation of the transient and also decreases the rate of change of pressure or velocity that it imparts to the flow it passes through. Both effects lead to a rounding in the predicted transient.

Taking the traditional waterhammer case as representative of interpolation techniques that velocity V and pressure p stand for the u_1 and u_2 variables in Tables 21.2 and 21.3 and referring to Fig. 21.2 as the general case where both c and V vary along the pipe and also where the possibility that $V > c$ is allowed to exist so that both C^+ and C^- characteristics slope downstream, a series of equations may be presented linking conditions at R and S (or S′) to conditions at the nodes A, C and B.

For the C^+ characteristic passing through R and linking conditions at R at time t to conditions at P at time $t + \Delta t$, consideration of the velocity variation yields

$$\frac{V_C - V_R}{V_C - V_A} = \frac{x_C - x_R}{\Delta x} = (V_R + c_R)\frac{\Delta t}{\Delta x},$$

since

$$x_C = x_P$$

and

$$\Delta x = x_C - x_A.$$

Similarly the wave speed terms yield

$$\frac{c_C - c_R}{c_C - c_A} = \frac{x_C - x_R}{\Delta x} = (V_R + c_R)\frac{\Delta t}{\Delta x}.$$

Simultaneous solution of these equations results in a series of interpolation relationships that allow the determination of base conditions at point R:

$$V_R = \frac{V_C + \theta(V_A c_C - V_C c_A)}{1 + \theta(V_C - V_A + c_C - c_A)}, \tag{21.43}$$

$$c_R = \frac{c_C + \theta V_R(c_A - c_C)}{[1 + \theta(c_C - c_A)]}, \tag{21.44}$$

where

$$\theta = \frac{\Delta t}{\Delta x},$$

and for the pressure or head terms

$$p_R = p_C - (p_C - p_A)\theta(V_R + c_R),\tag{21.45}$$

$$H_R = H_C - (H_C - H_A)\theta(V_R + c_R).\tag{21.46}$$

A number of points need to be stressed concerning the above equations:

1. If the wave speed is a constant, but still comparable to the local flow velocity, these equations will yield the interpolated values of flow velocity V with no modification.

2. In cases where the velocity is negligible with respect to a constant wave speed, so that

$$c_P = c_A = c_B,$$

these interpolation equations may be simplified as

$$\frac{V_C - V_R}{V_C - V_A} = \theta c_A,$$

thus

$$V_R = V_C - \theta c_A(V_C - V_A),$$

and pressure p, or head H, is given by

$$p_R = p_C - \theta c_A(p_C - p_A),$$

$$H_R = H_C - \theta c_A(H_C - H_A).$$

3. When the wave speed exceeds the local flow velocity, the C^- characteristic slopes upstream and the base conditions are found at point S in Fig. 21.2. By a similar series of substitutions to those above, the following expressions may be derived:

$$V_S = \frac{V_C - \theta(V_C c_B - V_B c_C)}{1 - \theta(V_C - V_B - c_C + c_B)},\tag{21.47}$$

$$c_S = \frac{c_C + \theta V_S(c_C - c_B)}{[1 + \theta(c_C - c_B)]};\tag{21.48}$$

and for the pressure or head terms

$$p_S = p_C + (p_C - p_B)\theta(V_S - c_S),\tag{21.49}$$

$$H_S = H_C + (H_C - H_B)\theta(V_S - c_S).\tag{21.50}$$

If $c \gg V$, and is assumed to be constant, then these equations reduce as before to

$$\frac{V_C - V_S}{V_C - V_B} = \theta c_B,$$

thus

$$V_S = V_C + \theta c_B (V_B - V_C),$$

and pressure p, or head H, is given by

$$p_S = p_C + \theta c_B (p_B - p_C),$$
$$H_S = H_C + \theta c_B (H_B - H_C).$$

4. If the local flow velocity exceeds the wave speed then the C^- characteristic also slopes downstream and the required base point, S', is to be found upstream of point P. This condition would not be expected in the analysis of pressure transients; however, it is the norm in many free surface flow applications of the method of characteristics as it represents the supercritical flow regime. In this case the interpolation equations are derived in the same manner as set out above, resulting in the following expressions

$$V_{S'} = \frac{V_C - \theta(V_A c_C + V_C c_A)}{1 + \theta(V_C - V_A + c_A - c_C)}, \tag{21.51}$$

$$c_{S'} = \frac{c_C + \theta V_{S'}(c_A - c_C)}{[1 + \theta(c_A - c_C)]}; \tag{21.52}$$

and for the pressure or head terms

$$p_{S'} = p_C - (p_C - p_A)\theta(V_{S'} - c_{S'}), \tag{21.53}$$

$$H_{S'} = H_C - (H_C - H_A)\theta(V_{S'} - c_{S'}). \tag{21.54}$$

EXAMPLE 21.1

Demonstrate that excessive interpolation can lead to rounding errors and artificial damping of a pressure transient.

Solution

The effect of interpolation, in the case where the wave speed c is constant along a pipelength, and where the value of c clearly exceeds the flow mean velocity, may be clearly demonstrated for a frictionless pipeline by application of surge program FM4SURG, Section 21.4. Figure 21.3 illustrates the predicted pressure variations at the exit valve of a 10 m long pipeline, appropriate wave speed 1000 m s^{-1}, following a valve closure in one pipe period. In the absence of friction and without interpolation, i.e. with a time-step of 0.002 seconds, equal to $\Delta x/c$, where Δx is 2 m, the pressure at the valve follows the periodic cycle discussed in Chapter 20, i.e. no attenuation with time due to the absence of frictional damping. When a reduced time-step is introduced with the same Δx, by interpolation, the effect of numerical damping becomes obvious; in the case illustrated a 50 per cent interpolation is introduced. While the effects of interpolation become less dramatic as the rate of change of the boundary conditions decreases, rapid changes can still occur within systems, particularly as a result of cavity collapse following column separation and, as the onset of such conditions is not always predictable from a study of a system design, it is advisable to limit interpolation.

FIGURE 21.3

Effect of a 50 per cent time-step reduction by interpolation on the pressure variation predicted at a rapidly closing valve in a frictionless pipeline (Swaffield and Boldy 1993)

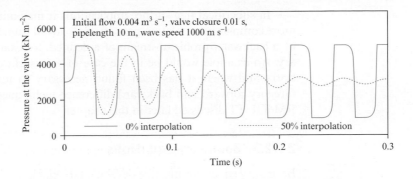

Therefore interpolation techniques must be used to yield values of the variables u_1, u_2 and x at each interpolation points, R, S or S', at the start of each time-step, Fig. 21.2. While Lister and later authors proposed a linear interpolation between nodes, as illustrated, in some cases this can lead to rounding errors sufficient to cause collapse of the solution. In these case a more sophisticated interpolation x-axis technique may be sufficient; however, in some cases there is an advantage in using the time-line interpolation illustrated in Fig. 21.4. While the timeline interpolation removes potential rounding errors associated with the linear interpolation technique, it can be difficult in cases where the flow conditions approach critical flow where the wave and fluid velocities are equal, implying a steep C^- characteristic and the necessity to store data at each node over very many time-steps. (Note that this is particular problem at the moving interface represented by a hydraulic jump upstream of an obstruction or pipe junction or at pipe entry. The critical depth at a free outfall is not, however, a problem as the C^+ slope tends to $1/2c$.)

It will be appreciated that the form of equations (21.38) and (21.40) remains unchanged except that the time-step, Δt in equation (21.38) becomes $(t_P - t_R)$ and in equation (21.40) either $(t_P - t_S)$ or, in free surface flows only, $(t_P - t_{S'})$ depending upon the relative values of fluid velocity and wave speed.

FIGURE 21.4

Schematic representation of time line interpolation for the general case when V and c are comparable and variable (Swaffield and Boldy 1993)

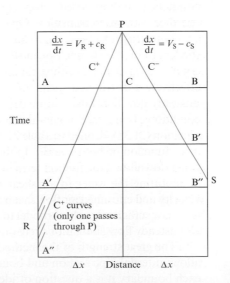

In subcritical free surface applications it may initially be necessary to introduce more complex interpolation procedures to avoid rounding errors leading to a collapse of a backwater profile upstream of a defined, for example critical depth, exit boundary. An example would be in the case of free surface rainwater gutters laid almost horizontal where it has been found necessary to utilize the Everett and Newton-Gregory forward and backward difference techniques, incorporated in the program FM4GUTT discussed later in this chapter.

21.2.3 Boundary equations

Boundary equations must be developed to enable solution at the entry and exit nodes for each conduit comprising the network. As will be appreciated from Fig. 21.2 there is only one available characteristic at each boundary, a C^+ at a downstream boundary and a C^- at the upstream boundary. The choice of appropriate boundary equations is fundamental to the application of the method of characteristics. In addition to providing a means of representing the conditions at the entry and exit from each conduit length, boundary conditions also allow the introduction of time-dependent events. Examples at the entry to a system are probably the easiest to discuss. In the case of free surface flows the entry conditions must represent the discharge profile to the network in terms of flowrate vs. time. The solution techniques described above then allow the simulation of the passage of this discharge along the channel to the next boundary – possibly a free outfall or a junction with a joining flow.

In the case of air pressure transients it will be necessary to model the boundary conditions by reference to either ambient pressure at the boundary or some expression linking pressure at the boundary to flow across it – for example, a fan characteristic or a damper loss coefficient. In particular venting systems, for example in underground structures, the use of air admittance valves would necessitate a boundary condition selected on the basis of local pressure, e.g. valve opens and shuts at some predetermined pressure level with a loss coefficient expression applicable during the open period – the value of the loss coefficient dependent upon the open percentage and hence the pressure differential across the relief valve, see Section 21.8.

In traditional waterhammer applications common boundaries include dead-ended conduits, including closed valves, constant pressure zones or pumps either at constant speed or with speed transition. Opening or closing valves may also be modelled.

In the case of free surface channel flows, where there is a distributed inflow along the length of the channel, or in the case of entrained air flow, where the moving air column is in contact with a falling annular film in a vertical pipe, where there is effectively a distributed shear force acting along each pipe length, the modelling of unsteady flow depends on time-dependent inputs defining these distributed unsteady conditions. In the case of lateral flow, referring to equation (21.2) and to coefficient C_3, equation (21.38), defined in Table 21.3, it will be appreciated that the lateral inflow term q is a function of both x and t. In the entrained air flow arising as a result of annular water downflows, the friction term in equation (21.27) may be reconsidered as a means of inputting the water to air shear force that depends on water downflow terminal velocity and entrained air flow mean velocity, Section 16.14. The method of characteristics is a sufficiently robust model to accommodate these alternative input paths – not all unsteady flow simulations rely on changes at entry to or exit from a system.

The great strength of the method of characteristics solution is that the simulation allows internal pipe section and boundary conditions to be considered separately. At each boundary it is a question of identifying the appropriate relationship to describe the fluid flow condition or the appliance response under unsteady flow conditions. Many of these boundary conditions have been fully developed within the international research programmes that underlie this portion of the text.

Boundary conditions may be subdivided into three broad categories:

1. *Passive boundary conditions* that arise as a result of the design of the system, for example junctions of two or more pipes, constant pressure reservoirs, dead-ended pipes, open discharges or changes in pipe cross-section, material or wall thickness.

2. *Active boundary conditions* that represent equipment connected to the system, for example valves, pumps or turbines.

3. Boundary conditions that arise as a result of the *propagation of transients* within the network and which are therefore not necessarily identified as being present at the design stage or by a simple examination of the system layout. In this category fall the boundary conditions needed to represent column separation, trapped air or gas release due to low system pressures. This category may also include 'moving' boundary conditions, for example the interface formed by the representation of a train moving through a tunnel or the discontinuity between free surface and full-bore flow in an initially partially filled conduit under surcharging conditions.

A further subcategory of boundary conditions represents the interface to pressure transient control and suppression devices, such as relief valves and air chambers, effectively a combination of categories (2) and (3).

In general suitable boundary equations will link either flow rate or pressure to time. In the case of some equipment boundary conditions it may be necessary to relate these variables via a monitoring of the equipment over the time period considered, for example valve position vs. time data will be linked to valve 'pressure loss' vs. position data in order to provide a suitable boundary condition to be solved with the pressure–velocity C^+ or C^- characteristic available at either pipe entry or exit. In other cases the required boundary equation will only be activated if the indicated conditions in the pipeline reach certain 'trigger' levels, for example the representation of a pressure relief valve will need to be present in the numerical model and will lie dormant until required by the solution.

Given the convention adopted in this text, namely that distance increases in the initial flow direction, it follows that a C^- characteristic will always be available at pipe entry and a C^+ characteristic will always be present at pipe exit. (It should be noted here that 'pipe entry' is assumed synonymous with inflow and a zero value of the distance measured along the pipe; however, this linkage is not necessary, as inflow to a system can be regarded as a normal negative flow, for example in vertical stacks carrying an annular water downflow that entrains an air flow that is often defined as negative.) While the development of suitable boundary conditions provides the main area of interest for the numerical modeller utilizing the method of characteristics, it has been necessary to provide a basic development of the method from the base equations of momentum and continuity and to demonstrate that the solutions generated are general in application. This has been the objective of this section, together with an emphasis upon the commonality of the solution across both full-bore liquid and gas flows and free surface conditions. Similarly this chapter has defined one of the main difficulties with the application of the method, namely the need to interpolate to obtain the base conditions for the characteristics, as well as outlining some possible improvements to limit the inherent rounding and dispersion errors.

Tables 21.4 to 21.6 illustrate the boundary conditions in each of these applications discussed.

TABLE 21.4   Boundary conditions: waterhammer applications of method of characteristics

BOUNDARY CONDITION	BOUNDARY EQUATION
Closed valve, dead end, non-return valve closed by −ve throughflow	$V = 0$
Open end with local loss coefficient K (note that K may be zero and that the absolute value of V ensures that the loss opposes motion)	$p = p_{atm} + 0.5\rho K V|V|$
Opening or closing valve at termination pipe, where K_t is valve loss coefficient at time t during either opening or closing valve motion and p_{out} is the prevailing pressure upstream or downstream of the valve. Note that as K is defined in terms of valve open area or some similar indicator it is also necessary to define the way this indicator varies with time to determine K_t	C^+/C^- solved with $p = p_{out} + 0.5\rho K_t V|V|$ If time $t >$ closing time, $V = 0$ If time $t >$ opening time, $K_t = K_{\text{fully open}}$
Opening or closing valve acting as a junction between two pipes. Note that the fully open version of this boundary equation covers the orifice between two pipes boundary	If $t <$ opening or closing time then C^+ upstream and C^- downstream solved with $p = p_{upstream} + 0.5\rho K_t V|V|$ and $\Sigma(AV) = 0$ If $t >$ closing time, $V = 0$ on upstream and downstream valve faces If $t >$ opening time, Δp across valve $= 0.5\rho K_{\text{fully open}} V_{upstream}\,|V_{upstream}|$
Junction of two or more pipes. (Note in this case it is necessary to define which pipes terminate at the junction, and are represented by a C^+ characteristic and which 'start' at the junction and are represented by a C^- characteristic. Loss coefficients may be zero. Note this boundary is still necessary if change at junction only affects wave speed through pipe material or wall thickness changes	$\Sigma(AV) = 0$ $p_1 = p_2 + 0.5\rho K_{(2-1)}V|V| = p_3 + \cdots$
Pump operating characteristic takes this form in the forward flow–forward rotation quadrant. It may also be necessary to represent the reverse flow–forward rotation quadrant prior to non-return valve closure	$\Delta p = C_1 + C_2 Q + C_3 Q^2 + \cdots$
Relief valves – inward or outward	$p <$ opening pressure, $Q_{valve} = 0$ $p >$ opening pressure, $Q_{valve} = (\Delta p_{valve}/K_{valve})^{0.5}$
Constant pressure zone	$p = p_{constant}$ solved with the available C^+ or C^- characteristic
Special case of column separation due to transient propagation through network, pressure remains at vapour level until cavity collapses	Volume cavity > 0, $p = p_{vapour}$ solved with C^+/C^- Volume cavity $= 0$, $V = 0$ solved with C^+/C^-
Trapped air	$p_{\text{air pocket}}\ \text{Vol}^n = $ constant solved with C^+/C^-

As shown by Fig. 21.2 only one characteristic can exist at entry or exit to a conduit. This equation must be solved with an appropriate boundary equation linking local pressure to flow rate or pressure or velocity through a time-dependent relationship.

TABLE 21.5 Boundary conditions: free surface wave propagation application of method of characteristics

BOUNDARY CONDITION	BOUNDARY EQUATION
Time-dependent inflow at channel entry.	
Subcritical flow	Solve available C^- equation with Q vs. time hydrograph
Supercritical flow	Inflow hydrograph solved with some assumption as to flow entry specific energy, e.g. critical or normal depth or some empirical relationship
Dead end, e.g. at rainwater gutter upstream termination	$V = 0$
Free outfall.	
Supercritical flows	Flow depth directly from solution C^+ and C^- equations
Subcritical flows, e.g. gutter discharge	Flow depth from solution C^+ equation and the critical depth equation (16.9)
Gutter discharge to a restricted outfall	Depth above critical determined by solution of C^- characteristic with an empirical relationship linking depth at the outlet to throughflow
Entry to a channel downstream of a junction	Flow depth equates to the critical depth for junction throughflow
Junctions	Solve C^+ for terminating channels with empirical junction relationship, $h_{\text{upstream}} = K_{\text{junction}} \sum (AV)^n_{\text{upstream}}$
Moving hydraulic jump, formed upstream of a junction, or other obstruction in channels where the flow is normally supercritical	C^+ and C^- equation in the supercritical flow upstream of the jump, solved with the C^- characteristic available downstream of the jump and the continuity of flow and momentum equations across the jump

TABLE 21.6   Boundary conditions: low-amplitude air pressure transient propagation application of method of characteristics

Boundary condition	Boundary equation
Dead end, closed damper or air admittance valve	$V = 0$
Constant pressure zone, including atmospheric pressure	$p = p_{constant}$
Open end, at entry or exit, with local loss coefficient K and connection to an air chamber at any known pressure	$p = p_{chamber} + 0.5\rho K V|V|$
Open-ended duct to atmosphere including wind shear where it would be necessary to have details of time-dependent wind effect	$p = p_{atm. \, at \, time \, t} + 0.5\rho K V|V|$
Control dampers where the damper loss coefficient will change with time as the damper is opened or closed. Note it is necessary to know loss coefficient vs. damper angle data as well as mode of damper motion	$p = p_{chamber} + 0.5\rho K_{time} V|V|$
Junction of two or more ducts, including change of cross-section in a single duct	Boundary condition as for waterhammer except that pipe material or wall thickness changes would not necessitate a boundary
Relief valves – inward or outward	$p <$ opening pressure, $Q_{valve} = 0$ $p >$ opening pressure, $Q_{valve} = (\Delta p_{valve} / K_{valve})^{0.5}$ $p >$ fully open pressure, $Q_{valve} = (\Delta p_{valve} / K_{fully \, open})^{0.5}$
Fan operating characteristic, including fan speed changes, in terms of flow and pressure coefficients, Section 25.4, where fan diameter and flow density are constant and the fan reference speed N_{ref} is known. The fan speed change case requires data on rate of change of speed with time	$\Delta p = (N/N_{ref})^2 [C_1 + C_2(N_{ref}/N)Q + C_3(N_{ref}/N)Q^2 + \cdots]$
Instantaneous flow stoppage, for example due to an open end or a chamber connection becoming closed	$V = 0$, Δp given by Joukowsky relationship $= \pm \rho c \Delta V$

As shown by Fig. 21.2 only one characteristic can exist at entry or exit to a conduit and must be solved with an appropriate boundary equation. As the characteristics link wave speed to local velocity so it will be necessary also to invoke equation (21.28) to determine pressure.

21.3 NETWORK SIMULATION

The computer programs linked to this chapter provide simple network analysis. More complex networks would naturally be addressed in practice but the methodology would be similar. Whether the flow condition is traditional waterhammer, low-amplitude air pressure transients or free surface wave attenuation, the simulation would have to include the following computing steps:

1. Ensure that the necessary constants are defined, e.g. density, atmospheric pressure, etc.

2. Define the network, input data for each pipe, duct or channel – diameter, length, wave speed in waterhammer applications, frictional data may include surface roughness for use in Colebrook–White or Manning n for open channels. Define internodal length for each pipe, duct or channel section.

3. Identify the boundary conditions at entry and exit from each specified pipe, duct or channel length. For junctions identify terminating pipes and those 'starting' at that junction in order to define whether the C^+ or C^- characteristic is used. Define any junction loss coefficients.

4. Identify boundary conditions that will generate the unsteady flow – opening or closing valves or dampers, inflow hydrographs – in terms of time.

5. Set up initial conditions at time zero throughout the network. In free surface flow cases this will require depth profile calculations for a 'trickle' flow.

6. Output data at time zero.

7. Increase time by Δt, where Δt is based on the internodal length and the maximum values of wave speed and local velocity. Choosing Δt in this way will lead to differential interpolation requirements in the various network sections.

8. Solve the C^+ and C^- equations at each node from 2 to n in each pipe, duct or channel length. Note that n internodal lengths Δx results in $n + 1$ nodes.

9. Solve the appropriate characteristic at each entry or exit node, or in the case of free surface flows also at any moving hydraulic jump locations. Note that the results should be scanned to ensure that new boundary equations are not required due to the transient conditions now prevailing, e.g. should any relief valves be treated as having opened, are there any vapour pockets to consider or have any moving valves or dampers completed their operation?

10. Output the data at this time-step and to save on storage space overwrite the base condition arrays to prepare for the next time-step calculations.

11. Check if simulation time complete, if not return to step (7), otherwise output data in some accessible form.

Clearly in practice this model would be more complex and could include a whole range of additional output information on the minimum or maximum pressure levels reached, whether column separation (vapour pockets) occurred or whether channels overflowed or junctions became surcharged. Figure 21.5 illustrates such an application to an analysis of the Concorde fuel system during refuelling.

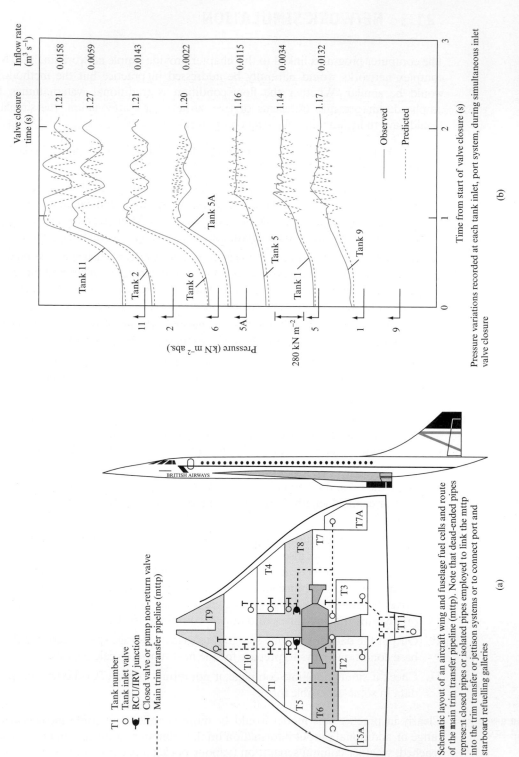

FIGURE 21.5   Illustration of the application of the closing valve boundary conditions to multiple valve closure during aircraft refuelling (Swaffield and Boldy 1993)

Schematic layout of an aircraft wing and fuselage fuel cells and route of the main trim transfer pipeline (mttp). Note that dead-ended pipes represent closed pipes or isolated pipes employed to link the mttp into the trim transfer or jettison systems or to connect port and starboard refuelling galleries

(a)

Pressure variations recorded at each tank inlet, port system, during simultaneous inlet valve closure

(b)

T1   Tank number
○   Tank inlet valve
◕   RCU/IRV junction
T   Closed valve or pump non-return valve
- - -   Main trim transfer pipeline (mttp)

21.4 COMPUTER PROGRAM FM4SURG. THE SIMULATION OF WATERHAMMER

A simple three-pipe system method of characteristics model is presented in order to allow the reader to investigate a range of boundary conditions to reflect individual interests.

The system presented consists of up to three pipes, one pipe being provided with an upstream supply while the other two feature downstream boundaries, including valves, dead ends, trapped air and constant pressure reservoirs. Both outwards and inwards relief valves are provided as optional boundaries, and a junction is provided for two- or three-pipe cases.

Column separation both upstream and downstream of an operated valve are included, the possibility of gas release to the downstream cavity case being included as an option. Variable modes of valve operation are automatically included by defining the valve closure in terms of the appropriate valve discharge coefficient at any time during closure. This option will allow the reader to experiment with two or more stage closures. Valve opening transients may also be considered, as may valves that are cycled through an opening/closing operation.

Table 21.7 illustrates the range of cases that may be investigated. Care should be taken in the simulation of pressure levels within a trapped air pocket following the opening of an upstream valve as the program will expect a two-pipe series configuration.

TABLE 21.7

Options possible with the three-pipe system model, FM4SURG

| | PIPE 1 | | PIPE 2 | | PIPE 3 | |
ENTRY	EXIT	ENTRY	EXIT	ENTRY	EXIT
Reservoir	Closing valve	–	–	–	–
Closing entry valve with/without gas release, with/without IRV	Reservoir	–	–	–	–
Closing entry valve closing, with/without gas release, with/without IRV	Junction 1–2	Junction 2–1	Reservoir	–	–
Reservoir	Junction 1–2	Junction 2–1	Exit valve, close/open ORV	–	–
Entry valve opening	Junction 1–2	Junction 2–1	Dead end with/without ORV or trapped gas	–	–
All the above combinations may then be repeated with three pipes					
Closing entry valve with/without gas release, with/without IRV	Junction 1–2–3	Junction 2–1–3	Reservoir	Junction 3–2–1	Dead end, trapped air with/without ORV

ORV – outwards relief valve; IRV – inwards relief valve, Chapter 6

21.4.1 Application examples

Program FM4SURG may be used to address each of the following examples 21.2 to 21.5.

EXAMPLE 21.2

Use FM4SURG to investigate the effect of valve closure rate on the pressure recorded by a pressure transducer located at a dead end remote from the flow-carrying pipe. Your investigation might include some or all the following parameters: branch length, branch diameter (both relative to the main pipes connecting a supply reservoir, pipe 1, to a discharge valve, pipe 2), wave speed in each pipe or the supply reservoir pressure.

Solution

Take as an example two pipes in series, 1 and 2, of 0.1 m diameter, 10 m length, with 2.5 m internodal computing sections and a wave speed of 1000 m s^{-1}. Pipe 1 is supplied from a reservoir at 1000 kN m^{-2} and pipe 2 terminates in a valve with a discharge loss at 0.1 m^3 s^{-1} throughflow of 200 kN m^{-2}. Take the flowing fluid as water with a density of 1000 kg m^{-3} and a vapour pressure of 6 kN m^{-2}. The effect of valve closure rates should be investigated by using linear closures in the time range 0.1 to 2.0 seconds. The branch, leading from the junction of pipes 1 and 2, has a length of 20 m and a diameter of 0.01 m, with a wave speed of 1200 m s^{-1}. The initial flow rate in pipes 1 and 2 is 0.002 m^3 s^{-1} and a friction factor of 0.01 is assumed for all pipes. The time-step is dictated by the wave speed in the branch, i.e. a value of $2.0833.10^{-4}$ (0.25 m/1200 m s^{-1}). The upstream reservoir pressure was 1000 kN m^{-2}.

FIGURE 21.6

Example 21.2(a) Valve closure in 2.0 seconds. (b) Valve closure in 0.1 seconds

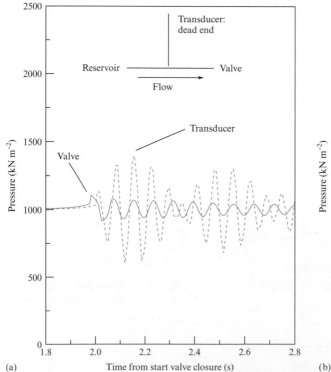

(a)

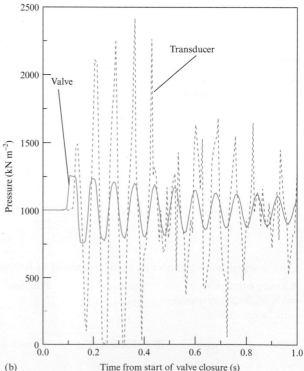

(b)

The pressure variation at the dead end of pipe 3 compared to the junction pressure at entry to pipe 3 is illustrated, Fig. 21.6(a) and (b), for valve closures of 0.1 and 2 seconds. In both cases the transducer is seen to accentuate the transient oscillation due to the +1 reflection coefficient at the dead end and due to the −1 reflection coefficient at the branch to main pipe junction, which traps transients in the branch. Similarly the relative lengths of the branch and the upstream and downstream pipe lengths introduce a complex set of transients into the branch which give, over a longer time scale, a bimodal oscillation form. Increasing valve closure time reduced the transients as expected but the overall form of the event is unchanged.

EXAMPLE 21.3

Discuss and investigate the pressure transient response of trapped air at a dead-ended pipe following the rapid changes in flow condition, for example the opening of an upstream valve – a situation found in practice in dry riser or sprinkler head fire protection systems – or the closure of a downstream valve as recorded by a remote transducer.

Solution

Trapped air at the end of a pipe can occur in a range of cases. The air pocket may be treated as the terminal boundary condition for the branch. This boundary arises as a result of flow conditions within the network and must, therefore, be included in a model as a boundary activated by the predicted flow conditions during transient propagation.

In the situation to be modelled, illustrated in Fig. 21.7, the volume of trapped gas is assumed known at some base pressure, conveniently atmospheric although the network steady flow pressure would also be suitable. The volume of gas is conveniently represented in pipe diameters. The unknowns are the fluid velocity at the fluid/gas interface and the pressure in the free gas volume. The volume of gas is assumed small compared to the pipe volume represented by one computing section Δx. The available equations are the C^+ characteristic at pipe exit, equation (21.38) and the gas law expressed as

$$P_{\text{gas}}^{t+\Delta t}(\text{Vol}_{\text{gas}}^{t+\Delta t})^n = P_{\text{gas}}^{t_0}(\text{Vol}_{\text{gas}}^{t_0})^n,$$

where $1 < n < 1.4$. (If $n > 1$ then the solution will require an iterative approach, the bisection method being efficient in such cases.) The volume of the gas at each time-step is given by

$$\text{Vol}_{\text{gas}}^{t+\Delta t} = \text{Vol}_{\text{gas}}^{t} - 0.5A(J)\Delta t\{V_{\text{p}}[J, n(J)+1] + v_{\text{p}}[J, n(J)+1]\},$$

where $v_{\text{p}}[J, n(J)+1]$ is the interface velocity at time t, i.e. one time-step earlier, an approach identical to that outlined for a column separation cavity summation, Example 21.4. It should be noted that the −ve sign in the gas volume summation ensures that the gas pocket volume decreases if the flow in the branch pipe is positive, i.e. towards the dead end. If $n = 1$ then these equations reduce to a quadratic in interface velocity.

FIGURE 21.7
Trapped air boundary with leakage

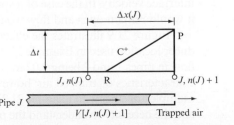

In cases where there is air leakage from the dead-ended pipe, illustrated by Fig. 21.7, as the gas pressure rises, for example in charged dry risers, the approach set out above will apply with a modification to the trapped gas volume summation. The leakage of gas, q, from the pipe will depend on the equivalent orifice equation, represented by

$$P_{gas}^{time} - P_{external} = k_{orifice}(q_{gas}^{time})^2,$$

and therefore

$$\mathrm{Vol}_{gas}^{t+\Delta t} = \mathrm{Vol}_{gas}^t$$

$$- 0.5\Delta t \left\| A(J)\{V_p[J, n(J)+1] + v_p[J, n(J)+1]\} \right.$$

$$\left. + \left(\frac{P_{gas}^{t+\Delta t} + P_{gas}^t - 2P_{external}}{k}\right)^{0.5} \right\|.$$

Consider a 20 m length of pipe, terminated by a closed end, fitted with an outwards pressure relief valve whose opening pressure may be set. The pipe is 0.1 m in diameter and is connected via a fast acting supply valve to a constant pressure reservoir, taken here to also represent a fast start up pump set. The reservoir pressure is set to 700 kN m^{-2} and the line pressure prior to valve opening to 100 kN m^{-2}. The trapped air volume at line pressure is taken as 20 pipe diameters. The supply valve opens in 0.1 seconds and has a loss of 200 kN m^{-2} at a throughflow of 0.1 m^3 s^{-1}.

Figure 21.8(a) illustrates the pressure and velocity variations expected at the interface between the trapped gas and the fluid column. The gas pressure rises sharply over the final stages of compression and the velocity of the fluid column falls to zero and then reverses as the trapped gas re-expands as a result of the reflection of the transient generated by the upstream boundary – in the case illustrated there is a constant pressure reservoir. This example also demonstrates that the fluid velocity reaches a maximum value just prior to the rapid pressure rise associated with the compression of the air pocket.

Experimental observations have shown that the peak pressure generated in the trapped gas may be considerably higher than the supply reservoir pressure depending upon the rate of opening of the supply valve illustrated. This example, and its extension to include gas leakage, has an important application in the study of pressure relief valve design for dry riser installations in buildings and other structures requiring fire protection. Figure 21.8(b) also illustrates the effect of introducing a relief valve set to open at a particular pressure – 750 kN m^{-2} in the case illustrated. Note that the pressure still rises above the opening pressure of the relief valve as air is forced out of the pipe. Once the air is exhausted the downstream boundary reverts to a zero velocity closed end and the pressure trace presented illustrates the effect this has on subsequent pressure oscillations in the pipe. Figure 21.8(c) illustrates the effect venting has on the interface velocity. In the case of a sprinkler head, assuming it survived the initial surge, it would remain open and flow would be continuous.

Figure 21.9 illustrates the effect of 30 pipe diameters of trapped air at the transducer location used in Example 21.2 following a valve closure in 0.1 seconds, the initial flow in pipes 1 and 2 being 0.02 m^3 s^{-1}. The form of the transducer output displays the characteristics of trapped air; however, the pressure recorded is severely in error. This example, together with Example 21.2, illustrates the importance of transient propagation and the necessity to understand the phenomena in the diagnosis of any system problem.

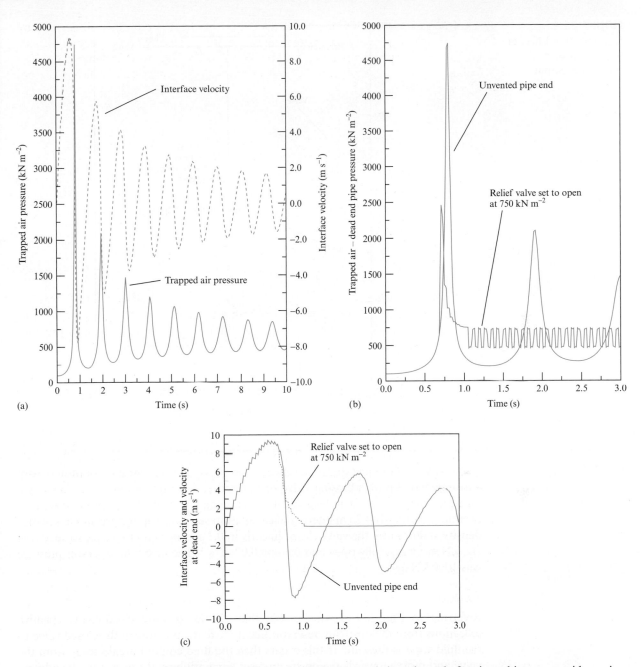

FIGURE 21.8 Example 21.3. (a) Trapped air at the end of a pipe subject to a rapid opening of an upstream valve – e.g. a sprinkler head or dry riser. (b) Relief valve effect in reducing peak pressure, venting trapped air and changing exit boundary condition to a dead end characterized by flow velocity = 0.0. (c) Effect of relief valve on the interface velocity. Note imposition of zero velocity boundary once trapped air removed

FIGURE 21.9

Example 21.3. Effect of
trapped air on a
transducer output

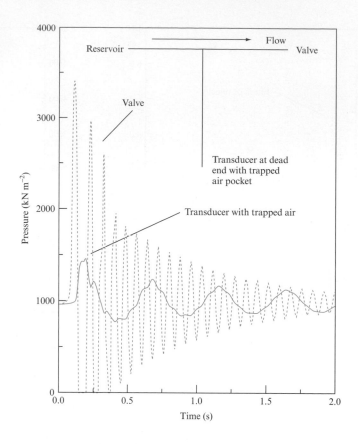

EXAMPLE 21.4

Use FM4SURG to investigate the effect known as column separation both upstream and downstream of a closing valve or at dead-ended pipe boundaries. The single pipe modelled was of 20 m length, 0.1 m diameter, applicable wave speed 1000 m s^{-1}, friction factor 0.01, 0.25 m internodal calculation sections, vapour pressure 6 kN m^{-2}, density 1000 kg m^{-3}, the valve closed linearly in 0.1 seconds and had a pressure loss of 200 kN m^{-2} when fully open and passing 0.02 m^3 s^{-1}. The upstream reservoir pressure was 1000 kN m^{-2}.

Solution

Referring to Fig. 20.1 it will be seen that, following valve closure, the negative reflections from the upstream reservoir may reduce the pressure at the closed valve to the fluid vapour pressure. If this occurs then the fluid column breaks away from the valve and moves towards the upstream reservoir until brought to rest by the adverse pressure gradient between the reservoir pressure and the vapour pocket at fluid vapour pressure. This pressure gradient then acts to accelerate the fluid column back towards the closed valve, collapsing the vapour pocket. The collapse of the vapour pocket brings the returning column to rest instantaneously as the flow 'hits' the closed valve, generating a Joukowsky pressure rise. The presence of any trapped air in the pocket will modify this, as shown in the previous examples.

The applicable boundary condition at the closed valve, $V = 0.0$, breaks down once the cavity forms, being replaced by the constant pressure at the closed valve, $p_{valve} = p_{vapour\ pressure}$, for the duration of the cavity. Cavity duration is normally determined by

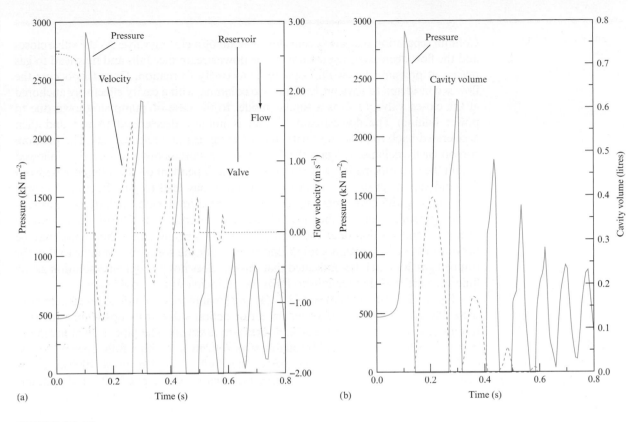

FIGURE 21.10

Example 21.4. Column separation upstream of a closed valve

summing the cavity volume over time based on the flow velocity at the interface determined by solving this boundary condition with the available C^+ characteristic:

$$\text{Vol}_{\text{time}=t} = \text{Vol}_{\text{time}=t-\Delta t} - 0.5A(V_{t,\,n+1} + V_{t-\Delta t,\,n+1})\Delta t,$$

where $n + 1$ denotes the final node of the pipe and the negative sign ensures that the cavity grows with reversed flow and vice versa.

Once the summed cavity volume becomes zero, or negative, at the end of a time-step the boundary condition reverts to zero velocity and the immediate effect is for the pressure on cavity collapse to rise to the Joukowsky pressure based on the final column velocity. As this is an instantaneous flow stoppage and as vapour pockets may form as the result of a 'slow' valve closure, the pressure generated may well exceed that on initial valve closure. The presence of other transient waves in the pipeline, in particular the reflected pressure wave generated on cavity formation, having a value equal to the difference between valve initial line pressure and fluid vapour pressure, will also increase the final pressure recorded at the valve during the first pipe period following cavity collapse. This pressure wave, being reflected between two constant pressure zones during column separation arrives at the closed valve to be reflected positively, so that the highest pressure at the closed valve following cavity collapse becomes

$$p_{\text{valve max}} = \rho c V_{\text{column final velocity}} + 2.0(p_{\text{valve at time}=0} - p_{\text{vapour}}).$$

Using a simple one-pipe model the effect of cavity formation and cavity collapse may be investigated, Fig. 21.10(a) and (b) confirms the discussion above.

EXAMPLE 21.5

Column separation may also occur downstream of a closing valve. As the valve closes and the flow decreases the pressure on the downstream face falls and may lead to gas release at pressures below atmosphere or to cavity formation. Once this occurs the flow is considered to have broken into two columns, with a cavity effectively anchored at the closed valve. (This is a similar model to the case of pump shutdown due to power failure.) The downstream column is initially decelerated to rest and then accelerated back to close the cavity and/or compress any released gas. The pressure rise on cavity collapse is again the Joukowsky instantaneous pressure rise, followed within the next pipe period with a secondary rise dependent on the difference between line and vapour pressure, and this combination may lead to pipe fracture. In some cases the negative transients can also lead to pipe implosion, making column separation potentially one of the most damaging transient events. Dissolved gas may be released from solution once the fluid pressure falls below atmospheric and may be present in the cavity region, and in that eventuality the solution is similar to that for trapped gas. Research has indicated that this gas does not go back into solution as the fluid pressure rises, but remains to form the boundary at the closed valve.

Figures 21.11(a) and (b) and 21.12(a) and (b) illustrate column separation in both these cases for a single pipe served by a fast action valve at its upstream entry and terminated downstream by a constant pressure reservoir. The pipe is 20 m in length, 0.1 m diameter with an initial throughflow of 0.02 m³ s⁻¹, with a fully open valve loss of 200 kN m⁻² at 0.1 m³ s⁻¹ throughflow, the fluid has a density of 1000 kg m⁻³, a vapour pressure of 6 kN m⁻² and a Bunsen solubility coefficient of 0.2. The applicable

FIGURE 21.11

Example 21.5. Column separation downstream of a closing valve, no gas release

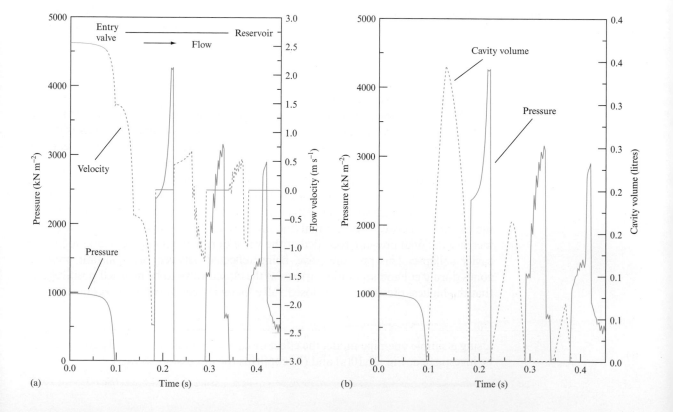

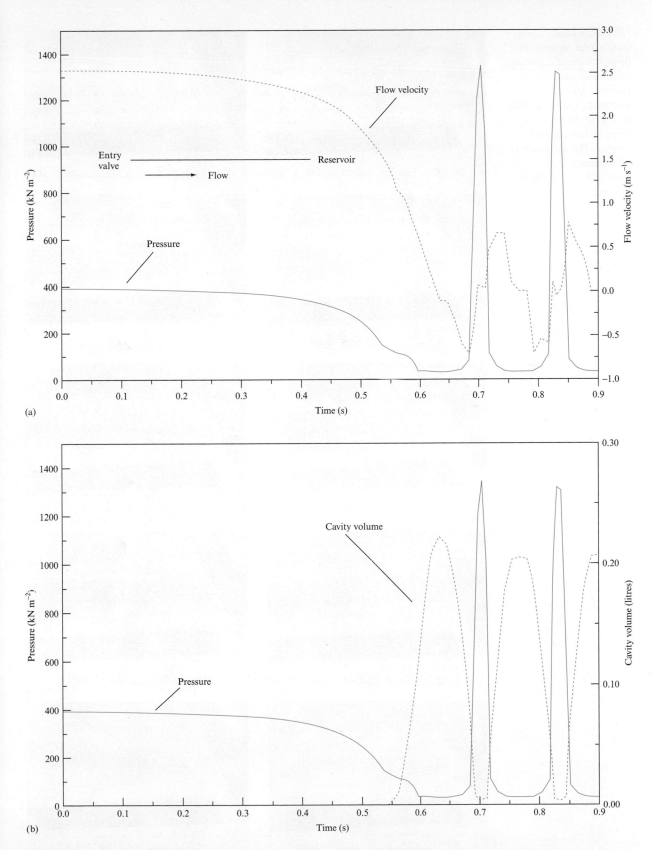

FIGURE 21.12 (a) and (b) Example 21.5. Column separation downstream of a closing valve with gas release

FIGURE 21.12

(c) Column separation downstream of a closing valve in an aviation kerosene 50 mm diameter pipeline as part of a study of pressure surge in aircraft fuel systems. Photos courtesy of Professor J. A. Swaffield, Heriot-Watt University

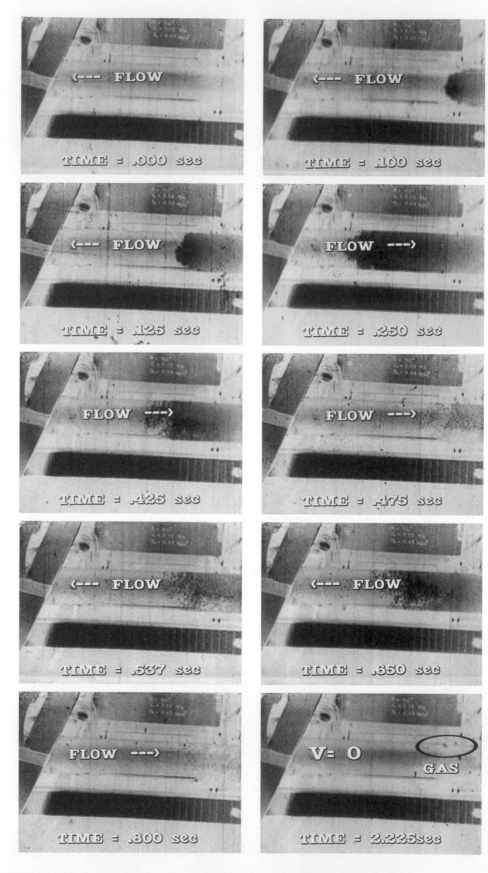

wave speed is $1000 \, \text{m s}^{-1}$ and the friction factor was taken as 0.01, the valve closed linearly in 0.1 seconds and the internodal computing sections were 0.25 m in length. In the 'no gas release' case the supply reservoir is at $1000 \, \text{kN m}^{-2}$ and in the gas release case $400 \, \text{kN m}^{-2}$ to accentuate the effect.

In the 'no gas release' case these results again illustrate the secondary pressure rise following cavity collapse due to the change in reflection coefficient at the closed valve from −1, when the cavity exists as a constant pressure zone, to +1, when it has closed and the closed valve forms the boundary. The importance of this secondary pressure rise is similar to line packing and could form an interesting investigation for the reader using FM4SURG. In the case of gas release, note the change in the form of the flow velocity reduction downstream of the valve once the pressure falls below atmospheric and dissolved gas starts to affect the calculations as the cavity forms.

The photographs, Fig. 21.12(c), taken downstream of a closing valve in a 50 mm diameter glass pipeline carrying aviation kerosene illustrate the gas release from solution at pressures below atmosphere and confirm the model discussed in this example. The cavity volume may be seen to grow and collapse cyclically until the pressure transient attenuates and separation no longer occurs. Gas remains out of solution throughout the process.

21.4.2 Additional investigations using FM4SURG

Computer program FM4SURG may also be used to investigate the following:

1. for a single pipeline, the effects of valve closure rate by, for example, introducing an initially rapid valve motion followed by a slower final closing period;

2. the regain of frictional pressure losses once flow is brought to rest by a downstream valve closure may be considerable in long pipelines with a high frictional loss. This effect may exceed the transient for slow valve closures. The effect is known as line packing and may be investigated by setting up a single pipe with a high friction loss, low velocity and a range of valve closure times and reservoir pressures;

3. the effect of line pressure on column separation in both upstream and downstream cases;

4. the effect of gas solubility on column separation downstream of a closing valve by varying the Bunsen solubility coefficient for the fluid used in Example 21.5;

5. the effect of introducing a relief valve to reduce pressure surge at a dead end; in particular the effect of valve opening pressure.

21.5 COMPUTER PROGRAMS FM4WAVE AND FM4GUTT. THE SIMULATION OF OPEN-CHANNEL FREE SURFACE AND PARTIALLY FILLED PIPE FLOW, WITH AND WITHOUT LATERAL INFLOW

Unsteady flows in open channels may be characterized by the changes in an inflow hydrograph shape as it propagates along the channel. In general an attenuation in the maximum depth and flow rate recorded at any downstream location is observed with the trailing edge of the wave flattening and a tendency for the leading edge to steepen.

These effects are illustrated in Fig. 21.13, where the hydrograph is thought of as a series of incremental waves each having an individual depth and wave speed, which increases with depth. For the trailing slope the deeper waves travel faster than the shallow trailing edge, thus 'stretching' the profile. Conversely, the leading edge steepens. Frictional effects and channel slope are also factors, so that not all waves become steep fronted automatically. The overall effect is one of depth attenuation as the wave propagates downstream; however, the phenomenon is complex and depends on such parameters as roughness, gradient and channel cross-sectional shape and the base flow over which the wave propagates. In general the deeper the base flow the less the attenuation, an observation easily explained in terms of the relative wave speeds. Cross-sectional shape also has a large part to play here as the cross-sectional profile will govern the increase in flow area with depth and hence the rate of change of wave speed with depth, equation (21.18). (Note that the term channel includes partially filled pipe flow as well as such channel examples as rainwater gutters.)

FIGURE 21.13

Unsteady flow wave attenuation in open channels or partially filled pipe flows

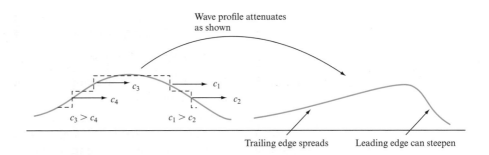

Accurate prediction of wave attenuation would be advantageous in pipe sizing for varying inflows, such as urban storm water systems, flood routing or building drainage schemes. Analytical solutions are limited but a numerical approach based on finite differences and the method of characteristics is practical.

The basic equations of motion and continuity defining unsteady free surface flows were developed in Section 21.1 and shown to be solvable via the method of characteristics. The need to cater for both subcritical and supercritical flow models led to the recognition, Fig. 21.2, that the two characteristic lines defining the conditions at a point P in the $x–t$ plane could originate upstream and downstream of P, i.e. subcritical flow, or both upstream of P, i.e. supercritical flow. In order to comply with the Courant stability criterion it was also shown that, as the free surface wave propagation speed and the likely flow velocity were comparable, interpolation would always be necessary to determine the base conditions for the characteristics, the necessary relationships being defined in Section 21.2.2.

The steady base conditions necessary prior to the initiation of an unsteady flow analysis may be determined from open-channel theory, as defined in Chapters 15 and 16. It may be assumed that the initial steady flow is at the normal depth appropriate to that channel and flow rate, equation (15.6). Conditions at entry and exit boundaries

TABLE 21.8 Program data options, FM4WAVE and FM4GUTT

DATA TYPE		INDIVIDUAL PARAMETER INPUT
Cross-sectional shape, (FM4WAVE and FM4GUTT)	Circular	Input diameter
	Rectangular	Input depth and width
	Trapezoidal	Input depth, base width and side slope
	Parabolic	Input diameter of enclosing circle and parabola/circle intersection depth
Channel data		Length, Manning n, slope, number of sections per m length
Simulation duration		Probably a good option to start with a short duration run to check data input
Output data interval		This may affect the degree of detail displayed on any subsequent data graphs and a balance may be needed between detail and storage depending on the computer used
Entry conditions, FM4WAVE		Inflow profile defined in terms of number of flow–time coordinate pairs followed by the time and flow rate. If the flow were to be supercritical input a factor, 0.7–1.0, to approximate effect of flow energy on entry, 1.0 indicates normal depth entry assumption and is safe for initial runs
	FM4GUTT	Lateral inflow defined in terms of flow rate or rainfall intensity, latter also requires roof area input
Exit conditions,	FM4WAVE	Program assigns critical depth boundary for subcritical flows
	FM4GUTT	Obstructions can result in gutter exit depth being above the expected critical depth. Option exists to introduce an empirical depth vs. flow rate curve

Note: Parabolic sections for both FM4WAVE and FM4GUTT depend on defining the depth at which the parabolic section perimeter intersects the perimeter of an imaginary enclosing circle, see Figure 21.14(a). This allows direct calculation of the parabola constant and allows subsequent program calculations of the flow depth, perimeter and free surface width required to allow the method of characteristics analysis to proceed.

may also be determined from the gradually varied steady flow treatment presented in Chapter 16. It must be noted that the solution requires a base condition at zero time and that zero flow, and therefore no flow depth, is not acceptable; this is a divergence from the full-bore transient cases where zero flow velocity is acceptable.

Thus the defining equations and the numerical techniques necessary to predict flow depth, velocity and surface wave speed have been developed, together with an acceptable approximation for frictional resistance, via either the Manning n approach or Colebrook–White applied to free surface flows, Section 15.2. In common with the pressure transient analysis the greatest challenge in the application of these techniques to the prediction of wave attenuation in partially filled pipe flows lies in the definition of boundary conditions.

Within the scope of this treatment channel exit conditions will be restricted to free discharge for either subcritical or supercritical flows. In the supercritical flow case the presence of the discharge cannot be communicated upstream, Fig. 21.2, hence the flow leaves the channel at a depth and velocity calculated from the C^+ and C^- characteristics for that node, both of which emanate from the upstream Δx.

The main differences between free surface wave attenuation analysis by the method of characteristics and that used for pressure transients have already been covered, namely that as flow velocity may, in supercritical flows, exceed the wave speed the characteristics may both emanate upstream of the calculation node. As the wave speed and the flow velocity may be similar in magnitude there may be interpolation difficulties. Supercritical flow conditions imply that information cannot travel upstream without the formation of a hydraulic jump to allow a moving transition from subcritical to supercritical flow conditions, a complication in building drainage system analysis upstream of channel junctions or obstructions to flow.

Computer programs FM4WAVE and FM4GUTT are available to allow wave attenuation in open channels or partially filled pipes, or gutters accepting lateral inflows, to be predicted. The range of program input options are illustrated in Table 21.8.

21.5.2 Application examples

FM4WAVE and FM4GUTT may be used to address each of the following examples.

EXAMPLE 21.6

Investigate wave attenuation in a parabolic cross-section channel subjected to a time-dependent inflow profile at a range of channel slopes.

Solution

Consider a 20 m long parabolic cross-section channel, enclosing diameter 0.1 m, intersection depth 0.08 m, of roughness represented by a 0.009 Manning n and subjected to an inflow hydrograph rising from 0.1 litres s^{-1} to 1.5 litres s^{-1} and returning to the 0.1 litres s^{-1} base flow.

The application of the method of characteristics to free surface flows depends upon the determination of wave speed at any particular flow depth. From equation (21.18) it will be realized that if flow cross-sectional area and surface width are known as functions of depth then this is possible. (Note that wetted perimeter will also be required as a function of depth to allow the determination of hydraulic mean depth and hence frictional resistance.) For circular, rectangular or any 'straight' sided cross-

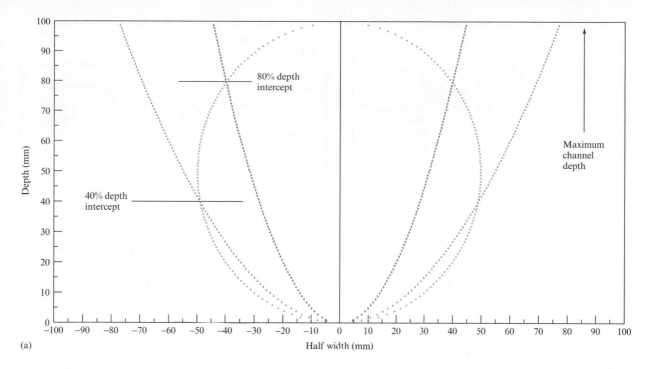

(a)

Half width (mm)

FIGURE 21.14

Example 21.6. (a) Definition of parabola based on defined parabola–circle intercept depth

section shape this is simply accomplished. For the parabolic sections suggested in this example, a shape found in sewers and some rainwater gutters, it is necessary to define the parabola coefficient. This is best done by imagining an enclosing circle and the intersection point with a parabola. For example, assume a circle of diameter d, with which a parabola intersects at a height h above the invert. The equation defining the parabola is thus given as the intersection point, which has coordinates h,t, where t is the half width at height h above the invert:

$$t = (h(d - h))^{0.5},$$

and the coefficient k in the parabola equation, $H = kT^2$, becomes

$$k = h/t^2,$$

which allows flow cross-sectional area and wetted perimeter to be calculated from the geometry of the parabola. Figure 21.14(a) to (c) illustrates these geometric considerations for an 80 per cent and 40 per cent parabola intercept height within a 0.1 m diameter circle.

Figure 21.14(d) illustrates the applied inflow hydrograph and the wave attenuation for a range of channel slopes. As slope decreases the wave attenuation increases. The use of Manning n is possibly not the best approximation to frictional resistance for small channels or partially filled pipe flows – Colebrook–White gives a slightly better representation of the wave attenuation but requires an iterative process omitted from these models for simplicity.

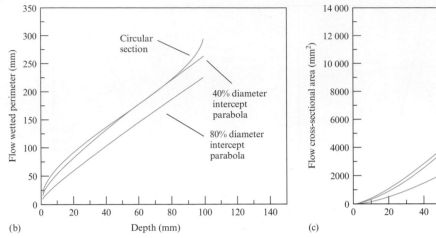

(b)

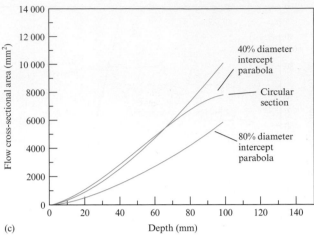

(c)

FIGURE 21.14
(b) Comparison of circle and parabolic wetted perimeters.
(c) Comparison of circle and parabolic cross-sectional areas.
(d) Wave attenuation in an 80 per cent diameter intecept parabolic open-channel, 0.1 m enclosing circle

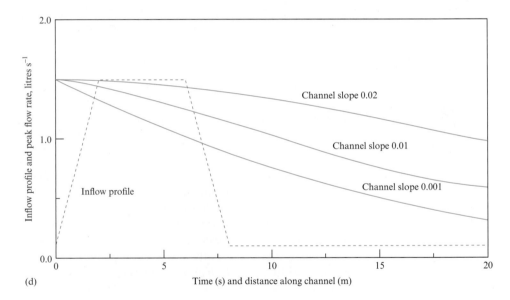

(d)

EXAMPLE 21.7

Investigate the gutter flow depths recorded in a trapezoidal gutter in response to a varying rainfall intensity

Solution

Consider a 10 m long open gutter, slope 0.001, Manning n 0.009, of trapezoidal section depth 0.2 m, base width 0.1 m and a side slope of 45°. The gutter accepts a rainfall equivalent to 0.1 litres $s^{-1}\,m^{-1}$ for 200 seconds and then a rising intensity to 0.3 litres $s^{-1}\,m^{-1}$ over the next 100 seconds. Rainfall returns to 0.1 litres $s^{-1}\,m^{-1}$ following 100 seconds at the peak level. In this case the exit depth is set to critical, equation (16.9). The flow depth along the gutter may be determined from FM4GUTT, as illustrated in Fig. 21.15(a) and (b), which shows the maximum depth along the gutter and the time varying depth at four locations as the rainfall intensity varies. Note that the upstream depth occurs at a zero flow velocity boundary.

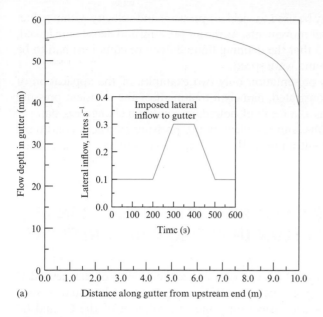

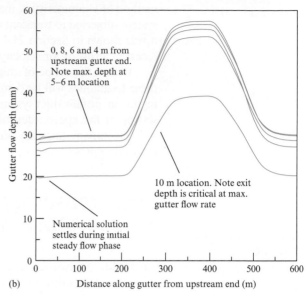

FIGURE 21.15
Example 21.7. (a)
Maximum flow depth
along the length of the
gutter. (b) Gutter flow
depths

As mentioned in the discussion of interpolation techniques the subcritical draw-down profile that exists upstream of the critical depth discharge presents problems in that small rounding errors inherent in a linear interpolation may lead to profile collapse. In the FM4GUTT simulation more accurate interpolation techniques, i.e. Everett and Newton–Gregory, are used that maintain the profile as the lateral inflow changes relatively slowly. For these stability reasons it is advisable to run FM4GUTT for a low intensity initial steady lateral inflow period to allow the gutter depths to stabilize, as illustrated in Fig. 21.15(b).

21.5.3 Additional investigations using FM4WAVE and FM4GUTT

The free surface flow simulations may be used to investigate:

1. for a single channel, the effects of cross-sectional shape on wave attenuation; for example, comparing channels with the same maximum depth cross-section could be interesting;

2. the effect of channel slope on rainwater gutter flow depth profiles, remembering that such gutters are normally set at a flat gradient leading to subcritical flow conditions.

21.6 SIMULATION OF LOW-AMPLITUDE AIR PRESSURE TRANSIENT PROPAGATION

It has already been shown that the propagation of low-amplitude air pressure transients may be defined by the same fundamental equations of motion and continuity developed in Section 21.1 and that the numerical solution of these equations, via the method of characteristics, may be used to yield air pressure and velocity within a duct system subjected to transient air movements. As the air pressure and density are linked, it was shown in Section 21.1.3 that the defining finite difference equations had to be recast in terms of air velocity and wave speed.

Within the scope of this presentation only two examples of the application of these techniques will be demonstrated, namely a treatment of the transient pressure regime in air flow duct systems as a result of boundary condition change, e.g. damper setting or fan speed alterations, and a review of the pressure transient conditions that accompany unsteady air entrainment by a falling water annulus, a case already referred to in Section 16.14.

21.7 COMPUTER PROGRAM FM4AIR. THE SIMULATION OF UNSTEADY AIR FLOW IN PIPE AND DUCT NETWORKS

TABLE 21.9
Program data options
FM4AIR
For each pipe, 1 to n, in the network make the following data input choices...

An air flow ductwork model, FM4AIR, can fully utilize the benefits of the method of characteristics solution technique. The connecting ducts between boundaries such as junctions, dampers or fans, are easily represented by solution of the C$^+$ and C$^-$

DATA TYPE	INDIVIDUAL PARAMETER INPUT
Duct cross-section, Circular	Input diameter
Rectangular	Input depth and width
Duct parameters	Length, surface roughness, in mm
Duct entry/exit condition	Atmospheric or defined zone pressure, entry/exit fixed grill or movable damper, dead end or originating/terminating at a junction
Junction type identification	Duct junction, two or more – zero loss, fan forms junction between two ducts only Fixed grill or movable damper forms junction between two ducts only
Duct entry/exit identification	In the flow direction identify number of ducts that terminate at a junction and the number that originate at that junction. Record each pipe identification number, 1 to n
Damper or grill loss coefficients	Loss coefficients K defined as Constant $\times$ Setting^n, where n may be chosen, eight suggested, the variation of K with setting is based on both fully open and 50% closed K values
Duct entry and exit data	Input detailed values based on the route choices already made
Junction data	Input fan characteristic data in terms of non-dimensional coefficients and reference speed Input any loss coefficients for grills or dampers Input fan speed change or damper setting vs. time data, this data in terms of number of data pairs followed by individual time and setting values
Calculation time-step, duration and output	Input number of Δx sections per m, simulation duration and output frequency. Flow output data in m^3 s^{-1} or litres s^{-1} may also be chosen

characteristics. As wave speed, i.e. the acoustic velocity in air, will exceed the flow velocity the characteristics emanate upstream and downstream of the calculation node. Frictional losses are represented by Colebrook–White and hence require duct roughness input. Boundary conditions may be written for all likely duct fittings, e.g. dampers, grills or fans. Similarly the ambient pressure in any space supplied to, or extracted from, may be included in the appropriate boundary condition. The flow direction is initially chosen and all boundaries may be identified as entry or exit conditions for the duct originating or terminating at that location. Table 21.5 has already illustrated these boundary conditions. Table 21.9 illustrates the data required to run the program FM4AIR.

The transients generated by damper motion or fan speed adjustment will be of low amplitude, however, these changes cannot be predicted by a steady flow analysis. The low-amplitude nature of the transients also simplifies the application of the method of characteristics as there is no need to incorporate any energy or thermal effects.

21.7.1 Application example

FM4AIR may be used to address the following example.

EXAMPLE 21.8

Utilize FM4AIR to investigate the variation in recirculation air content in a supply and extract two-fan network. For simplicity all ducts are circular of 0.2 m diameter and 4 m in length. The damper loss coefficient variation is shown in Fig. 21.16(a) as is the speed change data for both identical fans. The fan characteristics are identical and are assumed to have a polynomial fit equation having the following coefficients, as defined in Table 21.6,

$$C_1 = 430.0, \quad C_2 = -1.2, \quad C_3 = -0.001, \quad C_4 = 0.0,$$

with a reference speed of 500 rev min^{-1}.

Solution

Figure 21.16(a) and (b) illustrates the increasing flow rate as the fans speed up to their reference speed of 500 rev min^{-1} within the first 2 seconds of the simulation. At 2 seconds the recirculation control damper starts to close from its initial opening setting and the percentage of recirculation air flow drops. Between 4 and 8 seconds the damper is stationary and then reopens allowing the percentage recirculation air flow to rise. The continuity of flow condition is maintained throughout, as may be seen by a comparison of the flow rates in ducts 1, 2 and 7.

In using this program care must be taken to decide the positive flow direction for the network so that entry and exit boundary conditions may be correctly identified for each duct. In this case, for example, ducts 1 and 7 both terminate at the first junction on the supply system, duct 2 originates at that junction.

(Note that the time periods used are short in order to demonstrate the action of the simulation rather than to simulate actual rates of damper motion or fan speed change. However, due to the relationship defining duct period in terms of length and the wave speed, approximately 320 m s^{-1}, slower changes in boundary conditions would not result in simulation output much different from that presented.)

FIGURE 21.16
Example 21.8. (a) Air
duct network in study of
recirculation flow setting.
(b) Air flow continuity
at junction of inflow/
recirculation ducts

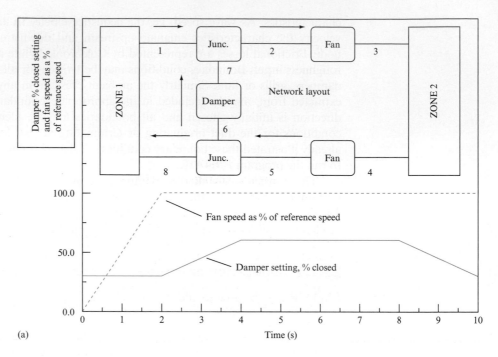

(a)

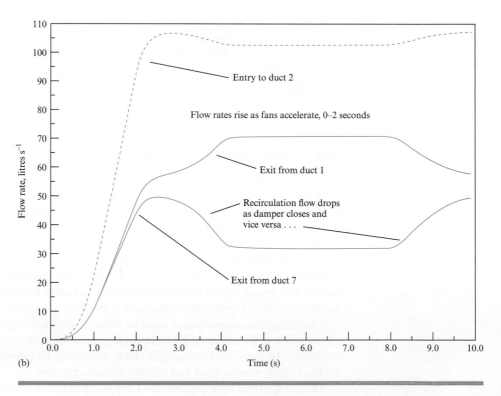

(b)

21.7.2 Additional investigations using FM4AIR

Program FM4AIR may be used to investigate the effects of:

1. damper pressure loss coefficient and control on the air flow distributions within a network;

2. zonal pressure level at system entry or exit;

3. multiple or manifold distribution downstream of a fan.

21.8 ENTRAINED AIR FLOW ANALYSIS REVIEW

Consider the flow condition illustrated in Fig. 21.17(a) and earlier in Fig. 16.19. Variable water flow, as an annulus falling at terminal velocity, Section 16.14, equation (16.28), entrains an air inflow to the network and consequently leads to subatmospheric pressures in the vertical air core due to the frictional pressure losses suffered in the upper dry stack and the separation losses due to the air flow passing through the inflowing water curtain at the discharging branch junction; see Section 16.14 and Fig. 16.18.

The frictional loss in the dry stack may be determined from an application of Darcy's equation for the entrained air flow and the stack parameters of pipe diameter and roughness. Separation losses at entry and due to passing through the discharging branch flow may also be determined so long as the appropriate loss coefficients are known. Figure 21.17(a) also illustrates the possible 'back' pressure at the base of the stack as the entrained air flow is forced through the water curtain formed at that location as the annular flow makes the transition to free surface flow in the horizontal drain. Thus these pressure losses may be combined as

$$\Delta p_{total} = \Delta p_{entry} + \Delta p_{dry\ pipe\ friction} + \Delta p_{branch\ junction} + \Delta p_{back\ pressure}. \tag{21.55}$$

The 'motive force' to entrain this air flow and compensate for these 'pressure losses' is derived from the shear force between the annular terminal velocity water layer and the air in the wet portion of the stack. Hence a 'negative' friction factor may be postulated that generates an equal pressure rise to that determined from equation (21.55) – the equivalent to a fan characteristic drawing air through the stack. Ongoing research has identified the format and relationships governing this shear force representation and allows the prediction of the transient response of the stack network to variations in applied water downflows.

Within the air core the conditions of air velocity and pressure may be found by solution of the available C^+ and C^- equations at each node and time-step. The upper boundary may be treated provided that a suitable boundary equation can be found and solved with the available C^+ characteristic. In the case illustrated the upper level boundary is provided by an air admittance valve. This form of valve allows air into the duct once the internal pressure has fallen to a preset level due to the suction provided by the falling water annulus. While the air pressure remains below this level, the valve continues to pass entrained air into the duct. However, should the air pressure in the duct increase above this opening value the valve will close and remain closed if the duct pressure exceeds atmosphere. Thus the upper boundary may be represented by a series of valve loss and boundary air velocity expressions, each solved as appropriate with the C^+ equation. The air admittance valve equations may be expressed as:

FIGURE 21.17

(a) Pressure variation down stack, 7 seconds into the applied annular water downflow, illustrating the component pressure losses and regain.
AAV = air admittance valve

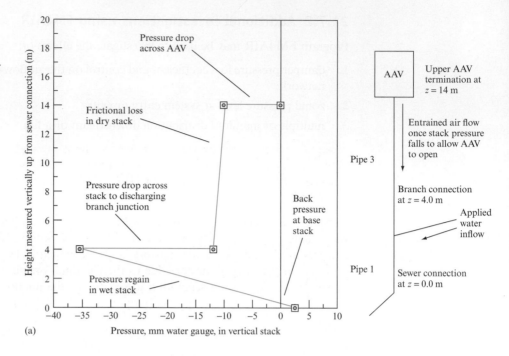

(a)

FIGURE 21.17

(b) Entrained air flow and associated pressure transients in a building drainage vertical stack carrying an annular water downflow

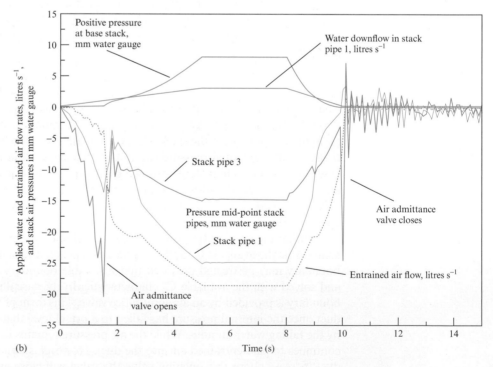

(b)

Closed valve, $u_{exit} = 0.0$,
(i.e. $p_{exit} > p_{open}$)

Partially open valve, $p_{atm} - p_{exit} = K0.5u_{exit}^2$,
(i.e. $p_{open} > p_{exit} > p_{max}$)

Fully open valve, $p_{atm} - p_{exit} = K_{min}0.5u_{exit}^2$,
(i.e. $p_{max} > p_{exit}$)

where p_{exit} refers to the air pressure in the duct immediately below the air admittance valve, i.e. at the duct entry node. (Note that for pressures below p_{max} the valve is fully open so that the loss coefficient becomes K_{min}.)

At the base of the stack the available C^- characteristic may be solved with an empirical loss coefficient expression for the water curtain. Similarly a complex empirical expression may be used to represent the loss coefficient at the discharging branch that takes into account the branch to stack geometry, illustrated in Fig. 16.18, as well as the volumetric flows in the branch and possibly the stack if there is another discharging branch above.

Figure 21.17(b) illustrates the variations in stack pressure as a result of a change in annular water downflow, in a system shown schematically in Fig. 21.17(a). The opening and closing sequence of the air admittance valve (AAV) in response to the changing stack air pressure is clearly identifiable at 1.5 and 10.0 seconds into the simulation. Note that as the AAV closes, a negative transient is propagated by exactly the same mechanism as that generating the pressure drop on valve closure in Example 21.5. As the driving empirical expressions for 'negative' friction factor and to a lesser extent discharging branch loss coefficient are the subject of continuing research, the simulation program behind these predictions is not included in the simulations available with this text.

If the air path down the duct is obstructed, as may be the case if the water downflow is sufficient to cause surcharging at the base of the stack, then the boundary condition at the base of the stack becomes zero air flow velocity. If this closure of the air path is deemed instantaneous, or less than the period calculated from stack height and air acoustic velocity, then a Joukowsky style pressure rise is experienced that travels up the stack and shuts the air admittance valve. Figure 21.18 illustrates this sequence of events, for the network shown in Fig. 21.17(a) and demonstrates further the identical nature of these low-amplitude air transients to the classical transient conditions normally treated by the method of characteristics. It is interesting to note that in this application the instantaneous stoppage of an airflow at 1 m s^{-1} will generate a pressure transient roughly equal to 40 mm of water gauge, an indication of the relative scale of the transient condition being studied. Figure 21.18 also indicates the difference in transient history at the mid-point of the upper dry stack if the AAV termination is replaced by a more conventional open end – normally protruding through a building roof. It will be appreciated that the AAV 'traps' the positive pressure transients within the stack, whereas the open end allows negative reflections to occur – identical to the transient case described in Example 21.4.

Overall the examples and demonstrations of the application of the method of characteristics to the simulation of low-amplitude air pressure transients has confirmed that this technique is applicable and that this category of transient propagation falls within the same family as the other cases considered in this chapter.

FIGURE 21.18

Comparison of the air pressure transient history at the mid-point of the dry stack following a stack base surcharge for AAV and open stack upper terminations

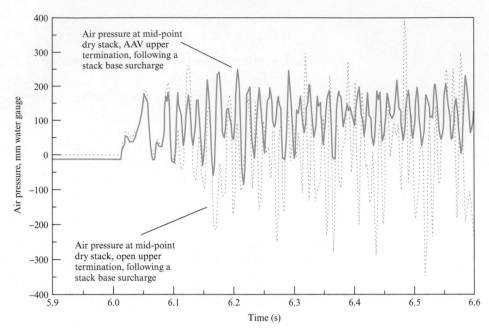

Concluding remarks

The analysis of unsteady flow phenomena depends upon the availability of fast computing and an understanding of the mechanisms to be modelled. The chapters included in Part VI of this text have sought to provide that understanding for a family of unsteady flow phenomena commonly met, from the traditional waterhammer cases to the more unusual application of the method of characteristics as the finite difference method of choice to free surface wave attenuation and low-amplitude air pressure transient propagation. The scope of the techniques presented is perhaps best illustrated by the ease with which non-standard cross-sections may be dealt with, as well as the ability to deal separately with new or unusual boundary conditions.

The examples given in this chapter use all the programs linked to this element of the text and further work is suggested following each example, so no further tutorial questions are presented. Clearly as these work suggestions are open-ended, solutions would be inappropriate, however, the examples presented in full should give sufficient confidence for the reader to proceed.

The simulation of unsteady flow phenomena will become increasingly important and applicable as access to computing power continues to rise exponentially. Design calculations and standards will incorporate these techniques as a matter of course so that the inclusion of such modelling in the future within fluid mechanics texts will no longer be seen as outside the scope of the text or the capabilities of the reader.

Summary of essential concepts and equations

1.   All applications of the method of characteristics depend upon adherence to the Courant criterion, equation (21.37), which links time-step to internodal length and wave propagation and fluid flow velocities:

$$\frac{\mathrm{d}x}{\mathrm{d}t} = u_1 \pm c,$$

where u_1 is defined and the value of c determined by Table 21.3.

2. The time-step throughout any network must be a constant at any time to allow continuity at junction boundaries. Thus the time-step used is the smallest consistent with the Courant criterion.

3. Interpolation is unavoidable in most single pipe/channel cases and always for networks. The most appropriate interpolation methodology must be used – linear for simple single pipe cases, time line where subcritical open-channel backwater profiles may collapse or more complex multi-point interpolation methods, such as the Newton–Gregory forward and backward difference methods.

4. The Joukowsky pressure change expression only applies to instantaneous changes in flow condition, or those occurring in less than one pipe period, defined to the nearest reflection source.

5. Pressures falling to vapour or gas release pressure will disrupt the expected periodic nature of transient response. It is necessary to include cavity formation or column separation boundary conditions within any surge program to lie dormant until triggered by local pressure conditions.

Further reading

General treatment of unsteady flow phenomena and the application of the method of characteristics:

Chaudhry, H. (1987). *Applied Hydraulic Transients*. Van Nostrand Reinhold, New York.

Fox, J. A. (1989). *Transient Flows in Pipes, Open Channels and Sewers*. Ellis Horwood, Chichester.

Lister, M. (1960). The numerical solution of hyperbolic partial differential equations by the method of characteristics. In *Mathematical Methods for Digital Computers*, ed. A. Ralston and H. S. Wilf. Wiley, New York.

Streeter, V. L. and Wylie, E. B. (1967). *Hydraulic Transients*. McGraw-Hill, New York.

Swaffield, J. A. and Boldy, A. P. (1994). *Pressure Surge in Pipe and Duct Systems*. Avebury Technical, Aldershot.

Swaffield, J. A. and Galowin, L. S. (1992). *The Engineered Design of Building Drainage Systems*. Ashgate, Aldershot.

Watters, G. Z. (1979). *Modern Analysis and Control of Unsteady Flow in Pipelines*. Ann Arbor Press, Michigan.

Wylie, E. B. and Streeter, V. L. (1983). *Fluid Transients*. FEB Press, Ann Arbor.

Papers with a particular emphasis on alternative interpolation techniques to improve stability of method of characteristics applications, particularly in free surface flow applications:

Goldberg, D. E. and Wylie, E. B. (1983). Characteristics method using time-line interpolations. *J. Hyd. Div. ASCE*, **109** (5), pp. 670–83.

Swaffield, J. A., Campbell, D. P. and Escarameia, M. (1999). Unsteady roof gutter flow simulation: development and application. *Proc. CIBSE Series A: BSER&T*, **20** (1), 29–40.

Swaffield, J. A. and Maxwell-Standing, K. (1986). Improvements in the application of the numerical method of characteristics to predict attenuation in unsteady partially filled pipe flow. *Journal of Research*, *NBS*, **91** (3), 389–93.

VII

Fluid Machinery. Theory, Performance and Application

Tunnel jet fans, image courtesy of Woods Air Movement Ltd

ENERGY MAY EXIST IN VARIOUS FORMS. HYDRAULIC energy is that which may be possessed by a fluid. It may be in the form of kinetic, pressure, potential, strain or thermal energy. Mechanical energy is that which is associated with moving or rotating parts of machines, usually transmitting power. It is thus the purpose of hydraulic machines to transform energy either from mechanical to hydraulic or from hydraulic to mechanical. This distinction, based on the 'direction' of energy transfer, forms the basis of grouping hydraulic machines into two distinct categories. All machines in which hydraulic energy forms the input and is transformed into mechanical energy, so that the output is in the form of a rotating shaft or a moving part of a machine, are known as turbines or motors. In the other category, the input is mechanical, the transfer is from mechanical into hydraulic energy and the output is in the form of a moving fluid, sometimes compressed and at elevated temperature. Such machines are called pumps, fans and compressors. Thus, in the first category, work is done by the fluid and energy is subtracted from it, whereas, in the second category of machines, the work is done on the fluid and energy is added to it.

However, sometimes fluids, because of their characteristic properties, are used by some machines as media to form a link in the energy transfer chain. In an hydraulic coupling, for example, mechanical energy is transformed into hydraulic, only to be changed back into mechanical in the other half of the coupling. There is no gain in mechanical advantage, but, because of the fluid properties and the type of fluid in the coupling, a smooth and gradual transfer of power is made possible.

The action of an hydraulic coupling is *rotodynamic*, as distinct from positive displacement, which is characteristic of, say, an hydraulic jack. Thus, the principle of machine operation affords further means of its classification and is quite independent of the direction of energy transfer.

In *positive displacement* machines, fluid is drawn or forced into a finite space bounded by mechanical parts and is then sealed in it by some mechanical means. The fluid is then forced out or allowed to flow out from the 'space' and the cycle is repeated. Thus, in positive displacement machines, the fluid flow is intermittent or fluctuating to a greater or lesser extent and the flow rate of the fluid is governed by the dimensions of the 'space' in the machine and by the frequency with which it is filled and emptied.

In rotodynamic machines, there is a free passage of fluid between the inlet and outlet of the machine without any intermittent 'sealing' taking place. All rotodynamic machines have a rotating part called a runner, impeller or rotor, which is able to rotate continuously and freely in the fluid, allowing an uninterrupted flow of fluid through it at the same time. Thus, the transfer of energy between the rotor and fluid is continuous and is a result of the rate of change of angular momentum.

These two criteria, namely the direction of energy transfer and the type of action, form the basis of classification of hydraulic machines, as shown above. From this, it will be seen that pumps and compressors increase the energy of the fluid and may be either rotodynamic or positive displacement. Fans are always rotodynamic. In turbines the work is done by the fluid and the action is rotodynamic, whereas motors are positive displacement machines also receiving energy from the fluid.

Note that in this part, all fluid velocities are, in fact, mean velocities. However, for convenience we have not included the bars over the symbols which have been conventional in the previous parts of this text.

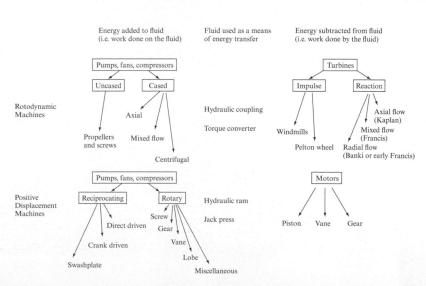

Classification of fluid machines

707

Theory of Rotodynamic Machines

THIS CHAPTER WILL CONSIDER THE FLOW THROUGH ROTODYNAMIC MACHINES, WHETHER fans, pumps or turbines, in terms of the energy transfers to, or from, the fluid, together with a detailed treatment of the fluid–blade relative velocity from inlet to exit. The contribution of each blade to the overall machine output will be considered in terms already introduced in the earlier treatment of the forces generated by fluid flow over an aerofoil surface. Losses inherent in the flow system formed within a rotodynamic machine will be considered and the resulting deviation from the idealized predictions of Euler's equation addressed. The effect of fluid compressibility is also considered for the case of rotodynamic compressors. ● ● ●

22.1 INTRODUCTION

This chapter is concerned with flow through rotodynamic machines and the relationships between the rate of fluid flow and the difference in total head across the impeller. Both are related to the type of machine under consideration and, hence, the geometric parameters of the impeller. However, certain fundamental relationships may be arrived at by considerations of angular momentum applied to a simplified, or idealized, impeller.

All rotodynamic machines, as previously stated, have a rotating part called the impeller, through which the fluid flow is continuous.

The direction of fluid flow in relation to the plane of impeller rotation distinguishes different classes of rotodynamic machines. One possibility is for the flow to be perpendicular to the impeller and, hence, along its axis of rotation, as shown in Fig. 22.1(a). Machines of this kind are called *axial flow* machines. In *centrifugal* machines (sometimes called 'radial flow'), although the fluid approaches the impeller axially, it turns at the machine's inlet so that the flow through the impeller is in the plane of the impeller rotation. This is shown in Fig. 22.1(b). *Mixed flow* machines constitute a third category. They derive their name from the fact that the flow through their impellers is partly axial and partly radial. Figure 22.2(a) shows a mixed flow fan impeller from the discharge side. It should be noted that the hub is conical; thus the direction of flow leaving the impeller is somewhere between the axial and radial. Figure 22.2(b) and (c) shows two other types of impeller.

Both pumps and turbines can be axial flow, mixed flow or radial flow. In the case of pumps, the last are normally referred to as centrifugal. All impellers consist of a supporting disc or cylinder and blades attached to it. It is the motion of the blades which is related to the motion of the fluid, one doing the work on the other or vice versa. In any case, there are forces exerted on the blades and, since they rotate with the impeller, torque is transmitted because of the rate of change of angular momentum.

It was shown in Section 5.7 that forces on moving vanes may be determined by considering the velocity triangles which represent, vectorially, the absolute velocity of the fluid, the velocity of the vane and the relative velocity between the two. However, it was justifiably assumed in Section 5.7 that the fluid velocity is the same across the jet. Also, the analysis was confined to the changes of kinetic energy of the fluid only, as the free jet is under atmospheric pressure throughout and, hence, there is no change

FIGURE 22.1

Axial flow and centrifugal impellers

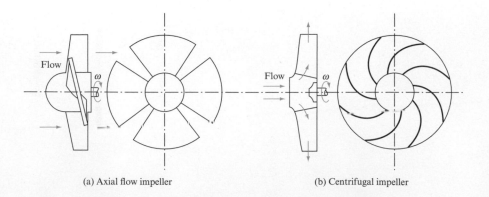

(a) Axial flow impeller   (b) Centrifugal impeller

FIGURE 22.2
(a) A mixed flow fan impeller. (By courtesy of Airscrew-Howden Ltd)
(b) A centrifugal pump impeller (shrouded). (By courtesy of Worthington-Simpson Ltd)

(a)

(b)

FIGURE 22.2
(c) A centrifugal pump impeller (unshrouded). (By courtesy of Worthington-Simpson Ltd)

(c)

in the pressure energy of the fluid. This is not the case when the fluid flows through the blade passages of an impeller. Nevertheless, an understanding of the analysis of Section 5.7 is a necessary prerequisite to the considerations of this chapter.

22.2 ONE-DIMENSIONAL THEORY

The real flow through an impeller is three dimensional, that is to say the velocity of the fluid is a function of three positional coordinates, say, in the cylindrical system, r, θ and z, as shown in Fig. 22.3. Thus, there is a variation of velocity not only along the radius but also across the blade passage in any plane parallel to the impeller rotation, say from the upper side of one blade to the underside of the adjacent blade, which constitutes an abrupt change – a discontinuity. Also, there is a variation of velocity in the meridional plane, i.e. along the axis of the impeller. The velocity distribution is,

FIGURE 22.3
A centrifugal impeller in relation to cylindrical cordinates

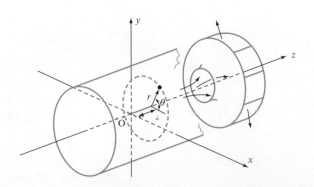

therefore, very complex and dependent upon the number of blades, their shapes and thicknesses, as well as on the width of the impeller and its variation with radius.

The one-dimensional theory simplifies the problem very considerably by making the following assumptions:

1. The blades are infinitely thin and the pressure difference across them is replaced by imaginary body forces acting on the fluid and producing torque.

2. The number of blades is infinitely large, so that the variation of velocity across blade passages is reduced and tends to zero. This assumption is equivalent to stipulating *axisymmetrical* flow, in which there is perfect symmetry with regard to the axis of impeller rotation. Thus,

$$\frac{\partial v}{\partial \theta} = 0.$$

3. Over that part of the impeller where transfer of energy takes place (blade passages) there is no variation of velocity in the meridional plane, i.e. across the width of the impeller. Thus,

$$\frac{\partial v}{\partial z} = 0.$$

The result of these assumptions is that whereas, in reality, $v = f(r, \theta, z)$, for the one-dimensional flow $v_\infty = f(r)$ only. Note that the suffix ∞ stipulates the assumption of an infinite number of blades and, hence, axisymmetry.

As a result, the flow through, say, a centrifugal impeller may be represented by a diagram such as Fig. 22.4. Although finite blades are shown, they are not taken into account in the theory. Furthermore, assumption (2) implies that the fluid streamlines are confined to infinitely narrow interblade passages and, hence, their paths are congruent with the shape of the interblade centreline, shown by a chain line. Thus, the flow of fluid through an impeller passage may be regarded as a flow of fluid particles along the centreline of the interblade passage.

The assumptions of the theory enable us to limit our analysis to changes of conditions which occur between impeller inlet and impeller outlet without reference to the space in between, where the real transfer of energy takes place. This space is treated as a 'black box' having an input in the form of an inlet velocity triangle and an output in the form of an outlet velocity triangle. Such velocity triangles for a centrifugal impeller rotating with a constant angular velocity ω are shown in Fig. 22.4.

At inlet, the fluid moving with an absolute velocity v_1 enters the impeller through a cylindrical surface of radius r_1 and may make an angle α_1 with the tangent at that radius. At outlet, the fluid leaves the impeller through a cylindrical surface of radius r_2, absolute velocity v_2 inclined to the tangent at outlet by the angle α_2.

The velocity triangles shown in Fig. 22.4 are obtained as follows. The inlet velocity triangle is constructed by first drawing the vector representing the absolute velocity v_1 at an angle α_1. The tangential velocity of the impeller, u_1, is then subtracted from it vectorially in order to obtain v_{r_1}, the relative velocity of the fluid with respect to the impeller blades at the radius r_1. In this basic velocity triangle, the absolute velocity v_1 is resolved into two components: one is the radial direction, called velocity of flow v_{f_1}, and the other, perpendicular to it and, hence, in the tangential direction, v_{w_1}, sometimes called velocity of whirl. These two components are useful in the analysis and, therefore, they are always shown as part of the velocity triangles.

FIGURE 22.4

One-dimensional flow through a centrifugal impeller

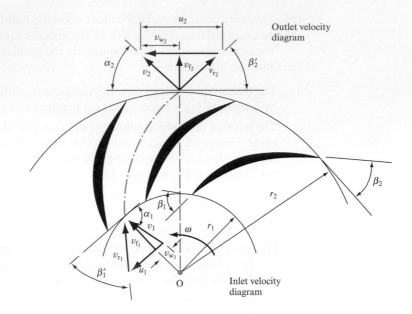

Similarly, the outlet velocity triangle consists of the absolute fluid velocity v_2 making an angle α_2 with the tangent at outlet, subtracted from which, vectorially, is the tangential blade velocity u_2 to give the relative velocity v_{r_2}. Here again, the absolute fluid velocity is resolved into radial (v_{f_2}) and tangential (v_{w_2}) components.

The general expression for the energy transfer between the impeller and the fluid, based on the one-dimensional theory and usually referred to as Euler's turbine equation, may be now derived as follows.

From Newton's second law applied to angular motion,

Torque = Rate of change of angular momentum.

Now,   Angular momentum = (Mass)(Tangential velocity)(Radius).

Therefore,

Angular momentum entering the impeller per second = $\dot{m} v_{w_1} r_1$,

Angular momentum leaving the impeller per second = $\dot{m} v_{w_2} r_2$,

in which $\dot{m}$ is the mass of fluid flowing per second. Therefore,

Rate of change of angular momentum = $\dot{m} v_{w_2} r_2 - \dot{m} v_{w_1} r_1$,

so that   Torque transmitted = $\dot{m}(v_{w_2} r_2 - v_{w_1} r_1)$.

Since the work done in unit time is given by the product of torque and angular velocity,

Work done per second = (Torque)$\omega = \dot{m}(v_{w_2} r_2 - v_{w_1} r_1)\omega$,

but $\omega = u/r$, so that $\omega r_2 = u_2$ and $\omega r_1 = u_1$. Hence, on substitution,

Work done per second, $E_t = \dot{m}(u_2 v_{w_2} - u_1 v_{w_1})$.   (22.1)

The SI units of the above expression are joules per second or watts.

Since the work done per second by the impeller on the fluid, such as in this case, is the rate of energy transfer, then:

Rate of energy transfer/Unit mass of fluid flowing, $Y = gE = E_t/\dot{m}$.

The product $gE = Y$, known as specific energy, is of significance in the case of pumps and fans. The units of Y are joules per kilogram.

From the specific energy, Euler's head E is given by

$$E = (1/g)(u_2 v_{w_2} - u_1 v_{w_1}). \tag{22.2}$$

The units of this equation are joules per kilogram divided by metres per second squared. This of course simplifies to metres and, therefore, is the same as all the terms of Bernoulli's equation and, consequently, E may be used in conjunction with it. Equation (22.2) is known as Euler's equation. From its mode of derivation it is apparent that Euler's equation applies to a pump (as derived) and to a turbine. In the latter case, however, since $u_1 v_{w_1} > u_2 v_{w_2}$, E would be negative, indicating the reversed direction of energy transfer. It is, therefore, common for a turbine to use the reversed order of terms in the brackets to yield positive E. Since the units of E reduce to metres of the fluid handled, it is often referred to as Euler's head and, in the case of pumps or fans, it represents the ideal theoretical head developed H_{th}.

It is useful to express Euler's head in terms of the absolute fluid velocities rather than their components. From the velocity triangles of Fig. 22.4,

$$v_{w_1} = v_1 \cos \alpha_1 \quad \text{and} \quad v_{w_2} = v_2 \cos \alpha_2,$$

so that $E = (1/g)(u_2 v_2 \cos \alpha_2 - u_1 v_1 \cos \alpha_1),$ \hfill (22.3)

but, using the cosine rule,

$$v_{r_1}^2 = u_1^2 + v_1^2 - 2u_1 v_1 \cos \alpha_1,$$

so that $u_1 v_1 \cos \alpha_1 = \frac{1}{2}(u_1^2 - v_{r_1}^2 + v_1^2).$

Similarly,

$$u_2 v_2 \cos \alpha_2 = \frac{1}{2}(u_2^2 - v_{r_2}^2 + v_2^2).$$

Substituting into (22.3),

$$E = (1/2g)(u_2^2 - u_1^2 + v_2^2 - v_1^2 + v_{r_1}^2 - v_{r_2}^2)$$

and $E = (v_2^2 - v_1^2)/2g + (u_2^2 - u_1^2)/2g + (v_{r_1}^2 - v_{r_2}^2)/2g.$ \hfill (22.4)

In this expression, the first term denotes the increase of the kinetic energy of the fluid in the impeller. The second term represents the energy used in setting the fluid in a circular motion about the impeller axis (forced vortex). The third term is the regain of static head due to a reduction of relative velocity in the fluid passing through the impeller.

Let us now consider the application of Euler's equation to centrifugal and axial flow machines.

In the former case, the velocity triangles are as shown in Fig. 22.4 and, in addition, the following relationships hold. Since, in general, $u = \omega r$, it follows that the tangential blade velocities at inlet and outlet are given by

$$u_1 = \omega r_1 \quad \text{and} \quad u_2 = \omega r_2. \tag{22.5}$$

Since the flow at inlet and outlet is through cylindrical surfaces and the velocity components normal to them are v_{f_1} and v_{f_2}, the continuity equation applied to inlet and outlet for the mass flow $\dot{m}$ and infinitely thin blades gives

$$\dot{m} = \rho_1 2\pi r_1 b_1 v_{f_1} = \rho_2 2\pi r_2 b_2 v_{f_2}, \tag{22.6}$$

where b_1 and b_2 are the impeller widths, as shown in Fig. 22.5, and ρ_1 and ρ_2 are the inlet and outlet densities, respectively. For incompressible flow, equation (22.6) simplifies to

$$r_1 b_1 v_{f_1} = r_2 b_2 v_{f_2}. \tag{22.7}$$

Now, assuming that m, ω, r_1 and r_2 are known, the following arguments are usually employed in order to draw the velocity triangles.

FIGURE 22.5

A centrifugal pump or fan impeller

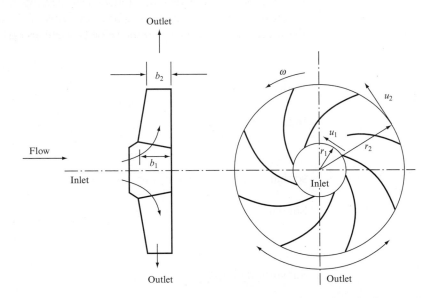

At inlet the usual assumptions are as follows:

1. The absolute velocity is radial. Therefore

$$v_1 = v_{f_1} \quad \text{and} \quad v_{w_1} = 0;$$

hence, v_1 is calculated from equation (22.6) and $\alpha_1 = 90°$. If this condition does not apply, which only occurs if there is a prewhirl (v_{w_1}) component present, perhaps due to inlet vanes or unfavourable inlet conditions, then v_{f_1} is calculated from equation (22.6) and α_1 can be determined only if v_{w_1} is known.

2. The blade angle at inlet β_1 is such that the blade meets the relative velocity tangentially. Thus $\beta_1 = \beta_1'$. This assumption is known as the 'no-shock' condition and is applied in determining the blade inlet angle during design in order to minimize the entry loss.

Thus the inlet triangles may be drawn.

For the outlet triangles, it is assumed that the fluid leaves the impeller with a relative velocity tangential to the blade at outlet. Thus, $\beta_2' = \beta_2$ and, in order to draw the outlet velocity triangles, β_2 must be known. The direction of v_{r_2} is then drawn, as well as the v_{f_2} vector, which is radial and whose magnitude is calculated from equation (22.6). It is, thus, possible to draw the u_2 ($= \omega r_2$) vector perpendicular to v_{f_2} and starting from the intersection with the direction of v_{r_2}. The absolute velocity v_2 is then obtained by completing the triangle, from which,

$$\cot \beta_2 = (u_2 - v_{w_2})/v_{f_2},$$

so that $\quad v_{w_2} = u_2 - v_{f_2} \cot \beta_2.$

Substituting this into Euler's equation, and remembering that $v_{w_1} = 0$, the following expression is obtained:

$$E = (u_2/g)(u_2 - v_{f_2} \cot \beta_2). \tag{22.8}$$

The total amount of energy transferred by the impeller is, thus,

$$E_t = \dot{m}gE = \dot{m}u_2(u_2 - v_{f_2} \cot \beta_2). \tag{22.9}$$

Consider now an axial flow machine, as shown in Fig. 22.6. The important difference between the axial flow machine and the centrifugal one is that since, in the former, the flow is axial, the changes from inlet to outlet take place at the same radius and, hence,

$$u_1 = u_2 = u = \omega r. \tag{22.10}$$

Also, since the flow area is the same at inlet and outlet,

$$v_{f_1} = v_{f_2} = v_f$$

and is obtained from

$$\dot{m} = \rho v_f \pi (R_2^2 - R_1^2). \tag{22.11}$$

The following assumptions are made with regard to the velocity triangles:

1. There is no prewhirl at inlet and, hence,

$$\alpha_1 = 90°, \quad v_{w_1} = 0, \quad v_1 = v_f.$$

2. For 'no-shock' condition, the blade is set at an angle such that it meets the relative fluid velocity tangentially or at an appropriate angle of incidence for an aerofoil section.

FIGURE 22.6
Axial flow impeller and
velocity triangles

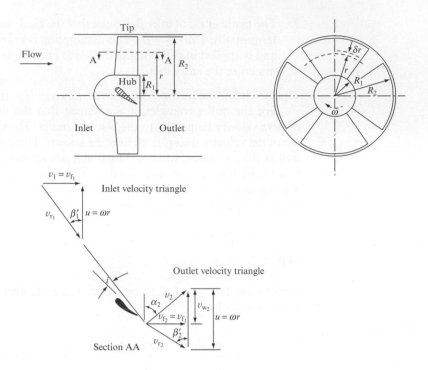

3. At outlet, the relative velocity leaves the blade tangentially and a similar
    procedure to that for a centrifugal impeller is used to complete the velocity
    triangles.

Here again, from the outlet triangle,

$$\cot \beta'_2 = (u - v_{w_2})/v_f,$$

so that $v_{w_2} = u - v_f \cot \beta'_2,$

which, on substitution into Euler's equation, gives

$$E = (u/g)(u - v_f \cot \beta'_2). \tag{22.12}$$

It is important, however, to realize that this equation applies to any particular
radius r and is not necessarily constant over the range from R_1 to R_2. For this
condition to apply, the increase of u with radius must be counterbalanced by an equal
decrease of $v_f \cot \beta'_2$. Since $v_f = $ constant, the blade must be twisted so that, for any
two radii r_a and r_b,

$$u_a^2 - u_a v_f \cot \beta'_{2a} = \text{constant} = u_b^2 - u_b v_f \cot \beta'_{2b}.$$

Rearranging,

$$v_f(u_b \cot \beta'_{2b} - u_a \cot \beta'_{2a}) = u_b^2 - u_a^2,$$

but $u = \omega r$, so that

$$v_f(\omega r_b \cot \beta'_{2b} - \omega r_a \cot \beta'_{2a}) = \omega^2(r_b^2 - r_a^2),$$

which gives

$$r_b \cot \beta'_{2b} - r_a \cot \beta'_{2a} = (\omega/v_f)(r_b^2 - r_a^2). \tag{22.13}$$

However, this condition, known as the 'free vortex' design, is not always met. It is then necessary to apply Euler's equation to an element dr and to integrate from R_1 to R_2 as follows.

The energy transfer for the element

$$dE_t = (u/g)(u - v_f \cot \beta'_2) \, dW.$$

However,

$$dW = 2\pi\rho g v_f r \, dr$$

and $\quad u = \omega r,$

so that $\quad E_t = 2\pi\rho\omega v_f \displaystyle\int_{R_1}^{R_2} r^2(\omega r - v_f \cot \beta'_2) \, dr, \tag{22.14}$

in which $\beta'_2 = f(r)$ must be known.

EXAMPLE 22.1

An axial flow fan has a hub diameter of 1.50 m and a tip diameter of 2.0 m. It rotates at 18 rad s^{-1} and, when handling 5.0 m^3 s^{-1} of air, develops a theoretical head equivalent to 17 mm of water. Determine the blade outlet and inlet angles at the hub and at the tip. Assume that the velocity of flow is independent of radius and that the energy transfer per unit length of blade (δr) is constant. Take the density of air as 1.2 kg m^{-3} and the density of water as 10^3 kg m^{-3}.

Solution

$$\text{Velocity of flow, } v_f = \frac{Q}{A} = \frac{Q}{\pi(R_2^2 - R_1^2)} = \frac{5}{\pi(1 - 0.5625)} = 3.64 \text{ m s}^{-1}.$$

Blade velocity at tip is given by

$$u_t = \omega R_2 = 18 \times 1 = 18 \text{ m s}^{-1}$$

and, at hub,

$$u_h = \omega R_1 = 18 \times 0.75 = 13.5 \text{ m s}^{-1}.$$

Since, for 'no-shock' condition, $\cot \beta_1 = u/v_f$, the inlet blade angle at tip is given by

$$\beta_{1t} = \cot^{-1}\left(\frac{18}{3.64}\right) = 11.4°$$

and, at hub,

$$\beta_{1h} = \cot^{-1}\left(\frac{13.5}{3.64}\right) = 15.1°.$$

Since the head generated by the tip and hub sections is the same, the outlet angles may be obtained by applying Euler's equation to these sections. From equation (22.12),

$$E = (u/g)(u - v_f \cot \beta_2),$$

but $\qquad E = H_{th} = 17$ mm of water $= 0.017 \times 10^3/1.2 = 14.16$ m of air.

Therefore at tip,

$$14.16 = (18/9.81)(18 - 3.64 \cot \beta_{2t}),$$

$$\beta_{2t} = \mathbf{19.5°}$$

and at hub,

$$14.16 = (13.5/9.81)(13.5 - 3.64 \cot \beta_{2h}),$$

$$\beta_{2h} = \mathbf{48.6°}.$$

22.3 ISOLATED BLADE AND CASCADE CONSIDERATIONS

The major assumption underlying the considerations of the previous section was that of an infinite number of blades in the impeller. In practice, of course, the number of blades is finite and their spacing depends on a particular impeller design and, therefore, may vary considerably. The distance between the adjacent blades, s, is known as *pitch*, whereas the ratio of blade chord to the pitch,

$$\sigma = c/s, \tag{22.15}$$

is called *blade solidity* and is a measure of the closeness of blades. If the blades are very close to each other, the passages between them may be treated as conduits and the flow through them treated accordingly as bounded flow. If, however, the blades are very far apart, they must be treated as bodies in an external flow, provided their mutual interference is negligible. Thus, the two extremes of high solidity and zero solidity provide well-defined situations requiring reasonably clear-cut treatment, but in real machines neither of them applies. Furthermore, what makes the treatment more difficult is the fact that in many impellers the blade solidity varies with radius. This is the case in axial flow impellers, where blades near the tip are much further apart than they are near the hub.

In this section we will first consider relationships which result from assuming that blades are far apart, so that $s \to \infty$ and, hence, $\sigma \to 0$. For such cases the theories expanded in Chapter 12 become directly applicable. We will then look at blades close together and arranged into cascades, for which a modified approach will be necessary.

It was shown in Chapter 12 that the lift is dependent upon circulation around the lifting surface or body and, also, that it is related to the pressure distribution around it:

$$L = \int_0^{2\pi} p \sin \theta \, d\theta = \tfrac{1}{2} C_L \rho U_0^2 A,$$

but also $L' = \rho U_0 \Gamma$ (from equation (7.67)), where L' is the lift per unit length of the body, so that

$$L = \rho U_0 \Gamma l,$$

where l is the length of the body. Thus, there is a direct relationship between the pressure distribution and, hence, the resultant force on the body, which may be an impeller blade, and the circulation around it, namely

$$\rho U_0 \Gamma l = \int_0^{2\pi} p \sin \theta \, d\theta,$$

which leads to

$$\Gamma = \tfrac{1}{2} C_L U_0 A / l. \tag{22.16}$$

It is, therefore, possible to relate Euler's equation to the circulation as follows. Consider the circulation around a single blade as shown in Fig. 22.7:

$$\Gamma_{ABCD} = \oint v \, ds = \int_A^B v \, ds + \int_B^C v \, dl + \int_C^D v \, ds + \int_D^A v \, dl.$$

However,

$$\int_B^C v \, dl = -\int_D^A v \, dl, \quad \int_A^B v \, ds = -v_{w_1} s_1 \quad \text{and} \quad \int_C^D v \, ds = v_{w_2} s_2.$$

Therefore, if we denote the circulation around the blade by $\Gamma_b = \Gamma_{ABCD}$, then,

$$\Gamma_b = s_2 v_{w_2} - s_1 v_{w_1}. \tag{22.17}$$

FIGURE 22.7
Circulation around
the blade

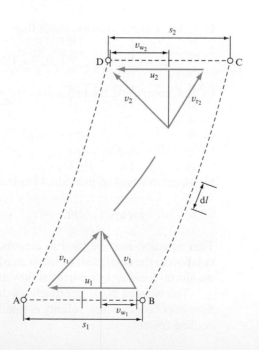

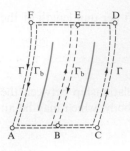

FIGURE 22.8
Circulation around two
blades

If we now consider two adjacent blades, the circulation around them may be obtained by considering Fig. 22.8:

$$\Gamma = \Gamma_{ACDF} = \int_A^B v\,ds + \int_B^C v\,ds + \int_C^D v\,dl + \int_D^E v\,ds + \int_E^F v\,ds + \int_F^A v\,dl.$$

However,

$$\int_C^D v\,dl = -\int_F^A v\,dl \quad \text{and} \quad \int_A^B v\,ds + \int_E^F v\,ds = \Gamma_b.$$

Also,

$$\int_B^C v\,ds + \int_D^E v\,ds = \Gamma_b.$$

Therefore,

$$\Gamma = \Gamma_b + \Gamma_b = 2\Gamma_b.$$

This result may be generalized for a number of blades z into

$$\Gamma = z\Gamma_b \tag{22.18}$$

and, substituting from equation (22.18), we obtain

$$\Gamma = z(s_2 v_{w_2} - s_1 v_{w_1}),$$

but $z s_1 = 2\pi r_1$ and $z s_2 = 2\pi r_2$, so that

$$\Gamma = 2\pi(r_2 v_{w_2} - r_1 v_{w_1}). \tag{22.19}$$

However, Euler's equation states that

$$E = (1/g)(u_2 v_{w_2} - u_1 v_{w_1}) = (\omega/g)(r_2 v_{w_2} - r_1 v_{w_1}),$$

so that, comparing the two equations, we obtain

$$\Gamma/2\pi = Eg/\omega$$

or $$E = (\omega/g)(\Gamma/2\pi). \tag{22.20}$$

However, in terms of individual blade circulation, this equation becomes

$$E = (\omega/g)(z\Gamma_b/2\pi). \tag{22.21}$$

This equation may be used in conjunction with equation (22.16) which relates circulation to the coefficient of lift in an ideal situation, since the latter equation is based on Kutta–Joukowsky's potential flow analysis.

The isolated blade approach, as described above, has some application to axial flow impellers in which solidity is small, such as propellers or axial flow fans used in cooling towers.

When solidity is significant, a different approach is required. An arrangement of geometrically identical blades such that they are at the same distance from one another and are positioned in the same way with respect to the direction of flow is called a *cascade*. If the blades are arranged along a straight line, the cascade is called *straight*, and if they are arranged around the circumference of a circle, they are referred to as *circular*. A development of an axial flow impeller constitutes a moving straight cascade, whereas a centrifugal impeller is a rotating circular cascade.

The main purpose of cascades is to deflect the flow. Hence, there is always a change of momentum across a cascade and a force associated with it. If the velocities upstream and downstream of a cascade are the same in magnitude, there will be change of momentum due to a change in direction, but it follows from Bernoulli's equation that there will be no pressure difference between the upstream and the downstream sides of the cascade. It is then known as an *impulse* cascade. If the pressure difference exists due to absolute velocities not being the same, the cascade is called *reaction*. Those reaction cascades in which fluid is accelerated (fall of pressure) are usually used in turbines, whereas those in which the fluid is decelerated and, hence, there is an increase of pressure are used in pumps and compressors.

Consider, now, a case of a stationary straight cascade of height Z. Let the upstream fluid velocity v_1 making an angle α_1 with the line of the cascade be deflected so that the downstream velocity v_2 makes an angle α_2, as shown in Fig. 22.9. Now the flow *deflection* is

$$\varepsilon = \alpha_2 - \alpha_1,$$

and it is an important characteristic of a cascade.

FIGURE 22.9

A straight cascade

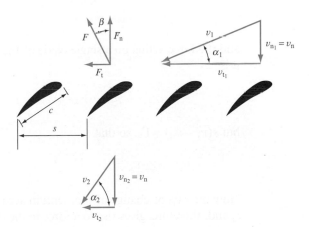

The fluid velocities v_1 and v_2 may be resolved into components parallel and normal to the cascade, v_t and v_n, respectively. Since there is a difference of velocity, the pressure change (assuming no loss and using Bernoulli's equation) is given by

$$p_1 - p_2 = \tfrac{1}{2}\rho(v_2^2 - v_1^2). \tag{22.22}$$

However, since, by the continuity equation, the mass flow through the cascade is

$$\dot{m} = sZ\rho_1 v_{n_1} = sZ\rho_2 v_{n_2},$$

where Z is the height of the cascade, it follows that for incompressible flow,

$$v_{n_1} = v_{n_2} = v_n.$$

Thus, the change in velocity is entirely due to the change of the tangential velocity component v_t, which also follows from substitution of the following relationships into equation (22.22):

$$v_2^2 = v_{t_2}^2 + v_n^2,$$
$$v_1^2 = v_{t_1}^2 + v_n^2,$$

from which

$$v_2^2 - v_1^2 = v_{t_2}^2 - v_{t_1}^2$$

and equation (22.22) becomes

$$p_1 - p_2 = \tfrac{1}{2}\rho(v_{t_2}^2 - v_{t_1}^2).$$

If we now introduce the mean tangential velocity,

$$v_t = \tfrac{1}{2}(v_{t_1} + v_{t_2}),$$

we obtain

$$p_1 - p_2 = \rho v_t(v_{t_2} + v_{t_1}) \tag{22.23}$$

and the force acting on a single blade of the cascade in the direction perpendicular to it is

$$F_n = sZ(p_1 - p_2) = sZ\rho v_t(v_{t_2} - v_{t_1}),$$

but $s(v_{t_2} - v_{t_1}) = \Gamma_b$, so that

$$F_n = \rho Z v_t \Gamma_b. \tag{22.24}$$

Now the rate of change of momentum across the cascade is again due to a change in v_t and, therefore, gives rise to a force in the direction of the cascade, F_t. So,

$$F_t = \dot{m}(v_{t_2} - v_{t_1}) = sZ\rho v_n(v_{t_2} - v_{t_1}),$$

from which

$$F_t = \rho Z v_n \Gamma_b. \tag{22.25}$$

The resultant force on the blade, therefore, is

$$F = \sqrt{(F_n^2 + F_t^2)} = \rho Z \Gamma_b \sqrt{(v_t^2 + v_n^2)}. \tag{22.26}$$

The direction of this force is given by the angle β such that

$$\cot \beta = F_n/F_t = \rho Z v_t \Gamma_b / \rho Z v_n \Gamma_b = v_t/v_n,$$

but $v_t = \frac{1}{2}(v_{t_1} + v_{t_2})$, which may be expressed in terms of inlet and outlet angles α_1 and α_2 by using the trigonometric relationships

$$v_{t_1} = v_n \cot \alpha_1 \quad \text{and} \quad v_{t_2} = v_n \cot \alpha_2.$$

Therefore,

$$v_t = \tfrac{1}{2} v_n(\cot \alpha_1 + \cot \alpha_2)$$

and $\qquad \cot \beta = \tfrac{1}{2}(\cot \alpha_1 + \cot \alpha_2).$ (22.27)

It also follows from simple geometry that β is equal to β_∞, defined as the mean direction of flow and obtained by superposition of the inlet and outlet velocity triangles, as shown in Fig. 22.10. Thus the force F being perpendicular to the mean direction of flow is the lift on the blade.

FIGURE 22.10

Combined inlet and outlet velocity triangles for a straight cascade

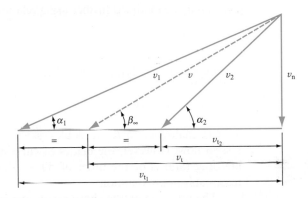

It is now possible to consider an axial flow impeller by selecting an annular element δr at a radius r (Fig. 22.6) and developing it into a moving cascade having velocity $u = \omega r$.

In this configuration the velocity with which the fluid approaches the cascade is the relative velocity v_{r_1} and takes the place of v_1 in the stationary cascade. Similarly

$$\left.\begin{aligned} &v_2 \text{ is replaced by } v_{r_2}, \\ &v_n \text{ is replaced by } v_f, \\ &v_{t_1} \text{ is replaced by } u, \\ &v_{t_2} \text{ is replaced by } (u - v_{w_2}). \end{aligned}\right\} \qquad (22.28)$$

The full velocity triangles for the usual case of $v_{w_1} = 0$ and, therefore, $v_1 = v_f$ are shown in Fig. 22.11. Note that the aerofoil blade is set at an angle of incidence i with respect to the mean flow direction. Thus the angle it makes with the direction of rotation, known as the blade *stagger* (or blade pitch) angle, is $(i + \beta_\infty)$.

FIGURE 22.11
Velocity triangles for
an axial flow impeller

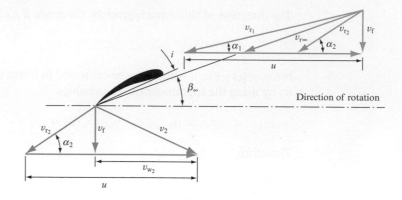

We can now obtain an expression for lift, which will then enable us to determine the energy transfer realized by a cascade formed by such blade elements. This may be obtained by first eliminating from equation (22.26) the term under the square root:

$$v_t^2 + v_n^2 = [(v_{t_1} + v_{t_2})/2]^2 + v_n^2,$$

which may be modified further using relationships (22.28):

$$v_t^2 + v_n^2 = (u - v_{w_2}/2)^2 + v_f^2 = v_{r\infty}^2.$$

The expression for lift, therefore, becomes

$$L = \rho Z \Gamma_b v_{r\infty}. \tag{22.29}$$

This is analogous to Kutta–Joukowsky's law and, likewise, applies to the ideal (no losses) flow. However, the blades of turbomachinery are generally efficient and the lift/drag ratio is of the order of 50, so that the above equation may be used as a reasonable approximation.

The lift given by the above expression is in terms of the blade circulation, which may be related to cascade data by equating equation (22.29) to equation (12.5),

$$\rho Z \Gamma_b v_{r\infty} = \tfrac{1}{2} C_L \rho A v_{r\infty}^2,$$

but, from Figure 22.9, $A = cZ$, so that

$$\Gamma_b = \tfrac{1}{2} C_L c v_{r\infty}. \tag{22.30}$$

It is now possible to relate the cascade lift coefficient C_L to the blade angle at inlet α_1, at outlet α_2 and the mean angle β_∞. From equation (22.17),

$$\Gamma_b = s(v_{w_2} - v_{w_1}),$$

but $v_{w_1} = 0$ and, from the outlet velocity triangle (Fig. 22.11),

$$v_{w_2} = u - v_f \cot \alpha_2,$$

so that $\Gamma_b = s(u - v_f \cot \alpha_2).$

Equating this expression to equation (22.30),

$$s(u - v_f \cot \alpha_2) = \tfrac{1}{2} C_L c v_{r\infty};$$

dividing by v_f,

$$s(u/v_f - \cot \alpha_2) = \tfrac{1}{2} C_L c v_{r\infty}/v_f.$$

But $u/v_f = \cot \alpha_1$ and $v_{r\infty}/v_f = 1/\sin \beta_\infty$; therefore,

$$s(\cot \alpha_1 - \cot \alpha_2) = C_L c/2 \sin \beta_\infty,$$

so that

$$C_L = 2(s/c)(\cot \alpha_1 - \cot \alpha_2) \sin \beta_\infty. \tag{22.31}$$

It is important to remember that the above equations apply to cascades and not to isolated aerofoils, because the theory is based on a change of momentum of the fluid stream due to a change of its direction (deflection), which only cascades achieve. Cascade data are obtained experimentally and a typical set may be as shown in Fig. 22.12.

FIGURE 22.12
Typical cascade data

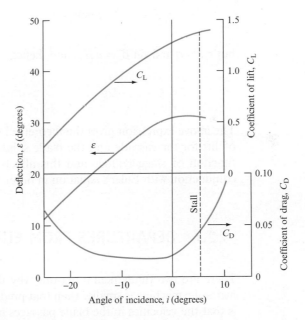

The alternative approach is to use the isolated aerofoil data but corrected by the use of a *cascade coefficient K*, defined as

$$K = \frac{\text{Cascade lift coefficient}}{\text{Aerofoil lift coefficient}}. \tag{22.32}$$

This approach is particularly appropriate for axial flow machines in which solidity varies considerably between hub and tip, but the values of K must be determined experimentally. Generally, K depends on the extent of overlap between adjacent blades, which is a function of solidity and the stagger angle.

The energy transfer occurring within a moving cascade may be obtained by considering the work done per second in the direction of motion, which, for a rotating impeller, is the plane of rotation:

$$\dot{W}_{d} = L \sin \beta_{\infty} u.$$

However, the lift for a radial element δr is

$$L = \tfrac{1}{2} C_{L} \rho c v_{r\infty}^{2} \, \delta r,$$

so that $\quad \dot{W}_{d} = \tfrac{1}{2} C_{L} \rho c u v_{r\infty}^{2} \sin \beta_{\infty} \, \delta r.$

Now, the weight of the fluid flowing per second through the elemental depth δr is

$$\rho g s v_{f} \, \delta r$$

and, hence, the energy transfer is

$$E = \tfrac{1}{2} C_{L} \rho c u v_{r\infty}^{2} \sin \beta_{\infty} \, \delta r / \rho g s v_{f} \, \delta r$$

$$= \frac{1}{2g} C_{L} \frac{c}{s} \frac{u}{v_{f}} v_{r\infty}^{2} \sin \beta_{\infty},$$

but $c/s = \sigma$ and $\sin \beta_{\infty} = v_{f}/v_{r\infty}$ and, hence,

$$E = C_{L} \sigma v_{r\infty} u / 2g. \qquad (22.33)$$

The above expression gives the theoretical energy transfer in terms of the coefficient of lift for the cascade and the blade solidity. However, since the lift coefficient is a function of stagger angle and this affects $v_{r\infty}$, equation (22.33) has to be used in conjunction with Euler's equation in order to obtain a design solution.

22.4   DEPARTURES FROM EULER'S THEORY AND LOSSES

There are two fundamental reasons why the actual energy transfer achieved by an hydraulic machine is smaller than that predicted by Euler's equation. The first reason is that the velocities in the blade passages and at the impeller outlet are not uniform owing to the presence of blades and the real flow being three dimensional. This results in a diminished velocity of whirl component and, hence, reduces Euler's head. This effect is not caused by friction and, therefore, does not represent a loss but follows from the ideal flow analysis of pressure and velocity distributions. The second reason is that in a real impeller there are losses of energy due to friction, separation and wakes associated with the development of boundary layers. We will consider these two effects separately.

In an impeller of a centrifugal pump, for example, the blades do work on the fluid by exerting an 'impelling' force on it. This is done by the upper or forward surface of

FIGURE 22.13

Effect of velocity
distribution on the outlet
velocity triangle

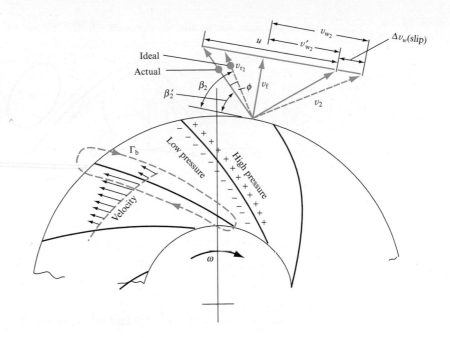

the blade. It follows that the pressure in the fluid on this side of the blade is greater
than that on the other side, as indicated in Fig. 22.13. Hence, the velocity near the
back side of the blade is greater than that near the forward side. This difference of
velocity on the two sides of the blade gives rise to the blade circulation Γ_b associated
with the lift. But the non-uniform velocity distribution is responsible for the mean
direction of flow leaving the impeller, being $\beta_2' = (\beta_2 - \phi)$ and not β_2, as assumed in
the ideal flow situation. This effect is responsible for *deviation* and results in the reduction of the all-important tangential velocity component (velocity of whirl) from v_{w_2} to
v_{w_2}', the reduction being Δv_w and called *slip*. Thus, the *slip factor* may be defined as

$$S_F = v_{w_2}'/v_{w_2}, \tag{22.34}$$

and the real velocity triangle is as shown by the full lines in Fig. 22.13, whereas that
corresponding to the ideal case is shown by the dashed lines.

There have been many attempts to predict the amount of slip and a number of
theories have been formulated. The earliest, due to Stodola, stipulates the existence of
a 'relative eddy' (shown in Fig. 22.14) which occurs between the adjacent blades. Since
a frictionless fluid passes through the impeller without rotation, it must also leave the
impeller without rotation. However, the impeller rotates at an angular velocity ω,
which means that the fluid must have a rotation relative to the impeller of $(-\omega)$, which
is the relative eddy. Stodola assumed that if the radius of a circle which may be
inscribed between the two adjacent blades at outlet is e, then the slip may be
considered to be the product of the relative eddy ω and e. Thus,

$$\Delta v_w = \omega e. \tag{22.35}$$

But the impeller circumference at outlet is $2\pi R_2$ and, hence, for z blades e may be
obtained approximately from

FIGURE 22.14
The 'relative eddy'

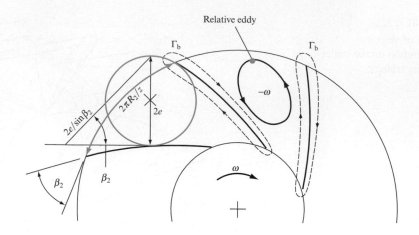

$$2e/\sin \beta_2 \cong 2\pi R_2/z,$$

$$e = (\pi R_2/z) \sin \beta_2.$$

Substituting into equation (22.35) and taking $\omega = u_2/R_2$,

$$\Delta v_w = \frac{u_2}{R_2} \times \frac{\pi R_2 \sin \beta_2}{z} = \frac{u_2 \pi}{z} \sin \beta_2.$$

Also, since

$$v_{w_2} = u_2 - v_f \cot \beta_2,$$

it follows that the slip factor may be expressed as

$$S_F = \frac{v_{w_2} - \Delta v_w}{v_{w_2}} = 1 - \frac{\Delta v_{w_2}}{v_{w_2}} = 1 - \frac{u_2 \pi \sin \beta_2}{z(u_2 - v_f \cot \beta_2)},$$

which becomes

$$S_F = 1 - \frac{\pi \sin \beta_2}{z[1 - (v_f/u_2) \cot \beta_2]}. \tag{22.36}$$

It is interesting to note that for an impeller with radial blades at the tip ($\beta_2 = 90°$), Stodola's slip factor reduces to

$$S_F = 1 - \pi/z. \tag{22.37}$$

The best known of the more exact theories is that due to Busemann, who considered the resultant flow as a superposition of flow through stationary vanes and a displacement due to rotation. This yields an expression for the slip factor of the form

$$S_F = [A - B(v_f/u_2) \cot \beta_2][1 - (v_f/u_2) \cot \beta_2], \tag{22.38}$$

in which both A and B are functions of R_2/R_1, β_2 and z.

Stanitz used relaxation methods (blade-to-blade solution) for impellers having $45° < \beta_2 < 90°$ and concluded that the slip velocity Δv_w is independent of β_2 and also that the slip factor was not affected by compressibility. His expression for the slip factor is

$$S_F = 1 - \frac{0.63\,\pi}{z[1 - (v_f/u_2)\cot\beta_2]},\qquad(22.39)$$

which, for radial blades, reduces to

$$S_F = 1 - 0.63\pi/z.\qquad(22.40)$$

On the whole, for pumps, the best agreement with experimental results is obtained in the following ranges of β_2:

$$20° < \beta_2 < 30°,\quad \text{Stodola's correction;}$$
$$30° < \beta_2 < 80°,\quad \text{Busemann's correction;}$$
$$80° < \beta_2 < 90°,\quad \text{Stanitz' correction.}$$

The effects of losses due to friction, separation, wakes, etc., are a completely different issue, although of equal importance. In a cascade these losses manifest themselves as a drop in pressure downstream of the cascade. Assuming that the velocity remains unchanged, the actual pressure difference across the cascade becomes

$$p_2 - p_1 = (\rho/2)(v_1^2 - v_2^2) - \Delta p,\qquad(22.41)$$

where Δp is the pressure loss in the cascade. This increased pressure difference affects the normal force component on the cascade and, hence, the resultant force, as shown in Fig. 22.15, which refers to a pump. In this case, assuming the same inlet pressure p_1 and the same inlet and outlet velocities, it is only the outlet pressure which is affected.

FIGURE 22.15

Effect of pressure losses on forces on a cascade

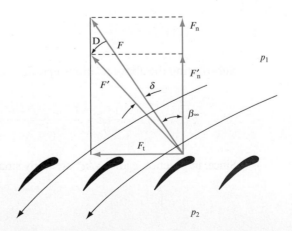

Without losses, the pump would generate greater pressure, so that the ideal pressure at outlet p_2' would be greater than p_2 actually achieved. Thus,

$$p_2 = p_2' - \Delta p.$$

It follows, therefore, that the actual force normal to the cascade is

$$F_n' = sZ(p_2 - p_1) = sZ(p_2' - p_1 - \Delta p),$$

whereas the theoretical or ideal force would be

$$F_n = sZ(p_2' - p_1).$$

These forces are shown in Fig. 22.15. The actual resultant force F' makes an angle $(\delta + \beta_\infty)$ with the normal. It is, therefore, no longer perpendicular to the mean direction of flow and, hence, is not equal to the lift. The angle δ is characteristic of the cascade efficiency, which may be defined as

$$\eta_c = \frac{p_2' - p_1 - \Delta p}{p_2' - p_1} = \frac{F_n'}{F_n}, \tag{22.42}$$

but $F_n' = F_t \cot(\beta_\infty + \delta)$ and $F_n = F_t \cot \beta_\infty$, so that

$$\eta_c = \frac{\cot(\beta_\infty + \delta)}{\cot \beta_\infty} = \frac{\cot \beta_\infty \cot \delta - 1}{\cot \beta_\infty (\cot \beta_\infty + \cot \delta)}.$$

This can be simplified to

$$\eta_c = \frac{1 - \tan \beta_\infty \tan \delta}{1 + \tan \delta \cot \beta_\infty}.$$

Now, let $\tan \delta = \varepsilon$,

$$\tan \beta_\infty = \frac{v_f}{u - v_{w_2}/2} = \frac{v_f}{u^*},$$

where $u^* = u - v_{w_2}/2$ and

$$\cot \beta_\infty = \frac{u - v_{w_2}/2}{v_f} = \frac{u^*}{v_f}.$$

Substituting the above relationships, the cascade (or blade) efficiency becomes

$$\eta_c = \frac{1 - \varepsilon \tan \beta_\infty}{1 + \varepsilon \cot \beta_\infty} = \frac{1 - \varepsilon v_f/u^*}{1 + \varepsilon u^*/v_f}. \tag{22.43}$$

Since, in practice, ε and v_f/u^* are fairly small, this expression approximates to

$$\eta_c = 1 - \varepsilon u^*/v_f. \tag{22.44}$$

An identical expression may be obtained for a turbine cascade, although the initial reasoning regarding pressure differences must relate to the reversed direction of energy transfer.

Referring back to Fig. 22.15, it will be seen that the actual force F' may be resolved into components perpendicular and parallel to the mean direction of flow, thus giving the actual lift and drag:

$$L = F' \sin \delta,$$

$$D = F' \cos \delta.$$

It is, thus, apparent that the losses in a cascade are related to the drag.

EXAMPLE 22.2

The impeller of a centrifugal pump has a diameter of 0.1 m and axial width at outlet of 15 mm. There are 16 blades swept backwards and inclined at 25° to the tangent to the periphery (Fig. 22.16). The flow rate through the impeller is 8.5 m³ h⁻¹ when it rotates at 750 rev min⁻¹. Calculate the head developed by the pump when handling water and assuming (a) one-dimensional ideal flow theory, (b) allowing for the relative eddy between the blades.

FIGURE 22.16

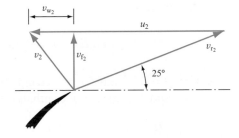

Solution

(a) Area at outlet, $A = \pi D t = \pi \times 0.1 \times 15 \times 10^{-3} = 4.71 \times 10^{-3}\,\mathrm{m^2}$.

Velocity of flow at outlet, $v_{f_2} = \dfrac{Q}{A} = \dfrac{8.5 \times 10^3}{3600 \times 4.71} = 0.501\,\mathrm{m\,s^{-1}}$.

Blade velocity at outlet, $u_2 = \dfrac{\pi N D}{60} = \dfrac{\pi \times 750 \times 0.1}{60} = 3.927\,\mathrm{m\,s^{-1}}$.

But, from the outlet velocity triangle,

$$\tan 25° = v_{f_2}/(u_2 - v_{w_2})$$

and so $v_{w_2} = u_2 - v_{f_2} \cot \beta_2 = 3.927 - 0.501 \cot 25°$

$$= 2.9\,\mathrm{m\,s^{-1}}.$$

The theoretical head is given by Euler's equation:

$$H_{th} = E_{th} = \frac{u_2 v_{w_2}}{g} = \frac{3.927 \times 2.9}{9.81} = \textbf{1.16 m of water}.$$

(b) Since $\beta_2 = 25°$, Stodola's slip factor may be used to allow for the relative eddy between the blades:

$$S_F = 1 - \frac{\pi \sin \beta_2}{z[1 - (v_{f_2}/u_2)\cot \beta_2]} = 0.886.$$

The head developed is

$$H = S_F H_{th} = 0.886 \times 1.16 = \mathbf{1.03 \ m \ of \ water.}$$

22.5 COMPRESSIBLE FLOW THROUGH ROTODYNAMIC MACHINES

Compressible flow involves appreciable changes of fluid density and, therefore, occurs most frequently in gases. Since, for gases, changes of density are related to changes of pressure and temperature by the equation of state, it is useful to make use of some of the fundamental thermodynamic concepts and relationships.

The first law of thermodynamics leads to the establishment of the steady flow energy equation:

$$\dot{Q} - \dot{W}_d = \dot{m}(H_2 - H_1) + \tfrac{1}{2}(v_2^2 - v_1^2) + g(z_2 - z_1), \tag{22.45}$$

where $\dot{Q}$ = heat transfer per second, $\dot{W}_d$ = work done per second, $\dot{m}$ = mass flow per second and H is the specific enthalpy. Hence, assuming constant specific heat c_p,

$$H_2 - H_1 = c_p(T_2 - T_1). \tag{22.46}$$

For gases the term $g(z_2 - z_1)$ is small and, therefore, may be ignored. Rearranging equation (22.45) we obtain

$$\dot{Q} - \dot{W}_d = \dot{m}[(H_2 + \tfrac{1}{2}v_2^2) - (H_1 + \tfrac{1}{2}v_1^2)].$$

Now, we define stagnation enthalpy by

$$H_T = H + \tfrac{1}{2}v^2, \tag{22.47}$$

where

$$H_{T_2} = H_2 + \tfrac{1}{2}v_2^2, \quad H_{T_1} = H_1 + \tfrac{1}{2}v_1^2 \quad \text{and} \quad H_{T_2} - H_{T_1} = c_p(T_{T_2} - T_{T_1}), \tag{22.48}$$

in which T_T is the stagnation (or total) temperature and c_p is assumed to be constant. However, for most turbomachinery the flow processes are very nearly adiabatic and, therefore, it is justifiable to write $Q = 0$. As a result,

$$\dot{W}_d = \dot{m}(H_{T_1} - H_{T_2}) \quad \text{for turbines} \tag{22.49}$$

and $\dot{W}_d = \dot{m}(H_{T_2} - H_{T_1}) \quad \text{for compressors.} \tag{22.50}$

FIGURE 22.17

Enthalpy changes in turbomachinery

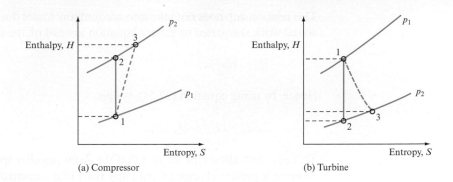

(a) Compressor   (b) Turbine

These processes may be most usefully represented in an enthalpy/entropy diagram (as shown in Fig. 22.17). For a compressor producing total pressure rise from p_1 to p_2, the increase of enthalpy corresponding to the assumption of no losses and, hence, of a reversible or isentropic process is represented by a straight line 1–2 ($S =$ constant). The actual process is not, however, isentropic because of losses due to friction and, therefore, the true increase of enthalpy is along line 1–3 ($S_3 > S_1$). Thus, it is possible to define isentropic efficiency as

$$\eta_i = \frac{\text{Isentropic work}}{\text{Actual work}} = \frac{H_{T_2} - H_{T_1}}{H_{T_3} - H_{T_1}}. \tag{22.51}$$

Similar reasoning applied to a turbine (Fig. 22.17(b)) leads to the isentropic efficiency being expressed as

$$\eta_i = \frac{H_{T_1} - H_{T_3}}{H_{T_1} - H_{T_2}}. \tag{22.52}$$

It is now possible to relate the change of enthalpy in the impeller to the velocities of the fluid flowing through it by making use of Euler's equation. From equation (22.50), the work done per unit mass flow is

$$\dot{W}_d / \dot{m} = H_{T_2} - H_{T_1} \quad \text{for a compressor.}$$

But from Euler's equation, the energy transfer per unit weight of fluid flowing is

$$E = (1/g)(u_2 v_{w_2} - u_1 v_{w_1})$$

and, therefore, the work done per unit mass of fluid flowing or specific energy is

$$y = \dot{W}_d / \dot{m} = Eg = (u_2 v_{w_2} - u_1 v_{w_1}).$$

Thus, equating the two expressions, we obtain

$$H_{T_2} - H_{T_1} = u_2 v_{w_2} - u_1 v_{w_1}.$$

Using the usual assumption of no whirl at inlet ($v_{w_1} = 0$), the above equation reduces to

$$H_{T_2} - H_{T_1} = u_2 v_{w_2}. \tag{22.53}$$

This relationship does not take into account any losses due to friction. If, however, the actual work is equated to Euler's equation instead of the isentropic work, we obtain

$$H_{T_3} - H_{T_1} = u_2 v_{w_2}.$$

Hence, by using equation (22.51), we get

$$u_2 v_{w_2} = (H_{T_2} - H_{T_1})/\eta_i. \tag{22.54}$$

This equation allows for the fact that the given impeller speed and fluid whirl velocity produce a greater change of enthalpy than that anticipated by the ideal frictionless conditions.

The problem now arises of how to account for changes in velocities through the impeller or around the impeller blades and in their lift and drag properties caused by compressibility. It is possible to use as an approximation the Prandtl–Glauert similarity rules, which apply to an isolated aerofoil at small angles of incidence. They state that the aerodynamic performance in compressible flow may be related to that in incompressible flow by the factor

$$\lambda = (1 - Ma_0^2)^{-1/2}, \tag{22.55}$$

where Ma_0 is the Mach number of the free stream. The factor λ is obtained by a transformation of a linearized, compressible flow, second-order differential equation in velocity potential to Laplace's equation for incompressible, two-dimensional, steady potential flow and is beyond the scope of this book.

Thus, for the same aerofoil, if u is the velocity at any point on its profile in an incompressible flow and the velocity at the same point in the compressible flow is u', then

$$u'/u = (1 - Ma_0^2)^{-1/2}$$

or $$u' = \lambda u. \tag{22.56}$$

Similarly,

$$C_L'/C_L = (1 - Ma_0^2)^{-1/2}$$

or $$C_L' = \lambda C_L. \tag{22.57}$$

It is also possible to consider two aerofoils, one in an incompressible flow (suffix i) and the other in a compressible flow (suffix c), such that their geometries are related by

$$(\text{Camber})_c = \lambda (\text{Camber})_i \tag{22.58}$$

and $$(\text{Maximum thickness})_c = \lambda (\text{Maximum thickness})_i. \tag{22.59}$$

For such related aerofoils, similarity rules due to Goethert are

$$\alpha_c = \lambda \alpha_i \quad \text{for angles of incidence,} \tag{22.60}$$

$$(C_L)_c = \lambda^2 (C_L)_i \quad \text{for lift coefficients.} \tag{22.61}$$

The Prandtl–Glauert and Goethert rules apply only to two-dimensional flow and may be used for Ma < 0.8. This limitation corresponds approximately to a likelihood of the Mach number becoming equal to unity at some point on the aerofoil. This will occur at a point of minimum pressure and, hence, maximum velocity. When the Mach number exceeds unity at such a point, a shock will be formed and, hence, the assumption of potential flow on which the Prandtl–Glauert similarity is based ceases to be valid.

The formation of a shock causes an immediate increase in losses because it constitutes an adverse pressure gradient and, therefore, induces boundary layer separation.

It is also possible for the velocity at some cross-section of the interblade passage to become equal to that of sound. When that happens, the flow becomes choked and it is not possible to increase the flow rate beyond the critical value corresponding to the choked flow.

EXAMPLE 22.3

A two-dimensional aerofoil has a ratio of maximum thickness to chord equal to 0.035 and a camber to chord ratio of 0.015. It is tested in a low-speed wind tunnel and gives a lift coefficient of 0.6 at an angle of incidence of 3°. What would be the lift coefficient of this aerofoil at Ma = 0.6? What would be the geometric characteristics and the lift coefficient of a related aerofoil at Ma = 0.6?

Solution

Assume that the conditions at the low-speed wind tunnel correspond to incompressible flow. For the same aerofoil, equation (22.57) holds and, therefore,

$$C'_L = \lambda C_L,$$

$$\lambda = (1 - Ma^2)^{-1/2} = [1 - (0.6)^2]^{-1/2} = 1.25.$$

Therefore,

$$C'_L = 1.25 \times 0.6 = \mathbf{0.75}.$$

A similar aerofoil would have a thickness to chord ratio of

$$0.035 \times [1 - (0.6)^2]^{-1/2} = \mathbf{0.043\,75}$$

and a camber to chord ratio of

$$0.015 \times [1 - (0.6)^2]^{-1/2} = \mathbf{0.018\,75}.$$

Its coefficient of lift would be

$$(C_L)_c = 0.6/(1 - Ma^2) = 0.6/0.64 = \mathbf{0.9375}$$

at an angle of incidence

$$\alpha_c = 3/(1 - Ma^2)^{1/2} = \mathbf{3.75°}.$$

Concluding remarks

The theory necessary to represent the performance of rotodynamic machines has been presented in this chapter. In doing so this treatment has been based upon earlier material, in particular the continuity of mass flow and the energy equation. The identification of losses within the machine derives from earlier work on aerofoil theory.

The remaining chapters will develop this theory to allow the performance of real machines, subject to inefficiencies, to be considered (Chapter 23) and the interface between the machine and the system to be modelled (Chapter 25).

Summary of important equations and concepts

1. This chapter emphasizes the importance of a full understanding of the flow and blade velocity relationships within a rotodynamic machine. The basic definition of torque, equation (22.1), is followed by the development of Euler's analysis of machine/fluid energy transfer, for both centrifugal and axial machines, equations (22.8) and (22.12).

2. The effect of blade pitch and solidity, defined in equation (22.15) as the ratio of blade chord to pitch, is addressed in Section 22.3, which introduces circulation, equation (22.16) and refers back to the Kutta–Joukowsky potential flow analysis of Chapter 7. Lift coefficient for the blade in terms of angle of attack is developed in equation (22.31) and a cascade coefficient is defined in equation (22.32) as the ratio of cascade to aerofoil lift coefficients, leading to the expression for theoretical energy transfer in terms of cascade lift coefficient and blade solidity, equation (22.33).

3. Real machines fail to deliver the theoretical energy transfer values developed due to the three-dimensional nature of flow in the blade passages reducing the whirl component, an effect not dependent on friction that follows from the ideal flow analysis of pressure and velocity distributions, and the friction, separation and wake phenomena associated with the boundary layers developed within the machine and interblade passages. These effects are investigated separately in Section 22.4 where slip factor is introduced, equation (22.34) and the zones of application, the Stodola, Busemann and Stanitz corrections are developed. Frictional losses are treated separately through cascade efficiency, equations (22.43) and (22.44).

4. Compressibility effects are discussed in Section 22.5 based on an application of the steady flow energy equation, equation (22.45), the development of an isentropic efficiency, equation (22.51) and reference to the Euler equation to determine the change in enthalpy generated in terms of blade and whirl velocities and isentropic efficiency, equation (22.54).

Further reading

Betz, A. (1966). *Introduction to the Theory of Flow Machines*. Pergamon, New York.

Dixon, S. L. (1966). *Fluid Mechanics, Thermodynamics of Machinery*. Pergamon, Oxford.

Ferguson, T. B. (1963). *The Centrifugal Compressor Stage*. Butterworths, London.

Lazarkiewicz, S. and Troskolanski, A. F. (1965). *Impeller Pumps*. Pergamon, Oxford.

Shepherd, D. G. (1956). *Principles of Turbomachinery*. Macmillan, London.

Problems

22.1 A centrifugal fan delivering $2\,m^3\,s^{-1}$ of air (density $1.2\,kg\,m^{-3}$) runs at 960 rev min^{-1}. The impeller outer diameter is 70 cm and the inner diameter is 48 cm. The impeller width at inlet is 16 cm and is designed for constant radial flow velocity. The blades are backward inclined making angles of 22.5° and 50° with the tangents at inlet and outlet, respectively. Draw the inlet and outlet velocity triangles and determine the theoretical head produced by the impeller.
[91.1 m of air]

22.2 A centrifugal pump delivers 0.3 m^3 s^{-1} of water at 1400 rev min^{-1}. The total effective head is 20 m. The impeller is 30 cm in diameter and 32 mm wide at exit, and is designed for constant velocity of flow. Both the suction and delivery pipes have the same bore. Calculate the following vane angles: (a) for the impeller vanes at exit; (b) for entry to the stationary guide vanes surrounding the impeller.
[(a) 37.28°, (b) 48.12°]

22.3 A centrifugal fan supplies air at a rate of 4.5 m^3 s^{-1} and total head of 100 mm of water. The outer diameter of the impeller is 50 cm and the outer width is 18 cm. The blades are backward inclined and of negligible thickness. If the fan runs at 1800 rev min^{-1} and assuming that the conversion of velocity head to pressure head in the volute is counterbalanced by the friction losses there and in the runner, determine the blade angle at outlet. Assume zero whirl at inlet and take air density as 1.23 kg m^{-3}. [27.8°]

22.4 When working at its best efficiency point, the blading at the mean radius, equal to 300 mm, of an axial flow pump deflects a stream approaching it at a relative angle of 60° to the axis through 15°, so that the water leaves it at a relative angle of 45°. Assuming that the water approaches it axially, and that the velocity of flow remains constant, draw the inlet and outlet velocity triangles under these conditions for a rotational speed of 600 rev min^{-1} and calculate the theoretical total head rise through the impeller. [15.3 m]

22.5 An axial flow pump operates at 500 rev min^{-1}. The outer diameter of the impeller is 750 mm and the hub diameter is 400 mm. At the mean blade radius, the inlet blade angle is 12° and the outlet blade angle is 15°, both measured with respect to the plane of impeller rotation. Sketch the corresponding velocity diagrams at inlet and outlet and estimate from them (a) the head generated by the pump, (b) the rate of flow through the pump, (c) the shaft power consumed by the pump. Assume hydraulic efficiency of 87 per cent and overall efficiency of 70 per cent.
[(a) 4.12 m, (b) 1.01 m^3 s^{-1}, (c) 58.3 kW]

22.6 Show that for a 'free vortex' flow through an axial flow impeller the circulation round the blade does not vary with radius. An axial flow fan delivers 20.0 m^3 s^{-1} of air and its major parameters are: rotational speed $N = 720$ rev min^{-1}; impeller diameter $D_2 = 1.00$ m; hub diameter D_1 = 0.45 m; number of blades $Z = 10$. The blades are of aerofoil cross-section which, for the optimum angle of incidence $i = 5°$, have the lift coefficient $C_L = 0.80$ and the chord at hub is $C_h = 70$ mm. Using the isolated blade theory and assuming 'free vortex' flow and constant velocity of flow, determine the total head rise across the impeller, the blade angle and chord length at tip and the blade angular twist between hub and tip. [12.38 m of air, 40.26°, 51 mm, 21.8°]

22.7 A downstream, guide vane, axial flow pump, 0.6 m in diameter and running at 950 rev min^{-1}, is to deliver 0.75 m^3 s^{-1} at a total head of 16 m. If the hub ratio is 0.6 and the blade solidity at hub and tip is 1.0 and 0.55, respectively, determine the blade angles at hub and tip and the guide vane inlet angles. Use the following aerofoil data:

Angle of incidence:	1.0°	4.0°	7.0°	10°	11°
Coefficient of lift:	0.46	0.87	1.16	1.39	stall

[11.35°, 24.85°, 51.8°, 64.7°]

22.8 The impeller of a centrifugal pump rotates at 1450 rev min^{-1} and is of 0.25 m diameter and 20 mm width at outlet. The blades are inclined backwards at 30° to the tangent at outlet and the whirl slip factor is 0.77. If the volumetric flow rate is 0.028 m^3 s^{-1} and neglecting shock losses and whirl at inlet, find the theoretical head developed by the impeller. Also, using Stodola's model of relative eddy, find the number of blades on the impeller.
[23.7 m, eight blades]

22.9 Show that for a centrifugal pump, neglecting losses, the condition for maximum efficiency is

$$u_2 = 2V_{f_2}/\tan\beta_2,$$

where u_2 is the blade peripheral speed at outlet, V_{f_2} is the outlet velocity of flow and β_2 is the blade angle at outlet measured with respect to the tangent.

A centrifugal pump with an impeller diameter of 10 cm and an axial width of 1.5 cm has swept-back blades inclined at 25° to the tangent to the periphery. If the impeller speed is 12.4 rev s^{-1} calculate the flow rate when the pump is operating at maximum efficiency. Assume zero swirl at inlet.
[0.0043 m^3 s^{-1}]

23

Performance of Rotodynamic Machines

THE OUTPUT FROM A ROTODYNAMIC MACHINE IS SHOWN TO BE DEPENDENT UPON A series of variables, including both machine and fluid properties. The concept of a machine characteristic is introduced, based upon a dimensional analysis, utilizing techniques already detailed in Chapter 8. Machine losses are identified and the variation of these losses with throughflow explained. The machine efficiency is introduced and shown to have a maximum value that may be used to guide machine selection. The dimensional analysis leads naturally to the application of the laws of similarity and the effect on machine performance of changes in speed and impeller diameter are discussed and demonstrated by means of the computer program introduced. Scale effects are considered in the application of the similarity laws. The effect of blade angle on the performance of fans and pumps is also illustrated, as is the relative performance of centrifugal, axial and mixed flow machines. The performance of turbines, including the Pelton wheel introduced in the earlier treatment of the momentum equation (Chapter 5), and centrifugal and axial machines is discussed in terms of performance characteristics and relative fluid velocities within the machine, together with the operation of hydraulic transmissions and fluid couplings. ● ● ●

23.1 THE CONCEPT OF PERFORMANCE CHARACTERISTICS

In the introduction to this part of the book a distinction was made between hydraulic machines in which work is done on the fluid (pumps, fans) and those machines in which work is done by the fluid and, therefore, energy is subtracted from it (turbines, motors). It was further stated that the process of energy transfer may be accomplished by either a 'positive displacement' action or by 'rotodynamic' action. In the first case, a volume of fluid fixed by the dimensions of the machine enters and leaves it at a frequency determined by the speed of operation of the machine. In the second case the flow is continuous through an impeller whose torque is equal to the rate of change of angular momentum of the fluid, as was shown in Chapter 22.

The fluid quantities involved in all hydraulic machines are the flow rate (Q) and the head (H), whereas the mechanical quantities associated with the machine itself are the power (P), speed (N), size (D) and efficiency (η). Although they are all of equal importance, the emphasis placed on certain of these quantities is different for pumps and for turbines. The output of a pump running at a given speed is the flow rate delivered by it and the head developed. Thus, a plot of head against flow rate at constant speed forms the fundamental performance characteristic of a pump. In order to achieve this performance, a power input is required which involves efficiency of energy transfer. Thus, it is useful to plot also the power P and the efficiency η against Q. Such a complete set of performance characteristics of a rotodynamic pump is shown in Fig. 23.1.

FIGURE 23.1

Typical pump characteristic

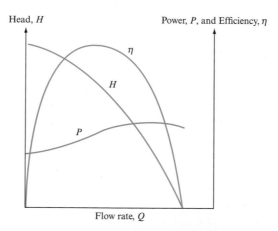

In the case of turbines, the output is the power developed at a given speed and, hence, the fundamental turbine characteristic consists of a plot of power against speed at constant head. The input in this case is the fluid flow rate and, therefore, this quantity as well as the efficiency is usually plotted against the speed to complete the set of turbine characteristics.

The performance characteristics, then, represent in a graphical form the relationships between the variables relevant to an hydraulic machine. Each and every hydraulic machine has its own set of characteristics which represent its performance.

For rotodynamic machines the concept of a characteristic follows directly from Euler's equation. It was shown in Chapter 22 (equations (22.8) and (22.12)) that the theoretical energy transfer per unit weight of fluid flowing through the machine, or the fluid head, may be given by

$$H = (u_2/g)(u_2 - v_{f_2} \cot \beta_2). \tag{23.1}$$

But $v_{f_2} = Q/A_2$, where A_2 is the impeller outlet area. Substituting,

$$H = (u_2/g)[u_2 - (Q/A_2)\cot \beta_2]$$

or $\quad H = u_2^2/g - (u_2/A_2g)Q \cot \beta_2.$

This equation may, for a constant speed of rotation and given impeller diameter so that $u_2 = $ constant and $A_2 = $ constant, be rewritten in a general form:

$$H = K_1 - K_2 Q \cot \beta_2. \tag{23.2}$$

It is, thus, seen that there is a definite functional relationship between the head and flow rate of a rotodynamic machine. This relationship constitutes the performance characteristic and is determined experimentally by performance tests. Thus, a pump, for example, will generate a head dependent upon the quantity of fluid it is handling. Furthermore, since for machines of different design features and different sizes the values of K_1, K_2 and β_2 will be different, their characteristics will also be different.

Equation (23.2) also shows the importance of the blade outlet angle β_2. This particular point will be discussed in greater detail in connection with centrifugal pumps in Section 23.7.

23.2 LOSSES AND EFFICIENCIES

All hydraulic machines convert energy from one form into another and it is a well-known fact that, in any energy conversion process, losses occur. Thus, hydraulic machines also suffer from losses of energy. How small these losses are or how good a machine is in converting energy is indicated by its efficiency. The efficiency of a machine is always defined as the ratio of the power output of the machine to the power input into it. However, hydraulic machines are complex, consisting of a number of parts through which the fluid moves and, thus, it is convenient for analytical and design purposes to consider component losses as well as their sum total and to express each component loss in the efficiency form.

Let us now consider these component losses one by one. First, the actual energy transfer in a rotodynamic machine occurs in its impeller. Here the fluid passes through the blade passages and either receives energy from the moving blades or imparts energy to them. In any case, there are two major sources of energy loss within the impeller. The inevitable contact between the fluid moving over solid surfaces gives rise to boundary layer development and, hence, to frictional losses, whereas the need for the fluid to change direction often results in separation and, hence, leads to separation (or shock) losses. Both these losses may be augmented by secondary flows which may occur within the impeller due to a pressure distribution across it and are usually prominent at off-design points of operation.

Thus if h_i is the head loss in the impeller and Q_i is the volumetric flow rate through the impeller, then the impeller power loss is

$$P_i = \rho g h_i Q_i. \tag{23.3}$$

Now, the flow rate through the impeller Q_i is usually not the same as that flowing through the machine, simply because some fluid passes through clearances between the impeller and the casing. In a pump, for example, of all the fluid passing through the impeller most flows into the discharge end but some passes through the inlet clearance and finds itself passing through the impeller again. Thus, the impeller always handles a greater volume than that discharged by the pump.

If we denote by q the volumetric flow rate leaking past the impeller and if H_i is the total head across the impeller, then the power loss due to the leakage may be expressed as

$$P_1 = \rho g H_i q. \tag{23.4}$$

In most machines the impeller is surrounded by a stationary casing so that the fluid passes through parts of the casing before it enters the impeller and after leaving it. Thus, losses due to friction (and possibly due to separation) occur in the casing as well. If the flow rate through the casing and, thus, through the machine is Q (greater or smaller than Q_i depending on whether it is a turbine or a pump, the difference being q) and the loss of head in the casing is h_c, then the power loss in the casing is

$$P_c = \rho g h_c Q. \tag{23.5}$$

Finally, there are mechanical losses of energy such as in the bearings and sealing glands which must be accounted for. It is normal practice in hydraulic machines to include within this category losses due to disc friction, sometimes referred to as 'windage' loss. This is the power required to spin the impeller at the required velocity without any work being done by the impeller or on the impeller by the fluid. This would be possible only if the impeller did not have any blades. Thus, windage loss accounts for the friction between the outer surfaces of the impeller rotating in a fluid surrounding it within the casing.

It is now possible to consider the energy balance for the whole machine, but here we must begin to distinguish between pumps and turbines because what represents the output of one is the input of the other and vice versa. For a pump:

$$
\underset{\substack{\text{Shaft}\\\text{power}\\\text{input}}}{P} = \underset{\substack{\text{Mechanical}\\\text{loss}}}{P_m} + \underset{\substack{\text{Impeller}\\\text{loss}}}{\rho g (h_i Q_i} + \underset{\substack{\text{Leakage}\\\text{loss}}}{H_i q} + \underset{\substack{\text{Casing}\\\text{loss}}}{h_c Q} + \underset{\substack{\text{Useful}\\\text{fluid}\\\text{power}}}{HQ).}
$$

$$\underbrace{\hphantom{\rho g (h_i Q_i + H_i q + h_c Q)}}_{\text{Hydraulic losses}} \tag{23.6}$$

For a turbine:

$$
\underset{\substack{\text{Fluid}\\\text{power}\\\text{input}}}{\rho g H Q} = \underset{\substack{\text{Mechanical}\\\text{loss}}}{P_m} + \underset{\substack{\text{Impeller}\\\text{loss}}}{\rho g (h_i Q_i} + \underset{\substack{\text{Casing}\\\text{loss}}}{h_c Q} + \underset{\substack{\text{Leakage}\\\text{loss}}}{H_i q)} + \underset{\substack{\text{Shaft}\\\text{power}\\\text{output}}}{P.}
$$

$$\underbrace{\hphantom{\rho g (h_i Q_i + h_c Q + H_i q)}}_{\text{Hydraulic losses}} \tag{23.7}$$

FIGURE 23.2

Energy balance for a
pump and summary of
efficiencies

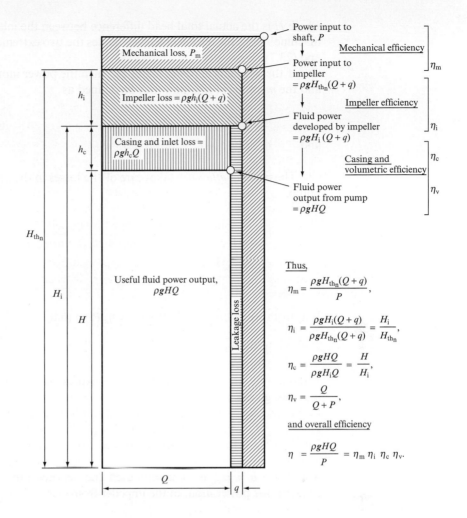

Figure 23.2 represents graphically the energy balance for a pump. A similar diagram may be constructed for a turbine.

Having discussed all the component losses in an hydraulic machine and the complete energy balance, it is now possible to define efficiencies.

The most important is the *overall efficiency*. It refers to the machine as a whole and is, therefore, always plotted as one of the performance characteristics. It is defined as:

$$\eta = \frac{\text{Power output of the machine}}{\text{Power input to the machine}}.$$

Hence, for a pump,

$$\eta = \frac{\text{Fluid power output}}{\text{Power input to shaft}} = \frac{\rho g H Q}{P} \tag{23.8}$$

and, for a turbine,

$$\eta = \frac{\text{Power output from shaft}}{\text{Fluid power input}} = \frac{P}{\rho g H Q}, \tag{23.9}$$

where H is the actual total head difference between the inlet and outlet flanges of the machine. Thus, the overall efficiency relates the two extreme terms of equations (23.6) and (23.7).

If the mechanical power loss is P_m, then the power input to the impeller is $P - P_m$ and the *mechanical efficiency* may be defined as

$$\eta_m = (P - P_m)/P \quad \text{for a pump} \tag{23.10}$$

and $\quad \eta_m = P/(P + P_m) \quad \text{for a turbine.} \tag{23.11}$

The *impeller efficiency* takes care of the losses in the impeller and, therefore, for a pump,

$$\eta_i = \frac{\text{Fluid power developed by impeller}}{\text{Mechanical power supplied to impeller}}$$

$$= \frac{\rho g H_i Q_i}{(P - P_m)}.$$

But, from equation (23.10), $P - P_m = \eta_m P$, so that

$$\eta_i = \rho g H_i Q_i / \eta_m P. \tag{23.12}$$

Alternatively, the denominator may be expressed in terms of the fluid loss in the impeller, giving

$$\eta_i = \frac{\rho g H_i Q_i}{\rho g H_i Q_i + \rho g h_i Q_i} = \frac{H_i}{H_i + h_i} = \frac{H_i}{H_{th_n}}, \tag{23.13}$$

where $H_i + h_i = H_{th_n}$ is sometimes used and denotes a theoretical total head deduced from the net power input to the impeller from

$$\eta_m P = \rho g Q_i (H_i + h_i) = \rho g Q_i H_{th_n}. \tag{23.14}$$

For a turbine,

$$\eta_i = \frac{\text{Mechanical power received by shaft}}{\text{Fluid power supplied to impeller}}$$

$$= (P + P_m)/\rho g H_i Q_i = P/\rho g H_i Q_i \eta_m = H_{th_n}/H_i. \tag{23.15}$$

The *volumetric efficiency* is not always appropriate, e.g. in axial flow machines, but is of importance in the case of centrifugal pumps and especially fans. In general, for pumps,

$$Q = Q_i - q \tag{23.16}$$

and, for turbines,

$$Q = Q_i + q, \tag{23.16a}$$

so that, for pumps, the volumetric efficiency is defined as

$$\eta_v = \frac{\text{Flow rate through machine}}{\text{Flow rate through impeller}}$$

$$= Q/Q_i = Q/(Q+q) \tag{23.17}$$

and, for turbines,

$$\eta_v = \frac{\text{Flow rate through impeller}}{\text{Flow rate through machine}}$$

$$= Q_i/Q = (Q-q)/Q. \tag{23.17a}$$

The *casing efficiency* accounts for the power loss in the casing. For a pump,

$$\eta_c = \frac{\text{Useful fluid power output}}{\text{Fluid power developed by impeller} - \text{Leakage loss}}$$

$$= \frac{\rho g H Q}{\rho g H_i Q_i - \rho g H_i q} = \frac{HQ}{H_i(Q_i - q)} = \frac{H}{H_i}. \tag{23.18}$$

For a turbine,

$$\eta_c = \frac{\text{Fluid power supplied to impeller} + \text{Leakage loss}}{\text{Fluid power received by casing}}$$

$$= \frac{\rho g H_i Q_i + \rho g H_i q}{\rho g H Q} = \frac{H_i(Q_i + q)}{HQ} = \frac{H_i}{H}. \tag{23.18a}$$

It is now possible to show that the overall efficiency is equal to the product of all the component efficiencies,

$$\eta = \eta_m \eta_i \eta_c \eta_v, \tag{23.19}$$

by substituting into the above the appropriate expression as follows:

$$\eta = \frac{P - P_m}{P} \times \frac{\rho g H_i Q_i}{P - P_m} \times \frac{H}{H_i} \times \frac{Q}{Q+q},$$

which simplifies to

$$\eta = \frac{\rho g H Q_i}{P} \times \frac{Q}{Q+q}.$$

But, since $Q + q = Q_i$, we obtain

$$\eta = \rho g H Q / P,$$

which is the expression (23.8) for the overall efficiency.

The internal losses of the machine, i.e. those occurring in the impeller and in the casing due to friction and separation, are sometimes called hydraulic losses, which give rise to the *hydraulic efficiency*, defined as

$$\eta_h = \frac{\text{Actual head}}{\text{Theoretical head}} = \frac{H}{H_{th_n}} = \eta_i \eta_c, \tag{23.20}$$

where H_{th_n} is the theoretical head calculated from the net power input to the impeller. Thus, equation (23.19) may be rewritten as

$$\eta = \eta_m \eta_h \eta_v. \tag{23.21}$$

The theoretical head calculated from Euler's equation (H_{th}) is not the same as the theoretical head calculated from the net power input (H_{th_n}) used in equation (23.14). The difference is due to 'head slip' discussed in Chapter 22. Thus, if the slip factor $S_F = v'_{w_2}/v_{w_2}$ (equation (22.34)) is introduced into Euler's equation,

$$H_{th} = u_2 v_{w_2}/g$$

(equation (22.2) for $v_{w_1} = 0$), then

$$H_{th_n} = u_2 v'_{w_2}/g = H_{th} S_F. \tag{23.22}$$

Thus, if the slip factor is known, the net theoretical head H_{th_n} may be calculated from Euler's head.

It is now possible to relate the theoretical characteristic obtained from Euler's equation to the actual characteristic by accounting for various losses responsible for the difference. The theoretical characteristic for a given blade angle at outlet β_2 is a straight line determined by equation (23.2), as shown in Fig. 23.3.

The use of the slip factor, which varies with flow rate, enables the H_{th_n} curve to be obtained. This represents the net head developed by the impeller, but, as discussed earlier, does not account for losses. These, for the machine as a whole, may be considered separately under the following categories:

1. *Shock losses*, which occur at the entry to the impeller, to the guide vanes, etc., especially at off-design operating conditions, may be simply expressed as

$$h_{sh} = k(Q - Q_N)^2, \tag{23.23}$$

where Q_N is the volumetric flow rate corresponding to the maximum efficiency point on the characteristic. Equation (23.23) assumes, therefore, that shock losses are zero at $Q = Q_N$. It is a parabola, which has a minimum at this point, as shown in Fig. 23.3.

2. *Friction losses*, which account for energy dissipation due to contact of the fluid with solid boundaries such as stationary vanes, impeller, casing, etc., are usually expressed in the form

$$h_f = k'Q^2, \tag{23.24}$$

where k' is a constant for a given machine. If it is assumed that when the machine operates at its maximum efficiency the shock losses are zero, then the hydraulic loss of head becomes equal to the friction losses at this point. Thus,

$$(H_{th_n} - H)_{h_{sh=0}} = (h_i + h_c)_{h_{sh=0}} = h_f.$$

FIGURE 23.3
Losses and characteristics
for a centrifugal fan

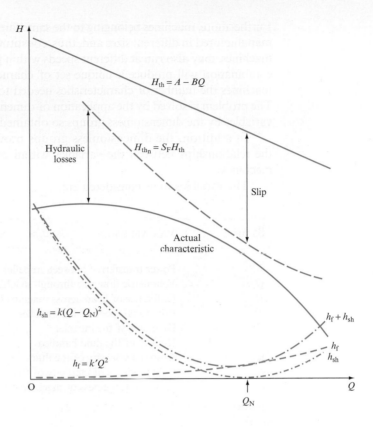

This relationship enables the value of k' to be established approximately. Equation (23.24) represents a parabola passing through the origin, also shown in Fig. 23.3.

The sum of the shock and friction losses when subtracted from the H_{th_n} curve gives the actual head/flow rate characteristic provided it is plotted against Q, the flow rate through the machine, and not $(Q + q)$, which represents the flow rate through the impeller. The disparity between these two quantities, defined by the volumetric efficiency, is usually dealt with during the determination of the H_{th_n} curve.

3. *The mechanical losses*, which usually include 'disc friction' or 'windage' loss due to the rotation of a 'bladeless' impeller, i.e. supporting discs or hub only, and bearing losses, do not affect the head/flow rate characteristic but only the power input and, hence, the overall efficiency.

Figure 23.3 shows the actual head/flow rate characteristic obtained by subtracting the hydraulic losses from the H_{th_n} curve.

23.3 DIMENSIONLESS COEFFICIENTS AND SIMILARITY LAWS

23.3.1 Dimensionless coefficients

The actual performance characteristics of rotodynamic machines have to be determined by experimental testing, and different machines have different characteristics.

Furthermore, machines belonging to the same family, i.e. being of the same design but manufactured in different sizes and, thus, constituting a series of geometrically similar machines, may also run at different speeds within practical limits. Each size and speed combination will produce a unique set of characteristics, so that for one family of machines the number of characteristics needed to be determined is impossibly large. The problem is solved by the application of dimensional analysis and by replacing the variables by the dimensionless groups so obtained.

In addition, the dimensionless groups provide the similarity laws governing the relationships between the variables within one family of geometrically similar machines.

The variables to be considered are:

SYMBOL	VARIABLE	DIMENSIONS
P	Power transferred between impeller and fluid	ML^2T^{-3}
Q	Volumetric flow rate through machine	L^3T^{-1}
H	Difference of head across machine ($= E$)	L
N	Rotational speed of the impeller	T^{-1}
D	Diameter of the impeller	L
ρ	Density of the fluid handled	ML^{-3}
μ	Absolute viscosity of the fluid	$ML^{-1}T^{-1}$
K	Bulk modulus of elasticity of the fluid	$ML^{-1}T^{-2}$
ε	Absolute roughness of machine's internal passages	L

Since the head H is the energy per unit weight of the fluid, it is appropriate to consider (gH) as a variable because it represents energy per unit mass, or specific energy y, which is more fundamental because it is independent of gravitational acceleration. Pumps, for example, develop the same specific energy (or work per unit mass of the fluid flowing) irrespective of the gravitational force.

Considering first the specific energy (head developed) as the dependent variable, the relationship between the variables may be written as

$$gH = \phi(Q, N, D, \rho, \mu, K, \varepsilon).$$

Using the indicial method, the power series reduces to

$$gH = kQ^a N^b D^c \rho^d \mu^e K^f \varepsilon^i$$

and substituting dimensions,

$$\frac{L^2}{T^2} = \left[\frac{L^3}{T}\right]^a \left[\frac{1}{T}\right]^b [L]^c \left[\frac{M}{L^3}\right]^d \left[\frac{M}{LT}\right]^e \left[\frac{M}{LT^2}\right]^f [L]^i.$$

Equating indices,

for [M]: $0 = d + e + f$,   therefore, $d = -e - f$;
for [T]: $-2 = -a - b - e - 2f$,   therefore, $b = 2 - a - e - 2f$;
for [L]: $2 = 3a + c - 3d - e - f + i$, therefore, $c = 2 - 3a - 2e - 2f - i$.

Substituting into the original equation,

$$gH = kQ^a N^{(2-a-e-2f)} D^{(2-3a-2e-2f-i)} \rho^{(-e-f)} \mu^e K^f \varepsilon^i$$

$$= kN^2 D^2 \left(\frac{Q}{ND^3}\right)^a \left(\frac{\mu}{ND^2\rho}\right)^e \left(\frac{K}{N^2 D^2 \rho}\right)^f \left(\frac{\varepsilon}{D}\right)^i.$$

Therefore,

$$\frac{gH}{N^2 D^2} = \phi\left[\left(\frac{Q}{ND^3}\right); \left(\frac{\mu}{ND^2\rho}\right); \left(\frac{K}{N^2 D^2 \rho}\right); \left(\frac{\varepsilon}{D}\right)\right]. \tag{23.25}$$

Now, $gH/N^2 D^2$ is the *head coefficient* K_H, and Q/ND^3 is the *flow coefficient*, K_Q. Also, since $ND \propto u$ it follows that

$$\frac{\mu}{ND^2\rho} \propto \frac{\mu}{uD\rho} \propto \frac{1}{\mathrm{Re}},$$

where Re is the Reynolds number based on impeller diameter. Also, since $\sqrt{(K/\rho)} = c$ (equation (5.30)) it follows that

$$\frac{K}{N^2 D^2 \rho} \propto \frac{c^2}{u^2} \propto \frac{1}{\mathrm{Ma}}.$$

Thus equation (23.25) may be rewritten as

$$K_H = \phi(K_Q, \mathrm{Re}, \mathrm{Ma}, \varepsilon/D), \tag{23.25a}$$

where ε/D is the relative roughness of the machine's internal passages.

Similarly, if the power is taken as the dependent variable the relationship between the variables may be written as

$$P = \phi(Q, N, D, \rho, \mu, K, \varepsilon),$$

and leads to

$$\frac{P}{N^3 D^5 \rho} = \phi\left[\left(\frac{Q}{ND^3}\right); \left(\frac{\mu}{ND^2\rho}\right); \left(\frac{K}{N^2 D^2 \rho}\right); \left(\frac{\varepsilon}{D}\right)\right]. \tag{23.26}$$

Now, $P/N^3 D^5 \rho$ is the *power coefficient*, K_P, and therefore the above equation may be restated as

$$K_P = \phi(K_Q, \mathrm{Re}, \mathrm{Ma}, \varepsilon/D). \tag{23.26a}$$

The functional relationships between K_H, K_P and K_Q are determinable by experiment and constitute a set of performance characteristics, which are of the same shape as the

H and P vs. Q characteristics, but represent the whole family of geometrically similar machines and are identical for all such machines if Re, Ma and relative roughness are the same.

23.3.2 Similarity laws

Since for all machines belonging to one family and operating under dynamically similar conditions the dimensionless coefficients are the same at corresponding points of their characteristics, it follows that the similarity laws governing the relationships between such corresponding points may be stated as follows:

$$\text{since } K_Q = Q/ND^3 = \text{constant}, \qquad Q \propto ND^3, \tag{23.27}$$

$$\text{since } K_H = gH/N^2D^2 = \text{constant}, \quad gH \propto N^2D^2, \tag{23.28}$$

$$\text{since } K_P = P/\rho N^3D^5 = \text{constant}, \quad P \propto \rho N^3D^5, \tag{23.29}$$

provided that Re, Ma and ε/D are also the same. It will also be shown, using a particular example, which follows, that

$$\eta = \text{constant}. \tag{23.30}$$

To illustrate the way in which the similarity laws are used in predicting the performance of a machine of a given size and running at a given speed from the known performance characteristics of a geometrically similar machine, consider a centrifugal pump whose characteristics when operating at a constant speed N_1 are as shown by the full lines in Fig. 23.4. Let it be required to establish the performance characteristics of the same pump but running at a faster speed N_2 (broken lines).

If, at N_1, the pump is operating at point X such that it delivers Q_X, generates head H_X and consumes power P_X at efficiency η_X, the corresponding point at speed N_2, marked X′, will be obtained by applying simultaneously the similarity laws as follows. From equation (23.27),

$$Q_X/N_1D^3 = Q_{X'}/N_2D^3,$$

but, since for the same pump the diameter D is the same,

$$Q_{X'} = Q_X N_2/N_1.$$

Similarly, from equation (23.28),

$$H_{X'} = H_X(N_2/N_1)^2.$$

Thus, plotting $H_{X'}$ against $Q_{X'}$, the point X′ on the new characteristic is obtained. The power required at N_2 follows from equation (23.29),

$$P_{X'} = P_X(N_2/N_1)^3,$$

which establishes a point on the power curve.

Now, the overall efficiency of the pump is defined as the ratio of the fluid power to the mechanical power supplied. The fluid power = $\rho g Q H$ (equation (6.30)), so that the efficiency,

$$\eta = \rho g H Q / P. \tag{23.31}$$

Let us now apply this expression to points X and X′ corresponding to speeds N_1 and N_2. At N_1,

$$\eta_X = \rho g Q_X H_X / P_X.$$

At N_2, $\eta_{X'} = \rho g Q_{X'} H_{X'} / P_{X'}.$

Dividing one expression by the other,

$$\frac{\eta_X}{\eta_{X'}} = \frac{Q_X}{Q_{X'}} \times \frac{H_X}{H_{X'}} \times \frac{P_{X'}}{P_X} = \frac{N_1}{N_2} \left(\frac{N_1}{N_2}\right)^2 \left(\frac{N_2}{N_1}\right)^3 = 1.$$

Thus, $\eta_X = \eta_{X'}$, which proves equation (23.30). However, although $\eta_{X'}$ is the same as η_X it is now plotted against $Q_{X'}$ so that its position on the graph is changed.

The procedure as outlined above for point X may be applied to other points on the characteristic curves such as Y and Z resulting in points Y′ and Z′, as shown in Fig. 23.4. Thus, a new set of characteristics corresponding to N_2 may be drawn by joining the 'primed' points. These new characteristic curves are shown dashed in the figure.

FIGURE 23.4

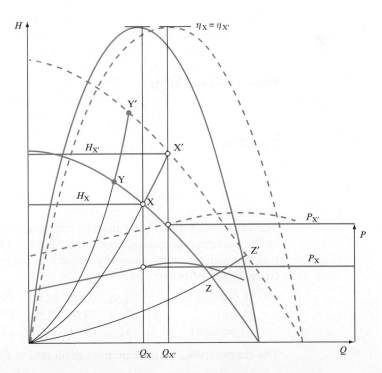

A change in pump size and, therefore, impeller diameter results also in a new set of characteristic curves, obtained by using equations (23.27) to (23.29), which yield the following relationships:

$$Q_{X''} = Q_X(D_2/D_1)^3, \quad H_{X''} = H_X(D_2/D_1)^2,$$

$$P_{X''} = P_X(D_2/D_1)^5, \quad \eta_{X''} = \eta_X.$$

Thus, the similarity laws enable us to obtain a set of characteristic curves for a machine from the known test data of a geometrically similar machine.

EXAMPLE 23.1

A centifugal pump, impeller diameter 0.50 m, when running at 750 rev min^{-1} gave on test the following performance characteristics:

Q (m^3 min^{-1})	0	7	14	21	28	35	42	49	56
H (m)		40.0	40.6	40.4	39.3	38.0	33.6	25.6	14.5	0
η (per cent)	0	41	60	74	83	83	74	51	0

Predict the performance of a geometrically similar pump of 0.35 m diameter and running at 1450 rev min^{-1}. Plot both sets of characteristics.

Solution

Let suffix 1 refer to the 0.5 m diameter pump and suffix 2 refer to the 0.35 m diameter pump. From equation (23.27),

$$Q_1/N_1 D_1^3 = Q_2/N_2 D_2^3.$$

Therefore,

$$Q_2 = Q_1(N_2/N_1)(D_2/D_1)^3$$
$$= Q_1(1450/750)(0.35/0.5)^3 = 0.663Q_1.$$

From equation (23.28),

$$H_1/N_1^2 D_1^2 = H_2/N_2^2 D_2^2.$$

Therefore,

$$H_2 = H_1(N_2/N_1)^2(D_2/D_1)^2$$
$$= H_1(1450/750)^2(0.35/0.50)^2 = 1.83H_1.$$

The values of Q_1 and H_1 are given by the table above. Therefore, by multiplying them by the multipliers calculated above, the values of Q_2 and H_2 may be tabulated. These, together with the same values of efficiency (equation (23.30)), constitute the predicted characteristic of pump 2 as follows:

Q_2 (m^3 min^{-1})	0	4.64	9.28	13.92	18.56	23.2	27.8	32.5	37.0
H_2 (m)		73.2	74.3	74.0	71.9	69.5	61.5	46.8	26.5	0
η (per cent)	0	41	60	74	83	83	74	51	0

The characteristics of both pumps are plotted in Fig. 23.5.

FIGURE 23.5

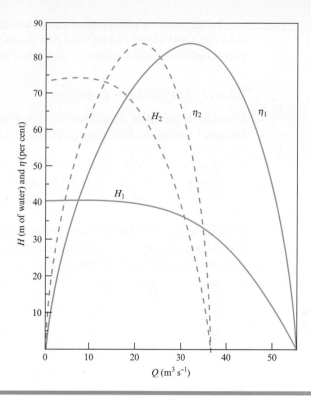

23.4 COMPUTER PROGRAM SIMPUMP

Equations (23.27) to (23.29) demonstrate the relationships that govern the flow and pressure characteristics for a family of geometrically similar fans or pumps when machine rotational speed or rotor diameter are changed. Thus if the machine performance is known at any one combination of speed and machine characteristic diameter, then its performance at another combination may be determined. Note that the pressure term is represented by the group gH/N^2D^2 but this may of course be recast for fans as $\Delta p/\rho N^2 D^2$, allowing the following procedure to be used for both fans and pumps, see also Sections 9.7 and 25.7, provided that the head term is expressed in metres of flowing fluid and that there is no air density change to be included between the given data condition and the condition at which the 'new' machine is to operate.

Computer program SIMPUMP accepts fan or pump characteristic data in one of the following forms:

1. flow rate Q (m³ s⁻¹), head H (m flowing fluid), power P (kW);
2. flow rate Q (m³ s⁻¹), head H (m flowing fluid), efficiency, η;
3. flow, head and power coefficient values – non-dimensional;
4. flow and head coefficient values – non-dimensional, and efficiency.

In addition the program requires the machine characteristic diameter and speed for which the base data above were known and the fluid density, assumed constant for both conditions.

In response to the input of the characteristic diameter and speed for the 'new' machine SIMPUMP displays the new characteristic in terms of both flow, head and power coefficients and efficiency and flow rate, head, power and efficiency.

23.4.1 Application example

Consider a pump of characteristic diameter $D = 0.4$ m and rotational speed 1500 rev min^{-1} having the following Q, H and η data:

Q (m^3 s^{-1})	0.05	0.10	0.15	0.2	0.25
H (m)	77.8	71.0	60.0	45.0	18.0
η (per cent)	66	79	78	60	12

For a geometrically similar pump under conditions of dynamic similarity, SIMPUMP determines that at a characteristic diameter of 0.75 m and 720 rev min^{-1} the pump performance characteristic is defined as:

KQ	KH	KP	η %
0.03	7.63	0.36	66
0.06	6.97	0.55	79
0.09	5.89	0.71	78
0.12	4.41	0.92	60
0.16	1.77	2.30	12

or

Q (m^3 s^{-1})	H (m)	P (kW)	η
0.16	63.02	148.19	66
0.32	57.51	225.96	79
0.47	48.60	290.10	78
0.63	36.45	377.13	60
0.79	14.58	942.82	12

23.4.2 Additional investigations using SIMPUMP

The simulation may be used to investigate:

1. data input in each of the four formats defined;
2. data input relevant to fans, remembering that the pressure rise across the fan should be input in metres of flowing fluid;
3. the application of the program to fans where there is a change in air density between the two machine conditions, hence introducing a density term into both the head and power coefficients.

23.5 SCALE EFFECTS

In the application of similarity laws as shown above, it was assumed that all criteria of dynamic similarity are satisfied, i.e. all the groups of equation (23.26) remain the same. This, however, is not true with regards to the dimensionless groups representing the Reynolds number, the Mach number and the relative roughness.

Consider first the Reynolds number, given by $\text{Re} = (ND^2\rho/\mu)$, which shows that a change of either speed or diameter alters the value of Re. Thus, in practice, $\text{Re} \neq$ constant. However, for water and air this effect is usually small because the values of Re are usually very high, the flow being fully turbulent.

A similar consideration of Mach number indicates that an increase of tip speed (by an increase either of N or of D) will make the Mach number higher. This not only means that one of the conditions of dynamic similarity is not satisfied but, in addition, may also mean that the compressibility effect (previously negligible) may now be of considerable importance. This second point must be watched carefully in the application of similarity laws to fans and compressors.

Consider now the effect of relative roughness. This again should be maintained constant, not only because it appears in equation (23.26) but also on account of geometrical similarity, which is the primary condition of any model laws to hold. Absolute roughness (ε) is the mean height of surface perturbances, which, therefore, remains the same for a given material and process used in the manufacture of the machine, irrespective of its size. Thus, any change of machine size involves a change of relative roughness (ε/D). On the whole, the larger the machine, the smaller the relative roughness will be. This will tend to make frictional losses relatively less important in larger machines.

In practice, it is also difficult to maintain geometrical similarity in clearances and some material thicknesses. The same gauge of sheet metal, for example, may be used for a range of sizes of fabricated impeller blades. Such deviations from geometrical similarity must obviously cause some departures from the idealized predictions based on the aforementioned similarity laws.

All such departures, which do occur in practice and which are due to the Reynolds number, Mach number, relative roughness or lack of strict geometrical similarity, are usually referred to as the *scale effect*. In general the scale effect tends to improve the performance of larger machines.

23.6 TYPE NUMBER

The performance of geometrically similar machines, i.e. machines belonging to one family, is governed by similarity laws and may be represented for the whole family by a single plot of dimensionless characteristics. Thus, the performance of machines belonging to different families may be compared by plotting their dimensionless characteristics on the same graph. Detailed comparison may then be achieved by analyzing the various aspects of the sets of curves. This method of comparison is satisfactory and often needed, but it lacks the brevity required in machine classification. This is obtained by the use of the *type number*, also known as the *specific speed*.

Every machine is designed to meet a specific duty, usually referred to as the *design point*. For a pump, for example, this would be stated in terms of the flow rate and the head developed and, thus, represents a particular point on its basic performance characteristic. The design point is normally associated with the maximum efficiency of the machine.

It is, thus, informative to compare machines by quoting the values of K_Q, K_H and K_P corresponding to their design points. However, since for pumps K_Q and K_H are the two most important parameters, their ratio would indicate the suitability of a particular pump for large or small volumes relative to the head developed. Furthermore, if the ratio is obtained in such a way that the impeller diameter is eliminated from it, then the comparison becomes independent of machine size. This is achieved by raising K_Q to the power 1/2 and K_H to the power 3/4. The result is the *type number*:

$$n_s = \frac{(K_Q)^{1/2}}{(K_H)^{3/4}} = \left(\frac{Q}{ND^3}\right)^{1/2}\left(\frac{N^2 D^2}{gH}\right)^{3/4}$$

$$= N\frac{Q^{1/2}}{(gH)^{3/4}} = N\frac{Q^{1/2}}{y^{3/4}}. \tag{23.32}$$

It must be realized that a value of type number can be calculated for any point on the characteristic curve. It will be equal to zero at point S in Fig. 23.6, because at that point the flow rate is zero, and will tend to infinity at point T because at large volumes head H tends to zero.

FIGURE 23.6

'Design point' on a pump characteristic

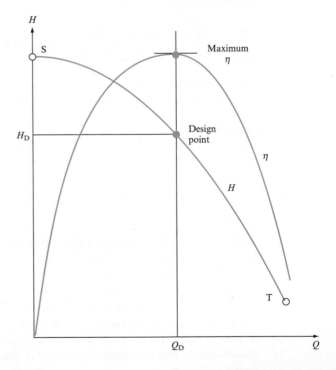

Such values are, however, of no practical interest and only the type number at the design point, usually referred to as *the* type number, is used for classification, comparison and design purposes.

The comparison of turbines is also achieved by the use of their type numbers. However, since for turbines the power developed is the most important variable, an alternative expression for type number in terms of power developed is obtained by eliminating D from the ratio of power and head coefficients. This is achieved by raising the power coefficient to the power of 1/2 and the head coefficient to the power of 5/4 and then taking their ratio:

$$n_s = \frac{(K_P)^{1/2}}{(K_H)^{5/4}} = \left(\frac{P}{\rho N^3 D^5}\right)^{1/2}\left(\frac{N^2 D^2}{gH}\right)^{5/4}$$

$$= NP^{1/2}/\rho^{1/2}(gH)^{5/4} = NP^{1/2}/(\rho^{1/2}y^{5/4}). \tag{23.33}$$

Equations (23.32) and (23.33) are fundamentally the same and are related by equation (23.31), which, for a turbine, takes the form

$$P_{\text{output}} = \eta\rho gHQ.$$

Substituting into equation (23.33),

$$n_s = N(\eta\rho gHQ)^{1/2}/\rho^{1/2}(gH)^{5/4} = N\eta^{1/2}Q^{1/2}/(gH)^{3/4}. \tag{23.34}$$

The type number, since it is obtained from dimensionless coefficients, is also a dimensionless quantity provided a consistent system of units, such as SI, is used. Unfortunately, the units used in practice are often not consistent (e.g. revolutions per minute for N, litres per hour for Q and metres for H) and when this is the case a symbol N_s is used. It is then essential to state the units used for all the relevant

FIGURE 23.7

Classification of pumps and turbines

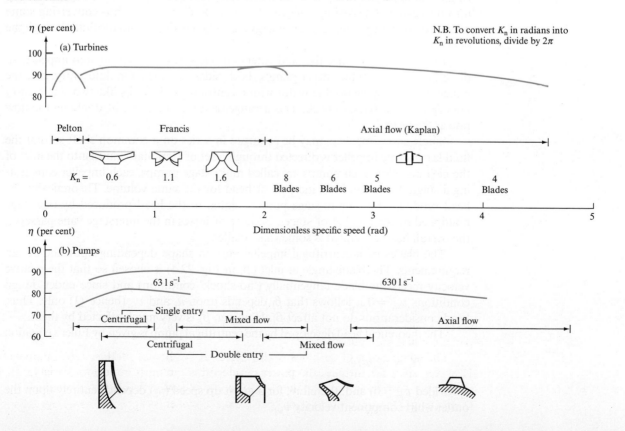

quantities. The rotational speed may be in either rev s^{-1} or rad s^{-1}. Thus two forms of the type number are possible:

$$n_s = NQ^{1/2}/(gH)^{3/4} \quad \text{or} \quad \omega_s = \omega Q^{1/2}/(gH)^{3/4} \quad \text{for pumps} \tag{23.35}$$

and

$$n_s = N\frac{(P/\rho)^{1/2}}{(gH)^{5/4}} \quad \text{or} \quad \omega_s = \omega \frac{(P/\rho)^{1/2}}{(gH)^{5/4}} \quad \text{for turbines.} \tag{23.36}$$

The relationship between the two forms is $\omega_s = 2\pi n_s$.

With the aid of type number the various types of pumps and turbines may be classified and compared, as shown in Fig. 23.7. Also, since the type number refers to the design point, it is used as the most important design parameter.

23.7 CENTRIFUGAL PUMPS AND FANS

Centrifugal pumps consist basically of an impeller rotating within a spiral casing. The fluid enters the pumps axially through the suction pipe via the eye of the impeller; it is discharged radially from the impeller around the entire circumference either into a ring of stationary diffuser vanes (and through them into the volute casing) or directly into the casing. The casing 'collects' the fluid, decelerates it – thus converting some of the kinetic energy into pressure energy – and finally discharges the fluid through the delivery flange.

In single-inlet pumps, the fluid enters on one side of the casing and impeller. In double-inlet (or double-entry) pumps, both sides are used for fluid entry and the impeller is usually of double width with a centre plate. It looks like two single-entry impellers placed back to back. This arrangement has the effect of doubling the flow rate at the head.

Single-entry impellers may be arranged in series on a common shaft so that the fluid leaving one impeller is directed through a set of stationary vanes into the inlet of the next impeller. Such pumps are called multi-stage pumps, each impeller constituting a stage. The effect is an increase of head for the same volume. Theoretically, the head produced by a multi-stage pump is equal to the head produced by one stage multiplied by the number of stages. Because of losses in the interstage vane passages, the overall head generated is somewhat smaller.

The blades of a centrifugal impeller vary in shape depending upon the design requirements. The blade angle at inlet (β_1 in Fig. 22.4) is chosen so that the relative velocity meets the blade tangentially ('no-shock' condition) and since under design conditions $v_{w_1} = 0$ it follows that β_1 depends upon u_1 and v_{f_1} (hence Q) only. Thus, head considerations do not affect β_1. However, β_2 is very much affected by them.

The theoretical head developed by the centrifugal pump is given by Euler's equation

$$H_{th} = v_{w_2}u_2/g$$

(provided $v_{w_1} = 0$) and, therefore, for a given tip speed (u_2) depends entirely upon the outlet whirl component velocity v_{w_2}.

FIGURE 23.8

Effect of blade outlet angle on the outlet velocity triangle

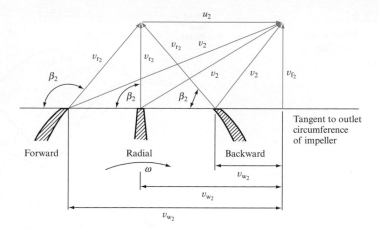

Let us now examine the dependence of this component upon the blade angle at outlet β_2. Figure 23.8 shows three different blade angles, often referred to as inclined backwards ($\beta_2 < 90°$), inclined forwards ($\beta_2 > 90°$) or radial ($\beta_2 = 90°$). The figure also shows a combined velocity diagram for the three types of blades.

The diagrams are drawn for the same u_2 and v_{f_2} (and, hence, the same speed of rotation, diameter and discharge) for the three blade types. It is clear that as β_2 increases, the absolute velocity v_2 (and hence the whirl component v_{w_2}) also increases. Therefore, the head developed depends upon β_2 and is larger for the forward-inclined blades. However, it must be remembered that the theoretical head given by Euler's equation is the total head developed by the impeller and, hence, embraces both the static and velocity head terms. Reference to Fig. 23.8 will show that the large head developed by the forward-inclined impeller blades includes a large proportion of velocity head since v_2 is very large. This presents practical difficulties in converting some of this kinetic energy into pressure energy. As shown in Section 10.8, the losses in a diffuser may be substantial, and are difficult to control.

The most common blade outlet angles for centrifugal pumps are from 15° to 90°, but, for fans, the range extends into forward-inclined blades (well-known multi-vane fans) with β_2 as large as 140°.

The effect of outlet blade angle on the performance characteristics is shown in Fig. 23.9. It is seen that the forward-bladed impeller generates a greater head at a given volume, but it must be remembered that a substantial part of this total head is in fact the velocity head.

The power characteristics also show fundamental differences, which are of considerable practical importance. For the backward-bladed impeller, the maximum occurs near the maximum efficiency point and any increase of flow rate beyond this point results in a decrease of power. Thus, an electric motor used to drive such a pump or fan may be safely rated at the maximum power. This type of power characteristic is called self-limiting.

This is not the case, however, for the radial- or forward-bladed impellers, for which the power is continuously rising. Choosing an appropriate motor, therefore, poses problems, because to have one rated for maximum power would mean overrating and an unnecessary expenditure if the pump will operate only near the maximum efficiency point. On the other hand, a small motor rated just for the operating point may be in danger of being overloaded should the pump be operated

FIGURE 23.9

The effect of blade outlet angle on performance characteristics

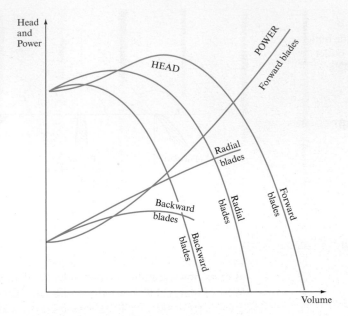

by mistake at a flow rate greater than the design value corresponding to the maximum efficiency point.

Centrifugal pumps and fans occupy the lower range of type numbers, up to approximately 1.8, as shown in Fig. 23.7. In general, the lower the type number of these machines the narrower is the impeller in relation to its diameter.

The overall efficiencies of centrifugal pumps are high, of the order of 90 per cent in the range of type numbers between 0.8 and 1.6. They tend to fall off rapidly at lower type numbers, mainly because of the increased frictional losses in the long interblade passages of these narrow impellers.

Also, the efficiencies depend upon the size of the machine and, hence, the capacity handled. The larger the machine, the higher is the efficiency (see Section 23.5).

For centrifugal fans, the highest efficiencies are realized by the 'aerofoil-bladed' fans. They are basically of the backward-bladed type, but the blades have an aerofoil profile rather than being of the same thickness. Their range of type numbers is from 0.5 to 1.6 and maximum efficiencies are of the order of 90 per cent.

23.8 AXIAL FLOW PUMPS AND FANS

Axial flow pumps and fans consist of an impeller rotating within a concentric cylindrical casing (Fig. 23.10). Thus, the direction of flow through the machine is axial throughout. For this reason axial flow pumps and fans have the smallest transverse dimensions of all rotodynamic machines.

It is usual for a set of stationary guide vanes to be present in all but the cheapest and least efficient machines. The guide vanes ensure that the flow at the outlet from the pump or fan is without a tangential component, that is to say is axial and without a swirl. This result may be achieved by either a set of downstream or a set of upstream vanes. The choice affects the construction and to some extent the type number.

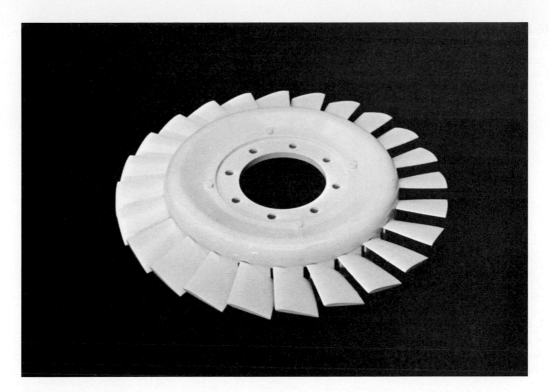

FIGURE 23.10

An impeller of an axial flow fan. (By courtesy of Airscrew-Howden Ltd)

Axial flow pumps and fans extend over the far end of the type number spectrum, starting from about $\omega_s = 2.8$ (see Fig. 23.7) to about 4.8. For fans, the range is somewhat wider, starting from about 1.4 to 2.4 for upstream guide vane fans, from 2.0 to 3.0 for downstream guide vane fans and from 2.5 to 4.5 for non-guide vane fans. Axial flow fans without a casing, called propeller fans, cover the range from 3.5 to 5.0, whereas at the other end of the range are the contrarotating fans, for which the type numbers are from 1.2 to 1.6. In these fans, the aim of having axial flow at outlet is achieved not by the guide vanes, which are omitted, but by having two impellers rotating in opposing directions, so that the whirl component produced by one is cancelled by the whirl component of the other (which is equal in magnitude but opposite in direction).

The disadvantages of axial flow pumps and fans are that they develop a low head (up to 20 m per stage) and have steeply descending efficiency curves and, hence, are only economical if operated at discharges corresponding to or very near to the design point. Also, the pressure/volume characteristic on the left of the maximum efficiency has a region of instability, as shown dashed in Fig. 23.11.

In addition, axial flow pumps have a limited suction capacity and, thus, are prone to cavitation, which considerably restricts their selection for some applications.

The blades of an axial flow impeller are fixed to a hub, usually permanently. In some cases, however, for special applications, the blades are adjustable so that the stagger angle may be varied and, thus, the performance altered. The effect of changing the stagger angle on the characteristics is shown in Fig. 23.12.

The relationship between the blade radial length and the size of the hub is expressed by the hub ratio, defined as

FIGURE 23.11
Typical flow
characteristics of an
axial flow pump

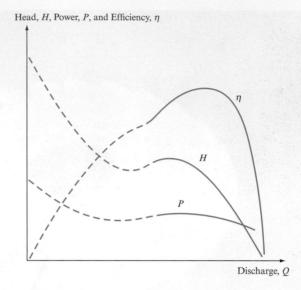

Head, *H*, Power, *P*, and Efficiency, η

η

H

P

Discharge, *Q*

FIGURE 23.12
Effect of stagger angle on
performance of an axial
flow pump

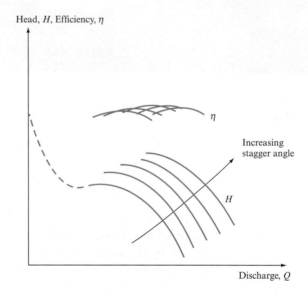

Head, *H*, Efficiency, η

η

Increasing
stagger angle

H

Discharge, *Q*

$$\text{Hub ratio} = \frac{\text{Hub diameter}}{\text{Impeller diameter}}. \tag{23.37}$$

This parameter affects the type number of the machine and, for pumps, varies from
0.3 to 0.6. A low hub ratio (long blades) is associated with high type numbers, whereas
a high ratio results in lower type numbers. Hence, the pumps and fans having low type
numbers develop relatively higher pressures. In the extreme, the axial flow compressors have a very high hub ratio (over 0.9) because their main purpose is to create high
gas pressure.

   The number of blades used depends to a large extent on the hub ratio. Generally,
the higher the hub ratio, the larger the number of blades used. For pumps, the usual
number is between 2 and 8, for fans between 2 and 16 and for compressors as many
as 32 blades are quite common.

23.9 MIXED FLOW PUMPS AND FANS

Mixed flow pumps occupy a position between the centrifugal and axial flow pumps. The impellers consist of a conical hub with blades attached in such a way that the flow into the impeller is axial, but through it the flow is partly axial and partly radial. On leaving the impeller, the fluid is usually diffused in the guide vanes, which lead into an axial outlet, as shown in Fig. 23.13. The alternative and less frequent solution is for the flow to be collected by a volute casing and then discharged in a plane normal to the axis of the impeller rotation. Such an arrangement is shown in Fig. 23.14. Both configurations cover the type number range from about 1.0 to 2.2, the latter type being more common at the lower end of this range.

One of the advantages of mixed flow pumps with axial discharge is that while they offer large discharges, they may be easily arranged in multi-stage units, thus providing high pressure.

FIGURE 23.13
Mixed flow pump with axial discharge

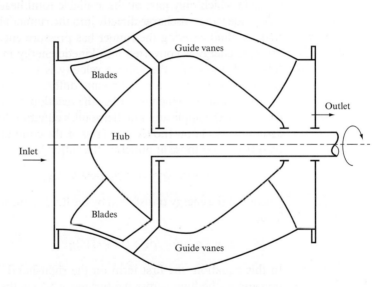

FIGURE 23.14
Mixed flow pump with volute casing

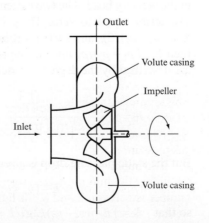

The efficiency of mixed flow pumps and fans is high, approaching the 90 per cent mark. Figure 22.2 shows a mixed flow impeller of a fan.

23.10 WATER TURBINES

Turbines are subdivided into *impulse* and *reaction* machines. In the impulse turbines, the total head available is first converted into the kinetic energy. This is usually accomplished in one or more nozzles. The jets issuing from the nozzles strike vanes attached to the periphery of a rotating wheel. Because of the rate of change of angular momentum and the motion of the vanes, work is done on the runner (impeller) by the fluid and, thus, energy is transferred. Since the fluid energy which is reduced on passing through the runner is entirely kinetic, it follows that the absolute velocity at outlet is smaller than the absolute velocity at inlet (jet velocity). Furthermore, the fluid pressure is atmospheric throughout and the relative velocity is constant except for a slight reduction due to friction.

In the reaction turbines, the fluid passes first through a ring of stationary guide vanes in which only part of the available total head is converted into kinetic energy. The guide vanes discharge directly into the runner along the whole of its periphery, so that the fluid entering the runner has pressure energy as well as kinetic energy. The pressure energy is converted into kinetic energy in the runner (the passage running full) and, therefore, the relative velocity is not constant but increases through the runner. There is, therefore, a pressure difference across the runner.

A parameter which describes the reaction turbines is the *degree of reaction*. It is derived by the application of Bernoulli's equation to the inlet and outlet of a turbine, assuming ideal flow (no losses). Thus, if the conditions at inlet are denoted by the use of suffix 1 and those at outlet by the suffix 2, then

$$p_1/\rho g + v_1^2/2g = E + p_2/\rho g + v_2^2/2g,$$

where E is the energy transferred by the fluid to the turbine per unit weight of the fluid. Thus,

$$E = (p_1 - p_2)/\rho g + (v_1^2 - v_2^2)/2g.$$

In this equation, the first term on the right-hand side represents the drop of static pressure in the fluid across the turbine, whereas the second term represents the drop in the velocity head. The two extreme solutions are obtained by making either of these two terms equal to zero. Thus, if the pressure is constant, so that $p_1 = p_2$, then $E = (v_1^2 - v_2^2)/2g$ and such a turbine is purely impulsive. If, on the other hand, $v_1 = v_2$, then $E = (p_1 - p_2)/\rho g$ and this represents pure reaction. The intermediate possibilities are described by the degree of reaction (R), defined as

$$R = \frac{\text{Static pressure drop}}{\text{Total energy transfer}}. \tag{23.38}$$

But the static pressure drop is given by

$$(p_1 - p_2)/\rho g = E - (v_1^2 - v_2^2)/2g,$$

so that $R = [E - (v_1^2 - v_2^2)/2g]/E = 1 - (v_1^2 - v_2^2)/2gE.$

Substituting now from Euler's equation for $E = v_{w_1} u_1/g$, we obtain

$$R = 1 - (v_1^2 - v_2^2)/2v_{w_1}u_1. \tag{23.39}$$

Water turbines are mainly used in power stations to drive electric generators. There are three well-known types which are used: the Pelton wheel, which is an impulse turbine, the Francis type and the axial flow (Kaplan) turbines, both being of the reaction type. Table 23.1 attempts to compare the three types.

TABLE 23.1

Comparison of water turbines

	PELTON WHEEL	FRANCIS	KAPLAN
Type number ω_s range (rad)	0.05–0.4	0.4–2.2	1.8–4.6
Operating total head (m)	100–1700	80–500	Up to 400
Maximum power output (MW)	55	40	30
Best efficiency (per cent)	93	94	94
Regulation mechanism	Spear nozzle and deflector plate	Guide vanes, surge tanks	Blade stagger

23.11 THE PELTON WHEEL

The Pelton wheel is an impulse turbine in which vanes, sometimes called 'buckets', of elliptical shape are attached to the periphery of a rotating wheel, as shown in Fig. 23.15. One or two nozzles project a jet of water tangentially to the vane pitch circle. The vanes are of double-outlet section, as shown in Fig. 23.16, so that the jet is split and leaves symmetrically on both sides of the vane. In this way the end thrust on the bearings and the shaft is eliminated.

The total head available at the nozzle is equal to the gross head less losses in the pipeline leading to the nozzle. If it is equal to H, then the velocity of jet issuing from the nozzle is

$$v = C_v\sqrt{(2gH)}, \tag{23.40}$$

FIGURE 23.15

Diagrammatic arrangement of a Pelton wheel

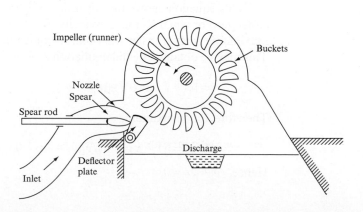

FIGURE 23.16

Velocity triangles for a
Pelton wheel

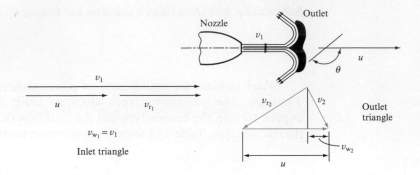

where C_v is the velocity coefficient and its value is between 0.97 and 0.99.

The total energy transferred to the wheel is given by Euler's equation:

$$E = (v_{w_1} u_1 - v_{w_2} u_2)/g,$$

but, as shown by the velocity triangles in Fig. 23.16, the peripheral vane velocity at outlet is the same as that at inlet,

$$u_1 = u_2 = u,$$

so that $E = (u/g)(v_{w_1} - v_{w_2}).$

However,

$$v_{w_2} = u - v_{r_2} \cos(180° - \theta) = u + v_{r_2} \cos \theta$$

and $v_{r_2} = k v_{r_1} = k(v - u),$

where k represents the reduction of the relative velocity due to friction. Thus,

$$v_{w_2} = u + k(v_1 - u) \cos \theta \quad \text{and} \quad v_{w_1} = v_1,$$

so that $E = (u/g)[v_1 - u - k(v_1 - u) \cos \theta]$

$$= (u/g)[v_1(1 - k \cos \theta) - u(1 - k \cos \theta)]$$

$$= (u/g)(v_1 - u)(1 - k \cos \theta). \tag{23.41}$$

This equation shows that there is no energy transfer when the vane velocity is either zero or equal to the jet velocity. It is reasonable to expect, therefore, that the maximum energy transfer will occur at some intermediate value of the vane velocity. This may be obtained by differentiation as follows:

$$E = [(1 - k \cos \theta)/g](v_1 u - u^2).$$

Therefore, for a maximum,

$$dE/du = [(1 - k \cos \theta)/g](v_1 - 2u) = 0.$$

Hence, $v_1 - 2u = 0,$

$$u = \tfrac{1}{2} v_1. \tag{23.42}$$

Substituting this value back into equation (23.41), the expression for maximum energy transfer is obtained:

$$E_{max} = (v_1/2g)(v_1 - \tfrac{1}{2}v_1)(1 - k \cos \theta)$$
$$= (v_1^2/4g)(1 - k \cos \theta).$$

Now, the energy input from the nozzle is the kinetic energy, which per unit weight of fluid flowing is

$$\text{Kinetic energy of the jet} = v_1^2/2g.$$

Thus, the maximum theoretical efficiency of the Pelton wheel becomes

$$\eta_{max} = E_{max}/\text{Kinetic energy of the jet}$$
$$= (v_1^2/4g)(1 - k \cos \theta)/(v_1^2/2g)$$

$$\eta_{max} = (1 - k \cos \theta)/2. \qquad (23.43)$$

In the ideal case, assuming no friction, there is no reduction of the relative velocity over the vane and, therefore, $k = 1$. Also, if $\theta = 180°$, the maximum efficiency becomes 100 per cent. In practice, however, friction exists and the value of k is in the region of 0.8 to 0.85. Also, the vane angle is usually 165°, to avoid the interference between the oncoming and outcoming jets. Thus, the ratio of the wheel velocity to the jet velocity becomes, in practice, somewhat smaller than the theoretical. Figure 23.17 shows the variation of the Pelton wheel efficiency with the speed ratio. It will be seen that, for the maximum efficiency, this ratio is about 0.46.

FIGURE 23.17

Pelton wheel efficiency as a function of speed ratio

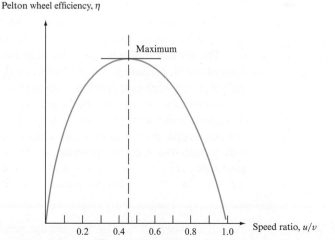

Pelton wheel efficiency, η

Maximum

0.2   0.4   0.6   0.8   1.0   Speed ratio, u/v

Since a Pelton wheel is usually employed to drive an electrical generator, it is required that its speed of rotation is constant, regardless of the load. Thus, u must be constant, but for maximum efficiency it is also important that the speed ratio is maintained constant as well. Since the jet velocity depends only upon the total head

H, which for a given installation is also constant, the velocity ratio may be kept constant provided there is no reduction of head at the nozzle. This means that a throttling process using a valve in the penstock is not suitable, since a valve reduces flow by reducing head and dissipating energy. It follows, then, that any alteration of the load on the turbine must be accompanied by a corresponding alteration of the water power, but with u/v remaining constant. Since $P = \rho gQH$, it follows that this requirement can only be achieved by alteration in Q such that H is unchanged. But

$$Q = Av = AC_v\sqrt{(2gH)}$$

and, therefore, to vary Q, the area of the jet must be changed. This is achieved by means of the needle (spear) shown in Fig. 23.15, which does not alter H.

Small changes in efficiency result because the nozzle loss represented by the value of C_v will be changed and, also, jet windage as well as bearing losses will change slightly. Figure 23.18 shows the typical variation of C_v with the jet opening.

FIGURE 23.18

Variation of C_v with jet opening

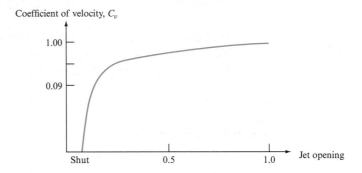

The movement of the needle may be controlled automatically by a governor operated by a servo-motor or some similar arrangement.

The provision of a needle valve does not, however, cater satisfactorily for sudden load removal, because it is not possible to shut the needle rapidly without the serious risk of a very high-pressure build-up in the pipe system (see Chapter 20). Surge tanks are not suitable because of the large heads involved in the Pelton wheel installations. Therefore, a deflector plate (as shown in Fig. 23.15) is used. The jet is deflected from the buckets partially or completely when the load is removed, the needle is then moved slowly into the required position and the deflector then returns to the original position.

EXAMPLE 23.2

A Pelton wheel driven by two similar jets transmits 3750 kW to the shaft when running at 375 rev min^{-1}. The head from the reservoir level to the nozzles is 200 m and the efficiency of power transmission through the pipelines and nozzles is 90 per cent. The jets are tangential to a 1.45 m diameter circle. The relative velocity decreases by 10 per cent as the water traverses the buckets, which are so shaped that they would, if stationary, deflect the jet through 165°. Neglecting windage losses, find (a) the efficiency of the runner and (b) the diameter of each jet.

Solution

(a) Efficiency of runner,

$$\eta = (2u/v_j^2)(v_j - u)(1 - k\cos\theta).$$

If the efficiency of the pipeline and the nozzle is 90 per cent, the head at the base of the nozzle convertible to jet velocity is

$$h = 0.9 \times 200 = 180 \text{ m},$$

and the jet velocity is

$$v_j = \sqrt{(2gh)} = \sqrt{(2 \times 9.81 \times 180)} = 59.5 \text{ m s}^{-1}.$$

Now, the bucket speed is

$$u = \pi DN/60 = \pi \times 1.45 \times 375/60 = 28.5 \text{ m s}^{-1}$$

and, hence, the runner efficiency is

$$\eta = [2 \times 28.5/(59.5)^2](59.5 - 28.5) \times 1.869 = \textbf{0.933 or 93.3 per cent}.$$

(b) If the runner is 93.3 per cent efficient, the total power of the jets is

$$P = 3750/0.933 = 4021 \text{ kW},$$

and the power per jet is $4021/2 = 2010.5$ kW. But, for a jet,

$$\text{Power} - \rho g A_j v_j(v_j^2/2g) = \rho(\pi d^2/8)v_j^3.$$

Therefore, equating,

$$2010.5 \times 10^3 = 10^3 \pi d^2(59.5)^3/8,$$

$$d^2 = 8 \times 2010.5/\pi(59.3)^3 = 0.024\,55$$

$$d = \textbf{0.157 m}.$$

23.12 FRANCIS TURBINES

A Francis turbine is a reaction machine, which means that during energy transfer in the runner (impeller) there is a drop in static pressure and a drop in velocity head. Only part of the total head presented to the machine is converted to velocity head before entering the runner. This is achieved in the stationary but adjustable guide vanes, shown in Fig. 23.19. It is important to realize that the machine is running full of water, which enters the impeller on its whole periphery. The guide vane ring may surround the runner on its outer periphery, in which case the flow of fluid is towards the runner centre. In such a case, the turbine is known as an *inward-flow* type. The alternative arrangement is for the fluid to enter the guide vanes at the centre and to

FIGURE 23.19

Francis turbine

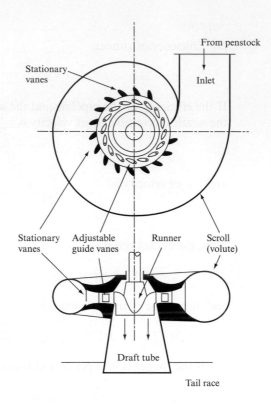

flow radially outwards into the runner which now surrounds the guide vanes. Such a turbine is known as an *outward-flow* type.

Consider an inward-flow Francis turbine, represented diagrammatically in Fig. 23.19. A section of the runner guide vane ring, showing the blades, vanes and velocity triangles, is given in Fig. 23.20.

The total head available to the machine is H and the water velocity on entering the guide vanes is v_0. The velocity leaving the guide vanes is v_1 and is related to v_0 by the continuity equation

$$v_0 A_0 = v_{f_1} A_1.$$

But $v_{f_1} = v_1 \sin \theta$, so that

$$v_0 A_0 = v_1 A_1 \sin \theta.$$

The direction of v_1 is governed by the guide vane angle θ. It is chosen in such a way that the relative velocity meets the runner blade tangentially, i.e. it makes an angle β_1 with the tangent at blade inlet. Thus,

$$\tan \theta = v_{f_1}/v_{w_1} \quad \text{and} \quad \tan \beta_1 = v_{f_1}/(u_1 - v_{w_1}).$$

Eliminating v_{w_1} from the two equations:

$$\tan \beta_1 = v_{f_1}/(u_1 - v_{f_1}/\tan \theta)$$

or

$$\cot \beta_1 = u_1/u_{f_1} - \cot \theta.$$

FIGURE 23.20

Section through part of
a Francis turbine

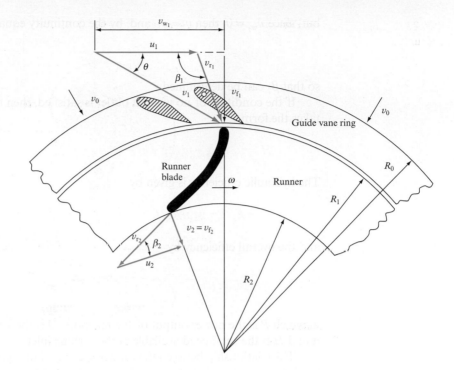

Therefore,

$$u_1/v_{f_1} = \cot \beta_1 + \cot \theta$$

or $$u_1 - v_{f_1}(\cot \beta_1 + \cot \theta),$$ (23.44)

The total energy at inlet to the runner consists of the velocity head $v_1^2/2g$ and pressure head H_1. In the runner, the fluid energy is decreased by E, which is transferred to the runner. Water leaves the impeller with kinetic energy $v_2^2/2g$. Thus, the following energy equations hold:

$$H = v_1^2/2g + H_1 + h_1'$$

and $$H = E + v_2^2/2g + h_1,$$

in which h_1' is the loss of head in the guide vane ring and h_1 is the loss in the whole turbine, including entry, guide vanes and runner.

The energy transferred E is given by Euler's equation, which, for the maximum energy transfer condition secured when $v_{w_2} = 0$, takes the form

$$E = v_{w_1} u_1/g.$$

The condition of no whirl component at outlet may be achieved by making the outlet blade angle β_2, such that the absolute velocity at outlet v_2 is radial, as shown in Fig. 23.20. From the outlet velocity diagram, then, it follows that

$$\tan \beta_2 = v_2/u_2,$$

but, since $v_{w_2} = 0$, then $v_2 = v_{f_2}$ and, by the continuity equation,

$$A_1 v_{f_1} = A_2 v_{f_2},$$

so that β_2 can be determined.

If the condition of no whirl at outlet is satisfied, then the second energy equation takes the form

$$H = v_{w_1} u_1 / g + v_2^2 / 2g + h_1. \tag{23.45}$$

The hydraulic efficiency is given by

$$\eta_h = E/H = v_{w_1} u_1 / gH, \tag{23.46}$$

and the overall efficiency by

$$\eta = P/\rho g Q H, \tag{23.47}$$

in which P is the power output of the machine, Q is the volumetric flow rate through it and H is the total head available at the turbine inlet.

The relationship between the runner speed and the spouting velocity, $\sqrt{(2gH)}$, for the Francis turbine is not so rigidly defined as for the Pelton wheel. In practice, the speed ratio $u_2/\sqrt{(2gH)}$ is contained within the limits 0.6 to 0.9.

Similarly to a Pelton wheel, a Francis turbine usually drives an alternator and, hence, its speed must be constant. Since the total head available is constant and dissipation of energy by throttling is undesirable, the regulation at part load is achieved by varying the guide vane angle θ, sometimes referred to as the gate. This is possible because there is no requirement for the speed ratio to remain constant. A change in θ results in a change in v_w and v_f. Thus, E is altered for given u. However, such changes mean a departure from the 'no-shock' conditions at inlet and also give rise to the whirl component at outlet. As a result, the efficiency at part load falls off more rapidly than in the case of the Pelton wheel. Also, vortex motion in the draft tube resulting from the whirl component may cause cavitation in the centre. Sudden load changes are catered for either by a bypass valve or by a surge tank.

EXAMPLE 23.3

In an inward-flow reaction turbine, the supply head is 12 m and the maximum discharge is 0.28 m³ s⁻¹. External diameter = 2 × (internal diameter) and the velocity of flow is constant and equal to $0.15\sqrt{(2gH)}$. The runner vanes are radial at inlet and the runner rotates at 300 rev min⁻¹.

Determine (a) the guide vane angles, (b) the vane angle at exit for radial discharge, (c) the widths of the runner at inlet and exit. The vanes occupy 10 per cent of the circumference and the hydraulic efficiency is 80 per cent.

Solution

(a) The velocity of flow is given by

$$v_{f_1} = v_{f_2} = 0.15\sqrt{(2gH)} = 0.15\sqrt{(2 \times 9.81 \times 12)} = 2.3\,\text{m s}^{-1}.$$

FIGURE 23.21

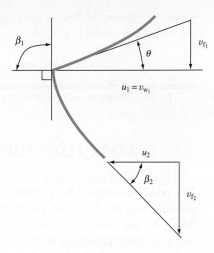

The efficiency is given by equation (23.46), but from Fig. 23.21

$$u_1 = v_{w_1} \quad \text{and} \quad v_{f_1}/v_{w_1} = \tan \theta.$$

Therefore,

$$\eta = v_{w_1}^2/gH,$$

but, since $H = 12$ and $\eta = 80$ per cent, it follows that

$$v_{w_1} = \sqrt{(0.8 \times 9.81 \times 12)} = 9.7 \text{ m s}^{-1}.$$

Therefore, $u_1 = 9.7 \text{ m s}^{-1}$ and

$$\tan \theta = v_{f_1}/u_1 = 2.3/9.7 = 0.237,$$
$$\theta = 13°20'.$$

(b) Since internal diameter $= \frac{1}{2} \times$ external diameter,

$$u_2 = \frac{1}{2}u_1 = 4.85 \text{ m s}^{-1}.$$

Therefore,

$$\tan \beta_2 = v_{f_2}/u_2 = 2.3/4.85 = 0.475,$$
$$\beta_2 = 25°20'.$$

(c) Now $u_1 = \omega r_1$, where $\omega = 300(2\pi/60)$ rad s^{-1}. Therefore,

$$r_1 = u_1/\omega = 9.7 \times 60/(300 \times 2\pi) = 0.31 \text{ m}.$$

Therefore, breadth at outlet is given by

$$b_1 = 0.28/(2.3 \times 0.9 \times 2\pi \times 0.31) = 0.0696 \text{ m}$$
$$= 69.6 \text{ mm}.$$

Since the velocity of flow is constant, and internal diameter = $\frac{1}{2}$ × external diameter, the breadth at outlet is given by

$$b_2 = 2 \times 69.6 = \mathbf{139.2\ mm}.$$

23.13 AXIAL FLOW TURBINES

The power developed by a turbine is proportional to the product of the total head available and the flow rate. Therefore, the power required from a turbine may, within limits, be obtained by a desired combination of these two quantities. For a Pelton wheel, in order to achieve high jet velocities, it is necessary that the total head is large and, consequently, the flow rate is usually small. However, the Pelton wheel becomes unsuitable if the head available is small, so that in order to achieve the desired power the quantity has to be greater. A Francis-type radial turbine is then used. Its proportions depend upon the flow rate which must pass through it. As in the case of pumps, for greater flow rate the size of the runner eye must be increased, the blade passages become shorter but wider, and a mixed-flow-type turbine results. If the process is carried further, an axial flow turbine is obtained because the maximum flow rate may be passed through when the flow is parallel to the axis.

Figure 23.22 shows that the arrangement of guide vanes for an axial flow turbine is similar to that for a Francis turbine. The guide vane ring is in a plane perpendicular to the shaft so that the flow through it is radial. The runner, however, is situated further downstream, so that between the guide vanes and the runner the fluid turns through a right angle into the axial direction. The purpose of the guide vanes is to impart whirl to the fluid so that when it approaches the runner it is essentially of a free vortex type, i.e. the tangential (whirl) velocity is inversely proportional to radius.

The runner blades must be long in order to accommodate the large flow rate and, consequently, considerations of strength required to transmit the tremendous torques involved impose the necessity for large blade chords. Thus, pitch/chord ratios of 1.0 to 1.5 are used and, hence, the number of blades is small, usually four, five or six.

The velocity of the blades is directly proportional to the radius whereas, as stated earlier, the fluid whirl velocity is inversely proportional to it. To cater for this difference, the runner blades are twisted so that the angle they make with the axis is greater at the tip than at the hub.

The blades may be cast as integral parts of the runner or may be welded to the hub. In such cases, the blade angles are fixed, resulting in a rapid fall of efficiency under part-load conditions because then the reduction of flow rate through the machine results in a mismatch between the direction of the fluid velocity relative to the runner and the blade angle. To offset this difficulty, runners may have adjustable or variable pitch blades, whereby they may be turned about their own axes, thus altering the stagger angle to meet the fluid tangentially. By this arrangement a very wide band of high efficiencies may be achieved. Axial flow turbines with variable pitch blades are known as Kaplan turbines. The efficiencies of Kaplan turbines are between 90 and 93 per cent and powers developed are up to 85 MW.

The velocity triangles shown in Fig. 23.22 are similar to those for the axial flow pump. The velocity of flow is axial at inlet and outlet and, of course, remains the same.

FIGURE 23.22
Axial flow turbines and
velocity triangles

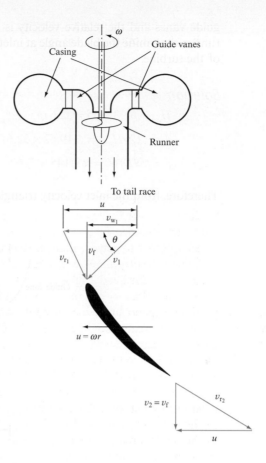

The whirl velocity is tangential. The blade velocity at inlet and outlet is the same, but varies along the blades with radius from hub to tip.

If the angular velocity of the runner is ω the blade velocity at radius r is given by $u = \omega r$; since at maximum efficiency $v_{w_2} = 0$ and $v_2 = v_f$, it follows that

$$E = uv_{w_1}/g,$$

in which $v_{w_1} = v_f \cot \theta$. Since E should be the same at the blade tip and at the hub, but u is greater at the tip, it follows that v_{w_1} must be reduced. Similarly, the velocity of flow v_f should remain constant along the blade and, therefore, $\cot \theta$ must be reduced towards the tip of the blade. Thus, θ has to be reduced and, consequently, the blade must be twisted so that it makes a greater angle with the axis at the tip than it does at the hub.

EXAMPLE 23.4

Water is supplied to an axial flow turbine under a total head of 35 m. The mean diameter of the runner is 2 m and it rotates at 145 rev min^{-1}. Water leaves the guide vanes at 30° to the direction of runner rotation and at mean radius the angle of the runner blade at outlet is 28°. If 7 per cent of the total head is lost in the casing and guide vanes and the relative velocity is reduced by 8 per cent due to friction in the runner, determine the blade angle at inlet (at mean radius) and the hydraulic efficiency of the turbine.

Solution

$$H_{net} = 0.93 \times 35 = 32.6 \, m,$$

$$v_1 = \sqrt{(2gH_{net})} = \sqrt{(19.62 \times 32.6)} = 25.3 \, m \, s^{-1},$$

$$u = \pi ND/60 = \pi \times 145 \times 2/60 = 15.2 \, m \, s^{-1}.$$

Therefore, from the inlet velocity triangle (Fig. 23.23),

FIGURE 23.23

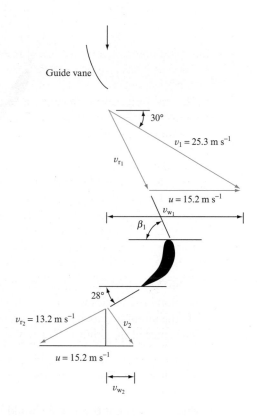

$$v_{r_1} = 14.3 \, m \, s^{-1}$$

and $$\beta_1 = \mathbf{62.2°}.$$

Therefore,

$$v_{r_2} = 0.92 \times 14.3 = 13.2 \, m \, s^{-1}$$

and, from the velocity triangles,

$$v_{w_1} = 21.9 \, m \, s^{-1}, \quad v_{w_2} = 3.6 \, m \, s^{-1}.$$

Therefore,

$$E = (u/g)(v_{w_1} - v_{w_2})$$

$$= 15.2 \times 18.3/9.81$$

$$= 28.4 \text{ m},$$

and so $\quad \eta_h = E/H = 28.4/35 = 0.811$

$$= \textbf{81.1 per cent.}$$

23.14 HYDRAULIC TRANSMISSIONS

At the beginning of this part, fluid machinery was primarily classified according to the direction in which energy is transferred. However, a class of machines exists in which the fluid is used as a means of energy transfer. It receives energy from a moving mechanical part only to give it up to another moving mechanical part. If the fluid action is rotodynamic, this class of machines constitutes hydrodynamic transmissions and includes two distinctly different types: fluid (or hydraulic) couplings and torque converters.

Fundamentally, hydrodynamic transmissions consist of two elements: a pump, usually referred to as the primary, and a turbine, known as the secondary. The pump is driven by a prime mover, such as an electric motor or an internal combustion engine; it gives energy to the fluid, usually oil of low viscosity, which then enters the turbine and transmits its acquired energy to it. The turbine shaft provides the mechanical energy output. No solid contact exists between the primary and the secondary. In a fluid coupling, shown diagrammatically in Fig. 23.24, two identical impellers are involved. They have radial blades within bowl-shaped shrouds. The space between the blades is full of oil. As the primary begins to rotate, the oil within its impeller moves towards the periphery and is discharged radially into the secondary at the outer radius. Within the secondary, it flows radially inwards towards the centre and is discharged back into the primary near the hub. For the flow to exist, the head produced by the primary must be greater than the centrifugal head of the secondary resisting the flow. This is only possible if the speed of the primary is greater than that of the secondary.

FIGURE 23.24

Fluid coupling

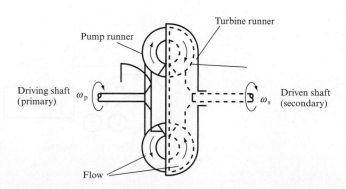

Thus, for torque transmission there must be a speed difference giving rise to the 'slip', defined as

$$s = (\omega_p - \omega_s)/\omega_p, \tag{23.48}$$

where ω_p = angular velocity of the primary, ω_s = angular velocity of the secondary.
The power input to the primary is

$$P_{in} = T_p \omega_p, \tag{23.49}$$

and the power output from the secondary is

$$P_{out} = T_s \omega_s. \tag{23.50}$$

Therefore, the efficiency of the transmission is

$$\eta = P_{out}/P_{in} = T_s \omega_s / T_p \omega_p. \tag{23.51}$$

However, in a fluid coupling, the input torque T_p and the output torque T_s must be the same, because there is no other element between the two parts to provide a torque reaction. Hence, for a coupling,

$$\eta = \omega_s/\omega_p = 1 - s. \tag{23.52}$$

The efficiency of hydraulic couplings is high, usually in excess of 94 per cent. The losses are due to friction and turbulence created when the fluid enters each impeller because the blades are not shaped to meet the flow without shock.

The great advantage of fluid couplings over other types of transmission lies in applications involving unsteady operation, because then torsional vibrations in either of the two halves of the coupling are not transmitted to the other. Also, since the torque is proportional to the speed, the full torque is not developed until the full speed is reached. Thus, the starting load is low, which makes starting of prime movers considerably easier.

Figure 23.25 shows velocity triangles for the primary and secondary at their inlets and outlets. The suffixes used are as follows: 1, inlet to the primary at radius r_i; 2, outlet from primary at radius r_o; 3, inlet to the secondary at r_o; 4, outlet from

FIGURE 23.25

Velocity triangles for a fluid coupling

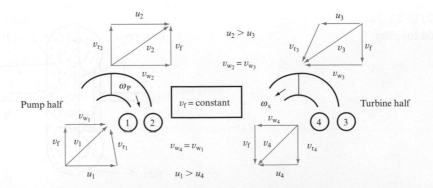

secondary at r_i. The assumption made is that of 'zero whirl slip', which means that the fluid leaves one impeller and enters the other with the same tangential velocity component, e.g.

$$v_{w_2} = v_{w_3} \quad \text{and} \quad v_{w_4} = v_{w_1}.$$

Now, by Euler's equation the work done by the primary per unit weight of the fluid is

$$E_p = (v_{w_2} u_2 - v_{w_1} u_1)/g.$$

But $u_2 = \omega_p r_o$ and $u_1 = \omega_p r_i$. Also, from the velocity triangles,

$$v_{w_2} = u_2 = \omega_p r_o \quad \text{and} \quad v_{w_1} = v_{w_4} = u_4 = \omega_s r_i.$$

Substituting into Euler's equation,

$$E_p = (\omega_p^2 r_o^2 - \omega_p \omega_s r_i^2)/g. \tag{23.53}$$

Similarly, the work done on the secondary per unit weight of fluid flowing is

$$E_s = (v_{w_3} u_3 - v_{w_4} u_4)/g.$$

But $\quad v_{w_3} = v_{w_2} = u_2 = \omega_p r_o, \quad u_3 = \omega_s r_o \quad \text{and} \quad v_{w_4} = u_4 = \omega_s r_i,$

so that $\quad E_s = (\omega_p \omega_s r_o^2 - \omega_s^2 r_i^2)/g. \tag{23.54}$

The energy dissipated may be obtained from the difference between equations (23.53) and (23.54):

$$\Delta E = E_p - E_s,$$

which, on substitution, gives

$$\Delta E = (\omega_p - \omega_s)(\omega_p r_o^2 - \omega_s r_i^2)/g. \tag{23.55}$$

Also, it is assumed that, in general,

$$\Delta E = kQ^2, \tag{23.56}$$

where Q is the flow rate through the coupling.

The above expressions are approximations to what actually happens in the coupling because of the difficulty in establishing radii r_o and r_i and because the flow rate varies with the radius.

Dimensional analysis may be applied to a fluid coupling as follows:

$$T = f(\rho, D, \omega_p, \omega_s, \mu, V),$$

where ρ = fluid density, μ = fluid viscosity, V = volume of fluid, D = diameter of impellers, ω_p = angular velocity of the primary, ω_s = angular velocity of the secondary. It leads to the establishment of the 'torque coefficient',

$$T/\rho \omega_p^2 D^5 = \phi(s, \rho \omega_p D^2/\mu, V/D^3). \tag{23.57}$$

For a given coupling, V/D^3 is constant, the second term is proportional to Reynolds number and, since the flow is very turbulent and its effect is insignificant, the torque coefficient may be approximated to

$$T/\rho\omega_p^2 D^5 = \phi'(s).\tag{23.58}$$

Similarly, since power $P = \omega T$, a power coefficient may be obtained:

$$P/\rho\omega_p^3 D^5 = \phi''(s).\tag{23.59}$$

The performance characteristics of an hydraulic coupling are shown in Fig. 23.26.

The difference between the fluid coupling and the torque converter is that while the former consists of only two runners, the latter also has a set of stationary vanes interposed between the two runners, as shown in Fig. 23.27.

FIGURE 23.26

Typical characteristics of a fluid coupling

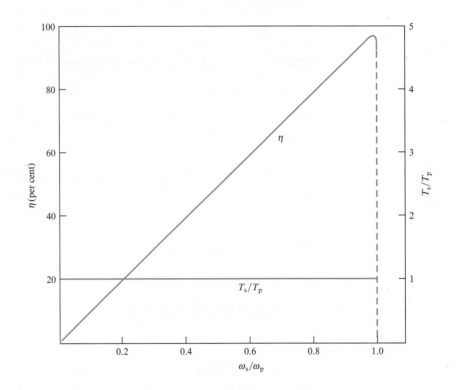

FIGURE 23.27

Torque converter

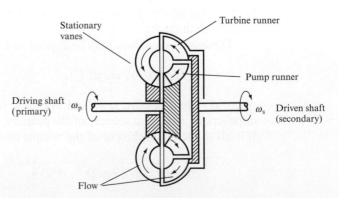

Since the stationary vanes change the angular momentum of the fluid passing through them, they are subjected to a torque. Also, since they do not rotate, an equal and opposite torque must be exerted on them from the housing. Thus, the existence of this additional torque to the system means that the torque on the secondary runner is not the same as that on the primary. The relationship between the torques is now

$$T_s = T_p + T_v, \tag{23.60}$$

where T_s = torque on the secondary, T_p = torque on the primary, T_v = torque on the stationary vanes. The efficiency of the torque converter, therefore, is as given by equation (23.51), namely

$$\eta = T_s \omega_s / T_p \omega_p.$$

It is possible for T_v to be either positive or negative. If the vanes' design is such that they receive torque from the fluid which is in the opposite direction to that exerted on the driven shaft, T_v is positive, signifying an increased output torque. It may in fact be as much as five times the input torque of the primary runner. If, on the other hand, the vanes by virtue of their design receive the torque which is in the same direction as that of the driven shaft, T_v is negative and the secondary torque is reduced.

The dimensional analysis for a torque converter yields the following relationship after rejecting the Re and the V/D^3 terms:

$$T/\rho \omega_p^2 D^5 = \phi(s, T_s/T_p). \tag{23.61}$$

The maximum efficiency of a torque converter is smaller than that of a fluid coupling because of the additional losses in the guide vanes. Figure 23.28 shows typical performance characteristics of a torque converter.

FIGURE 23.28

Typical characteristics of a torque converter

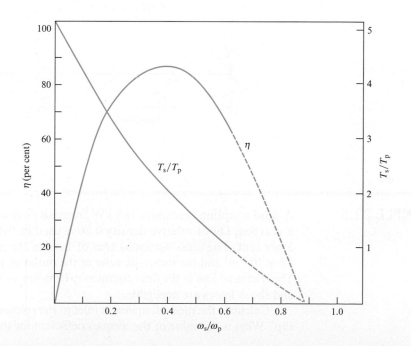

If a large speed reduction is required, torque converters usually incorporate more than one set of stationary guide vanes and secondary runners.

Figures 23.26 and 23.28 show that when $\omega_s \rightarrow \omega_p$ (or $\omega_s/\omega_p \rightarrow 1$) the efficiency of a torque converter is low, while that of a coupling is high. It is advantageous, therefore, to design the guide vanes so that they are allowed to rotate. Under normal operation, when $\omega_p > \omega_s$, the stator is held in a fixed position by the torque on the stator vanes, but when the speed of the primary approaches that of the secondary, the vanes are allowed to freewheel. Thus, they cease to influence the torque transmitted, T_v becomes zero and the torque converter performs as a fluid coupling. Typical performance characteristics of such a torque converter are shown in Fig. 23.29.

FIGURE 23.29

Typical characteristics of a torque converter capable of operation as a fluid coupling

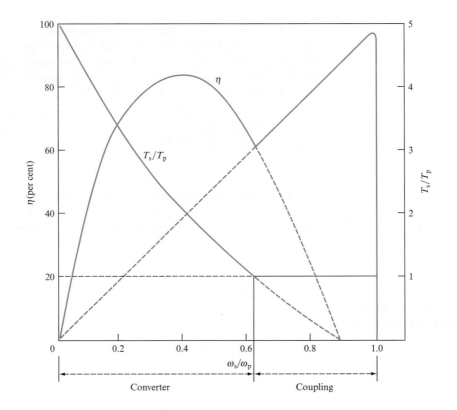

EXAMPLE 23.5

A fluid coupling transmits 185 kW from an engine running at 2250 rev min^{-1} to a gearbox. Oil of relative density 0.86 is used in the coupling, which has a slip of 3 per cent. The cross-sectional area of flow in the primary member is constant at 2.8×10^{-2} m^2 and the mean diameter at the outlet of the primary member is 460 mm. The frictional loss in the fluid circuit may be taken as 3.5 times the mean velocity head and shock losses are negligible.

Calculate the mean diameter at inlet to the primary member assuming zero 'whirl slip'. What is the value of the torque coefficient for this coupling?

Solution

Slip is given by

$$s = (\omega_p - \omega_s)/\omega_p = 0.03.$$

Therefore,

$$1 - \omega_s/\omega_p = 0.03$$

and so the efficiency is

$$\eta = \omega_s/\omega_p = 1 - 0.03 = 0.97.$$

Power lost in friction is $0.03 \times 185 = 5.55$ kW. But power lost in friction is also given by $\rho g Q h_f$ and

$$h_f = 3.5v^2/2g = 3.5\,(Q/A)^2/2g = 3.5[Q^2/(2.8 \times 10^{-2})^2]/2g.$$

Therefore,

$$5.55 \times 10^3 = 0.86 \times 10^3 g Q \times 3.5[Q^2/(2.8 \times 10^{-2})^2]/2g$$

$$Q^3 = 5.55 \times 2 \times (2.8)^2 \times 10^{-4}/(0.86 \times 3.5) = 29 \times 10^{-4}$$

$$Q = \mathbf{0.143\ m^3\ s^{-1}}.$$

Energy transfer to the primary is

$$E = (u_o v_{wo} - u_i v_{wi})/g,$$

where the suffixes o and i refer to the outlet and inlet, respectively. Therefore,

$$u_o = \omega_p r_o, \quad u_i = \omega_p r_i, \quad v_{wo} = \omega_p r_o, \quad v_{wi} = v_{w_4} = v_{w_1} = \omega_s r_i$$

and, hence,

$$E = (\omega_p^2 r_o^2 - \omega_p \omega_s r_i^2)/g = \omega_p^2[r_o^2 - (\omega_s/\omega_p)r_i^2]/g.$$

The power transmitted is

$$P = \rho g Q \omega_p^2[r_o^2 - (\omega_s/\omega_p)r_i^2]/g = \rho Q \omega_p^2[r_o^2 - (\omega_s/\omega_p)r_i^2].$$

Now, $\quad \omega_p = 2\pi N/60 = 2\pi \times 2250/60 = 236$ rad s^{-1}

and $\quad \omega_s/\omega_p = 0.97.$

Therefore,

$$185 = 0.86 \times 0.143 \times (236)^2[(0.46/2)^2 - 0.97r_i^2],$$

$$(0.23)^2 - 0.97r_i^2 = 185/[0.86 \times 0.143 \times (236)^2]$$

$$0.97r_i^2 = 0.053 - 0.0272$$

$$r_i^2 = 0.0265$$

$$r_i = 0.163\,\text{m} = 163\,\text{mm}.$$

Hence, $\quad\quad\quad D_i = \mathbf{326\,mm}.$

Torque coefficient is calculated from

$$\text{Torque coefficient} = T/\rho\omega_p^2 D^5 = P/\rho\omega_p^3 D^5$$
$$= 185/[0.86 \times (236)^3 \times (0.46)^5]$$
$$= \mathbf{0.0008}.$$

23.15 WIND ENERGY EXTRACTION, HORIZONTAL AND VERTICAL AXIS GENERATORS

In the search for renewable sources of energy the development of the windmill, a device with a centuries old tradition, has attracted considerable attention. Studies of wind energy availability have led to the identification of suitable sites for 'wind farms', notably in the USA but also in Europe, for example the installations in the Netherlands along the sea coast barrages of the Zuider Zee. In order to discuss the potential of wind energy it is first necessary to analyze the potential of the wind turbine by the application of the principles of momentum and continuity introduced earlier, Chapter 5.

FIGURE 23.30
(a) Air flow over a rotating propeller, illustrating flow acceleration and slipstream boundary.
(b) Air flow over a wind turbine, illustrating flow deceleration and slipstream boundary

Consider a propeller in an undisturbed steady flow, Fig. 23.30(a), approach velocity u_1. The applied torque to the propeller is used to generate thrust and hence a propulsive force to the attached craft. The approaching air flow is accelerated through the rotor, the velocity rising from u_2 to u_3 across the rotor itself and accelerating downstream to a velocity u_4. The airstream affected reduces in area downstream. While the pressure rises across the rotor, it must return to its upstream steady flow value downstream of the rotor location. Hence all the components for a momentum equation analysis are present, as the diameter and rotational speed of the rotor would be set and known.

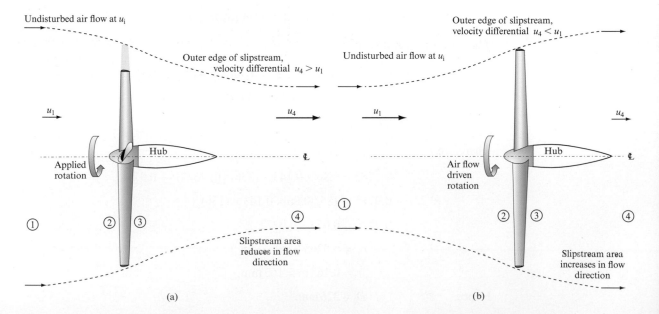

(a)   (b)

Similarly, consider a wind turbine mounted in an air flow. Here the objective is to extract energy from the air flow and hence the airstream decelerates as it passes the rotor section and so the downstream affected flow area increases rather than decreasing as is the case for the propeller. However, the application of the momentum equation remains the same in principle, Fig. 23.30(b).

For either case consider the force acting on the fluid. Applying Bernoulli's equation from section 1 to 2 upstream of the rotor and from section 3 to 4 downstream of the rotor yields

$$p_1 + 0.5\rho u_1^2 = p_2 + 0.5\rho u_2^2 \tag{23.62}$$

and

$$p_3 + 0.5\rho u_3^2 = p_4 + 0.5\rho u_4^2, \tag{23.63}$$

which may be combined to relate conditions across the rotor as the pressures at the upstream and downstream sections 1 and 4 are the same and the velocities at 2 and 3, on either side of the rotor, may be taken as identical by continuity if the rotor is assumed to have minimal flow direction thickness. Hence,

$$p_2 - p_3 = 0.5\rho(u_1^2 - u_4^2). \tag{23.64}$$

Note that these equations will apply to both cases, for the turbine the pressure drops across the turbine rotor, the opposite being true for the propeller.

The thrust on the wind turbine, or the thrust applied to the air flow in the propeller case, may be expressed as

$$F = \rho Q(u_1 - u_4), \tag{23.65}$$

where Q is the throughflow, $Q = A_i u_i$, where $i = 1$ to 4. The force may also be expressed in terms of the pressures on either side of the rotor disc as

$$F = A(p_2 - p_3), \tag{23.66}$$

where the sign again reflects the application and A is the area of the rotor. Combining equations (23.64) and (23.66) with $Q = u_2 A$, defines the velocity at the rotor as the mean of the upstream and downstream values, a result initially due to Froude,

$$u_2 = u_3 = 0.5(u_1 + u_4). \tag{23.67}$$

For a propeller application the rate of doing useful work, i.e. the power output, may be expressed for motion at velocity u_1 through a stationary fluid as

$$P = u_1 \rho Q(u_1 - u_4), \tag{23.68}$$

to which must be added the wasted kinetic energy imparted to the surrounding fluid that is set in motion relative to ground at a velocity $(u_4 - u_1)$, allowing an expression for propeller efficiency to be expressed as

$$\eta = [u_1 \rho Q (u_1 - u_4)]/[u_1 \rho Q(u_1 - u_4) + 0.5\rho Q(u_4 - u_1)^2], \tag{23.69}$$

which reduces to

$$\eta = u_1/[u_1 + 0.5\,(u_4 - u_1)]. \tag{23.70}$$

It should be noted that this is very much a theoretical upper limit to propeller efficiency as no account is taken of friction or the swirl components of velocity imparted by the propeller action.

For the wind turbine a similar efficiency may be developed. The definition applied is to express efficiency as the ratio of the loss of kinetic energy suffered by the airstream, i.e. the opposite of the kinetic energy loss term for the propeller analysis, to the undisturbed power of the airstream passing through the rotor disc area, hence

$$\eta = [0.5\rho A u_2(u_1^2 - u_4^2)]/(0.5\rho A u_1 u_1^2), \tag{23.71}$$

which reduces to

$$\eta = 0.5(u_1 + u_4)(u_1^2 - u_4^2)/u_1^3. \tag{23.72}$$

The maximum efficiency can then be determined for a frictionless system with no account of swirl components as

$$\eta = 0.5(1 + u_r - u_r^2 - u_r^3), \tag{23.73}$$

where $u_r = u_4/u_1$ and the maximum efficiency is obtained when $\mathrm{d}\eta/\mathrm{d}u_r = 0$,

$$\mathrm{d}\eta/\mathrm{d}u_r = 0.5\,(1 - 2u_r - 3u_r^2) = 0, \tag{23.74}$$

$$u_r = 1/3 \tag{23.75}$$

and

$$\eta = 0.5\,(1 + 1/3 - 1/9 - 1/27) = 16/27 = 59.3\%. \tag{23.76}$$

The above derivation of η, also referred to in some texts as the power coefficient, makes no acknowledgement of the disturbed nature of the airstream due to the presence of the wind turbine. Referring to Fig. 23.30(a) and (b), it will be appreciated that the actual velocity profiles downstream of the rotor merge to zero remote from the rotor centreline, rather than having the values indicated. Experimental studies have suggested that the efficiency may be slightly higher than the theoretical value derived due to the depression effect behind the turbine drawing in a mass of air into a vortex thereby increasing in momentum and power. However, the theoretical efficiency does not take into account the aerodynamic losses, or the mechanical and electrical power losses which will overall severely limit the efficiency of the machine. Wind energy does offer substantial advantages, however, within a renewable energy policy, as already demonstrated by the installation of wind farms, particularly in California where, in 1995, 15 per cent of the state electrical power generation was by this process.

23.16 APPLICATION OF WIND TURBINES TO ELECTRICAL POWER GENERATION

In 1996 there were more than 20 000 wind turbines worldwide, with an installed power generation capacity in excess of 2200 MW. The windmill, however, has a long and fruitful history, dating from 200 BC in China and the Middle East. Europe had windmills from the twelfth century, their numbers reaching a peak in Holland and the UK in the 1700s. The advent of the steam engine obviously reduced this form of natural energy extraction; however, the windmill continued to be used for the traditional tasks of grinding and water pumping up to the present day in remote locations and developing country applications, particularly when linked to water supply provision. The 'modern' wind turbine era may be roughly dated from the oil supply problems of the 1970s and while the more dire predictions of oil shortage and energy deficits have failed to materialize, the growth in sustainability research and the constraints on energy utilization and generation make wind energy a highly promising area for further development of known and well-understood technologies.

It has been shown, equation (23.68), that the power output of the wind turbine depends upon the cube of airstream velocity and the square of the diameter of the rotor. Thus a three-fold increase in wind speed will generate 27 times the power output for a given rotor diameter. Therefore siting of the wind turbine is of the utmost sensitivity. As an example of the likely power output from horizontal axis wind turbines, predicted maximum theoretical power values rise from 0.4 kW to 23 kW as rotor diameter rises from 3.8 m to 30 m in a 10 mph wind and from 46 kW to 2950 kW for the same rotors with a wind speed of 50 mph, results in line with the airstream velocity and rotor diameter dependencies mentioned above. These results assume that the wind turbine converts some 60 per cent of the wind energy into usable output; however, aerodynamic, mechanical and electrical losses further reduce the actual output to between 50 and 70 per cent of these values.

In the UK the Meteorological Office has developed data to represent the annual mean wind speeds over open sites as a guide to siting and sizing of wind turbines. In recent studies the possibility of offshore wind farms has been investigated, as such a siting avoids the objections on grounds of visual intrusion and noise, let alone disruption to wildlife. A sea-borne site would, however, have other obvious problems, particularly weather and safety related issues.

Generally, the most efficient design remains the horizontal axis two-bladed wind turbine, Fig. 23.31, with rotor blades historically up to 30 m, shown here in comparison with other 'landscape intrusions', such as telecommunications repeater masts and the rarer vertical axis wind turbine. In strong winds load is avoided by feathering the blades, identical to that process on propeller-driven aircraft when an engine is taken out of use. Modern horizontal axis wind turbines feature blade control. The unsteady nature of the power generation leads to the need for storage via battery systems as well as the possibility of direct supply to an a.c. load, as shown in Fig. 23.32, along with the traditional 'Dutch windmill' and the multi-vane unit conventionally used in a rural or developing world application for water pumping.

A disadvantage of the horizontal axis turbine is access to the generator in the blade hub and in recent years some interest has been shown in vertical axis wind turbines, Fig. 23.33, where the generator and all associated controls may be mounted

FIGURE 23.31
Comparison of wind
turbine heights with other
windmills and pylons

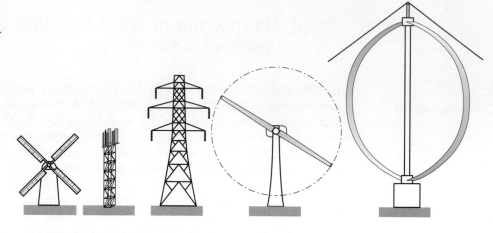

FIGURE 23.32
Conventional
horizontal axis wind
turbine compared with
the traditional 'Dutch'
windmill and the multi-
vane windmill much used
in developing country
applications

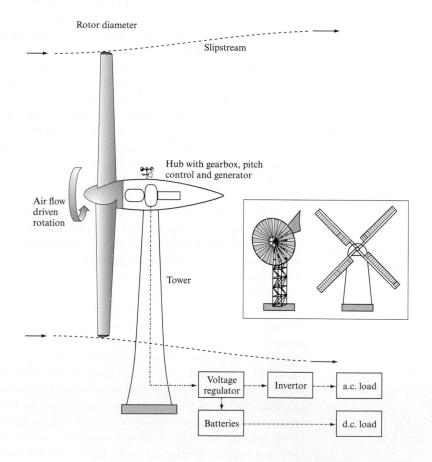

FIGURE 23.33

Vertical axis wind turbine
– Darrieus rotor.
Efficiency can equal
horizontal axis machines,
compared with the less
efficient Savonius rotor
principle (top left)

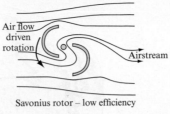

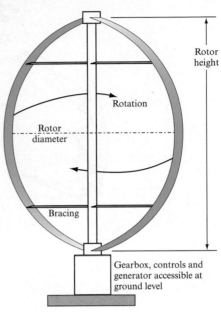

FIGURE 23.34

Power coefficient
variation with tip speed
and pitch for a given
typical horizontal axis
wind turbine

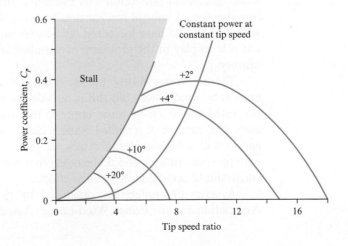

FIGURE 23.35
Wind contour map of
the UK

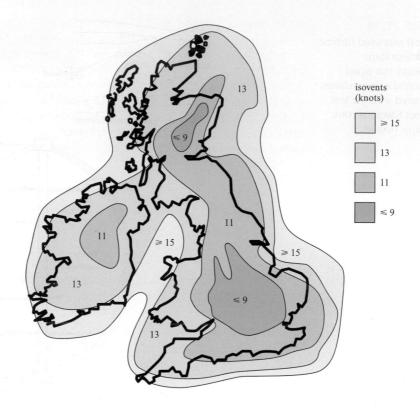

isovents
(knots)

≥ 15

13

11

≤ 9

at the base of the vertical axis. Vertical axis turbines are generally less efficient than the more traditional horizontal axis machines. Improvements in blade design and the introduction of control mechanisms to allow the rotor and generator speed to vary with wind speed have increased efficiencies markedly. Wind energy has an important role to play in the provision of sustainable energy generation in the twenty-first century.

There is now a substantial literature that addresses the sustainable energy generation by wind power, but this is outside the scope of this text. The aerofoil theory presented here does, however, apply to the study of wind turbine design and efficiency. The concept of a stalled blade is clearly applicable. Figure 23.34 illustrates the earlier stall at high angles of attack, or pitch, and the variation of power coefficient with tip speed ratio. It should be noted that holding the tip speed subsonic is a design constraint to avoid compressibility effects.

A series of conferences organized by the British Hydromechanics Research Association and the British Wind Energy Association over the past 25 years provides an interesting perspective on the changing research objectives. Similarly, research methodologies designed to aid siting and sizing of wind turbines may be accessed through the Meteorological Office and UK government publications, again over the past 30 years. Figure 23.35 illustrates the wind contours, or isovents, for the UK and Eire zones, explaining the attraction of offshore wind turbine installations. International interest in wind power generation remains strong with substantial publications and guidance available from the US Department of Energy, as well as a wide range of texts designed to cover the whole spectrum of interest in this topic, in particular addressing methodologies for energy storage, necessary as the wind is an unsteady source of energy.

EXAMPLE 23.6

A wind turbine used to generate electricity differs in a number of important ways from a water turbine operating within a conduit. While the air pressure remotely upstream and downstream of the rotor remains at atmospheric pressure, the air flow velocity drops across the rotor while the cross-section of the affected air flow increases.

Using dimensional analysis determine the dependence of generated power on wind speed and turbine rotor diameter, and hence determine the increase in wind velocity or the increase in rotor diameter required to double the power generated by a wind turbine belonging to a dimensionally similar family of machines. Detail the factors determining wind turbine power generation.

Solution

Identify the variables involved, neglecting for simplicity the geometric terms such as hub diameter, blade chord and roughness, which are all non-dimensionalized with respect to rotor diameter D.

VARIABLE	M	L	T
Power, P	1	2	-3
Diameter, D	0	1	0
Rotational speed, N	0	0	-1
Upstream wind velocity, V_1	0	1	-1
Air density, ρ	1	-3	0
Air dynamic viscosity, μ	1	-1	-1

Six variables in three dimensions require three groups with ρ, D and N as the repeating variables.

By inspection, the groups are power coefficient $P/\rho N^3 D^5$, tip speed ratio V_1/ND, and Reynolds number $\rho V_1 D/\mu$. Note that tip speed ratio replaces the flow rate coefficient for water turbines and pumps/fans due to the absence of an enclosing conduit. (ND is the tip speed.)

Hence P depends on D^2 and $(ND)^3$, and therefore to double power output the wind speed would have to rise by a factor $2^{0.33} = 1.28$, or the rotor diameter would have to increase by $2^{0.5} = 1.41$.

In summary the output of wind generators depends upon:

1. the cube of the wind velocity through the rotor;

2. the square of rotor diameter;

3. the nature of the wind in terms of its inherent unsteady nature and its gust frequency;

4. the overall efficiency of the mechanical and electrical components of the turbine, which when compounded with the maximum possible theoretical efficiency derived can reduce overall efficiencies to below 30 per cent.

Concluding remarks

In defining the performance of rotodynamic machines, this chapter has utilized the theory presented in Chapter 22, together with earlier material, particularly in terms of dimensional analysis and similarity (Chapters 8 and 9). It has been shown that the general approach is equally applicable to pumps, fans and turbines. The application of dimensional analysis and similarity has been shown to be essential if the results of machine testing are to be translated into operational units whose likely performance in any particular network may be predicted at the network design stage. Again it may be stressed that the necessary equations to allow the development of machine performance prediction are firmly rooted in the fundamental principles of continuity, energy and momentum, together with detailed knowledge of the flow patterns within the machine that is informed by work on aerofoils.

This chapter also defined the likely machine design to cope with particular flow conditions, for both pumps/fans and turbines.

The machine characteristics developed will be utilized in Chapter 25 to determine the operating point for a machine, or multiple machines, introduced into a fluid network.

Summary of important equations and concepts

1. This chapter introduces the fundamental relationships that define pump, fan or turbine performance in terms of flow rate, pressure change, machine speed and characteristic dimensions. These machine characteristics form the basis for any system analysis to determine machine/system operating points, equations (23.25) and (23.26).

2. The loss terms contributing to the determination of machine efficiency are identified in Section 23.2, which also stresses that machine characteristics have to be experimentally determined prior to the application of the similarity laws, Section 23.3.2, to determine the characteristics of 'families' of machines obeying the laws of geometric and dynamic similarity.

3. The application of the similarity laws is stressed in Section 23.3 and through the computer program SIMPUMP, Section 23.4.

4. The significance of type number or specific speed is demonstrated for both pumps and turbines, and the expected values for centrifugal (0.5–1.6) and axial (2.8–4.8) pumps and fans (1.2–5.0 for various designs) are suggested.

5. Turbines are discussed in Section 23.10, including the Pelton wheel, Francis and Kaplan designs covering reaction, centrifugal and axial machines. In each case the machine efficiency is developed and the zone of appropriate application defined drawing on the techniques developed in Chapter 22 to determine machine/fluid velocity relationships.

6. The application of wind turbines is introduced and dimensional analysis employed to identify the dependencies on wind velocity and rotor diameter. The inefficiencies inherent in the system are also investigated and quantified.

Problems

23.1 A 0.4 m diameter fan running at 970 rev min^{-1} is tested when the air temperature is 10 °C and the barometric pressure is 772 mm Hg and the following data are observed: $Q = 0.7 \text{ m}^3\text{s}^{-1}$; fan total pressure = 25 mm of water; shaft power = 250 W.

Find the corresponding volume flow, fan total pressure and shaft power of a geometrically similar fan of 1 m diameter, running at 500 rev min^{-1} when the air temperature is 16 °C and the barometric pressure is 760 mm Hg. Assume that the fan efficiency is unchanged.

[5.64 m^{3}s^{-1}, 40 mm of water, 3.22 kW]

23.2 A centrifugal pump will operate at 300 rev min^{-1} delivering 6 m^{3}s^{-1} against 100 m head.

Laboratory facilities for a model are: maximum flow 0.28 m^{3}s^{-1} and maximum power available 225 kW. Using water and assuming that the efficiencies of model and

prototype are the same, find the speed of the model and the scale ratio. Also calculate the specific speed.

[1196 rev min^{-1}, 4.4, 0.439 (rad)]

23.3 A centrifugal fan delivers 2.0 m^{3}s^{-1} when running at 960 rev min^{-1}. The impeller diameter is 70 cm and the diameter at blade inlet is 48 cm. The air enters the impeller with a small whirl component in the direction of impeller rotation, but the relative velocity meets the blade tangentially. The impeller width at inlet is 16 cm and at outlet, 11.5 cm. The blades are backward inclined making angles of 22.5° and 50° with the tangents at inlet and outlet, respectively.

Draw to scale the inlet and outlet velocity triangles and from them determine the theoretical total head produced by the impeller.

Assuming that the losses at inlet in the impeller and in the casing amount to 70 per cent of the velocity head at impeller outlet and that the velocity head at fan discharge is 0.1 of the velocity head at impeller outlet, calculate the fan static pressure in millimetres of water column. Assume the air density to be 1.2 kg m^{-3} and neglect the effect of blade thickness and interblade circulation.　　[66.6 mm]

23.4 A centrifugal pump running at 2950 rev min^{-1} gave the following results at peak efficiency when pumping water during a laboratory test:

Effective head:　　　$H = 75$ m of water
Flow rate:　　　　　$Q = 0.05 \, \text{m}^3 \text{s}^{-1}$
Overall efficiency:　$\eta = 76$ per cent.

(a) Calculate the specific speed of the pump in dimensionless terms and based on rotational speed in revolutions per second.
(b) A dynamically similar pump is to operate at a corresponding point of its characteristic when delivering 0.45 m^3 s^{-1} of water against a total head of 117 m. Determine the rotational speed at which the pump should run to meet the duty and the ratio of its impeller diameter to that of the model pump tested in the laboratory, stating any assumptions made. What will be the power consumed by the pump?
[(a) 0.078; (b) 1379 rev min^{-1}, 2.68, 679 kW]

23.5 An axial flow pump handling oil of specific gravity 0.8 at a rate of 1.0 m^3 s^{-1} is fitted with downstream guide vanes and its impeller rotates at 250 rev min^{-1}. The oil approaches the impeller axially, and the velocity of flow, which may be assumed constant from the hub to the tip, is 3.0 m s^{-1}. The pump consumes 60 kW, the overall efficiency being 77 per cent and the hydraulic efficiency (including guide vanes) is 86 per cent. If the impeller diameter is 0.8 m and the hub diameter is 0.4 m find the inlet and outlet blade angles and the guide vane inlet angles at the hub and at the tip. Assume that the total head generated at the hub and at the tip is the same.
[at hub: 29.8°, 159.06°, 13.2°; at tip: 16°, 36.5°, 25°]

23.6 In a Pelton wheel the diameter of the bucket circle is 2 m and the deflecting angle of the bucket is 162°. The jet is of 165 mm diameter, the pressure behind the nozzle is 1000 kN m^{-2} and the wheel rotates at 320 rev min^{-1}. Neglecting friction, find the power developed by the wheel and the hydraulic efficiency.　　[701 kW, 73.3 per cent]

23.7 A Pelton wheel nozzle, for which the coefficient of velocity is 0.97, is 400 m below the water surface of a lake. The jet diameter is 80 mm, the pipe diameter is 0.6 m, its length is 4 km and $f = 0.008$. The buckets deflect the jet through 165° and they run at 0.48 jet speed, bucket friction reducing the relative velocity at outlet by 15 per cent of the

relative velocity at inlet. The mechanical efficiency of the turbine is 90 per cent. Find the flow rate and the shaft power developed by the turbine.　　[0.42 m^3 s^{-1}, 1189 kW]

23.8 Three identical, double-jet Pelton wheels operate under a gross head of 400 m. The nozzles are of 75 mm diameter with a coefficient of velocity 0.97. The pitch circle of the buckets is 1.2 m diameter and the bucket speed is 0.46 × (jet velocity). The buckets deflect the jet by 165° and owing to friction the relative velocity is reduced by 18 per cent. The mechanical efficiency is 96 per cent. The water from the reservoir is supplied to the turbines by means of two parallel pipes, each of 0.5 m diameter and 450 m long, having a friction factor $f = 0.0075$.
　　If the quantity of water supplied to each turbine is 0.65 m^3 s^{-1} calculate the shaft power developed by it and its rotational speed.　　[1905 kW, 620 rev min^{-1}]

23.9 A vertical shaft Francis turbine has an overall efficiency of 90 per cent and runs at 428 rev min^{-1} with a water discharge of 15.5 m^3 s^{-1}. The velocity at the inlet of the spiral casing is 9 m s^{-1} and the pressure head at this point is 260 m of water, the centreline of the spiral casing inlet being 3.3 m above the tail water level. The diameter of the runner at inlet is 2.4 m and the width at inlet is 0.3 m. The hydraulic efficiency is 93 per cent. Determine the output power, the specific speed, the guide vane angle and the runner vane angle at inlet.　　[36 MW, 0.073, 8.7°, 37°]

23.10 An axial flow turbine operates under a head of 21.8 m and develops 21 MW when running at 140 rev min^{-1}. The external runner diameter is 4.5 m and the hub diameter is 2.0 m. If the hydraulic efficiency is 94 per cent and the overall efficiency is 88 per cent, determine the inlet and outlet blade angles at the mean radius.　　[30°, 20° 20′]

23.11 An axial flow turbine, with fixed stator blades upstream of the rotor, running at 250 rev min^{-1} has an outer diameter of 1.8 m and an inner diameter of 0.75 m. At the mean diameter the outlet angle is 140° in the stator and the rotor blade angle at inlet is 30°, both measured from the direction of blade velocity. Determine (a) the flow rate for which the angle of incidence for the rotor blades is zero assuming that the axial velocity is uniform, (b) the rotor blade angle at outlet if the whirl component there is zero, (c) the theoretical power output if the change of whirl is independent of the radius.
[(a) 12 m^3 s^{-1}, (b) 18.9°, (c) 1362 kW]

23.12 A fluid coupling has a mean diameter at inlet equal to 0.6 of the mean diameter at outlet which is 0.38 m. The efficiency of the coupling is 96.5 per cent and the specific gravity of the oil used is 0.85. Assuming that the cross-sectional area of the flow passage is constant and equal to 0.026 m^2 and that the loss round the fluid circuit is four times the mean velocity head, calculate the power transmitted by the coupling from an engine running at 2400 rev min^{-1}.

[167.7 kW]

23.13 The torque coefficient of a particular design of a hydraulic coupling is given approximately by

$$\frac{T}{\rho \omega_p^2 D^5} = 0.25\left(1 - \frac{\omega_s}{\omega_p}\right).$$

A 250 mm diameter coupling using a working fluid of specific gravity 0.85 is driven by a motor running at 1450 rev min^{-1}. What is the maximum torque which could be transmitted if the 'slip' is not to exceed 4 per cent?

If the power output is to be increased by 20 per cent to what value should the input speed be increased in order to maintain the same slip? [191 N m, 1540 rev min^{-1}]

Positive Displacement Machines

THIS CHAPTER TREATS THE RECIPROCATING MACHINE IN MUCH THE SAME MANNER AS the rotodynamic machines covered in Chapters 22 and 23. The essential differences in machine performance are highlighted and the momentum equation utilized, as in Chapter 19, to determine the operating pressures of a machine–network combination, where the system losses are represented by application of the steady flow energy equation. Rotary vane and gear pumps are discussed, together with hydraulic motors. ● ● ●

24.1 RECIPROCATING PUMPS

A reciprocating pump consists essentially of a piston moving to and fro in a cylinder. The piston is driven by a crank powered by some prime mover such as an electric motor, IC engine or stream engine. Small portable reciprocating pumps may be hand operated.

When a piston moves away from the valve end of the cylinder, i.e. to the right in Fig. 24.1, pressure is reduced in the cylinder. This enables atmospheric pressure p_a acting on the free surface of the liquid in the lower reservoir to force the liquid up the suction pipe and into the cylinder. The suction valve (2) is a one-way valve and opens when the liquid is moving into the cylinder. Thus the outward motion of the piston constitutes a suction stroke. It is then followed by a delivery stroke during which the liquid in the cylinder is pushed out through the delivery valve (3) and into the upper reservoir. During the delivery stroke, valve (2) is closed because of the fluid pressure exerted on it. The whole cycle is then repeated at a frequency dependent upon the rotational speed of the crank ω. Each cycle may be represented by a plot of pressure in the cylinder against the volume of the liquid, as shown in Fig. 24.2. During the suction stroke the pressure in the cylinder is below atmospheric as represented by line ab in the diagram. On reversal of the direction of motion of the piston, i.e. at the end of the suction stroke and beginning of the delivery stroke, the pressure rises abruptly along the line bc while the volume remains the same. The delivery stroke follows, during which the high delivery pressure is maintained. This is represented by the line cd. At the end of the delivery stroke the pressure falls along da and the cycle starts again.

However, the simplified analysis above does not take into account the effects that are actually present, namely that of the inertia of the liquid in the pipes, which opposes any changes in velocity, and that of the frictional losses in the pipes.

At the end of each stroke the liquid in the cylinder and in the relevant pipe must be brought to rest, i.e. decelerated. Immediately afterwards at the beginning of the following stroke, the fluid in the cylinder and in the associated pipe must be accelerated. These accelerations and decelerations result in additional pressures being involved. It was shown in Chapter 19 that the inertia pressure is given by

$$p_i = \rho g h_i = \rho l \frac{\mathrm{d}v}{\mathrm{d}t},$$

where l is the length of pipe and $\mathrm{d}v/\mathrm{d}t$ is the acceleration of the fluid. If the cross-sectional area of the cylinder is A and that of the pipe is a then by continuity

$$av = Au,$$

if u is the velocity of the piston, so that

$$\frac{\mathrm{d}v}{\mathrm{d}t} = \frac{A}{a}\frac{\mathrm{d}u}{\mathrm{d}t},$$

and

FIGURE 24.1
Reciprocating pump
installation

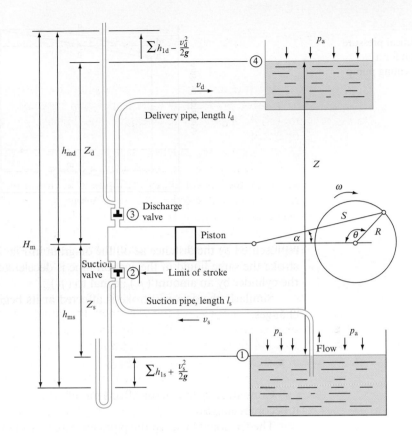

FIGURE 24.2
Basic pressure diagram
for a reciprocating pump

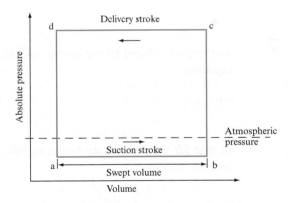

$$p_i = \rho l \frac{A}{a} \frac{du}{dt}.$$ (24.1)

Thus at the beginning of the suction stroke (point a) the liquid in the suction pipe
must be accelerated so that the pressure in the cylinder must be lowered by an amount

$$(p_i)_{ae} = \rho l_s \frac{A}{a_s} \frac{du}{dt},$$

FIGURE 24.3

Theoretical pressure
diagram for a
reciprocating pump

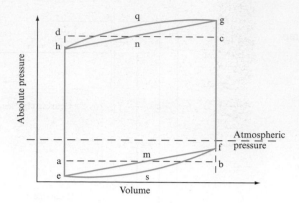

represented by the distance ae on the diagram in Fig. 24.3. At the end of the suction
stroke the same liquid in the suction pipe is decelerated so that it exerts pressure on
the cylinder by an amount $(p_i)_{fb}$ equal to $(p_i)_{ae}$.

Similarly the delivery stroke is affected at its beginning and its end by pressure
changes:

$$(p_i)_{cg} = -(p_i)_{dh} = \rho l_d \frac{A}{a_d} \frac{du}{dt}.$$

Consequently the inertia effects modify the simple pressure diagram abcd so that
it becomes emfgnh.

The frictional losses in the pipes are given by the Darcy equation

$$h_f = \frac{4fl}{d} \frac{v^2}{2g},$$

and may be related to the piston velocity by substitution of v from the continuity
equation

$$v = \frac{A}{a} u,$$

so that for the delivery stroke during which the frictional losses in the delivery pipe
become relevant,

$$h_{fd} = \frac{4fl_d}{d_d} \left(\frac{A}{a_d}\right)^2 \frac{u^2}{2g}. \tag{24.2}$$

Now, the piston velocity may be obtained from the displacement equation,
assuming simple harmonic motion, in terms of the crank radius r and crank angle θ
and ultimately in terms of the angular velocity ω and time t, giving

$$u = \omega r \sin \theta = \omega r \sin \omega t.$$

The equation shows that when $\theta = 0$, i.e. at the beginning and at the end of each
stroke, $u = 0$ and therefore the frictional effects are zero, but the acceleration (or

deceleration) $\mathrm{d}u/\mathrm{d}t$ is a maximum and hence the inertia effect are maximum. When $\theta = 90°$ or $270°$, i.e. at the middle of each stroke, the velocity is maximum and hence the frictional effects reach a maximum, whereas the acceleration (or deceleration) is zero and therefore the inertia effects vanish. Also, it follows from the Darcy equation that $h_f \propto v^2$, so that curves esf and gqh representing frictional effects on the diagram are parabolae superimposed on lines emf and gnh. Thus the theoretical pressure diagram becomes esfgqh.

The work done by the piston may be obtained as follows. If the instantaneous pressure in the cylinder is p then the force exerted by the fluid on the piston is Ap. If now the piston moves a distance δx then the instantaneous work done is $Ap\delta x$. The total work done during the cycle is therefore

$$\mathrm{WD/cycle} = A \int p\,\mathrm{d}x, \tag{24.3}$$

and is represented by the area of the pressure diagram.

The rate at which the liquid is delivered by the pump clearly depends upon the pump speed, since

$$\text{Volume delivered in one stroke} = AS = V,$$

where S = piston stroke and V = swept volume.

Thus, if the pump speed is N (rev s^{-1}), then the theoretical volume delivered in 1 s is

$$Q_{\mathrm{th}} = ASN,$$

and since $N = \omega/2\pi$,

$$Q_{\mathrm{th}} = AS\omega/2\pi = V\omega/2\pi. \tag{24.4}$$

Thus the pump discharge is directly proportional to the rotational speed and is entirely independent of the pressure against which the pump is delivering.

Because of the leakage of liquid through glands the actual pump discharge is smaller than the theoretical. If the leakage is q then the actual delivery,

$$Q = Q_{\mathrm{th}} - q \tag{24.5}$$

and the volumetric efficiency,

$$\eta_v = Q/Q_{\mathrm{th}} = Q/(Q + q). \tag{24.6}$$

Sometimes an expression known as *slip* is used:

$$\mathrm{Slip} = \frac{Q_{\mathrm{th}} - Q}{Q_{\mathrm{th}}} = 1 - \eta_v. \tag{24.7}$$

The pump pressure depends upon the system against which it is working, as shown in Fig. 24.1, and may be obtained as follows. Applying the energy equation to points (1) and (4) and using the liquid level in the lower reservoir as datum,

$$\frac{\text{Total energy}}{\text{at (1)}} + \text{WD by pump} = \frac{\text{Total energy}}{\text{at (4)}} + \frac{\text{Losses in}}{\text{the system,}}$$

but

Total energy at (1) per unit mass of fluid flowing $= p_a/\rho$,

WD by the pump per unit mass of fluid flowing $= gH$ (where $H =$ pump head),

Total energy at (4) per unit mass of fluid flowing $= gZ + p_a/\rho$,

Losses in the system $= g\Sigma h_{1s} + g\Sigma h_{1d}$.

Substituting,

$$p_a/\rho + gH = gZ + p_a/\rho + g\Sigma h_{1s} + g\Sigma h_{1d}.$$

Thus, rearranging, the pump head,

$$H = Z + \Sigma h_{1s} + \Sigma h_{1d}. \tag{24.8}$$

For reasons explained later it is sometimes useful to consider the pump suction side and delivery side separately and this is why the static lift Z (the difference between the water levels if the reservoirs are open to atmosphere) is considered to be the sum of the suction lift Z_s and delivery lift Z_d. If the difference in levels between the inlet and outlet to the pump is neglected (between (2) and (3) in the diagram) because it is usually very small compared with the rest of the system, the two sides of the pump may be analyzed separately as follows.

Applying the steady flow energy equation between the lower reservoir (assuming constant water level) and the pipe at pump inlet (2),

$$\frac{p_a}{\rho g} = \frac{p_s}{\rho g} + Z_s + \frac{v_s^2}{2g} + \Sigma h_{1s},$$

from which the static pressure at pump inlet

$$\frac{p_s}{\rho g} = \frac{p_a}{\rho g} - \left(Z_s + \frac{v_s^2}{2g} + \Sigma h_{1s} \right). \tag{24.9}$$

The expression in brackets is known as the manometric suction head, h_{ms}, because it represents the negative gauge pressure shown by a manometer attached to the pump inlet.

Similarly, applying the steady flow energy equation between the delivery pipe at the pump outlet (3) and the water level in the upper reservoir,

$$\frac{p_d}{\rho g} + \frac{v_d^2}{2g} = \frac{p_a}{\rho g} + Z_d + \Sigma h_{1d},$$

so that the static pressure at pump outlet,

$$\frac{p_d}{\rho g} = \frac{p_a}{\rho g} + \left(Z_d + \Sigma h_{1d} - \frac{v_d^2}{2g} \right). \tag{24.10}$$

Here the expression in brackets is the manometric delivery head, h_{md}, because it represents the gauge pressure shown by a manometer connected to the pump outlet. The pump manometric head H_m is defined by

$$H_m = h_{md} + h_{ms}. \tag{24.11}$$

Therefore, substituting,

$$H_m = Z + \Sigma h_{1s} + \Sigma h_{1d} - (v_d^2 - v_s^2)/2g. \tag{24.12}$$

Comparing equations (24.8) and (24.12),

$$H - H_m = (v_d^2 - v_s^2)/2g.$$

If the delivery and suction pipes are of the same diameter, then $v_d = v_s$ and $H = H_m$.
The internal head, H_i, generated by the pump is greater than the pump head, H, the difference accounting for the internal losses within the pump, h_{1p}. Thus,

$$H_i = H + h_{1p}. \tag{24.13}$$

The internal fluid power generated by the pump is

$$P_i = \rho g H_i Q_{th}, \tag{24.14}$$

and the actual power output of the pump is

$$P = \rho g H Q. \tag{24.15}$$

If the power input to the pump from the prime mover is P_o, then the overall pump efficiency is:

$$\eta = P/P_o = \rho g H Q/P_o. \tag{24.16}$$

Component efficiencies are also useful and they are as follows:

Hydraulic efficiency, $\eta_h = H_m/H_i$. $\tag{24.17}$

Mechanical efficiency, $\eta_m = P_i/P_o$. $\tag{24.18}$

It follows therefore that the overall pump efficiency,

$$\eta = \frac{P}{P_o} = \frac{P_i}{P_o} \times \frac{P}{P_i} = \frac{P_i}{P_o} \times \frac{\rho g H Q}{\rho g H_i Q_{th}} = \frac{P_i}{P_o} \times \frac{H}{H_i} \times \frac{Q}{Q_{th}},$$

so that

$$\eta = \eta_m \eta_h \eta_v. \tag{24.19}$$

When considering a pump installation an important limitation on the location of the pump, in relation to the level of the lower reservoir, is the manometric suction head. It represents the lowest pressure in the system (at pump inlet), in particular during the beginning of the suction stroke represented by point e in Fig. 24.3. If this pressure falls to the value of the liquid vapour pressure, cavitation will occur and delivery will cease. This phenomenon is discussed fully in Section 25.10.

One major disadvantage of reciprocating pumps is the fluctuating flow. It can be reduced by fitting air cylinders to either the suction pipe or the delivery pipe or both. Air cylinders are closed vessels which act similarly to surge tanks. The decelerating liquid moves into the cylinder compressing the enclosed air and thus storing energy in it. When the fluid is accelerated the energy in the air is released, thus augmenting the accelerating force. By this process the fluctuations in the flow are smoothed out to an extent dependent upon the size of the air vessels. Figure 24.4 shows an actual indicator diagram of a pump fitted with air cylinders. It shows that the effects of inertia have been largely eliminated.

FIGURE 24.4

Indicator diagram for a reciprocating pump with air cylinders

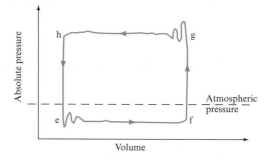

The provision of an air chamber reduces the total friction loss to be overcome in the system by maintaining a steady flow from the pump. Without the air chamber the work done against friction would vary with piston position and, as the flow–friction loss relationship is parabolic, the mean frictional resistance is two-thirds of the maximum, i.e. at mid-discharge stroke,

$$\text{Mean friction loss} = \frac{2}{3} \times \frac{4fL}{2dg}\left(\frac{A\omega r}{a}\right)^2.$$

With an air chamber close to the pump, so that the inertia effects in the short connecting pipe may be ignored, then

$$\text{Mean friction loss} = \text{Constant flow-based loss}$$

$$= \frac{4fL}{2dg}\left(\frac{A}{a}\frac{\omega r}{\pi}\right)^2.$$

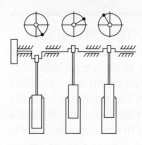

FIGURE 24.5

Three-cylinder, single-acting ram pump

Therefore,

$$\frac{\text{Work done with air vessel}}{\text{Work done without air vessel}} = \frac{1/\pi^2}{2/3} = \frac{3}{2\pi^2}.$$

Another method of dealing with the fluctuations in the flow is the use of multi-cyclinder pumps. In these several cylinders act in parallel and out of phase as shown in Fig. 24.5.

A less effective solution is provided by a double-acting pump in which both sides of the piston are connected to the suction and delivery pipes and both sides of the piston work on the fluid. When one side is on suction stroke the other side at the same time is on the delivery stroke.

Figure 24.6 compares the flow rate fluctuations and the mean deliveries of single-acting, double-acting and two-cylinder, double-acting pumps.

FIGURE 24.6

Variation of discharge with crank angle θ for:
(a) single-cylinder, single-acting pump;
(b) single-cylinder, double-acting pump;
(c) two-cylinder, double-acting pump

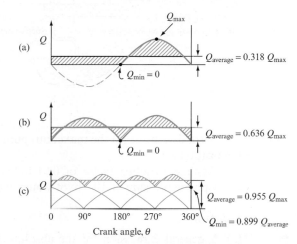

The theoretical plot of the pump head H against flow rate Q at a constant speed is a vertical straight line as shown in Fig. 24.7. However, as the head against which the pump is working is increased the flow rate is in practice slightly reduced because of internal leakage.

FIGURE 24.7

Characteristic of a reciprocating pump

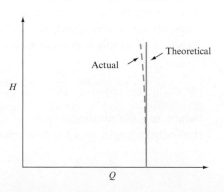

EXAMPLE 24.1

A single-acting, single-cylinder, positive displacement pump (Fig. 24.8) is used to drain an excavation. The pump has a bore of 150 mm and a stroke of 400 mm. The suction and discharge pipes are both of 50 mm diameter, the suction pipe being 2 m long and the discharge pipe 15 m along. The suction lift to the pump is 1.5 m while the discharge is 6 m above the level of the water surface in the excavation. In the absence of any air chambers on either (a) pump suction or (b) discharge, calculate for (a, b) the absolute pressure head in the cylinder at the (i) start, (ii) end and (iii) middle of each stroke if the pump drive is at 0.2 rev s⁻¹ and may be assumed to be simple harmonic.

FIGURE 24.8

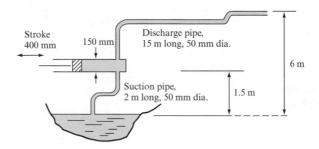

Also, determine (c) the maximum pump speed if separation is to be avoided on the piston face.

Assume a friction factor of 0.01 for both pipes, a pump slip of 4 per cent, an atmospheric pressure of 10.3 m of water, and a fluid vapour pressure of 2.4 m.

Solution

(a) A general expression for the absolute head in the cylinder during the suction stroke may be written as

$$H = H_{at} \qquad - H_s \qquad - H_{si} \qquad - H_{sf}.$$

| Atmosphere pressure | Suction lift | Suction acceleration | Friction in suction pipe |

While the suction lift term H_s remains a constant during the stroke the values of the inertia head, H_{si}, associated with the acceleration of the fluid along the suction pipe vary during the stroke depending on piston speed. Similarly, as the inflow to the pump depends on piston speed, the friction loss in the suction pipe depends on piston position. If simple harmonic motion is assumed then the piston speed is given by

$$u = \omega r \sin \omega t, \quad \frac{du}{dt} = w^2 r \cos wt, \quad \frac{du}{dt} = w^2 r \text{ when } t = 0$$

where ω is the rotational speed of the pump drive in radians per second and r is the half-stroke length, and the flow velocity in a pipe of area a is therefore

$$v = u\frac{A}{a},$$

where A is the cylinder cross-sectional area. At the beginning and end of the stroke $u = 0$, but the acceleration or inertia pressure necessary to move fluid is a maximum as dv/dt is a maximum at these times:

$$h_{si} = \pm \frac{L}{g} \frac{A}{a} \omega^2 r,$$

where L is the length of the pipe.

At mid-stroke time $dv/dt = 0$ but the piston velocity is a maximum, thus giving a maximum friction loss along the pipe:

$$h_{sf} = \frac{4fL}{2dg} \left(\frac{vA}{a} \right)^2,$$

where d is the diameter of the suction pipe.

Therefore the general equation is

$$H = H_{at} - H_s \pm \frac{L}{g} \frac{A\omega^2 r}{a} - \frac{4fL}{2dg} \left(\frac{vA}{a} \right)^2.$$

(i) Cylinder head at start of suction stroke, $u = 0$, so $h_{sf} = 0$. Therefore from the general equation above

$$H = 10.3 - 1.5 - \frac{2}{9.81} \left(\frac{150}{50} \right)^2 (0.2 \times 2\pi)^2 \frac{200}{1000}$$

$$= 10.3 - 1.5 - 0.58 = \textbf{8.22 m of water.}$$

(Note inertia head is negative as flow is accelerated.)

(ii) Cylinder head at end of suction stroke,

$$H = H_{at} - H_s + \frac{L}{g} \frac{A\omega^2}{a} r - \frac{4fL}{2dg} \left(\frac{vA}{a} \right)^2.$$

Again $u = 0$; note inertia head is positive as flow is decelerated to rest.

$$H = 10.3 - 1.5 + 0.58 = \textbf{9.38 m of water.}$$

(iii) Cylinder head at mid-suction stroke,

$$H = H_{at} - H_s - \frac{L}{g} \frac{du}{dt} - \frac{4fL}{2dg} \left(\frac{vA}{a} \right)^2,$$

but $du/dt = 0$ and $u = \omega r$.

$$H = 10.3 - 1.5 - \frac{4 \times 0.01 \times 2 \times 1000}{2 \times 50 \times 9.81} \times \left[0.2 \times 2\pi \times \frac{200}{1000} \left(\frac{150}{50} \right)^2 \right]^2$$

$$= \textbf{8.38 m.}$$

(b) During discharge the same form of the equation may be employed:

$$H = H_{at} + H_d \qquad + H_{di} \qquad + H_{df}.$$

Delivery   Delivery   Delivery
lift   acceleration   pipe friction

At the start and end of the stroke, $u = 0$; therefore $H_{df} = 0$ and at mid-stroke $du/dt = 0$ so that $H_{di} = 0$.

(i) Cylinder head at start of discharge stroke,

$$H = 10.3 + 4.5 + \frac{L}{g}\frac{A}{a}\omega^2 r$$

$$= 14.8 + \frac{15}{9.81}\left(\frac{150}{50}\right)^2 (0.2 \times 2\pi)^2 \frac{200}{1000}$$

$$H = \textbf{19.15 m of water.}$$

(ii) Cylinder head at end of discharge stroke,

$$H = 10.3 + 4.5 - \frac{L}{g}\frac{A}{a}\omega^2 r = \textbf{10.45 m of water.}$$

(iii) Cylinder head at mid-discharge stroke,

$$H = 10.3 + 4.5 + \frac{L}{g}\frac{du}{dt} + \frac{4fL}{2dg}\left(\frac{uA}{a}\right)^2.$$

Now $du/dt = 0$, $u = \omega r$, and as slip $= 4$ per cent,

$$(uA)_{actual} = 0.96\,(uA)_{theory}.$$

Therefore,

$$H = 10.3 + 4.5 + \frac{4 \times 0.01 \times 15 \times 1000}{2 \times 50 \times 9.81} \times \left[0.96 \times 0.2 \times 2\pi \left(\frac{150}{50}\right)^2 \frac{200}{1000}\right]^2$$

$$= \textbf{17.68 m of water.}$$

(c) Separation may be defined as the production of vapour in the cylinder on pump suction (see Section 25.10) and this requires the absolute pump pressure to fall to 2.4 m absolute. The minimum pressure occurs at the start of the suction stroke; thus

$$2.4 = 10.3 - 1.5 - \frac{L}{g}\omega^2 r \frac{A}{a}$$

$$-6.4 = -\frac{L}{g}\frac{A}{a}\omega^2 r$$

$$\omega^2 = 6.4 \times \frac{9.81}{2} \times \left(\frac{50}{150}\right)^2 \frac{1000}{200} = 17.44$$

$$\omega = 4.176 \text{ rad s}^{-1} = 4.176/2\pi \text{ rev s}^{-1} = 0.665 \text{ rev s}^{-1}.$$

The drive speed for separation, and hence maximum pump rotation, is **40 rev min**$^{-1}$.

EXAMPLE 24.2

A single-acting, single-cylinder, positive displacement pump, driven at 0.4 rev s⁻¹, has a bore of 200 mm and a stroke of 500 mm (Fig. 24.9). The suction and discharge pipes are both 100 mm in diameter. The suction lift is 0.4 m and the suction pipe is 3 m long. The water is discharged at a point 20 m above the pump level by means of a pipe 200 m long, fitted with a large air chamber 20 m from the pump. Calculate the absolute pump cylinder pressures at the (i) start, (ii) end and (iii) mid-stroke times for both (a) suction and (b) discharge assuming no slip at the pump and a friction factor of 0.008 for both pipes.

FIGURE 24.9

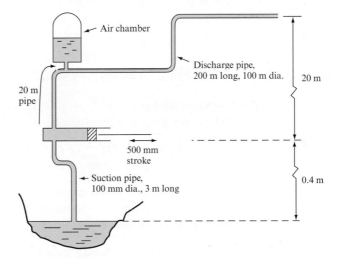

Take atmospheric pressure as 10.3 m.

Further assess the effect of the introduction of the air chamber.

Solution

(a) Suction stroke:

(i) Cylinder head at start of suction stroke,

$$H = H_{at} - H_{\substack{static \\ lift}} - H_{\substack{suction \\ acceleration}} - H_{\substack{suction \\ friction\ loss}}$$

$$= 10.3 - 0.4 - \frac{L}{g}\omega^2 r \frac{A}{a}, \text{ as } H_{sf} = 0,$$

i.e. as $u = 0$, $du/dt = $ maximum at start of stroke,

$$H = 10.3 - 0.4 - \frac{3}{9.81}(0.4 \times 2\pi)^2 \left(\frac{200}{100}\right)^2 \frac{250}{1000}$$

$$H = 9.9 - 1.93 = \textbf{7.97 m of water.}$$

(ii) Cylinder head at end of suction stroke,

$$H = H_{at} - H_{\substack{static \\ lift}} + H_{\substack{suction \\ deceleration}},$$

i.e. as $u = 0$ at end stroke, friction loss is zero,

$$H = 10.3 - 0.4 + 1.93 = \textbf{11.83 m of water.}$$

(iii) Cylinder head at mid-suction stroke,

$$H = H_{at} - H_{\substack{suction \\ lift}} - H_{\substack{friction, \\ loss}}$$

i.e. as $du/dt = 0$ at mid-stroke, inertia head is zero,

$$H = 10.3 - 0.4 - \frac{4fL}{2dg}(\omega r)^2 \left(\frac{A}{a}\right)^2$$

$$= 10.3 - 0.4 - \frac{4 \times 0.008 \times 3}{2 \times 100 \times 9.81} \times 1000 \times \left(0.4 \times 2\pi \times \frac{250}{1000}\right)^2 \left(\frac{200}{100}\right)^4$$

$$= 10.3 - 0.4 - 0.31 = \textbf{9.59 m.}$$

(b) Discharge stroke:

(i) Cylinder head at start of stroke,

$$H = H_{at} + H_{\substack{static \\ lift}} + H_{\substack{inertia \\ head}} + H_{friction}.$$

As $u = 0$ at start of cycle, friction loss might at first sight appear to be zero; however, the action of the air chamber is to maintain a steady discharge from the system. Thus the discharge is a constant, at a value of

$$Q = \text{Piston area} \times \text{Stroke} \times \text{rev s}^{-1}$$

$$= A \times 2r \times n = 2rA(\omega/2\pi) = rA(\omega/\pi).$$

Thus the friction loss in the system is a constant at

$$h_f = 4f(L-l)\left(r\frac{A}{a}\frac{\omega}{\pi}\right)^2,$$

where L is the total length of pipe and l is the length of the pipe from the pump to the air chamber. Therefore,

$$H = 10.3 + 20 + \frac{l}{g}\omega^2 r \frac{A}{a} + \frac{4f}{2dg}(L-l)\left(r\frac{A}{a}\frac{\omega}{\pi}\right)^2$$

$$= 30.3 + \frac{20}{9.81}(0.4 \times 2\pi)^2 \frac{250}{1000}\left(\frac{200}{100}\right)^2$$

$$+ \frac{4 \times 0.008 \times 1000}{2 \times 100 \times 9.81}(180)\left(r\frac{A}{a}\frac{\omega}{\pi}\right)^2,$$

where

$$r\frac{A}{a}\frac{\omega}{\pi} = \frac{250}{1000}\left(\frac{200}{100}\right)^2 \frac{0.4 \times 2\pi}{\pi} = 0.8.$$

$$H = 30.3 + 12.87 + 1.88 = \textbf{45.05 m of water.}$$

(ii) Cylinder head at end of discharge stroke,

$$H = H_{at} + \underset{\substack{\text{static} \\ \text{lift}}}{H_{static}} - \underset{\substack{\text{deceleration} \\ \text{inertia head}}}{H_{deceleration}} + \underset{\substack{\text{steady} \\ \text{friction loss}}}{H_{steady}}$$

$$= 30.3 - 12.87 + 1.93 = \textbf{19.36 m of water.}$$

(iii) Cylinder head at mid-discharge stroke,

$$H = H_{at} + \underset{\substack{\text{static} \\ \text{lift}}}{H_{static}} + \underset{\substack{\text{inertia} \\ \text{head}=0}}{H_{inertia}} + \underset{\substack{\text{friction} \\ \text{in pipe 1}}}{H_{friction}} + \underset{\substack{\text{steady} \\ \text{friction}}}{H_{steady}}$$

$$= 30.3 + \frac{4fl}{2dg}\left(\omega r \frac{A}{a}\right) + 1.93$$

$$= 30.3 + \frac{4}{2} \times \frac{0.008}{100} \times \frac{20}{9.81} \times 1000 \left[0.4 \times 2\pi \times \left(\frac{250}{1000}\right)\left(\frac{200}{100}\right)^2\right]^2 + 1.93$$

$$H = 30.3 + 2.06 + 1.93 = \textbf{34.3 m of water.}$$

EXAMPLE 24.3

For the reciprocating pump described in Example 24.1, calculate the increase in pump speed in rev min^{-1} if a large air chamber were fitted close to the pump suction valve.

Solution

The effect of introducing an air chamber would be to remove the inertia head effects calculated previously.

Therefore, as the minimum head occurs at the start of the suction cycle, and to avoid separation, it must be at least 2.4 m absolute, it follows that

$$H = H_{at} - \underset{\substack{\text{suction} \\ \text{lift}}}{H_{suction}} - \underset{\substack{\text{inertia} \\ \text{head}}}{H_{inertia}} - \underset{\substack{\text{friction} \\ \text{loss}}}{H_{friction}}$$

$$= \textbf{Vapour head of 2.4 m of water.}$$

Now $H_{inertia} = 0$, $H_{friction\ loss} = $ constant value based on the mean flow rate in the suction pipe, $(A/a)\,\omega\,(r/\pi)$:

$$H_{friction} = \frac{4fL}{2dg}\left(\frac{A}{a}\frac{\omega r}{\pi}\right)^2$$

$$2.4 = 10.3 - 1.5 - \frac{4fL}{2dg}\left(\frac{A}{a}\frac{\omega r}{\pi}\right)^2$$

$$6.4 = \frac{4 \times 0.01 \times 2 \times 1000}{2 \times 50 \times 9.81}\left(\frac{150}{50}\right)^4\left(\frac{\omega}{\pi} \times \frac{200}{1000}\right)^2$$

$$\omega = 15.4 \text{ rad s}^{-1} = 140 \text{ rev min}^{-1}.$$

Increase in speed $= 140 - 40 = \textbf{100 rev min}^{-1}.$

24.2 ROTARY PUMPS

In rotary pumps there is at least one rotating element which displaces a finite volume of fluid during each revolution. The most common are the gear pump, vane pump, screw pump and the rotary piston pump. All rotary pumps are distinguished by the following features.

1. There is usually more than one rotating chamber, so that the effects of inertia are minimized and the fluctuations in flow rate are small or effectively eliminated.

2. The chambers rotate so that in turn they come into direct contact with the inlet and outlet ports making valves unnecessary.

3. The rotational speed is usually quite high, so the pumps can be coupled directly to high-speed prime movers.

In general, rotary pumps, like all positive displacement pumps, are suitable for applications calling for large heads and small volumes, or specific speeds less than 0.2 (see Fig. 23.7). Rotary pumps are particularly used in oil hydraulics applications, not only because their 'positive' action is desirable, but also because of the relatively large-pressure and small-volume requirements of such applications.

Rotary hydraulic pumps may be divided into two basic types:

1. fixed capacity, in which the swept volume per revolution is fixed;

2. variable capacity, in which the swept volume per revolution may be varied within certain limits depending upon the type of pump.

24.3 ROTARY GEAR PUMPS

Gear pumps belong to the fixed capacity category. A rotary gear pump consists essentially of two intermeshing spur gears which are identical and which are surrounded by a closely fitting casing as shown in Fig. 24.10. One of the pinions is driven directly by the prime mover while the other is allowed to rotate freely. The fluid enters the spaces between the teeth and the casing, and moves with the teeth along the outer periphery until it reaches the outlet where it is expelled from the pump.

If the area enclosed by two adjacent teeth and the casing is a and the axial length of the pinion is l, then the volume of fluid enclosed between adjacent teeth is al and the total volume carried round by one pinion in one revolution is aln, where n is the

FIGURE 24.10
Gear pump

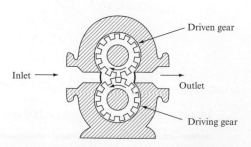

number of teeth. If, further, the volumetric efficiency is η_v, then the pump discharge is given by

$$Q = 2\eta_v a l n N, \tag{24.20}$$

where N is the speed of rotation and a may be expressed in terms of the geometric parameters of the gears.

Gear pumps are used for flow rates up to about $400 \, \text{m}^3 \, \text{h}^{-1}$ working against pressures as high as $17 \, \text{MN} \, \text{m}^{-2}$. The volumetric efficiency of gear pumps is in the order of 96 per cent at pressures of about $4 \, \text{MN} \, \text{m}^{-2}$ but decreases as the pressure rises.

The overall efficiency, which takes into account the mechanical losses, is given by

$$\eta_o = \frac{\rho g H Q}{P} = \frac{pQ}{P} = \eta_m \eta_v, \tag{24.21}$$

where η_m is the mechanical efficiency.

Figure 24.11 shows typical performance curves for a gear pump at different speeds.

FIGURE 24.11

Typical performance curves of a gear pump

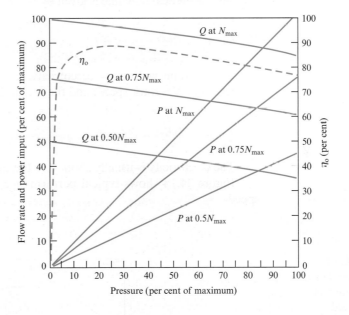

24.4 **ROTARY VANE PUMPS**

A vane pump consists of a rotor, which is fitted with vanes (blades) which are free to slide radially in the slots within the rotor, as shown in Fig. 24.12. The rotor is positioned eccentrically within a circular sleeve which is part of the casing. Thus as the rotor revolves the vanes subjected to a centrifugal force move radially outwards and their tips follow the contour of the casing sleeve. Fluid enters through the inlet port,

FIGURE 24.12
Vane pump

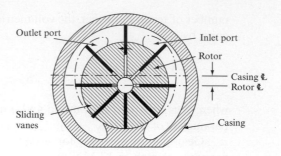

is trapped between the moving vanes and pushed towards the outlet port through which it leaves the pump. Ten to 12 vanes are usually employed. The slots and hence the vanes may be slightly inclined backwards with respect to the direction of rotation to minimize friction and wear at the vanes' tips.

This type of pump either can be of the constant capacity type, in which case the rotor eccentricity is fixed, or the casing may be allowed to move in order to alter the eccentricity, in which case the pump is of the variable capacity type. Clearly the amount of eccentricity controls the volume swept by the vanes and hence the flow rate of the fluid handled.

The theoretical flow rate is given by

$$Q_{th} = 2LenDN \sin(\pi/n), \tag{24.22}$$

(where L = vane width, e = eccentricity, n = number of vanes, D = casing inner diameter and N = rotor speed) and the actual flow rate is

$$Q = Q_{th}\eta_v. \tag{24.23}$$

The eccentricity is usually about $0.3r$ but depends upon the number of vanes.

Figure 24.13 shows typical performance curves of a vane pump at constant speed.

FIGURE 24.13
Typical characteristics
of a vane pump at
constant speed

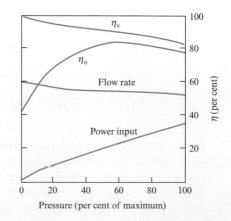

24.5 ROTARY PISTON PUMPS

There are two kinds of rotary piston pumps: the radial piston and the swash-plate (axial piston) pumps.

In the former the cylinders are arranged radially in a rotor (cylinder block) which is mounted eccentrically within a circular outer casing as shown in Fig. 24.14. As the rotor moves round, the pistons reciprocate in and out in the cylinders, their stroke being determined by the eccentricity e. One-half of the central stationary opening acts as the fluid inlet whereas the other half constitutes the outlet.

FIGURE 24.14

Rotary radial piston pump

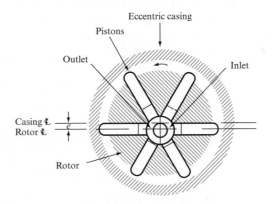

Since the stroke is equal to $2e$ it follows that the theoretical discharge is given by

$$Q_{th} = \tfrac{1}{2} e\pi d^2 nN, \tag{24.24}$$

where e = eccentricity, d = cylinder diameter, n = number of cylinders and N = rotational speed.

Such pumps are not suitable for high pressures because of excessive leakage and, hence, experience a considerable drop in volumetric efficiency.

An alternative arrangement for a radial piston pump is to have a stationary cylinder block and the pistons operated by an eccentrically mounted shaft. Since in both cases it is possible to design these pumps in such a way that the eccentricity may be varied, they then become variable discharge pumps.

In a swash-plate type of pump, shown diagrammatically in Fig. 24.15, the cylinders and pistons are arranged axially in a circle of a stationary casing. The driving shaft rotates an inclined swash plate, which actuates the pistons.

FIGURE 24.15
Rotary swash-plate pump

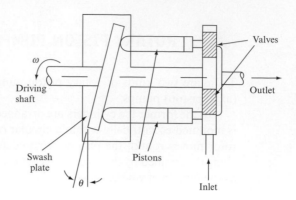

For these pumps the theoretical discharge is given by:

$$Q_{th} = \tfrac{1}{2}\pi d^2 rnN \tan\theta, \hspace{3cm} (24.25)$$

where d = piston diameter, r = cylinder centreline radius, n = number of cylinders, N = rotational speed and θ = swash-plate angle.

If the swash plate is of a variable angle type, the pump discharge may be altered and therefore this type of pump may be designed as a variable delivery pump.

24.6 HYDRAULIC MOTORS

Hydraulic and pneumatic motors are machines in which the fluid energy is converted into mechanical energy. The fluid energy is provided either by suitable oil supplied at high pressure or by compressed air. Most rotary positive displacement pumps, as described in the previous sections, can be used as motors.

For example, gear motors are very similar in construction to gear pumps and many may be used as either pumps or motors depending on whether the sealing arrangements are adequate for both modes of operation. Usually pumps tend to have plain bearings, whereas motors are fitted with ball or roller bearings. This is to facilitate starting by reducing the initial torque required to overcome friction.

Similarly vane pumps may be run as motors, again subject to suitable sealing provisions.

Figure 24.16 shows typical performance curves of a swash-plate-type hydraulic motor operating at constant pressure.

Hydraulic units which combine a pump driving a motor constitute self-sufficient hydraulic drives. For example, a vane-type pump may drive an oil motor, both operating within a common housing. The advantage of such an arrangement is that by varying the pump eccentricity the speed of the motor may be varied. Further-

FIGURE 24.16

Typical characteristics
of a swash-plate-type
hydraulic motor at
constant pressure

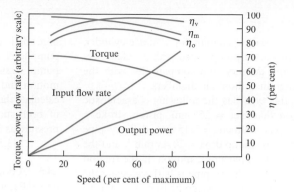

more, by reversing the eccentricity the direction of flow is reversed and hence the motor is put into reverse. Such an arrangement therefore offers great flexibility in operation.

Concluding remarks

The treatment of positive displacement machines presented has been entirely based on the momentum, energy and continuity equations developed earlier in the text. Comparisons between the pump performance characteristics for rotodynamic and positive displacement machines have demonstrated the insensitivity of the latter to delivery flow, a considerable divergence from the rotodynamic case.

Summary of important equations and concepts

1. For reciprocating pumps equation (24.4) confirms that pump discharge is directly proportional to rotational speed and is independent of the pressure against which the pump delivers.

2. Volumetric efficiency and slip are defined, equations (24.6) and (24.7), and the application of the steady flow energy equation is demonstrated in the determination of system pressure levels, e.g. equation (24.9). Machine overall efficiency is determined based on mechanical and hydraulic efficiencies, equation (24.19).

3. The introduction of air chambers to smooth the flow is introduced, a similar application to that suggested in surge suppression, Chapter 20.

4. Theoretical pump discharge rates are developed for a range of machines, including rotary gear pumps, equation (24.20), rotary vane pumps, equation (24.22), rotary piston pumps, equation (24.24) and swash-plate pumps, equation (24.25).

5. The importance of cavitation in the definition of operating conditions is reinforced by Examples 24.1 through 24.3.

Problems

24.1 Sketch an indicator diagram for a single-cylinder, single-acting reciprocating pump, fitted with air chambers on both suction and delivery, and compare this with the corresponding diagram without air chambers.

A pump of this type has the following characteristic dimensions: piston diameter = 255 mm, piston stroke = 460 mm, delivery pipe diameter = 115 mm, delivery pipe length = 50 m, speed of pump drive = 20 rev min⁻¹, and the piston moves with simple harmonic motion. If the friction factor applicable to the delivery pipe is 0.01, calculate the reduction in power required to overcome friction if a large air chamber is fitted to the pump discharge.

[0.215 kW]

24.2 Define percentage slip and separation with reference to positive displacement pumps.

Determine the maximum speeds at which a double-acting reciprocating pump can be run under the following conditions: (*a*) no air vessel on the suction side, (*b*) a large air vessel on the suction side close to the pump inlet. The suction lift is 3.65 m, the suction pipe length is 6 m of 100 mm diameter tubing. The cylinder diameter is 100 mm and the stroke is 460 mm. Assume simple harmonic motion for the piston movement and take the limit of pump operation at 2.4 m absolute at an atmospheric pressure of 10.3 m.

[(*a*) 53 rev min⁻¹, (*b*) 769 rev min⁻¹]

24.3 In a reciprocating pump the velocity of the water in the suction pipe during the suction stroke varies between zero and *V*, the displacement of the piston being simple harmonic. Prove that the mean friction head during the stroke is

$$4fLV^2/3gD,$$

where *L*, *D* and *f* are the length, diameter and friction factor for the suction pipe.

Further show that, if a large air chamber is added to the suction pipe close to the pump, the mean friction during the stroke reduces to

$$4fL\left(\frac{A\omega r}{a\pi}\right)\!\bigg/2gD,$$

where *A* and *a* are the cylinder and suction pipe areas, *ω* is the drive speed in radians per second and *r* is the half-stroke length.

24.4 A single-acting reciprocating pump, having a bore and stroke of 200 mm and 400 mm, respectively, runs at 20 rev min⁻¹. The suction pipe, which has a diameter of 100 mm and a length of 9.1 m, has no air chamber fitted. The suction lift is 3.6 m. The discharge pipe is also 100 mm

bore and is 470 m long, the discharge being 15.2 m above the pump level, and is fitted with an air chamber 15 m from the pump. Assuming simple harmonic motion for the piston motion and taking the friction factor as 0.01 for all pipes, calculate the cylinder pressures at the start, middle and end of both suction and discharge strokes. Take atmospheric pressure as 10.34 m.

[3.5 m, 6.22 m, 9.99 m, 33.54 m, 29.04 m, 22.82 m]

24.5 A double-acting reciprocating pump, of 180 mm bore and 360 mm stroke, draws water from a level 3 m below the pump and delivers it to a point 48 m above the water level. Both suction and delivery pipes are of 100 mm diameter and of 6 and 76 m length, respectively. The pump piston has simple harmonic motion and makes 40 double strokes per minute. Large air chambers are fitted to both pump suction and delivery, 1.5 m away on the suction side and 4.5 m away on the discharge side. The friction factor for both pipes may be taken as 0.008. Determine the pressure difference across the pump at the start of the stroke.   [55.01 m]

24.6 A double-acting positive displacement pump has a bore of 150 mm and a stroke of 300 mm. The suction pipe has a bore of 100 mm and is fitted with an air chamber. Calculate the rate of flow into or out of the air chamber when the crank driving the piston makes angles of 30°, 90° and 120° with inner dead centre.

Determine also the crank angles at which there is no flow to or from the air vessel. Assume drive speed to be 2 rev s⁻¹ and that the piston has simple harmonic motion.

[4.5 litres s⁻¹, 12.2 litres s⁻¹, 7.7 litres s⁻¹, 39.4°, 140.6°]

24.7 A double-acting, single-cylinder, positive displacement pump of 190 mm bore and 380 mm stroke runs at 36 double strokes per minute. The suction head is 3.65 m and the discharge static lift is 30.5 m. The suction pipe is 9 m long and the discharge pipe is 61 m long. Large air chambers are provided 3 m away on the suction side and 6 m away on the discharge side. Take the friction factor for both pipes as 0.008. Calculate at the beginning of the stroke the head at the two ends of the cylinder and the load on the piston rod, neglecting its size and assuming simple harmonic motion.

[3.6 m, 47.37 m, 12.2 kN]

24.8 A double-acting, single-cylinder reciprocating pump has a cylinder of 90 mm bore and a stroke of 150 mm. The ratio of length of crank to connecting rod is 1 : 4. The suction lift is 2.43 m, the suction pipe being of 64 mm diameter and 4.25 m length. The separation losses at the suction side valves may be taken as the equivalent of an extra 2.4 m of suction pipe length. Calculate the speed at which separation occurs, assuming a vapour level of 1.2 m absolute and an atmospheric pressure of 10.34 m absolute, and the maximum friction loss at this speed.

[97.6 rev min^{-1}, 0.39 m]

24.9 An hydraulic motor provides a shaft power of 90 kW when running at 2500 rev min^{-1}. The actual capacity of the motor is 38×10^{-6} m^3 rad^{-1}. What must be the pressure drop in the motor if the overall efficiency is 83 per cent?

If the flow through the motor is to be provided by a pump with an actual capacity of 65×10^{-6} m^3 rad^{-1}, at what speed should the pump operate?

[7.51 MN m^{-2}, 1462 rev min^{-1}]

Machine—Network Interactions

THIS CHAPTER ADDRESSES THE MATCHING OF FANS AND PUMPS TO THE NETWORK and as such utilizes both the machine performance characteristics, introduced in Chapter 23, and the steady flow energy equation, introduced in Chapter 6 and applied to pipe and duct networks in Chapter 14. The operation of multiple fans and pumps, either in series or in parallel in gas or liquid systems, is introduced and particular problems of machine stability addressed. The simultaneous solution of an appropriate machine pressure vs. flow characteristic with the system frictional and separation loss equation, that may also include pressure difference and static lift terms from the steady flow energy equation, is demonstrated, either graphically or via the use of the computer program introduced in this chapter. For pumps the possibility of cavitation constraining performance is also addressed.

While fans and pumps are normally expected to provide flows within duct or pipe systems, the application of fans to tunnel ventilation depends upon air entrainment by the fan throughflow. The ventilation of tunnels by jet fans is discussed, the treatment drawing upon the earlier development of the momentum equation in Chapter 5 and the definition of flow separation losses in Chapter 10. ● ● ●

It has already been shown that the flow of liquids and gases through pipe and duct systems depends upon the availability of the necessary energy transfers to overcome the frictional and separation 'losses' inherent in the system. These energy inputs may be provided simply by reference to the elevation of the system, i.e. the transfer of potential energy, or may require the input of mechanical energy to the system via fans or pumps. Similarly, the extraction of available energy through turbine installations depends upon the availability of a potential energy differential to generate both the flow and the turbine operation. The treatment of the steady flow energy equation in Chapter 6 and the subsequent applications of this approach to the flow of liquids and gases in Chapter 14 have laid a framework for the consideration of the interaction between the fan, pump and turbine characteristics and the system characteristic. The conduit network may be defined in terms of its physical layout, generating the static lift term, and the properties of both the fluid and the pipes or ducts comprising the system, generating the frictional and separation loss terms. A system will operate so that the available energy equals the demands of the system characteristic. This chapter will develop this relationship further for both fan and pump applications.

25.1　FANS, PUMPS AND FLUID NETWORKS

Chapter 14 illustrated the application of the steady flow energy equation to determine the flow of fluid through a pipe network as a result of an elevation difference between two reservoirs, or the pressure rise available as a result of introducing a fan or pump into the conduit. It was shown that the overall loss along a particular pipe or duct length could be expressed as

$$\Delta h = KQ^2/2g, \quad \Delta p = K\rho Q^2/2, \tag{25.1}$$

where K is the equivalent loss coefficient, incorporating both the frictional and separation losses, and by definition the pipe length, diameter and area terms appropriate to that pipe or duct length. This equation is frequently called the system resistance; Δh is the head loss in the pipe due to the flow Q, and Δp is the analogous pressure loss. (Note that historically Δh was used as this subject area was predominantly concerned with water and it was understood that the head value referred to was in metres, or earlier feet, of *water*. However, if the fluid changes then the units of h also change to metres of the flowing fluid. This is potentially confusing and the use of pressure, p, measured in newtons per square metre is to be preferred.)

　　Thus to maintain flow Q at this rate, energy E must be measured in appropriate head or pressure units must be supplied. In the two-reservoir problem, if the difference in reservoir levels is ΔZ, it is this potential energy difference that causes the flow. Thus

$$\rho g\Delta Z = 0.5\rho KQ^2. \tag{25.2}$$

In such a case there is no need to introduce a pump as the flow is maintained by the difference in elevation between the reservoirs. However, if the flow direction were to be reversed, as shown in Fig. 25.1, then it would become necessary to introduce a pump to overcome both the frictional (or system) resistance and the elevation difference (or static lift) between the new flow direction entry and exit. Thus the total energy required to maintain flow and raise it against the gravitational force a distance ΔZ is given by

FIGURE 25.1

Pump and pipe system

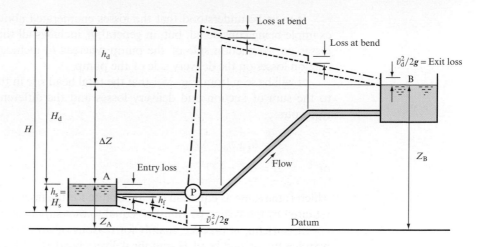

$$E = \rho g \Delta Z + 0.5 K \rho Q^2 \tag{25.3}$$

in pressure terms, or, obviously, in head terms as

$$E = \Delta Z + K Q^2 / 2g.$$

This equation is known as the system characteristic and the $\rho g \Delta Z$ term is referred to as the static lift.

Utilizing head as appropriate to a water-based pumping system, the network illustrated in Fig. 25.1 may be analyzed further to indicate the total energy line and the hydraulic grade line for the system. In pressure terms these lines are known as the total pressure and static pressure lines. The downward slope of both lines represents the frictional losses along the pipe length, while the vertical drops represent separation losses assumed to be concentrated at fittings. The system loss coefficient, K, introduced above incorporates the following terms:

1.   On the suction side of the pump:

Entry loss + Friction loss in the suction pipe,

so that

$$h_s = K_1 \frac{v_s^2}{2g} + \frac{4 f l_s}{d_c} \frac{v_s^2}{2g}.$$

2.   On the delivery side of the pump:

Losses due to bends + Frictional losses + Exit loss,

so that

$$h_d = K_2 \frac{v_d^2}{2g} + \frac{4 f l_d}{d_d} \frac{v_d^2}{2g} + \frac{v_d^2}{2g}$$

and $$\frac{K Q^2}{2g} = h_s + h_d. \tag{25.4}$$

It must be understood that the losses enumerated above refer to the particular example being considered, but, in general, h_s includes all the separation and friction losses on the suction side of the pump whereas h_d includes all the separation and friction losses on the delivery side of the pump.

It will be seen from Fig. 25.1 that the total head rise in the pump H is, thus, equal to the sum of suction and delivery losses and the difference in levels between the reservoirs:

$$H = h_s + h_d + \Delta Z$$

or $\qquad H = \Delta Z + KQ^2/2g,$ $\hfill$ (25.5)

which is the same as equation (25.3) because for the flow to be maintained the energy required by the system (E) must be equal to that supplied to it by the pump (H).

It is useful, in practice, mainly for reasons of avoiding cavitation, to distinguish between the suction head H_s and the delivery head H_d of the pump. The suction head includes losses h_s and that part of ΔZ which happens to be on the suction side of the pump. The delivery head includes losses h_d and that part of ΔZ which is on the delivery side (see Fig. 25.1). Thus,

$$H = H_s + H_d$$ $\hfill$ (25.6)

and in our example (Fig. 25.1),

$$H_s = h_s \quad \text{and} \quad H_d = h_d + \Delta Z.$$

Clearly H_d and H_s depend upon the location of the pump in relation to the levels of points A and B, but their sum H is independent of it.

Returning to equation (25.5), the system resistance is a parabola, which when plotted offers a simple means of determining head loss for any given value of Q. It must be remembered, though, that if the pipe system is in any way modified or additional losses are introduced (such as partially closing a valve), a new and different parabola will result because the value of K will be changed.

The system characteristic takes into account the difference in elevation, ΔZ, in addition to the head loss. A typical system characteristic is shown in Fig. 25.2. In a system handling air, the Z term is usually negligible because of the low value of air density.

FIGURE 25.2
System characteristic

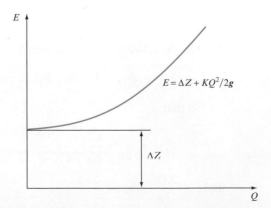

For a rotodynamic pump, the head generated is not constant but is a function of discharge (as discussed in Chapter 23), the relationship between the two being the pump characteristic

$$H = f(Q).$$

Clearly, then, if a rotodynamic pump operates in conjunction with a pipe system, the two must handle the same volume and, at the same time, the head generated by the pump must be equal to the system energy requirement at that flow rate. Therefore,

$$H = E$$

or $$f(Q) = \Delta Z + KQ^2/2g. \tag{25.7}$$

The solution of this equation may be obtained graphically or by computer because, clearly, it is the intersection of the pump and system characteristics. This is shown in Fig. 25.3, and the point where the two characteristics cross is known as the *operating*

FIGURE 25.3

Pump and system characteristics

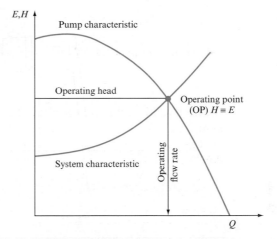

FIGURE 25.4

Fan operating against estimated and actual system characteristics

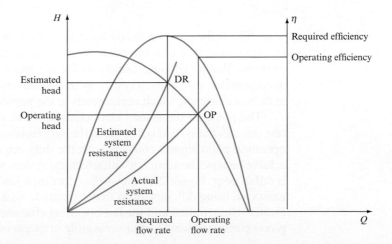

FIGURE 25.5
Positive displacement
pump and system
characteristic

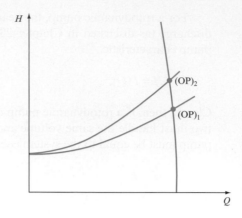

FIGURE 25.6
Total, static and kinetic
pressure variations along
a fan and duct system
(Woods Air Movement
Ltd)

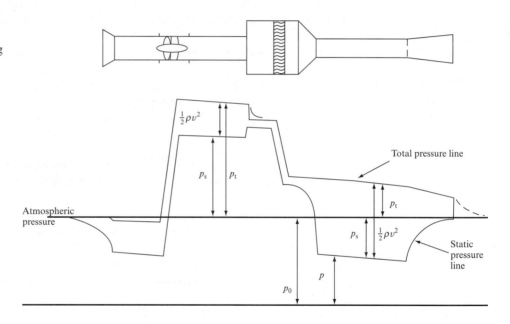

point. This is the point on the pump characteristic at which the pump operates and, at the same time, it is also the point on the system characteristic at which the system operates. 'Pump matching' usually means the process of selecting a pump to operate in conjunction with a given system so that it delivers the required flow rate, operating at its best efficiency, which corresponds to the pump's design point.

The point on the system characteristic which corresponds to the required flow rate through the system is known as the *duty required*. Thus, for correct matching, the operating point should coincide with the duty required. This is not always easy to achieve because the accuracy with which the system resistance is estimated, in practice, is rather poor. Figure 25.4 shows the effect on a fan application of the actual system resistance being different from that estimated. As a result, the flow rate delivered is greater than required and the fan operating efficiency is lower and, consequently, the power consumed would be unnecessarily in excess of that expected.

FIGURE 25.7

Fan and system matching, emphasizing the influence of the flow-independent room pressures

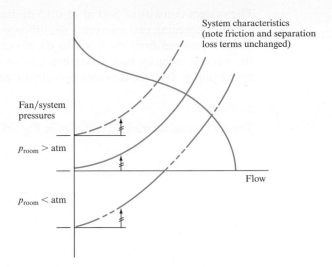

Figure 25.5 shows the negligible effect of system characteristic on the flow rate delivered by a positive displacement pump. Because of its almost vertical characteristic it is, for practical purposes, independent of the head requirement of the system. Since the efficiency is also almost constant, positive displacement pumps are selected on a flow rate basis only, subject to maximum pressure and power limitations for a given pump.

Figure 25.6 illustrates the pressure relationships along a ducted fan system, possibly utilized for the provision of ventilation or fume extraction. The fan operating point lies at a pressure equal to the sum of the frictional and separation losses for the system at the operating flow rate. In this example, and in most air duct cases, the static lift term may be ignored, but there may be an extra load if the flow is delivered to a space held above atmosphere, or, conversely, if the fan extracts air from such a space there will be some 'free' flow. These effects are in many ways identical to the static lift term for pumps as the effect is independent of the flow through the duct. Figure 25.7 illustrates the resulting system characteristic for both cases and the consequent effect upon the operating point.

(It should be noted that, with reference to Fig. 25.1, if the water flow were to be delivered to a pressurized reservoir, or drawn from a pressurized source, then the application of the steady flow energy equation would have included a pressure difference term independent of flow rate that would have had to be either added to or subtracted from the static lift term. Similarly, if the pump in Fig. 25.1 were to have been pumping downhill then a 'free' flow effect would have operated which would have ensured flow even if the pump was not driven. Simple examples of this effect for pumps can be found in the delivery of fuel from mobile tankers to underground storage tanks or the operation of aircraft pumps during air-to-air refuelling operations.)

EXAMPLE 25.1

The characteristics of a centrifugal pump handling water are:

Q (m³ s⁻¹)	0.010	0.014	0.017	0.019	0.024
H (m)	9.5	8.7	7.4	6.1	0.9
η (per cent)	65	81	78	68	12

The system consists of 840 m of 0.15 m diameter pipes with absolute roughness 6×10^{-6} m joining two reservoirs, the difference between water levels being 3 m. The water is pumped from the lower to the upper reservoir. Neglecting all losses except friction determine the rate of flow between the two reservoirs and the power consumed by the pump. Take the dynamic viscosity of water as $\mu = 1.14 \times 10^{-3}$ N s m^{-2}.

Solution

The pump characteristics are drawn in Fig. 25.8. In order to use equation (14.2),

FIGURE 25.8

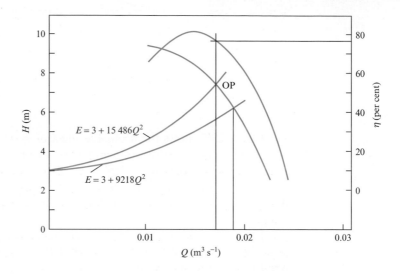

$$h_f = flQ^2/3d^5,$$

to determine the frictional loss in the pipe, it is first necessary to establish the value of the friction factor f. The relative roughness of the pipe is given by

$$\varepsilon/d = 6 \times 10^{-6}/0.15 = 40 \times 10^{-6}.$$

Now, from the Moody chart (Fig. 10.7), since Re is not known, take as a first approximation $f = 0.0025$. This gives

$$h_f = [0.0025 \times 840/3 \times (0.15)^5]Q^2 = 9218Q^2.$$

Therefore, the system characteristic becomes

$$E = 3 + 9218Q^2.$$

Plotting this, the intersection with the pump characteristic gives

$$Q = 0.0188 \text{ m}^3 \text{ s}^{-1},$$

and so the mean velocity in the pipe is

$$\bar{v} = 4Q/\pi d^2 = 4 \times 0.0188/\pi(0.15)^2 = 1.06 \text{ m s}^{-1}.$$

Hence, $\text{Re} = \rho\bar{v}d/\mu = 10^3 \times 1.06 \times 0.15/1.14 \times 10^{-3} = 0.14 \times 10^6.$

Referring to the Moody chart again, $f = 0.0042$, so that the system characteristic becomes

$$E' = 3 + 15\,486Q^2.$$

Plotting this, the intersection with the pump characteristic gives

$$Q = 0.017 \text{ m}^3 \text{ s}^{-1}.$$

Checking on Reynolds number gives $\text{Re} = 0.13 \times 10^6$, which is close enough to the previous value to accept as the operating point: $Q = 0.017$ m^3 s^{-1}; $H = 7.45$ m; $\eta = 78$ per cent; and so, power consumed is given by

$$P = \rho g H Q/\eta = 9.81 \times 10^3 \times 7.45 \times 0.017/0.78 = \textbf{1.59 kW.}$$

25.2 PARALLEL AND SERIES PUMP OPERATION

It is sometimes necessary to use more than one pump in conjunction with a given system. The pumps may be used 'in series' or 'in parallel'. In the first case, the inlet of the second pump is connected to the outlet of the first pump so that the same flow rate passes through each pump, but the heads generated by the two pumps are added together for a given flow rate. In the parallel operation each pump handles part of the flow rate because the inlets of the pumps as well as the outlets are coupled together. Thus the total flow rate passing through the system is equal to the sum of the flow rates passing through the individual pumps at a given head, which is the same for each pump. Figure 25.9 shows the combined characteristics for two identical pumps

FIGURE 25.9

Parallel and series operation of two identical rotodynamic pumps

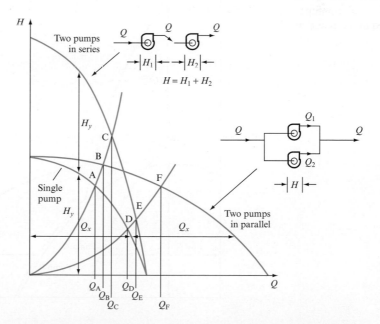

operating in parallel and in series against two resistances R_1 and R_2. For clarity, a system characteristic for which $\Delta Z = 0$ has been used. From the single-pump characteristic the combined characteristics of the pumps operating in parallel are obtained by the horizontal addition of values of Q_x for a given head to the characteristic of the single pump. Similarly, the combined characteristics of the two pumps connected in series are obtained by vertical addition of values of H_y for every value of Q.

Having obtained the combined characteristics it will be seen that at the system resistance R_1, for example, the single pump will operate at point A, the two pumps connected in parallel will operate at B and when connected in series at point C. Similarly, at R_2 the corresponding operating points will be D for a single pump, E for series operation and F for parallel operation.

It is also possible to use two or more dissimilar pumps either in series or in parallel. The procedure for obtaining the combined characteristics is the same as described above. Figure 25.10 shows a case of two dissimilar pumps and it should be noticed that certain parts of the combined characteristics are identical with that of the single pump and therefore no benefit whatsoever is achieved by the addition of the second pump if the system characteristic crosses the pump characteristic in these regions.

The cases represented by Figs 25.9 and 25.10 are only general demonstrations that each case should be studied on its own, and the answer will obviously depend upon the shape of the pump characteristic and upon the system characteristic.

The shape of the pump characteristic is especially important in the case of parallel operation as it may, for certain system characteristics, lead to unstable operation. This occurs when the system characteristic crosses the combined 'pumps in parallel' characteristic at more than one point or is coincidental with it. Axial flow pumps and fans are particularly vulnerable to this danger because they have initially rising head/flow rate characteristics.

FIGURE 25.10

Combined characteristics of two dissimilar pumps connected in series and in parallel

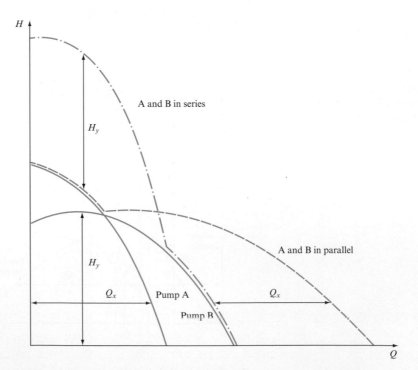

EXAMPLE 25.2

The characteristics of an axial flow pump delivering water are as follows:

Q (m³ s⁻¹)	0	0.040	0.069	0.092	0.115	0.138	0.180
H (m)	5.6	4.20	4.35	4.03	3.38	2.42	0

When two such pumps are connected in parallel, the flow rate through the system is the same as when they are connected in series. Determine the flow rate that a single pump would deliver if connected to the same system. Assume the system characteristic to be purely resistive (no static lift).

Solution

The single-pump characteristic is plotted in Fig. 25.11. From it, parts of the combined characteristics for two such pumps connected in series and in parallel are drawn. Since the volume delivered in each case is the same, the system resistance must pass through point A, which is the intersection of the 'parallel' and the 'series' characteristics. This is

FIGURE 25.11

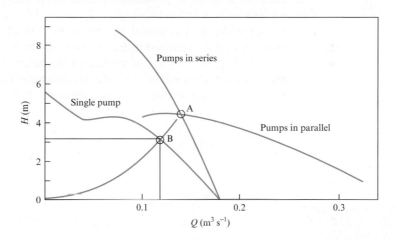

$Q = 0.14 \text{ m}^3 \text{ s}^{-1}$ and $H = 4.45$ m.

Now, a parabola representing the system characteristic is drawn through this point and through the origin ($\Delta Z = 0$). It crosses the single-pump characteristic at B, which would be the operating point for the single pump. It is

$Q = 0.12 \text{ m}^3 \text{ s}^{-1}$ and $H = 3.15$ m.

25.3 FANS IN SERIES AND PARALLEL

As fans and pumps are both examples of rotodynamic machinery, the combined characteristics of fans in series or parallel operation may be treated in the same way as described for pumps. Series or parallel operation may be utilized as a means of controlling the operating point of the combination. Similarly, there are maintenance advantages in being able to switch load between units without the necessity to close down the system.

Fans in parallel can be used to control air flow delivered by switching off one or more units, possibly in response to the output of a space contamination sensor (Chapter 26). It is often advisable to allow the parallel fans to deliver to a plenum space and then arrange for onward flow into a duct system or via entry grilles into a ventilated space. Reversed flow must be avoided by fitting dampers that close as the individual fan slows down to rest. The characteristics of parallel fans are shown in Fig. 25.12 for three identical units. The total volume delivered against any system resistance is dependent on the number of units operating. It will be appreciated that the maximum benefit of parallel fan operation in increasing delivered flow rate lies in their use with low-resistance systems; this result is also true for pumps in parallel.

The combination of parallel fan characteristics is simple if the individual fan characteristics do not include a maximum. Figure 25.13 illustrates the technique for both identical and different fans; the characteristics are simply added. This approach encounters difficulties when the individual fan characteristics include a peak, and becomes even more prone to fan instability when a maximum and minimum are present.

Consider first potential instability for fan characteristics with a peak in the pressure vs. flow relationship. Figure 25.14 illustrates the combination of two identical fan characteristics. The new combined characteristic is formed by adding the flow rates indicated by each at any constant pressure. Thus 'below' the peak, points D to E, a single combined characteristic appears. However, for points from A to D two

FIGURE 25.12

Use of a number of identical parallel fans to control the operating point for a given constant system characteristic (Woods Air Movement Ltd)

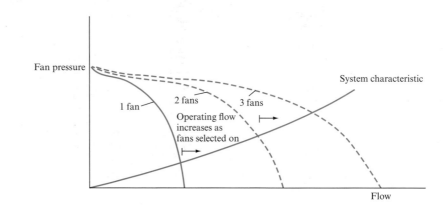

FIGURE 25.13

Addition of fan characteristics for the parallel operation of identical and non-identical fans

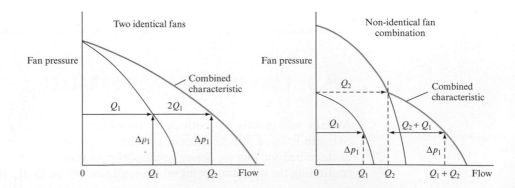

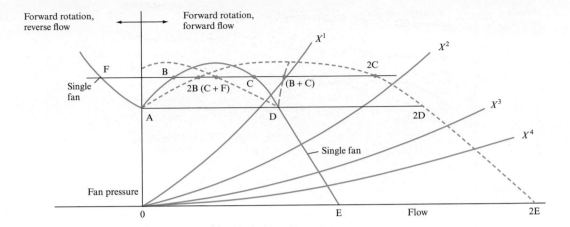

FIGURE 25.14

Summation of identical fan characteristics for parallel operation. The presence of a maximum in the characteristic leads to a range of possible combined characteristics and instability problems

possible characteristics appear, namely that formed by doubling values at, for example, B and C, and that formed by adding values at B and C, a correct process under this technique as both points are at the same pressure. For a particular system characteristic, labelled X^1, there are therefore three possible intersection points on the combined fan characteristic. (Further complications involving the second quadrant of the fan characteristic, i.e. its reverse flow, forward rotation operation, have been ignored within the scope of this presentation.) The fans can oscillate between these operating points.

System characteristics that intersect the combined fan characteristic to the right of point D will be stable (curves X^{2-4} in Fig. 25.14).

In the case of two identical fans with both a maximum and minimum present in the fan characteristic, then the combined characteristic has the form illustrated in Fig. 25.15. Again a particular system characteristic may intersect the fan curve at up to three points as shown, while another, lower, system resistance curve located to the right of the unstable region will enjoy stable fan operation.

In the case where one fan is much smaller than the other it is necessary to introduce the fan characteristic in its second quadrant, i.e. forward rotation but reversed flow. In this case air may be forced back through the smaller fan although its drive is still operating. (This effect may also be encountered in pumps and may be a

FIGURE 25.15

Combined characteristic possible when the two identical fan curves feature both a maximum and a minimum. Note multiple intersections for system characteristic X^1

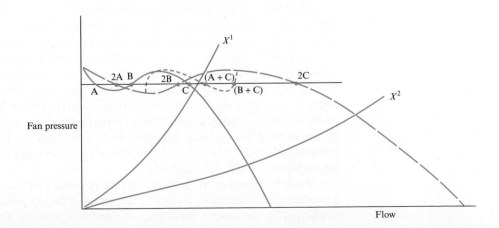

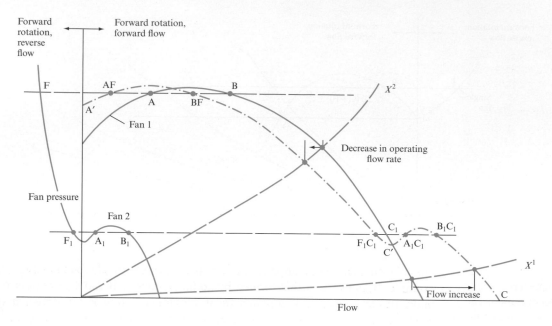

FIGURE 25.16

Combination of two dissimilar fan characteristics for parallel operation can lead to reduced operating flow owing to reverse flow through the 'weaker' fan

problem during the generation of transients in pumped networks where the pumps are not adequately protected by non-return valves.)

In the second quadrant the fan characteristic becomes a loss curve, the pressure necessary to force a reversed air flow through the fan rising as that flow increases, as shown in Fig. 25.16. Combination of the fan characteristics results in the curve ABC; this is illustrated where the smaller fan curve has effectively been added to the lower right-hand portion of the larger fan characteristic. However, for points to the left of C_1 the combined fan characteristic must take into account the smaller fan's second quadrant loss curve; thus the combined characteristic at flows less than that at C_1 is along curve A'C'C. Consideration of two different system resistance curves confirms the expected result that the larger fan forces air back through the smaller. For a low-resistance system, characteristic X^1, the flow delivered by the combined fans increases. However, for a higher-resistance system, characteristic X^2, the intersection point indicates less air delivered to the system by both fans together than by the larger fan on its own, the difference being the flow forced back through the smaller fan.

Thus the additive technique described may be used, when combined with both a knowledge of the system resistance and the fan characteristic in the reverse flow, forward rotation quadrant, to predict quite complex flow interactions.

Contrarotating axial flow fans in series are useful in controlling and increasing delivered flow in higher-resistance systems, as shown in Fig. 25.17. Successive fans should be contrarotating. When one or more fans are switched off they will idle, i.e. rotate as a result of the flow passing through from the driven fans, at about two-thirds of their driven rotation speed, with a consequent additional system loss of around 50 per cent of flow kinetic pressure. Axial flow fans are suitable for this type of installation; centrifugal fans and axial fans with guide vanes are not suitable owing to the excessive idling losses incurred.

While it is convenient simply to add the pressure vs. flow characteristics to obtain a combined characteristic, it must be appreciated that in practice each successive fan is affected by the output conditions of the fan upstream, and therefore the overall

FIGURE 25.17

Series fan operation. Note that flow interaction between successive fans leads to the combined characteristics deviating from the simple summation model (Woods Air Movement Ltd)

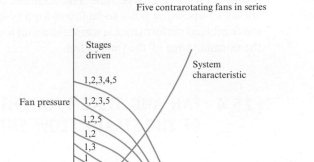

Five contrarotating fans in series

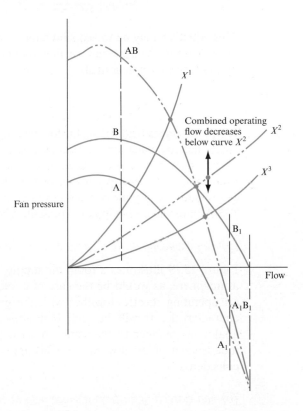

FIGURE 25.18

Series fan operation with dissimilar fans

performance is not strictly proportional to the number of fans in operation. In addition, if fans are to be switched off to control the overall delivered flow, it is necessary to choose the correct order to avoid excessive swirl effects.

In cases where the series fans are not identical, complications similar to those described for parallel operation can arise. Figure 25.18 illustrates one such difficulty where one fan is obviously larger and has a greater range of delivered flow rates. In this case it is necessary to determine the smaller fan characteristic for these flow rates, equivalent to the smaller fan's operation being assisted by a reduced, or suction, pressure on its discharge face. The addition of the individual characteristics then allows an operating point to be found for any system resistance. It will be noted that the combined performance is worse than that for the single larger fan for flows beyond the normal range of the smaller fan.

25.4   FAN AND SYSTEM MATCHING. AN APPLICATION OF THE STEADY FLOW ENERGY EQUATION

The most appropriate form of the steady flow energy equation in the matching of fan and system characteristics may be expressed as

$$\Delta p_{\text{fan}} = (p_2 - p_1) + [(\rho g Z)_2 - (\rho g Z)_1] + [(0.5 \rho V^2)_2 - (0.5 \rho V^2)_1]$$
$$+ \Sigma(\Delta p_{\text{friction}} + \Delta p_{\text{separation}}).$$

This equation may be solved simultaneously, either graphically or by computer program, with the fan pressure vs. flow rate characteristic, which may be approximated by a third-order polynomial:

$$\Delta p_{\text{fan}} = C_1 + C_2 Q_{\text{fan}} + C_3 Q_{\text{fan}}^2 + C_4 Q_{\text{fan}}^3.$$

It will be shown below that this formulation of the steady flow energy equation may be applied to a wide range of fan installation and operating conditions. It will be seen that this form of the equation, as stressed in Chapter 6, is expressed in terms of pressure units, or energy per unit volume. In the treatment of fans this is appropriate as pressure is easily measured and values are independent of the flowing fluid.

A number of examples are presented below.

(a) Fan supplying air to a space held above atmospheric pressure

Figure 25.19 illustrates a room air supply, the room to be held at a pressure above atmosphere, as would be the case in a 'clean' room manufacturing application or an operating theatre, together with the graph of the appropriate fan and system characteristic. It will be seen that above-atmosphere room pressure acts in an analogous manner to the static lift term for pumps discussed earlier, in that it is independent of air flow rate. In this application the use of fans in series could be considered.

(b) Fan extracting air from a space held above atmospheric pressure

Figure 25.20 illustrates a room ventilation extract system, the room held at a pressure above atmosphere, again as would be the case in a 'clean' room manufacturing application or an operating theatre, together with the graph of the appropriate fan and system characteristic. It will be seen that above-atmosphere room pressure acts to increase the air flow, similar to the case of a pump operating 'downhill'. In this application the use of fans in parallel could be considered as the system resistance is effectively reduced.

FIGURE 25.19

Solution of the fan and system characteristics when the fan operates against a room pressure above atmosphere

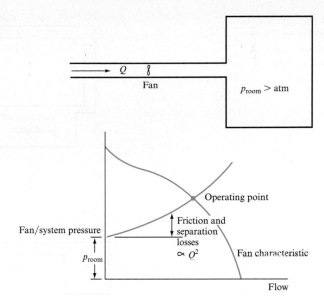

FIGURE 25.20

Solution of the fan and system characteristics when the fan operates to extract air from a room held above atmospheric pressure

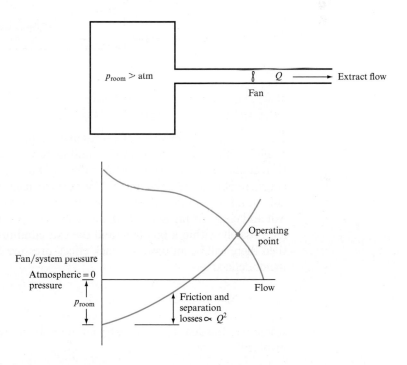

(c) Closed loop ventilation

Figure 25.21 illustrates a closed loop ventilation system where the extracted air is reprocessed for reuse. Applications clearly include aircraft and submarine systems. The steady flow energy equation may be applied to this process by 'opening up' the loop as shown. A system characteristic may then be drawn and the system operating point found as illustrated.

FIGURE 25.21

Closed loop ventilation
treated by 'breaking' into
loop and determining loss
characteristic

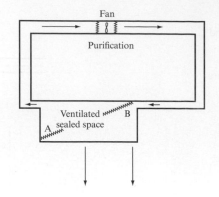

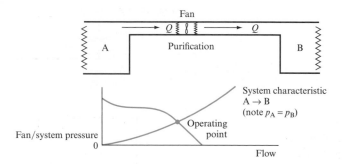

(d) Effect of density variation

Large barometric pressure differences may occur in fan/ductwork applications, e.g. in
dealing with skyscrapers, chimneys and mine ventilation. The fan characteristic may
be redefined in terms of the new air density, but there is also the effect on the system
characteristic to be considered. For vertical height differences of less than about
200 m, it is acceptable to introduce a mean air density, calculated from vertical stack
exit and entry air pressures. If there exists a temperature difference between the entry
air, possibly within a building, and the exit conditions, possibly at a roof discharge,
then there will be an upward 'stack effect' pressure difference that will aid air movement, defined as

$$\Delta p = \rho_1 g H \Delta t / (273 + t_1),$$

where ρ_1 is the density of the entry air at temperature t_1 and H is the height difference
involved.

EXAMPLE 25.3

A louvre in an air extract duct of 0.5 m diameter is controlled by a carbon dioxide
sensor mounted in a university library and is used to control the air flow rate
generated by an axial flow fan whose pressure–flow characteristics are shown below.

   If the duct is 20 m long and has a friction factor of 0.008, with entry and exit grilles
having a joint separation loss coefficient of 0.2, determine the air extract rate when the
louvre is set to 60° closed and the increase in flow if the louvre is opened to 30°.

Assume an air density of 1.2 kg m^{-3}. The louvre separation loss coefficient, K, may be calculated from

$$K = 1.05 + 1.25^{180°/(90°-\theta)},$$

where θ is measured in degrees.

The fan characteristics are as shown below:

Δp (N m^{-2})	4.90	5.20	4.75	3.90	2.10	1.00
Q (m^3 s^{-1})	0.0	1.0	2.0	3.0	4.0	5.0

Solution

Application of the steady flow energy equation

$$\Delta p_{\text{fan}} = (p_2 - p_1) + [(\rho gZ)_2 - (\rho gZ)_1] + [(0.5\rho V^2)_2 - (0.5\rho V^2)_1]$$
$$+ \Sigma(\Delta p_{\text{friction}} + \Delta p_{\text{separation}})$$

to the extract air duct, in the absence of any static lift or room pressurization terms, yields

$$\Delta p_{\text{fan}} = \Sigma(\Delta p_{\text{friction}} + \Delta p_{\text{separation}}).$$

Frictional losses may be expressed as

$$\Delta p_f = 4fL\rho Q^2/(2DA^2)$$
$$= 4 \times 1.2 \times 0.008 \times 20.0 \times Q^2/[2 \times 0.5 \times (\pi \times D^2/4)^2]$$
$$= 19.9Q^2.$$

Separation losses may be expressed, with a louvre angle of 60° closed, as

$$\Delta p_s = 0.5\rho KQ^2/A^2$$
$$= 0.5 \times 1.2 \times (1.25 + 1.25^{180°/(90°-60°)}) \times Q^2/(\pi \times D^2/4)^2$$
$$= 15.6 \times (1.25 + 1.25^{180°/(90°-60°)}) \times Q^2$$
$$= 15.6 \times 5.06 \times Q^2$$
$$= 79.0Q^2.$$

Thus the system characteristic in this case becomes

$$\Delta p_{\text{system}} = 98.9Q^2.$$

Figure 25.22 illustrates the resulting system operating point at a flow rate of 2.15 m^3 s^{-1}.

If the louvre is opened to 30°, as the carbon dioxide content rises the separation loss coefficient drops to 3.20 and the system resistance coefficient reduces to 69.8. The system operating point therefore moves to a higher flow rate of 2.5 m^3 s^{-1}, as shown in Fig. 25.22.

It has already been stressed that control in this way is not the most energy efficient. The steady flow energy equation clearly shows that flow control is only

FIGURE 25.22

Fan and system matching for Examples 25.3, 25.4 and 25.6

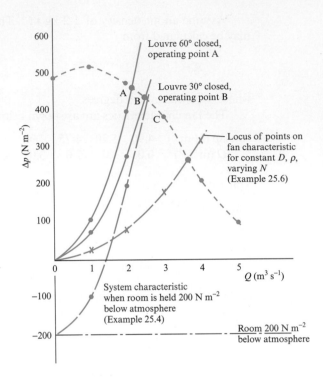

achieved by altering the loss in the system, rather than acting on the energy input to the system. A possibly better approach is to use fan speed control.

EXAMPLE 25.4

For the case given in Example 25.3 determine the system operating point at a 60° louvre setting if the ventilated space is held at 200 N m^{-2} below atmosphere.

Solution

The system characteristic becomes

$$\Delta p_{system} = (p_{atmosphere} - p_1) + \Sigma(\Delta p_{friction} + \Delta p_{separation})$$

or

$$\Delta p_{system} = -200 + 98.9Q^2$$

and the operating point from Fig. 25.22 becomes **2.55 m^3 s^{-1}**.

25.5 CHANGE IN THE PUMP SPEED AND THE SYSTEM

Pump characteristics, as shown in previous examples, refer to a given speed of pump operation. A change in pump speed will result in a different characteristic which may be predicted, subject to small-scale effects, by the use of similarity laws, discussed in Section 23.3 and Chapter 9. These give the following relationships:

$$Q/ND^3 = \text{constant}, \quad gH/N^2D^2 = \text{constant}, \quad \eta = \text{constant}.$$

For a given pump size, the impeller diameter is constant and, therefore, the above relationships reduce to:

$$Q/N = \text{constant}, \tag{25.8}$$

$$gH/N^2 = \text{constant}. \tag{25.9}$$

Thus, if a pump characteristic at speed N_1 (Fig. 25.23) is known and it is required to establish the pump's characteristic at speed N_2, the above relationships may be applied to any point on the 'N_1' characteristic – such as point A – giving values of Q' and H', which establish point A' on the 'N_2' characteristic. This point is the *corresponding* point to A, and the pump efficiency at A' is the same as at A. Thus,

FIGURE 25.23

Effect of pump speed change on matching with the system without static lift

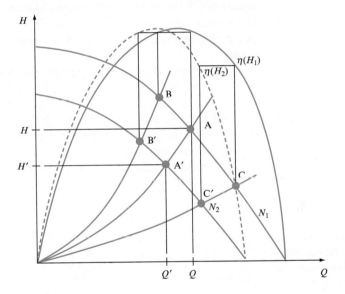

$$Q' = Q(N_2/N_1) \tag{25.10}$$

and $\quad H' = H(N_2/N_1)^2. \tag{25.11}$

A similar procedure is applied to other, arbitrarily chosen, points on the N_1 characteristic, such as points B or C, and the corresponding points B' and C' are established. Through them, the predicted characteristic at N_2 is drawn together with the efficiency curve, remembering that, since $\eta' = \eta$ at corresponding points, no calculations are required but merely a replot of efficiency values for the corresponding points at N_2.

It is interesting to note that the corresponding points lie on a parabola passing through the origin. This may be shown as follows. From equations (25.8) and (25.9),

$$Q = cN \quad \text{and} \quad H = kN^2;$$

eliminating N,

$$(Q/c)^2 = H/k \quad \text{or} \quad H = (k/c^2)Q^2,$$

which, in general, may be written as

$$H = CQ^2. \tag{25.12}$$

This equation is of the same form as equation (25.1) for the system resistance. Thus, if the system characteristic is purely resistive ($\Delta Z = 0$), the change in pump speed results in the corresponding points lying on the system characteristic. This means that, for any purely resistive system characteristic, the operating points at different pump speeds will be the corresponding points and, hence, the application of similarity laws does not necessitate replotting the pump characteristic.

This is not the case if $\Delta Z \neq 0$, as illustrated in Fig. 25.24. Here the operating point at N_1 has its corresponding point A' at N_2, but the system characteristic crosses the new pump characteristic (at N_2) at point B' and not at A'. Hence, the flow rate delivered will be $Q_{(N_2)}$ and not that corresponding to A'. Thus, the application of equation (25.10) to $Q_{(N_1)}$ will not give the correct result. Similarly, the efficiency will be that at C' and not C. In such cases it is, therefore, necessary to replot part of the pump characteristic at the required new speed in order to establish the new operating point.

For positive displacement pumps, the flow rate is directly proportional to the pump speed and, hence, equation (25.8) holds. Since the pump characteristic is a vertical line (Fig. 25.25), a change in speed will result in a proportional change in flow rate and a change of head against which the pump will operate, this change depending entirely upon the system characteristic. However, this is of little consequence unless it exceeds the maximum for the system or the pump. The power consumed by the pump will, of course, be affected, but the efficiency may be assumed constant.

FIGURE 25.24

Effect of pump speed change on matching with system which contains static lift

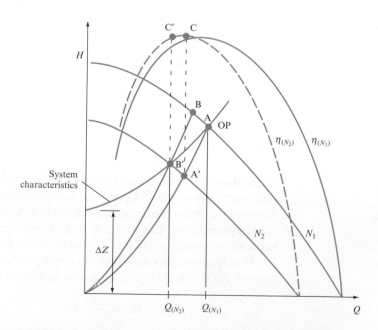

FIGURE 25.25
Matching of a positive
displacement pump with
a system characteristic

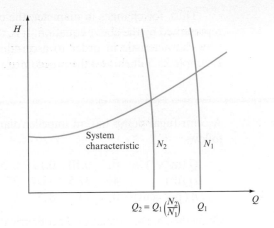

25.6 CHANGE IN THE PUMP SIZE AND THE SYSTEM

Geometrically similar pumps, i.e. of the same design, are made at different sizes. The impeller diameters are, consequently, different and it is again possible to apply the similarity laws in order to predict the performance of a pump having diameter D_2 from the known characteristic of a similar pump having diameter D_1, both pumps running at the same speed. Thus, if $N =$ constant, the similarity laws give the following relationships:

$$Q/D^3 = \text{constant},\tag{25.13}$$

$$gH/D^2 = \text{constant}.\tag{25.14}$$

Hence, selecting arbitrary points on the known characteristic (for D_1) it is possible to calculate the corresponding points on the required characteristic for D_2. If, for example, at point A the flow rate is Q and the head is H, then at A',

$$Q' = Q(D_2/D_1)^3,\tag{25.15}$$

$$H' = H(D_2/D_1)^2.\tag{25.16}$$

It is important to observe, however, that these corresponding points do not lie on a square parabola such as the non-resistive system characteristic. This may be shown by eliminating D from equations (25.13) and (25.14):

$$D = Q^{1/3}/c = H^{1/2}/k,$$

so that

$$H = CQ^{2/3}.\tag{25.17}$$

Thus, for changes in diameter, the corresponding points ($\eta = \eta'$) lie on a curve represented by the above equation. It is, therefore, always necessary to plot part of the new characteristic in order to ascertain where it crosses the system characteristic. Example 25.5 illustrates the procedure.

EXAMPLE 25.5

A centrifugal pump has an impeller diameter of 0.5 m and its characteristics are as follows:

Q (m³ s⁻¹)	0	0.10	0.15	0.20	0.25	0.30
H (m)		40	37.5	33.0	27.5	20.0	12.0
η (per cent)	0	73	82	81	71	48

Draw the characteristics of a geometrically similar pump having an impeller diameter of 0.562 m and running at the same speed.

If the two pumps operate against a system which includes a static lift of 10 m and is such that the smaller pump delivers 0.22 m³ s⁻¹, establish the operating point of the larger pump and the operating efficiencies of both pumps. Show on your graph some lines connecting the corresponding points on the two characteristics.

Solution

The pump characteristic for $D_1 = 0.5$ m is drawn in Fig. 25.26. In order to obtain the pump characteristic for $D_2 = 0.562$ m, points A, B, C and D are selected and the similarity laws are applied as follows:

$$Q_2 = Q_1(0.562/0.5)^3 = 1.42Q_1,$$

$$H_2 = H_1(0.562/0.5)^2 = 1.263H_1.$$

FIGURE 25.26

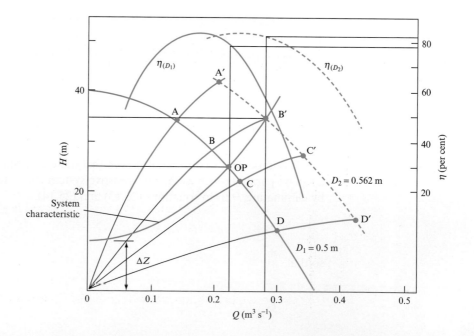

Using these relationships, points A′, B′, C′ and D′ are obtained. Through them is plotted the new characteristic for $D_2 = 0.562$ m. The efficiency values corresponding to A, B, C and D are moved horizontally to correspond to points A′, B′, C′, D′, respectively.

The system resistance is obtained by drawing a parabola through the operating point and $\Delta Z = 10$ m. Its intersections with the two pump characteristics give

$$Q = 0.22 \text{ m}^3 \text{ s}^{-1}, \ H = 25 \text{ m}, \ \eta = 78 \text{ per cent}, \quad \text{for } D_1 = 0.5 \text{ m};$$

$$Q = 0.28 \text{ m}^3 \text{ s}^{-1}, \ H = 35 \text{ m}, \ \eta = 82 \text{ per cent}, \quad \text{for } D_2 = 0.562 \text{ m}.$$

The corresponding points are joined by lines AA′, BB′, CC′ and DD′, all of which pass through the origin.

25.7 CHANGES IN FAN SPEED, DIAMETER AND AIR DENSITY

It has been shown (Chapter 22) that fan power, pressure generated and throughflow may be expressed by a series of dimensionless groupings

$$P/N^3D^5\rho = \phi(Q/ND^3, \mu/\rho ND^2, k/D, a/D, \ldots),$$

$$\Delta p/N^2D^2\rho = \phi_2(Q/ND^3, \mu/\rho ND^2, k/D, a/D, \ldots),$$

where P is the shaft power, Q is the throughflow of air of density ρ, viscosity μ, for a fan of diameter D rotating at speed N with fan blades having representative characteristic dimensions a and roughness k. Under conditions of geometric and dynamic similarity these functions may be utilized to generate 'new' fan characteristics for the same machine operating at different speeds or with differing flow densities, or for a family of machines the effect of fan impeller diameter may be determined. Under conditions of strict geometric and dynamic similarity it therefore follows that

$$(\Delta p/N^2D^2\rho)_1 = (\Delta p/N^2D^2\rho)_2, \quad (Q/ND^3)_1 = (Q/ND^3)_2.$$

For a constant density and fan diameter it then follows that

$$(\Delta p/N^2)_1 = (\Delta p/N^2)_2, \quad (Q/N)_1 = (Q/N)_2$$

and, hence, every point on the pressure–throughflow characteristic for a constant diameter and flow density and a particular fan speed lies on a locus defined by the relationship

$$(\Delta p_1/\Delta p_2) = (Q_1/Q_2)^2$$

between pressure and throughflow. Figure 25.27 illustrates the family of fan characteristics so formed.

FIGURE 25.27
Schematic of the effect of
density, speed and
diameter change for a
family of fan
characteristics under
conditions of geometric
and dynamic similarity

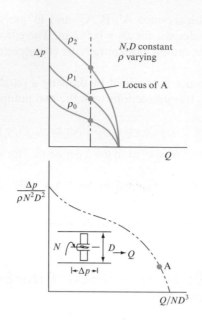

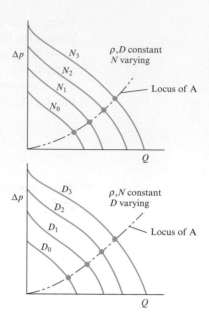

A similar reasoning for constant speed and flow density yields complementary equations:

$$(\Delta p/D^2)_1 = (\Delta p/D^2)_2, \quad (Q/D^3)_1 = (Q/D^3)_2,$$

$$(\Delta p_1/\Delta p_2) = (Q_1/Q_2)^{2/3},$$

also illustrated in Fig. 25.27.

For constant speed and diameter but varying flow density it will be seen that only the pressure group value changes, the flow coefficient being independent of flow density. Thus the fan characteristics for varying density may be deduced directly from the pressure coefficient

$$(\Delta p/\rho)_1 = (\Delta p/\rho)_2, \quad (Q)_1 = (Q)_2,$$

$$(\Delta p_1/\Delta p_2) = (\rho_1/\rho_2), \quad (Q)_1 = (Q)_2,$$

leading to a new family of fan characteristics having the same range of flow coordinates (Fig. 25.27).

EXAMPLE 25.6

For the air extract duct discussed in Example 25.3 determine the fan speeds that would have been necessary to generate the operating points for both louvre settings, i.e. 60 and 30 degrees closed. Assume that the original fan speed was 2800 rev min⁻¹, but note that this information is strictly superfluous as speed ratios may be determined. Also assume that the system loss characteristics remain the same except for the deletion of the louvre and entry/exit grille loss term.

Solution

Application of the steady flow energy equation

$$\Delta p_{\text{fan}} = (p_2 - p_1) + [(\rho g Z)_2 - (\rho g Z)_1] + [(0.5\rho V^2)_2 - (0.5\rho V^2)_1]$$
$$+ \Sigma(\Delta p_{\text{friction}} + \Delta p_{\text{separation}})$$

to the extract air duct, in the absence of any static lift, room pressurization or fitting separation loss terms, yields

$$\Delta p_{\text{fan}} = \Sigma \Delta p_{\text{friction}}.$$

Frictional losses may be expressed as

$$\Delta p_f = 4fL\rho Q^2/(2DA^2)$$
$$= 4 \times 1.2 \times 0.008 \times 20.0 \times Q^2/[2 \times 0.5 \times (\pi \times D^2/4)^2]$$
$$= 19.9 Q^2,$$

which in this case becomes the system characteristic.

From Fig. 25.22 the system operating point with an initial louvre setting of 60° was at a flow rate of 2.15 m³ s⁻¹ and a pressure of 462 N m⁻².

Without the louvre the system loss at this flow becomes 92 N m⁻², and it is therefore necessary to determine the fan speed that would have generated this operating point as the intersection of the system and modified fan characteristics.

The locus of pressure–flow for varying fan speed that passes through this point (Fig. 25.22) may be determined as

$$\Delta p_2 = (92/2.15^2)Q_2^2 = 19.89 Q_2^2,$$

which intersects the original fan characteristic at coordinates 3.7 m³ s⁻¹ and 272.3 N m⁻², these coordinates being the correct values to utilize subsequently in calculating the fan speed at the new 'louvre-less' system operating point. From the speed relationships

$$(\Delta p/N^2)_1 = (\Delta p/N^2)_2, \quad (Q/N)_1 = (Q/N)_2,$$

it follows that the speed reduction ratio becomes, from the pressure coordinates,

$$N_2/N_1 = (92/272.3)^{0.5} = 0.58$$

and, merely as a check, from the flow coordinates

$$N_2/N_1 = (2.15/3.7) = 0.58.$$

Thus the new fan speed corresponding to the 60° louvre setting would be **1627 rev min⁻¹**.

It should be stressed that while the curve in this case overlays the system characteristic, this is only because the example chosen did not include any static lift or pressurization terms in the system characteristic. This expression is related only to the fan characteristic and is independent of the system. The earlier treatment of the pump characteristic also makes this distinction clear. This example has therefore been included in this form to emphasize that point.

As the carbon dioxide level rose in Example 25.3, the louvre angle was decreased to 30° with a consequent rise in flow to 2.5 m³ s⁻¹. The fan speed ratio at this flow may thus be calculated as

$$N_2/N_1 = (2.5/3.7) = 0.676,$$

a new fan speed of **1891 rev min⁻¹**.

25.8 JET FANS

Jet fans are axial flow fans designed to produce a high-velocity jet at outlet. Unlike ordinary fans, jet fans are not used in conjunction with closed ducting which then constitutes the system resistance against which the fan operates. Instead jet fans are mounted in road tunnels, normally suspended from the tunnel roof, and, by virtue of the high-velocity jet so generated, a secondary air flow is induced along the length of the tunnel.

Road tunnels require ventilation to dilute the carbon monoxide and oxides of nitrogen emitted by road vehicles and to meet the acceptable health requirements on pollution.

The traditional way of ventilating a road tunnel, transverse ventilation, consists of a fully ducted system with a fresh air supply duct and a polluted air extract duct, each running the full length of the tunnel, or tunnel section served. Figure 25.28 illustrates this and other commonly employed ventilation systems. Closely spaced inlet grilles connect the supply duct with the tunnel, usually at road level, and similarly spaced outlet grilles connect into the exhaust duct at high level. Both supply and exhaust ducts are fitted with fans at both ends of the ventilated section, these fans selected in accordance with the system/machine matching criteria already discussed. For short tunnels the semi-transverse system of ventilation may be used. In this modification the supply duct arrangements remain unchanged but the exhaust, polluted air, is allowed to flow along the length of the tunnel and exit via the tunnel portals.

The longitudinal system of tunnel ventilation does not require any ducting. The whole required air volume moves through the tunnel at constant velocity from entry to exit. This flow is induced by jet fans suspended from the tunnel roof and blowing in the same direction as the traffic, assuming a single-direction traffic flow as would probably be the case in longer-tunnel applications. The number and size of these fans depend upon the maximum resistance to air flow and the total air flow through the tunnel.

Consider an idealized axisymmetric case of a number of jet fans mounted centrally along the length, L, of a tunnel, as shown in Fig. 25.29. The tunnel is assumed to be of constant circular cross-sectional shape. The equal spacing between the fans is X, the jet velocity is u_0 and the fully developed, tunnel air flow velocity is v_t. The fan jets induce a secondary air flow. At the fan location this flow is assumed to have a velocity v_0, as shown in Fig. 25.30. In each case the velocities are assumed constant across any tunnel section.

Applying the continuity equation between sections 0–0 and 2–2 in Fig. 25.29,

$$au_0 + (A - a)v_0 = Av_t,$$

FIGURE 25.28
Road tunnel ventilation
systems; below, examples
of road tunnel cross-
sections (Woods Air
Movement Ltd)

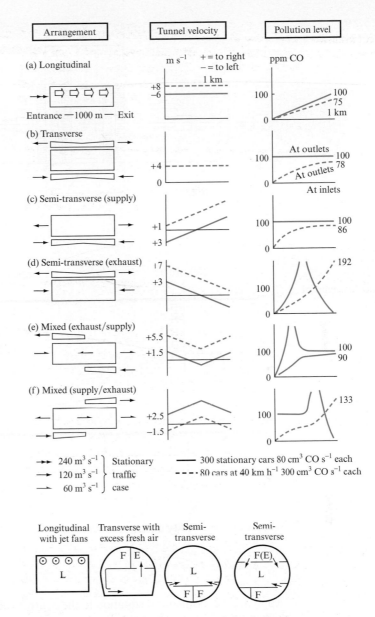

F: fresh air supply airway.   E: exhaust airway.   L: longitudinal flow.
F(E): Fresh air supply, reversible for emergency extraction

where A is the tunnel cross-sectional area and a is the fan outlet area. Let the velocity
ratio be

$$\omega = u_0/v_t$$

and the area ratio

$$\alpha = a/A,$$

FIGURE 25.29

Series of jet fans in a
tunnel

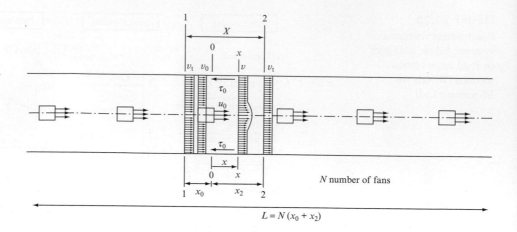

FIGURE 25.30

Velocities downstream of
a jet fan in a tunnel

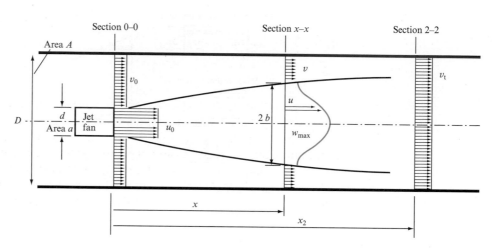

so that the continuity equation becomes, when divided by av_t,

$$\alpha\omega + (1-\alpha)(v_0/v_t) = 1.$$

Applying the momentum equation to the control volume defined by sections 0–0 and
2–2 and the tunnel walls yields the expression

Rate of change of momentum	=	Pressure force (on sections 0–0 and 2–2)	−	Shear stress on tunnel wall	−	Force to maintain flow kinetic energy

$$\rho Av_t^2 - \rho[(A-a)v_0^2 + au_0^2] = (p_0 - p_2)A - \tau_0 \pi D x_2 - \rho A(v_t^2/2),$$

where

$$\tau_0 = \rho f v_m^2/2 = \rho f(v_0 + v_t)^2/8 = \rho f(E+1)^2 v_t^2/8$$

as

$$v_m = \tfrac{1}{2}(v_0 + v_t), \quad v_m^2 = \tfrac{1}{4}v_t^2(E+1)^2, \quad E = v_0/v_t.$$

Substituting and dividing both sides by $\rho A v_t^2$,

$$1 - [(1-\alpha)E^2 + \alpha\omega^2] = \frac{p_0 - p_2}{\rho v_t^2} - \left[\frac{f(E+1)^2 \pi D x_2}{8\pi D^2/4} + \frac{1}{2}\right]$$

$$= \frac{p_0 - p_2}{\rho v_t^2} - \left[\frac{f(E+1)^2 x_2}{2D} + \frac{1}{2}\right],$$

but from the steady flow energy equation applied between sections 1–1 and 0–0

$$p_0 + \rho(v_0^2/2) = p_1 + \rho(v_t^2/2) - \Delta p_{\text{friction}_{1-0}} - p_{\text{loss}},$$

where, from the Darcy equation,

$$\Delta p_{\text{friction}_{1-0}} = 4\rho f x_0 v_m^2/2D.$$

Referring to Figures 25.29 and 25.30 it will be seen that the flow approaching the fan at velocity v_t splits into two components, one passing through the fan and the other bypassing it. The bypass flow velocity is reduced relative to its approach value owing to the fan effect. Thus there exists a pressure loss term that may be seen as the equivalent to the sudden expansion loss that would have been incurred by such a reduction in flow velocity. From Section 10.8.1 it will be seen that this equivalent pressure loss may be expressed as follows, recognizing that only a part of the flow is involved:

$$p_{\text{loss}} = \frac{m_t - m_j}{m_t}\rho\frac{(v_t - v_0)^2}{2} = \frac{Av_t - auu_0}{Av_t}\rho\frac{(v_t - v_0)^2}{2},$$

where m_t and m_j are the mass flow rates appropriate to the downstream tunnel and the fan jet.

The energy equation between 1–1 and 0–0 may therefore be developed into

$$\frac{p_0 - p_1}{\rho v_t^2} = -\frac{f x_0 (E+1)^2}{2D} - (1-\alpha\omega)\frac{(1-E)^2}{2} + \frac{(1-E^2)}{2},$$

but

$$p_0 - p_1 = p_0 - p_2$$

since by longitudinal symmetry $p_1 = p_2$ as the pressures at tunnel entry and exit are atmospheric.

Let

$$C = (1 - \alpha\omega)(1-E)^2 + E^2 - 1,$$

thereby combining the kinetic energy and p_{loss} terms from the energy equation. Thus by substitution

$$1 - (1-\alpha)E^2 - \alpha\omega^2 = -\frac{fx_0(E+1)^2}{2D} - \frac{C}{2} - \frac{fx_2(E+1)^2}{2D} - \frac{1}{2}$$

$$= -\frac{f(x_0+x_2)(E+1)^2}{2D} - \frac{C}{2} - \frac{1}{2}$$

and therefore, finally,

$$2[\alpha\omega^2 + (1-\alpha)E^2 - 1] = \frac{fx(E+1)^2}{D} + C + 1. \qquad (25.18)$$

This equation indicates that for a given tunnel diameter, length and friction factor, there is only one value of the velocity ratio. Thus, if the area ratio is retained, an increase in jet velocity will increase the tunnel flow proportionally to maintain the same velocity ratio.

If n fans are used in a tunnel, as shown in Fig. 25.30, then each additional fan adds to the rate of change of momentum a force equivalent to the thrust developed by each fan, reduced by the pressure force required to overcome the loss due to the sudden enlargement and the kinetic energy difference at each fan entry.

Applying the momentum equation between the fan inlet and outlet, the thrust, T, developed by the fan may be expressed as

$$T = \rho a u_0 (u_0 - v_t)$$

and

$$T/\rho A v_t^2 = \alpha\omega(\omega - 1).$$

Thus the addition to the rate of change of momentum is

$$T/\rho A v_t^2 = \alpha\omega(\omega - 1) - \tfrac{1}{2}C.$$

Hence for n fans, equation (25.18) becomes

$$2\left\{[\alpha\omega^2 + (1-\alpha)E^2 - 1] + (n-1)\left[\alpha\omega(\omega - 1) - \frac{C}{2}\right]\right\}$$

$$= \frac{fx(E+1)^2}{D} + C + 1. \qquad (25.19)$$

The theory developed above does not take into account the fact that, in practice, fans are not mounted centrally but near the tunnel roof. This affects the flow pattern in the tunnel and the induced air flow. To compensate for this, a correction factor, K_1, is introduced with a value less than unity:

$$K_1 \left[2 \left\{ [\alpha\omega^2 + (1-\alpha)E^2 - 1] + (n-1)\left[\alpha\omega(\omega-1) - \frac{C}{2} \right] \right\} \right]$$

$$= \frac{fx(E+1)^2}{D} + C + 1. \tag{25.20}$$

An alternative approach to this problem is to consider the fan thrust divided by the tunnel cross-sectional area as the pressure difference promoting the tunnel flow and thus equal to the sum of the tunnel losses. Thus

$$T/A = \Sigma \text{losses},$$

where the losses include wall friction and the tunnel entry and exit separation losses, the latter terms expressed by loss coefficient values of 0.5 for the entry and 1.0 for the exit condition.

Hence, if there are n fans supplying the thrust then

$$n\frac{T}{A} = \rho \left(\frac{4fLv_t^2}{2D} + 1.5\frac{v_t^2}{2} \right). \tag{25.21}$$

In practice this method is preferred as fan thrust can be determined experimentally in a laboratory, whereas u_0 is far more difficult to establish.

If the laboratory-measured value of fan thrust is used it must be corrected for the fact that, in a tunnel, the fan encounters at inlet a velocity v_t and not still air, as would be the case under laboratory conditions. Thus it will develop a reduced thrust in the tunnel. To compensate for this a factor K_2 is introduced, along with the K_1 factor introduced to correct for the roof mounting of the fans. Thus

$$n\frac{T}{A}K_1K_2 = \rho \left(\frac{4fLv_t^2}{2D} + 1.5\frac{v_t^2}{2} \right). \tag{25.22}$$

The longitudinal system of road tunnel ventilation has several advantages over the transverse system: the civil and mechanical costs are reduced as there are no additional ducts and no plant room is necessary; the installation is straightforward; air flow control is simple by fan switching; and generally the power consumption is lower.

EXAMPLE 25.7

A road tunnel, 1 km long, has a friction factor of 0.008 and mean hydraulic diameter of 10 m. The required tunnel air flow is 300 m³ s⁻¹. It is proposed to use jet fans having 1 m diameter, outlet velocity 25.3 m s⁻¹, developing a thrust of 590 N and absorbing power of 12.6 kW.

Determine the number of fans required and the total power consumption. Take the values of K_1 and K_2 to be 0.96 and 0.9, respectively.

Solution

The area ratio and the required tunnel air flow velocity are

$$\alpha = \left(\frac{1}{10}\right)^2 = 0.01, \quad v_t = \frac{4 \times 300}{\pi \times 10^2} = 3.82,$$

$$\omega = \left(\frac{25.3}{3.82}\right) = 6.62, \quad E = \frac{1 - \alpha\omega}{1 - \alpha} = 0.943,$$

$$C = (1 - \alpha\omega)(1 - E)^2 + E^2 - 1$$

$$= (1 - 0.01 \times 6.62)(1 - 0.943)^2 + 0.943^2 - 1$$

$$= 0.9338 \times 0.0033 + 0.8892 - 1 = -0.1077.$$

From equation (25.20), substituting into

$$\frac{fX(E+1)^2}{D} + C + 1$$

$$= K_1[\![2\{[\alpha\omega^2 + (1 - \alpha)E^2 - 1] + (n - 1)[\alpha\omega(\omega - 1) - C/2]\}]\!]$$

yields

$$[0.008 \times 1000 \times (0.943 + 1)^2/10 + (-0.1077) + 1]$$

$$= 0.96\{2[0.01 \times 6.62^2 + (1 - 0.01)0.943^2 - 1]\}$$

$$+ 2.0\{0.96(n - 1)[0.01 \times 6.62(6.62 - 1) - (-0.1077)/2]\}.$$

Thus

$$3.02 - 0.1077 + 1 = 0.96[2(0.438 + 0.88 - 1) + 2(n - 1)(0.372 + 0.0538)]$$

$$3.91 = 0.96[0.636 + (n - 1)0.8516]$$

$$4.07 - 0.636 = 0.8516(n - 1)$$

$$n = 5.04$$

and, hence, the number of fans is five or six. (Clearly a design decision and not a simple rounding up or down is required.)

Utilizing the alternative approach (equation (25.22)),

$$\text{Friction} = \rho\frac{4fLv_t^2}{2D} = \frac{1.2 \times 4 \times 0.008 \times 1000}{2 \times 10}v_t^2 = 1.92v_t^2.$$

The losses at entry and exit reduce to

$$\rho 1.5\frac{v_t^2}{2} = 1.2 \times 1.5\frac{v_t^2}{2} = 0.9v_t^2.$$

Substituting into equation (25.22),

$$n \frac{590 \times 4}{\pi \times 100} 0.96 \times 0.9 = 2.82 v_t^2 = 41.15$$

$$6.49n = 41.15$$

$$n = 6.34.$$

Therefore the number of fans predicted is **six**.
The total power consumed is $6 \times 12.6 = \mathbf{75.6\,kW}$.

EXAMPLE 25.8

To ventilate a tunnel 750 m long, equivalent diameter 9 m and friction factor 0.008, it is proposed to use ten jet fans, each of diameter 630 mm, outlet velocity 27.9 m s^{-1} and developing a thrust of 290 N.

Determine the resulting tunnel air flow velocity. Take $K_1 = 0.98$ and $K_2 = 0.92$.

Solution

From equation (25.20) assume that $E = 1$ as a starting value, i.e. the tunnel friction loss is based on the tunnel air flow velocity rather than on the mean of velocities v_0 and v_t.

Equation (25.20) becomes, as $C = 0$

$$\{2[\alpha\omega^2 + (n-1)\alpha\omega(\omega - 1)]\} K_1 = 4fL/D + 1.0,$$

$$\alpha = (0.63/9)^2 = 0.0049,$$

$$4fL/2D = 4 \times 0.008 \times 750/9 = 2.666.$$

$$2\{[0.0049\omega^2 + 9 \times 0.0049\omega(\omega - 1)]\}0.98 = 2.666 + 1,$$

$$0.049\omega^2 - 0.0441\omega - 1.87 = 0.$$

This quadratic is solvable for ω:

$$\omega = 6.7,$$

and the tunnel velocity becomes

$$v_t = 27.9/6.7 = 4.16 \text{ m s}^{-1}.$$

Alternatively, utilizing equation (25.22),

$$n\frac{T}{A} K_1 K_2 = \rho \left(\frac{4fL v_t^2}{2D} + 1.5\frac{v_t^2}{2} \right)$$

$$10 \left(\frac{290}{\pi \times 9^2/4} \right) 0.98 \times 0.92 = 1.2 \left(\frac{4 \times 0.008 \times 750}{2 \times 9} v_t^2 + \frac{1.5 v_t^2}{2} \right)$$

$$41.09 = 2.5 v_t^2$$

$$16.44 = v_t^2$$

and, hence, $v_t = \mathbf{4.05 \text{ m s}^{-1}}$.

25.9 COMPUTER PROGRAM MATCH

The determination of the operating point for a fan or pump connected into a given pipe or duct system has been discussed in Section 25.1 and demonstrated by Fig. 25.3. The system characteristic is developed from the steady flow energy equation and hence contains frictional and separation loss terms dependent upon Q^2 and flow-independent terms dependent upon the static lift and/or pressure differential to be compensated for by the machine operation, Section 25.4. Similarly, the fan or pump characteristic at any particular machine speed and diameter may be seen to relate delivered flow Q to generated head or pressure difference across the machine, H or Δp. Normally machine data are known as an experimental set of data points and hence machine characteristics in equation form are normally represented by a polynomial fit; generally second order (quadratic equations) are sufficient to match the accuracy of the known data.

Hence, any fan or pump operation within a system may be seen as a mechanical analogue computer solution to two simultaneous equations in Q and H or Δp, graphically illustrated by Figs 25.19 and 25.20.

Computer program MATCH allows this solution to be undertaken numerically, replacing the graphical solution common prior to ready access to computing power. Note that this technique is embedded within the FM4AIR program introduced in Chapter 21. The program accepts fan or pump data in a number of possible formats listed on the screen, namely:

1. *'Quadratic' option*: here machine diameter and speed may be changed and the head vs. flow quadratic coefficients change automatically dependent upon the values chosen for the coefficients in a head coefficient vs. flow coefficient equation. Default values are given to aid the user.

2. *'Dimensionless' option*: here data are requested in the form of a table of flow and head coefficients, similar to that generated by program SIMPUMP, Section 23.4. The program then fits a quadratic and displays the resulting equation for H vs. Q, where the coefficients change as the declared machine diameter and speed are changed.

3. *'Known' option*: here tabular data in terms of flow Q and head H are required. Operation of the program then fits a head vs. flow quadratic to these data. There is no provision for machine speed or diameter input.

In options (1) and (2) the machine characteristic quadratic may be seen to change as changes are made to either machine diameter and/or speed.

In each of the above options the system characteristic is required and the two coefficients necessary may be input directly by changing the default values displayed in the system quadratic, $H = (Z \text{ constant}) + (K \text{ constant}) Q^2$.

25.9.1 Application example

Option 1. Set pump speed to 408 rev min^{-1} and diameter to 0.5 m with the machine characteristic set at $KH = 10 + 10Q + 5Q^2$ (i.e. the default values). Set the system characteristic by changing the Z constant to 1.5 and the k constant to 12. The operating point is shown to be at $Q = 0.448$ m^3 s^{-1} and $H = 3.97$ m. Changing the k

constant to 48, i.e. increasing frictional and separation losses reduces the flow rate to $0.32 \text{ m}^3 \text{ s}^{-1}$.

Option 2. Using the default tabular data for KH and KQ, setting the machine diameter to 1.0 m and speed to 500 for system loss curve $H = 10 + 40Q^2$ yields an operating point of $0.852 \text{ m}^3 \text{ s}^{-1}$ and 38.9 m. Increasing the static lift to 35 m reduces the flow to $0.57 \text{ m}^3 \text{ s}^{-1}$.

Option 3. Here the machine characteristic is determined from the tabular input data for Q and H. Adopting the following table of Q and H values

Q (m³ s⁻¹)	0.2	1.6	3.0	4.0
H (m)	10.0	9.5	7.0	6.0

generates a machine quadratic with coefficients 10.2, −0.3 and −0.1, and setting the system characteristic to $h = 5 + 1.2Q^2$ yields an operating point at $1.8 \text{ m}^3 \text{ s}^{-1}$ and 8.924 m.

25.9.2 Additional investigation using MATCH

The program may be used to investigate:

1. the relative importance of static lift and frictional and separation losses in determining system operating point;

2. the influence of machine speed and diameter on the machine characteristic;

3. the change in operating point achieved by increasing systematically either machine speed or diameter in order to confirm the relationships presented in Section 25.7 and Fig. 25.27.

25.10 CAVITATION IN PUMPS AND TURBINES

Cavitation is the name given to a phenomenon which consists, basically, of local vaporization of a liquid. When the absolute pressure falls to a value equal to or lower than the vapour pressure of the liquid at the given temperature, small bubbles of vapour are formed and boiling occurs. Since liquids normally have air dissolved in them, the lowering of pressure to a value near to the vapour pressure releases this air first. The combination of air release and vaporization is known as cavitation.

In practice, cavitation starts at pressures somewhat higher than the vapour pressure of the liquid. The actual mechanism of cavitation inception is not yet known, but it appears to be associated with the existence of microscopic gas nuclei which cause cavitation. One theory suggests that these nuclei are present in the pores of the solid material at the fluid boundary. It is because of the presence of these nuclei that a fluid cannot withstand tension. It is estimated that, in their absence, a tension of 10 000 atm could be transmitted by water.

The nuclei give rise to the formation of bubbles during cavitation inception. These bubbles grow and collapse, producing pressure waves of high intensity, only to be followed by the formation of successive bubbles. Each cycle lasts only a few milliseconds, but the local pressures are enormous (maybe up to 4000 atm). Similarly, local temperatures may increase by as much as 800 °C.

The occurrence of cavitation is accompanied by a crackling noise and weak emission of light.

In a flowing system, the liquid may be subjected to changes in velocity and, consequently, changes in pressure. When the velocity increases, the pressure falls and, if it falls to a sufficiently low level, cavitation may occur. The bubbles may subsequently flow with the fluid into the region of higher pressure, where they collapse. Thus, cavitation may occur not only at pump inlets or draft tubes of turbines but also on hydrofoils, propellers, in venturi meters or syphons. In general, the effects of cavitation are noise, erosion of metal surfaces and the vibration of the system.

The most general and very useful cavitation parameter is the *cavitation coefficient*, σ, defined as

$$\sigma = (p_1 - p_c)/\tfrac{1}{2}\rho\bar{v}_1^2, \tag{25.23}$$

where p_1 = upstream or ambient static pressure, p_c = critical pressure at which cavitation occurs (usually taken as vapour pressure), and $\bar{v}_1$ = mean upstream fluid velocity.

The value of σ at which cavitation starts is called the critical cavitation coefficient σ_{crit} and is referred to as the *inception point*. Theoretically, cavitation starts when the pressure falls to the value of the vapour pressure of the liquid, but the latter is a function of temperature. Thus, a system which will operate satisfactorily without cavitation during winter may give cavitation trouble in the summer when the temperature is higher and, therefore, the vapour pressure of the liquid is also higher. Figure 25.31 gives the values of vapour pressure of water as a function of temperature.

FIGURE 25.31
Variation of water vapour pressure with temperature

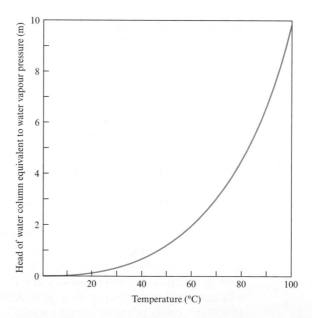

In rotodynamic pumps, cavitation occurs at the pump inlet, where the pressure is lowest, as demonstrated in Fig. 25.32, which shows the hydraulic gradient for a simplified pump system.

If the absolute static pressure at the pump inlet is p_i, then,

$$p_i = p_{atm} - \rho g H_s, \tag{25.24}$$

FIGURE 25.32

Total energy gradient for the suction side of a pump system

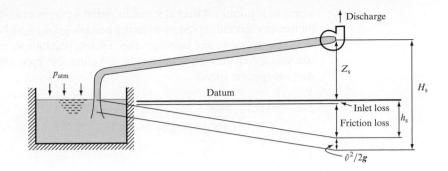

where p_{atm} is the atmospheric pressure and H_s is the suction head, which includes not only the suction lift Z_s but also the sum of the losses in the inlet pipe h_s and the velocity head, so that

$$H_s = Z_s + h_s + \bar{v}^2/2g. \tag{25.25}$$

Now, if the vapour pressure is p_{vap} then theoretically cavitation starts when

$$p_i = p_{vap}$$

and the difference

$$H = (p_i - p_{vap})/\rho g$$

is a measure of the absolute head available at the pump inlet above the vapour pressure (above cavitation inception) and is known as the *net positive suction head* or simply NPSH. Thus,

$$\text{NPSH} = (p_i - p_{vap})/\rho g = p_{atm}/\rho g - H_s - p_{vap}/\rho g. \tag{25.26}$$

In SI, NPSH is replaced by the *net positive suction energy* (NPSE), defined as

$$\text{NPSE} = g\text{NPSH} = p_{atm}/\rho - gH_s - p_{vap}/\rho. \tag{25.27}$$

Thus, NPSE or NPSH represents the difference between the total energy at the inlet flange of the pump and the vapour energy: it is the energy available to the pump on its suction side. If the pump total head is H, no more than the NPSH part of it should be used at the suction side if cavitation is to be avoided.

It was suggested by Thoma that NPSH is proportional to the pump total head H and he defined a cavitation coefficient,

$$\sigma_{Th} = \text{NPSH}/H = \text{NPSE}/gH. \tag{25.28}$$

Another useful parameter is the *suction specific speed*, analogous to the type number or specific speed,

$$K_s = \omega Q^{1/2}/(\text{NPSE})^{3/4} = \omega Q^{1/2}/(g\text{NPSH})^{3/4}, \tag{25.29}$$

where ω = pump rotational speed in radians per second, Q = pump flow rate in cubic metres per second, NPSE is in joules per kilogram, and NPSH is in metres.

The relationship between the Thoma cavitation coefficient and the suction specific speed may be obtained by dividing the type number of the pump by the suction specific speed:

$$\frac{\omega_s}{K_s} = \frac{\omega Q^{1/2}}{(gH)^{3/4}} \Big/ \frac{\omega Q^{1/2}}{(g\mathrm{NPSH})^{3/4}} = \frac{(\mathrm{NPSH})^{3/4}}{H^{3/4}} = \sigma_{\mathrm{Th}}^{3/4}.$$

Thus,

$$\sigma_{\mathrm{Th}} = (\omega_s/K_s)^{4/3}. \tag{25.30}$$

For geometrically similar pumps the scaling laws may be obtained from

$$\frac{\mathrm{NPSH}_1}{\mathrm{NPSH}_2} = \left(\frac{N_1}{N_2}\right)^2 \left(\frac{D_1}{D_2}\right)^2. \tag{25.31}$$

The main effect of cavitation on pumps, besides erosion and vibration, is the possibility of performance failure. Figure 25.33 shows the characteristic curve of a centrifugal pump with performance failure due to the cavitation indicated.

FIGURE 25.33
Cavitation effects on pump characteristic

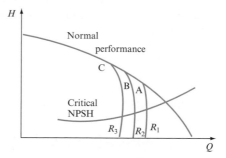

If the pump on test is throttled at discharge, the normal characteristic is obtained. If, however, the inlet valve is partially closed, so that the inlet pressure is lowered by a resistance R_1, and then the pump is tested by opening the discharge valve (starting from shut-off), it will perform normally as far as point A, but a further opening of the discharge valve will no longer produce any increase in flow. Repeating the test with a greater inlet resistance, say R_2, will cause the falling off of performance earlier, namely at some point such as B. By testing a pump in the manner described, it is possible to determine the absolute pressure at inlet at which cavitation occurs (performance failure) and, hence, to calculate the corresponding NPSH (or NPSE). If, now, the critical NPSH is defined as the point at which the head falls by an arbitrary percentage, usually 2 or 3 per cent, below the normal non-cavitating performance, then the critical NPSH may be plotted against the flow rate, as shown in Fig. 25.33, alongside the other pump characteristics. It then shows the minimum NPSH required by the pump in order to avoid cavitation.

In positive displacement pumps, the manometric suction head must be always greater than the vapour pressure. The vapour pressure increases with temperature, as shown in Fig. 25.31. For example, at 90 °C it is equivalent to a column of 7.14 m of water. Supposing the atmospheric pressure is equivalent to 740 mm of mercury, which is $0.74 \times 13.6 = 10.06$ m of water, then the available difference for the manometric suction head is $10.06 - 7.14 = 2.92$ m of water. However, allowing for the drop of pressure due to inertia at the beginning of the suction stroke and losses in the valve, the manometric suction head must be considerably smaller than that figure. For pumps handling cold water, the maximum manometric head is, in practice, between 6 and 6.5 metres. Figure 25.34 gives the relationship between the manometric suction head and temperature for reciprocating pumps handling water, from which it is seen that, if the water temperature is above 70 °C, the pump must be below the lower reservoir to ensure positive water pressure.

FIGURE 25.34

Maximum manometric suction head for reciprocating pumps handling water

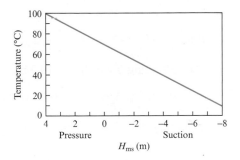

In turbines, the areas susceptible to cavitation are the blade trailing edge and the draft tube, since in these places the pressure is likely to be the lowest. It is possible to avoid cavitation in turbines altogether by submerging them to a low level, but this usually means excavation work, which in view of the large size of average water turbines tends to be very costly. Therefore, cavitation is often accepted and provisions are made for the periodic repair of the damage caused by it. Because cavitation occurs on the downstream side of a turbine it has very little, if any, effect on its performance.

Turbine cavitation is usually defined by the Thoma coefficient

$$\sigma_{\text{Th}} = (H_{\text{atm}} - Z - h_{\text{vap}})/H, \tag{25.32}$$

where H_{atm} = atmospheric head, Z = height of centreline of turbine above tailrace, h_{vap} = vapour pressure head, and H = net head across turbine.

As with pumps, a suction specific speed K_s is also used. An empirical relationship between $(K_s)_{\text{crit}}$ and ω_s, suggested by Noskievic, is of the form

$$(K_s)_{\text{crit}} = a/\sqrt{\omega_s}, \tag{25.33}$$

where the value of a, a constant, is between 4.5 and 5.8.

EXAMPLE 25.9

A centrifugal pump having dimensionless specific speed, based on rotational speed in radians per second equal to 0.45, is to pump 0.85 m³ s⁻¹ at a total head of 152 m. The pump will take water with a vapour pressure of 350 N m^{-2} from a storage basin at sea level. For the pump speed consistent with the above requirements, calculate the elevation of the pump inlet relative to the water level based on an acceptable value of suction specific speed equal to 3.2.

Solution

$$\omega_s = \omega Q^{1/2}/(gH)^{3/4} = 0.45,$$

$$K_s = \omega Q^{1/2}/(gH_{t_1})^{3/4} = 3.2,$$

where $H_{t_1} = \text{NPSH} = H_{atm} - Z - h - H_{vap}$. Therefore,

$$\omega_s/K_s = (H_{t_1}/H)^{3/4} = 0.45/3.2,$$

$$H_{t_1} = H(0.45/3.2)^{4/3} = H/13.6$$

$$= 152/13.6 = 11.2 \text{ m}.$$

But, $H_{t_1} = H_{atm} - Z - h - H_{vap}$

and, assuming $h = 0$, $H_{atm} = 10.3$ m,

$$H_{vap} = 350/9.81 \times 10^3 = 0.036 \text{ m}.$$

Therefore,

$$Z = H_{atm} - H_{t_1} - H_{vap} = 10.3 - 11.2 - 0.036$$

$$= -0.936 \text{ m},$$

and so the pump must be submerged below the water level.

25.11 FAN AND PUMP SELECTION

Usually, pump selection starts with the required flow rate and head being specified. These are the two essential quantities which the pump, when operating in conjunction with the given system, must deliver. In addition some constraints on the pump selected may be present. These can be:

1. pump speed (which may be specified if the prime mover and its speed are known);
2. minimum operating efficiency (which may be specified and must be guaranteed by the manufacturer);
3. static lift (which may affect the location of the pump with regard to cavitation hazards);
4. type of pump (i.e. centrifugal or axial), because of ease of installation in a particular system;
5. space available for the pump and the type of drive;
6. type of fluid to be handled (not only its density and viscosity, but also whether it carries abrasive particles, solid matter, acids, etc.);

7. minimum noise level (in the case of fans);

8. non-overloading power characteristic (which may be essential).

All these aspects will influence the type of pump or fan chosen. However, in the absence of all the above restrictions, in theory any pump or fan type may be selected for any duty. The difference will be in the size of the machine and its speed of operation. Also, some machine types are more efficient than others. Generally speaking, if a low-type-number pump or fan is selected, it will be large and its speed of operation will be low. Conversely, if a high-type-number pump or fan is chosen, it will be smaller but it will have to run faster. This is well illustrated in Example 25.10 following.

If the only specification given is the flow rate and the operating head, usually referred to as the *duty required* (DR), the first step is to decide the pump speed. Since it is desirable for the pump to operate at its best efficiency and, therefore, at its design point (DP) the pump type number is used because, by definition, it describes uniquely the design point. Since

$$n_s = NQ^{1/2}/(gH)^{3/4},$$

and Q as well as H are given, it is possible to establish the numerical relationship

$$n_s = AN. \tag{25.34}$$

This equation demonstrates clearly that the decision with regards to the pump speed (N) immediately specifies the pump type number, and, hence, the pump type which has to be used.

The most common pump and fan drives are a.c. electric motors, which run at speeds governed by the a.c. frequency and the number of poles. Thus, the synchronous speed of an a.c. motor is given by

$$N_{\text{synch}} = (2f/n)60 \text{ rev min}^{-1}, \tag{25.35}$$

where f = frequency and n = number of poles. However, motors run at speeds less than synchronous because of the required slip. The table below gives the synchronous and nominal motor speeds for 50 Hz supply:

No. of poles	N_{SYNCH}	N_{NOMINAL}
2	3000	2900
4	1500	1450
6	1000	960
8	750	720
10	600	575
12	500	480
14	430	410
16	375	360

If, therefore, an electric motor is to be used, the different values of nominal speeds may be substituted into equation (25.34), thus giving a choice of pump types that may be used. Some may have to be rejected because of any of the restrictions listed above. The final choice, therefore, will be based on the balance between the initial and the running costs of the machine chosen. A high-type-number machine will be small and, hence, initially cheaper. It may, however, be less efficient and, hence, its running costs may be higher. Generally, for small pumps and fans requiring small powers, the capital expenditure is usually considered more important. It is certainly not the case for large pumps and fans consuming substantial amounts of power. In such cases it is essential to select a machine which will have the highest possible efficiency, thus reducing its running costs to a minimum.

Having established the pump speed and, hence, the type, it is next necessary to determine the pump size, i.e. the diameter of its impeller. For selection and comparison purposes, pump and fan characteristics are usually drawn in terms of dimensionless coefficients, namely the flow coefficient

$$K_Q = Q/ND^3, \tag{25.36}$$

and the head coefficient

$$K_H = gH/N^2D^2, \tag{25.37}$$

which were discussed in Chapter 23.

The value of either of the above two coefficients corresponding to the design point (hence n_s chosen) may be used to calculate the diameter required for the speed chosen. Alternatively, for each pump type it is useful to have, in addition to the value of the type number, the corresponding value of the *specific diameter*, defined as

$$D_s = D(gH)^{1/4}/Q^{1/2} = K_H^{1/4}/K_Q^{1/2}. \tag{25.38}$$

If this is known for the particular pump selected, its diameter may be calculated immediately.

A common difficulty arises when, for the speed chosen, the calculated type number does not correspond to any of the type numbers of the pumps available. The ideal answer would be to design a new pump to suit the specification exactly. This very costly procedure may only be justified in the rare cases of very large pumps and/or if a substantial number of them is required.

In all other cases, it is necessary to choose a pump whose type number is closest to and greater than that required. It will mean that the pump will not operate at its design point but somewhat to the right of it. If the pump type number is less than that required, it will operate somewhere to the left of the design point, which for some types may lead to an unstable operation, depending upon the shape of the head/flow rate characteristic. In any case the operating efficiency will be less than maximum.

A procedure for pump selection is illustrated by Example 25.10.

The selection of positive displacement pumps presents fewer problems because their operation is (within limits) independent of the system characteristic. It is precisely this feature which makes them eminently suitable for applications where the system resistance is difficult to estimate or, more commonly, when it is subject to

FIGURE 25.35

Comparative fan designs at equal output power (Woods Air Movement Ltd)

A Backward-curved Halfwidth 630 mm
42 rev s⁻¹
13.5 kW at Δ
17 kW at φ
Sp. speed 0.31

B Backward-curved Fullwidth 630 mm
36 rev s⁻¹
12 kW at Δ
14 kW at φ
Sp. speed 1.56

C Axial 50 per cent hub 630 mm
48 rev s⁻¹
13.5 kW at Δ
15 kW at φ
Sp. speed 2.95

D Forward-curved centrifugal 700 mm
18 rev s⁻¹
15 kW at Δ
30 kW at φ
Sp. speed 1.50

E Multi-vane centrifugal 850 mm
9 rev s⁻¹
15 kW at Δ
30 kW at φ
Sp. speed 1.23

F Axial 35 per cent hub 1000 mm
24 rev s⁻¹
13 kW at Δ
15 kW at φ
Sp. speed 5.0

G Axial 25 per cent hub 2000 mm
12 rev s⁻¹
12.5 kW at Δ
14 kW at φ
Sp. speed 6.5

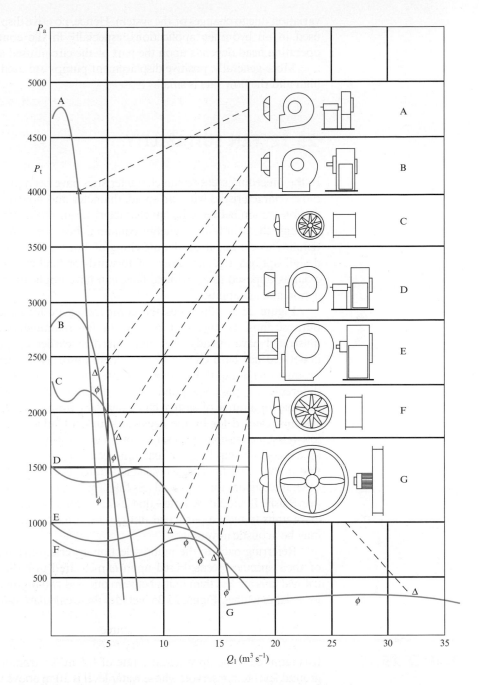

Each fan, operating at top speed and best efficiency point Δ, is chosen for an output
$Q \times P_t = 10$ kW
Peak input power is taken at φ
Drawings are to a uniform scale of 1:120

variation due to changes of the system. Hence, positive displacement rotary pumps are used in oil hydraulic applications, especially in the control systems in which the operating head depends upon the part of the circuit used at any time.

More generally, positive displacement pumps are used when the required head is high and the flow rate is small.

25.12 FAN SUITABILITY

In the preceding sections fan characteristics have been presented and the variation of those characteristics with fan speed, diameter and flow density considered. Similarly, the interaction between the fan characteristic and the system requirements, as defined through the steady flow energy equation, has been investigated. It therefore only remains to comment on the matching of fan type to the system requirements. Fan design options, e.g. the choice of forward- or backward-swept blades or the use of axial as opposed to centrifugal fans, can have implications for the system–fan interaction and suitability.

Figure 25.35 illustrates the comparative performance of a range of general purpose fans, as suggested by a major fan manufacturer. The fan design options illustrated have already been introduced in earlier chapters. In each case the fan illustrated has been chosen to provide an absorbed power output, calculated as the product of the fan throughflow and pressure rise, of 10 kW. It may be seen that the different fan types correspond to quite different fan characteristics and thus would be suitable for a range of system characteristics, from relatively low-flow–high-resistance systems, catered for by the backward-curved-bladed centrifugal unit, to high-flow–low-resistance systems best served by an axial fan with a small hub blockage. There is an area of overlap between axial and centrifugal units at moderate pressures. Note that in each case the characteristic illustrated applies to the fan's top speed. A family of characteristics for any speed less than this could therefore be generated as described previously. It should be noted that using a fan at less than top speed will necessitate extra units to meet a particular duty and will have space implications; however, there may be acoustic advantages.

Referring back to the pump selection criteria, fans may also be defined in terms of their specific speeds, based upon ω measured in radians. Typical values for centrifugal fans are less than unity, between one and two for mixed flow fans and above two for axial machines. Figure 25.35 includes these calculations for the fan types illustrated.

EXAMPLE 25.10

It is required to pump water at a rate of 0.5 m³ s⁻¹ from a sump which is 7 m below ground level to a reservoir whose water level is 20 m above the water level in the sump. The calculated losses due to friction and separation in the proposed pipeline amount to 52 m of water.

Select a suitable pump, which must be direct driven by an a.c. synchronous electric motor, to meet the above duty. Pumps A, B and C are available for selection and their characteristics are given in Fig. 25.36. For the selected pump, specify its size, speed, efficiency, power consumed and any requirements regarding its location with respect to the water level in the sump. Assume that the cavitation characteristics given obey similarity laws and refer to a saturated water vapour head of 0.2 m and

FIGURE 25.36

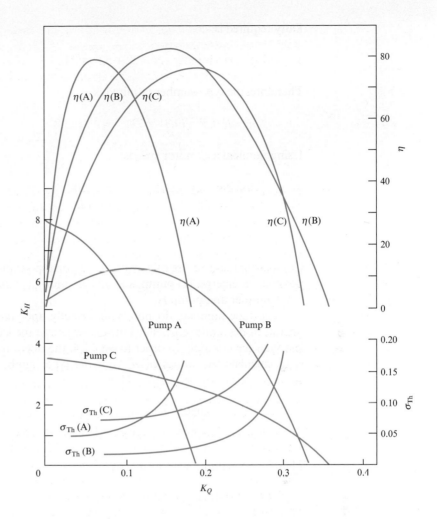

barometric pressure of 750 mm of mercury, the conditions at which the pump will operate.

Solution

From the characteristics given, the values of K_Q and K_H corresponding to the maximum efficiency of each pump are established and the values of type number n_s calculated from

$$n_s = Q^{1/2}/K_H^{3/4},$$

which gives:

	K_Q	K_H	n_s	σ_{Th}
Pump A	0.07	6.75	0.0631	0.055
Pump B	0.20	5.50	0.125	0.035
Pump C	0.16	2.80	0.185	0.085

Duty required is

$$Q = 0.5 \text{ m}^3 \text{ s}^{-1}, \quad H = 20 + 52 = 72 \text{ m}.$$

Therefore, the type number required is

$$n_s = N \, 0.5^{1/2}/(9.81 \times 72)^{3/4} = N/193.8.$$

Using nominal a.c. motor speeds:

for 2900 rev min^{-1}	$n_s = 0.250$
1450	0.125
960	0.0826
720	0.0619

Comparing these values of n_s with those corresponding to the pumps available, two possibilities emerge: (1) pump A at 720 rev min^{-1}; (2) pump B at 1450 rev min^{-1}.

Consider first pump A.

The type numbers do not match exactly and, therefore, the pump would not operate at maximum efficiency but somewhere to the left of it, since the required n_s is less than the pump n_s. In order to establish the operating point, one possible method is to calculate the values of n_s' (not *the* type number n_s) at points on the pump characteristic to the left of K_N. Thus,

$$\text{at } K_Q = 0.060, \quad K_H = 7.00 \quad \text{and} \quad n_s' = 0.057;$$

$$\text{at } K_Q = 0.065, \quad K_H = 6.85 \quad \text{and} \quad n_s' = 0.062.$$

Therefore, the pump will operate at

$$K_Q = 0.065 \quad \text{and} \quad \eta = 79 \text{ per cent,}$$

from which, since $K_Q = Q/ND^3$,

$$D = (Q/NK_Q)^{1/3} = (0.5 \times 60/720 \times 0.065)^{1/3} = 0.862 \text{ m.}$$

Power consumed is given by

$$P = 9.81 \times 0.5 \times 72/0.79 = 447 \text{ kW.}$$

Cavitation restrictions are

$$\sigma_{\text{Th}} = \text{NPSH}/H = 0.050$$

at the operating point, and so

$$\text{NPSH} = 72 \times 0.050 = 3.6 \text{ m.}$$

Now, $\text{NPSH} = p_{\text{atm}}/\rho g - H_s - p_{\text{vap}}/\rho g,$

$$p_{\text{atm}}/\rho g = 0.750 \times 13.6 = 10.2 \text{ m.}$$

Therefore,

$$H_s = 10.2 - 0.2 - 3.6 = 6.4 \text{ m}.$$

Thus, the maximum suction head including all pipe losses must not exceed (say) 6 m in order to avoid cavitation. Since ground level is 7 m above the sump, the pump would have to be lowered below ground level by at least 1 m, preferably more to account for pipe losses.

Consider now pump B.

The type number matches exactly and, therefore, the pump would operate at the maximum efficiency of 77 per cent.

$$D = (Q/NK_Q)^{1/3} = (0.5 \times 60/1450 \times 0.2)^{1/3} = 0.47 \text{ m}.$$

Power consumed is given by

$$P = 9.81 \times 0.5 \times 72/0.77 = 458.6 \text{ kW},$$

cavitation by

$$\text{NPSH} = \sigma_{Th}H = 0.035 \times 72 = 2.52 \text{ m}.$$

Therefore,

$$H_s = 10.2 - 0.2 - 2.52 = 7.48 \text{ m}.$$

Since, in this case, H_s is greater than the ground level above the sump, it will be possible to use this pump at ground level, provided it is close to the sump so that all pipe losses are on the delivery side of the pump. Also, pump B will be smaller than pump A. Conclusion: select pump B as follows:

$$D = 0.47 \text{ m}, \quad N = 1450 \text{ rev min}^{-1}, \quad \eta = 77 \text{ per cent}, \quad P = 458.6 \text{ kW}.$$

Concluding remarks

The study of rotodynamic machine performance and characteristics developed in Chapter 23 may be seen to be directed towards the interaction studies presented in this chapter. The solution of the relevant characteristic with the system loss curve allows the operating point of the system to be developed and the most suitable machine, or combination of machines, to be selected. This chapter has also presented the techniques necessary to determine the combined characteristics of fans and pumps operating in series and parallel, including the avoidance of machine instability. While these techniques are readily understood graphically, it is clear that computer simulation would be advantageous as it would allow rapid identification of unsuitable combinations.

The problems inherent in the pumping of fluids subject to cavitation were also raised in this chapter. This is a design limitation, but the careful selection of pump delivery and suction pipe diameters and location of the machine can minimize the risk of cavitation.

The chapter also introduced the use of jet fans in the ventilation of road tunnels. This example serves to emphasize again the solution of a wide range of fluid flow problems by reference to the fundamental equations stressed throughout this text.

Summary of important equations and concepts

1. The matching of fans or pumps to particular systems is a fundamental application of the steady flow energy equation and the concepts of machine characteristics introduced in Chapter 23. The operating point is defined as the intersection point of these two relationships, traditionally found by graphical techniques but now numerically by computing.

2. The series and parallel operation of fans and pumps is discussed, Sections 25.2 and 25.3, and the derivation of combined characteristics introduced and demonstrated.

3. Combinations of non-identical fan characteristics in parallel or series are shown to lead to a possible degradation in overall performance, Fig. 25.16, and 'hunting', or instability leading to multiple possible operating points, Figs 25.14 and 25.15.

4. The 'fan laws' are introduced, Section 25.7, and shown to be merely an alternative presentation of the dimensional analysis presented in Section 23.3.

5. Fans for tunnel ventilation are considered and equations developed to determine the number of units required to meet a particular ventilation demand, Section 25.8.

6. Cavitation as a limit on pump operation and as a determinant of pump location and attached pipe sizing is discussed in Section 25.10, with the cavitation constant defined by equation (25.28) and the suction specific speed by equation (25.29).

Problems

25.1 The characteristics of a fan are as follows:

Volume flow ($m^3 h^{-1}$)	0	2000	4000	6000
Fan total pressure (mm of water)	50	54.5	56	54.5
Fan input power (kW)	0.4	0.63	0.90	1.20

Volume flow ($m^3 h^{-1}$)	8000	10000	12000
Fan total pressure (mm of water)	50	42.5	32
Fan input power (kW)	1.53	1.70	1.75

If the system resistance is 60 mm of total water column at 7000 $m^3 h^{-1}$ determine the fan operating point, the power consumed and fan total efficiency.

[6600 $m^3 s^{-1}$, 53 mm, 953 W, 73 per cent]

25.2 A centrifugal pump has the following characteristics:

Q ($m^3 h^{-1}$)	0	23	46	69	92	115
h (m)	17	16	13.5	10.5	6.6	2
η (per cent)	0	49.5	61	63.5	53	10

The pump is used to pump water from a low reservoir to a high reservoir through a total length of 800 m of pipe 15 cm in diameter. The difference between the water levels in the reservoirs is 8 m. Neglecting all losses except friction and assuming $f = 0.004$, find the rate of flow between the reservoirs. Also determine the power input to the pump.

[60 $m^3 h^{-1}$, 3.07 kW]

25.3 Performance figures for a centrifugal fan are tabulated below. Plot these and superimpose a shaft power curve. From this determine the shaft power at the operating point if the system resistance is 100 mm of water at 40 $m^3 s^{-1}$ and also the power if the output is reduced to 25 $m^3 s^{-1}$ by damper regulation.

Q ($m^3 s^{-1}$)	0	10	20	30	40
h (mm of water)	85	92.5	95	90	80
η (per cent)	0	46	66	70	67

Q (m³ s⁻¹)	50	60	70
h (mm of water)	65	47.5	25
η (per cent)	60	48	32

[44 kW, 33 kW]

25.4 The characteristics of two rotodynamic pumps at constant speed are as follows:

Pump A:

Q (m³ s⁻¹)	0	0.006	0.012	0.018
H (m)	22.6	21.9	20.3	17.7
η (per cent)	0	32	74	86

Q (m³ s⁻¹)	0.024	0.030	0.036
H (m)	14.2	9.7	3.9
η (per cent)	85	66	28

Pump B:

Q (m³ s⁻¹)	0	0.006	0.012	0.018
H (m)	16.2	13.6	11.9	11.6
η (per cent)	0	14	34	60

Q (m³ s⁻¹)	0.024	0.030	0.036
H (m)	10.7	9.0	6.4
η (per cent)	80	80	60

One of the above pumps is required to lift water continuously through 3.2 m of vertical lift and the pipe to be used is 21 m long, 10 cm in diameter, and the friction coefficient is 0.005.

Select the more suitable pump for this duty and justify your selection. What power input will be required by the selected pump?   [Pump B, 3.49 kW]

25.5 A centrifugal pump is used to circulate water in a closed loop experimental rig, consisting of: two vertical pipes, one 4 m long and the other 3 m long, two horizontal pipes, each 1.3 m long, three 90° bends and a vertical 'working section' 1 m long. The pump is situated in one of the two low-level corners of the circuit. The pipes and the bends are 7.5 cm in diameter and the working section has a cross-sectional area of 125 cm². The friction factor for all pipes is 0.006 and the loss in each bend may be taken as $0.1\, v^2/2g$, where v is the mean velocity in metres per second. The loss in the 'working section' may be taken as equivalent to a frictional loss in a 1 m long pipe of 7.5 cm diameter.

Determine the mean velocity in the 'working section' if the pump characteristic is as follows:

Q (m³ s⁻¹)	0	0.006	0.012	0.018	0.024	0.027
H (m)	3.20	3.13	2.90	2.42	1.62	0.98

[1.32 m s⁻¹]

25.6 A 60 cm diameter, impeller centrifugal pump has the following characteristic at 750 rev min⁻¹:

Q (m³ min⁻¹)	H (m)	η (per cent)
0	40.0	0
7	40.6	41
14	40.4	60
21	39.3	74
28	38.0	83
35	33.6	83
42	25.6	74
49	14.5	51
56	0	0

(a) If the system resistance is purely frictional and is 40 m at 42 m³ min⁻¹ determine the pump operating point and power absorbed.

(b) The pump is used to pump water from one reservoir to another, the difference between levels being 13 m. The pipeline is 45 cm in diameter, 130 m long, $f = 0.005$ and it contains two gate valves ($K = 0.2$) and ten 90° bends ($K = 0.35$). Obtain the volume delivered by the pump and power absorbed.

(c) If, for the system at (b) a geometrically similar pump but of 50 cm diameter is used running at 900 rev min⁻¹, determine the volume delivered and power consumed.

[(a) 37 m³ min⁻¹, 232 kW; (b) 43.5 m³ min⁻¹, 234 kW; (c) 32.5 m³ min⁻¹, 163 kW]

25.7 The characteristic of an axial flow pump running at 1450 rev min⁻¹ is as follows:

Q (m³ s⁻¹)	0	0.046	0.069	0.092	0.115
H (m)	5.6	4.2	4.35	4.03	3.38

Q (m³ s⁻¹)	0.138	0.180
H (m)	2.42	0

When two such pumps are connected in parallel the flow rate through the system is the same as when they are connected in series. At what speed should a single pump run in order to deliver the same volume?

Assume the system characteristic to be purely resistive (no static lift).   [1700 rev min⁻¹]

25.8 The characteristics of a centrifugal pump at constant speed are as follows:

Q (m³ s⁻¹)	0	0.012	0.018	0.024
H (m)	22.6	21.3	19.4	16.2
η (per cent)	0	74	86	85

Q (m³ s⁻¹)	0.030	0.036	0.042
H (m)	11.6	6.5	0.6
η (per cent)	70	46	8

The pump is used to lift water over a vertical distance of 6.5 m by means of a 10 cm diameter pipe, 65 m long, for which the friction coefficient $f = 0.005$.

(a) Determine the rate of flow and the power supplied to the pump.

(b) If it is required to increase the rate of flow, and this may be achieved only by the addition of a second, identical pump (running at the same speed), investigate whether it should be connected in series or in parallel with the original pump. Justify your answer by determining the increased rate of flow and power consumed by both pumps.

[(a) 0.0268 m³ s⁻¹, 4.73 kW;
(b) parallel 13.2 kW, series 9.9 kW]

25.9   The characteristic of a pump in terms of dimensionless coefficients may be approximated to $K_H = 8 - 2K_Q - 210K_Q^2$. Such a pump having an impeller diameter of 0.4 m and running at 1450 rev min⁻¹ operates against a system characteristic represented by $h = 20 + 300Q^2$. Determine the flow rate delivered and the pump operating head. What would be the flow rate through the system if two such pumps were connected (a) in series and (b) in parallel?

[0.217 m³ s⁻¹, 34.12 m; (a) 0.253 m³ s⁻¹, (b) 0.326 m³ s⁻¹]

25.10   A pump has the following characteristic when running at 1450 rev min⁻¹:

Q (m³ s⁻¹)	0	0.225	0.335	0.425	0.545
H (m)	20	17	15	13	10

Q (m³ s⁻¹)	0.650	0.750	0.800
H (m)	7	3	0

A system is designed where the static lift is 5 m and the operating point is $H = 11.1$ m and $Q = 0.5$ m³ s⁻¹ using the pump as above. The system is redesigned, the static lift being 5 m as before, but the frictional and other losses increase by 40 per cent. Find the new pump speed such that the flow rate of 0.5 m³ s⁻¹ can be maintained.   [1576 rev min⁻¹]

25.11   Cavitation tests were performed on a pump giving the following results: $Q = 0.05$ m³ s⁻¹; $H = 37$ m; barometric pressure 760 mm of mercury; ambient temperature 25 °C. Cavitation began when the total head at pump inlet was 4 m. Calculate the value of Thoma cavitation coefficient and the NPSH.

What would be the maximum height of this pump above water level if it is to operate at the same point on its characteristic in the ambient conditions of a barometric pressure of 640 mm of mercury and a temperature of 10 °C?

[0.162, 6.086 m, 2.55 m]

25.12   The critical Thoma number for a certain type of turbine varies in the following manner:

N_s (rev min⁻¹, kW, m units)	0	50	100	150
σ_{Th}	0	0.04	0.1	0.18

N_s (rev min⁻¹, kW, m units)	200	250
σ_{Th}	0.28	0.41

A turbine runs at 300 rev min⁻¹ under a net head of 50 m and produces 2 MW of power. The runner outlet velocity of fluid is 10.4 m s⁻¹ and this point is 4.7 m above the tail race. The atmospheric pressure is equivalent to 10.3 m of water and the saturation pressure for water is 0.04 bar. Determine whether cavitation is likely to occur and find the head loss between runner outlet and tail race.

[No cavitation; 10.2 m]

25.13   A centrifugal pump, having four stages in parallel, delivers 218 litres s⁻¹ of liquid against a head of 26 m, the diameter of the impellers being 229 mm and the speed 1700 rev min⁻¹. A pump is to be made up with a number of identical stages in series of similar construction to those in the first pump to run at 1250 rev min⁻¹ and to deliver 282 litres s⁻¹ against a head of 265 m. Find the diameter of the impellers and the number of stages required.

[439 mm, five stages]

25.14   A louvre in a 0.25 m diameter extract fan and duct system is used to control air flow rates. If the flow rate is 2.0 m³ s⁻¹ with the louvre angle θ set to 60° open, calculate:

(a) the louvre separation loss coefficient at this setting and the pressure drop across the louvre,

(b) the reduction in flow rate if the louvre is shut down to 30° open if the pressure drop across the louvre remains constant.

Assume air density to be 1.2 kg m⁻³ and that the relationship between louvre loss coefficient K and louvre open angle θ is

$$\log_e K = 2 \times (90 - \theta)/180 - 1.25,$$

where θ is measured in degrees.

[0.398, 396 N m⁻², 0.31 m³ s⁻¹]

25.15   A ventilation duct includes a fan and control louvre. Determine the pressure rise across the fan if, at a louvre setting of 35° to the horizontal, the ventilation rate is 6 changes per hour. The room has a volume of 300 m³ and is held at 250 N m⁻² above atmosphere. The duct is 0.25 m in diameter and has a 15 m overall length. Entry and exit losses are equivalent to 33.3 pipe diameters and the louvre has a loss coefficient defined below,

K	2.0	3.2	6.4	12.8	30.6
Louvre angle	0	20	40	60	80

Take air density to be 1.2 kg m⁻³ and the duct friction factor as 0.009.   [542.5 N m⁻²]

25.16   The speed of a 1.0 m diameter extract fan in a library is controlled by the level of CO_2 sensed within the space. Assume that the combined frictional and separation losses in the fan extract ductwork may be expressed as

$\Delta p\,\mathrm{N\,m^{-2}} = 850Q^2$, where Q is measured in $\mathrm{m^3\,s^{-1}}$. If the relationship between fan speed N and the contamination level c is defined by $N = kc^{0.5}$, determine the change in extract flow consequent upon a halving of the contamination within the space from a level at which the appropriate fan speed was 1200 rev min^{-1}. The fan characteristics are given below:

$\Delta p\,10^6/(\rho N^2 D^2)$	520	475	390	210	100
$Q\,10^3/(ND^3)$	0	2.0	3.0	4.0	5.0

Take air density as $1.2\ \mathrm{kg\,m^{-3}}$.   [0.32 m^3 s^{-1}]

25.17   The speed of a 1.2 m diameter extract fan in a process area is controlled by the level of airborne contamination sensed within the space. Assume that the combined frictional and separation losses in the fan extract ductwork may be expressed as $\Delta p\,\mathrm{N\,m^{-2}} = 120\,Q^2$, where Q is measured in $\mathrm{m^3\,s^{-1}}$. If the relationship between fan speed N and the contamination level c is defined by $N = kc^3$, determine system operating point and the change in contamination resulting in a halving of the extract air flow rate from a level at which the appropriate fan speed was 1200 rev min^{-1}. The fan characteristics are given below:

$\Delta p\,10^6/(\rho N^2 D^2)$	520	475	390	210	100
$Q\,10^3/(ND^3)$	0	2.0	3.0	4.0	5.0

Take air density as $1.2\ \mathrm{kg\,m^{-1}}$.

[0.52 m^3 s^{-1}, 84.8 N m^{-2}, 87.5%]

25.18   An extract fan/duct system consists of 10 m of 0.5 m diameter ductwork, friction factor 0.008, incorporating an axial flow fan. The duct entry and exit grilles may be represented by separation losses equivalent to 30 pipe diameters of ductwork. The design flow rate is 2.6 m^3 s^{-1}, the design setting of the control louvre being 45° at that flow.

The damper louvre loss coefficient may be expressed as $k = c_1 + c_2[\pi/(\pi - 2\alpha)]^3$, where c_1 and c_2 are empirical coefficients and α is the damper angle to the horizontal, with $\alpha = 0$ when the damper is fully open and $\alpha = 90°$ when the damper is shut. Laboratory experiments have indicated that $k = 1.2$ when $\alpha = 15°$ and 3.6 when $\alpha = 60°$.

Determine the pressure rise across the fan at the design condition. Assume air density at 1.3 kg m^{-3}.

[153.87 N m^{-2}]

26

Ventilation

THIS CHAPTER ADDRESSES ONE OF THE MOST COMMON APPLICATIONS OF FANS, NAMELY the ventilation of habitable or process space. The treatment includes a presentation of the contamination decay equation as applied to fully mixed air/contaminant systems, together with a program that may be used to predict multiple, non-reacting, contaminant levels within a space over a time period that includes changes in either the contamination generation rate or the rate of ventilation air flow provided. The processes of natural ventilation are briefly introduced to provide a basis for the decision to introduce mechanical ventilation. ● ● ●

26.1 VENTILATION AND AIRBORNE CONTAMINATION IN SPACES

The provision of a ventilation air supply to both occupied and unoccupied spaces within buildings and other inhabited structures, such as forms of transport, is necessary in order to replenish the oxygen supply; to act as a dilutant to carbon dioxide, odours, process emissions; to prevent the build-up of potentially explosive vapour mixtures in unoccupied plant spaces; to provide air movement, as a constituent part of comfort; and to control airborne contamination in industrial ventilation. While the detailed treatment of ventilation is outside the scope of this text, the basic techniques and mechanisms are clearly applications of fluid mechanics principles.

There are two techniques available to provide space ventilation, namely natural ventilation, relying upon differential pressures or stack effects acting on 'gaps', either intentional, such as window openings, or unintentional, such as cracks in structures that allow an air infiltration path, and mechanical ventilation, relying on fan-driven systems operating within sealed structures, e.g. aircraft at altitude, submarines or buildings constructed to have low infiltration characteristics. It is clear that the first is suitable for domestic buildings and structures where neither close energy management nor process-generated contamination is considered a problem. The second, mechanical, technique is suitable for energy-controlled buildings, those where sealing the building becomes necessary for other reasons, e.g. acoustic pollution, and buildings incorporating processes generating contaminants. Mechanical ventilation is naturally the only alternative for the other examples quoted above.

26.1.1 Minimum ventilation rates

Minimum ventilation rates are provided as guidance in the design of building services systems and are normally based on the need to dilute carbon dioxide and odours, provide air movement and/or reduce heat loads. Specialist minimum rates for the prevention of process contamination or explosive mixture build-up are also available. Some general guidelines applicable to habitable spaces are indicated in Tables 26.1 and 26.2. The supply of ventilation air flows to achieve these levels can be either natural, i.e. relying upon cracks, windows, doors, etc., or mechanical involving a fan and ductwork network.

TABLE 26.1
Guidelines on minimum ventilation rates

Air space per person (m^3)	Fresh air supply per person (litres s^{-1}) (air changes per hour)		
	Minimum	Non-smoking	Smoking
3	11.3	17.0 (2.0)	22.6
6	7.1	10.7 (6.5)	14.2
9	5.2	7.8 (3.2)	10.4
12	4.0	6.0 (1.8)	8.0

TABLE 26.2

Levels of ventilation
required in various
building applications

APPLICATION	LITRES PER SECOND PER PERSON	Litres s^{-1} m^{-2}
Domestic	8–12	–
Board rooms	18–25	–
Bars	12–18	–
Dept. stores	5–8	–
Factories	–	0.8
Garages	–	8.0
Op. theatres	–	16.0
Hospital wards	8–12	–
General offices	5–8	1.3–2.0
Private offices	8–12	1.3–2.0
Restaurants	12–18	–
Theatres	5–8	–

Note: lavatory spaces, particularly those without external openable windows, are a special case where the recommendations are that mechanical ventilation is provided at a rate equal to 15 air changes per hour, 80 litres s^{-1} per WC bowl installed or 16 litres $s^{-1} m^{-2}$, whichever is the greatest.

In many cases these ventilation levels are based on the need to remove carbon dioxide. The effects of carbon dioxide contamination by volume are as follows:

1–2 per cent	Continuous exposure leads to headaches and dyspnoea (breathlessness)
3 per cent	Severe headaches
5 per cent	Mental depression
6 per cent	Visual impairment
10 per cent	Unconsciousness.

As shown above, the rate of ventilation air supply may be expressed as a number of air changes per hour, i.e. based on the total volume of the space served, a rate based on the number of litres per second supplied per person or per square metre of floor space. It is essential to remember in using the equations for concentration build-up or decay that the SI system of units must be applied and so time must be expressed in seconds.

The use of litres per second per person is best if the occupation of the space is likely to be high or varying. The use of litres per second per square metres of floor space is best for situations where the occupancy is fixed, e.g. an office space. The use of air changes per hour can be misleading in spaces with a high ceiling and hence a large volume.

Carbon dioxide is a common contaminant whose concentration build-up in a space is monitored. Carbon dioxide is present in ambient air at a level of 0.03 per cent and is generated by an occupant of a space at a rate of 4.72×10^{-3} litres s^{-1}.

The concentration equation to be developed will allow the changes in concentration level in a space of a contaminant to be predicted. It will be shown that this equation can be applied separately to any number of contaminants present in a space provided that these contaminants do not react with each other.

26.2 CALCULATION OF VENTILATION REQUIREMENTS

Figure 26.1 illustrates the general case of ventilation and contamination growth, or decay, within a space. Processes or occupation of the space will cause the contamination levels to rise unless the space is adequately ventilated. It is also necessary to consider the possibility of contamination entering the space from outside by means of the ventilation air flow. It is possible to develop a general expression to encompass these effects, the non-relevant terms being dropped if the effect they represent is not present.

FIGURE 26.1

Definition of the terms constituting the decay equation for contamination within ventilated spaces, for either naturally or mechanically ventilated conditions

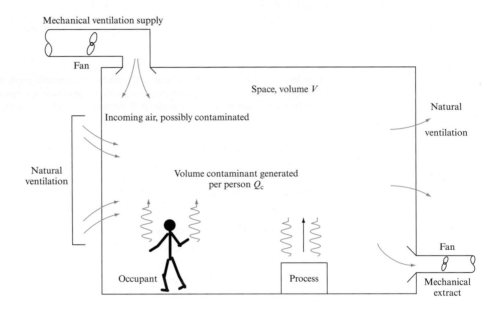

In the derivation of the general contamination equation the following terms may be defined:

c_i initial concentration of contaminant in the space at time zero;

c concentration at time t;

c_0 concentration of the contaminant in the incoming air supply;

Q incoming ventilation air supply expressed as a volumetric flow per second per person;

Q_c volume of contaminant produced per person within the space per second;

V volume of the space per person in occupation;

n number of air changes per hour for the whole space.

It is usual to refer to the contamination levels as parts per 10 000. It is thus possible to write down a balanced equation across a small time increment dt as follows:

In a time increment dt the contamination in the space increases due to possible inflow via the ventilation air and due to generation within the space

$$= (Qc_0/10\,000 + Q_c)\, dt \text{ per person in the space.}$$

Contamination leaves the space with the extract air. Here it is assumed that there is no build-up of pressure in the space so that no 'air storage' term exists in the continuity of air flow equation. This removed contamination has a value

$$= (Qc/10\,000)\, \mathrm{d}t \text{ per person in the space.}$$

(It will be seen that the extract air flow is assumed to be equal to the inflow rate; it is the level of contamination carried by this flow that differs.)

The net change in contamination in the space over the small time increment $\mathrm{d}t$ is thus

$$\mathrm{d}c = [(Qc_0/10\,000 + Q_\mathrm{c}) - Q_\mathrm{c}/10\,000]\, \mathrm{d}t \text{ per person.}$$

Expressed as a concentration $\mathrm{d}c$ in parts per $10\,000$ of air/unit volume of room space,

$$\mathrm{d}c/10\,000 = (1/V)[(Qc_0/10\,000 + Q_\mathrm{c}) - Qc/10\,000]\, \mathrm{d}t.$$

Rearranging

$$\mathrm{d}c/\mathrm{d}t + (Qc/V) = (Qc_0 + 10\,000 Q_\mathrm{c})/V.$$

This differential equation may be solved by the use of an integrating factor as follows. Combine the first two terms to form a function capable of integration,

$$\mathrm{d}(c\ \mathrm{e}^{Qt/V})/\mathrm{d}t = \mathrm{e}^{Qt/V}\mathrm{d}c/\mathrm{d}t + \mathrm{e}^{Qt/V}Qc/V,$$

i.e. using the product rule gives

$$\mathrm{e}^{Qt/V}(\mathrm{d}c/\mathrm{d}t + Qc/V).$$

Hence, multiplying both sides by an integrating factor $\mathrm{e}^{Qt/V}$ allows the solution of the differential equation

$$\mathrm{e}^{Qt/V}(\mathrm{d}c/\mathrm{d}t + Qc/V) = \mathrm{d}(c\ \mathrm{e}^{Qt/V})/\mathrm{d}t$$
$$= \mathrm{e}^{Qt/V}(Qc_0 + 10\,000 Q_\mathrm{c})/V.$$

Integrating both sides with respect to t yields

$$c\ \mathrm{e}^{Qt/V} = (V/Q)\mathrm{e}^{Qt/V}(Qc_0 + 10\,000 Q_\mathrm{c})/V + A.$$

Note that the V term cancels at this stage and that A is a constant of integration whose value must be determined from known conditions at time zero.

At $t = 0$, $c = c_\mathrm{i}$, A is therefore calculated as

$$A = c_\mathrm{i} - (c_0 + 10\,000 Q_\mathrm{c}/Q).$$

Thus a general expression for the contamination within the space at any time t is given by

$$c = (c_0 + 10\,000 Q_\mathrm{c}/Q)(1 - \mathrm{e}^{-Qt/V}) + c_\mathrm{i}\ \mathrm{e}^{-Qt/V}.$$

Two points are worth reinforcing at this stage:

1. The contamination equation is *only valid* during continuous processes. If there is a change in the applied ventilation rate or in the rate of production of contaminant by a process within the space, then the analysis must be restarted with a 'new' time zero. Naturally, the final contamination concentration becomes the initial value for the new application of the concentration equation.

2. The equation may be applied simultaneously and in parallel to two or more contaminants within the same space so long as these contaminants are separate entities and do not react with each other.

The general expression for contamination at time t, independent of the number of occupants, within a continuous process for any individual contaminant is thus

$$c = (c_0 + 10\,000Q_c/Q)(1 - e^{-Qt/V}) + c_i e^{-Qt/V}, \qquad (26.1)$$

A number of special cases may be considered:

1. Fresh air supplied to the space contains no contaminant, e.g. if the contaminant is only present in the room as the result of a process or as a means of monitoring ventilation, and there are no people or processes active in the space, then

$$c = c_i\, e^{-Qt/V} = c_i\, e^{-nt}. \qquad (26.2)$$

This is the standard decay equation used to determine the natural ventilation of a space. If a known concentration of tracer gas is introduced into the space at time zero and the level of this contaminant is then monitored for a period of time, it is possible to determine the value of the 'number of air changes per hour', n. Plotting c vs. t in a natural log field results in an equation

$$\log_e(c) = \log_e(c_i) - nt, \qquad (26.3)$$

which gives n as the gradient of the resulting straight line. (Note that t is always in seconds to maintain SI units.) In utilizing this technique to determine the natural ventilation rate it is worth stressing that efficient mixing of the tracer gas with the air within the space is necessary to obtain a realistic measurement of 'air changes per hour'.

2. If the supply is contaminated with a contaminant not initially present in the space then the level of this contamination will rise:

$$c = c_0(1 - e^{-Qt/V}) = c_0(1 - e^{-nt}). \qquad (26.4)$$

3. If the contamination is only a direct result of the people or processes active in the space then the initial contamination is zero as is the contamination carried in from outside:

$$c = (10\,000Q_c/Q)(1 - e^{-nt}). \qquad (26.5)$$

It follows that any combination of the above processes may be considered provided that the 'clock is stopped' at each time that the overall process changes, e.g. a change in the extract rate or the cessation of a process that generated a contaminant.

FIGURE 26.2
Contamination decay and
growth under a range of
conditions. Note special
case for determination of
natural ventilation air
change per hour

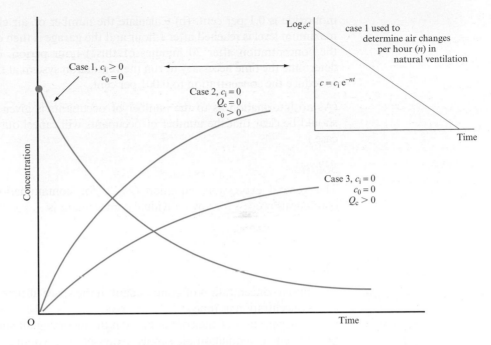

Case 1, $c_i > 0$
$c_0 = 0$

Case 2, $c_i = 0$
$Q_c = 0$
$c_0 > 0$

Case 3, $c_i = 0$
$c_0 = 0$
$Q_c > 0$

$\log_e c$

case 1 used to
determine air changes
per hour (n) in
natural ventilation

$c = c_i\, e^{-nt}$

Time

Concentration

Time

O

FIGURE 26.3
Variation of contaminant
concentration over a
24-hour period owing to
an intermittent process.
Zones A, C:
contamination due to
process, so $Q_c > 0$, $c_i \geqslant 0$,
$c_0 = 0$; fan-assisted
ventilation. Zones B, D:
no process, $Q_c = 0$, $c_i > 0$,
$c_0 = 0$; fan-assisted
ventilation. Zone E:
overnight, $Q_c = 0$, $c_i > 0$,
$c_0 = 0$; natural ventilation
only

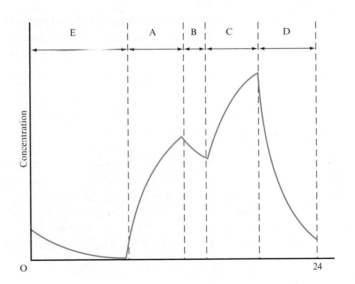

E   A   B   C   D

Concentration

O   24

Figure 26.2 illustrates each of the cases above, while Figure 26.3 illustrates the
change in contamination level due to a process that is intermittent over a 24-hour
period, e.g. paint spraying.

EXAMPLE 26.1

A garage has a volume of 60 m by 30 m by 3 m and contains cars that generate
$0.0024\ \mathrm{m^3\,s^{-1}}$ of carbon monoxide. (a) Calculate the number of air changes per hour if
the garage is in continuous use and the maximum permissible concentration of carbon

monoxide is 0.1 per cent. (b) Calculate the number of air changes per hour if this maximum level is reached after 1 hour and the garage is then out of use. (c) Calculate the concentration after 20 minutes of this 1-hour period. (d) For case (b) above determine the time necessary to run the ventilation system at the rate calculated in (b) to reduce the concentration to 0.001 per cent.

(As no information as to the number of occupants is given assume one person. It should be clear that the number of occupants will cancel out of the equation in any case.)

Solution

The general expression, equation (26.1), for contamination at time t within a continuous process for any individual contaminant is

$$c = (c_0 + 10\,000 Q_c/Q)(1 - e^{-Qt/V}) + c_i\, e^{-Qt/V},$$

where

c_i initial concentration of contaminant in the space at time zero;
c concentration at time t;
c_0 concentration of the contaminant in the incoming air supply;
Q incoming ventilation air supply expressed as a volumetric flow per person per second;
Q_c volume of contaminant produced per person within the space per second;
V volume of the space per person in occupation;
n number of air changes per hour for the whole space.

(a) In case (a):

$c_0 = 0$, $c_i = 0.1$ per cent or 10 parts in 10 000 as the garage is in continuous use.

$V = 5400\ \text{m}^3$.

$Q_c = 0.0024\ \text{m}^3\,\text{s}^{-1}$.

$Q = Vn/3600$ to retain seconds as the time unit.

Thus

$$10 = [0 + 10\,000 \times 0.0024/(5400n/3600)](1 - e^{-nt}) + 10\,e^{-nt}$$

$$10 = (16/n)(1 - e^{-nt}) + 10\,e^{-nt}$$

$n = 1.6$ as in this case of continuous use t would be infinite.

(b) In case (b):

$c_i = 0$, $c_0 = 0$ and $c = 0.1$ per cent at time $t = (1 \times 3600)$ s

$$c = (c_0 + 10\,000 Q_c/Q)(1 - e^{-Qt/V}) + c_i\, e^{-Qt/V}$$

$$10 = [0 + 10\,000 \times 0.0024/(5400n/3600)](1 - e^{-Z}) + 0,$$

where $Z = Qt/V$, $t = 1 \times 3600$ s

$$10 = (16/n)(1 - e^{-Z}),$$

where $Z = nt$, with t measured in hours

$$n = 1.6(1 - e^{-nt}).$$

If $n = 1.5$, $\quad 1.5 = 1.6(1 - 0.22) = 1.24$, $\quad$ error $= -0.26$.

If $n = 1.2$, $\quad 1.2 = 1.6(1 - 0.30) = 1.112$, $\quad$ error $= -0.088$.

If $n = 0.9$, $\quad 0.9 = 1.6(1 - 0.41) = 0.95$, $\quad$ error $= +0.05$.

If $n = 1.0$, $\quad 1.0 = 1.6(1 - 0.37) = 1.01$, $\quad$ error $= +0.01$.

By trial therefore the necessary value of $n = 1.0$ air changes per hour. (A graphical representation of this equation would give a better approximation.) Note that this is less than case (a) as the limit is reached earlier.

(c) In case (c) the problem is to calculate c after 20 minutes, 1200 seconds or 0.33 hours. Apply the general equation (26.1), as above with $t = 1200$:

$$c = (c_0 + 10\,000 Q_c/Q)(1 - e^{-Qt/V}) + c_i\, e^{-Qt/V}$$

$$= [0 + 10\,000 \times 0.0024/(5400 \times 1.0/3600)](1 - e^{-nt})$$

$$= 16(1 - e^{-1.0 \times 0.33})$$

$$= 16(1 - 0.72) = 4.53 \text{ parts in } 10\,000.$$

(d) In case (d) the initial concentration is 0.1 per cent and no new contamination is generated, hence $Q_c = 0$ and as no carbon monoxide is carried in by the vent system, it follows that $c_0 = 0$. The target c is 0.001 per cent, or 0.1 parts in 10 000.

Applying the general equation (26.1),

$$c = c_i\, e^{-nt}$$

results in

$$0.1 = 10 \times e^{-1.0t}$$

as the final contamination level aimed at is 0.001 per cent and the air change rate is 1.0 per hour.

Thus, with t in hours

$$1.0t = \log_e(100) = 4.61$$

$t = 4.61$ hours or 276.3 minutes.

26.3 COMPUTER PROGRAM CONTAM

Program CONTAM predicts the contamination level in a ventilated, and possibly occupied, space as a result of possibly both the infiltration of contaminant in the ventilation supply and the generation of contaminant within the space, either by virtue of its occupation or the processes carried on within it. The prediction is based on the general form of the decay equation, equation (26.1), and therefore assumes uniform mixing within the space. The program will accept up to ten parallel contaminants, each being treated for a maximum of ten time periods. In this context a time period is defined as the time between changes in condition that necessitate the 'restarting' of the decay equation application, e.g. a change in room occupation, a change in the rate of contamination generation or a change in the applied ventilation rate.

The program requires as data, in SI units with the exception of time which is input in hours and minutes as appropriate and contamination in parts per 10 000, the following input:

1. the number of contaminant cases to be considered (note this facility may also be used as means of comparing the effect of changing ventilation rate for a particular example);
2. the number of sample periods;
3. the initial contamination in the space at the start of the analysis;
4. the volume of the space to be considered;
5. the start time for the analysis, 24-hour clock format.

The following data are repeated for each sample period, entered in a screen table:

6.1 the duration of the time period, hours;
6.2 the interval at which contamination levels are required in this time period, minutes;
6.3 the concentration of the contaminant in the supply air (note this can be zero), parts per 10 000;
6.4 the ventilation air supply rate, $m^3 s^{-1}$;
6.5 the number of people in the space (note zero acceptable) during this period;
6.6 the volume of contaminant produced per person in the space, $m^3 s^{-1}$, during that period (note if no occupants, then zero is acceptable; however, if the space is unoccupied with a source of contaminant active, then input total rate of contaminant generation).

26.3.1 Application example

Program CONTAM was used to investigate the changing carbon dioxide levels in a lecture hall over a 24-hour period. The data fields entered were as follows:

Contamination episode	0 (note only one contaminant to be considered)				
Sample periods	5				
Initial conc. (pp10 000)	3.0				
Total volume (m^3)	200				
Start time	9.0				
Period durations (hours)	3.0	2.0	4.0	3.0	12.0
Sample intervals (minutes)	5.0	5.0	5.0	5.0	10.0
Air supply conc. (pp10 000)	3.0	3.0	3.0	3.0	3.0
Supply airflow ($m^3 s^{-1}$)	0.2	0.05	0.2	0.1	0.05
Number of occupants	40	5	60	20	0
Contam./person/s ($m^3 s^{-1}$)	5e–6	5e–6	5e–6	5e–6	0.0

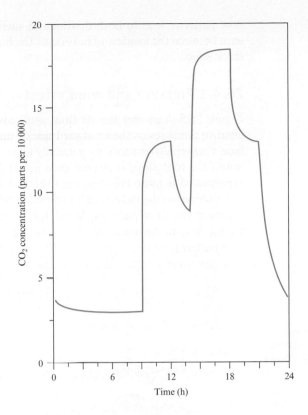

FIGURE 26.4
Carbon dioxide
concentration in a $200\,m^3$
space over a 24-hour
period, predicted by use
of program CONTAM

The variation in CO_2 level over the 24-hour period is illustrated in Fig. 26.4, and is displayed on screen by the program.

26.3.2 Additional investigations using CONTAM

The simulation may be used to investigate:

1. multiple contamination sources to determine which governs the occupation rate of the space or the duration of any process;
2. the effect of externally generated contamination levels on usability of a space – e.g. carbon monoxide or other contaminant infiltration from adjacent road activity. In this case the initial concentration in the space would be zero and the internal generation of contamination would be zero;
3. the use of an initial concentration of contaminant to determine the natural ventilation rate for an enclosed space.

26.4 PRINCIPLES OF NATURAL VENTILATION

Natural ventilation relies on the presence of openings in the ventilated structure and the effect of two mechanisms that cause air exchange through these openings, based on temperature and pressure effects. Changes in pressure over the surface of the building, due to the effect of the building and surrounding structures on the local air

flow patterns, lead to both driving and suction effects, while the temperature difference between the inside and outside of the building leads to density variations and the
stack effect.

26.4.1 Pressure and wind effects

Figure 26.5 illustrates the air flow over a building form. The interaction results in
positive pressures on the windward side of the building and suction pressures on its lee
face. Pressures over roofs are generally lower than those in the undisturbed airstream,
unless the roof slope is steeper than about 45°. In practice, the actual pressure field
experienced is more complex than this, as the presence of other structures, trees or
obstructions in the vicinity of the building being considered leads to localized vortices
affecting the flow patterns. Wind tunnel testing has been utilized to allow these
interactions to be predicted. The acoustic effects that accompany the generated air
flow patterns are also of interest to the designer, as well as localized loading effects that
may determine construction decisions.

FIGURE 26.5

(a) Air pressures on a
building façade as a result
of prevailing winds. Note
downstream vortices.
(b) Interaction between
buildings situated in the
air flow patterns
downstream of adjacent
structures

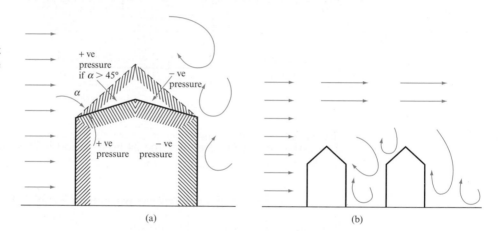

Based on the orifice flow expressions detailed in Chapter 6, it follows that the
volumetric flow through an orifice assumed in the building windward façade to
represent an infiltration air path crack may be expressed as

$$Q_{in} = C_d A u,$$

where u is the air flow velocity through the orifice, area A and discharge coefficient C_d.
    The kinetic energy of the air flow may be expressed as

$$\Delta p = 0.5 \rho u^2,$$

where Δp is the pressure difference across the orifice, at this stage either in or out of
the enclosure, and ρ is the appropriate air density, so that the summation of the
inflows to the building due to infiltration cracks on its windward façade may be
expressed as (Fig. 26.6)

$$\sum Q_{in} = C_d \sum A_{in} (2 \Delta p_{in} / \rho)^{0.5} \tag{26.6}$$

FIGURE 26.6

Infiltration due to
pressure and wind effects

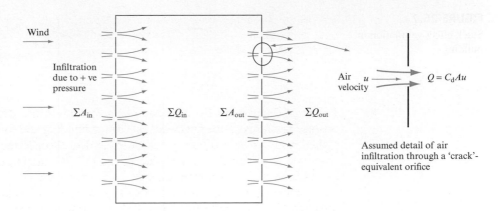

if it is assumed that all the orifices have the same discharge coefficient. The outflow
from the enclosure due to the difference between the internal building pressure and the
suction pressures acting on the leeward façade may similarly be expressed as

$$\Sigma Q_{out} = C_d \Sigma A_{out}(2\Delta p_{out}/\rho)^{0.5},$$ (26.7)

again assuming that all the infiltration cracks have the same discharge coefficient. If it
is further assumed that steady flow continuity exists, such that

$$\Sigma Q_{in} = \Sigma Q_{out} = \Sigma Q_{inf},$$

then it follows that summing the pressure differences as

$$\Delta p_{in} + \Delta p_{out} = \Delta p_{wl},$$ (26.8)

where Δp_{wl} is the pressure difference between the windward and leeward façades of the
building, yields an expression for infiltration air flow of the form

$$\Delta p_{wl} = (\rho \Sigma Q_{inf}^2/2C_d^2)[1/(\Sigma A_{in})^2 + 1/(\Sigma A_{out})^2]$$ (26.9)

$$\Sigma Q_{inf}^2 = \Delta p_{wl} 2C_d^2/\rho[1/(\Sigma A_{in})^2 + 1/(\Sigma A_{out})^2].$$ (26.10)

It is clear that there are a wide range of assumptions inherent in this formulation,
including the problem of assessing crack size and discharge coefficient, summing the
infiltration paths and the fundamentally unsteady nature of the wind-generated
pressures. However, the combination of air infiltration by design, i.e. by providing
openable windows, and the seemingly unavoidable infiltration paths left in construction provides the basis for natural ventilation. As such the natural ventilation rate
provided may be determined by the tracer gas decay methodology mentioned in
Section 26.2.

26.4.2 Temperature and stack effects

Differences in temperature between the air in a building and that outside result in the
establishment of a density difference across the building fabric that causes the internal

FIGURE 26.7

Stack effect ventilation in buildings

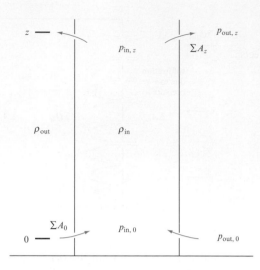

air, if warmer, to rise and exit the building through high-level vents, while at the same time cooler air from outside enters the enclosure via lower-level openings. (Note that this process is reversed if the outside air is warmer than the building air, an effect that may be found in heavyweight structures in hot climates.) Figure 26.7 illustrates this effect and the notation adopted for the relative pressure, temperature and density values.

Over a building height z, the relationship between pressure and level in the outside air may be expressed as

$$p_{out,0} = p_{out,z} + g\rho_{out}z \tag{26.11}$$

and, correspondingly, for the air column within the building,

$$p_{in,0} = p_{in,z} + g\rho_{in}z. \tag{26.12}$$

Therefore the pressure differences between the internal and external air at the base of the building and at a height z may be introduced by subtracting these two expressions and introducing the term $\Delta\rho$, the density difference between the internal and external air:

$$p_{out,0} - p_{in,0} = gz(\rho_{out} - \rho_{in}) - (p_{in,z} - p_{out,z}). \tag{26.13}$$

Under steady conditions the mass flow entering the building must equal the mass flow leaving at the higher level. Note that as the density changes between these flows, it is mass and not volumetric flow that must be equated. Thus from the previous treatment of orifice flow it follows that mass continuity may be expressed as

$$\rho_{out}C_d\Sigma A_{z=0}[2(p_{out,0} - p_{in,0})/\rho_{out}]^{0.5}$$
$$= \rho_{in}C_d\Sigma A_z[2(p_{in,z} - p_{out,z})/\rho_{in}]^{0.5}.$$

Assuming as before that the orifice discharge coefficients have a single value, this mass flow continuity equation reduces to

$$p_{in,z} - p_{out,z} = (p_{out,0} - p_{in,0})(\rho_{out}A_{z=0}^2)/(\rho_{in}A_z^2),$$

so that $p_{out,0} - p_{in,0} = gz(\rho_{out} - \rho_{in}) - (p_{out,0} - p_{in,0})(\rho_{out}A_{z=0}^2)/(\rho_{in}A_z^2)$

and, hence,

$$p_{out,0} - p_{in,0} = [gz(\rho_{out} - \rho_{in})]/[1 + (\rho_{out}A_{z=0}^2)/(\rho_{in}A_z^2)].$$

As the ventilation air flow entering the building at low level may be expressed as

$$\Sigma Q_{inf} = C_d \Sigma A_{z=0}[2(p_{out,0} - p_{in,0})/\rho_{out}]^{0.5},$$

it follows by substitution that

$$\Sigma Q_{inf} = C_d \Sigma A_{z=0}[\![2gz(\rho_{out} - \rho_{in})/\{\rho_{out}[1 + (\rho_{out}A_{z=0}^2)/(\rho_{in}A_z^2)]\}]\!]^{0.5}.$$

$$(26.14)$$

This expression may be simplified and expressed in terms of the more easily measured air temperatures if it is assumed that the term ρ_{out}/ρ_{in} tends to unity and that the term $(\rho_{out} - \rho_{in})/\rho_{out}$ may be replaced by $\Delta T/T$, where T is the mean of the inside and outside air temperatures and ΔT is the difference between the mean air temperatures, with respect to height z inside and outside the building. The expression for infiltration volumetric flow thus becomes

$$\Sigma Q_{inf} = [C_d/(1/\Sigma A_{z=0}^2 + 1/\Sigma A_z^2)](2gz\Delta T/T)^{0.5}.$$

$$(26.15)$$

In practice, both pressure- and temperature-driven natural ventilation infiltration air flows will exist together. It is possible for the pressure effect to aid and enhance the stack air movement if the pressure is increased on the orifices contributing an inflow to the stack, or temperature-driven, flows. However, the situations met in practice are likely to be far more complex. These uncertainties add to those already identified above. While the treatment described here has concentrated on internal to external building gradients, it must also be remembered that gradients also exist between spaces within a building. The importance of such models has been highlighted by the studies of air movement between rooms with interconnecting doors and the implications of such infiltration on passive-smoking-generated ailments.

26.5 MECHANICAL VENTILATION

Mechanical ventilation allows the control of contamination levels within a space as well as contributing to the energy management by determining the temperature levels at which 'used' air is returned to atmosphere. In this connection it may well be used in conjunction with a heat exchanger system to recover energy and then utilized to preheat incoming fresh air. Air drawn from outside by a fan and duct system is distributed to all zones within the building. The design of the necessary ductwork and the determination of the operating points of the fans may be undertaken based upon the techniques discussed elsewhere in this text.

A range of air supply and mixing techniques are used to ensure that the spaces served by the system are adequately ventilated and that contamination levels are held to satisfactory levels. The simplest technique is the local exhaust system, as demonstrated by a laboratory fume cupboard or a commercial kitchen oven hood, that is intended to remove the contamination at source before it can infiltrate into the rest of the space. Here the fan mounted in the ductwork draws air over, or through, the space to be ventilated. It is stressed that this system is only efficient locally.

The need to provide 'clean' environments for both medical (i.e. operating theatres) and electronic manufacturing purposes has led to considerable research aimed at the prevention of contamination spread in these spaces. Known as piston ventilation, such systems rely on low-turbulence air flows that may be introduced to the space via ceiling- or wall-mounted grille and filter systems. The downflow, or crossflow, of ventilation air at flow velocities in the region of $0.5\,\mathrm{m\,s^{-1}}$ effectively drives any contamination generated towards low-level-mounted extract ducts.

The introduction of mechanical ventilation or air-conditioning systems into large buildings prompted a range of techniques to ensure that satisfactory mixing takes place. Without a full mixing process the predictions of the contamination equation developed in Section 26.2 become doubtful. Although outside the scope of this treatment, techniques to ensure mixing are now well understood. However, the movement of air within a space does depend upon the obstructions present and therefore the utilization of a space, in terms of the arrangement of furniture, partitions and localized process and heat sources, can become important on a micro level. Generally, such systems are referred to as mixing ventilation. This technique has many properties in common with displacement ventilation where, in modern systems, ventilation air is introduced at low level, to maximize the benefit to the occupant, and rises due to buoyancy effects caused by the local heat sources within the space, thereby displacing previously contaminated air through high-level extract ducts. Figure 26.8 illustrates these modes of mechanical ventilation.

FIGURE 26.8
(a) Localized contamination extract system – fume cupboards or process extract.
(b) Schematic of piston ventilation. Crossflow also utilized. (c) Air distribution in a mixing ventilation system

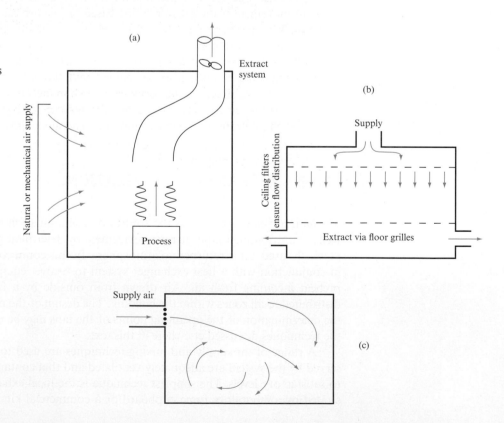

Concluding remarks

While the earlier chapters in this part of the text have concentrated on fan and pump performance, characteristics and matching, this chapter has introduced one of the main reasons for the introduction of a fan and duct network, necessitating much of the analysis presented. The provision of mechanical ventilation to habitable and process spaces is a major application for fan technology and therefore this chapter has been included to provide a basis for understanding the mechanisms of contamination decay, either by natural or by mechanical means.

The decay equation introduced during this treatment is restricted to fully mixed contamination and is also only applicable so long as there is no change in the contamination generation rate or ventilation rate during the time period covered by the integration. However, the expression is general in that it may be applied sequentially to time periods having constant ventilation or contamination generation rates, or constant occupation levels. Similarly, it may be applied in parallel to any number of non-reacting contaminants that may be present together within the space studied. It will again be noted in the derivation of the decay equation that the only necessary equation was the conservation of mass, including the storage terms, again reinforcing the view that fluid mechanics is a subject area capable of analysis by the fundamental equations of continuity, momentum and energy.

Summary of important equations and concepts

1. The requirement for ventilation is defined and tabular data, Tables 26.1 and 26.2, provided to indicate the required levels dependent upon space usage. The effects of CO_2 deprivation are also discussed.

2. The decay equation is developed, Section 26.2, equation (26.1), and its limitations are that it only applies to steady state conditions, i.e. no changes to the parameters, e.g. no change in occupancy, ventilation rate, contamination generation or infiltration rate. Any change in any of these parameters necessitates a restarting of the application of the equation.

3. As long as there are no reacting contaminants present, the equation may be employed in parallel for a range of contaminants.

4. The use of the decay equation to measure natural ventilation rates is discussed, Section 26.2.

5. A computer program encompassing all the points above is presented in Section 26.3.

6. The principles of natural ventilation are discussed in Section 26.4.

Further reading

Awbi, H. B. (1991). *Ventilation of Buildings*. E & F Spon, London.

CIBSE (1986). *Guide to Current Practice*, Volume A. Chartered Institution of Building Services Engineers, London.

Croome, D. J. and Roberts, B. M. (1981). *Air Conditioning and Ventilating of Buildings*, 2nd Edition, Volume 1. Pergamon, Oxford.

Eastop, T. D. and Watson, W. E. (1992). *Mechanical Services for Buildings.* Longman, Harlow.

Goodfellow, H. D. (1985). *Advanced Design of Ventilating Systems.* Elsevier Science, Amsterdam.

Jones, W. P. (1989). *Air Conditioning Engineering.* Edward Arnold, London.

Problems

26.1 A classroom having a volume of $283\,m^3$ undergoes 1.5 air changes per hour from natural ventilation. The concentration of carbon dioxide in the outside air is 0.03 per cent and the production of carbon dioxide per person is $4.72 \times 10^{-6}\,m^3\,s^{-1}$.

(*a*) What is the maximum occupancy of the space if the carbon dioxide concentration is to be less than 0.1 per cent at the end of the first hour, assuming an initial concentration equal to the ambient outside conditions?
(*b*) What is the maximum occupancy if the space is continuously used and the concentration must never exceed 0.1 per cent?   [(*a*) 22, (*b*) 17]

26.2 If regulations stipulate that the minimum amount of fresh air to be supplied to a cinema is 8 litres s^{-1} per person and that the minimum amount of space allowable is $12\,m^3$ per person, calculate the concentration of carbon dioxide present after one hour as a percentage. Assume that fresh air contains contaminant at 0.03 per cent and that contaminant generation within the space is 4.72×10^{-3} litres s^{-1} per person.   [0.084 per cent]

26.3 Use the general equation for the contamination level in an enclosure to indicate how the natural ventilation for an unoccupied space may be measured.
   Write down the form of the contamination equation that would apply to the following cases, illustrating your answer with appropriate contamination vs. time curves:

(**i**)   Generation of contaminant within a space that is mechanically ventilated.

(**ii**) Decay of an initial contamination level within a space by mechanical ventilation, followed by a period of natural ventilation.

(Indicate any assumptions in your answers.)
   A lecture theatre has a volume of $1500\,m^3$. The maximum level of carbon dioxide at the end of an hour is 0.1 per cent, assuming an initial concentration of 0.03 per equal, equal to that in the outside air used for ventilation. Natural ventilation can provide 1.5 air changes per hour. If each occupant generates carbon dioxide at a rate of $5.0 \times 10^{-6}\,m^3\,s^{-1}$, determine:

(*a*) The maximum number of occupants who can use the space for an hour.
(*b*) The air change rate that would be necessary to double this occupancy for an hour.   [(*a*) 113, (*b*) 3.75]

26.4 The general equation for contamination decay within a space may be expressed as

$$c = (c_0 + 10\,000 Q_c/Q)(1 - e^{-Qt/V}) + c_i\,e^{-Qt/V}.$$

(*a*) Use this general equation to determine the number of air changes per hour that result from natural ventilation in an unoccupied space.
(*b*) Write down the special forms of this equation for the following cases, illustrating the resulting expressions graphically:

(**i**)   Following a zero initial contamination, a process is instituted that generates a steady contaminant input to the ventilated space, incoming air being free of contamination.

(**ii**) On completion of a process, contamination is allowed to decay by steady ventilation of the space, incoming air again being free of contamination, initially by mechanical ventilation and later by natural ventilation.

(*c*) Paint spraying takes place in a ventilated space from 9 a.m. to 12 noon and from 2 p.m. to 5 p.m. Sketch the variation of paint solvent fume contamination over a 24-hour period, assuming a zero initial contamination at 9 a.m. and a constant ventilation rate from 9 a.m. to 9 p.m. (Note that no calculations are required but your answer should reflect changes in conditions in the space during the 24 hours considered.)

26.5 Five operatives are employed in an aircraft hangar, volume $5000\,m^3$, to spray a camouflage scheme. An extract fan system removes air at a rate of $4\,m^3\,s^{-1}$, with the hangar remaining at atmospheric pressure. Paint is sprayed continuously by each operative at a steady rate of 15 litres h^{-1}, the paint having the following specification:

Density	$1.2\,kg\,l^{-1}$
Solvent content	25 per cent by mass
Drying time (i.e. solvent evaporation time)	assumed instantaneous
Specific volume solvent vapour	$0.5\,m^3\,kg^{-1}$
Lower explosive limit for solvent vapour in air	2.0 per cent by volume.

(a) Determine the maximum duration that the operatives may work before the contamination level exceeds 1 per cent of the lower explosive limit.

(b) Once the process has been stopped, determine the time necessary for the contamination level to fall to 1 per cent of the lower explosive limit with the ventilation fans operating at a reduced rate of extraction of $2\,m^3\,s^{-1}$.

[66.7 mins, 3.19 hours]

Appendix 1
Some Properties of
Common Fluids

TABLE A1.1 Variation of some properties of water with temperature

TEMPERATURE (°C)	DENSITY, ρ (kg m^{-3})	DYNAMIC VISCOSITY, μ (kg m^{-1} s^{-1})	KINEMATIC VISCOSITY, ν (m^2 s^{-1})	SURFACE TENSION, σ (N m^{-1})	VAPOUR PRESSURE HEAD, $p_v/\rho g$ (m)	BULK MODULUS OF ELASTICITY, K (MN m^{-2})
0	999.9	1.792 ($\times 10^{-3}$)	1.792 ($\times 10^{-6}$)	7.62 ($\times 10^{-2}$)	0.06	2040
5	1000.0	1.519	1.519	7.54	0.09	2060
10	999.7	1.308	1.308	7.48	0.12	2110
15	999.1	1.140	1.141	7.41	0.17	2140
20	998.2	1.005	1.007	7.36	0.25	2200
25	997.1	0.894	0.897	7.26	0.33	2220
30	995.7	0.801	0.804	7.18	0.44	2230
35	994.1	0.723	0.727	7.10	0.58	2240
40	992.2	0.656	0.661	7.01	0.76	2270
45	990.2	0.599	0.605	6.92	0.98	2290
50	988.1	0.549	0.556	6.82	1.26	2300
55	985.7	0.506	0.513	6.74	1.61	2310
60	983.2	0.469	0.477	6.68	2.03	2280
65	980.6	0.436	0.444	6.58	2.56	2260
70	977.8	0.406	0.415	6.50	3.20	2250
75	974.9	0.380	0.390	6.40	3.96	2230
80	971.8	0.357	0.367	6.30	4.86	2210
85	968.6	0.336	0.347	6.20	5.93	2170
90	965.3	0.317	2.328	6.12	7.18	2160
95	961.9	0.299	0.311	6.02	8.62	2110
100	958.4	0.284	0.296	5.94	10.33	2070

TABLE A1.2
Variation of bulk
modulus of elasticity of
water with temperature
and pressure

	BULK MODULUS, K (MN m^{-2})			
PRESSURE (bar)	0 °C	20 °C	49 °C	93 °C
1	2040	2200	2289	2130
100	2068	2275	2358	2199
300	2186	2399	2496	2330
1000	2620	2827	2937	2792

TABLE A1.3
Variation of some
properties of air with
temperature at
atmospheric pressure

TEMPERATURE, T (°C)	DENSITY, ρ (kg m^{-3})	DYNAMIC VISCOSITY, μ (kg m^{-1} s^{-1})	KINEMATIC VISCOSITY, v (m^2 s^{-1})
−40	1.52	14.94 ($\times 10^{-6}$)	9.83 ($\times 10^{-6}$)
−20	1.40	15.92	11.37
0	1.29	17.05	13.22
20	1.20	18.15	15.13
40	1.12	19.05	17.01
60	1.06	19.82	18.70
80	0.99	20.65	20.86
100	0.94	21.85	23.24
120	0.90	23.20	25.78

TABLE A1.4
Some properties of
common liquids

LIQUID	DENSITY, ρ (kg m^{-3})	SURFACE TENSION, σ (N m^{-1})	DYNAMIC VISCOSITY, μ (kg m^{-1} s^{-1})	BULK MODULUS, K (GN m^{-2})
Temperature, 20 °C				
Water, fresh	998	72.7 ($\times 10^{-3}$)	1.00 ($\times 10^{-3}$)	2.05
sea	1 025			
Alcohol, ethyl	789	22.3	1.197	1.32
Benzene	879	28.9	0.647	1.10
Carbon tetrachloride	1 632	26.8	0.972	1.12
Glycerol	1 262	63	620	4.03
Mercury	13 546	472	1.552	26.2
Paraffin oil	800	26	1.9	1.62
Temperature, 38 °C				
Oil, SAE 10	880–950	30	29	
SAE 30	880–950	30	96	

TABLE A1.5   Some properties of common gases (at $p = 1$ atm, $T = 273$ K)

Gas	Molecular weight	Density, ρ (kg m^{-3})	Gas constant, R (J kg^{-1} K^{-1})	Specific heats c_p (J kg^{-1} K^{-1})	c_v (J kg^{-1} K^{-1})	Specific heat ratio, γ ($= c_p/c_v$)	Dynamic viscosity, μ (kg m^{-1} s^{-1})
Air		1.293	287	993	708	1.402	17.05 ($\times 10^{-6}$)
Carbon monoxide	28.0	1.250	297	1 050	748	1.404	16.6
Carbon dioxide	44.0	1.977	189	834	640	1.304	14
Helium	4.0	0.179	2077	5 240	3 157	1.66	18.6
Hydrogen	2.02	0.090	4121	14 300	10 140	1.41	8.35
Methane	16.04	0.717		2 200	1 676	1.313	10.3
Nitrogen	28.0	1.250	297	1 040	741	1.404	16.7
Oxygen	32.0	1.429	260	913	652	1.40	19.2
Water vapour	18.0	0.800	462	2 020 (373 K)	1 519	1.33	8.7

TABLE A1.6
International Standard Atmosphere

Altitude above sea level (m)	Absolute pressure (bar)	Absolute temperature (K)	Mass density (kg m^{-3})	Kinematic viscosity (m^2 s^{-1})	Velocity of sound (m s^{-1})
0	1.013 2	288.15	1.225 0	1.461 ($\times 10^{-5}$)	340.3
1 000	0.898 8	281.7	1.111 7	1.581	336.4
2 000	0.795 0	275.2	1.006 6	1.715	332.5
4 000	0.616 6	262.2	0.819 4	2.028	324.6
6 000	0.472 2	249.2	0.660 2	2.416	316.5
8 000	0.356 5	236.2	0.525 8	2.904	308.1
10 000	0.265 0	223.3	0.413 4	3.525	299.5
11 500	0.209 8	216.7	0.337 5	4.213	295.1
14 000	0.141 7	216.7	0.227 9	6.239	295.1
16 000	0.103 5	216.7	0.166 5	8.540	295.1
18 000	0.075 65	216.7	0.121 6	11.686	295.1
20 000	0.055 29	216.7	0.088 92	15.989	295.1
22 000	0.040 47	218.6	0.064 51	22.201	296.4
24 000	0.029 72	220.6	0.046 94	30.743	297.7
26 000	0.021 88	222.5	0.034 26	42.439	299.1
28 000	0.016 16	224.5	0.025 08	58.405	300.4
30 000	0.011 97	226.5	0.018 41	80.134	301.7
32 000	0.008 89	228.5	0.013 56	109.62	303.0

TABLE A1.7
Solubility of air in pure water at various temperatures

TEMPERATURE (°C)	VOLUME OF AIR DISSOLVED* (m³)
0	0.029
10	0.023
30	0.016
70	0.012
100	0.011

* Measured at 0 °C and 1 atm pressure per unit volume of water under a pressure of 1 atm.

TABLE A1.8   Absolute viscosity of some common fluids

FLUID	ABSOLUTE VISCOSITY $\times 10^6$ (kg m^{-1} s^{-1})							
	−20 °C	0 °C	20 °C	40 °C	60 °C	80 °C	100 °C	120 °C
Carbon dioxide	13.4	14.6	15.8	16.8	17.7	18.7	19.4	20.1
Helium	18.7	19.4	20.5	21.6	22.5	23.4	23.9	24.4
Hydrogen	8.5	8.9	9.6	9.9	10.2	10.5	10.6	10.7
Alcohol, ethyl	2530	1 720	1200	770	570	470	–	–
Castor oil	–	–	–	249 000	81 000	34 000	17 000	–
Crude oil (relative density 0.86)	–	17 000	8100	5 300	3 800	3 000	2 400	2000
Glycerine	–	–	–	163 000	43 000	21 000	–	–
Mercury	1820	1 670	1580	1 440	1 380	1 330	1 260	1150
Paraffin	–	3 160	1910	1 290	–	–	–	–
Petrol	–	390	300	240	210	–	–	–

Appendix 2
Values of Drag Coefficient C_D for Various Body Shapes

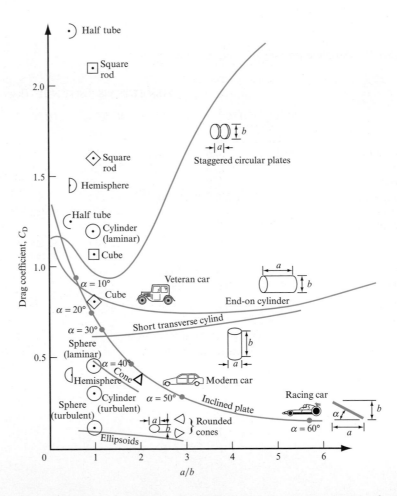

Figure A2.1 Values of drag coefficient C_D for various body shapes at $Re \simeq 10^5$, based on frontal area, except for the inclined plate, where $A = L \times W$.

$$C_D = 2(\text{Drag})/(\rho v^2 A).$$

Flow is from left to right with respect to the body shape indicated. Where not shown, a is the body dimension in the direction of flow and b is perpendicular to it.

Index

IMPORTANT: READ CAREFULLY
WARNING: BY OPENING THE PACKAGE YOU AGREE TO BE BOUND BY THE TERMS OF THE LICENCE AGREEMENT BELOW.

This is a legally binding agreement between You (the user or purchaser) and Pearson Education Limited. By retaining this licence, any software media or accompanying written materials or carrying out any of the permitted activities You agree to be bound by the terms of the licence agreement below.

If You do not agree to these terms then promptly return the entire publication (this licence and all software, written materials, packaging and any other components received with it) with Your sales receipt to Your supplier for a full refund.

SINGLE USER LICENCE AGREEMENT

❐ YOU ARE PERMITTED TO:

- Use (load into temporary memory or permanent storage) a single copy of the software on only one computer at a time. If this computer is linked to a network then the software may only be installed in a manner such that it is not accessible to other machines on the network.

- Make one copy of the software solely for backup purposes or copy it to a single hard disk, provided you keep the original solely for back up purposes.

- Transfer the software from one computer to another provided that you only use it on one computer at a time.

❐ YOU MAY NOT:

- Rent or lease the software or any part of the publication.

- Copy any part of the documentation, except where specifically indicated otherwise.

- Make copies of the software, other than for backup purposes.

- Reverse engineer, decompile or disassemble the software.

- Use the software on more than one computer at a time.

- Install the software on any networked computer in a way that could allow access to it from more than one machine on the network.

- Use the software in any way not specified above without the prior written consent of Pearson Education Limited.

ONE COPY ONLY

This licence is for a single user copy of the software
PEARSON EDUCATION LIMITED RESERVES THE RIGHT TO TERMINATE THIS LICENCE BY WRITTEN NOTICE AND TO TAKE ACTION TO RECOVER ANY DAMAGES SUFFERED BY PEARSON EDUCATION LIMITED IF YOU BREACH ANY PROVISION OF THIS AGREEMENT.

Pearson Education Limited owns the software You only own the disk on which the software is supplied.

LIMITED WARRANTY

Pearson Education Limited warrants that the diskette or CD rom on which the software is supplied are free from defects in materials and workmanship under normal use for ninety (90) days from the date You receive them. This warranty is limited to You and is not transferable. Pearson Education Limited does not warrant that the functions of the software meet Your requirements or that the media is compatible with any computer system on which it is used or that the operation of the software will be unlimited or error free.

You assume responsibility for selecting the software to achieve Your intended results and for the installation of, the use of and the results obtained from the software. The entire liability of Pearson Education Limited and its suppliers and your only remedy shall be replacement of the components that do not meet this warranty free of charge.

This limited warranty is void if any damage has resulted from accident, abuse, misapplication, service or modification by someone other than Pearson Education Limited. In no event shall Pearson Education Limited or its suppliers by liable for any damages whatsoever arising out of installation of the software, even if advised of the possibility of such damages. Pearson Education Limited will not be liable for any loss or damage of any nature suffered by any party as a result of reliance upon or reproduction of or any errors in the content of the publication.

Pearson Education Limited does not limit its liability for death or personal injury caused by its negligence.

This licence agreement shall be governed by and interpreted and construed in accordance with English law.